DARK MATTER

AIP CONFERENCE PROCEEDINGS 336

DARK MATTER

COLLEGE PARK, MD OCTOBER, 1994

EDITOR: STEPHEN S. HOLT
CHARLES L. BENNETT
NASA/GODDARD
SPACE FLIGHT CENTER

AIP PRESS

American Institute of Physics **New York**

L.C. Catalog Card No. 95-76538
ISBN 1-56396-438-4
DOE CONF-9410287

Printed in the United States of America.

CONTENTS

PREFACE .... **xi**
S. S. Holt

INTRODUCTION TO DARK MATTER

Introductory Survey **3**
M. J. Rees
Dark Matter—A Personal View **10**
B. Paczyński
From the Standard Model to Dark Matter **18**
F. Wilczek
Neutrino Mass and Dark Matter **44**
T. J. Bowles
Balloon-Borne Direct Search for Ionizing Massive Particles as a Component of the Galactic Halo Dark Matter **53**
P. C. McGuire, T. Bowen, D. L. Barker, P. G. Halverson, K. R. Kendall, T. S. Metcalfe, R. S. Norton, A. E. Pifer, L. M. Barbier, E. R. Christian, K. E. Krombel, J. W. Mitchell, J. F. Ormes, R. E. Streitmatter, A. J. Davis, A. W. Labrador, R. A. Mewaldt, S. M. Schindler, R. L. Golden, S. J. Stochaj, and W. R. Webber
Looking Backward, Darkly **57**
V. Trimble

LOW MASS STARS AS DARK MATTER

Very Low Mass Stars As Dark Matter **71**
J. Leibert
Recent Developments in Gravitational Microlensing and the Latest MACHO Results: Microlensing Towards the Galactic Bulge **77**
D. P. Bennett, C. Alcock, R. A. Allsman, T. S. Axelrod, K. H. Cook, K. C. Freeman, K. Griest, S. L. Marshall, S. Perlmutter, B. A. Peterson, M. R. Pratt, P. J. Quinn, A. W. Rodgers, C. W. Stubbs, and W. Sutherland
The Results of Deep CCD Field Surveys: Very Low Mass Halo Population Stars as Dark Matter **91**
P. C. Boeshaar, J. A. Tyson, and G. M. Bernstein
Direct Detection of Dark Matter from TwQSO Microlensing **95**
R. Schild and D. J. Thomson

DARK MATTER IN GALAXIES

Dark Matter in the Milky Way 101
L. Blitz

Dark Matter in External Galaxies—HI Observations 111
A. Bosma

The Shape of the Dark Matter Halo of NGC 4244 121
R. P. Olling and J. H. van Gorkom

Dark Matter and the Hubble Sequence of Disk Galaxies 125
A. Broeils

Disks and Dark Halos of Dwarf Spirals 129
V. Friese, B. Fuchs, and R. Wielen

The Nature of Dark Matter and Galactic Rotation Curves 133
I. Khan

Rotation Curves of 967 Spiral Galaxies: Implications for Dark Matter 137
M. Persic, P. Salucci, and F. Stel

Another Flattened Dark Halo: Polar Ring Galaxy A0136-0801 141
P. D. Sackett and R. W. Pogge

Dark Matter in Elliptical Galaxies 145
F. Bertola and A. Pizzella

Mapping Dark Matter Around Early Type Galaxies 149
B. Tremblay, T. B. Williams, and D. Merritt

Probing the Mass Distribution and Formation History of Galaxies with Globular Clusters 153
S. E. Zepf

A Faint Luminous Halo that May Trace the Dark Matter Around Spiral Galaxy NGC 5907 157
H. Morrison, P. Sackett, P. Harding, and T. Boroson

Cold Molecular Gas as Dark Matter Candidate in Galaxies and Clusters 161
D. Pfenniger and F. Combes

Preliminary Evidence for Dust in Galactic Halos 165
D. Zaritsky

Galaxy Halos and the Nature of Dark Matter 168
B. Moore

X-Ray (ASCA) Measurements of Elliptical Galaxies 172
K. Makishima

X-Ray Constraints on the Intrinsic Shape of the Lenticular Galaxy NGC 1332 182
D. A. Buote and C. R. Canizares

Constraints on Dark Matter in Clusters and Galaxies from the Shapes of X-ray Images 186
C. R. Canizares and D. A. Buote

Dark Matter in NGC 4636 190
M. Loewenstein, R. Mushotzky, H. Awaki, H. Matsumoto, K. Makishima, and K. Matsushita

The Edge of Darkness: Measurements of the cD Galaxy in A665 194
E. J. Gaidos

DARK MATTER IN CLUSTERS

Clusters and Dark Matter 201
N. A. Bahcall

Gas Mass and Total Mass in Clusters of Galaxies 208
L. M. Lubin

Photometric Properties of Clusters of Galaxies 212
B. Garilli, D. Maccagni, L. Carrasco, and E. Recillas

A Merger of Subclusters in the Galaxy Cluster Abell 754 216
A. I. Zabludoff and D. Zaritsky

Kinematics and Dynamics of the Morgan Poor Clusters 221
C. M. Bird, T. C. Beers, J. R. Kriessler, and J. P. Huchra

Determination of the Mass Distribution in Rich Clusters from Imaging X-Ray Spectroscopy 231
R. Mushotzky, M. Loewenstein, K. Arnaud, and Y. Fukazawa

X-Ray Constraints on Dark Matter in Groups of Galaxies 243
J. S. Mulchaey

The Velocity Dispersion-Temperature Correlation from a Limited Cluster Sample 251
C. M. Bird and R. F. Mushotzky

ASCA Measurements of X-Rays from Poor Clusters 255
T. Ohashi

Mapping the Properties of Galaxy Clusters with ROSAT 264
D. Cannon and T. Ponman

ASCA Observation of the NGC 2300 Group of Galaxies 268
Y. Tawara, Y. Sakima, K. Yamashita, and R. Mushotzky

A Comparison of X-ray and Optical Wavelet Maps in the Central Region of Coma 272
D. Gerbal, A. Biviano, F. Durret, O. Le Fevre, F. Le Peintre, C. Lobo, A. Mazure, and E. Slezak

A Hierarchical Clustering Method for the Description of Structures in Galaxy Clusters 276
D. Gerbal and A. Serna

The Effect of a Partially Virialized Gas on the Baryon Fraction Derived from X-ray Measurements 280
M. J. Henriksen

GRAVITATIONAL LENSING AS A PROBE OF DARK MATTER

Dark Matter Mapping by Gravitational Lens Tomography 287
J. A. Tyson

Gravitational Lens Constraints on Cosmology and Cosmogony 297
C. S. Kochanek

Gravitational Lensing and the Mass Distribution of Clusters 307
M. Bartelmann and R. Narayan

Mass Distributions of Clusters from Gravitational Magnification 320
T. Broadhurst

Evidence from Gravitational Lensing for a Non-Thermal Pressure Support in the Cluster of Galaxies A2218 330
A. Loeb and S. Mao

Gravitational Lensing of Variability Selected QSOs by Galaxy Clusters 331
L. L. Rodrigues-Williams and M. R. S. Hawkins

Measuring the Bias Factor With the Weak Lensing Effect 335
M. Bartelmann

Detection of Lens Candidates for the Double QSO Q2345+007 339
P. Fischer, J. A. Tyson, G. M. Bernstein, and P. Guhathakurta

Gravitational Lensing of Quasar 0957+561 and the Determination of H_0 343
G. Rhee, G. Bernstein, T. Tyson, and P. Fischer

Testing Cosmogonic Models With Gravitational Lensing 347
J. Wambsganss, R. Cen, J. P. Ostriker, and E. L. Turner

Reconstructing Cluster Lensing Mass Distributions from Gravitational Arclets 351
G. Wilson, S. Cole, and C. Frenk

Arc Statistics With Realistic Cluster Potentials 355
M. Bartelmann, M. Steinmetz, and A. Weiss

THE LARGE SCALE DISTRIBUTION AND MOTIONS OF GALAXIES

Indications of Dark Matter Derived from Large Scale Flows 361
M. Davis and A. Nusser

Using Galaxy Clusters to Trace the Large Scale Velocity Field 371
M. Postman

A Spherical Harmonic Approach to Redshift Distortions 381
A. N. Taylor and A. F. Heavens

Galaxy Bias and the Mean Density of the Universe 386
A. N. Taylor and A. F. Heavens

Spatial Clustering of Radio Sources in the 1987 Green Banks Survey 390
H. Sicotte and P. J. E. Peebles

COLD AND HOT DARK MATTER

The Second Coming of Cold Dark Matter? 397
W. H. Zurek, B. C. Bromley, and M. S. Warren

The Power Spectrum of Dark Matter Halos in a CDM Universe 407
T. G. Brainerd and J. V. Villumsen

Cold+Hot Dark Matter Cosmology with $m(\nu_\mu)\approx m(\nu_\tau)\approx 2.4$ eV 411
A. Klypin, J. R. Primack, J. Holtzman, and D. O. Caldwell

Massive Neutrinos and Galaxy Formation 420
C.-P. Ma

The Influence of CDM on the Dynamical Estimate of Ω in the Local Neighborhood 425
A. M. Dunn and R. Laflamme

Dark Matter in Voids 429
R. Fong, A. G. Doroshkevich, and V. I. Turchaninov

Rich Cluster Simulation: Dynamics and Mass Estimates 433
B. C. Bromley, M. S. Warren, W. H. Zurek, and P. J. Quinn

Hydrodynamic Simulations of Galaxy Formation 437
A. A. Thoul

Large-Scale Structure Formation and the Segregation of Dark and Baryonic Matter 441
H. Martel, P. R. Shapiro, A. Valinia, and J. V. Villumsen

The Effect of Reionization of the Intergalactic Medium on Galaxy Formation 446
P. R. Shapiro and H. Martel

The Little Bang: Peculiarities in the Local Group and Some Dynamical Explanations 450
M. J. Valtonen, M. L. McCall, K. A. Innanen, J. Q. Zheng, and G. G. Byrd

COSMIC MICROWAVE BACKGROUND ANISOTROPY AND DARK MATTER

What Can We Learn from Future CMBR Experiments? 457
D. N. Spergel

Dark Matter-Cosmic Microwave Background Connection 467
K. M. Górski

COBE-DMR Two-Year Large Scale Anisotropy Results 477
C. L. Bennett, A. Banday, K. Górski, G. Hinshaw, A. Kogut, and E. L. Wright

On the RMS Anisotropy at 7 and 10 Degrees Observed in the COBE-DMR Two Year Sky Maps 481
A. J. Banday, K. M. Górski, A. Kogut, G. Hinshaw, C. L. Bennett, C. H. Lineweaver, G. F. Smoot, and L. Tenorio

Simulated CMB Maps at 0.5 Degree Resolution: Basic Results **485**
G. Hinshaw, C. L. Bennett, and A. Kogut
CMB Anisotropy at 0.5 Degree Angular Scale II. Analysis of Peaks **489**
A. Kogut, G. Hinshaw, and C. L. Bennett

ALTERNATIVE APPROACHES TO THE DARK MATTER PROBLEM

Alternatives to Dark Matter **495**
D. Kazanas
Microlensing, Newton-Einstein Gravity, and Conformal Gravity **505**
P. D. Mannheim
Magnetogravitodynamics Model of Galaxy Formation **509**
H. D. Greyber
Evidence for a Possible Common Origin of Baryonic and Non-Baryonic Dark Matter **513**
E. J. Sternglass
Dark Matter and Cosmic Plasma Physical Processes **517**
P. Ammiraju
Baryon Fraction in the Universe, Antiprotron Lifetime, and the Diffuse Radiation **520**
S. A. Stephens and N. Finetti

SUMMARY

Summary **527**
J. Silk
Appendix A—Conference Program **533**
Appendix B—List of Attendees **537**
Appendix C—Physical Constants **547**
Author Index **551**
Subject Index **555**

PREFACE

This is the fifth in a series of annual October Astrophysics Conferences in Maryland. These conferences are organized by astrophysicists at the Goddard Space Flight Center and the University of Maryland. The topic for each conference is selected by a permanent committee of senior scientific staff with the help of an International Advisory Committee, the current membership of which is:

Roger Blandford, Pasadena
Arnon Dar, Haifa
Alan Dressler, Pasadena
Riccardo Giacconi, Munich
Guenther Hasinger, Potsdam
Steve Holt, Greenbelt
Frank Kerr, College Park
Dick McCray, Boulder
Jim Peebles, Princeton
Sir Martin Rees, Cambridge
Vera Rubin, Washington
Joseph Silk, Berkeley
Rashid Sunyaev, Moscow
Alex Szalay, Budapest
Yasuo Tanaka, Tokyo
Scott Tremaine, Toronto

The subject chosen for this conference was "Dark Matter," with its program developed by the Scientific Organizing Committee:

Chuck Bennett
Leo Blitz
Kris Górski
Demos Kazanas
John Mather
Richard Mushotzky
Alex Szalay
Virginia Trimble

"Dark Matter" is probably the consensus choice as the most important problem to be addressed in astrophysics. The inconsistencies between dynamically-inferred and luminosity-inferred mass densities on various distance scales are somewhat problematic, but much more serious difficulties are encountered when comparisons are made with the densities demanded by our "best" theories. Lots of dark baryonic mass is required to achieve a density consistent with the observed light element abundances in order to satisfy big-bang nucleosynthesis, and an order of magnitude more non-baryonic matter is needed to achieve the critical density demanded by inflation. The conference was designed to confront the problem with new results from both "traditional" dynamical analyses and from the emerging gravitational lensing activities.

As usual, the conference began with two invited summaries reviewing the general subject; this year's were delivered by *Martin Ress* (to introduce the problem and its ramifications) and *Bohdan Paczinski* (to introduce the potential of gravitational lensing in addressing the problem). The conference then proceeded through the next two days with a series of non-parallelled sessions, each devoted to a specific topic and led by a distinguished session chair. Each of these sessions featured two or three invited talks and an extensive discussion period that may have also included one or two short contributions "promoted" from the poster papers by the session chair.

All the chairs should be commended for keeping the activities lively and the speakers to their allotted times:

Marek Abramowicz
Neta Bahcall
Chuck Bennett
Lawrence Krauss
John Mather
Vera Rubin
Alex Szalay
Michael Turner
Ned Wright

At previous meetings in this series, the banquet at the conclusion of the second day has usually featured a speaker who has been one of the early heroes of the discipline. This year we broke with tradition by inviting a speaker who is still one of the younger and more active researchers in the field, because *Michael Turner* is one of the most entertaining speakers in any scientific discipline. He didn't disappoint us.

The concluding session after lunch on the third day consisted of a remarkable rapporteur presentation by *Joe Silk*. That it was a thoughtful and accurate representation of the conference and the status of the discipline was not the remarkable part—we know that we can expect such a presentation from him. What was truly remarkable was that it was prepared after he sat through each of the sessions with a freshly broken arm, unable to make more than childish scrawls in lieu of detailed notes.

I like to think that the Proceedings from these conferences have been very successful insofar as they have constituted close-to-complete current summaries of the subject disciplines, which have been published and available to the community well before the next annual October meeting. This volume is particularly interesting insofar as it is the first which has substantial overlap with the subject of a previous conference in this series. It is a tribute to the organizers of this meeting that these Proceedings truly update and complement (with virtually no duplication) those of the first October meeting, *"After the First Three Minutes."*

Steve Holt
January 1995

INTRODUCTION TO DARK MATTER

INTRODUCTORY SURVEY

Martin J. Rees
Institute of Astronomy, Madingley Road, Cambridge, CB3 0HA, UK.
mjr@mail.ast.cam.ac.uk

1. INTRODUCTION

It would be hard to imagine a bigger subject for a conference than this one. Our three-day meeting will cover 90 per cent, and maybe even 99 per cent, of what is in the Universe. My purpose in this introductory talk will be to explain why most astronomers are perplexed about how much dark matter there is, and what it might be. These comments will serve as warm-up remarks for Dr Pacynzski, who will give a more comprehensive survey.

Intimations of dark matter date back, of course, to studies of motions in clusters of galaxies in the 1930s. These were later supplemented by analyses of motions within the local group, particularly the classic 1959 paper of Kahn and Woltjer [1]; and by radio and optical study of rotation curves in the outer parts of disc galaxies. This is not going to be primarily a historical review, but I would like nevertheless to start off by going back 20 years, to 1974, because that was when a consensus about the dynamical dominance of dark matter was crystalised, particularly in two important papers.

One of the classic 1974 papers, by three Estonian astronomers, Einasto, Kaasik, and Saar [2], stated that"the mass of galactic coronae exceeds the mass of populations of known stars by 1 order of magnitude, as do the effective dimensions. The mass luminosity ratio rises to $f = 100$ for spiral and $f = 120$ for elliptical galaxies. With $H = 50$ km per sec per Mpc this ratio for the Coma cluster is 170. In the second paper, by Ostriker, Peebles, and Yahil [3] at Princeton, it was stated that "currently available observations strongly indicate that the mass of spiral galaxies increases almost linearly with radius to nearly 1 Mpc, and that the ratio of this mass to the light within the Holmberg radius is 200 M/L".

These particular inferences have been buttressed enormously by progress in the last 20 years, and we shall hear from many other speakers during this meeting about the current evidence. But it is remarkable that the conclusions have not drastically changed.

Another indirect constraint on the amount of dark matter in baryonic form comes from the abundances of light elements predicted by cosmic nucleosynthesis. As has been well-known since the 1960s, these abundances depend on the baryon density during the few minutes over which the universe cools through the temperature range from 1 MeV to 100 keV, and therefore (since the present background temperature is known) can be related directly to the present baryon density. The predicted helium abundance increases only slowly with density, but the measurements of helium are now precise enough to provide a significant upper limit. Deuterium, however, is a more sensitive diagnostic of the primordial baryon density. Since it is an intermediate product in the production of helium, more deuterium survives in a universe of low baryon density. Moreover, it is now much more clearer than it was in the 1960s that deuterium is best explained as a relic of the early universe [4]. 1974, plainly a vintage year for this subject, also saw the publication of a review by Gott, Gunn, Schramm, and Tinsley [5]. These authors adopted a synoptic approach, and tried to seek consistent ranges for the density parameter Ω and the Hubble constant. They considered three constraints. The first was the requirement that the universe (whose age depends on the Hubble constant and, in Friedman models, on Ω) should be older than the oldest stars. The second was an upper limit on the baryon density from deuterium. And the third was a lower limit to Ω, of order 0.1, set by the amount of dark matter that was reliably established by dynamical arguments. They claimed that there was a very small window, with Ω of order 0.1 and a low Hubble constant, such that the age constraints could be satisfied and all the reliably-established dark matter could be baryonic.

2. UPDATE SINCE 1974

What has changed regarding Gott *et al.*'s arguments in the last twenty years? The uncertainties about the Hubble constant and stellar ages are unfortunately still with us. So let us confront that problem squarely and pass on.

There is firmer evidence on the dark matter density in clusters of galaxies, though its net import does not substantially change the old estimates.

However, the issue of extra dark matter between clusters, maybe even sufficient to provide the critical density, is now a more lively one, and I shall return to it later.

Constraints from nucleosynthesis

Estimates of deuterium as a measure of baryon density have improved, particularly through a better understanding of the relationship of deuterium and helium[3]. Mike Turner is a coauthor (with Copi and Schramm) of the most up-to-date review of primordial nucleosynthesis [6], and he should have ample chance in his after-dinner speech tomorrow to discuss these issues at length.

There has recently been a flurry of interest in cosmic deuterium, stimulated by the claim of a high relative abundance of deuterium to hydrogen, of order 3×10^{-4}, in a high redshift damped Lyman-alpha absorption system along the line of sight to a quasar [7]. If this result were to stand up, it would push down the permitted baryon density, completely ruling out the possibility that most halo dark matter could be baryonic unless one abandons other standard assumptions. However, it would be wise to suspend judgements on this issue. The alleged deuterium line is a weak satellite of a very strong feature attributed to high-column-density HI. It is indeed displaced from the centre of a strong HI feature by 80 kilometres per second, equivalent to the expected isotropic shift, and there is only a few per cent chance of finding a random weak line in the Lyman forest in this position. But there may very well be an excess of weak 'satellite' lines close to any damped Lyman alpha system (due to gas associated with the same 'protogalaxy'). Until we are sure that there are more systems displaced by 80 km/s than by, say, 60 or 100 km/s the significance of this claim for high deuterium must remain in doubt. Further data from Keck ought, fortunately, to settle this question within the next couple of years.

Baryonic dark matter

The possibility of baryonic dark matter in stars or stellar remnants was addressed with particular thoroughness in a 1984 paper by Carr, Bond, and

Arnett [8]. These authors showed, through a variety of arguments that are now well-known, that there were two possible mass ranges. Dark matter could exist in black holes in the mass range between a few hundred and 10^6 solar masses, which could be a remnant of a population of early massive stars that ended their lives collapsing via the pair production instability. Some constraints on high mass objects in our Galaxy are set by the lack of evidence for accretion onto those passing through the Disc, and so forth. The other possibility is brown dwarf or planetary mass objects, similar to stars except they are below the threshold of around 0.07 solar masses needed to trigger hydrogen fusion. The interesting recent work, involving gravitational microlensing and the search for evidence of lensing in our own galaxy, will feature strongly later in this conference. I shall therefore say no more about it, especially as Bohdan Pacynzski is better qualified to introduce it. Here again, we can be optimistic that the total amount of mass in the lensing objects (and perhaps their typical masses) will be clarified within the next couple of years.

Clusters of galaxies

I turn now to the evidence on dark matter in clusters. Traditionally, this has been inferred from application of the virial theorem to galaxy motions. But there are now two other lines of attack. Maps of the X-ray brightness profile and temperature are now good enough to allow estimates of the depth (and even the profile) of the gravitational well confining the hot X-ray emitting gas. And the detection of large numbers of very faint background galaxies whose shapes are distorted, often into conspicuous arcs, by the effects of light bending due to the clusters gravitational field, will soon offer very direct information about the total mass distribution, whatever that mass may be. One of the early highlights of the data from the post-refurbishment HST is a superb picture of the cluster Abell 2218, with a redshift 0.18, by Ellis, Kneib, and Smail [9], which shows very large numbers of obvious background arcs.

It will soon seem natural to discuss the structure and dynamics of cluster masses in an order different from the traditional one. We shall first infer the shape of the potential well directly by reconstructing it from the observed distortion by gravitational lensing of background galaxies. It will then be possible to decide whether the observed spatial distribution of galaxies, and the spread in their velocities, is consistent with an isotropic equilibrium in that particular potential; if it isn't, the angular distribution of velocities must be more complex, or the system must be out of equilibrium. X-ray maps will reveal whether the gas has a temperature and density profile consistent with that potential. Any discrepancy would motivate us to consider whether the gas is partially supported by rotation, macroscopic bulk motion, magnetic pressure, relativistic particles, etc. (The gas can of course be somewhat inhomogeneous, but the clumping factor is constrained because gas confined in the potential well cannot be on a very much higher adiabat than the gas that dominates the X-ray emission.)

Clustering must be envisaged in the more general context of overall cosmic structure formation. Numerical simulations of this are now a heavy industry, and an increasingly sophisticated one. Most of these simulations are based on the assumption that the dominant gravitating stuff is nonbaryonic. So let us briefly consider this option.

Non-baryonic dark matter

One of the main changes since Gott *et al.*wrote their 1974 paper has been the much greater willingness to invoke nonbaryonic matter. Non-zero neutrino masses are no longer thought theoretically unacceptable, and there is a willingness to invoke new kinds of particles, particularly those predicted by supersymmetric theories. One topic which does not figure much on the programme for this conference is *direct detection* of nonbaryonic matter. Neutrinos seem impossible to detect by feasible current techniques, and axions present a very severe experimental challenge. But there is substantial interest in detecting heavy neutral particles, such as the lightest stable supersymmetric particles. These techniques involve detecting the recoil in the rare event when one of these particles, which would pervade the entire halo moving with speeds of about $10^{-3}c$, interacts with a nucleus in an experimental detector. We should certainly spare a thought, and give every possible encouragement, to those of our colleagues, mainly working down mineshafts, who have accepted the challenge to detect dark matter. Even the optimist cannot predict success with great confidence, but the attainable upper limits are themselves becoming significant, and detection of such particles would tell us what 90 per cent of the Universe is made of, as well as perhaps discovering an entirely new class of particles that cannot be produced terrestrially.

Of course, progress in fundamental physics may sometime tell us whether supersymmetric particles actually exist, and what their masses and cross-sections are. Calculating how many should survive from the first millisecond, and what they contribute to the dark matter density, would then be as straightforward in principle as calculating how much helium and deuterium survives from the first few minutes. Experiments may also tell us that neutrinos have a mass of a few eV (in which case they would definitely contribute to the dark matter, unless we abandon our extrapolation even back to times of order one second).

Simulations of structure formation

But even in the absence of such basic physical knowledge, we can still explore the cosmogonic consequences of different non-baryonic candidates. As is well known, the essential distinction here is between those, such as light neutrinos, which move fast enough that they homogenise on small scales, and those that are sufficiently slow-moving that they can be treated as 'cold'.

The most intensively studied model for structure formation involves the latter assumption. In particular, a benchmark for comparison of observations has been the so-called 'standard' CDM model. This involves a package of five assumptions.

1. The primordial spectrum has the Harrison-Zeldovich form (index n=1 in the conventional notation), and the fluctuations are Gaussian.
2. The universe is dynamically dominated by cold nonbaryonic matter which interacts only gravitationally with everything else.
3. The density is taken to be equal to the critical value: in other words, $\Omega = 1$.
4. Galaxies are related to dark matter by a simple biasing prescription.
5. Neutrino masses are taken to be zero.

The outcome of these simulations is tested against the data by comparing the relative amplitude of clustering on different scales with what is actually observed at the current epoch, and (on large scales) by fitting the amplitude of the microwave background fluctuations. The z-dependence of the structure offers another test. There is limited evidence on how the observed large-scale structure has evolved, but on galactic scales there are constrains back to redshifts of 5 from the existence of quasars, neutral hydrogen clouds, etc.

It is now well-known that this 5-item package, the 'standard' CDM model, has difficulty reconciling small- and large-scale structure and the microwave background fluctuation amplitude. However, there are several modifications which are physically motivated and by no means simply 'ad hoc'. First, the primordial fluctuation spectrum could be tilted, so that the amplitude increases slowly with scale. Indeed most inflationary models predict that this should occur. Also one could consider models where Ω is different from unity: these are either open, or else flat with the extra curvature made up by a non-zero cosmological constant. (Indeed, a model where 80 per cent of the critical density comes from the cosmological constant and only 20 per cent from dark matter and baryons gives an excellent fit to all scales of structure; such a model also eases the 'age problem' that will face us if the Hubble constant settles down at a high value)

An uncertainty in comparing such models with the real universe concerns the relation between the galaxies and the dark matter. This involves complicated gas dynamics, feedback from star formation, etc, and is too complicated to be adequately simulated. The simple scheme which relates the distribution of galaxies to that of the dark matter via a single 'biasing parameter' is plainly over-simplified: the galaxy formation efficiency may depend on environment, etc., in many ways. A cleaner test of such models is possible if we can directly probe the distribution of the dark matter, either by determining the deviations from Hubble flow that it induces in the galaxies, or by detecting weak lensing due to inhomogeneities on supercluster scales. Several speakers will be telling us about the current status of these challenging projects.

It may turn out that neutrino masses are not exactly zero. The so-called 'hybrid' or 'mixed' dark matter models, in which a neutrino has a mass of a few eV, surmount some of the difficulties of standard CDM. If experimentalists find such evidence for neutrino masses, believers in CDM would delightedly incorporate it in their existing models, ending up with a better fit.

3. IS OMEGA REALLY 1?

The other change since Gott *et al.*'s classic paper is that there is now a strong theoretical prejudice in favour of $\Omega = 1$, stemming from the attractiveness of the general concept of an inflationary universe. Many people seem happy to accept whatever non-baryonic matter and biasing would be needed to meet this requirement. Inflationary models naturally predict that the universe expands enough to stretch the universe flat, in the sense that the Robertson-Walker curvature radius would become enormously larger than the present Hubble scale. Anything different from a flat universe would, as is well known, involve fine tuning in the expansion factor. This tuning is implausible at the level of a few per cent, even in the more optimistically contrived scenarios. However, most variants of inflation provide an even stronger argument in favour of $\Omega = 1$. In these models, if the universe had inflated only enough to make the present Robertson-Walker curvature of order the Hubble radius, there would be quadrupole or dipole effects in the microwave background of order unity. Some recently developed models, however, manage to avoid this latter constraint.

What, then, is the observational case for or against a critical density? Until a few years ago, there was essentially no evidence for more than 20 per cent of the critical density. I think everyone would agree that any such evidence is still tentative. However, we shall hear from several speakers about the attempts to measure and interpret large-scale streaming motions. We shall also hear about the attempts to detect 'clumping' in the dark matter on large scales by searching for correlated distortions in distant galaxies that might be due to 'weak' gravitational lensing.

Some of the classical 'geometrical' methods should soon become more helpful. The Hubble diagram for supernovae may be extended to high enough redshifts to reveal the deceleration parameter; further studies may firm up the earlier tentative evidence [10] from the angular diameters of high redshift sources in favour of a high density. On the other hand, if the Hubble constant error bars are reduced, and the Hubble time becomes less than 15 billion years, this will obviously argue against a standard Friedman model with the critical density, in which the time since the big bang is only two thirds of the Hubble time.

Finally, there are some other rather less direct lines of evidence on Ω which will will attract increasing attention, even if they do not figure prominently on the programme for this meeting.

Clusters of galaxies offer two such arguments. The first is an inference from the irregular shapes of most clusters, indicating that they have undergone recent mergers of subcomponents each comprising a substantial fraction of the total mass. In a low density universe, structure forms early and is thereafter frozen in. On the other hand, formation continues if Ω is high. Therefore the prevalence of conspicuous substructure points towards a high Ω, though I do not yet think this is quantitative enough to be able to say that it requires the full critical density.

But a quite distinct argument, again based on clusters, suggests a low Omega. X-ray data show that the baryon fraction in a cluster, mainly in hot gas, is typically between 10 and 20 per cent. (The exact fraction depends, of course, on the Hubble constant). This has been inferred from detailed study of the Coma cluster [11], and also, by D. White and Fabian [12], from X-ray data on a sample of 19 clusters. When the baryonic fraction of the mass in the cluster is compared with the baryonic fraction in the universe allowed by standard big bang nucleosynthesis, there is a contradiction if Ω is more than about 0.3. A high Ω requires either abandonment of standard nucleosynthesis (despite the evidence that this is in good shape), or a mechanism that can concentrate baryons (by a factor of about 3) by pushing them relative to dark matter through distances as large as the turnaround radius of a cluster. Resolution of this dilemma may come from a combination of small effects and uncertainties, but at the moment it seems a serious argument against $\Omega = 1$.

Another quite different estimate of Ω will soon come from studies of microwave background fluctuations. The COBE data refer to angular scales of 10°. However, several other groups are now reporting fluctuations on angular scales of order 1°. These latter scales seem to display a larger amplitude than found by COBE. This is precisely what is expected on scales smaller than the horizon at recombination, because there is then a contribution from Doppler motions, etc. This horizon subtends about 2° in a flat model, but the angle scales as $\Omega^{\frac{1}{2}}$. Firm evidence for an upturn in the background fluctuation amplitude on angular scales of one or two degrees, would be hard to reconcile with a low Ω, where any doppler contribution would be restricted to angular scales below one degree.

The pace of advance in the subject is clear from a glance at our conference programme, where the emphasis is on topics that barely existed a few years ago. We probably won't get definitive answers this week to the basic questions – what is the dark matter? and does it provide the critical density? – but we can look forward to a fascinating and wide-ranging meeting.

REFERENCES

1 Kahn, F.D. and Woltjer, L.W. 1959, Ap.J., 130, 703
2 Einasto, J., Kaasik, A. and Saar, E. 1974, Nature, 250, 309
3 Ostriker, J.P., Peebles, P.J.E. and Yahil, A. 1994, Ap.J. (Lett.), 193, L1
4 Reeves, H. 1994, Rev. Mod. Phys., 66, 193
5 Gott, J.R., Gunn, J.E., Schramm, D.N. and Tinsley, B. 1974, Ap.J., 194, 543
6 Copi, E., Schramm, D.N. and Turner, M.S. 1994, Science (in press)
7 Songalia, A., Cowie, L.L., Hogan, C.J. and Rugers, M. 1994, Nature, 368, 599
8 Carr, B.J., Bond, J.R. and Arnett, W.D. 1984, Ap.J., 277, 455
9 Ellis, R.S., Kneib, J.P. and Smail, I. 1994, Ap.J. (in press)
10 Kellermann, K.I. 1993, Nature, 361, 134
11 White, S.D.M. *etal* 1993, Nature, 366, 429
12 White, D.A. and Fabian, A.C. 1994, MNRAS (in press)

DARK MATTER – A PERSONAL VIEW

Bohdan Paczyński
Princeton University Observatory, Princeton, NJ 08544-1001
bp@astro.princeton.edu

ABSTRACT

Traditional evidence for large amount of dark matter is based on dynamical consideration for systems with $t_{dyn} \gg t_{obs}$. Recent observational and theoretical developments in gravitational lensing offer a much more robust determination of the mass distribution in some galaxies and their clusters, with the precision comparable to that obtainable for double stars for which $t_{dyn} < t_{obs}$, and offer independent and direct evidence for the presence of dark matter.

Gravitational microlensing and femtolensing offer a possibility to detect MACHOs with masses in excess of $\sim 10^{-15}\ M_{\odot}$. The recent detections of microlensing events by the EROS, MACHO and OGLE teams do not require any dark lenses as ordinary low mass stars are compatible with the observations. However, these searches will soon either detect genuine MACHOs, or they will place stringent upper limits on their number density.

1. INTRODUCTION

The presence of dark matter is inferred through its gravitational effect on the luminous matter. The same basic concept led John C. Adams and Urbain Le Verrier to the prediction of the existence of a dark object which was responsible, through its gravity, for the perturbations of the orbit of Uranus. The object was discovered by Johann Galle in 1846, and was named Neptune. More recently accurate masses of the dark matter binary system PSR 1913+16 were determined on the basis of dynamical considerations (Taylor & Weisberg 1989, and references therein). For Neptune and for PSR 1913+16 the mass to light ratio (M/L) is enormous, much higher than for any galaxy or a cluster of galaxies. Presumably, these two types of dark objects, and their M/L, are not typical for the universe at large.

Less precise dynamical consideration are the basis for the claim that there is plenty of dark matter in clusters of galaxies (Zwicky 1933) as well as in galaxies (Oort 1932, Freeman 1970). The propagation of light from distant sources is a very powerful tracer of mass distribution through the phenomenon called gravitational lensing.

Various inferences have a different degree of reliability. My personal rating of the various methods is as follows:

1) dynamical, with $t_{dyn} < t_{obs}$ – the best
2) gravitational lensing – good
3) dynamical, with $t_{dyn} > t_{obs}$ – fair
4) mass to light ratio – poor
5) cosmological principles – meaningless

where t_{dyn} is the dynamical time scale for the system, like the binary period, or the virialization time, and t_{obs} is the length of time over which the observations were carried out.

There is no doubt that the first technique provides a very robust mass measurements. This is the only truly fundamental method of measuring masses in astronomy. In practice it is applicable only to binary stars and to planets. The third method, with $t_{dyn} > t_{obs}$, or rather $t_{dyn} \gg t_{obs}$, is by far the most commonly used in modern inferences for the presence of dark matter. Its reliability varies a lot from case to case. The second method, which only recently developed to the extent that it is practical under a range of conditions, is the main one to be discussed in this review. It is very direct, using light rays as tracers of space-time geometry, and it does not require the mass distribution to be in any kind of equilibrium. Its main practical weakness is that in most cases the model of mass distribution is not unique. Fortunately, there are many cases in which models are reasonably accurate, and the observational as well as modeling techniques are developing rapidly.

The last two methods are far less reliable. For some astronomical object, be it a galaxy or a cluster of galaxies, the mass is estimated with the method (3). The luminosity of the system is measured directly, and the M/L is calculated. The estimates of M/L are available for many galaxies, groups and clusters of galaxies, and some average (median?) value of M/L is guessed to be representative for the whole universe. This value being of the order of $\sim 300\ M_\odot/L_\odot$ implies that most matter in the universe is dark.

It is important to remember that we have no fundamental theory that would allow us to calculate the efficiency of star formation from diffuse gas or the stellar mass function. It is not known how the efficiency of star formation and the shape of the mass function depend on the initial gas pressure, magnetic fields, chemical composition, ambient radiation field, local energy density in cosmic rays, etc. Our knowledge in this field is empirical only, and even this is incomplete and often unreliable. Therefore, my personal rating of the estimates of the amount of dark matter in the universe as based on the method (4) is 'poor'.

My personal rating of the method (5) is 'meaningless' even though there is a strong theoretical bias towards cosmology with $\Omega = 1$. The common justification for this bias comes from the theory of inflation, which in its original form 'predicted' $\Omega = 1$. However, the modern theory of inflation can accommodate almost anything: certainly $\Omega = 0.3$ (Kamionkowski et al. 1994, Bucher et al. 1994), and a 'tilted' (i.e. non-Zeldovich) spectrum of primordial perturbations (Adams et al. 1992) are among the many 'predictions'. Therefore, even though the inflation remains a wonderful concept, and even though in some distant future it may become rigorous, it currently provides no realistic estimate for the amount of dark matter. It is listed as a method number 5 for historical reasons only.

There is no hope to use method (1) on cosmological scale, as typically we may have $t_{dyn} \sim 10^8$ years, or even more, orders of magnitude above t_{obs}. Hence the method (3) is by far the most common. However, just as on the stellar scale the method (1) is the bedrock of all secondary methods, so the method (2) should become the bedrock for mass estimates on the galactic scale, even though its range of applicability is rather limited so far. There are many excellent recent reviews and books on gravitational lensing: Blandford & Narayan (1992), Refsdal & Surdej (1994), Schneider, Ehlers, & Falco (1992), Surdej at al. (1993). As all important references can be found in these reviews, and the page limit imposed on this paper is very strict, I make no attempt to provide a complete or even fair reference to all important contributions to this rapidly growing field, and I apologize for any discomfort this may create.

2. GRAVITATIONAL MACROLENSING

Strong gravitational lensing makes multiple images of a single source. If the separation between the images is large enough (arcseconds in optical domain, milli arcseconds in the radio) so that they can be seen separately, the phenomenon is referred to as macrolensing. The gravitational lensing is called weak if only one distorted image of a source is present.

The largest scale on which the strong gravitational lensing was detected and on which robust quantitative models of the lens mass distribution are available are clusters of galaxies, in which 'luminous arcs' were discovered a few years ago (Lynds & Petrosian 1989, and references therein). Later, less spectacular but much more common 'arclets' were found in many clusters. These are caused by the weak gravitational lensing and are so numerous that they proved to be very useful for the studies of mass distribution (Tyson et al. 1990, and references therein). The 'arcs' and the 'arclets' are the highly distorted images of galaxies which are at a larger redshift than the cluster itself, and so they are easily recognized being much bluer than the cluster galaxies.

The models of cluster mass distributions will be discussed later during this symposium, so I would like to point out only a few obvious results. First, the 'arcs' demonstrate that the density of matter increases strongly towards cluster centers and exceed the critical value needed for strong gravitational lensing. In other words the core radii of the mass distribution are much smaller than believed only a decade ago. Second, the column mass density within a cluster can be measured with the lensing model while the surface brightness can be measured directly. Hence, the value of M/L can be determined with a fairly high accuracy.

Many gravitational lenses are caused by galaxies, which form a double or a quadruple image of a distant quasar. The total mass contained between the images can be measured with a very high precision (Kochanek 1991), and so can be the light of the lensing galaxy, thereby allowing a precise M/L estimate to be made. In some cases a model gives a column mass density along the lines of sight towards the individual images (cf. Kochanek 1994).

An interesting case is the double quasar 0957+561 (Walsh et al. 1979), which has a VLBI radio jet some 40 milli arcseconds long. Two different images of the jet are observed. The matter along the two lines of sight is mostly dark and the millilensing properties are different. Therefore, a detailed comparison of the two radio images offers an opportunity to either detect the presence of $\sim 10^6\ M_\odot$ compact objects (black holes? dark clusters?) or demonstrate that such objects do not exist (Wambsganss & Paczyński 1992, Garrett et al. 1994).

3. GRAVITATIONAL MICROLENSING

A lensing galaxy is made of stars and some dark matter. We know nothing about the form of dark matter, but the stars can be treated as point masses. Each image of a lensed quasar is seen through the lensing galaxy, and in turn is lensed by the stars close to the line of sight, which split each image into a number of sub-images. The sub-images are so close to each other that they cannot be seen separately and the phenomenon is called gravitational microlensing. It can be detected because the relative motion of the lens, the source, and the observer, or the motions within the lens make the combined intensity of all micro-images vary. There are two extreme types of microlensing, corresponding to a very low and large 'optical depth', respectively. In the very low optical depth limit there is either no, or at most one star close to the line of sight, and there are either one or just two bright micro-images. In cass of a large optical depth there are many stars close to the line of sight and many bright micro-images are formed. The properties of lensing in these two regimes are very different.

The optical depth to gravitational microlensing is very low when we look at the stars in our galaxy, or at the stars in the Magellanic Clouds. Along the lines of sight towards these stars there may be other stars, brown dwarfs, planets and black holes acting as gravitational microlenses. The probability that the brightness of a star is magnified by more than ~ 0.3 mag is very low, $\sim (V_{rot}/c)^2 \sim 10^{-6}$, where V_{rot} is the rotational velocity of our galaxy (Paczyński 1986, 1991, Griest 1991, Griest et al. 1991). Three groups are conducting the search for such events: EROS, MACHO, and OGLE, and all three reported the detection of candidate events towards the LMC and/or the galactic bulge (Alcock et al. 1993, 1995, Aubourg et al. 1993, Udalski et al 1993, Udalski et al. 1994a, 1994b, 1994c, Bennett 1994)

This is a field at its infancy, but some reasonably firm conclusions are already available. First, it has been demonstrated that it is possible to detect the very rare events of gravitational microlensing at a rate of a few per year (OGLE) or even a few dozens per year (MACHO). Second, it is possible to detect the events in real time with the OGLE 'early warning system' EWS (Paczyński 1994a, Udalski et al. 1994c), and with the MACHO 'alert system' (Bennett et al. 1994). This is important, as the follow-up observations are possible while the event unfolds. Third, the rate of events towards the galactic bulge turned out to be higher than expected by a factor of 2-4, indicating that our understanding of the galactic structure is inadequate. It follows, that microlensing offers a new way to study the galactic structure. Four, even though the search for microlensing events was undertaken with the dark matter in mind, the results available so far neither prove nor disprove the hypothesis that dark matter is made of massive compact objects (MACHOs). All events detected so far are compatible (within observational and theoretical errors) with the lenses being ordinary stars. My personal conclusion is that MACHOs, if they exist, will be detected within the next few years.

There is one major misconception which invalidates many theoretical studies: the incompleteness of our knowledge about the galactic structure means that currently there is no meaningful way to relate the observed time scales of microlensing events to the lens masses, as it is not clear if the lenses detected towards the galactic bulge are in the galactic disk (Alcock et al. 1995), or are they in the galactic bar (Paczyński et al. 1994), while the lenses detected towards the LMC may be in the disk or the halo of our galaxy, or they may be in the bulge or the halo of the LMC. The inferences about the lens masses will become realistic when the location of lenses is established with the future observations.

There are also misconceptions about the statistical properties expected of microlensing events. As the events are so rare it is commonly believed that they should not repeat. In fact we do know that the stars are commonly double, and the same presumably holds for the lenses. Mao & Paczyński (1991) estimated that $\sim 10\%$ of all events may have a strong signature of a double lens. Indeed, in a small sample of 13 events detected by the OGLE there is one very dramatic case of a double lens, the OGLE #7 (Udalski et al. 1994d, Bennett et al 1994), in which the two lensing stars were separated by approximately one Einstein ring radius. Another possible OGLE double lens was analyzed by Mao & DiStefano (1995). The distribution of binary separations is uniform in the logarithm of separation. Therefore, there should also be sources microlensed by the two components of a binary with the separation exceeding the Einstein ring radius. In this case there should be two separate single microlensing events, separated by a few months or a few years. A similar dual event might be observed if the source is double (Griest 1992, Griest & Hu 1993). I roughly estimate that a few percent of all events should be like that, i.e. they should repeat. If such repeating cases are not found then either binary stars are less common than we think, or the detection criteria discriminate against them.

A very different regime of microlensing and a very different type of variability is expected when the optical depth is modest or even large. This is the case of a quasar lensed by a galaxy at a cosmological distance. The first clear case of such microlensing was reported for 2237+0305, i.e. Huchra's lens (Huchra et al. 1985, Irwin et al. 1989). The most dramatic event was reported by Pen et al. (1994): the luminosity of one of the four quasar images increased by 1.5 mag during a time interval of ~ 2 months, and declined by the same amount during a few days. This is compatible with a theoretical picture in which the magnification pattern is a maze of caustics produced by the stars randomly distributed in the lensing galaxy (Wambsganss et al. 1990). The most common light variation is caused by the source (the quasar) crossing one of the many caustics. The variation on one side is relatively slow, with the time scale proportional to the square root of a typical stellar mass. The variation on the other side would be instantaneous for a point source, and has a finite rate for an extended source, with the duration proportional to the source size. Still, there is a surprise in the Pen et al. (1994) result: it implies the quasar is as small as $\sim 10^{14}$ cm in the continuum light, much smaller than expected in the conventional accretion disk models (Rauch & Blandford 1991, Jaroszyński et al. 1992).

There are also misconceptions about the microlensing in the large optical depth regime. Contrary to common belief it is not possible to relate the observed light variation to a specific star, as the caustics are formed by many stars acting together, gravity being a long range force, Also, the relation between the time scale of the light variability and the microlens masses has not been worked out in any paper published so far. Even though the time scale has to be proportional to a square root of some average microlens mass, the dimensionless coefficient of proportionality is not the same as it is in the optically thin case. Therefore, it is not possible (at this time) to estimate the microlens masses in the Huchra's lens to better than an order of magnitude.

4. GRAVITATIONAL FEMTOLENSING

Gravitational microlensing, as well as macrolensing, operates within the approximation of geometrical optics. Yet, under certain conditions the diffraction effects might be important. It is required that the photon wavelength must be longer than the Schwarzschild radius of the lensing mass, and the source has to be smaller than the Fresnel length. Gould (1992) was the first to point out that this effect may be used to search for dark matter made of objects of $\sim 10^{-15}\ M_{\odot}$, as they might show up as anomalous spectral features in gamma-ray bursts at cosmological distance. Gould also invented the name 'femtolensing' to describe the phenomenon. The theory was farther elaborated by Stanek et al. (1993) and by Ulmer & Goodman (1994). Jaroszyński (1994) found that the diffraction effects can be present even when the photon wavelength is much smaller than the lens Schwarzschil radius, and in fact they limit the resolving power of caustics in Huchra's lens to $\sim 10^{-12}$ seconds of arc. No case of gravitational femtolensing has been detected as yet.

5. THE VALUES OF H_0 , Ω and Λ – GRAVITATIONAL MEGALENSING

The three most important cosmological parameters: H_0, Ω, and Λ may be determined using gravitational lensing effects. When two well separated images of the source can be observed, and the source is variable, then the time delay between the variations as observed in the two images is proportional to the Hubble time, i.e. to H_0^{-1} (Kayser 1993, and references therein). The other parameters may also be estimated using the lens statistics (King 1993, and references therein). However, the classical method of estimating Ω and Λ is based on the redshift - angular diameter, or equivalently redshift - luminosity relation. Fundamentally this is the phenomenon of gravitational lensing by the universe as a whole, and it might be called megalensing. Currently, the best prospect to apply this classical method may be offered by supernovae of Type Ia (Colgate 1979, Branch & Tammann 1992, and references therein, Phillips 1993).

6. CONCLUSIONS

Strong gravitational macrolensing offers the most direct, and perhaps the most accurate determination of the mass of cosmological objects like galaxies and their clusters. Unfortunately, this method is applicable only to the objects with near critical column mass density. The weak gravitational macrolensing can be applied in more general, lower density conditions, but it is also less direct and hence less reliable. In some cases it is possible to measure directly the column mass density within the beam towards the images. Gravitational macrolensing can be used (and it is used) to establish the presence of dark matter of some kind, though so far it does not distinguish between the MACHOs and the WIMPS.

Gravitational microlensing (and femtolensing) can provide the most direct determination of the presence of MACHOs in a very large range of masses, from 10^{-15} $M_\odot$ to 10^6 $M_\odot$ (Nemiroff 1993, and references therein), but so far there is no definite evidence for or against MACHOs in any mass range. However, the recent successes of the MACHO and OGLE teams in detecting microlensing within our galaxy suggest that genuine MACHOs will be detected within a few years.

It is possible that supernovae Type Ia will allow the determination of Ω and Λ using the classical redshift - luminosity relations, which are related to gravitational lensing by all matter within the beam of radiation - gravitational megalensing. The Hubble constant may be measured, at least in principle. with the time delay effect.

Clearly, gravitational lensing is not the only way to study cosmology and to search for dark matter. However, it is one of the most useful and versatile approaches, and in some cases (mass determination) it is the most reliable.

ACKNOWLEDGEMENTS & APOLOGIES

It is a great pleasure to acknowledge that many of the ideas presented in this review are the result of numerous conversations and discussions with many friends, too many to list them all. I apologize for not being more specific. I also apologize for not providing more complete references and for not mentioning many important contributions.

This project was supported with the NSF grant AST 93-13620 and NASA grant NAG5-1901.

REFERENCES

Adams, F. C. et al. 1992, Phys. Rev. D, 47, 426
Alcock, C., et al. 1993, Nature, 365, 621
Alcock, C., et al. 1995, ApJ, in press
Aubourg, E., et al. 1993, Nature, 365, 623
Bennett, D. P. 1994, this conference
Blandford, R. D., & Narayan, R. 1992, AAR&A, 30, 311
Branch, D., & Tammann, G. A. 1992, ARA&A, 30, 359
Bucher, M., Goldhaber, A. S., & Turok, N. 1994, preprint iassns-hep-94-81, PUPT-94-1507
Colgate, S. A. 1979, ApJ, 232, 404
Freeman, K. C. 1970, ApJ, 161, 802
Garrett, M. A. et al. 1994, MNRAS, 270, 457
Griest, K. 1991, ApJ, 366, 412
Griest, K., et al. 1991, ApJ, 372, L79
Griest, K. 1992, ApJ, 397, 362
Griest, K., & Hu, W. 1993, ApJ, 407, 440
Huchra, J., et al 1985, AJ, 90, 691
Irwin, M. J., et al. 1989, AJ, 98, 1989
Jaroszyński, M., Wambsganss, J., & Paczyński, B. 1992, ApJ, 396, L65
Jaroszyński, M. 1994, in preparation
Kamionkowski, M., Ratra, B., Spergel, D. N., & Sugiyama, N. 1994, ApJ, 434, L1
Kayser, R. 1993, in Gravitational Lenses in the Universe, p. 5, Surdej, J. at al. eds., Proc. 31st Liège Coll.
King, P. 1993, in Gravitational Lenses in the Universe, p. 53, Surdej, J. at al. eds., Proc. 31st Liège Coll.
Kochanek, C. S. 1991, ApJ, 373, 354
Kochanek, C. S. 1991, ApJ, submitted = Harvard - Smithsonian CfA Preprint No. 3933
Lynds, R., & Petrosian, V. 1989, ApJ, 336, 1
Mao, S., & Paczyński, B. 1991, ApJ, 374, L37
Mao, S., & DiStefano, R. 1995, ApJ, in press
Nemiroff, 1993, in Gravitational Lenses in the Universe, p. 53, Surdej, J. at al. eds., Proc. 31st Liège Coll.
Oort, J. H. 1932, BAN, 6, 249
Paczyński, B. 1986, ApJ, 304, 1
Paczyński, B. 1991, ApJ, 371, L63
Paczyński, B. 1994a, IAU Circ. No. 5997
Paczyński, B., et al. 1994, ApJ, 435, L113
Pen, Ue-Li, et al. 1993, in Gravitational Lenses in the Universe, p. 111, Surdej, J. at al. eds., Proc. 31st Liège Coll.
Phillips, M. M. 1993, ApJ, 413, L105
Rauch, K. P., & Blandford, R. D. 1991, ApJ, 381, L39

Refsdal, S. & Surdej, J. 1994, Rep. Prog. Phys., 56, 117
Schneider, P., Ehlers, J., & Falco, E. E. 1992, Gravitational Lenses (Springer Verlag)
Stanek, K. Z., Paczyński, B., & Goodman, J. 1993, ApJ, 413, L7
Surdej, J. at al. eds. 1993, Gravitational Lenses in the Universe, Proc. 31st Liège Coll.
Taylor, J. H., & Weisberg, J. M. 1989, ApJ, 345, 434
Tyson, J. A., Valdes, F., & Wenk, R. A. 1990, ApJ, 349, L1
Udalski, A., et al. 1994a, ApJ, 426, L69
Udalski, A., et al. 1994b, Acta Astron., 44, 165
Udalski, A., et al. 1994c, Acta Astron., 44, 227
Udalski, A., et al. 1994d, ApJ, 436, L103
Ulmer, A., & Goodman, J. 1994, ApJ, submitted = Princeton Observatory Preprint No. 569
Walsh, D., Carswell, R. F., & Weyman, R. J. 1979, Nature, 279, 381
Wambsganss, J., & Paczyński, B. 1992, ApJ, 397, L1
Wambsganss, J., Paczyński, B., & Schneider, P. 1990, ApJ, 358, L33
Zwicky, Z. 1933, Helv. Phys. Acta, 6, 110

IASSNS-HEP-94/112
October 1994

From the Standard Model to Dark Matter*

FRANK WILCZEK†

School of Natural Sciences
Institute for Advanced Study
Olden Lane
Princeton, N.J. 08540

ABSTRACT

The standard model of particle physics is marvelously successful. However, it is obviously not a complete or final theory. I shall argue here that the structure of the standard model gives some quite concrete, compelling hints regarding what lies beyond. Taking these hints seriously, one is led to predict the existence of new types of very weakly interacting matter, stable on cosmological time scales and produced with cosmologically interesting densities–that is, "dark matter".

* Invited talk at 5th Annual October Maryland Astrophysics Conference

† Research supported in part by DOE grant DE-FG02-90ER40542. WILCZEK@IASSNS.BITNET

I have been asked to discuss particle physics candidates for dark matter. There are many ways one could go with such an assignment. I have made a very particular choice [1]. I will discuss the converging lines of thought and evidence leading from the consolidation and recent precision testing of the standard model to compelling ideas about unification of forces and the possibility of low-energy supersymmetry. These ideas produce as an important bonus a very attractive dark-matter candidate, the lightest supersymmetric particle – specified more precisely, using a concept to be defined below, as the lightest particle with odd R-parity.

Particle physics provides several other interesting and attractive dark matter candidates, notably including axions and massive neutrinos. It is entirely possible that one or both of these species provides a significant component of the mass density of the Universe. It is important vigorously to pursue experimental programs to detect each of them. Given the available time I had to make a choice, however. One pedagogical argument for presenting the supersymmetric option to an audience outside particle physics is that it is organically linked to the absolutely central theoretical ideas for going beyond the standard model, so I'll be able to lead into supersymmetric dark matter through presentation of these central ideas. Axions and neutrino masses are tied to important, but more peripheral ideas.

Critique of the Standard Model

The standard model of particle physics is based upon the gauge groups SU(3)×SU(2)×U(1) of strong, electromagnetic and weak interactions acting on the quark and lepton multiplets as shown in Figure 1.

In this Figure I have depicted only one family (u,d,e,ν_e) of quarks and leptons; in reality there seem to be three families which are mere copies of one another as far as their interactions with the gauge bosons are concerned, but differ in mass. Actually in the Figure I have ignored masses altogether, and allowed myself the convenient fiction of pretending that the quarks and leptons have a definite chirality – right- or left-handed – as they would if they were massless. (The more

three colors, and the weak by transformations between two others, what could be more natural than to embed both theories into a larger theory of transformations among all five colors? This idea has the additional attraction that an extra U(1) symmetry commuting with the strong SU(3) and weak SU(2) symmetries automatically appears, which we can attempt to identify with the remaining gauge symmetry of the standard model, that is hypercharge. For while in the separate SU(3) and SU(2) theories we must throw out the two gauge bosons which couple respectively to the color combinations R+W+B and G+P, in the SU(5) theory we only project out R+W+B+G+P, while the orthogonal combination (R+W+B)-$\frac{3}{2}$(G+P) remains.

Georgi and Glashow [4] originated this line of thought, and showed how it could be used to bring some order to the quark and lepton sector, and in particular to supply a satisfying explanation of the weird hypercharge assignments in the standard model. As shown in Figure 2, the five scattered SU(3)×SU(2)×U(1) multiplets get organized into just two representations of SU(5).

In making this unification it is necessary to allow transformations between (what were previously thought to be) particles and antiparticles of the same chirality, and also between quarks and leptons. It is convenient to work with left-handed fields only; since the conjugate of a right-handed field is left-handed, we don't lose track of anything by doing so, once we disabuse ourselves of the idea that a given field is intrinsically either genuine or "anti".

As shown in Figure 2, there is one group of ten left-handed fermions that have all possible combinations of one unit of each of two different colors, and another group of five left-handed fermions that each carry just one negative unit of some color. (These are the ten-dimensional antisymmetric tensor and the complex conjugate of the five-dimensional vector representation, commonly referred to as the "five-bar".) What is important for you to take away from this discussion is not so much the precise details of the scheme, but the idea that *the structure of the standard model, with the particle assignments gleaned from decades of experimental effort and theoretical interpretation, is perfectly reproduced by a simple abstract set of rules for manipulating symmetrical symbols.* Thus for example the object RB in this Figure has just the strong, electromagnetic, and weak interactions we expect of the complex conjugate of the right-handed up-quark, without our having to instruct the theory further. If you've never done it I heartily recommend to you the simple exercise of working out the hypercharges of the objects in Figure 2 and checking against what you need in the standard model – after doing it, you'll find it's impossible ever to look at the standard model in quite the same way again.

SU(5): 5 colors RWBGP

$\underline{10}$: 2 different color labels (antisymmetric tensor)

$$\begin{array}{ll} u_L: & RP,\ \ WP,\ \ BP \\ d_L: & RG,\ \ WG,\ \ BG \\ u^c_L: & RW,\ \ WB,\ \ BR \\ & (\bar{B})\ \ (\bar{R})\ \ (\bar{W}) \\ e^c_L: & GP \\ & (\ ) \end{array} \qquad \begin{pmatrix} 0 & u^c & u^c & u & d \\ & 0 & u^c & u & d \\ & & 0 & u & d \\ & * & & 0 & e \\ & & & & 0 \end{pmatrix}$$

$\underline{\bar{5}}$: 1 anticolor label

$$\begin{array}{ll} d^c_L: & \bar{R},\ \ \bar{W},\ \ \bar{B} \\ e_L: & \bar{P} \\ \nu_L: & \bar{G} \end{array} \qquad (d^c \ \ d^c \ \ d^c \ \ e \ \ \nu)$$

$$\boxed{Y = -\tfrac{1}{3}\,(R+W+B) + \tfrac{1}{2}\,(G+P)}$$

FIGURE 2

Figure 2 - Unification of fermions in SU(5).

There is a beautiful extension of SU(5) to a slightly larger group, SO(10), which permits one to unite all the fermions of a family into a single multiplet. The relevant representations for the fermions is a 16-dimensional spinor representation. Some of its features are depicted in Figure 3, as I shall now explain.

$$(\pm\pm\pm\pm\pm) \;:\; \underline{\text{even}}\ \#\ of\ -$$

$$\begin{array}{rlcc}
 & (++-|+-) & 6 & (\mathrm{u_L}, \mathrm{d_L}) \\
10: & (+--|++) & 3 & \mathrm{u_L^c} \\
 & (+++|--) & 1 & \mathrm{e_L^c} \\
 & & & \\
\bar{5}: & (+--|--) & \bar{3} & \mathrm{d_L^c} \\
 & (---|+-) & \bar{2} & (\mathrm{e_L}, \nu_L) \\
 & & & \\
1: & (+++|++) & 1 & \mathrm{N_R}
\end{array}$$

FIGURE 3

Figure 3 - Unification of fermions in SO(10). The rule is that all possible combinations of 5 + and - signs occur, subject to the constraint that the total number of - signs is even. The SU(5) gauge bosons within SO(10) do not change the numbers of signs, and one sees the SU(5) multiplets emerging. However there are additional transformations in SO(10) but not in SU(5), which allow any fermion to be transformed into any other. Permutations of signs within the first three slots or within the last three slots are not indicated. The numbers in the left-hand column indicates the SU(5) quantum multiplets – to be compared with Figure 2; the numbers in the third column indicates the multiplicity of standard model multiplets – to be compared with Figure 1.

Spinor representations are most easily constructed iteratively. To construct spinors for rotations in n-dimensional space, one needs an algebra of γ-matrices obeying the anticommutation relations

$$\{\gamma_i^{(n)}, \gamma_j^{(n)}\} = 2\delta_{ij} \tag{1}$$

for $i, j \leq n$. For $n = 2$ one can use the first two Pauli matrices: $\gamma_1^{(2)} = \tau_1 = \begin{pmatrix} 0 & 1 \\ 1 & 0 \end{pmatrix}$, $\gamma_2^{(2)} = \tau_2 = \begin{pmatrix} 0 & -i \\ i & 0 \end{pmatrix}$. Then to move from $n-2$ to n dimensions one uses the set

$$\gamma_i^{(n)} = \gamma_i^{(n-2)} \otimes \tau_3 \tag{2}$$

for $i \leq n-2$ and

$$\begin{aligned} \gamma_{n-1}^{(n)} &= 1 \otimes \tau_1 \\ \gamma_n^{(n)} &= 1 \otimes \tau_2 \ , \end{aligned} \tag{3}$$

where of course in (3) 1 denotes the $2^{\frac{n}{2}} \times 2^{\frac{n}{2}}$ dimensional unit matrix. One can easily unfold the $\gamma^{(n)}$ into n-fold tensor products, or alternatively write down explicit ordinary matrix representations. Thus for example

$$\gamma_1^{(4)} = \begin{pmatrix} 0 & 1 & 0 & 0 \\ 1 & 0 & 0 & 0 \\ 0 & 0 & 0 & -1 \\ 0 & 0 & -1 & 0 \end{pmatrix} \qquad \gamma_2^{(4)} = \begin{pmatrix} 0 & -i & 0 & 0 \\ i & 0 & 0 & 0 \\ 0 & 0 & 0 & i \\ 0 & 0 & -i & 0 \end{pmatrix}$$

$$\gamma_3^{(4)} = \begin{pmatrix} 0 & 0 & 1 & 0 \\ 0 & 0 & 0 & 1 \\ 1 & 0 & 0 & 0 \\ 0 & 1 & 0 & 0 \end{pmatrix} \qquad \gamma_4^{(4)} = \begin{pmatrix} 0 & 0 & -i & 0 \\ 0 & 0 & 0 & -i \\ i & 0 & 0 & 0 \\ 0 & i & 0 & 0 \end{pmatrix} \tag{4}$$

and

$$\gamma_1^{(6)} = \begin{pmatrix} 0 & 1 & 0 & 0 & 0 & 0 & 0 & 0 \\ 1 & 0 & 0 & 0 & 0 & 0 & 0 & 0 \\ 0 & 0 & 0 & -1 & 0 & 0 & 0 & 0 \\ 0 & 0 & -1 & 0 & 0 & 0 & 0 & 0 \\ 0 & 0 & 0 & 0 & 0 & -1 & 0 & 0 \\ 0 & 0 & 0 & 0 & -1 & 0 & 0 & 0 \\ 0 & 0 & 0 & 0 & 0 & 0 & 0 & 1 \\ 0 & 0 & 0 & 0 & 0 & 0 & 1 & 0 \end{pmatrix} \quad \gamma_2^{(6)} = \begin{pmatrix} 0 & -i & 0 & 0 & 0 & 0 & 0 & 0 \\ i & 0 & 0 & 0 & 0 & 0 & 0 & 0 \\ 0 & 0 & 0 & i & 0 & 0 & 0 & 0 \\ 0 & 0 & -i & 0 & 0 & 0 & 0 & 0 \\ 0 & 0 & 0 & 0 & 0 & i & 0 & 0 \\ 0 & 0 & 0 & 0 & -i & 0 & 0 & 0 \\ 0 & 0 & 0 & 0 & 0 & 0 & 0 & -i \\ 0 & 0 & 0 & 0 & 0 & 0 & i & 0 \end{pmatrix}$$

$$\gamma_3^{(6)} = \begin{pmatrix} 0 & 0 & 1 & 0 & 0 & 0 & 0 & 0 \\ 0 & 0 & 0 & 1 & 0 & 0 & 0 & 0 \\ 1 & 0 & 0 & 0 & 0 & 0 & 0 & 0 \\ 0 & 1 & 0 & 0 & 0 & 0 & 0 & 0 \\ 0 & 0 & 0 & 0 & 0 & 0 & -1 & 0 \\ 0 & 0 & 0 & 0 & 0 & 0 & 0 & -1 \\ 0 & 0 & 0 & 0 & -1 & 0 & 0 & 0 \\ 0 & 0 & 0 & 0 & 0 & -1 & 0 & 0 \end{pmatrix} \quad \gamma_4^{(6)} = \begin{pmatrix} 0 & 0 & -i & 0 & 0 & 0 & 0 & 0 \\ 0 & 0 & 0 & -i & 0 & 0 & 0 & 0 \\ i & 0 & 0 & 0 & 0 & 0 & 0 & 0 \\ 0 & i & 0 & 0 & 0 & 0 & 0 & 0 \\ 0 & 0 & 0 & 0 & 0 & 0 & i & 0 \\ 0 & 0 & 0 & 0 & 0 & 0 & 0 & i \\ 0 & 0 & 0 & 0 & -i & 0 & 0 & 0 \\ 0 & 0 & 0 & 0 & 0 & -i & 0 & 0 \end{pmatrix}$$

$$\gamma_5^{(6)} = \begin{pmatrix} 0 & 0 & 0 & 0 & 1 & 0 & 0 & 0 \\ 0 & 0 & 0 & 0 & 0 & 1 & 0 & 0 \\ 0 & 0 & 0 & 0 & 0 & 0 & 1 & 0 \\ 0 & 0 & 0 & 0 & 0 & 0 & 0 & 1 \\ 1 & 0 & 0 & 0 & 0 & 0 & 0 & 0 \\ 0 & 1 & 0 & 0 & 0 & 0 & 0 & 0 \\ 0 & 0 & 1 & 0 & 0 & 0 & 0 & 0 \\ 0 & 0 & 0 & 1 & 0 & 0 & 0 & 0 \end{pmatrix} \quad \gamma_6^{(6)} = \begin{pmatrix} 0 & 0 & 0 & 0 & -i & 0 & 0 & 0 \\ 0 & 0 & 0 & 0 & 0 & -i & 0 & 0 \\ 0 & 0 & 0 & 0 & 0 & 0 & -i & 0 \\ 0 & 0 & 0 & 0 & 0 & 0 & 0 & -i \\ i & 0 & 0 & 0 & 0 & 0 & 0 & 0 \\ 0 & i & 0 & 0 & 0 & 0 & 0 & 0 \\ 0 & 0 & i & 0 & 0 & 0 & 0 & 0 \\ 0 & 0 & 0 & i & 0 & 0 & 0 & 0 \end{pmatrix} \tag{5}$$

Clearly the dimension of the $\gamma^{(n)}$ matrices is doubling each time n increases by two. (We will not need to be concerned with odd n here, but of course if n is odd we can satisfy (1) by using the first n $\gamma^{(n+1)}$s.)

Now it is very easy to verify from (1) that the matrices

$$\sigma_{ij}^{(n)} \equiv \frac{-i}{4}[\gamma_i^{(n)}, \gamma_j^{(n)}]$$

satisfy the commutation relations

$$i[\sigma_{ij}^{(n)}, \sigma_{kl}^{(n)}] = \delta_{ik}\sigma_{jl}^{(n)} - \delta_{il}\sigma_{jk}^{(n)} . \tag{6}$$

These are the same commutation relations one has for the infinitesimal rotations around the ij and kl axes. Thus we have represented the Lie algebra of rotations in n dimensions by a set of $2^{\frac{n}{2}} \times 2^{\frac{n}{2}}$ matrices; that is, we have a $2^{\frac{n}{2}}$ dimensional representation of $SO(n)$. This representation is not quite irreducible, however. One easily verifies that the product

$$\Gamma^{(n)} = (-i)^{\frac{n}{2}}\gamma_1^{(n)}\gamma_2^{(n)}\cdots\gamma_n^{(n)} \tag{7}$$

anticommutes with all the $\gamma_i^{(n)}$, and therefore commutes with all the $\sigma_{ij}^{(n)}$. Indeed, in the specific representation we have chosen $\Gamma^{(n)}$ unfolds into the n-fold tensor product

$$\Gamma^{(n)} = \tau_3 \otimes \tau_3 \cdots \otimes \tau_3 . \tag{8}$$

We also have

$$(\Gamma^{(n)})^2 = 1 . \tag{9}$$

Thus by projecting onto the eigenspaces of $\Gamma^{(n)}$, using the projection operators $(1 \pm \Gamma^{(n)})/2$, we find two representations of dimension $2^{\frac{n}{2}-1}$. These representations turn out to be irreducible. The representation we shall be most interested in is the 16 dimensional representation of SO(10).

Now at last we are in position to make contact with the notations of Figure 3. One can label states in the spinor representation, as we have constructed it, very conveniently in terms of the eigenvalues of the operators

$$\begin{aligned} 2\sigma_{12}^{(10)} &= \tau_3 \otimes 1 \otimes 1 \otimes 1 \otimes 1 \\ 2\sigma_{34}^{(10)} &= 1 \otimes \tau_3 \otimes 1 \otimes 1 \otimes 1 \\ &\cdots . \end{aligned} \tag{10}$$

Indeed, these form a complete set of commuting observables, so their simultaneous eigenvalues can be used to label the states. In this way, we see that each state in the spinor representation, as we have constructed it, corresponds to an ordered choice of 5 $\pm$ signs. Projection with $(1 + \Gamma^{(10)})/2$ corresponds to imposing the constraint that the product of these signs is positive, *i.e.* that there should be an even number of - signs. Thus we arrive at the 16 states depicted in Figure 3.

For the application to physics it is important to identify explicitly the quantum numbers of the standard model within the abstract realization of the postulated unified gauge symmetry group. How does the symmetry of the standard model sit within $SO(10)$, and how do particles in the spinor 16 representation transform under it? Well, the group $SU(n)$ forms a subgroup of $SO(2n)$ in a very canonical way. Indeed, $SU(n)$ is the group that preserves a Hermitean inner product between vectors in a n-complex dimensional vector space. But the real part of this inner product is just the ordinary real inner product of the 2n-real dimensional vectors formed from the real and imaginary parts of n-complex dimensional vectors, so $SU(n)$ is the subgroup of $SO(2n)$, which leaves the imaginary part as well as the real part of the inner product invariant.

Thus the $SU(3) \times SU(2)$ part of the standard model is easy to locate within $SO(6) \times SO(4)$ inside $SO(10)$. If we declare that the components 2,4,6,8,10 are the imaginary parts of the complex vectors whose real parts are the components 1,3,5,7,9 respectively, then $SU(3)$ will naturally act on the first six components – that is, it will consist of suitable combinations of the $\sigma_{ij}^{(10)}$ with $i, j \leq 6$; and $SU(2)$ will consist of suitable combinations of the $\sigma_{ij}^{(10)}$ with $7 \leq i, j$.

With this way of organizing things, it becomes easy to identify the physical meaning of the operators in (10) and the $\pm$ signs used to label the states. Indeed, each operator in (10) becomes the generator of an infinitesimal *phase rotation* of one of the five complex components of vectors in $SU(5)$. The first three of these correspond to generators of QCD color charges, say red, white, and blue. It is natural to call the final two the generators weak color charges, say green and purple. The generators

$$\Sigma_{\mathrm{QCD}} = \sigma_{12}^{(10)} + \sigma_{34}^{(10)} + \sigma_{56}^{(10)} \tag{11}$$

and

$$\Sigma_{\mathrm{weak}} = \sigma_{78}^{(10)} + \sigma_{9,10}^{(10)} \tag{12}$$

represent phase rotations that commute with $SU(3)\times SU(2)$. Indeed $\Sigma_{\mathrm{QCD}}+\Sigma_{\mathrm{weak}}$ generates a common phase rotation of all 5 complex vector components. It is not properly in $SU(5)$ at all; that's the difference between $SU(5)$ and just $U(5)$. The hypercharge generator in $SU(5)$ is proportional to the 'traceless' combination $2\Sigma_{\mathrm{QCD}} - 3\Sigma_{\mathrm{weak}}$.

By the way, $U(5)$ can be located within $SO(10)$ by the condition that $U(5)$ generators are the combinations of $SO(10)$ generators that commute with $\Sigma_{\mathrm{QCD}} + \Sigma_{\mathrm{weak}} \equiv J$. Indeed J implements, as we have seen, an overall phase rotation. The condition that a linear transformation which already leaves the real part of the inner product invariant should also leave the imaginary part invariant is exactly that it respect such a phase rotation.

All the concepts used in constructing Figure 3 have now been spelled out, and at this point it ought to be a pleasant exercise for you to verify that each "particle" constructed in this abstract mathematical way has just the quantum numbers to be identified with one of the fundamental fermions (with one interesting exception, as we shall discuss immediately below). In this accounting, one represents each fermion using a left-handed chiral field, by taking charge conjugates if necessary. Thus if one wishes to find the right-handed up quark, for instance, one should look for the left-handed anti-up antiquark. For an example, consider $(++-+-)$. It forms part of a QCD triplet together with $(+-++-)$ and $(-+++-)$, and part of a weak doublet together with $(++--+)$. $(+++--)$, on the contrary, is a singlet both for QCD and for weak interactions. The hypercharges of these two objects are in the ratio 2:12. These are exactly the quantum numbers one wants for $(++-+-)$ to be a component of a left-handed quark field, and $(+++--)$ the charge conjugate of e_R, the right-handed component of the electron.

Proceeding along these lines, one finds an uncanny fit between the abstract quantum numbers of $SO(10)$ spinors and the ones observed for particles in the real world: the charge spectrum, and all the strong, electromagnetic, and weak interactions of the standard model, are incorporated. There is also one extra state in the model, however, namely $(+++++)$. This state is a singlet under all the interactions of the standard model. Thus it is not surprising that the particle corresponding to this state would escape easy detection, even were it to exist. One can make very good use of this state in constructing models of massive neutrinos, but I promised not to get into that subject.

What I have shown you here is very old and standard mathematics. Whatever small novelty there is, is in the packaging. That said, I must admit that I find this way of presenting things, culminating in the labeling of fermion states by ordered bits as displayed in Figure 3, very appealing and seductive. I beg indulgence to mention two sorts of fantasies it suggests. First, of course, one might speculate that there are additional colors, thus being led to $SO(10+x)$ theories. This large symmetry group can be put to use in attempting to address the question why there are multiple families of fermions with identical $SO(10)$ (or at least standard model) quantum numbers. It is a remarkable fact, quite transparent in our construction of the spinor representation, that a spinor of $SO(10+n)$ breaks up into several spinors (and an equal number of antispinors) under $SO(10)$, with no other representations appearing. For some adventures in trying to exploit these ideas see [6] . Alternatively, one might speculate that the representation of particles as bit-structures is a profound feature of the physical world, conceivably more fundamental than any particular gauge group or even than gauge theory altogether. A relatively modest (but conversely, relatively concrete) idea in this vein is that the different particles should be described as some sort of soliton or magnetic monopole, and the various signs indicate occupation (or not) of a set of zero modes.

Unification: coupling values

We have seen that simple unification schemes are successful at the level of *classification*; but new questions arise when we consider the dynamics which underlies them.

Part of the power of gauge symmetry is that it fully dictates the interactions of the gauge bosons, once an overall coupling constant is specified. Thus if SU(5) or some higher symmetry were exact, then the fundamental strengths of the different color-changing interactions would have to be equal, as would the (properly normalized) hypercharge coupling strength. In reality the coupling strengths of the gauge bosons in SU(3)×SU(2)×U(1) are not observed to be equal, but rather follow the pattern $g_3 \gg g_2 > g_1$.

Fortunately, experience with QCD emphasizes that couplings "run". The physical mechanism of this effect is that in quantum field theory the vacuum must be regarded as a polarizable medium, since virtual particle-anti-particle pairs can screen charge. Thus one might expect that effective charges measured at shorter distances, or equivalently at larger energy-momentum or mass scales, could be different from what they appear at longer distances. If one had only screening then the effective couplings would grow at shorter distances, as one penetrated deeper insider the screening cloud. However it is a famous fact [7] that due to paramagnetic spin-spin attraction of like charge vector gluons [8], these particles tend to *antiscreen* color charge, thus giving rise to the opposite effect – asymptotic freedom – that the effective coupling tends to shrink at short distances. This effect is the basis of all perturbative QCD phenomenology, which is a vast and vastly successful enterprise. For our present purpose of understanding the disparity of the observed couplings, it is just what the doctor ordered. As was first pointed out by Georgi,

Quinn, and Weinberg [9], if a gauge symmetry such as SU(5) is spontaneously broken at some very short distance then we should not expect that the effective couplings probed at much larger distances, such as are actually measured at practical accelerators, will be equal. Rather they will all have have been affected to a greater or lesser extent by vacuum screening and anti-screening, starting from a common value at the unification scale but then diverging from one another. The pattern $g_3 \gg g_2 > g_1$ is just what one should expect, since the antiscreening or asymptotic freedom effect is more pronounced for larger gauge groups, which have more types of virtual gluons.

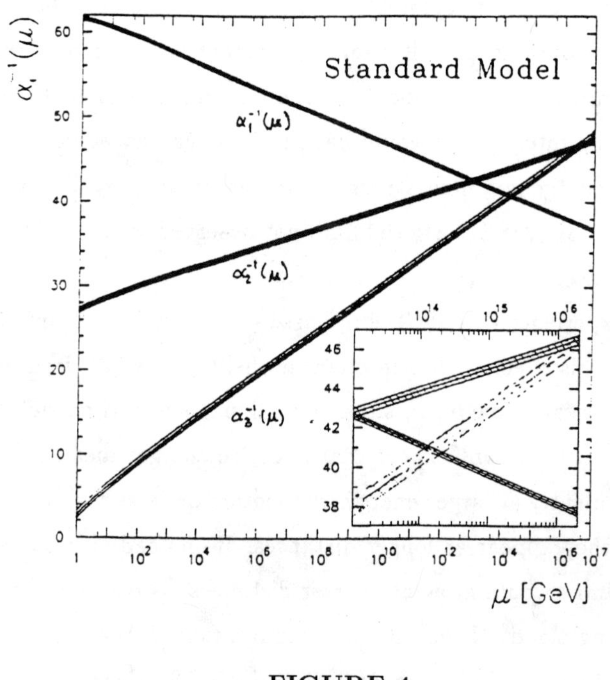

FIGURE 4

Figure 4 - The failure of the running couplings, normalized according to SU(5) and extrapolated taking into account only the virtual exchange of the "known" particles of the standard model (including the top quark and Higgs boson) to meet. Note that only with quite recent experiments, which greatly improved the precision of the determination of low-energy couplings, did the discrepancy become significant.

The marvelous thing is that the running of the couplings gives us a truly quantitative handle on the ideas of unification, for the following reason. To fix the relevant aspects of unification, one basically needs only to fix two parameters: the scale at which the couplings unite, which is essentially the scale at which the unified symmetry breaks; and their value when then unite. Given these, one calculates three outputs: the three *a priori* independent couplings for the gauge groups in SU(3)×SU(2)×U(1). Thus the framework is eminently falsifiable. The miraculous thing is, how close it comes to working (Figure 4).

The unification of couplings occurs at a very large mass scale, $M_{\rm un.} \sim 10^{15}$ Gev. In the simplest version, this is the magnitude of the scalar field vacuum expectation value that spontaneously breaks SU(5) down to the standard model symmetry SU(3)×SU(2)×U(1), and is analogous to the scale $v \approx 250$ Gev for electroweak symmetry breaking. The largeness of this large scale mass scale is important in several ways.

• It explains why the exchange of gauge bosons that are in SU(5) but not in SU(3)×SU(2)×U(1), which reshuffles strong into weak colors and generically violates the conservation of baryon number, does not lead to a catastrophically quick decay of nucleons. The rate of decay goes as the inverse fourth power of the mass of the exchanged gauge particle, so the baryon-number violating processes are predicted to be far slower than ordinary weak processes, as they had better be.

• $M_{\rm un.}$ is significantly smaller than the Planck scale $M_{\rm Planck} \sim 10^{19}$ Gev at which exchange of gravitons competes quantitatively with the other interactions, but not ridiculously so. This indicates that while the unification of couplings calculation itself is probably safe from gravitational corrections, the unavoidable logical next step in unification must be to bring gravity into the mix.

• Finally one must ask how the tiny ratio of symmetry-breaking mass scales $v/M_{\rm un.} \sim 10^{-13}$ required arises dynamically, and whether it is stable. This is the so-called gauge hierarchy problem, which we shall discuss in a more concrete form a little later.

The success of the GQW calculation in explaining the observed hierarchy $g_3 \gg$

$g_2 > g_1$ of couplings and the approximate stability of the proton is quite striking. In performing it, we assumed that the known and confidently expected particles of the standard model exhaust the spectrum up to the unification scale, and that the rules of quantum field theory could be extrapolated without alteration up to this mass scale – thirteen orders of magnitude beyond the domain they were designed to describe. It is a triumph for minimalism, both existential and conceptual.

However, on further examination it is not quite good enough. Accurate modern measurements of the couplings show a small but definite discrepancy between the couplings, as appears in Figure 4. And heroic dedicated experiments to search for proton decay did not find it [10]; they currently exclude the minimal SU(5) prediction $\tau_p \sim 10^{31}$ yrs. by about two orders of magnitude.

Given the magnitude of the extrapolation involved, perhaps we should not have hoped for more. There are several perfectly plausible bits of physics that could upset the calculation, such as the existence of particles with masses much higher than the electroweak but much smaller than the unification scale. As virtual particles these would affect the running of the couplings, and yet one certainly cannot exclude their existence on direct experimental grounds. If we just add particles in some haphazard way things will only get worse: minimal SU(5) nearly works, so the generic perturbation from it will be deleterious. This is a major difficulty for so-called technicolor models, which postulate many new particles in complex patterns. Even if some *ad hoc* prescription could be made to work, that would be a disappointing outcome from what appeared to be one of our most precious, elegantly straightforward clues regarding physics well beyond the standard model.

Virtual supersymmetry?

Fortunately, there is a theoretical idea which is attractive in many other ways, and seems to point a way out from this impasse. That is the idea of supersymmetry [11]. Supersymmetry is a symmetry that extends the Poincare symmetry of special relativity (there is also a general relativistic version). In a supersymmetric theory one has not only transformations among particle states with different energy-momentum but also between particle states of different *spin*. Thus spin 0 particles can be put in multiplets together with spin $\frac{1}{2}$ particles, or spin $\frac{1}{2}$ with spin 1, and so forth.

Supersymmetry is certainly not a symmetry in nature: for example, there is certainly no bosonic particle with the mass and charge of the electron. More generally if one defines the R-parity quantum number

$$R \equiv (-)^{3B+L+2S} ,$$

which should be accurate to the extent that baryon and lepton number are conserved, then one finds that all currently known particles are R even whereas their supersymmetric partners would be R odd. Nevertheless there are many reasons to be interested in supersymmetry, of which I shall mention three.

- You will notice that we have made progress in uniting the gauge bosons with each other, and the various quarks and leptons with each other, but not the gauge bosons with the quarks and leptons. It takes supersymmetry – perhaps spontaneously broken – to make this feasible.

- Supersymmetry was invented in the context of string theory, and seems to be necessary for constructing consistent string theories containing gravity (critical string theories) that are at all realistic.

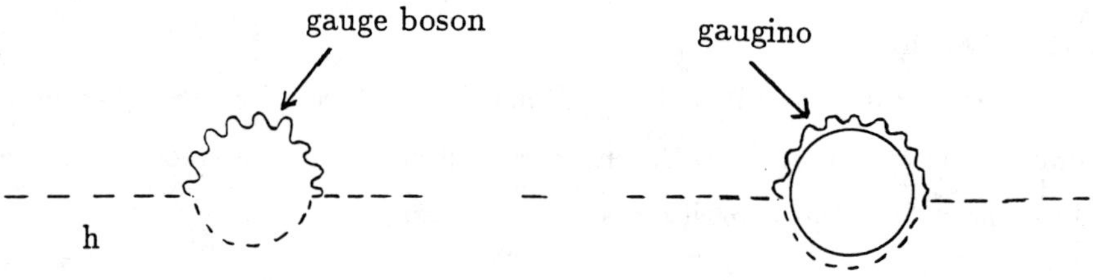

FIGURE 5

Figure 5 - A typical quadratically divergent contribution to the (mass)^2 of the Higgs boson, and the supersymmetric contribution which, as long as supersymmetry is not too badly broken, will largely cancel it.

• Most important for our purposes, supersymmetry can help us to understand the vast disparity between weak and unified symmetry breaking scales mentioned above. This disparity is known as the gauge hierarchy problem. It actually raises several distinct problems, including the following. In calculating radiative corrections to the (mass)^2 of the Higgs particle from diagrams of the type shown in Figure 5 one finds an infinite, and also large, contribution. By this I mean that the divergence is quadratic in the ultraviolet cutoff. No ordinary symmetry will make its coefficient vanish. If we imagine that the unification scale provides the cutoff, we find that the radiative correction to the (mass)^2 is much larger than the final value we want. (If the Higgs field were composite, with a soft form factor, this problem might be ameliorated. Following that road leads to technicolor, which as mentioned before seems to lead us far away from our best source of inspiration.) As a formal matter one can simply cancel the radiative correction against a large bare contribution of the opposite sign, but in the absence of some deeper motivating principle this seems to be a horribly ugly procedure. Now in a supersymmetric theory for any set of virtual particles circulating in the loop there will also be another graph with their supersymmetric partners circulating. If the partners were accurately degenerate, the contributions would cancel. Otherwise, the threatened quadratic divergence will be cut off only at virtual momenta such that the difference in $(\text{mass})^2$ between the virtual particle and its supersymmetric partner is negligible. Thus we will be assured adequate cancelation if and only if supersymmetric partners are not too far split in mass – in the present context, if the splitting is not much greater than the weak scale. This is (a crude version of) the most important *quantitative* argument which suggests the relevance of "low-energy" supersymmetry.

The effect of low-energy supersymmetry on the running of the couplings was first considered long ago [12], well before the crisis described at the end of the previous section was evident. One might fear that such a huge expansion of the theory, which essentially doubles the spectrum, would utterly destroy the approximate success of the minimal SU(5) calculation. This is not true, however. To a first approximation since supersymmetry is a space-time rather than an internal symmetry it does not affect the group-theoretic structure of the calculation. Thus to a first approximation the absolute rate at which the couplings run with momentum is affected, but not the relative rates. The main effect is that the supersymmetric partners of the color gluons, the gluinos, weaken the asymptotic freedom of the strong interaction. Thus they tend to make its effective coupling decrease and approach the others more slowly. Thus their merger requires a longer lever arm, and the scale at which the couplings meet increases by an order of magnitude or so, to about 10^{16} Gev. Also the common value of the effective couplings at unification is slightly larger than in conventional unification ($\frac{g^2_{\text{un.}}}{4\pi} \approx \frac{1}{25}$ *versus* $\frac{1}{40}$). This increase in unification scale significantly reduces the predicted rate for proton decay through exchange of the dangerous color-changing gauge bosons, so that it no longer conflicts with existing experimental limits.

Upon more careful examination there is another effect of low-energy supersymmetry on the running of the couplings, which although quantitatively small has become of prime interest. There is an important exception to the general rule that adding supersymmetric partners does not immediately (at the one loop level) affect the relative rates at which the couplings run. This rule works for particles that come in complete SU(5) multiplets, such as the quarks and leptons (which, since they don't upset the full SU(5) symmetry, have basically no effect) or for the supersymmetric partners of the gauge bosons, because they just renormalize the existing, dominant effect of the gauge bosons themselves. However there is one peculiar additional contribution, from the supersymmetric partner of the Higgs doublet. It affects only the weak SU(2) and hypercharge U(1) couplings. (On phenomenological grounds the SU(5) color triplet partner of the Higgs doublet must be extremely massive, so its virtual exchange is not important below the unification scale. *Why* that should be so, is another aspect of the hierarchy problem.) Moreover, for slightly technical reasons even in the minimal supersymmetric model it is necessary to have two different Higgs doublets with opposite hypercharges. The net affect of doubling the number of Higgs fields and including their supersymmetric partners is a sixfold enhancement of the asymmetric Higgs field contribution to the running of weak and hypercharge couplings. This causes a small, accurately calculable change in the calculation. From Figure 6 you see that it is a most welcome one. Indeed, in the minimal implementation of supersymmetric unification, it puts the running of couplings calculation right back on the money [13].

Since the running of the couplings with scale is logarithmic the unification of couplings calculation is not terribly sensitive to the exact scale at which supersymmetry is broken, say between 100 Gev and 10 Tev. There have been attempts to push the calculation further, in order to address this question of the supersymmetry breaking scale, but they are controversial. It is not obvious to me that such calculations will ever achieve the resolution of interest.

For example, comparable uncertainties arise from the splittings among the very large number of particles with masses of order the unification scale, whose theory is poorly developed and unreliable.

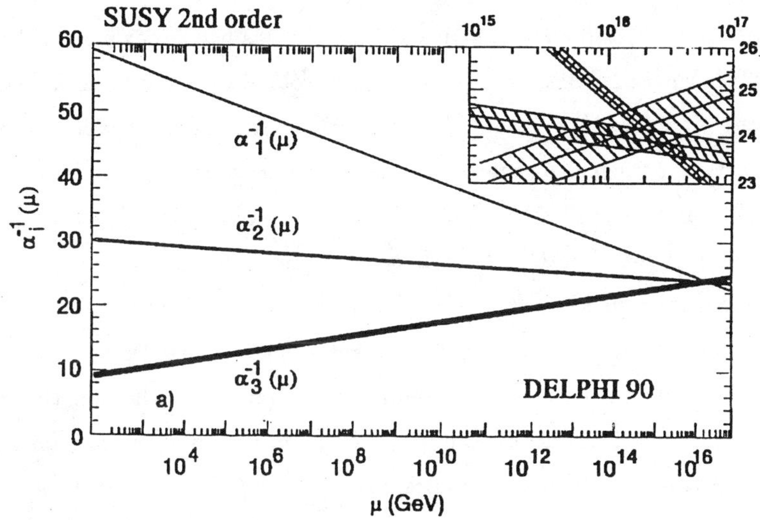

FIGURE 6

Figure 6 - When the exchange of the virtual particles necessary to implement low-energy supersymmetry, a calculation along the lines of Figure 4 comes into adequate agreement with experiment.

In any case, if we are not too greedy the main points still shine through:

• If supersymmetry is to fulfill its destiny of elucidating the hierarchy problem in any straightforward way, then the supersymmetric partners of the known particles cannot be much heavier than the SU(2)×U(1) electroweak breaking scale, *i.e.* they should not be beyond the expected reach of LHC.

• If we assume this to be the case then the meeting of the couplings takes place in the simplest minimal models of unification, without further assumption – a most remarkable and non-trivial fact.

To the extent R-parity is valid, the lightest R-odd particle is forbidden to decay. Since baryon number and lepton number, not to mention spin, are rather accurately conserved this particle has a chance to be extremely stable. Since it can annihilate in pairs, there is no possibility of an "intrinsic asymmetry" – thus the density of relic particles left over from the big bang can be calculated in a straightforward fashion, in the given a model of particle physics. It has been known for a long time that particles with annihilation cross-section of roughly weak interaction strength and masses of a few Gev. would be produced with cosmologically interesting densities*. Given this encouragement from big bang cosmology, in the context of the foregoing discussion, I think you will agree that the lightest R-odd particle is a most interesting candidate to provide cosmological dark matter.

In the conservative kinds of models that try to make minimal additions to the standard model, while incorporating the advantages for unification of couplings and stability against large radiative corrections that I mentioned before, the lightest R-odd particle usually turns out to be a linear combination of the supersymmetric partners of the neutral Higgs particles, and the photon and Z bosons. These various partners are called the higgsino, photino, and bino, respectively, and the combination of definite mass is called the neutralino. It is a spin-1/2 electrically neutral particle.

There is considerable uncertainty in the predictions for the mass and interaction properties of the lightest supersymmetric particle, since at the present stage of knowledge many parameters in the models must be taken as free variables. For example, in some ranges of parameters the lightest supersymmetric particle turns out to be charged. This is difficult to reconcile with the idea that it is cosmologically stable, since there are extremely powerful experimental constraints on such particles. Their cosmology is also problematic – they would presumably dissipate, fall into the disc, find one another and annihilate efficiently giving a high-energy photon background, ... Dark matter really ought to be dark (or, more accurately,

---

⋆ The principles for such calculations were first discussed, mainly in the context of searching for relic quarks, by Zeldovich. Another seminal contribution was a very clear and influential paper on heavy (that is, several Gev.) neutrinos by Lee and Weinberg [14].

transparent). In other ranges of parameters the lightest R-odd particle turns out to be the supersymmetric partner of the neutrino. This raises many interesting additional issues (including the possibility of an intrinsic asymmetry!), but resembles the neutralino scenario in broad outline.

There is a sizable literature devoted to discussions of the best methods for detecting supersymmetric particles, including excellent reviews [15, 16]. Both more-or-less conventional techniques of particle physics to produce and sense the superparticles at higher energy accelerators, and extraordinary techniques to sense the cosmological background itself have been contemplated. Supersymmetry also provides new mechanisms for CP violation, proton decay, and flavor-changing processes that could come in at experimentally detectable levels.

I hope I've been able to convey to you a few core ideas for physics beyond the standard model that can be understood fairly simply and that appear likely to be of permanent value. They provide, in my opinion, very good specific reasons to be hopeful about the future of experimental particle physics, and related domains of cosmology, if we can summon up the national or international will to pursue it. One hears the distant rumbling of big game afoot.

REFERENCES

1. The reader may wish to compare S. Dimopoulos, S. Raby, F. Wilczek *Physics Today* Oct. 1991, p. 25ff for a related discussion and additional references. The present text is cloned from F. Wilczek, *Annals N.Y. Acad. Sci.* **V688** 94 . The principal differences are that an extensive discussion SO(10) unification, from a somewhat unorthodox pont of view, is included here; and that the implications for dark matter are spelled out a bit.
2. The best source for up-to-date information and references is undoubtedly review talks at the standard international conferences. The Aachen conference *Twenty Years of QCD*, ed. Zerwas (World Scientific) is especially useful.
3. Recently there has been rather spectacular progress in lattice QCD, so anything other than the latest preprints does not represent the state of the art. See especially C. Davies, K. Hornbostel, G. Lepage, A. Lidsey, J. Shigemitsu, J. Sloan *A Precise Determination of* α_s *From Lattice QCD*, Cornell preprint (August 1994).
4. H. Georgi, S. Glashow *Phys. Rev. Lett.* **32**, 438 (1974).
5. H. Georgi, in *AIP Conference Proceedings* **23**, ed. C. Carlson (AIP Press, 1975).
6. M. Gell-Mann, P. Raymond, and R. Slansky in *Supergravity*, ed. P. van Nieuwenhuizen and D. Freedman (North Holland, 1979); F. Wilczek and A. Zee *Phys. Rev.* **D 25**, 553 (1982).
7. D. Gross, F. Wilczek *Phys. Rev. Lett.* **30**, 1343 (1973); *Phys. Rev.* **D8**, 3633 (1973); H. D. Politzer *Phys. Rev. Lett.* **30**, 1346 (1973).
8. R. Hughes *Phys. Lett.* **B97**, 246 (1980); *Nucl. Phys.* **B186**, 376 (1981); N. K. Nielsen *Am. J. Physics* **49**, 1171 (1981).
9. H. Georgi, H. Quinn, S. Weinberg *Phys. Rev. Lett.* **33**, 451 (1974).
10. R. Becker-Szendy *et al.*, *Phys. Rev.* **D42**, 2974 (1990); Particle Data Group, *Phys. Lett.* **B239**, 1 (1990).
11. A very useful reprint collection with some historical commentary is S. Ferrara, ed. *Supersymmetry* (2 vols.) (North-Holland/World Scientific, Singapore

1987).

12. S. Dimopoulos, S. Raby, F. Wilczek *Phys. Rev.* **D24**, 1681 (1981); W. Marciano and G. Senjanovic *Phys. Rev.* **D25**, 3092 (1982); M. Einhorn and D. Jones *Nucl. Phys.* **B196**, 475 (1982).
13. U. Amaldi, W. de Boer, H. Furstenau *Phys. Lett.* **B260**, 447 (1991); P. Langacker, M. Luo *Phys. Rev.* **D44**, 817 (1991). See also A. Sirlin *Phys. Lett.* **B232**, 123 (1989); J. Ellis, S. Kelley, D. Nanopoulos *Phys. Lett.* **B249**, 441 (1990).
14. B. Lee, S. Weinberg *Phys. Rev. Lett.* **39**, 165 (1977).
15. For supersymmetric particle physics: H. Haber and G. Kane, *Phys. Rep.* **117**, 75 (1985).
16. For supersymmetric dark matter: G. Jungman, K. Griest, and M. Kamionkowski, to appear shortly.

NEUTRINO MASS AND DARK MATTER

Thomas J. Bowles
Physics Division, MS D449
Los Alamos National Laboratory
Los Alamos, New Mexico 87545

Abstract

The existence of a finite neutrino mass would have important consequences in particle physics, astrophysics, and cosmology. Experimental sensitivities have continued to be pushed down without any confirmed evidence for a finite neutrino mass. Yet there are several observations which might be possible indications of a finite neutrino mass. If these hints are confirmed, then it is likely that at least one of the neutrinos has a mass in the eV range. As such, it would be a prime candidate for the hot dark matter that seems to be required to explain a number of observations of the large-scale structure of the Universe.

1. INTRODUCTION

Neutrinos play a dominant role in both particle physics, astrophysics, and cosmology. In our present understanding of the strong, weak, and electromagnetic forces, the group structure of the Standard Model is $SU(3)_C \otimes SU(2)_L \otimes U(1)_{EM}$. In the Weinberg-Salam-Glashow Standard Electroweak Model, left-handed neutrinos sit in a doublet, while right-handed neutrinos are in a singlet, and therefore do not interact with the other known particles. Also in this model, the neutrinos are intrinsically massless. However, while the W-S-G model provides an amazingly accurate picture of our present cold Universe, it has a number of deficits. The Standard Model does not explain the origin of the group structure, it does not reduce the number of coupling constants required, nor does it offer any prediction for the physical masses of the particles. Thus, it is generally assumed that the Standard Model is but a subset of some larger gauge theory. A wide variety of Grand Unified field Theories (GUTs), Super Symmetric Models (SUSY), and Superstring models have been proposed as the model for this larger structure. In general, these models predict nonzero neutrino masses and contain mechanisms that provide for lepton-number violation. Thus, a variety of new phenomena are predicted, including finite neutrino masses and the possibility that neutrinos can oscillate from one type to another.

In astrophysics, neutrinos play direct roles in the energy-producing reactions in stars, as well as being the primary driving force in supernova explosions. As the cross sections for neutrinos are so small (typically 10^{-40} cm^2), they generally do not interact as they exit a star and so they can provide a direct means of probing the stellar interior. In contrast, the energy we observe radiated from stars takes of order 10^6 years to reach the surface of the star and suffers so many interactions that any direct information is lost.

Finally, in Standard Big Bang cosmology, the number density of relic neutrinos is comparable to that of photons, and is therefore of order 10^9 times higher than that of baryons. Thus, if neutrinos have even a very small mass (of order a few tens of eV), they will be the dominant form of matter in the Universe and may be an explanation for dark matter. Even with masses of only a few eV, they will strongly influence the large scale structure of the Universe.

2. BETA DECAY

There are three known flavors of neutrinos: electron, muon, and tau (together with their corresponding antineutrinos). Neutrinos may also be classified as Dirac, in which case the right- and left-handed neutrinos ν_R and ν_L and their antiparticles ν_R and ν_L are distinct states, or Majorana, in which case $\nu \equiv \nu$, so that there are only two states, ν_R and ν_L. The electron and muon neutrinos have been directly observed in many experiments. The tau neutrino has not yet been directly observed, but is assumed to exist, as it is required in the Standard Model and from LEP results. By observing the beta decay spectra of nuclei, muons, and tau particles, one can kinematically reconstruct the events, and the existence of a finite neutrino mass will produce a well-defined distortion in the spectra.

2.1 Tau Neutrino

The best limit on the ν_τ mass comes from the ARGUS collaboration at DESY. By observing the decay of a τ to 5-π final states, one can constrain the ν_τ mass by looking for events very close to the endpoint. A total of 20 events has been observed, and coupled with recent new measurements of the τ mass, provides an upper limit on the τ mass of $m_{\nu_\tau} < 31$ MeV (95% CL).[1] With higher statistics and improved resolution, one might hope to achieve limits of 10 MeV in the future.

2.2 Muon Neutrino

The most stringent limit on the ν_μ mass comes from measurements of the μ momentum following the decay of stopped pions. The most recent determination of m_{π^+}, coupled with the muon data of Abela et al.,[2] gives $m_{\nu_\mu} < -0.097(72)$ MeV, which, using the Bayesian prescription,[3] yields an upper limit of 270 keV (90% CL).

However, new measurements at the Paul Scherrer Institute (PSI) of m_{π^+} have resulted in a reevaluation[4] of the Abela et al. data. This results in a new value of the mass of ν_μ of $m_{\nu_\mu} < -0.127(25)$ MeV, resulting in a 5σ negative central value. The difficulty is that m_{π^+} is determined using pionic X-rays, and the precision is limited principally by theoretical uncertainties such as electron screening and strong interactions (e.g., absorption from the 3d state). Thus, the measurement using stopped pions has a severe systematic problem and cannot be used.

The next best method of determining m_{ν_μ} mass is using π decay in flight in which both the π and μ momenta are directly measured. This method is relatively insensitive to m_π and m_μ. The measurement of Anderhub et al.[5] at PSI thus provides the most reliable limit on m_{ν_μ} with $m_{\nu_\mu}{}^2 = -0.14(20)$ MeV^2, resulting in an upper limit on the mass of ν_μ of $m_{\nu_\mu} < 500$ keV (90% CL).

2.3 Electron Neutrino

Tritium beta decay offers an almost ideal means of searching for a neutrino mass. It is a superallowed decay, the endpoint energy is quite low (18.6 keV), and the atomic final-state effects can be well understood. In these experiments, the beta decay spectrum is measured over a wide region far below the endpoint and then the expected spectral shape in the endpoint region is extrapolated assuming a zero neutrino mass and including all effects (energy loss, spectrometer resolution, decay to different atomic final states, backscattering, background, variation of spectrometer acceptance efficiency with energy) that can distort the spectrum. One then measures the spectrum in the endpoint region and compares it with the extrapolated spectrum. A deviation between the extrapolated and measured spectra can be indicative of a finite neutrino mass.

Originally, a Russian group initially claimed to see evidence for a ν_e mass of 35 eV, which was later reduced to 26 eV.[6] Of particular concern in these measurements was the use of a tritiated amino acid (valine) as the source material, which required extensive (and somewhat uncertain) theoretical calculations to take into account the atomic and molecular final-state effects. At present, six other experiments have reported results using much simpler source materials, ranging from pure molecular tritium to tritiated molecular compounds. As shown in Table 1, all six experiments rule out the ITEP result. However, it must also be noted that all five experiments find a best fit for $m_{\nu_e}^2$ that is negative. In fact, the weighted average of the five experiments is $m_{\nu_e}^2 = -121 \pm 19$ eV^2 (not including the INR result, as no systematic uncertainty has yet been given). While the spread in results between the different experiments, in outside of statistical fluctuations, this is due to the selection of the region over which the fit is made. Fundamentally, all of the experiments see the same effect with the same magnitude when fits over the same extended region are made. This effect appears as a several percent branch in the beta spectrum begining about 120 eV down from the endpoint. Physically, this corresponds to an observed excess of events in the endpoint region, rather than a deficit, which would be indicative of a finite neutrino mass. Using the Bayesian method, these combined data result in a limit on m_{ν_e} of 5 eV (95% CL). However, as the result is 6.4σ negative, this limit must be viewed with some suspicion.

Table 1. Limits on the mass of ν_e.

Group	$m_{\nu_e}^2$ (eV^2)	95% CL Limit (eV)
Los Alamos[7]	$-147 \pm 68 \pm 41$	9.3
Zurich[8]	$-24 \pm 48 \pm 61$	11.5
INS Tokyo[9]	$-65 \pm 85 \pm 65$	13
Livermore[10]	$-130 \pm 20 \pm 15$	8.0
Mainz[11]	$-39 \pm 34 \pm 15$	7.2
INR[12]	$-18 \pm 6 \pm$?	4.5

The origin of the negative central value could be due to either 1) a statistical fluke, 2) an undetermined systematic error in the experiments, or 3) difficulties with the theoretical description of the spectrum. The chance that it is a statistical fluctuation is only 1.2%, assuming that m_ν is actually zero. An independent check for possible systematic problems has been made by comparing the endpoint measurements from the experiments with the known T-^{3}He mass difference. One finds very good agreement at the few-eV level, making it less likely (but not ruling out) that the explanation is a systematic error. However, clearly additional studies of possible effects are warranted and are under way. Finally, there are two possible uncertainties in the theoretical description of the spectrum. The first is that the effect of decays populating different atomic and molecular final states comes entirely from theory. In the case of tritium, it is believed that these final-state distributions can be calculated with high accuracy, and in fact several different calculations agree quite well. The other possibility is that some new physics is involved; for example, tachyonic neutrinos, capture of relic neutrinos from the Big Bang, the existence of new particles, etc. While most such ideas are ruled out by other data, one should not preclude such a possibility, and further theoretical work is under way.

Nonetheless, the use of the Bayesian method provides a relatively stable limit for m_{ν_e}, and it is unlikely that m_{ν_e} can exceed 10 eV. Such a limit not only rules out the ITEP claim, but also eliminates ν_e (or ν_e) as the dominant component of dark matter.

3. DOUBLE BETA DECAY

The pairing force in nuclei results in a number of cases in which beta decay from a nucleus with (Z, A) is not energetically allowed to a daughter with (Z-1, A), but double beta ($\beta\beta$) decay from the (Z, A) parent is energetically allowed to a daughter with (Z-2, A). This can proceed by three possible mechanisms: two-ν $\beta\beta$ decay, no-ν $\beta\beta$ decay, and no-ν $\beta\beta$ decay with the emission of a Majoron. The two-ν $\beta\beta$ decay is an allowed second-order process requiring no new physics. No-ν $\beta\beta$ decay requires that the neutrino have a mass, that lepton number be violated, and that the neutrino is a Majorana neutrino. No-ν $\beta\beta$ decay with Majoron emission has, in addition, the requirement that a new particle, the Majoron (a massless Goldstone boson) also exists.

The most stringent limits on the neutrino mass in no-ν $\beta\beta$ decay come from the experiments using Ge solid-state detectors, which are highly enriched in ^{76}Ge. In this case, the detector is also the sample. The best limit to date comes from a Moscow-Heidelberg collaboration, which has set a limit on no-ν $\beta\beta$ decay of $< 5.1 \times 10^{24}$ yrs (90% CL).[17] Using calculated nuclear matrix elements, one can then use this result to set a limit on the mass of a Majorana electron neutrino of < 0.68 eV (90% CL). Larger isotopic ^{76}Ge detectors with lower backgrounds are under construction and one might hope to reach sensitivities of 0.1 eV, ultimately.

4. NEUTRINO OSCILLATIONS

If the neutrino has a nonzero mass and lepton number is also violated, then the physical neutrinos that we observe (e, μ, and τ) are not mass eigenstates. Instead there exist three mass eigenstates (ν_1, ν_2, and ν_3) with masses m_{ν_1}, m_{ν_2}, and m_{ν_3}. The three physical neutrinos are linear combinations of these mass eigenstates. For simplicity, one can consider the case of only two neutrinos, ν_e and ν_μ. Then, ν_e is predominantly composed of ν_1 with a small admixture (determined by a mixing angle θ between ν_1 and ν_2) of ν_2. It is then possible for ν_e to oscillate into ν_μ as it propagates. The probability for oscillation to occur is:

$$P(\nu_e \to \nu_\mu) = \sin^2 2\theta \;\; \sin^2\{1.27 \times \Delta m^2(eV^2) \times L(m)/E_\nu(MeV)\}$$

where $\Delta m^2 = |m_{\nu_2}{}^2 - m_{\nu_1}{}^2|$.

A variety of searches, both terrestrial and non terrestrial, have been carried out and no definitive evidence for neutrino oscillations has been found in terrestrial experiments.[20]

The LSND experiment currently being carried out at LAMPF utilizes a 200 ton detector filled with mineral oil and a small admixture of scintillant.[15] Thus, the detector simultaneously detects Cherenkov and scintillation light (in roughly equal amounts) using 1200 8" photomultiplier tubes. This allows one to detemine postion, angle, and total energy of the detected particle. The goal of the experiment is to search for neutrino oscillations in the $\nu_\mu \rightarrow \nu_e$ channel and to measure neutrino-proton elastic scattering. The signature for $\nu_\mu \rightarrow \nu_e$ is ν_e p -> e^+n followed by np $\rightarrow$ dγ (2.2 MeV). In this channel, LSND observed 8 events after the first year of running with an expected background of 0.9 events. One possible interpretation of this result is consistent with oscillations with $\Delta m2 \approx 6$ eV^2 and $\sin^2 2\theta \approx 0.01$. This is just barely not ruled out byu other existing limits. LSND will continue to take additional data and to study systematics. If the data hold up after further running, only then might it be possible to claim possible evidence for neutrino oscillations.

5. ATMOSPHERIC NEUTRINOS

Cosmic rays (primarily protons and gammas) striking the upper atmosphere produce showers of pions and muons that decay, yielding ν_e and ν_μ with typical energies of around 1 GeV. The flux of these neutrinos can be calculated to about 30% accuracy and is sufficiently large that they can be observed in large underground detectors. By measuring the ratio of electron- to muon-neutrinos, one can search for neutrino oscillations in a manner that is relatively free of the individual flux uncertainties, as the ratio can be calculated to about 5% accuracy. As the distance traveled by the neutrinos ranges from 10 to 10,000 km, one has sensitivity to small values of Δm^2 that are otherwise inaccessible.

Measurements of atmospheric neutrinos have been carried out by a number of large underground detectors: Kamiokande and IMB (which are both large water Cherenkov detectors); Frejus, NUSEX, Baksan, and Soudan II (which are either scintillator-based detectors or tracking calorimeters). In their analyses, they compare the observed ratio of ν_μ/ν_e divided by the Monte Carlo (MC) calculated ratio of ν_μ/ν_e. The results are given in Table 2. The most sensitive of the detectors, Kamiokande and IMB, observe a significant deficit of the relative number of ν_μ compared to ν_e. A possible explanation of this deficit may be attributed to neutrino oscillations. IMB does not make any claim to observe neutrino oscillations, due to its much larger systematic uncertainties.

Table 2. Atmospheric neutrino data.

Group	ν_μ/ν_e(Data)/ν_μ/ν_e(MC)
Kamiokande[16]	$0.60 +0.07 -0.06 \pm 0.05$
IMB-3[17]	$0.54 \pm 0.05 \pm 0.12$
	$1.01 \pm 0.03 \pm 0.11^a$
Frejus[18]	$1.06 \pm 0.18 \pm 0.15$
	$0.87 \pm 0.16 \pm 0.08^b$
NUSEX[19]	$0.99 +0.35/-0.25 \pm$?

^{a} Stopping/through muons
^{b} Fully contained events

However, it may well be that the deficit is due to a systematic effect rather than new physics. Two major concerns have been raised. First, it is quite difficult to separate electrons from muons at low energies (a few hundred MeV) based on the differences in the observed Cherenkov rings. IMB and Kamiokande jointly plan to check this by building a 1-kiloton water Cherenkov detector at KEK by using beams of charged particles (π, μ, and e) to check the accuracy of the identification.

Preliminary results from a test run of the Kamiokande PMTs indicates that the particle identification at low energies is very close to that expected. Thus, it appears unlikely that the deficit is due to particle misidentification. It is also interesting that recently Soudan II, which is a tracking calorimeter, observed the same deficit, albeit with large statistical uncertainties at present. Second, there is possible concern that the cross sections used in the Monte Carlo calculations may be incorrect. These cross sections are calculated using the Fermi Gas Model (FGM), as the momentum transfer is low at these energies, and nuclear effects are important. However, data recently published of measurements made at LAMPF by LSND of ν_μ cross sections on ^{12}C are at some variance with the FGM predictions.[20] Preliminary analysis of the first LSND data at LAMPF is also in agreement with a lower-than-expected cross section.[15] Preliminary theoretical work apparently does not find any large differences with improved nuclear models. It may be possible to check the cross sections by exposing the KEK test detector to a beam of low-energy neutrinos. This possibility is being investigated, but it will require upgrades of the accelerator and construction of a neutrino beam line, and so is unlikely to happen soon.

6. SOLAR NEUTRINOS

Perhaps the most outstanding discrepancy between prediction and measurements in current particle physics comes from the solar neutrino problem, in which the radiochemical chlorine experiment of Davis et al. observes a deficit of high-energy neutrinos of a factor of four.[21] This experiment is sensitive primarily to the ^8B and ^7Be neutrinos produced by fusion reactions in the Sun. The deficit of the ^8B neutrinos has been confirmed by the Kamiokande (KIII) experiment, which is a large real-time water Cherenkov detector that observes a factor of two deficit.[22] The observed deficits appear to be statistically quite significant with small systematic errors, so that the ratio of the measured rate compared to the Bahcall-Pinnsenault Standard Solar Model (SSM)[23] predictions are: $\phi(Cl)/SSM = 0.23 \pm 0.02 \pm 0.03$ and $\phi(KIII)/SSM = 0.49 \pm 0.04 \pm 0.06$. Different SSM give somewhat different predictions. For example, using the Turck-Chieze et al.[24] SSM, $\phi(Cl)/SSM = 0.29 \pm 0.02 \pm 0.03$ and $\phi(KIII)/SSM = 0.63 \pm 0.05 \pm 0.08$. While the differences are appreciable, they are still insufficient to resolve the problem. However, the predicted fluxes are extremely temperature dependent, as the ^8B (^7Be) flux scales with the core temperature of the Sun (T_C) as $\phi(^8B) \propto T_C{}^{18}$ [$\phi(^7Be) \propto T_C{}^8$]. Thus, a 5% decrease in T_C can lower the fluxes to be in agreement with the observations. Many Nonstandard Solar Models have been invoked to try to do just this, incorporating effects such as turbulent diffusion, massive mass loss, strong magnetic fields, a burnt-out core, etc. However, all of these models have run into problems in trying to reproduce other measured parameters (e.g., the luminosity) of the Sun. Other explanations involving new physics such as WIMPs (Weakly Interacting Massive Particles), neutrino magnetic moments, neutrino decay, neutrinos oscillations, etc., have also been proffered. But again, most of these explanations have been ruled out by either laboratory or astrophysical measurements. It appears that perhaps the most likely particle physics solution is that of matter-enhanced neutrino oscillation, the Mikheyev-Smirnov-Wolfenstein (MSW) oscillations. In order to try to resolve the source of the deficit, two radiochemical gallium experiments, SAGE (Soviet-American Gallium Experiment) and GALLEX (GALLium EXperiment) were mounted. The threshold of the gallium experiments is sufficiently low that it is sensitive to the dominant flux of low-energy p-p neutrinos, which are produced in the primary energy-producing reaction of the Sun, the fusion of two protons. The flux of the p-p neutrinos is determined to about 2% accuracy in a relatively model-independent manner and is directly coupled to the measured solar luminosity. The gallium experiments have now announced their first results, and also find a significant deficit. SAGE reports[25] the deficit (compared to the Bahcall-Pinnsenault SSM) to be ϕ(SAGE)/SSM = 0.52 ± 0.08 +0.05/-0.07 while GALLEX observes[26] ϕ(GALLEX)/SSM = 0.60 ± 0.08 ± 0.05, with a weighted average to obtain ϕ(Ga)/SSM = 0.56 ± 0.08. If one assumes only that fusion reactions power the Sun and that the Sun is in thermal equilibrium, one expects a minimal rate in the Ga experiments of 0.60 SSM. A rate significantly lower than this would require one to invoke new physics as the solution to the solar neutrino problem. Thus, while the Ga experiments seem to favor a particle physics solution, they are not yet precise enough to definitively rule out an astrophysics solution. In addition, it will be important to check the quoted efficiencies of the gallium experiments using an artificial neutrino source. Both SAGE and GALLEX are planning to do this with intense ^{51}Cr sources. However, using the results of the four experiments, a recent analysis[27] indicates that any model invoking a cooler Sun is ruled out at the 99.99% confidence level. All four experiments are consistent with the hypothesis of MSW oscillations, and the regions in the parameter space of Δm^2 versus $\sin^2 2\theta$ are shown[27] in Figure 1.

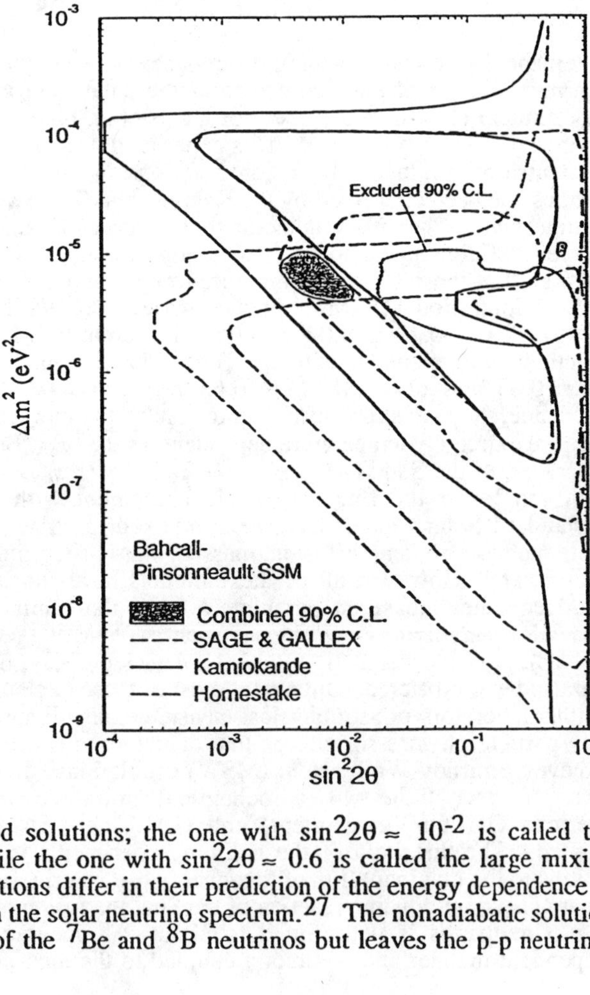

Figure 1. Allowed regions of neutrino oscillation parameters for the results from the four solar neutrino experiments.

There are two allowed solutions; the one with $\sin^2 2\theta \approx 10^{-2}$ is called the nonadiabatic solution, while the one with $\sin^2 2\theta \approx 0.6$ is called the large mixing angle solution. These solutions differ in their prediction of the energy dependence of the effect of oscillations on the solar neutrino spectrum.[27] The nonadiabatic solution gives strong suppression of the ^{7}Be and ^{8}B neutrinos but leaves the p-p neutrinos essentially unaffected. The large mixing-angle solution provides for a roughly equal suppression of neutrinos at all energies. In order to determine the origin of the solar neutrino problem, it now appears that model-independent tests must be made. This could done by either directly measuring the solar neutrino spectrum and looking for the distortions predicted by the MSW oscillations, or to measure the flux of ν_e and compare it to measurements of the total flux of $\nu_e + \nu_\mu + \nu_\tau$. Three experiments[28] to do this are under way. In the Sudbury Neutrino Observatory (SNO) experiment, which is a large water Cherenkov detector using 1 kiloton of heavy water (D_2O), one will be able to directly measure the ^8B neutrino spectrum by observing the energetic electron emitted in the charged current (CC) reaction $\nu_e + d \rightarrow p + p + e^-$. In addition, one can measure the total flux of all neutrinos above a 2.2-MeV threshold by observing the neutron created in the neutral current (NC) reaction $\nu_e + d \rightarrow p + n$. Finally, in SNO one can also observe the elastic scattering (ES) reaction $\nu_e + e^- \rightarrow \nu_e' + e^{-'}$. In the Super-Kamiokande project, a 50-kiloton water Cherenkov detector will be built to study the ES reaction This experiment will give rates almost 100 times higher than Kamiokande and will be able to measure the ^8B neutrino energy spectrum from the recoil electron spectrum. The SNO and Super-Kamiokande projects expect to come on the air in 1996. Finally, the Borexino project would use 300 tons of scintillator viewed by an array of photomultipliers to directly observe the ^7Be neutrinos via the ES reaction. This experiment is under development and hopes to come on line about 1996. These experiments will provide real-time information with better resolutions and count rates almost 100 times higher than previous solar neutrino experiments. One expects that they should be able to determine the origin of the solar neutrino problem in a model-independent manner.

The nonadiabatic solution of the solar neutrino problem provides values of Δm^2 and $\sin^2 2\theta$ that are consistent with the expectations of SO(10) grand unified field theories. If one then invokes the "see-saw" mechanism, in which the neutrino masses scale as the square of either the lepton masses, then one finds that the mass of the electron neutrino is very small (of order μeV), the mass of muon neutrino is about 3 meV, and the mass of the tau neutrino is in the eV range. This would make the tau neutrino a prime candidate for hot dark matter.

7. CONCLUSIONS

At present, laboratory measurements are all consistent with the hypothesis that neutrinos are massless. However, extensions beyond the Standard Model generically predict nonzero neutrino masses. There are however several tantalizing hints of nonzero neutrino mass in data from LSND, atmospheric neutrinos, and solar neutrinos. In the case of LSND, other experiments place constraints on the possible oscillation parameters, but do not completely exclude the LSND result. Additional data acquisition and studies of systematics are now under way. In the case of atmospheric neutrinos, only one experiment has the sensitivity to claim consistency with neutrino oscillations but further studies are under way. In the case of solar neutrinos, four experiments all see significant deficits below the rates predicted. Various analyses indicate it is unlikely that changes to the Standard Solar Models can accommodate the results. The most likely consistent explanation is that we are observing matter-enhanced MSW neutrino oscillations. Yet, given the extreme dependence of the neutrino fluxes on the core temperature of the Sun, it is difficult to definitively rule out astrophysical solutions. A new round of experiments will provide model-independent tests, which should provide the solution to the solar neutrino problem in the next five to seven years.

If either the LSND, atmospheric neutrino, or solar neutrino results are confirmed to be due to neutrino oscillations, then it is likely that the mass of at least one of the neutrinos is in the eV range. As such, it would dramatically change our view of cosmology. Neutrinos would then comprise a significant fraction of the mass of the Universe and would likely be the primary source of the hot dark matter which seems to be required to explain a number of observations of the large-scale structure of the Universe.

8. REFERENCES

1. Albrecht A et al., Phys. Lett. B **292** (1992) 221.
2. Abela R et al., Phys. Lett. B **146** (1984) 431.
3. Particle Data Group, Phys. Lett. B **170** (1986) 1.
4. Frosch R 1992. (private communication.)
5. Anderhub HB et al., Phys. Lett. B **114** (1982) 76.
6. Boris S et al., JETP Lett. **45** (1987) 333.
7. Robertson RGH et al., Phys. Rev. Lett. **67** (1991) 957.
8. Holzschuh E, Fritschi M, Kundig W, Phys. Lett. B **287** (1992) 381.
9. Kawakami H. et al., Phys. Lett. B **256** (1991) 105.
10. Stoeffl W and Decman D, to be published in Proc. of the 1st Intl. Conf. on Non Accelerator Particle Physics, Bangalore, India, Jan. 1994.
11. Weinheimer C et al., Phys. Lett. B **300** (1993) 210.
12. Belesev AI et al., INR preprint 862/94, July 1994. Also to be published in the Proc. of the 27th Intl. Conf. on High Energy Physics, Glasgow, Scotland, July 1994.
13. Balysh, A et al., to be published in the Proc. of the 27th Intl. Conf. on High Energy Physics, Glasgow, Scotland, July 1994.
14. Schneps J, Proc. XV Intl. Conf. on Neutrino Physics and Astrophysics, Granada, Spain (1992) 207.
15. Louis W et al., to be published in Proc. XVI Intl. Conf. on Neutrino Physics and Astrophysics, Eilat, Israel, June 1994.
16. Kajita T et al., Proc 26th Intl. Conf. on High Energy Physics, Dallas (1992) 1187.
17. IMB Collaboration, Phys. Rev. Lett. **66** (1991) 2561, Phys. Rev. Lett. **69** (1992) 1010.
18. Frejus Collaboration, Phys. Lett. B **245** (1991) 305.
19. NUSEX Collaboration, Europhys. Lett. **8** (1989) 611.
20. Koetke DD et al., Phys. Rev. C **46** (1992) 2554.
21. Bowles T and Gavrin V, Ann. Rev. Nucl. Part. Sci. **43** (1993) 117.
22. Totsuka Y et al., to be published in Proc. XVI Intl. Conf. on Neutrino Physics and Astrophysics, Eilat, Israel, June 1994.
23. Bahcall JN and Pinnsenault MH, Rev. Mod. Phys. **64** (1992) 885.
24. Turck-Chieze S and Lopes I, Astroph. J. **408** (1993) 347.
25. Abdurashitov JN et al., Phys. Lett. B **328** (1994) 234.
26. Anselmann P et al., Phys. Lett B327 (1994) 377.
27. Bludman SA et al., Phys. Rev. D47 (1993) 117.
28. McDonald AB, Proc. of the Texas/PASCOS 92 Intl. Conf. on Relativistic Astrophysics and Particle Cosmology, Berkeley, CA, Annals of the NY Acad. of Sciences **688** (1993) 605.

BALLOON-BORNE DIRECT SEARCH FOR IONIZING MASSIVE PARTICLES AS A COMPONENT OF THE GALACTIC HALO DARK MATTER

(The Arizona-IMAX Collaboration)

P.C. McGuire[1], T. Bowen, D.L. Barker, P.G. Halverson[2], K.R. Kendall, T.S. Metcalfe, R.S. Norton, A.E. Pifer:

Department of Physics, University of Arizona, Tucson AZ 85721

L.M.Barbier, E.R.Christian, K.E.Krombel, J.W. Mitchell, J.F. Ormes, R.E. Streitmatter:

Laboratory for High Energy Astrophysics, NASA/GSFC, Greenbelt, MD 20771

A.J. Davis, A.W. Labrador, R.A. Mewaldt, S.M. Schindler:

Space Radiation Laboratory, California Institute of Technology, Pasadena, CA 91125

R.L. Golden, S.J. Stochaj, W.R. Webber[3] :

Particle Astrophysics Laboratory, New Mexico State University, Las Cruces, NM 88003

[1]EMAIL: mcguire@soliton.physics.arizona.edu or DECnet – UAZHEP::MCGUIRE

[2]Physics Department, University of California at Irvine, Irvine, CA 92717

[3]Astronomy Department., New Mexico State University, Las Cruces, NM 88003

ABSTRACT

A dark matter (DM) search experiment was flown on the IMAX balloon payload to search for a possible minor component of the dark matter in the Galactic halo: ionizing massive particles (IMPs) ($m_x > 10^4$ GeV/c^2) that cannot penetrate the atmosphere due to their low-velocities and high energy-loss. The DM search experiment consisted of a delayed coincidence between four large plastic scintillation detectors arranged in a vertical stack. In order to search for ultra-slow particles which do not slow down in the IMAX telescope, the experiment contained TDCs which measured the time-delays $T_{i,i+1} \in [0.3, 14.0]$ μs between hits in successive counters to $\sim 2\%$ precision. We present IMP flux limits for non-slowing IMPs and also for IMPs which slow down significantly within the IMAX telescope. This experiment effectively closes much of a previously unconstrained 'window' in the mass/cross-section joint parameter space for massive particles as the dominant halo DM.

1. INTRODUCTION & MOTIVATION

A whole host of different 'elementary' particles have been hypothesized as solutions to the dark matter problem[1](e.g. a small neutrino rest mass, WIMPs, monopoles, CHAMPs, very massive neutrinos, SIMPs, strange quark nuggets, axions); and novel particle detectors are required to detect each different dark matter candidate. Weakly Interacting Massive Particles (WIMPs) are considered to be a likely dark matter particle candidate. In the early universe, the low WIMP cross-section might allow enough WIMPs to survive annihilation from anti-WIMPs, so as to be abundant enough to solve the missing matter problem today. However, Ionizing Massive Particles (IMPS), which presently have a much higher interaction cross-section with ordinary matter, might have a primodial annihilation cross-section that is similar to the WIMP primordial annihilation cross-section. Therefore, IMPs *might* have a relic abundance that is large enough to constitute all the dark matter in the universe and be unnoticed in past experiments[2]. Examples of IMPs include: CHAMPs (electrically CHArged Massive Particles), SIMPs (Strongly Interacting Massive Particles), monopoles and "strange" quark nuggets. If IMPs have antiparticles, the range of IMP masses for which IMPs could be the dominant (halo) dark matter (DDM) includes $m_x \in [6 \times 10^2, 3 \times 10^5]h$ GeV, where $h \sim 1$ is the Hubble constant[2]. Since the high interaction cross-section of IMPs with ordinary nuclei should have easily observable consequences, many experiments have been completed[3] or clever arguments invented[2] to rule out different hypothetical IMPs as the DDM, within the theoretically most favored ranges of IMP mass, m_x, number density, n_x, and interaction cross-section with ordinary nuclei σ_{xN}. However, Starkman *et al.* have found that for a significant range of the joint mass and interaction cross-section parameter space, generic IMPs have *not* been ruled out as the dominant halo dark matter, as previously thought.

Our work was unprecedented for several reasons. First and foremost, dark matter hunters have rarely searched for dark matter particles at balloon or satellite altitude. Second, usually when a search for dark matter particles at balloon or satellite altitude has been performed, the experimenters have been unable to reject the background with flux $\sim 1\ \mathrm{cm^{-2}s^{-1}sr^{-1}}$ from ordinary cosmic rays, which deposit energies of $> 2\ \mathrm{MeV/g/cm^2}$. Our search for IMPs with a four-fold delayed coincidence between the four scintillation detectors in the IMAX stack was the first dark matter particle search experiment flown at balloon altitude which could reject the cosmic ray background to a flux level of $\sim 10^{-5}\ \mathrm{cm^{-2}s^{-1}sr^{-1}}$ and have a relatively small energy-loss threshold ($dE/dx\,(\mathrm{THR}) \sim 3.5\ \mathrm{MeV/g/cm^2}$). This article briefly discusses the results of our balloon-borne dark matter search — for more details, see Ref.[4].

2. THE IMP SEARCH EXPERIMENT & RESULTS

The parameters of the IMAX balloon flight opportunity are listed in Table 1. In order to search for IMPs over a wide range of velocities, $\beta = (0.33 \text{ to } 3.3) \times 10^{-3}$, triggering is accomplished in a specially designed detector module with a series of delayed coincidence gates. All 4 time intervals and 3 pulse heights are read out by an ORTEC 811 ADC CAMAC module.

Number of scintillation counters	4
Total thickness of counters	$5\ \mathrm{g/cm^2}$
Total thickness of material above counter telescope (including $5\ \mathrm{g/cm^2}$ atmosphere)	$6\ \mathrm{g/cm^2}$
Total thickness of material between counters	$8\ \mathrm{g/cm^2}$
Total thickness to reach and traverse experiment	$19\ \mathrm{g/cm^2}$
Geometry factor	$\Omega = 100\ \mathrm{cm^2 sr}$
Time at float altitude (for high quality DM data)	$t = 1.8 \times 10^4$ seconds
IMP velocity range	$\beta = (0.33 \text{ to } 3.3) \times 10^{-3}$
Trigger thresholds in 1 cm plastic scintillators: $\beta \approx 1, Z = \pm 1$, 35% of Landau peak of MIPs in 1 cm; $dL/dx = 0.018\mathrm{MeV/cm}$; $Z = +1$ IMPs (e.g. X^+ and $X^- - \alpha$ CHAMPs), monopoles, dyons,(1/5)e superstring particles are all above threshold over full IMP velocity range listed above	

Table 1: Parameters of IMAX Balloon-Borne Experiment

The wide coincidence timing gates (μs instead of ns) ensured a steady stream of accidental coincidences uniformly distributed in each time gate, the principal background for the IMP search. The accidental rate $A(n)$ for n counters is proportional to each counter's singles rate and to the gatewidths for each of the delayed coincidences. At float altitude, $R_i \approx 3500$ Hz for each counter, and with the gate widths as given in Ref. [4], $A(3) \approx 4$ Hz; about 44×10^3 events were recorded (reduced somewhat by deadtime) during the five hours of excellent IMP search data at float altitude. The time-interval distributions are uniform, as expected, except for the peaks in T_{12} and T_{23} which are due to anomalous events, of unknown origin, which are eliminated from the data set when the velocity limits, $v_{\min} = 0.1\mathrm{m}/\mu\mathrm{s}$, $v_{\max} = 0.67\mathrm{m}/\mu\mathrm{s}$, corresponding to the limits of the T_{34} distribution, are applied to T_{12} and T_{23}. Since $A(4) \approx 0.055$ Hz, 653 four-fold accidental coincidences were observed during the five hour observation period. However, if the three independent time measurements must be consistent in the data analysis with the time delays of an IMP traversal within, say, $\pm 1\sigma$, where $\sigma = 2\%$, then two of the timing gates are effectively narrowed; one of the gates is not narrowed since the IMP is allowed to have any velocity consistent with the gate widths. Then the effective accidental rate is $A(4) = 2.6 \times 10^{-4}$ Hz, corresponding to $N_{\mathrm{IMP}} \sim 5$ accidental events which mimic IMP traversals in the 5 hours of analysed data. Thus the IMAX time-of-flight IMP search has a minimum detectable flux, $\phi_{\mathrm{IMP,\ min.\ det.}}$, at the 4σ level using $N_{\mathrm{IMP}}+N_{\mathrm{ACCID}} \geq 14$ as well as the geometry factor Ω and time t from Table 1, of $\phi_{\mathrm{IMP,\ min.\ det.}} = 5 \times 10^{-6}\ \mathrm{cm^{-2}s^{-1}sr^{-1}}$(see Tables 2 and 3 for our actual results).

When we assume a specific IMP-nucleus interaction model to parametrize the different IMP scattering cross-sections with different nuclei, we indeed find that our IMP search does place useful *new* constraints on IMP parameter space, closing a wide-open window in parameter space ($10^6 < m_x < 10^8$ GeV; see Figure 1). The windows shown here are from the Starkman *et al.* interpretation, adapted from and including the results of the recent BPRS publication. We eliminate a large fraction of the remaining window W_2 for the coherent IMP-nucleus interactions. The remaining portion of the window W_2 can be constrained by a sea level IMP search with low thresholds and a significant background rejection capability.

This work was supported at Arizona by the NSF and at the other institutions by NASA.

Search	$\frac{dE}{dx}(THR)$	$\sigma_{\min}(\mathrm{cm}^2)$	$a_1^{\max}$ $(\frac{\mathrm{cm}^2}{g})$	$(\frac{\sigma}{m_x})^{\max}$ $(\frac{\mathrm{cm}^2}{\mathrm{GeV}})$	$\mathrm{A}\Omega$ $(\mathrm{cm}^2\mathrm{sr})$
$\Delta v \approx 0$	3.0	6.2×10^{-21}	0.013	2.31×10^{-26}	100
small Δv	3.0	6.2×10^{-21}	0.07	1.2×10^{-25}	100
large Δv	3.0	6.2×10^{-21}	0.123	2.1×10^{-25}	100
Triples	3.0	6.2×10^{-21}	0.123	2.1×10^{-25}	135
Singles Rate	0.9	1.6×10^{-21}	0.38	3.3×10^{-25}	9120

Table 2: Summary of IMAX Results, part 1. For each IMP search, we tabulate the energy loss detection threshold ($dE/dx(THR)$($\mathbf{MeV/g/cm^2}$)), the corresponding threshold cross-section including the β^2-dependence of dE/dx ($\sigma_{\min}$), the maximum slowing down rate ($a_1^{\max}$), the corresponding maximum elastic cross-section to mass ratio ($(\sigma/m_x)^{\max}$), and the geometry factor ($\mathbf{A}\Omega$) for the search.

Search	N	B	$n_{s,\max}$	$\phi_{\max}(\frac{1}{\mathrm{cm^2 s\ sr}})$	$m_{\max}$ (GeV)
$\Delta v \approx 0$	5	4.0 ± 0.3	9.1	6.5×10^{-6}	1.5×10^{11}
small-Δv	5	7.5 ± 0.1	6.9	4.9×10^{-6}	2.0×10^{11}
large-Δv	16	18.3 ± 2.1	11.1	7.9×10^{-6}	1.3×10^{11}
Triples	4 Hz	0 Hz	4 Hz	3×10^{-2}	3.3×10^{7}
Singles	4000 Hz	0 Hz	4000 Hz	0.44	2.27×10^{6}

Table 3: Summary of IMAX Results, part 2. For each IMP search, we tabulate the number of signal (**N**) and background (**B**) events observed, the 95% C.L. Poisson upper limit ($n_{s\max}$) on the number of IMP events, the upper limit on the flux ($\phi_{\max}$), and the maximum mass detectable ($m_{\max}$) (assuming IMPs are all the dark matter).

REFERENCES

[1] J.R. Primack et al., Ann. Rev. Nucl. Part. Sci. 38, 751 (1988).

[2]A. De Rujula et al., Nucl. Phys. B333, 173 (1990). M.W. Goodman and E. Witten, Phys. Rev. D31, 3059 (1985). H. Goldberg and L.J. Hall, Phys. Lett. B174, 151 (1988).G.D. Starkman et al., Phys. Rev. D41, 3594 (1990).S. Dimopoulos et al., Phys. Rev. D41, 2388 (1990). A. Gould et al.,Phys. Lett. B238, 337 (1990). R.S. Chivukula et al., Phys. Rev. L65, 957 (1990). E. Witten, Phys. Rev. D30, 272 (1984).

[3] J. Rich et al., Phys. Lett. B194, 173 (1987). Snowden-Ifft et al., Ap. J. L364, 25 (1990). B. Barish et al., Phys. Rev. D36, 2641 (1987).C. Bacci et al. (BPRS Collab.), to appear in Astroparticle Physics (1994).

[4]P. McGuire and T. Bowen, Proc. 23rd ICRC, Vol.4, p.726, Calgary (1993). J.W. Mitchell et al., Ibid, Vol.1, p.519, Calgary (1993). McGuire, Bowen, L.M. Barbier, et al., Ibid, Vol.4, p.621, Calgary (1993). McGuire, Ph. D. Diss., U. of Arizona, unpublished (1994).

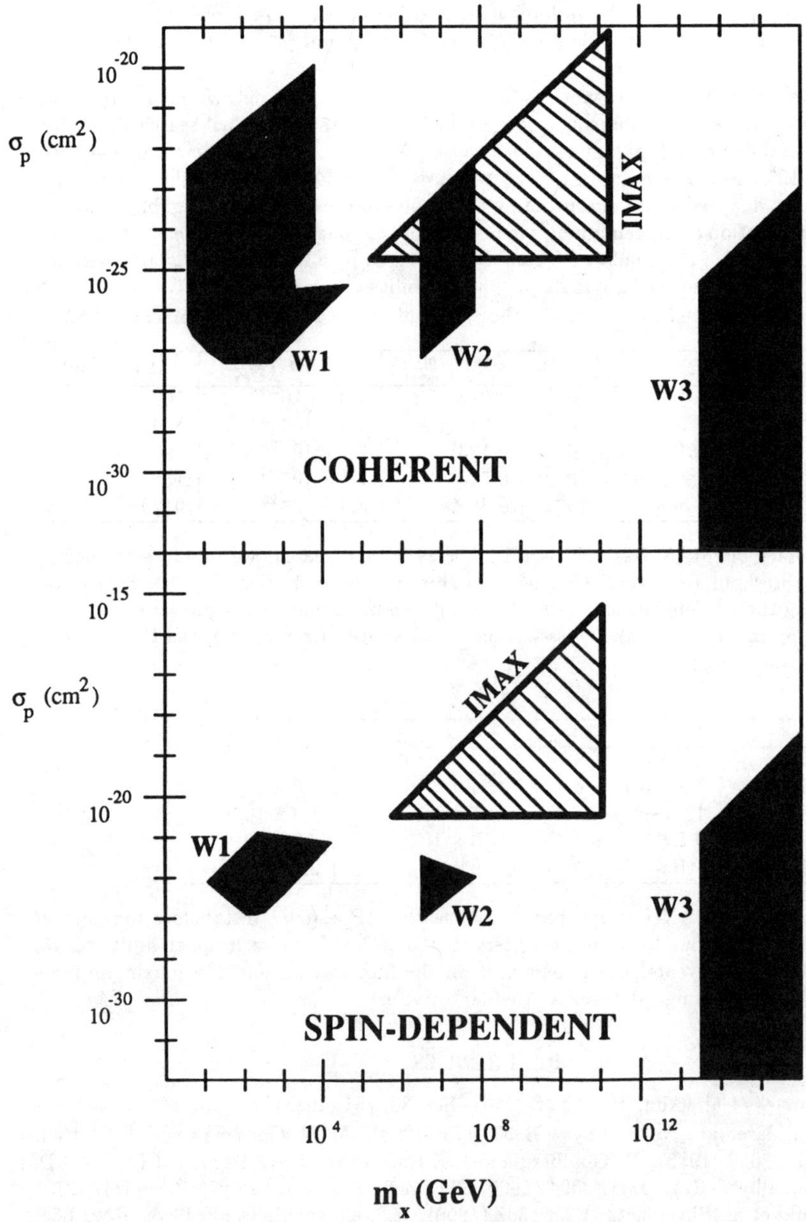

Figure 1: We show the IMAX constraints on the SIMP mass/cross-section parameter space for coherent interactions and for spin-dependent interactions. Without the IMAX results, there are three different non-excluded windows in each parameter space W_1, W_2, and W_3. The hatched areas are the IMAX excluded regions, assuming that SIMPs are all the galactic halo dark matter ($f_d = 1$).

LOOKING BACKWARD, DARKLY

Virginia Trimble
Astronomy Department, University of Maryland, College Park MD 20742
and
Department of Physics, University of California, Irvine CA 92717-4575

ABSTRACT

Dark matter, both as a concept and as a name, has a surprisingly long history. We here trace its story from 19th century origins in the Newtonian two- and three-body problems through early 20th century applications of velocity dispersions of stars and galaxies to the first considerations of the most promising modern candidates.

1. INTRODUCTION

The physics and chemistry textbooks of our fathers' era nearly always began by explaining what the Ancient Greeks had thought about the subject. This has largely gone out of fashion, and a good thing, too. Where it persists, in astronomy texts that often begin with Ptolemaic cosmology, students with short attention spans quite typically come out of our courses firmly convinced (a) that the sun goes around the earth and (b) that they learned it in Astro 100. But an historical viewpoint is rather appropriate for the present subject, because Aristotle and his colleagues seem to have taught that all terrestrial phenomena were attributable to four elements -- earth, air, fire, and water (nearly always in that order) -- while celestial phenomena were the realm of a fifth element, or quintescence.

The Greek point of view went gradually out of fashion when Galileo and Newton showed that celestial objects obey the same laws of gravitation as terrestrial apples and pendula. Its death knell came with the 19th century spectroscopists, Bunsen, Kirchhof, and others, who showed that the spectra of the sun and stars display absorption lines due to sodium and, eventually, a wide range of other normal chemical elements.

There is today no doubt that the planets, stars, interstellar gas, galaxies, and, indeed, anything capable of emitting and absorbing any form of electromagnetic radiation, consist of the elements found on earth, though in rather different proportions, strongly dominated by hydrogen and helium, and collectively called baryonic matter. But a large body of evidence now indicates that 90% or more of the matter that contributes to gravitational forces and potentials in the universe does not emit its fair share of light, and this dark matter may well be non-baryonic.

The history of dark matter, in this restricted sense of non-luminous or subluminous material whose existence we deduce from its gravitational effect on brighter things, can then be divided into five periods, roughly analogous to the periods of human history, but with somewhat different dates. These are the Neolithic (before 1900) the Classical (1920-39), the Mediaeval (1940-60), the Rennaissance (1961-

1973), and the Soviet and American Revolutions (1974). A shift in astronomical conventional ideas in favor of dark matter was largely complete by the time of the 1977 Yale conference on evolution of galaxies and stellar populations, where Ivan King (1977), in his introductory talk, accepts as given that 90% of the mass present in galaxies is not accounted for, and expresses some surprise that the effect of this stuff on chemical evolution seems to be so small.

2. NEOLITHIC DARK MATTER (BEFORE 1900)

The late 18th and 19th centuries saw three applications of Newtonian mechanics that implied the existence of non-luminous objects. Curiously, each of the three corresponds closely to a well-defined modern class of astronomical entity that is still a viable dark matter candidate on some scale.

First, in purely theoretical terms, John Michell (1784), followed by Pierre Simon, Marquis de Laplace (1799), considered objects sufficiently compact to have escape velocities in excess of the speed of light. Michell remarked that "all light emitted from such a body could be made to return to it by its own proper gravity," and Laplace concluded that "the largest luminous bodies in the universe may, through this cause, be invisible." The assertions are generally regarded as the first predictions of black holes, whose observational confirmation came with the demonstration that the X-ray emitting component in Cyg X-1 has a mass in excess of the maximum possible for neutron stars (Bolton 1972) and, more gradually, with the understanding of the central enegines in active galaxies (Salpeter 1964, Zeldovich 1964; Holt et al. 1992).

On the observational front, Wilhelm Friedrich Bessel (1845) analyzed irregularities in the motions of Sirius and Procyon shortly before Urbain J.J. Leverrier (1845) and John Couch Adams (1847) looked at the orbit of Uranus. Bessel concluded that each star had an invisible companion of roughly its own mass, while Adams and Leverrier predicted a new planet. The latter (Neptune) remained unseen for less than a year (Clerke 1885, and many more recent histories), and subsequent applications of the method led to the discovery of Pluto (Tombaugh and Moore 1980) and the non-discovery of Planet X (Standish 1993). Neptune (along with Jupiter, Saturn, and Uranus) is an example of a gas giant or brown dwarf, substellar bodies with at least roughly the cosmic abundance of hydrogen and helium.

The stars counted as dark until 1862 and 1896 respectively, when they were seen by Alvan G. Clark, testing an 18½" objective, and J.M. Schaeberle (1896), using the Lick 36". Schaeberle seems to have been rather a cantankerous character, given to writing about the "worthless telescopes" and "strange articles" of his contemporaries. Clark is widely advertized as having initially mistaken Sirius B for a reflection in his new telescope, but I have no idea whether the story, which I heard from a modern amateur telescope builder, is true, or whether documentary evidence exists. Sirius B (initially called Sirius *comes*) and Procyon B were the first two of a very numerous stellar class, the white dwarfs.

In what sense are the three Neolithic classes still possible contributors to dark matter? Black holes must be either primordial ones of 10^{15-30} grams or pre-galactic ones of 10^{4-6} $M_\odot$. The intermediate masses (and neutron stars as galactic dark matter) are ruled out by limits on the diffuse galactic X-ray background, which they would contribute to by accreting interstellar gas when they passed through the galactic plane. Masses below 10^{15} g will already have died in bursts of Hawking-Bekenstein radiation. BH numbers close to that mass are a bit constrained by such bursts not having been seen; they are, anyhow, not the known gamma ray bursters. Gas giants or brown dwarfs must be separate objects in space, not companions, which are already included in the inventory of mass associated with stars and are known to be rare. And an important contribution from white dwarfs is possible only if they are old enough to have faded below 10^{-4} $L_\odot$.

3. CLASSICAL DARK MATTER (1920-39)

Moving into the 20th century, we find both Sir James H. Jeans (1922) and Johannes C. Kapteyn (1922) determining the local mass density from stellar distributions and velocity dispersions above the galactic plane. Their results (0.143 and 0.099 $M_\odot/pc^3$) roughly bracket the modern range of values, as do their conclusions, based on comparing the mass per luminous star implied by their numbers with the average mass of known binary systems (1.6 $M_\odot$ at that time). Jeans opined that "there must be about three dark stars in the universe for every bright star," while Kapteyn, noting that "we therefore have the means of estimating the mass of dark matter in the universe," concluded that "as matters stand at present it appears at once that this mass cannot be excessive." Writing before Hubble had shown the spiral nebulae to be separate stellar systems, they said "universe" where we would say "galaxy."

The definitive early study of this kind by Jan H. Oort (1932) gives us the usual name, Oort limit, for the dynamically determined mass density in the plane of our galaxy. He found 0.08 to 0.11 $M_\odot/pc^3$ and said that "one unit of light corresponds to 1.8 solar masses." Stars brighter than M_V = 13.5 contributed 0.038 $M_\odot/pc^3$ of his total. Oort's opinion was that stars between M_V = 13.5 and 18.5 would account for most of the rest. This was based heavily on apparent (and erroneous) gravitational redshifts near 240 km/s for the white dwarfs Procyon B and van Maanen 2, which (using R.H. Fowler's non-relativistic equations for degenerate matter) would have required them to have masses much larger than 1 $M_\odot$. By the time of Oort's (1965) article in the Kuiper compendium, his estimate had risen to 0.150 $M_\odot/pc^3$, with "invisible stars and gas" responsible for 40% of the total.

Whether the galactic disk really has a dark matter component separate from the well-established halo one remains a topic of dispute, with Bahcall (1986) and his colleagues arguing the affirmative and Gilmore and Kuijken (1983) among those arguing the negative. The measurement is a difficult one both because the disk mass you get scales with logarithmic gradient of numbers of stars vs. height from the plane and because the stellar distances must be found by assuming that their luminosities are the same as those of nearby stars of the same spcetral type (which can be wrong if they differ in either chemical composition or in evolutionary stage).

A second classical pillar of dark matter studies is the investigation of velocity dispersions in rich clusters of galaxies. Fritz Zwicky (1933) looked at the Coma cluster and concluded "dass dunkle Materie in sehr viel groesserer Dichte vorhanden is als leuchtende Materie." Incidentally, this sentence is not an illustration of the widespread story that Zwicky spoke seven languages -- all badly. "Dass" functions as a conjunction, not an article. A few years later Sinclair Smith (1936) examined the velocities of Virgo cluster galaxies and found a total mass of 10^{14} M_o (for a distance of 2 Mpc, thus 5-10 X 10^{14} M_o at a modern distance). He concluded "that the mass per nebula is 2 X 10^{11} M_o ... far larger than Hubble's value of 10^9 M_o for the mass of an average nebula. It is possible that both figures are correct and that the difference represents a great mass of internebular material within the cluster."

Zwicky (1937, 1939, 1951, 1961) continued to take an active interest in the subject for a number of years, advocating a combination of dwarf galaxies and intergalactic material to make up the dominant dark component at the time of his 1959 Handbuch article, and intergalactic stars and pygmy galaxies at the 1961 IAU Symposium (by which time a 21 cm survey of Coma by Muller [1959] had ruled out neutral hydrogen as a candidate). Among his other original contributions to astrophysics were supernovae, neutron stars, gravitational lensing by galaxies, a rest mass of 10^{-63} g for the graviton (based on the non-existence of superclusters of galaxies), and object Hades (which he would probably now claim as a prediction of quark stars). Dying in 1974, Zwicky just missed the chance to say "I told you so" to his converted colleagues on the subject of dark matter.

Smith, whose 1936 paper explicitly cites Zwicky, died in 1938 and does not seem to have been an "I told you so" sort of person anyhow. His Virgo study was done with a spectrograph of his own design and construction and was his only foray into galactic spectroscopy. His other devices included a high time resolution spectrograph to study emission by exploding wires, a vertical seismometer, and optical oscillograph, a 1932 photo-electric photometer, a radiometer (incorporated into a microphotometer built by Olin C. Wilson), and much of the control instrumentation for the 200" telescope.

Holmberg (1937), in the only classical period examination of binary galaxies, found an "average mass of the galaxies = 1.0 X 10^{11} M_o" (5 X 10^{11} on a modern distance scale) from a handful of pairs. Remarkably, given the large errors and limited statistics of all determinations comfortably between Hubble's single-galaxy and Zwicky's rich-cluster results, as we now think it should. Holmberg himself (citing Zwicky and Smith as well as Hubble and Shapley) felt that his value was a satisfactory one, "in very good agreement with the values earlier derived." He remained an outstanding extragalactic astronomer throughout his career.

Horace Babcock (1939), who measured the first spiral galaxy rotation curve, in contrast, immediately left the field and is now primarily known for contributions to solar physics, especially the theory of the 22 year activity cycle. He explained who in a handwritten letter responding to my inquiry about why no-one (including him) followed up on his work for such a long time. First, his AAS talk on the subject was heavily criticized (presumably at the August, 1939 Berkeley meeting, but I haven't checked this). Second, most unusually for an astronomy thesis completed at Lick at the time, he was not encouraged to submit his work to the Astrophysical Journal, then as now the most prestigious journal in the field. It appeared only as a Lick Observatory Bulletin. This was an experience I could identify with. Though most of the astronomy theses written at Caltech when I was a student there were published in ApJ, I was instructed to send mine to a lower-profile publication. But if you live long enough, these things take care of themselves. I am now an associate editor of the Astrophysical Journal, and anything Babcock sends we will find room for.

It is probably significant that Babcock's published thesis does not cite Zwicky, Smith, or Holmberg, and, according to an oral history interview published elsewhere, he was very possibly not aware of their work at the time. As a result, he was not prepared to find or to defend a total mass as large as the 10^{11} M_o (5×10^{11} for modern distances) implied by the rotation speed he measured for his outermost point in the galaxy or to find "a ratio of mass to luminosity...much greater than that...in the vicinity of the sun." That last point, 380 km/sec nearly 100' out from the center of M31, came from a multi-night exposure and is now known to be erroneous (though his conclusion was not so far off from current ideas). Vera Rubin has wondered whether he might perhaps have had doubts about it even at the time.

The second rotation curve ever measured was that of M33, published in 1940 (and so leading naturally into the next section) by Mayall and Aller (1940). M33 is now known to be tidally distorted and truncated, so that the small mass and M/L they found was inevitable. It is hard to say in retrospect whether this was good or bad luck.

4. MEDIAEVAL DARK MATTER (1940-60)

Astronomical paper production fell precipitously during the war, and the immediate post-war flowering focussed more on studies of stellar structure and evolution than on galaxies. Nevertheless, to this period belong a number of studies of velocity dispersions in clusters and groups of galaxies, hints that single galaxies might have larger masses than at first evident, and the advent of steady state cosmology, with its firm prediction of significant intergalactic material.

First, the cluster velocity dispersions were persistently large, independent of the philosophical predispositions of the observers. Mass-to-luminosity ratios (in solar units and assuming gravitationally bound systems) of 100 to 1000 were found for the clusters in Virgo (Ambartsumian 1958, Oort 1959, van den Bergh 1960), in Coma (Schwarzschild 1954, de Vaucouleurs 1959, van den Bergh 1960), in Hercules (the Burbidges 1959), and in Canum Venaticorum (van den Bergh 1960) and for a number of small groups (Holmberg 1950, Zwicky 1956, the Burbidges 1959, de Vaucouleurs 1959a, 1960, and Limber and Mathews 1960)

Next, several lines of evidnece gradually indicated that, in general, masses of galaxies got larger when measured on larger scales. Kunth (1952) and Lohman (1956) found, for instance, a mass for the Milky Way from the velocity dispersion of its globular clusters that was two or three times that implied by the circular velocity in the solar neighborhood. The rapid rotation of the outskirts of M31, Babcock's nemesis, was confirmed by van de Hulst's (1957) 21 cm data, implying a global M/L near 20, vs. 2 in the nucleus (Lalleman et al. 1960). Elliptical galaxies also showed larger global M/L's than nuclear ones (Osterbrock 1962). And many galaxy pairs had smaller M/L's than whole clusters, but larger ones than the galaxies considered individually (van den Bergh 1960; Kahn and Woltjer 1959 on the Local Group). Page's (1952, 1956) much smaller masses for binary galaxies complicated the issue considerably and seem to have resulted from too-stringent criteria for regarding pairs as truly bound.

Schwarzschild (1954) suggested that the dark matter, at least in elliptical galaxies, might be contributed by old white dwarfs that had had M_V near +1 on the main sequence, an interesting foreshadowing of the idea later called bimodal star formation. The bimodality is in the mass distribution N(M) of the stars formed and is required so as not to exceed the very tight limits on numbers of old, low metallicity stars in the solar neighborhood now. Whatever hypothetical process made the $M_v = 1$, 2-3 M_o stars ancestral to the postulated white dwarfs is not allowed also to make 0.4 - 0.9 M_o stars that would still be around.

Finally, our medieaval period was the era of Bondi and Gold's (1948) and Hoyle's (1948) Steady State universes, which necessarily have a good deal of pregalactic dark matter around at all epochs. In Steady State with a creation field, the average density is exactly $3H^2/8\pi G$, a hundred times or so larger than the density contributed by luminous stars and galaxies. Hoyle (1959) predicted that the missing 99% would be largely in the form of intergalactic hydrogen at $T > 10^5$K. Other members of the astronomical community thought this large preponderance of invisible stuff considerably less likely, in language ranging from the gentle words of Oort (1959) and Bonnor (1960) to the scornful ones of McVittie (1959). We now know that intergalactic hydrogen comprises very little of the cosmic mass density, lower temperatures being ruled out by absence of absorption in the Lyman alpha line (Gunn-Peterson effect) and higher ones by the absence of distortion of the microwave background spectrum by scattering (related to the Sunyaev-Zeldovich effect).

Despite (or because of??) the large number of uncoordinated observational results, this period was marked by a profound lack of concensus on what the data meant and even on whether they needed to be taken terribly seriously.

5. THE RENAISSANCE OF DARK MATTER (1961-73)

The first American General Assembly of the International Astronomical Union (Berkeley 1961) had among its fringe activities a pair of conferences in Santa Barbara, one addressing problems of extragalatic research in general (IAU Symposium 15, proceedings edited by Mc Vittie 1961) and one focussed more specifically on the issue of stability of clusters of galaxies with large velocity dispersions (proceedings edit ed by Newman, Page, and Scott 1961). Under the circumstances, you might have expected Zwicky to be the keynote speaker at both. In fact, he appears briefly in the proceedings of IAUS 15 and not at all in the written record of the second meeting. According to participant Vera Rubin, he was actually forced to arrange an unofficial, after dinner session in order to be heard, but then devoted most of his talk to attacting the existence of superclusters of galaxies. Although we would now say he had been wrong about this, his typical "cell size" for a cluster was 40 h^{-1} Mpc, and many of his clusters had complex substructure and would probably now be called superclusters, sheets, filaments, and so forth.

Opinions at the two meetings had crystallized around two strongly opposing views. The rich clusters must either be bound by dark matter, associated more with the clusters as a whole than with the individual galaxies, or the clusters must be short lived and currently expanding out of some fairly violent explosion.

Binding was advocated by Zwicky (1961), Minkowski (1961), Limber (1961), Baum (1961), who, with Zwicky, thought there might be enough intracluster starlight to bind the clusters with luminous material, and Abell (1961). Abell went further to propose "second-order clustering of galaxies and interactions between clusters of galaxies" which implied superclusters of 50 h^{-1} Mpc diameter and masses of 10^{16}-10^{17} M_o (H = 100). In this matter, he was specifically opposed to Zwicky, and I remember in the mid-60's Abell at a Caltech colloquium offering "political asylum at UCLA" to one of Zwicky's associates who had had the misfortune to find evidence for superclustering.

Van Albada (1961) and von Hoerner (1961) presented results of analytic and numerical models for the dynamical evolution of clusters whose implication was that, for bound clusters, Virial masses were likely to be wrong, but not by factors of more than two or three. Today, we would describe von Hoerner's calculation as an n-body simulation, with n = 16 (and also with roughly the correct result).

Unstable expanding clusters were first proposed by Ambartsumian (1958, 1961) by analogy with known expanding clusters of young stars like the Trapezium (now thought to have formed from bound gas clouds, but to have been liberated when stellar winds and supernovae blew away large amounts of still-uncondensed gas). Supporters of this view included Vorontsov-Velyaminov (1961), Kalloghjian (1961), Markarian (1961), de Vaucouleurs (1961), and the Burbidges (1961). The analogy is a false one, since many star clusters, and no galaxy ones, are demonstrably bound by their luminous material. In addition, we see no 10^{17} M_o gas clouds that could be the immediate precursors of unbound clusters to form in the near future. In fairness to Ambarsumian, he never supposed we did. His image was more nearly akin to that of Jeans, who once described the spiral nebulae as places where material was pouring into the universe from elsewhere.

The dissenters from the two main streams included Holmberg (1961) who believed that observational errors, foreground/background interlopers, and substructure in clusters could account for the large apparent velocity dispersions with no need for either dark matter or explosions, and Lemaitre (1961), who proposed that the rich clusters might might be constantly exchanging galaxies with the field, so that the configurations were permanent, but the individual members not gravitationally bound. Van den Bergh (1961) considered, but rejected a contribution from non-gravitational forces that might bind the clusters without requiring additional mass. Shortly after these meetings, Finzi (1963) summarized the evidence for M/L increasing with radius and suggested the closely related idea of "assuming a law of gravitation that implies a much stronger attraction at a long distance than that prediced by the law of Newton."

Virtually all these points of view have modern adherents or analogs. Ideas related to Ambartsumian's have been re-explored under the name "lagging cores" by Arp. et al. (1990). Binney (1971) has revived the suggestion of Smith (1936), Zwicky (1961), and Baum (1961) that galaxies might have faint luminous extensions, sufficient to keep M/L roughly constant with radius. Valentijn's (1990) claim that spiral disks are luminous but opaque has the same effect of binding things without requiring large values of M/L. The importance of substructure, interlopers, and uncertainties in the observations and their interpretation continues to be emphasized by Valtonen and Byrd (1986). Lemaitre's proposal seems to have no current supporters, but the density wave model of spiral structure has some similar features as does the model for Hickson compact groups of galaxies in which they are continuously re-formed from among the membership of looser groups.

Non-gravitational forces resurface frequently in the papers of people new to the field who shout, "but you have forgotten about magnetic fields." Actually we haven't. Magnetic fields are relativistic fluids and, unanchored, will expand at the speed of light. If they are to confine anything, they must be tied to non-zero-rest-mass material or have it to push against, and you come out behind on M/L because you need to hang on to the fields as well as to the large velocities (Katz 1994). Finally, Milgrom (1986) and Sanders (1986), among others, have presented testable versions of non-Newtonian gravity, in the form either of a lower limit to gravitational acceleration or of an additional Yukawa-form gravitational potential. Both experience (probably fatal) difficulties in accounting simultaneously for rotation of very small galaxies and velocity dispersions of very large clusters.

6. THE ~~SOVIET~~ AND AMERICAN REVOLUTIONS (1974)

The winner was, however, undoubtedly cosmological dark matter. The turning point came not through any one particular new sort of observation (though X-ray detection of rich clusters, measuring both the amount of intracluster gas and the depth of the potential confining it eventually brought some laggards aboard the bandwagon). Rather, it was the publication of two short summaries of existing data, indicative of M/L increasing monotonically with the length scale on which it was measured that persuaded most of the astronomical community, first, that the observations should be taken seriously and, second, that non-luminous gravitating material, or dark matter, was much the most likely explanation. Einasto, Kaasik, and Saar (1974) said, "The mass of galactic coronas exceeds that mass of populations of known stars by one order of magnitude, as do the effective dimensions" and, "The mass luminosity ratio rises to $f \gtrsim 100$ for spiral and $f \gtrsim 120$ for elliptical galaxies. With H = 50 km/s/Mpc this ratio for the Coma cluster is 170." Ostriker, Peebles, and Yahil (1974) summarized the situation by saying, "Currently available observations strongly indicate that the mass of spiral galaxies increases almost linearly with radius to nearly 1 Mpc...and that the ratio of this mass to the light within the Holmberg radius, f, is $\sim 200\ (M/L)_o$." A paper by Ozernoi (1974), which cites the Einasto et al. work (in preprint form) but not that by Ostriker et al, was perhaps the first to emphasize that the dark matter must be largely communal in rich clusters of galaxies and not associated with the individual objects.

Since then, it has been the non-believers who have found themselves out of step and, sometimes, out of print. There was an accompanying shift in nomenclature from "missing mass" (still found in many elementary textbooks) back to Kapteyn's (1922) original "dark matter," presumably because most of us felt that it is the light that is missing, not the mass. My views on the post-revolutionary status of dark matter have been summarized in two now rapidly-aging reviews (Trimble 1987, 1988).

7. THE DARK MATTER CANDIDATES

The inventory of baryonic dark matter candidates has not expanded beyond the original Neolithic trio of black holes, brown dwarfs, and old white dwarfs (nor have large numbers of any of these turned up in contexts indicating major contributions to the mass supply). Most of specific non-baryonic candidates for dark matter discussed at this meeting and elsewhere are relatively recent inventions and belong to the realm of current events rather than history. Exceptions are: Einstein's cosmological constant (even in its modern use to stretch out the age of the universe, which popped up last in about 1970, when there seemed to be an excess of quasars with redshifts near 1.955, as you would expect from a coasting epoch): magnetic monopoles, which, of course, go back to Dirac (though I don't suppose he planned to close the universe with them); and a non-zero-rest-mass for ordinary neutrinos, considered by Gershtein and Zeldovich (1966). On the assumptions that the density of the universe is not more than 10 times the closure density, that only one species (electron or muon neutrinos) need be considered at a time, and that neutrinos had been in thermal equilibrium with photons at some temperature well above 1 MeV, they derived a maximum possible mass of 400 eV/c^2 for either species. This was above the laboratory limit at the time for electron neutrinos, but well below it for muon neutrinos. Cowsik and McClelland (1972) brought the limit down within striking distance of lab values for the electron neutrino by considering both species at once and using a less generous limit for the density of the universe.

Since then, there has been a monotonic increase in the cuteness of the names given to the candidates. The one of the two authors of the first WIMP paper (Steigman and Turner 1984) who is currently speaking to me indicates that the motivation of the paper was as much the coinage of the name as the tracing out of the particle properties. And about MACHOs, Winos, and hedgehogs, perhaps the less said the better.

ACKNOWLEDGEMENTS

We can, just barely, still reach directly much of the history of the major ideas in dark matter astrophysics. I am particularly indebted to Horace W. Babcock, Daniel Magnus Popper, Vera Cooper Rubin, Lawrence Hugh Aller, and the late Olin C. Wilson for sharing their first-hand memories of critical events in the area.

REFERENCES

Abell, G.O. 1961. AJ 66. 607
Adams, J.C. 1847 Mem. RAS 16
van Albada, C.B. 1961. AJ 66, 590
Ambartsumian, V.A. 1958. in Solvay Conference: La Structure et l'Evolution de l'Universe (Sovay, Inst. de Physique) p. 163
..... 1961. AJ 66, 536
Arp, H.C., G.R. Burbidge, F. Hoyle, J.V. Narlikar, and N.C. Wickramasinghe 1990. Nature 346, 807
Babcock, H.W. 1939. Lick Obs. Bull. 19, 41
Bahcall, J.N. 1986. ARA&A 24, 577
Baum, W.A. in IAU Symp. 15, McVittie 1961, p. 255
van den Bergh, S. 1960. MNRAS 121, 387 and ApJ 131, 558
..... 1961. AJ 66, 566
Bessell, W.F. 1845. AN 22, 145, 169 & 185; MNRAS 6, 136
Binney, J. 1986. Phil. Trans. Roy. Soc. A320, 431
Bondi, H. and T. Gold 1948. MNRAS 108, 252
Bonnor, W.B. 1960. MNRAS 121, 475
Bolton, C.T. 1972. Nature 235, 271
Burbidge, E.M. and G.R. Burbidge 1959. ApJ 130, 15, 23, & 629
..... 1961. AJ 66, 541
Clerke, Agnes Mary 1885. A Popular History of Astronomy During the Nineteenth Century (Edingburgh, Adam & Charles Black)
Cowsik, R. & J. McClelland 1972. PRL 29, 669
Einasto, J., A. Kaasik, and E. Saar 1974. Nature 250, 309
Finzi, A. 1963. MNRAS 127, 21
Gershtein, S.S. and Ya. B. Zeldovich 1966. JETP Lett. 4, 175
Gilmore, G. and K. Kuijken 1983. MNRAS 202, 1025
von Hoerner, S. 1961. AJ 66, 580
Holmberg, E. 1937. Lund Annals 6, 1
..... 1950. Meddeilung Lund Astron. Bos. Ser. 2, No. 128
..... 1961. AJ 66,
Holt, S.S. et al. eds. 1992. Testing the AGN Paradigm (AIP, New York)
Hoyle, F. 1948. MNRAS 108, 372
..... 1959. in Paris Symposium on Radio Astronomy (IAU Symp. 9) ed. R.N. Bracewell (Stanford Univ. Press) pl 529
van de Hulst, H.C. et al. 1957. BAN 14, 1
Jeans, J.H. 1922. MNRAS 82, 130
Kahn, F.D. and L. Woltjer 1959. ApJ 180, 105
Kalloghjian, A. 1961. AJ 66, 554

Kapteyn, J.C. 1922. ApJ 55, 302
Katz, J.I. 1994. ASS 213, 155
King. I.R. 1977. in The Evolution of Galaxies and Stellar Populations, ed. B.M. Tinsley & R.B. Larson (Yale University Observatory) p. 1
Kunth, R. 1952. Zs. f. Ap. 28, 234
Lalleman, A. 1960. PASP 72, 76
de Laplace, P.S. 1796. Exposition du Systeme du Monde, Vol. 2, p. 305 (Paris: J.B.M. Duprat)
Lemaitre, G. 1961. AJ 66, 603
Leverrier, U.J.J. 25, No. 580
Limber, D.N. 1961. AJ 66, 572
Limber, D.N. and W.G. Mathews 1960. ApJ 132, 286
Lohman, W. 1956. Zs. f. Phys. 144, 66
Markarian, B.E. 1961. AJ 66, 555
McVittie, G.C. 1961. ed. Problems in Extragalactic Research (IAU Symp. 15) New York: McMillan
McVittie, G.C. 1961. op cit p. 449
Michell, J. 1784. Phil. Trans. Roy. Soc. 74, 35
Milgrom, M. 1986. ApJ 306, 9
Minkowski, R. 1961. AJ 66, 558
Muller, C.A. 1959. BAN 14, 339
Newman, J. T.L. Page, and E. Scott, eds. 1961 AJ 66
Oort, J.H. 1932. BAN 6, 249
..... 1958. in Solvay Conference: La Structure et l'Evolution de l'Universe (Solvay, Inst. de Physique) p. 163
..... 1965. in A. Blaauw & M. Schmidt eds. Galactic Structure (Univ. of Chicago Press) p. 455
Osterbrock. D.E. 1962. in Interstellar Matter in Galaxies (New York: Benjamin) p. 117
Oort, J.H. 1959 in Paris Symposium on Radio Astronomy (IAU Symp. 9), ed. R.N. Bracewell (Stanford Univ. Press_ p. 353
Ostriker, J.P., P.J.E. Peebles, and A. Yahil 1974. ApJ 193, L1
Ozernoi, L.M. 1974. Sov. Astron. AJ
Mayall, N.U. and L.H. Aller 1940. PASP 52, 278
Page, T.L. 1952. ApJ 116, 63
..... 1956. ApJ 136, 685
Salpeter, E.E. 1964. ApJ 140, 794
Sanders, R.H. 1986. MNRAS 223, 539
Schaeberle, J.M. 1896. AJ 17, 37
Schwarzschild, M. 1954. AJ 59, 273
Smith, S. 1937. ApJ 83, 23
Standish, E.M. 1993. AJ 105, 2900
Steigman, G. and M. Turner 1984. Nucl. Phys. B253, 73
Tombaugh, C.W. and P.W. Moore 1980. Out of the Darkness: The Planet Pluto (Harrisburg: Stackpole)
Trimble, V. 1987. ARA&A 25, 425
..... 1988. Contemp. Phys. 29, 373
Valentijn, E.A. 1990. Nature 346, 152
Valtonen, M.J. and G.C. Byrd 1986. ApJ 303, 523
de Vaucouleurs, G. 1959. Observatory 79, 113 and ApJ 130, 718
..... 1960. ApJ 131, 585
..... 1961. AJ 66, 629
Vorontsov-Velyaminov. V. 1961. AJ 66, 551
Zeldovich, Ya. B. 1964. Sov. Phys. Doklady 9, 195
Zwicky, F. 1933. Helvetica Physica Acta 6, 110
..... 1937. ApJ 86, 217 and Proc. NAS 23, 251

Zwicky, F. 1939, Proc. NAS 25, 605
..... 1951, PASP 63, 61
..... 1959, Handbuch der Physik 53, 373
..... 1956, Ergebnisse der Exakten Naturwissenschaften (Springer, Tracts in Modern Physics) 29, 344
..... 1961, in G.C. McVittie, 1961, p. 347

LOW MASS STARS AS DARK MATTER

Very Low Mass Stars As Dark Matter

James Liebert
Steward Observatory, Univ. of Arizona, Tucson AZ 85721
liebert@as.arizona.edu

ABSTRACT

1. INTRODUCTION

It is now generally believed that a substantial fraction of matter in the Galaxy is not accounted for in visible form, but its existence is inferred from measurements of the velocities of gas, stars and stellar systems far from the center of the Galaxy (*cf.*Fich & Tremaine 1991). This unseen matter is at least comparable to and could be several times greater than the mass accounted for by stars, gas and dust. With the recent discovery of microlensing events by objects which may be in the halo of our galaxy (Alcock *et al.* 1993 – the MACHO collaboration; Aubourg *et al.* 1993 – the EROS collaboration), there has been a resurgence of interest in examining low mass stellar objects as the source of an important fraction of the unseen matter in this component of our Galaxy.

Measurements of the stellar velocity and density distribution perpendicular to the Galactic disk provide an estimate of the total disk density. This turns out be to be at least 0.1 $M_{\odot}$pc^{-}3, and it has long been suspected (Oort 1932) that this exceeds the density of visible stars, gas and dust. However, the evidence is controversial (Bahcall 1984, Kuijken and Gilmore 1989, Gould 1990, Bahcall et al. 1992), though the situation allows at least a moderate amount of additional mass in the form of baryonic entities of low luminosity. In fact it is likely that any such additional disk material is baryonic in nature, since it would be dissipational; on the other hand, it would contribute little to the missing matter of the Galaxy as a whole.

The candidate stellar objects for halo and disk dark matter traditionally include low mass objects near or below the hydrogen-burning mass limit and stellar remnants, especially white dwarfs. For completeness, we discuss the latter first.

2. WHITE DWARFS OF THE DISK AND HALO

A reasonably representative census of the low luminosity stars in the solar neighborhood is displayed in Fig.1, an HR Diagram of stars with accurate trigonometric parallaxes measured in the U.S. Naval Observatory program (Monet *et al.* 1992). The sequence on the left are the degenerate dwarfs, the vast majority of which show kinematics indicative of membership in the disk population. It may be noted that the white dwarf sequence ends just past $M_V \sim +16$, while the disk main sequence on the right extends about three magnitudes fainter.

It has been argued for more than a decade (Liebert *et al.* 1979; Winget *et al.* 1987) that the observed white dwarf sequence cuts off because of the finite age of the Galactic disk, some 9–10 Gyr according to Winget *et al.* . The luminosity function used in the preceding analysis is published as Liebert, Dahn & Monet (1988). Some confirmation of this functional form has come from the independent studies of proper motion samples by Evans (1992) – see also von Hippel, Gilmore and Jones (1993) – and of a sample of wide binaries with cool white dwarf components by Oswalt and Smith (1994). Ruiz, Anguita and Maza (1989) have an ongoing proper motion survey of the largely unexplored southern sky (at faint magnitudes) and have discovered one white dwarf fainter than $M_V \sim +17$ (with a good trigonometric parallax measurement by these authors). The age assignable to the Galactic disk, and its uncertainty, remain subject to considerable discussion, since these depend on such assumed parameters as the thicknesses of outer hydrogen and helium layer masses and on the possible phase separation in the crystallizing C/O core (Wood 1994; D'Antona and Mazzitelli 1990; Winget and Van Horn 1987).

Recently there has been theoretical exploration of the white dwarf luminosity function as a possible probe of the age of the Galactic halo (Tamanaha *et al.* 1990; Mochkovitch *et al.* — 1990), though the observational data do not yet reach to faint enough magnitudes (Liebert, Dahn & Monet 1989). It is unlikely that white dwarfs dominate the dark halo mass, however, since evidence for this would normally be expected in the observed LF of halo stars (from their counterparts of lower initial mass) and/or the Galactic chemical enrichment history (from their more massive counterparts).

3. LOW MASS STELLAR OBJECTS OF THE DISK

Regardless of the fact that any unaccounted-for disk material will not satisfy the Galactic halo missing mass problem, there has been considerable interest in the possibility that substellar objects might exist in the solar neighborhood, perhaps in substantial numbers. There are two key parts to determining an accurate mass function near the stellar mass limit: (1) an accurate LF, and (2) an accurate conversion from luminosity to mass in the regime where the functional form is changing dramatically.

The statistical determination of the LF at the faint limit has improved dramatically in the last several years, especially due to the comprehensive photographic, photometric and astrometric survey of Tinney, Reid & Mould (1993) and the photometric survey of Kirkpatrick *et al.* (1994). There is no doubt that the mass function flattens in slope below $0.5 M_\odot$. There is some evidence for a peak near $0.25 M_\odot$, followed by a gentle decline. However, the LF appears to rise steeply near $M_v > +17$ ($M_{bol} > +13$), which suggests that the mass function could be rising abruptly as one crosses the boundary for stable hydrogen–burning (Tinney 1993). This leaves open the possibility that brown dwarfs are numerically important in the solar neighborhood, though unimportant to the mass budget of the disk. On the other hand, it must be acknowledged that not a single, unambiguous brown dwarf has yet been identified from *any* survey; in particular, such objects cannot be numerous as companions to nearby stars (see the review by Burrows & Liebert 1993).

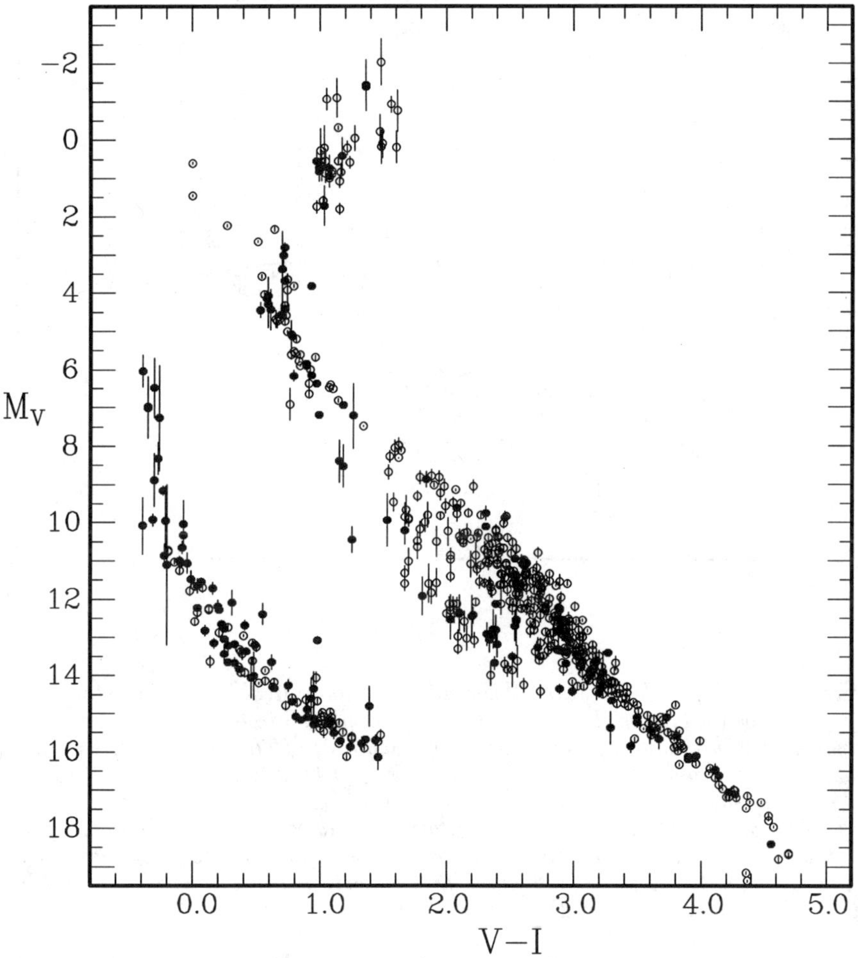

Figure 1. *The M_v vs. $V-I$ diagram for a selection of 680 stars with USNO trigonometric parallaxes either published (open circles) or unpublished but nearly completed (filled circles) – courtesy of C.C. Dahn. A detailed discussion of the USNO parallax results is found in Monet et al. (1992).*

4. LOW MASS STELLAR OBJECTS OF THE HALO

As indicated in the first paragraph, there has been great interest in the possibility that low mass stellar and substellar objects in the halo might explain (1) the large mass inferred from the Galactic rotation curve far from the Galactic Center, and (2) the gravitational microlensing events detected by the MACHO and EROS projects in lines of sight through the halo.

Dahn, Liebert, Harris, & Guetter (1994) have estimated the LF of a kinematically-selected sample of Pop II (visible spheroid) stars in the solar neighborhood, following closely the technique of Bahcall & Casertano (1986). This

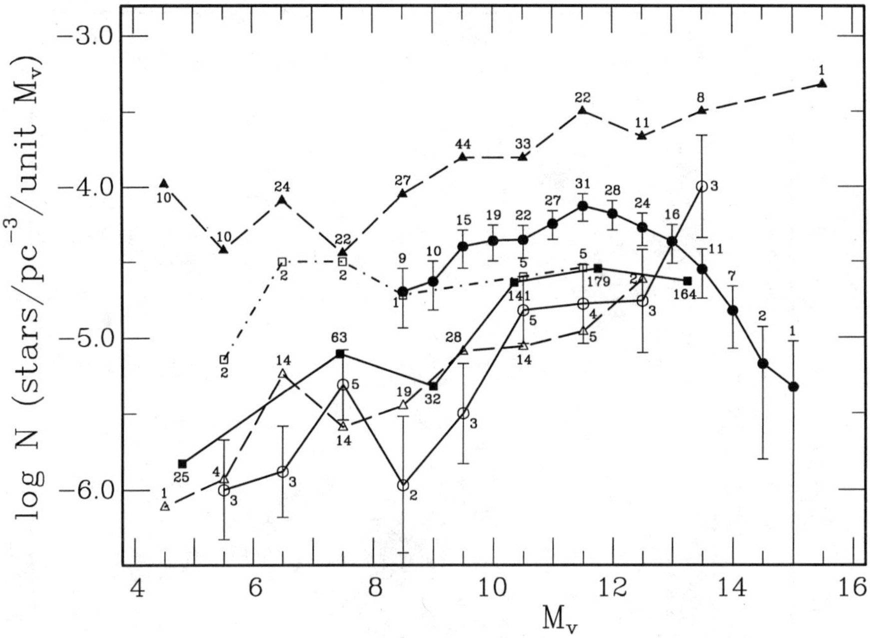

Figure 2. *The Pop. II LF from the study of kinematically-selected stars in the solar neighborhood from Dahn et al. (1994, filled circles, with error bars) is compared with previous determinations by Schmidt (1975; open squares), Eggen (1983; filled triangles), Bahcall & Casertano (1986; open triangles), Dawson (1986; open squares), and Richer & Fahlman (1992; open circles, with error bars). The small numbers near each plotted point indicate the number of stars contributing to the derived value. The present sample includes only objects with tangential velocities* $v_{tan} > 220$ *km/s and a derived correction factor of* $1/\chi = 2.46$ *has been applied (from Dahn et al. 1994).*

distribution is shown as the filled circles in Fig.2, along with several earlier determinations. Most of the stars in the new sample had trigonometric parallax measurements, a feature that deep pencil-beam surveys lack. None of the determinations yields a mass density of visible Pop. II stars that can account for the "dark halo." Moreover, the Dahn *et al.* function peaks at $M_V \sim +12$ and appears to turn over; thus, the expected mass density in the substellar regime at $M_V \geq +15$ is modest. The other determinations have little sensitivity at the lowest luminosities of visible Pop. II stars, but only that of Richer and Fahlman (1992) suggests a rising slope to the LF.

Now Bahcall *et al.* (1994) using the Wide-Field Camera of the recently repaired Hubble Space Telescope have searched for visible stars to the greater galactocentric distances at which the "dark halo" material should be better represented. Still they find no evidence for any excessive number of stars; in fact, if one simply extends the Dahn *et al.* LF to the appropriate distance assuming an $R^{-3.5}$ dependence for these Pop. II stars, then the numbers predicted are consistent with what Bahcall *et al.* found.

Deep ground-based pencil-beam surveys have pushed the CCD detector state-of-the-art to fainter magnitudes, using telescopes larger in aperture than the HST and covering larger areas of the sky. Particularly important have been the surveys of Boeshaar *et al.* (1986), Tyson(1988), Hu *et al.* (1994), and Boeshaar, Tyson, & Bernstein (1995, this volume). Again, no evidence has been found for any population of low luminosity stars forming a Galactic corona or "dark halo." On the other hand, it must be kept in mind that the spatial distributions of the visible Population II and the inferred dark halo matter are quite different. These results constrain any stellar population accounting for the dark halo to have a primarily substellar mass function, very different from the visible Pop. II LF.

The author is grateful to the National Science Foundation for support through grant AST 92–17961. I thank Conard Dahn and Hugh Harris for the use of figures from work in press and for many fruitful discussions. I thank John Bahcall and Don Schneider for hospitality and many important discussions at the Institute for Advanced Study.

REFERENCES

Alcock, C. *et al.* 1993, Nature, 365, 621
Aubourg, E. *et al.* 1993, Nature, 365, 623
Bahcall, J.N. 1984, ApJ 276, 169
Bahcall, J.N., & Casertano, S. 1986, ApJ, 308, 347
Bahcall, J., Flynn, C., Gould, A., & Kirhakos, S. 1994, ApJL, 435, L51
Bahcall, J., Flynn, C., & Gould, A. 1992, ApJ 389, 234
Boeshaar, P.C., Tyson, J.A., & Bernstein, G.M. 1995, this volume
Boeshaar, P.C., Tyson, J.A., Seitzer, P. 1986, in *Astrophysics of Brown Dwarfs*, eds. Kafatos, M.C., Harrington, R.S., Maran, S.P., Cambridge: Cambridge U. Press, p. 76
Burrows, A., & Liebert, J. 1993, Rev. Mod. Phys., 65, 301
D'Antona, F., and Mazzitelli, I. 1990, ARAA, 28, 139
Dawson, P.C. 1986, ApJ 311, 984
Dahn, C.C. and Liebert, J., Harris, H.C., and Guetter, H.H. 1994, in Proceedings of the ESO Workshop *The Bottom of the Main Sequence And Beyond*, ed. C.G. Tinney, Berlin: Springer-Verlag, in press
Eggen, O.J. 1983, ApJS, 51, 183
Evans, D.W. 1992, MNRAS, 255, 521
Fich, M., & Tremaine, S. 1991, ARAA, 29, 409
Gould, A. 1990, MNRAS, 244, 25
Hu, E. *et al.* 1994, Nature, 371, 493
Kirkpatrick, D., McGraw, J.T., Hess, T.R., Liebert, J., & McCarthy, Jr., D.W. 1994, ApJS 94, 749
Kuijken, K., & Gilmore, G. 1989, MNRAS, 239, 571
Liebert, J., Dahn, C.C., Gresham, M., & Strittmatter, P.A. 1979, ApJ 233, 226
Liebert, J., Dahn, C.C., & Monet, D.G. 1988, ApJ 332, 891
Liebert, J., Dahn, C.C., & Monet, D.G. 1989 in *White Dwarfs*, Proc. IAU Coll. 114, ed. G. Wegner, Berlin: Springer-Verlag, p. 15
Mochkovitch, R., Garcia-Barro, E., Hernanz, M., Isern, J., & Panis, J.F. 1990, AstrAp, 233, 456
Monet, D.G., Dahn, C.C., Vrba, F.J., Harris, H.C., Pier, J.R., Luginbuhl, C.B., Ables, H.D. 1992, AJ 103, 638
Oort, J.H. 1960, Bull. Astron. Inst. Netherlands 6, 249

Oswalt, T.D., & Smith, J.A. 1994, in *Proc. 9th European Workshop on White Dwarfs*, eds. K. Werner & D. Koester, Kiel, Germany, in press

Richer, H.B., and Fahlman, G.G. 1992, Nature 358, 383

Ruiz, M.T., Anguita, C., & Maza, J. 1989, in *White Dwarfs*, Proc. IAU Coll. 114, ed. G. Wegner, Berlin: Springer-Verlag, p. 122

Tamanaha, C.M., Silk, J., Wood, M.A., & Winget, D.E. 1990, ApJ, 358, 164

Tinney, C.G. 1993, ApJ, 414, 279

Tinney, C.G., Reid, I.N., & Mould, J.R. 1993, ApJ, 414, 254

Tyson, J.A. 1988, AJ 96, 1

Von Hippel, T., Gilmore, G., & Jones, D.H.P. 1993, in *White Dwarfs: Advances in Theory and Observation*, NATO ASI Series, ed. M.A. Barstow, Berlin: Kluwer Acad. Publ., p. 7

Winget, D.E., Hansen, C.J., Liebert, J., Van Horn, H.M., Fontaine, G., Nather, R.E., Kepler, S.O., & Lamb, D.Q. 1987, ApJ Lett., 315, L77

Winget, D.E., & Van Horn, H.M. 1987, in *The Second Conference on Faint Blue Stars*, IAU Coll. 87, eds. A.G.D. Philip, D.S. Hayes, and J. Liebert, Schenectady NY: L. Davis Press, p. 363

Recent Developments in Gravitational Microlensing and the Latest MACHO Results: Microlensing Towards the Galactic Bulge

D.P. Bennett[†,*], C. Alcock[*,†], R.A. Allsman[*],
T.S. Axelrod[*], K.H. Cook[*,†], K.C. Freeman[‡],
K. Griest[†,‖], S.L. Marshall[†,♭], S. Perlmutter[†],
B.A. Peterson[‡], M.R. Pratt[†,♭], P.J. Quinn[‡], A.W. Rodgers[‡],
C.W. Stubbs[†,♣], W. Sutherland[♠]
(The MACHO Collaboration)

[*] Lawrence Livermore National Laboratory, Livermore, CA 94550

[†] Center for Particle Astrophysics, University of California, Berkeley, CA 94720

[‡] Mt. Stromlo and Siding Spring Observatories, Australian National University, Weston, ACT 2611, Australia

[‖] Department of Physics, University of California, San Diego, CA 92039

[♭] Department of Physics, University of California, Santa Barbara, CA 93106

[♣] Department of Astronomy, University of Washington, Seattle, WA 98195

[♠] Department of Physics, University of Oxford, Oxford OX1 3RH, U.K.

ABSTRACT

We review recent gravitational microlensing results from the EROS, MACHO, and OGLE collaborations, and present some details of the very latest MACHO results toward the Galactic Bulge. The MACHO collaboration has now discovered in excess of 40 microlensing events toward the Galactic Bulge during the 1993 observing season. A preliminary analysis of this data suggests a much higher microlensing optical depth than predicted by standard galactic models suggesting that these models will have to be revised. This may have important implications for the structure of the Galaxy and its dark halo. Also shown are MACHO data of the first microlensing event ever detected substantially before peak amplification, the first detection of parallax effects in a microlensing event, and the first caustic crossing to be resolved in a microlensing event.

Introduction

The suggestion by Paczyński (1986) that compact objects in the halo of our Galaxy might be detected by means of gravitational microlensing was confirmed in a rather spectacular fashion last September when the EROS, MACHO, and OGLE collaborations each discovered their first candidate microlensing events in the course of less than a month (Alcock, *et al.*, 1993, Aubourg, *et al.*, 1993, and Udalski, *et al.*, 1993). Since then, the total number of microlensing events discovered has grown to about 60, and the implications of the microlensing results toward both the Large Magellanic Cloud and the Galactic bulge remain somewhat mysterious.

The basic physics of microlensing is quite simple. If a compact object passes very close to the line of sight to a background star, the light will be deflected to produce two images of the star. In the case of perfect alignment, the star will appear as an 'Einstein ring' with a radius of

$$r_E = \sqrt{\frac{4GMLx(1-x)}{c^2}},$$

where M is the lens mass, L is the observer-source distance and x is the ratio of the observer-lens and observer-source distances. In a typical situation of imperfect alignment, the image appears as two arcs. On cosmological scales, many cases of quasars multiply imaged by foreground galaxies are known ; however, for stellar lensing the angular separation of the lensed images is much smaller. For a source distance of 50 kpc $\approx 10^{10}$ AU and a lens distance of 10 kpc, the Einstein radius is $r_E \approx 8\sqrt{M/\mathrm{M}_\odot}$ AU ; this yields an angular separation of ~ 0.001 arcsecond which is well below even the Space Telescope resolution, hence the term 'microlensing'.

However, in the point source approximation, the lensing produces a net amplification of the source by a factor

$$A = \frac{u^2+2}{u\sqrt{u^2+4}}$$

where $u = b/r_E$ and b is the distance of the lens from the direct line of sight. (The above scale of 8 AU is much larger than the radius of a star, hence the point source approximation is accurate for lens masses $\gtrsim 10^{-5}\mathrm{M}_\odot$). The amplification is approximately u^{-1} for $u \lesssim 0.5$, and $1+2u^{-4}$ for $u \gg 1$, hence the amplification can be very large but falls rapidly for $u \gtrsim 1$. Since objects in the Galaxy are in relative motion, this amplification will be time-dependent; for a typical lens transverse velocity of 200 km/s, the duration is

$$\hat{t} \equiv \frac{2r_E}{v_\perp} \approx 140\sqrt{M/\mathrm{M}_\odot} \text{ days}.$$

This is a convenient timescale for astronomical observations, and thus by sampling on a range of timescales the microlensing searches may be sensitive to a wide mass range from small planets of $\sim 10^{-6}\mathrm{M}_\odot$ to black holes of $\sim 100\mathrm{M}_\odot$, covering most of the plausible candidates.

Optical Depth

The 'optical depth' τ for gravitational microlensing is defined as the probability that a given star is lensed with $u < 1$ or $A > 1.34$ at any given time, and is

$$\tau = \pi \int_0^L \frac{\rho(l)}{M} r_E^2(l)\, dl$$

where l is the distance along the line-of-sight and ρ is the dark matter density. Since $r_E \propto \sqrt{M}$, while for a given ρ the number density of lenses $\propto M^{-1}$, the optical depth is independent of the individual MACHO masses. Using the virial theorem, it is found that $\tau \sim (v/c)^2$, where v is the rotation speed of the Galaxy. More detailed calculations (Griest, 1991) give an optical depth for lensing by halo dark matter of stars in the Large Magellanic Cloud (LMC) of

$$\tau_{\rm LMC} \approx 5 \times 10^{-7}$$

This very low value is the main difficulty of the experiment; only one star in two million will be amplified by $A > 1.34$ at any given time, while the fraction of variable stars is much higher, $\sim 3 \times 10^{-3}$.

Microlensing Signatures

Fortunately, microlensing has many strong signatures which can discriminate it from stellar variability :

1) Since the optical depth is so low, only one event should be seen in any given star.
2) The deflection of light is wavelength-independent, hence the star should not change color during the amplification.
3) The accelerations of galactic objects are negligible on timescales of these events, hence the events should be symmetrical in time, and have a shape derived from (2) with $u(t) = \sqrt{u_{min}^2 + (v_\perp(t - t_{max})/r_E)^2}$. Examples of such light-curves are shown in Figure 2.

All these characteristics are distinct from known types of intrinsic variable stars; most variable stars are periodic or semi-regular, and do not remain constant for long durations. They usually change temperature and hence color as they vary, and they usually have asymmetrical lightcurves with a rapid rise and slower fall.

In addition to these individual criteria, if many candidate microlensing events are detected, there are further statistical tests that can be applied:

4) The events should occur equally in stars of different colors and luminosities.
5) The distribution of impact parameter u_{min} should be uniform from 0 to the experimental cutoff $u(A_{min})$.
6) The event timescales and peak amplifications should be uncorrelated. (In practice these distributions will be modulated by the detection efficiencies, which can be computed from simulations).

It is worth clarifying two common misconceptions: first, this experiment does *not* provide an estimate of Ω, since the amount of dark matter within 50 kpc of spiral galaxies is approximately known from rotation curve data. This represents $\Omega \sim 0.05$: if $\Omega = 1$ either the halos must extend far beyond 50 kpc or there must be intergalactic dark matter. Secondly, the lenses are not required to be dark, merely significantly fainter than the source stars. However, since the LMC is located 30° from the galactic plane, the optical depth from known stars is only $\lesssim 10^{-8}$ (Bahcall, *et al.*, 1994).

Microlensing Searches

Due to the low optical depth, a very large number of stars must be monitored over a long period to achieve a significant detection rate. The optimal targets when searching for Machos in the galactic halo are the Large and Small Magellanic Clouds, the largest of the Milky Way's satellite galaxies, since they have a high surface density of stars and are distant enough at 50 and 60 kpc to provide a good path length through the dark halo. This requires a Southern hemisphere observatory.

Another target for microlensing seaches is the Galactic bulge where a higher microlensing optical depth is predicted due to lensing by faint stars in the Galactic disk (Griest, *et al.*, 1991, Paczyński, 1991). Naively, one might expect that microlensing toward the bulge would not be of great relevance for the halo dark matter problem, but as we shall see, this is not the case.

There are curently three microlensing surveys that are taking data and have reported results: the EROS, MACHO, and OGLE collaborations. The EROS collaboration operates two seperate experiments which observe the LMC. They search for events lasting from days to months using photographic plates taken at the ESO Schmidt and for events lasting less than a day from a 40-cm telescope with a large format CCD array camera also located at the ESO site in La Silla, Chile. They plan to upgrade to a dedicated 1-m telescope with a very large format two color CCD camera system (64 or 128 Mega-pixels) in late 1995 or 1996. The OGLE collaboration now observes the Galactic bulge with ~ 75 nights per year on the 1-m Swope telescope at Las Campanas, Chile, but they also plan to upgrade to a dedicated 1.3-m telescope in 1995 or 1996.

The MACHO collaboration has full-time use of the 1.27-m telescope at Mt.Stromlo Observatory near Canberra, Australia, which is used to image 0.5 square degrees of sky simultaneously in two color bands. The two foci are equipped with large CCD cameras (Stubbs, *et al.*, 1993), each containing 4 Loral CCD chips of 2048×2048 pixels. Most of the observing time is divided between the LMC and bulge, with a small amount of time being spent on the SMC.

Microlensing Towards the LMC

The first Galactic microlensing events every discovered were announced simultaneously by the EROS and MACHO collaboration in the fall of 1993 (Alcock, *et al.*, 1993, Aubourg, *et al.*, 1993). These data initiated a flurry of speculation as to whether the the halo might be entirely baryonic or whether some non-halo population might be responsible for the microlensing toward the LMC. Although the EROS and MACHO collaborations have not yet announced definitive results on the microlensing optical depth toward the LMC, preliminary results given in a number of conferences suggest that the optical depth is both significantly below the prediction ($\tau \approx 5\times 10^{-6}$) for a "standard halo" composed entirely of Machos and significantly above the maximum possible value ($\tau < 10^{-8}$) due to known distributions of stars (Bahcall *et al.*, 1994). The possibility that these events are not microlensing but rather a new variable star phenomena now seems rather unlikely as spectral follow-up has confirmed that they appear to be normal stars (Della Valle, 1994, Beaulieu, *et al.*, 1994, Giraud, 1994, and Alcock, *et al.*, 1994a).

One possibility that remains viable is that the microlensing may be due to faint stars in the LMC (Sahu, 1994), but this seems unlikely for several reasons. First, to get the required event rate, one need to assume that the mass of the LMC is almost entirely baryonic while the Milky Way is apparently dominated by non-baryonic matter. It is difficult to see how this could come about. Also, although microlensing by a self-consistent model of the LMC which is massive enough to explain the observed events has never been investigated, naive arguments suggest that the timescales of the observed events could most easily be explained by microlensing by objects of substellar mass.

Thus, the most reasonable conclusion is that the microlensing is due to brown dwarfs in a halo or perhaps spheroidal population (Giudice, Mollerach, & Roulet, 1993), but that the these objects do not provide the dominant mass for a "standard halo." It is tempting to conclude that there must be another component that dominates the halo, except that, as we shall see, the microlensing optical depth toward the galactic bulge is much higher than expected. Thus, another viable possibility is that the dynamics of the inner galaxy are dominated by a massive disk and bulge, and that Machos are the dominant component of a much smaller than "standard" halo.

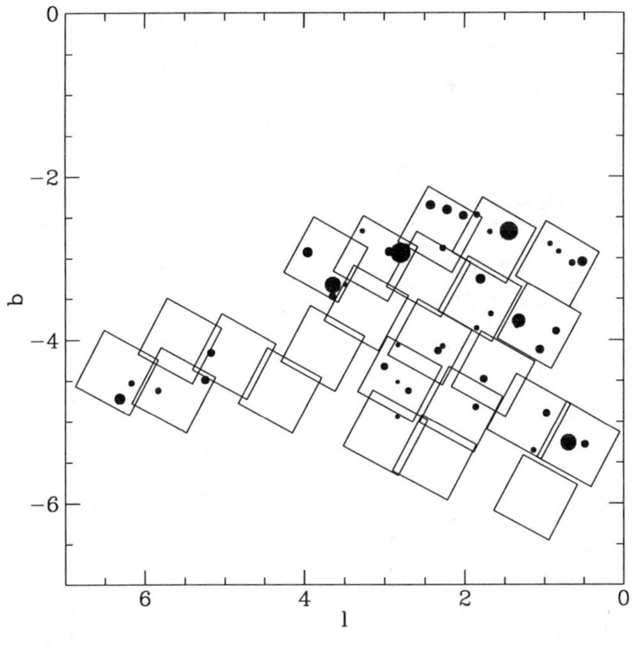

Figure 1. The location of the 24 bulge fields that were observed by the MACHO group and analyzed for this paper. The spots indicate the locations of detected events, and the spot area is proportional to the $\hat{t}$ value for the event.

Microlensing Events Toward the Galactic Bulge

Within a week after the EROS and MACHO teams announced the first microlensing events detected toward the LMC, the OGLE collaboration had discovered the first microlensing event ever seen toward the Galactic bulge (Udalski, *et al.*, 1993). Subsequent analyses by the OGLE group revealed about 10 additional events, and revealed a very high optical depth towards Baade's window (Udalski, *et al.*, 1994a). These results are fully consistent with the MACHO results that will be discussed in the remainder of this paper.

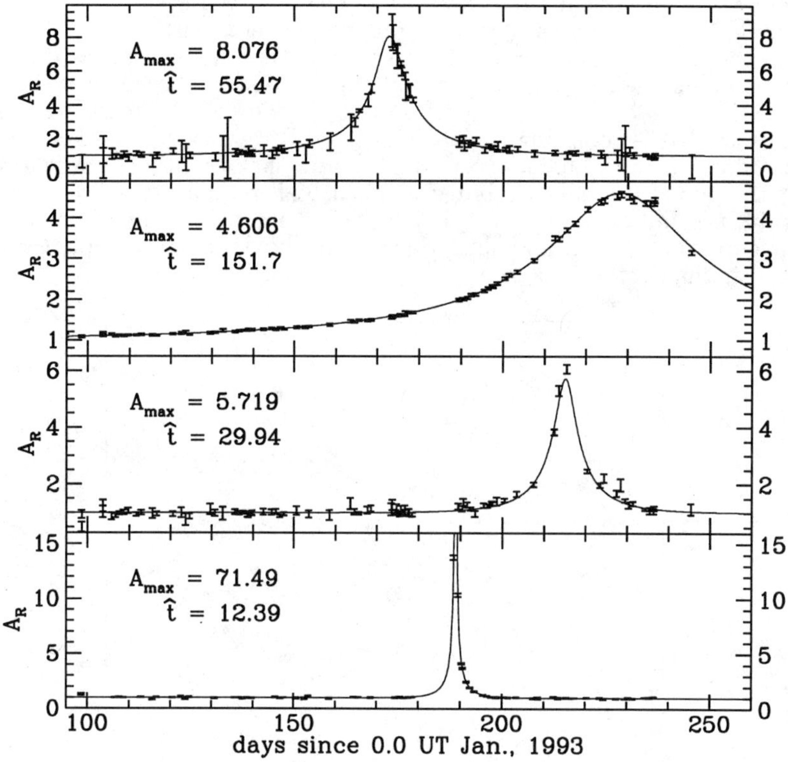

Figure 2. MACHO-Red band lightcurves of 4 events from field 101 located at $\ell \approx 3.8$, $b \approx -3.1$.

We have recently analyzed the 2-color light curve data for about 10 million in 24 fields toward the Galactic bulge. The location of these fields in galactic coordinates (as well as the location of the events) in shown in Fig. 1. Before starting the microlensing search, measurements with questionable PSF chi square, crowding, missing pixel, or cosmic ray flags are flagged as bad measurements and removed from further consideration. Then, we run matched filters through the data in both colors, and 37,000 stars passed the filter trigger threshold and were fit jointly in both colors with a microlensing light curve. These stars form our "level 1" candidate list and are subject to further cuts on the difference between the χ^2 values for the microlensing and flux=const. fit, and the microlensing fit χ^2 per d.o.f. Further cuts are made on the number of data points near the peak and the average estimated error. Four of the 45 candidates which pass these cuts are shown in Fig. 2. A few of the 45 events may be variable stars, but the majority are almost certainly microlensing events. A color magnitude diagram showing the locations of these 45 stars as well as 7% of the stars within 2.5' of these stars is shown in Fig. 3.

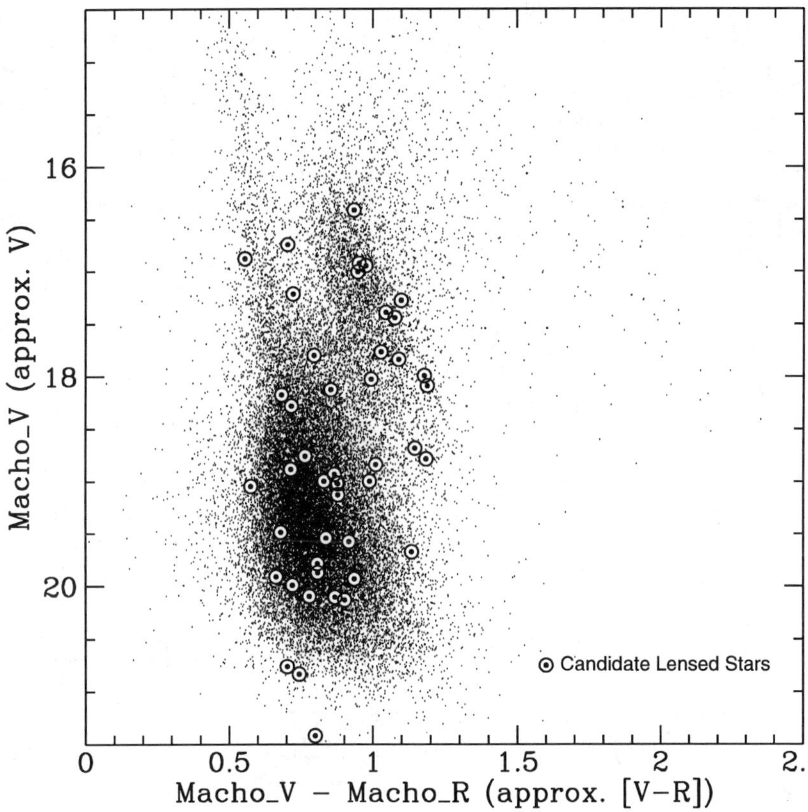

Figure 3. A color magnitude diagram for the 45 microlensing candidates and 7% of the stars that fall within 2.5' of the candidates.

Preliminary Optical Depth Estimates

The primary measurable quantity in a gravitational microlensing experiment is the microlensing optical depth which is defined to be the instantaneous probability that a random star is magnified by a lensing object by more than a factor of 1.34. This is a measure of the mass in microlensing objects along the line of sight to the source stars. Experimentally, one can define the measured optical depth as the amount of time that the microlensed stars spend inside the Einstein ring (*i.e.*are amplified by more than 1.34).

$$\tau_{\rm meas} = \frac{1}{E}\frac{\pi}{4}\sum_i \frac{\hat{t}_i}{\epsilon(\hat{t}_i)} \ .$$

where E is the total exposure (in star-years), $\hat{t}_i$ is the Einstein ring diameter crossing time, and $\epsilon(\hat{t}_i)$ is the detection efficiency.

The detection efficiency ϵ is properly determined by adding "fake" stars with the same luminosity function as the real stars to the raw images. The brightnesses of these "fake" stars can be modulated according to randomly selected microlensing light curves, and then the analysis can be run to see how many of these simulated events are recovered. A much simpler procedure would be to add simulated microlensing light curves to the light curve database, but this yields ϵ values that are generally too large because it neglects the effects of systematic errors in the photometry and the blending of stellar images which tends to make lensing events more difficult to see. The ϵ values calculated in this way are known as sampling efficiencies, and these are the only efficiencies that we will present here. $\epsilon(\hat{t})$ is plotted for several different data cuts in Fig. 4.

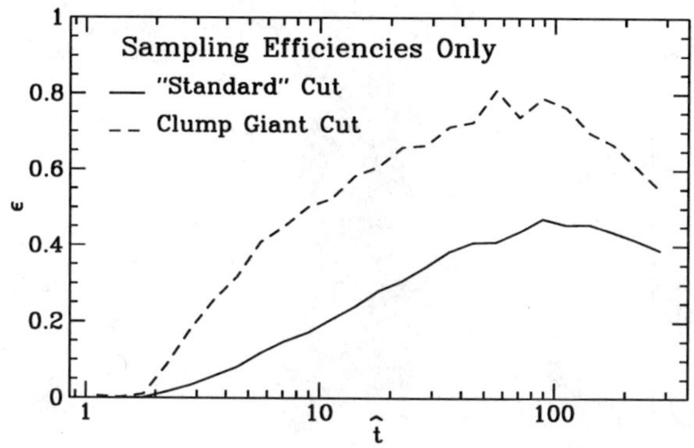

Figure 4. Sampling Efficiencies are plotted for our standard and clump giant cuts. The efficiencies are higher for clump giants because they are much brighter than the average star.

Microlensing Optical Depth Towards the Galactic Bulge $\times 10^6$

		Confidence Levels						
cut	N	0.025	0.05	0.16	**measured**	0.84	0.95	0.975
standard	43	1.02	1.11	1.29	**1.58**	1.93	2.16	2.28
clump giant	13	1.25	1.48	2.06	**3.03**	4.47	5.46	5.96
$\|b\| < 3.1$	18	1.81	2.11	2.72	**3.79**	5.27	6.26	6.83
clump giant w/ $\|b\| < 3.1$	9	2.04	2.57	3.91	**6.32**	10.27	13.07	14.48

Table 1. The opticcal depth τ is listed in units of 10^{-6} for the different cuts discussed in the text. Note that the $|b| < 3.1$ cuts have selection effects that have not been included in the confidence level estimation.

Another problem that interferes with the determination of the microlensing optical depth toward the Galactic Bulge is that many of the stars that we see when we look towards the galactic bulge are foreground disk stars. If we do not correct for this, then our estimates of the optical depth τ will be underestimates. One way to avoid both the problem of foreground stars and the complications due to stellar blends is to concentrate on a class of stars that is both bright and "known" to be in the galactic bulge: the "clump giant" stars. These are relatively low mass core helium burning giants–the horizontal branch of a metal rich population. Their location is marked on the color magnitude diagram in Fig. 3.

The measured optical depths for a number of different cuts on the data are shown in Table 1. This table displays the measured optical depth and various confidence level limits on the microlensing optical depth for several different cuts on the bulge data sample. The optical depth τ corresponding to a confidence level, P, is the τ value such that a fraction P of simulated data sets have a measured optical depth $> \tau$. The simulated data sets are constructed assuming Poisson statistics and an actual $\hat{t}$ distribution that is identical to the observed $\hat{t}$ distribution.

The standard cut referred to in Table 1 is quite similar to the one which generated the 45 events shown in Fig. 2, but a few more cuts had to be added for consistency with the monte carlo events used for the efficiency estimation. Two additional types of cuts are referred to in Table 1. The first is the clump giant cut alluded to above. The fact that this cut seems to imply a noticeably higher optical depth ($\tau = 3.0^{+1.5}_{-0.9} \times 10^{-6}$) than the standard cut ($\tau = 1.58^{+0.35}_{-0.28} \times 10^{-6}$) suggests that the full efficiency corrections and/or the foreground star corrections are important for the full sample.

The other cut referred to in Table 1 is a cut to include only the 5 fields which have a central galactic latitude of $|b| < 3.1$. This cut is intended to emphasize the contribution of microlensing by objects in the galactic disk as opposed to the galactic bulge or bar. A galactic bar elongated along the line of sight has been proposed as the source of the large microlensing optical depth seen toward the galactic bulge by the OGLE group (Paczyński, *et al.*, 1994) and by Zhou, Spergel, and Rich (1994). We (Alcock, *et.al.*, 1994b) and Gould (1994b) have suggested that a heavy, "maximal" disk might account for the large optical depth seen toward the bulge. The results shown in Table 1 do seem to suggest that microlensing by the disk is contributing a large fraction of the total optical depth, but we should caution that the cut at $|b| < 3.1$ has been selected after the distribution of events with b was known. So, this cut can be considered to have been made in a biased way. A more sophisticated analysis is now under way to determine if, in fact, these data indicate that the "maximal disk" model is preferred.

An "Alert" Event

Fig. 5 shows the light curve of the first gravitational microlensing event ever detected substantially before its peak amplification. The time of discovery is indicated. An IAU circular (IAUC 6068) was submitted about 24 hours after the event was discovered, and a great deal of photometric and spectroscopic data was collected during the course of the event. The team of Benetti, Pasquini, West, and de Lapparent obtained spectra on 6 different nights from the NTT and 3.6m telescopes at ESO, and have confirmed that the spectral features of the star did not change as the star rose to a peak amplification of $A = 3.5$ and then dropped back down to its usual brightness (IAUC 6069 and 6071). Similar data was obtained at a number of large telescopes including a spectrum obtained by Dick Hunstead at the AAT on Sept. 2. The ESO and AAT data indicate that star is classified as a K0 class II-IV star with a radial velocity of ~ -200km/sec.

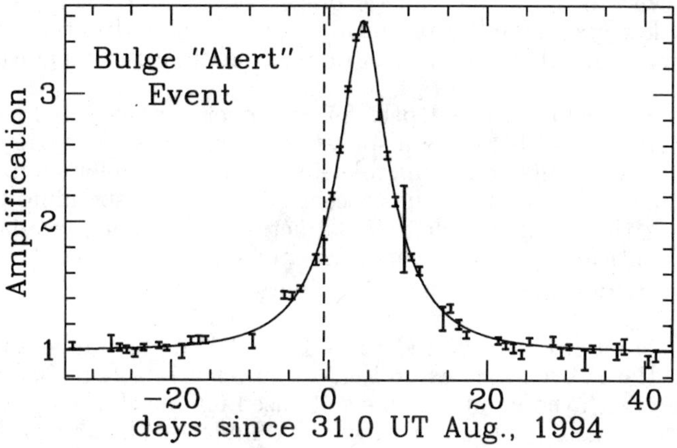

Figure 5. The light curve of the first microlensing event discovered substantially before peak amplification. The vertical dashed line is the time of discovery.

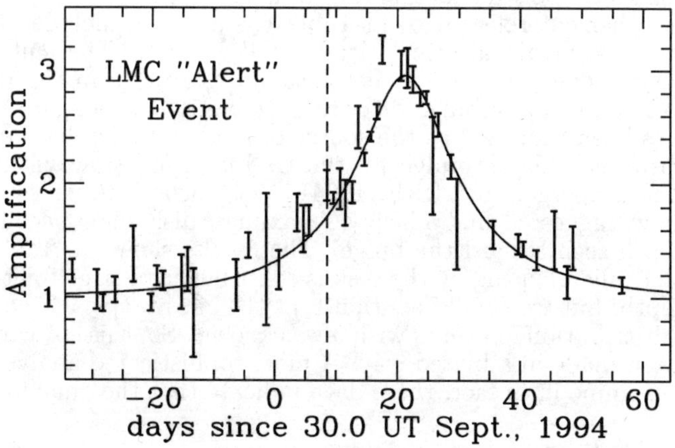

Figure 6. The light curve of the first microlensing event discovered in progress towards the LMC. The vertical dashed line is the time of discovery.

An alert system has also been demonstrated by the OGLE collaboration (Udalski, *et.al.*, 1994c) who have detected two moderate amplification events in progress.

At the time of this conference, the first microlensing event ever detected in real time toward the LMC appears to be in progress!! The light curve of this event is shown in Fig. 6. Giraud (private communication) obtained a spectrum of this star near peak amplifiction, and confirmed that it is a normal F star.

A Parallax Event

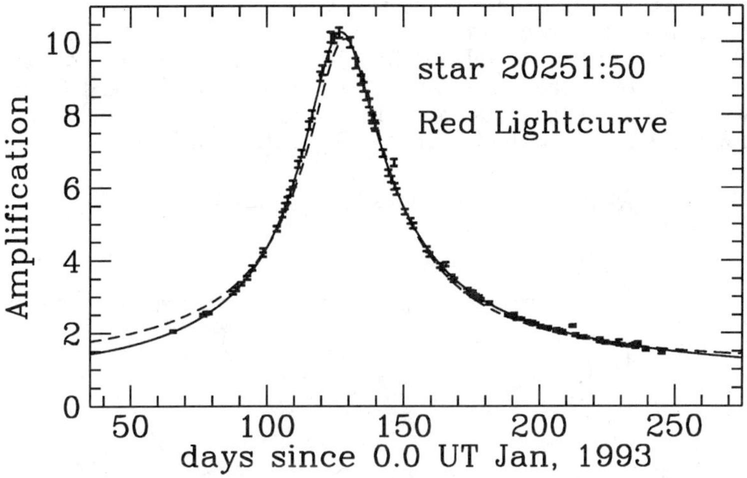

Figure 7. The MACHO-Red band light curve for the longest event yet detected by the MACHO project. The solid curve is the fit light curve which includes the effect of the earth's motion, while the dashed curve is the best fit ignoring the motion of the earth.

Fig. 7 shows the light curve of the longest timescale microlensing event that we have detected, and the first to show a significant asymmetry in it's light curve. This asymmetry can be explained as a parallax effect–the change in the earth's velocity over the hundreds of days that this event has lasted has caused a deviation from the simple 3-parameter light curve that describes most microlensing events. This light curve is well fit by a model which includes the orbital motion of the earth with the transverse velocity of the microlensing object projecting to a transverse velocity of 54 km/sec at the position of the earth. This gives us the information we need to determine the mass of the microlensing object as a function of the earth-lens distance. The implied mass is shown in Fig. 8. It seems clear from Fig. 8 that the lensing object must very likely be a low mass star in the Galactic disk or a brown dwarf in the galactic bulge.

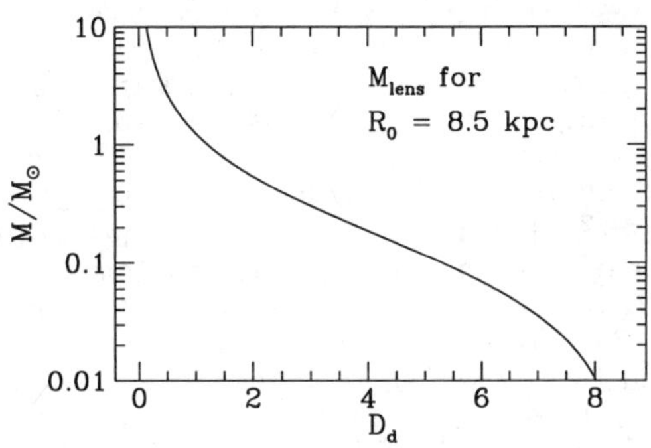

Figure 8. The implied mass is plotted as a function of the distance to the lens for the parallax event.

Much progress could be made toward resolving the mysteries of microlensing if the parallax effect could be observed for a large number of microlensing events. However, it is only for the very long events that the earth will move enough during the event for such an effect to be detectable. This situation could be remedied if a small satellite could observe events from a solar orbit, as suggested by Gould (1994a-b). The satellite would need to be alerted to events detected in progress from the ground, but this capability has already been demonstrated by the OGLE and MACHO groups.

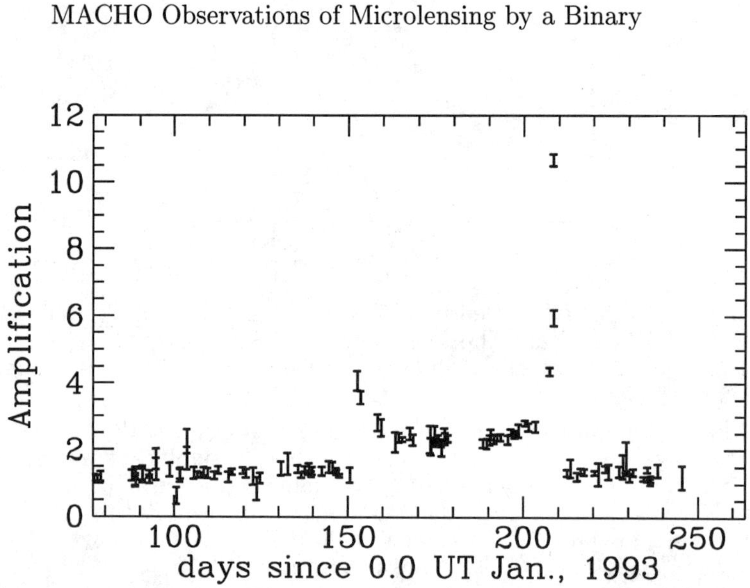

Figure 9. The MACHO-Blue band light curve of the binary lens event first seen by OGLE.

Fig. 9 shows light curves of a microlensing event that clearly shows characteristics of a binary lens with "caustic crossings" where the source star crosses into (or out of) a region where two extra images are created (making a total of 5). This event was first discovered by the OGLE collaboration (Udalski, *et.al.*, 1994c), but the MACHO data provides a nice confirmation with very good coverage of the second caustic crossing which was largely missed by OGLE. In fact, the MACHO coverage of the second caustic crossing is so good, that the finite angular size of the source star is resolved in the MACHO data. The divergent amplification of the 4th and 5th images near a caustic can be shown to have the form $A \approx d^{-1/2}$ where d is the distance to the caustic on the interior side. The modification of this formula for a finite size source star with a realistic limb darkening law has been derived by Schneider and Wagoner (1987). Fig. 10 shows a fit using the 4 MACHO Blue band data points closest to the caustic using a 3 parameter model with the Schneider and Wagoner profile. The fit $\chi^2 = 0.009$ for 1 d.o.f. This fit fixes the time for the caustic to cross the diameter of the star to be 10 hours. Using the OGLE fit value for the angle of the caustic crossing, we find that the time for the lens position to cross the diameter of the star is about 8 hours. This implies a lens velocity of $v_t = x(48\text{km/sec})R_{\text{star}}/R_{\odot}$, where x is the fractional distance between us and the lensing object. This will lead to a contraint on the mass of and distance to the lens.

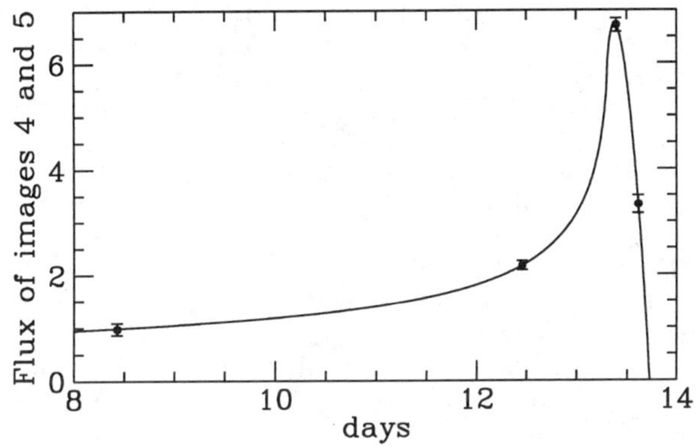

Figure 10. A fit to the 2nd caustic crossing of the binary lens event using the MACHO-Blue band data.

4. Conclusions

The microlensing surveys toward the galactic bulge have yielded a number of surprizes including exotic events such as the binary and parallax events. Although they now have the status of curiosities that are interesting in their own right, they also have the virtue that we know more about the properties of the lenses in these events than the lenses in the other microlensing events. With the help of the recently developed "Alert" systems, we will soon be obtaining higher quality data on events that are discovered in progress which should increase the chances of detecting these exotic events. With higher quality data coupled with more observations covering a larger range of ℓ and b, we hope to shed some light on the major bulge microlensing mystery, which is: Why is the optical depth so high? The resolution of this puzzle will probably help to determine whether the Galactic halo is made predominantly of baryonic or non-baryonic matter.

ACKNOWLEDGEMENTS

This research was supported by the National Science Foundation through the Center for Particle Astrophysics at UC, Berkeley and by the Department of Energy through the LDRD program at the Lawrence Livermore National Laboratory.

REFERENCES

Alcock, C., Akerlof, C.W., Allsman, R.A., Axelrod, T.S., Bennett, D.P., Chan, S., Cook, K.H., Freeman, K.C., Griest, K., Marshall, S.L., Park, H.-S., Perlmutter, S., Peterson, B.A., Pratt, M.R., Quinn, P.J., Rodgers, A.W., Stubbs, C.W., & Sutherland, W., 1993, *Nature*, **365**, 621.

Alcock, C., Allsman, R.A., Axelrod, T.S., Bennett, D.P., Cook, K.H., Freeman, K.C., Griest, K., Marshall, S.L., Perlmutter, S., Peterson, B.A., Pratt, M.R., Quinn, P.J., Rodgers, A.W., Stubbs, C.W., & Sutherland, W., 1994a, in preparation.

Alcock, C., Allsman, R.A., Axelrod, T.S., Bennett, D.P., Cook, K.H., Freeman, K.C., Griest, K., Marshall, S.L., Perlmutter, S., Peterson, B.A., Pratt, M.R., Quinn, P.J., Rodgers, A.W., Stubbs, C.W., & Sutherland, W., 1994b, *ApJ*, in press.

Aubourg, E., Bareyre, P., Brehin, S., Gros, M., Lachieze-Rey, M., Laurent, B., Lesquoy, E., Magneville, C., Milsztajn, A., Moscosco, L., Queinnec, F., Rich, J., Spiro, M., Vigroux, L., Zylberajch, S., Ansari, R., Cavalier, F., Moniez, M., Beaulieu, J.-P., Ferlet, R., Grison, Ph., Vidal-Madjar, A., Guibert, J., Moreau, O., Tajahmady, F., Maurice, E., Prevot, L., & Gry, C., 1993, *Nature*, **365**, 623

Bahcall, J.N. Flynn, C., Gould, A., & Kirhakos, S., 1994, *Nature*, in press.

Beaulieu J.P., Ferlet R., Grison Ph., Vidal-Madjar A., Kneib JP., Maurice E., Prevot L., Gry C., Guibert J., Moreau O., Tajahmady F., Aubourg E., Bareyre P., Brehin S., Gros M., Lachieze-Rey M., Laurent B., Lesquoy E., Magneville C., Milsztajn A., Moscoso L., Queinnec F., Rich J., Spiro M., Vigroux L., Zylberajch S., Ansari R., Cavalier F., & Moniez M., 1994, preprint.

Della Valle, M., 1994 *A. & A.*, **287**, L31

Giudice, G.F., Mollerach, S., & Roulet, E., 1993, (preprint: CERN-TH.7127/93)

Gould, A., 1994a, *ApJ*, **421**, L75

Gould, A., 1994b, preprints.

Griest, K., 1991, *ApJ*, **366**, 412

Griest, K., Alcock, C., Axelrod, T.S., Bennett, D.P., Cook, K.H., Freeman, K.C., Park, H.-S., Perlmutter, S., Peterson, B.A., Quinn, P.J., Rodgers, A.W., & Stubbs, C.W., 1991, *ApJ Lett.*, **372**, L79

Kiraga M., and Paczyński, B. 1994, *ApJ Lett.*, in press.

Paczyński, B, 1986, *ApJ*, **304**, 1

Paczyński, B. 1991, *ApJ Lett.*, **371**, L63

Paczyński, B., Stanek, K.Z., Udalski, A., Szymanski, M., Kaluzny, J., Kubiak, M., Mateo, M., & Krzeminski W., 1994 *ApJ Lett.*, **435**, L113

Sahu, K.C., 1994, *Nature*, **370**, 275.

Schneider, P., & Wagoner, R.V., 1987, *ApJ*, **314**, 154

Stubbs, C., Marshall, S.L., Cook, K.H., Hills, R., Noonan, J., Akerlof, C.W., Axelrod, T.S., Bennett, D.P., Dagley, K., Freeman, K.C., Griest, K., Park, H.-S., Perlmutter, S., Peterson, B.A., Quinn, P.J., Rodgers, A.W., Sosin, C., & Sutherland, W., 1993, *SPIE Proceedings*, **1900**, 192

Udalski, A., Szymanski, M., Kaluzny, J., Kubiak, M., Krzeminski, W., Mateo, M., Preston, G.W., & Paczynski, B., 1993, *Acta Astronomica*, **43**, 289

Udalski, A., Szymanski, M., Stanek, K.Z., Kaluzny, J., Kubiak, M., Mateo, M., Krzeminski, W., Paczynski, B., & Venkat, R., 1994a, *Acta Astronomica*, **44**, 165

Udalski, A., Szymanski, M., Kaluzny, J., Kubiak, M., Mateo, M., Krzeminski, W., Paczynski, B., & Venkat, R., 1994b, *Acta Astronomica*, **44**, 227

Udalski, A., Szymanski, M., S. Mao, R. Di Stefano, J. Kaluzny, M. Kubiak, M. Mateo, & W. Krzeminski, 1994b, *ApJ Lett.*, in press.

Zhou, H.S., Spergel, D.N. and Rich, R.M., 1994, preprint.

The Results of Deep CCD Field Surveys: Very Low Mass Halo Population Stars As Dark Matter

P. C. Boeshaar
Drew University, Madison, NJ 07940
boeshaa@drew.drew.edu

J. A. Tyson
AT&T Bell Laboratories, Murray Hill, NJ 07974
tyson@physics.att.com

G. M. Bernstein
University of Michigan, Ann Arbor, MI 48109
garyb@astro.lsa.umich.edu

ABSTRACT

Halo and disk M dwarfs differ significantly in metallicity. Multi-color deep CCD surveys are uniquely capable of detecting these separate populations of stars with differing metallicity and therefore colors. Analyzing very deep three-band CCD images covering 192 arcmin^2 at high galactic latitude we find no evidence for a population of extreme low mass M subdwarfs sufficient to account for the halo dark matter. These observations covering a volume of 2×10^5 pc^3 are consistent with extrapolations of a halo luminosity function determined using low metallicity stars in the solar neighborhood.

1. INTRODUCTION

With the recent discovery of microlensing events by objects in the halo of our galaxy (Alcock *et al.* 1993), there has been a resurgence of interest in examining low mass stellar objects as the source of an important fraction of the unseen matter in our Galaxy. While direct constraints on the stellar mass function may be obtained by microlensing, a significant excess of low mass objects with $M \leq 0.1 M_{\odot}$ would have observable consequences for the faint tail of the stellar luminosity function (LF). Our knowledge of the shape of the stellar LF has been limited until recently to the local neighborhood within 10-20 pc of the sun, and covers a mixture of population types.

The recent results for the halo M subdwarf population in the solar neighborhood by Dahn *et al.* (1994) using measured trigonometric parallaxes are generally consistent with the earlier results of Dawson (1986) based on a statistical analysis of proper motions. Both of these studies suggest a flat or falling LF at the faint end where $M_V = 14 - 15$, close to the theoretical estimate for the hydrogen burning limit for stars with halo metallicities (Monet *et al.* 1992). This appears to be at odds with one recent survey for halo stars; deep V, R and I-band CCD observations were used by Richer and Fahlman (1992) to estimate photometric parallaxes of halo stars more than 1200 pc above the galactic plane. In contrast to the solar neighborhood results, Richer and Fahlman's analysis suggests that the faint end halo LF and mass functions are steeply rising.

2. OBSERVATIONS

Very deep multicolor CCD surveys provide a rich database which can be exploited to sample the low mass stellar halo population out to several kpc, provided that proper color calibration is available to separate low mass stars of differing metallicities as well as morphologically misclassified galaxies. Deep three-band ($B_J < 27.5$, $R < 26.4$, $I < 25$ mag) CCD images of 12 widely spaced high galactic latitude fields covering a total of 144 arcmin^2 on the sky have been obtained as part of a 4-m survey done at CTIO originally to study the faint galaxy count problem. Together with a single 1024^2 CCD field covering 48 arcmin^2 on the sky obtained at KPNO, these data covering 192 arcmin^2 have been analyzed to search for M dwarfs near the halo and disk hydrogen burning limits. A detailed discussion of the observations, image processing, and star-galaxy separation techniques can be found in Boeshaar *et al.* (1986) and Tyson (1988).

Our color data have been carefully calibrated by observing standard stars of different temperatures and luminosities which have spectroscopically determined metallicities. Many of these observations were necessary in order to show how to separate out the different population types in our color-color plane (Gullixon *et al.* 1995). Metal deficient halo M stars have redder $B_J - R$ colors than their disk counterparts due to decreased molecular band blanketing that dominates the V through I passbands. As can be seen in Figure 1, the M subdwarfs of varying metal deficiency lie above their disk population counterparts in our color-color plane, with the objects of lowest metallicity having the reddest colors in $B_J - R$. The M subdwarfs in this diagram have been chosen from the sample of Dahn *et al.* (1994), and the object with the reddest R-I color in Figure 1 is LHS 1742a, one of the least luminous and most metal poor M subdwarf discovered to date. The extreme halo population M stars plotted cover $M_V \approx 13 - 14.5$. The disk M dwarfs encompass the complete spectral range dM0 - dM9, or $M_V \approx 8 - 19.5$.

A comparison with color-color data for galaxies found in our fields, most of which are very blue, indicates that no galaxies are found redward of R-I = 3. This is the region where the latest M dwarfs are found. Furthermore, minimal galaxy contamination is expected in the region where the halo population M subdwarfs are located.

3. RESULTS

If LHS 1742a is representative of the least luminous halo M subdwarfs near the hydrogen burning limit, then we might expect such objects to have $M_V \approx 15$. At this absolute magnitude, our survey is complete out to 3000 pc, covering a volume of approximately 205,000 pc^3. There are six objects in our deep CCD survey data having colors consistent with M subdwarfs of $M_V \approx 13.5 - 15$. A comparative analysis of our galaxy data in the same fields at similar apparent magnitudes suggests that this subdwarf sample may be contaminated by at most 1-2 misclassified compact high redshift galaxies of similar color which appear stellar. However pollution of our ground-based stellar counts by misclassified very red galaxies is not a major problem. There appear to be few very red galaxies which are faint and compact: recent HST WFPC2 Medium Deep Survey data reveal no new population of faint red compact galaxies (Griffiths *et al.* 1994).

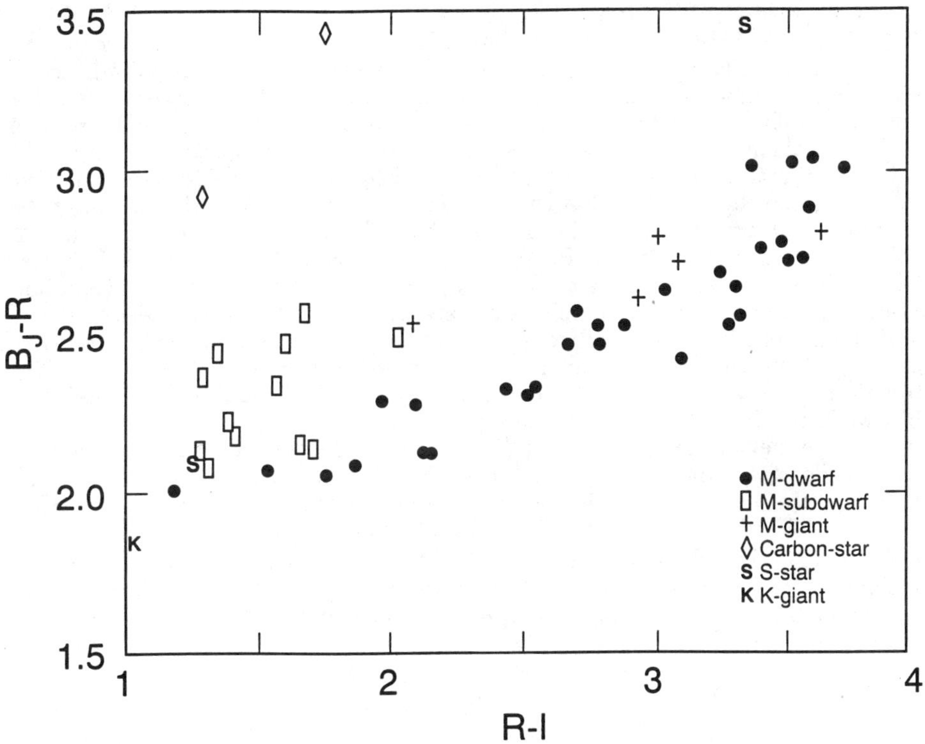

Figure 1. *Colors of red standard stars with spectral classification observed on the B_J, R, I system. Disk M dwarfs cover the complete spectral range dM0-dM9, while the halo M subdwarfs have been selected from the list of Dahn et al. (1994), representative of the complete observed range in metal deficiency and temperature.*

Our subdwarf counts agree quite well with the predictions made using the luminosity function of Dahn *et al.* (1994), i.e. that we should find 5 ± 3 of the least luminous subdwarfs within our volume. By comparison, the halo luminosity function of Richer and Fahlman (1992) would predict over five times as many low mass M subdwarfs than we detect in our surveys. This is clearly not the case, even though we go at least as deep their survey. If the dark matter halo were composed of metal poor halo stars near the hydrogen burning limit with color similar to the reddest subdwarfs observed in Figure 1, then we should have detected over 2000 such objects. Our data also eliminates the possibility that 0.092 $M_\odot$ zero metallicity stars at the hydrogen burning limit as modeled by Saumon *et al.* (1994) constitute a major portion of the halo dark matter. Such objects would be found approximately 1 mag bluer in R-I and 0.5 mag bluer in $B_J - R$ than our reddest halo population stars, a region with three or four objects. We suspect that many of these latter objects are misclassified galaxies.

In time, parallel HST WFC2 very deep multicolor imaging should improve on these results because of its superior angular resolution. While multicolor data is crucial for studying these dwarf stellar populations, Bahcall *et al.* (1994)

recently reported HST observations in one color of a single 4.4 arcmin^2 field, in F606W and F814W bands. Using their I limit of 25.3 mag and an M_I of 11.5 (Monet *et al.* 1992) as representative of the latest halo extreme M subdwarfs identified spectroscopically, we estimate that Bahcall *et al.* have surveyed a volume one fifth of that for which we have accurate deep CCD color data. Two objects in their data have $V - I \leq 3$ colors and distances consistent with the M subdwarfs defined above, i.e. results statistically comparable to ours. In their interpretation they assume that that the least luminous halo and disk stars have similar V-I colors. However, it is known that faint halo and disk stars differ in color. It is not surprising that Bahcall *et al.* discovered no stars with V-I greater than 3, if such objects belong to the disk population as current theory (Saumon *et al.* 1994, and references therein) suggests. Assuming an average disk M dwarf scale height of 350 pc, Bahcall *et al.* sampled a few tens of cubic parsecs for disk dwarfs. Using the best current measurements of the disk M dwarf luminosity function (Kirkpatrick *et al.* 1994), at most one $V - I > 3$ and no V-I = 4-5 disk M dwarfs would be expected in that HST field.

Recent evidence (Kirkpatrick *et al.* 1994) suggests that the latest M dwarfs with V-I's of 4-5 may have a smaller scale height than earlier disk M dwarfs and that their distribution is nonuniform. Our survey has detected six stellar objects with colors indicative of M dwarfs having $V - I \approx 4 - 5$ and $M_I \approx 14$ ($M_V \approx 19$), all located in our single 1024^2 field. Assuming these objects are among the latest M dwarfs, their I magnitudes indicate that they all lie within 250 pc of the galactic plane, though our survey is complete to 300 pc for such objects.

REFERENCES

Alcock, C. *et al.* 1993, Nature, 365, 621

Bahcall, J., Flynn, C., Gould, A., & Kirhakos, S. 1994, ApJ 435, L51

Boeshaar, P.C., Tyson, J.A., Seitzer, P. 1986, in Astrophysics of Brown Dwarfs, eds. Kafatos, M.C., Harrington, R.S., Maran, S.P. (Cambridge U. Press, NY), p. 76

Dawson, P.C. 1986, ApJ 311, 984

Dahn, C.C. and Liebert, J., Harris, H.C., and Guetter, H.H. 1994, in Proceedings of the ESO Workshop "The Bottom of the Main Sequence And Beyond", ed. C.G. Tinney (Springer-Verlag, Heidelberg), in press

Griffiths, R.E. *et al.* 1994, ApJ Lett., 435, L19

Gullixon, C.A., Boeshaar, P.C., Tyson, J.A., Seitzer, P. 1995, ApJS, in press

Kirkpatrick, D. *et al.* 1994, ApJS 94, 749

Monet, D.G., Dahn, C.C., Vrba, F.J., Harris, H.C., Pier, J.R., Luginbuhl, C.B., Ables, H.D. 1992, AJ 103, 638

Richer, H.B., and Fahlman, G.G. 1992, Nature 358, 383

Saumon, D., Bergeron, P., Lumine, J.I., Hubbard, W.B., Burrows, A. 1994, ApJ 424, 333

Tyson, J.A. 1988, AJ 96, 1

DIRECT DETECTION OF DARK MATTER FROM TWQSO MICROLENSING

Rudy Schild
Center for Astrophysics, 60 Garden St. Cambridge MA 02138
rschild@cfa.harvard.edu

David J. Thomson
AT&T Bell Labs 2C-360, Murray Hill, NJ 07974-2070
djt@research.att.com

ABSTRACT

Gravitational microlensing in the direction of Q0957+561 A,B (TwQSO) is expected to show the spectrum of mass objects in the $z = 0.37$ lens galaxy, because the existence of a widely separated double image implies a microlensing optical depth near unity. In our 15 year record of brightness of the two quasar images, we find evidence for a range of masses from approximately $0.1 M_o$ to $10^{-5} M_o$. These are evidenced by continuous power in the Fourier power spectrum of the observed microlensing over a range of frequencies from 0.1 to 4 cycles per year. A sharp high frequency cutoff (lower mass bound) is found.

1. INTRODUCTION

Microlensing by a solar mass star in the $z = 0.37$ lens galaxy G1 of the Q0957 system should cause a brightness enhancement lasting about 30 years. The observed microlensing record shows an overall drift with this time scale and 0.3 magnitude amplitude, but also a complex network of much faster events which are as yet unexplained. Because the speed of a mass object in the lens galaxy should be independent of its mass, and because the duration of a microlensing event should vary as the square-root of mass, the mapping from event duration to microlens mass should be simple.

2. DATA AND ANALYSIS

The data set now comprises 835 data points, where each point is the average for a night's observing. The interpolated data are shown in Fig. 1 for a time delay of 1.1 years, and one-sided error bars are shown for every 10'th data point. The record is interrupted in 1989, when the 24" telescope was being replaced with a 48". The upper plot is the A component, and a "rug plot" at the bottom shows the dates when observations were actually made. Errors are smallest for the most recent data, and the jitter that at first looks like noise in fact is a network of overlapping cusps. A wavelet analysis of this data shows that the cusp network does not line up for any value of time delay, including the 404 day Schild (1991) and 536 day Press et al (1992) values.

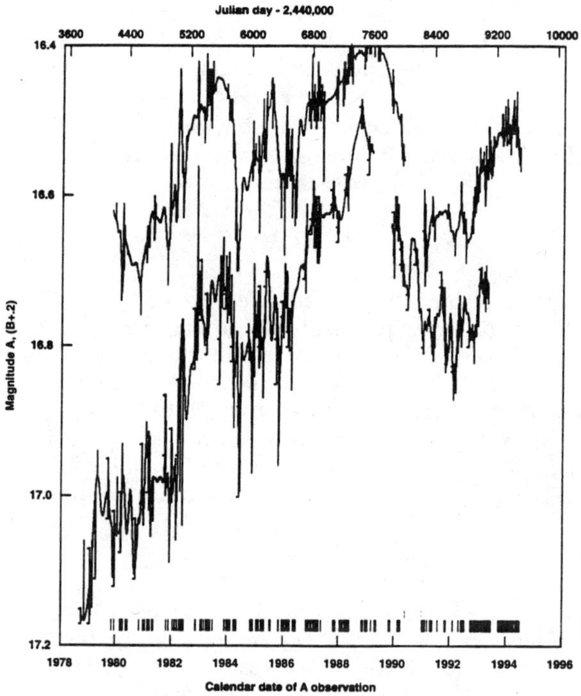

Fig. 1. The Data Set

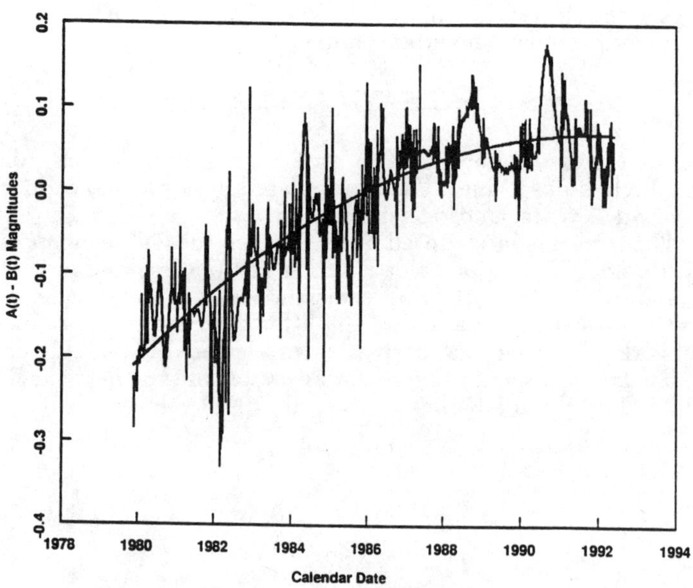

Fig. 2. The Microlensing Record.

The difference between the two quasar brightness records, phased by the 1.1 year time delay, is shown as magnitudes vs. the calendar date of the A observation in Fig. 2. A cubic fit is shown as a solid line to represent the probable long-term drift signalling ingress of a solar mass microlensing star. The elaborate network of cusps on time scales of several months must signal the presence of microlensing mass objects of only $10^{-5} M_o$. These cusps may at first seem to be observational noise, but recall that with approximately 50 cusps observed, each must be seen in many data points. The profile of an individual cusp is shown in Fig. 3. The elaborate cusp network does not phase up between the A and B images for any value of time delay, which is one of the principal reasons why the time delay has been so difficult to determine.

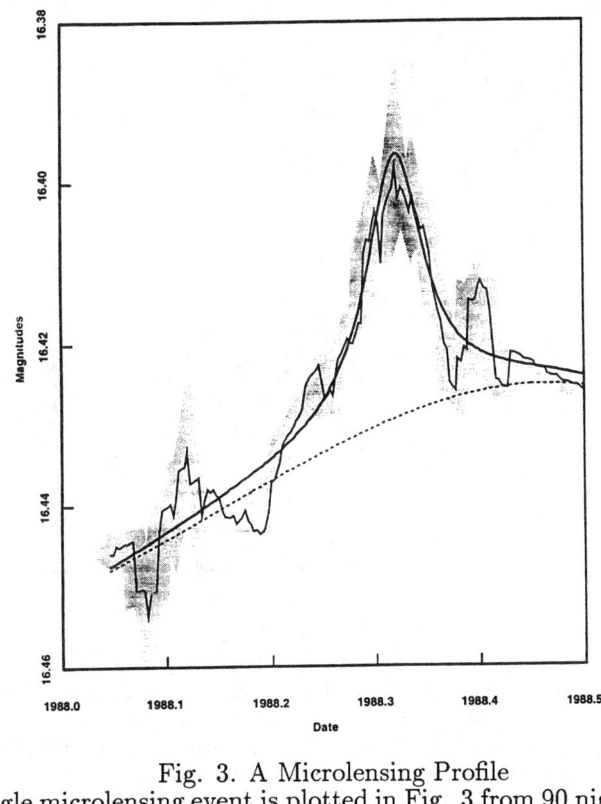

Fig. 3. A Microlensing Profile

A single microlensing event is plotted in Fig. 3 from 90 nights of A component data from 1988. The one sigma error interval is shown as a grey band. The total duration of this event was approximately 75 days, and the amplitude only 0.04 magnitudes. This event may be seen in the A data record of Fig. 1. Note that in addition to the obvious main peak, numerous secondary peaks are seen with amplitudes of only 0.01 magnitudes and event durations of only a week. These rapid events appear to be both positive and negative (ie, magnifying and demagnifying) and probably originate in a different population of microlenses. Such events are seen throughout the data record, except that for some months they seem to vanish.

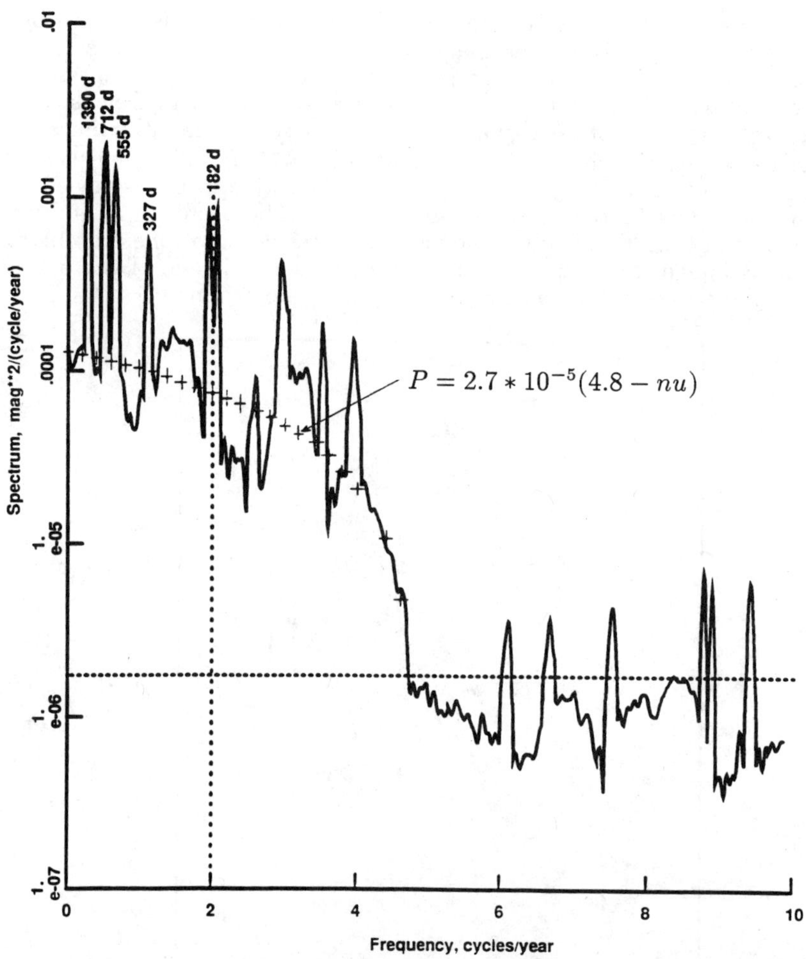

Fig. 4. The Fourier Power Spectrum

The Fourier power spectrum of the microlensing cusps, with the cubic fit to the long-term (30 year) drift due to solar mass star ingress subtracted off, is shown with logarithmic scaling of the power. The peaks near 1 cycle/year show x10 power enhancements over the continuum. Five discrete peaks are labeled, and they may reflect orbital motions of the microlensing masses. The expected noise level is shown as a horizontal dashed line. A continuum is observed and well fit by the analytical expression $P = A(4.8 - nu)$ where $A = 2.7 * 10^{-5} mag^2(cy/yr)^{-2}$.

REFERENCES

Press, W., Rybicki, G.B. and Hewitt, J.N., 1992, Ap. J. 385,416.
Schild, R., 1990, A. J. 100, 1771.

DARK MATTER IN GALAXIES

Dark Matter in the Milky Way

Leo Blitz
Astronomy Department, University of Maryland, College Park, MD, 20742

Email ID
blitz@astro.umd.edu

ABSTRACT

The current state of the search for dark matter in the Milky Way is reviewed starting with the current state of rotation curve studies. From the dynamical evidence, there continues to be strong evidence for a flat rotation curve to at least 2.5 $R_\circ$, and the various studies seem to be in reasonably good agreement. Other dynamical evidence for Galactic dark matter does not appear to be particularly compelling. The many searches for brown dwarfs and very low mass stars are now at a level where it appears reasonable to conclude that these objects are probably not responsible for the flat rotation curve of the Milky Way. Another baryonic candidate, very cold H_2 clouds, also do not appear to be able to explain the flat rotation curves. Although the dynamical evidence for dark matter in the Milky Way appears to be as strong as that for other Galaxies, we appear to be little closer to identifying what the dark matter actually is than we were ten years ago.

1. INTRODUCTION

The most compelling evidence for the existence of dark matter has come from the measurement of flat galaxy rotation curves. The period 1978 – 1980 brought a number of measurements of the Milky Way rotation curve out to large enough distances to suggest that the Galaxy too probably contains large quantities of dark matter. Other suggestions for the existence of dark matter have come from Oort analyses of the total midplane density of the Galactic disk near the Sun, though the interpretation of these has been controversial. The present state of dynamical measurements to detect the presence of Galactic dark matter are discussed in §2. Also included in this section is a discussion of the radial HI scale height distribution and how it constrains the Galactic dark matter to be in the halo rather than the disk. If the Milky Way does contain dark matter, and if that dark matter is baryonic, the most likely candidate has been thought to be brown dwarf stars: objects like the planet Jupiter of standard stellar compostion that contain too little mass to burn hydrogen. The most recent observational evidence regarding the existence of brown dwarfs is discussed in §3. Also discussed in §3 is a more recent suggestion that the dark matter may be the result of very cold molecular hydrogen clouds that have so far escaped detection. In §4 a search for dark matter is briefly reviewed in which the signature of a triaxial dark matter halo might be detected through the non-circular motions of the outer Galactic disk. The results from the gravitational microlensing experiments are not discussed in this review; they are well described elsewhere in this volume.

2. DYNAMICAL EVIDENCE FOR DARK MATTER IN THE GALAXY

The Rotation Curve

The existence of flat rotation curves has provided the strongest evidence for the existence of dark matter (Rubin *et al.* 1978; Bosma 1978). Because the light distribution of spiral galaxies is always found to be exponentially decreasing, if the mass-to-light ratio is constant, as is generally thought to be the case in galactic disks, then the rotation curves should be decreasing in the outer parts of most spiral galaxies. Perhaps even more compelling are the rotation curves determined from the 21 cm HI line which often remain flat well beyond where there are *any* stars visible, suggesting that some non-luminous matter must be present to make the galaxies rotate as fast as they do in their outer reaches.

Measurements of the Galactic rotation curve in its outer parts requires the determination of independent distances and velocities to a set of objects, or collections of gas. The measurements can be divided into three basic kinds, each of which has its own particular limitations: 1) luminous main sequence stars, 2) standard candles, and 3) HI profiles.

For the luminous main sequence stars, generally OB stars, either the exciting stars of HII regions or reflection nebulae are identified, and their distances are determined from spectroscopic parallaxes (Jackson, FitzGerald & Moffatt 1979; Blitz 1979; Chini and Wink 1984; Fich *et al.* 1989; Brand & Blitz 1993). A variant of this approach is to find open clusters and determine their distances by main sequence fitting (Hron 1987). In the former case velocities are found quite accurately from associated CO clouds, or less accurately from the velocities of the stars themselves, which is usually the only way to get velocities for the clusters.

The standard candles that have been successfully used to get the good rotation curves are planetary nebulae (Schneider & Terzian 1982), carbon stars (Schechter *et al.* 1988, Aaronson *et al.* 1989, 1990), and cepheid variables (Caldwell & Coulson 1987; (Welch 1988; Pont *et al.* 1995). For these stars, it is necessary to calibrate the standard candle (in the case of planetary nebulae, it is to the exciting star), or use the period-luminosity-color relation for cepheid variables. Velocities are measured either from radio recombination lines (for planetary nebulae) or directly from the stellar velocities. In general, the velocities of these objects are not as good as the velocities determined from the CO line profiles. However, this can be offset by the larger number of objects often found at great distances from the Galactic center, as is the case for carbon stars, for example.

In order to determine the rotation curve directly from the HI, the velocity is measured directly from the doppler shift and has a very small uncertainty. It is necessary, however, to establish that particular parcels of gas are coherent and share similar distances from the center. This was successfully accomplished by Petrovskaya & Teerikorpi (1986), and also by Merrifield (1992). Merrifield's method makes the assumption that the Galaxy is axisymmetric; he uses the radial velocity of the HI to find gas in a ring of constant Galactic radius. The angular scale height of the gas is determined at different longitudes and the distance to the ring of HI is found by the variation of its angular scale height across the sky. The method is akin to finding the distance of a crater by seeing how the angle its rim makes along various lines of sight varies in azimuth for an observer inside it but not at its center. Because the Galactic HI layer extends beyond the optical disk, the rotation curve measured in this way is determined to a somewhat larger distance than from the others.

The results are all largely consistent with one another. None of the rotation curves shows any sign of falling off at the last measured point. There remains a discrepancy between the CO and planetary nebula rotation curves, and those measured from carbon stars and cepheid variables, the former showing a small rise beyond the Perseus arm. Hron (1989) has argued that this may be due to distant metallicity effects which would tend to overestimate the diestances to the OB stars.

In any event, the rotation curve is not declining even though one would expect it to well before $2R_o$ if the mass of the Milky Way came only from stars. The scale length of the disk is measured from COBE data to be 3 kpc (Blitz & Spergel in preparation), and 2 R_o is nearly seven scale lengths. The evidence for dark matter based on the rotation curve of the Milky Way remains as strong as ever.

In addition to the evidence for dark matter from the rotation curve itself, there is additional, but less compelling evidence that the rotation curve remains flat to even larger distances. It is beyond the scope of this article to discuss the relevant studies, but the interested reader is referred to a good review of the mass of the Galaxy by Fich & Tremaine (1991).

Dark Matter in the Disk

Oort (1932) pioneered the method of determining the total quantity of matter in the midplane of the Milky Way near the Sun by measuring the velocity dispersion and vertical stellar density gradient of a population of stars that trace the gravitational potential. By carefully summing up the contributions of the known luminous components and comparing the sum to the gravitationally determined value, it is possible in principle to determine whether any of the matter in the midplane is not luminous. Oort pointed out a factor of two discrepancy between the luminous matter and the dynamically determined mass, but it was unclear at the time how much was due to unobserved interstellar gas. Oort's method was modernized by Bahcall (1984a) who developed a method of self-consistent modeling and applied it to archival data (Bahcall, 1984b,c). Bahcall concluded that there was still a factor of two discrepancy between the two estimates of midplane density, even though the interstellar matter was now well accounted for, and he attributed the discrepancy to dark matter in the disk.

Since Bahcall's work in 1984, there have been a number of other studies that have led to different conclusions. A good review of the major points made by different authors has been written by Gould (1995). The most well known of the dissenting studies was by Kuijken & Gilmore (1989, 1991) who found, from a new sample of K dwarfs sensititive to mass between about 300 pc – 2000 pc from the Galactic midplane that the dynamically determined disk surface density $\Sigma_0 = 46 \pm 9\ M_\odot pc^{-2}$ is consistent with the sum of the luminous contributions. Kuijken & Gilmore also found a halo dark matter component of $\Sigma_0 =\sim 25$ $M_\odot pc^{-2}$ below $z < 1.1$ kpc. Gould (1995) has criticized the Kuijken & Gilmore interpretation because they assume a spherical dark matter halo and then asked, given this assumption, what is the best fit value for the disk? Their data, Gould claims, is more consistent with no halo dark matter and either a large amount of disk dark matter, (or alternatively and extremely flattened halo). Subsequent studies by Bahcall *et al.* (1992) and by Flynn & Fuchs (1995), have found $\Sigma_0 = 85 \pm 25\ M_\odot pc^{-2}$, and $\Sigma_0 = 56 \pm 12\ M_\odot pc^{-2}$ not including a dark halo component, for tracer components closer to the plane than Kuijken & Gilmore.

It seems that the controversy is still not resolved, but that there is not yet a compelling case for the existence of disk dark matter. If there were, it would be highly dissipational, and low mass stars or brown dwarfs are strongly implicated. However, the inability to detect such objects from a number of different recent studies, suggests that the systematics of doing the Oort problem are so difficult, that a convincing detection or non-detection of disk dark matter has not yet been made.

The HI Scale Height

A variant of the Oort analysis is to use the outer Galaxy atomic hydrogen to determine the scale height of the gas as a function of Galactic radius. Because extragalactic observations suggest that the velocity dispersion of the HI is constant as far out as it can be measured, the assumption of a constant gas velocity dispersion allows one to determine the run of the surface mass density as a function of radius as well. If we make assumptions about what is known about the radial mass distribution from the rotation curve, then we can get some idea whether the mass responsible for keeping the rotation curve flat is in the disk, or whether it is in a more spheroidal distribution. Such an analysis was carried out by (Kulkarni *et al.* 1982).

The rms scale height of the HI layer was found from the Weaver and Williams (1974) survey to rise linearly from a value of about 200 pc at $R_\circ$, to a value of 1700 pc at $3R_\circ$ Kulkarni *et al.* (1982). If we assume that the stellar scale height $h_* >> h_g$, the scale height of the gas, it can be shown that

$$< h_g^2(R) > = \frac{Q^2}{4\pi G\rho_*(R)}$$

where Q^2 is the sum of the thermal, terbulent, magnetic, and cosmic ray pressure terms,

$$Q^2 =< V_{gz}^2 > + \frac{{B_0}^2}{8\pi\rho_g} + \frac{P_{CR}}{\rho_g}.$$

If we assume from observations of other galaxies that $Q(R) = const$, and that a flat rotation curve implies that the mass surface density,

$$\sigma(R) = \rho_*(R)h_*(R) \propto R^{-1},$$

then it can be shown that

$$\frac{h_g^2(R_2)}{h_g^2(R_1)} = \frac{R_2 h_*(R_2)}{R_1 h_*(R_1)},$$

where the numerical subscripts refer to material at different Galactic radii.

Now, if we take, $h_g(R_0) = 200$ pc (Falgarone and Lequeux 1980), and $h_g(3R_0) = 1.7$ kpc (Kulkarni *et al.* 1982), then

$$h_*(3R_0) = 24h_*(R_0).$$

that is, if the stellar scale height is about 300 pc in the vicinity of the Sun, and if the flat rotation curve is entirely the result of matter in a thin disk, then the scale height of the self-gravitating component would be about 7 kpc, for a full width at half maximum of about 17 kpc, at a radius of about 24 kpc. That is, the thickness of the layer responsible for confining the gas would have to be about equal to the radius of the galaxy, in contradiction to the assumption that the self-gravitating material is in a thin disk. Another way of saying this is that the scale height of the gas rises too quickly for the matter giving rise to a flat rotation curve to be confined to a thin disk. This was early observational evidence that the dark matter in the Milky Way had to be largely distributed in a spheroidal componenet.

3. THE SEARCH FOR BARYONIC DARK MATTER IN THE GALAXY

The prime suspect in baryonic dark matter searches are brown dwarfs, objects without enough mass to initiate stellar burning, and thus with very low luminosities. The primary reason they are thought to be common is that the initial mass function (IMF) apparently has a power law form even for low mass stars, and if the extrapolated power laws continue into the range of brown dwarf masses, < 0.08 $M_\odot$, then there might be enough of them to produce flat rotation curves. Because there is nothing to suggest that the process of fragmentation to produce stars has any connection to the nuclear burning process, it seems plausible that the IMF could continue down to very low masses. There have been three recent attempts to detect brown dwarfs and the luminous stars very close to the hydrogen burning limit that are sensitive enough to place interesting limits on the the presence of brown dwarfs. The inability to find even a single confirmed brown dwarf suggests that the reasons that the rotation curve of the Galaxy is flat lies elsewhere.

Brown Dwarfs in Clusters

The initial luminosity function (as opposed to the IMF) can, in principle be measured directly from young clusters. With the advent of sensitive infrared array cameras, it has become possible to probe clusters still in the process of formation, and simply count stars in different magnitude ranges. Infrared arrays make it possible to peer through the dusty shrouds that make the star forming regions so opaque. At 2μm, the extinction due to dust is only 10% of what it is in the visible, making it possible to photograph all but the most heavily obscured regions.

Two studies are especially noteworthy, a review by Zinnecker *et al.*(1993), and by Lada & Lada (1995). In both studies, there have been images made of very young embedded star clusters, the Trapezium cluster and the cluster IC348 respectively. In addition to the decrease in opacity, the infrared is particularly advantageous for determining what the low end of the IMF is like because many stars in very young embedded clusters are still on their Hayashi tracks and are overluminous compared to their main sequence counterparts.

In both clusters, there is a turnover in the luminosity function at 12 and 11 magnitudes respectively. These turnovers occur 2 mag in the case of Orion, and 4 mag in the case of IC348 brighter than the faintest measured stars. Although turning the luminosity function into a mass function is dependent on uncertain stellar modeling, the observed turnover suggests that there is some characteristic mass, probably close to 0.3 $M_\odot$, below which the numbers dwindle significantly. The details of the observed luminosity functions for both clusters can be explained by effects of evolution and non-coeval star formation within each cluster. However, in the Zinnecker *et al.* study, there is a complete absence of stars 1 mag brighter than the completeness limit of 15 magnitudes.

The Trapezium is thought to be 1 $\times 10^6$ y old (Prosser *et al.* 1994), and evolutionary effects are thought to be minimal. For this age, and for the distance of Orion, the hydrogen burning limit has a K magnitude of 12.8. Thus, although stars are detected to K = 13.5, there is no evidence that Brown Dwarfs exist in any significant numbers in the Trapezium cluster. Lada & Lada find a similar result for the IC 348 cluster, but their completeness limit is closer to the hydrogen burning limit than is true for the Trapezium. If one may generalize from these two clusters, few if any brown dwarfs are currently being formed in the disk of the Galaxy.

Proto-brown Dwarfs

One of the problems of detecting brown dwarfs directly is their very low inferred luminosities. Even in the infrared, unless brown dwarfs go through a deuterium burning phase similar to that of low mass pre-main sequence stars, they might not be visible in existing searches. Brown dwarfs are faint not so much because the intensity of radiation from their surfaces are so small, but because they are so little. If one could only look for *large* brown dwarfs, it would make detections so much easier. But in their early collapse phase, they are quite distended, and therefore are expected to be relatively luminous.

If we imagine that brown dwarfs form in the same way as ordinary stars (which is likely if they are going to contribute significantly to the dark matter), then the objects that become brown dwarfs, the proto-brown dwarfs, will form in giant molecular clouds (GMCs). Because stars are believed to form by means of a Jeans instability, one may convert the Jeans criterion into a *density* criterion. We ask, what is the density required to form an object with a mass < 0.08 $M_\odot$ if the temperature of the cloud in which it forms is within the range of 5 – 20 K? These densities are in the range of $10^6 - 10^7$ cm^{-3}, and by looking for molecular transitions whose excitation is sensitive to these densitites, one should be able to find the condensations that will ultimately form brown dwarfs.

Molecular clouds are very inhomogeneous and the clumps within them are often easy to disentangle because of the doppler shifts associated with their relative motions (see *e.g.* Williams *et al.* 1994). Many of these clumps contain molecular cores with embedded young stellar objects (*e.g.* Myers and Benson 1983), and the cores seem to have masses comparable to the stars that form within them. Therefore, the signature of a proto-brown dwarf is an object, distinct from the background emission, with a mass < 0.08 $M_\odot$ that is gravitationally bound. Using such tracers as DCO^+, it is possible to survey dense regions of active star formation for proto-brown dwarfs with a mass sensitivity down to several Jupiter masses in a reasonable amount of telescope time.

In order to show that a clump is gravitationally bound, it is necessary to demonstrate that the gravitational mass, RV^2/G is equal to the luminous mass and that both are less than a stellar mass. This is because an object with a luminous mass < 0.08 M/solar does not insure that it is gravitationally bound. There are, in fact, many molecular clouds that are not gravitationally bound (Magnani *et al.*1985), and many of the clumps within a GMC are also not gravitationally bound (Williams & Blitz 1995). Another way of saying this is that the luminous mass of a cloud is a good measure of its total mass, and the gravitational mass is a measure of whether or not it is gravitationally bound. If both are equal (to insure that no systematic errors are bing made) and less than 0.08 M/solar, it should be a good sign that the cloud will form a brown dwarf. Unlike current infrared techniques, this method can determine the masses of candidate objects directly, and thus avboids the uncertain stellar modeling.

Pound & Blitz (1993, 1995) used this technique to look for proto-brown dwarfs in several regions, including the dense molecular cloud B18 in Taurus, and in the Oph B2 molecular cloud. They identified numerous objects with masses < 0.08 $M_\odot$, demonstrating that the method does indeed work, and were able to identify many gravitationally bound clumps. However, nearly all of the bound objects had masses > 0.08 $M_\odot$; only four had masses less than the hydrogen burning limit, one in Taurus, and three in Ophiuchus. However, even these four were suspect because they were found at the edges of the mapped regions. Nevertheless, in Oph B2, the star formation rate is known to some degree, and the results could be used to determine an upper limit to the slope of the IMF at brown dwarf masses. Pound & Blitz concluded that even if the candidate objects are real, the stellar mass function is sharply declining even into the brown dwarf mass range. Using the low mass IMF of Tinney (1993), Pound & Blitz expected to find 8 objects even if the power-law slope of the IMF is as low as 1.0 (compared to 2.3 for a Salpeter IMF) in the mass range 0.04 – 0.08 $M_\odot$ in Oph B2. At most, they found three objects, and more than likely none. Evidently, molecular clouds at the present epoch are not forming very many objets with less than a stellar mass.

Faint Red Stars

Using deep HST imaging, Bahcall *et al.*(1995), tried to find whether faint red stars near the hydrogen burning limit could contribute significantly to the mass of the Galaxy. They acquired two images in V′ and I′ toward $l = 241°$, $b = -51°$, down to a limiting $I < 25.3$ for $2 < V\text{-}I < 5$. If the dark halo were comprised of these faint stars, Bahcall *et al.* expected to find 64 in their field, but they found no stars with $V\text{-}I > 3.0$, and only 5 stars for $2 < V\text{-}I < 3$.

They concluded that stars near the hydrogen burning limit account for less than 5% of the mass of the dark halo of the Galaxy. They also find that these stars make up a similarly small fraction of the mass of the Galactic disk. Late type M dwrafs, they conclude are dynamically unimportant in the Milky Way. Finally, in agreement with the infrared studies of embedded stellar clusters, and with the search for proto-brown dwarfs, they aregue that the disk luminosity function actually falls for $15 > M_V > 19$, the hydrogen burning limit.

The preponderance of evidence suggests that there are few, if any, brown dwarfs in either the disk or the halo of the Milky Way, and suggests that the search for Milky Way dark matter should look elsewhere. However, the failure to detect the brown dwarfs raises yet another question. How is it that stars know that they shouldn't form unless they can actually burn hydrogen?

Cold Hydrogen Clouds

There have been two sets of suggestions that the dark matter in the Milky Way is composed of small, cold gravitationally bound clouds of hydrogen gas (Pfenniger Combes 1994a,b; Gerhard & Silk 1995). The properties of the clouds in both studies are similar, but in the Pfenniger & Combes work, the clouds are in a thickish disk. The HI scale height distribution with radius seems to rule out a thin dark matter contribution, so I will concentrate on the Gerhard & Silk idea.

Gerhard & Silk hypothesize that to survive for a Hubble time, the collision time between clouds must be $> {H_0}^{-1}$, and that to avoid detection, they must be small enough to have masses close to the Jeans mass. Furthermore, the evaporation time must also be $> H_0{}^{-1}$. They find that cold clouds of molecular hydrogen that satisfy these criteria have diameters D $\sim$ 0.02 pc, M $\sim$ 85 $M_\odot$, and n $\sim 10^7\ cm^{-3}$. Such clouds, if primordial, would be very difficult to detect because their surface covering fraction on the sky is less than 1%, and if they are primordial, contain little CO. Furthermore, their small expected HI envelopes require sensitivities that could only be achieved with a dedicated search. Furthermore, these clouds are surprisingly robust, and would not be affected very much, on average, by a passage through the disk of the Milky Way.

Nevertheless, such clouds would have been detected, if they existed because they will pass through HII regions in the vicinity of the Sun, and should have bright cometary morphologies because they are traveling throught the HII regions at velocities of $\sim$ 150 km s^{-1}. In an HII region like the Rosette Nebula, the ionized envelopes of these passing H_2 balls is about 0.1 pc making them easily detectable on archival photographic plates, with an emission measure of $\sim 10^5$ cm^{-6} pc, about 10 times the background brightness. Furthermore, in a volume the size of the Rosette Nebula, one expects to find about 4 such objects, and there should be hundreds visible against the catalogued optical HII regions in the solar vicinity. Surely observers perusing the POSS plates would have noticed these bright streaks in the HII regions, if they were to exist.

4. THE SHAPE OF THE DISK

If the Galaxy does contain a dark matter halo, and the halo is primordial, what will its shape be? Detailed numerical simulations suggest that the halo will not be round, but will instead be somewhat triaxial (Binney 1995). The degree of triaxiality of the halo can be affected by the initial conditions at the time of formation, including the degree to which gas is present, but any triaxial dark matter potential will affect the shape of the disk. That is, even if the underlying mass distribution of the disk is axisymmetric, a triaxial dark matter potential will cause deviations in the distribution and velocity fields of the disk.

In an attempt to understand certain irregularities in the distribution of the outer Galaxy HI distribution, Blitz & Spergel (1991) postulated that the gravitational potential in which the disk finds itself is not axisymmetric, rather it responds to a triaxial distribution of spheroid stars. They argued that the evidence is consistent with a disk that has elliptical stellar orbits near the solar circle, and gets rounder with increasing distance from the Galactic Center. With their model, Blitz & Spergel were able to explain a large number of disparate kinematic phenomena in the Milky Way.

Kuijken & Tremaine (1992, 1994) argued that the effects of such a disk triaxiality should be apparent in the motions of the stars in the solar vicinity, and developed a particularly useful parameterization of the motions of stars in a non-axisymmetric disk. They found that the measured vertex deviation of stars differs from the prediction of Blitz & Spergel by about 3σ, whereas the uncertainties on most of the other predictions were too large to serve as a useful test. However, Kuijken & Tremaine (1994) postulated a simple model for the Galaxy in which the entire disk is elliptical, and oriented at some angle relative to the sun-center line of the Milky Way. They decomposed the ellipticity into a symmetric and antisymmetric part, and looked hard for signatures of the antisymmetric part, to which the previous study of Blitz & Spergel was insensitive. Kuijken & Tremaine found that their best solution did require the disk to be non-axisymmetric, but their favored model had the disk increasingly elliptical with distance form the center with one of the principal axes closely aligned to the sun-center line of the Milky Way.

These two non-axisymmetric models of the Galaxy have not yet been reconciled, although there is some hope that COBE observations of the near-infrared radiation of the stars in the disk of the Galaxy may help to distinguish between them. Additional evidence may come from very distant HI possibly detected by the new Burton & Hartmann HI survey of the Milky Way (Hartmann 1994).

The search for dark matter in the Milky Way, has mode some progress in the last 10 years, but we seem to be little closer to identifying its nature than we were in 1985. Perhaps the main advance is that faint stars and brown dwarfs seem to be present in insufficient numbers to be dynamically interesting, but little more can be said of the nature of the dark matter in the Galaxy.

ACKNOWLEDGEMENTS

Partial support for this work comes from a grant from the National Science Foundation. I would like to thank the Leiden Observatory, and the Institute of Astronomy in Cambridge for their hospitality during the time this paper was written.

REFERENCES

Aaronson, M., Blanco, V., Cook, K.H., & Schechter, P.L., 1989, ApJS, 70, 637
Aaronson, M., Blanco, V., Cook, K.H., Olszewski, E.W., & Schechter, P.L., 1990, ApJS, 73, 841
Bahcall, J.N., 1984a, ApJ, 276, 156
Bahcall, J.N., 1984b, ApJ, 276, 169
Bahcall, J.N., 1984c, ApJ, 287, 926
Bahcall, J.N., Flynn, C., & Gould, A., 1992, ApJ, 389, 234
Bahcall, J.N., Flynn, C., Gould, A., & Kirhakos, S., 1994, ApJL, 435, L51
Binney, J., 1995, in *Unsolved Problems of the Milky Way*, Blitz & Teuben, eds., Kluwer: Dordrecht, in press
Blitz L., 1979, ApJ, 231, L115
Blitz, L. & Spergel, D., 1991, ApJ 370, 205
Bosma A., 1978, Ph. D. Thesis, University of Groningen
Brand, J., & Blitz, L., 1993, A&A, 275, 67
Caldwell, J.A.R., and Coulson, I.M., 1987, AJ, 93, 1090
Chini R., Wink J.E., 1984, A&A 139, L5
Fich M., Blitz L., & Stark A.A., 1989, ApJ, 342, 272
Fich, M., & Tremaine, S., 1991, ARAA, 29, 409
Flynn, C., & Fuchs, B., 1994, A&A, in press
Gerhard, O.E., & Silk, J., 1994, Nature, submitted
Gould, A., 1995, in *Unsolved Problems of the Milky Way*, Blitz & Teuben, eds., Kluwer: Dordrecht, in press
Hartmann, D., 1994, Ph.D. thesis, Leiden University
Hron J., 1987, A&A, 176, 34
Hron J., 1989, A&A, 222, 85
Jackson P.D., FitzGerald M.P., & Moffat A.F.J., 1979, in *The Large-Scale Characteristics of the Galaxy*, Burton W.B. (ed.), Reidel: Dordrecht, 221

Kuijken, K., & Gilmore, G., 1989, MNRAS, 239, 605
Kuijken, K., & Gilmore, G., 1991, ApJ, 367, L9
Kuijken, K., & Tremaine 1992, in *The Dynamics of Disk Galaxies*, B. Sundelius, ed., Goteborg: Chalmers University of Technology, 71
Kuijken, K., & Tremaine 1994, ApJ, 421, 178
Kulkarni, S.R., Blitz, L., and Heiles, C., 1982, ApJL, 259, L63
Lada, E.A., & Lada, C.J., 1995, ApJ, in press
Magnani, L., Blitz, L., and Mundy, L. 1985, ApJ, 295, 402
Merrifield M.R., 1992, AJ, 103, 1552
Myers, P.C., & Benson, P.J., 1983, ApJ, 266, 309
Oort, J.H., 1932, BAN, 6, 249
Petrovskaya I.V., Teerikorpi P., 1986, A&A, 163, 39
Pfenniger, D., Combes, F., & Martinet, L., 1994, A&A, 275, 79
Pfenniger, D., & Combes, F., 1994, A&A, 275, 94
Pont, F., Queloz, D., Mayor, M., & Bratschi, P., 1995, in *Unsolved Problems of the Milky Way*, Blitz & Teuben, eds., Kluwer: Dordrecht, in press
Pound, M., & Blitz, L., 1993, ApJ, 418, 328
Pound, M., & Blitz, L., 1995, ApJ, in press
Prosser, C.F., *et al.*, 1994, ApJ, 421, 517
Rubin, V.C., Ford, K., and Thonnard, N., 1978, ApJL, 225, L107
Schechter, P., Aaranson, M., Cook, K.H., & Blanco, V., 1988, in *The Outer Galaxy*, Blitz & Lockman, eds., Springer: New York, 31
Schneider S.E., & Terzian Y., 1983, ApJ, 274, L61
Tinney, C.G., 1993, Ph.D. thesis, California Institute of Technology
Welch, D., 1988, in *The Mass of the Galaxy*, M. Fich, ed., CITA: Toronto, 68
Williams, J., de Geus, E.J., & Blitz, L., 1994, ApJ, 428, 693
Williams, J., & Blitz, L., 1995, ApJ, submitted
Zinnecker, H., McCaughrean, M.J., & Wilking, B.A., 1993, in *Protostars and Planets III*, eds. Levy, Lunine, and Mathews, University of Arizona Press: Tucson, p.429

DARK MATTER IN EXTERNAL GALAXIES - HI OBSERVATIONS

Albert Bosma
Observatoire de Marseille
2 Place Le Verrier, 13248 Marseille Cedex 4, FRANCE
bosma@batis.cnrs-mrs.fr

ABSTRACT

HI observations of spiral galaxies provide some of the strongest arguments for the existence of dark matter. I will review here the main efforts to quantify the amount of dark matter and its distribution, and indicate some constraints on its nature.

1. "STANDARD" ROTATION CURVE ANALYSIS

It is now customary to construct composite bulge/disk/halo mass models of spiral galaxies, using as input an observed HI rotation curve, a radial luminosity distribution in some colour band, and a radial distribution of the gas density (usually that of HI multiplied by ~1.36 to account for the He contribution). For a typical galaxy like NGC 3198 (e.g. Begeman, 1989) the rotation curves for the disk and gas components do not add up to match the observed rotation curve, and the remaining matter is attributed to a dark halo, which has a distribution conveniently described with a few parameters, such as a core radius and a central density.

Such an analysis rests on several assumptions, and 15 - 20 years ago was not evident at all. Several remarks are in order : 1) the need for dark matter comes from the mismatch between the observed rotation curve and the one calculated for the "visible" material alone, under the assumption of a constant mass-to-light ratio for each of the components bulge (if present) and disk, 2) it is clear that this mismatch is larger when the rotation curve extends out further, and 3) the attribution of the dark matter to a halo is essentially based on the theoretical work on disk stability by Ostriker and Peebles (1973), who introduced the concept of a stabilizing halo interior to the outer disk radius to prevent bar instabilities in numerical simulations of spiral galaxies. Further work on disk stability has essentially removed the requirement for a spherical halo (e.g. Athanassoula & Sellwood, 1986), provided the disk itself is dynamically hot enough.

The rotation curve extent focusses attention to 21-cm line data, since these can extend out to 2 - 3 times the optical radius (e.g. Bosma 1978, 1981, Begeman 1987); also Kalnajs (1983) and Kent (1986) have shown that most optical rotation curves can be fitted without the need for a dark matter component. The extent of HI rotation curves seems to be set by the ionization of the outer HI by the extragalactic background (cf. Maloney 1993).

Early work on mass modelling seldom took the step of computing mass models from the observed light distributions : instead the rotation curve was inverted to yield a mass surface density denstribution. Thus Bosma (1978, 1981) finds that the local mass-to-light ratio increases as function of radius (cf. Bosma

& Van der Kruit 1979), and that the local mass-to-gas ratio becomes more or less constant in the outer parts. The latter result can be recast in another way : one can scale the rotation curve of the gas with a constant factor in order to obtain the rotation curve of the halo (e.g. Carignan et al. 1990). This has led to arguments that the HI traces the dark matter rather well, that the dark matter might be baryonic, and that its distribution might be flat.

However, the increase in local mass-to-light ratio outwards does not say much more about the properties of the matter responsible for it, while with a composite mass model, which reflects identifiable physical components, the analysis can proceed further. The most debatable step in the analysis is actually which value to adopt for the mass-to-light ratio of the disk, and whether the assumption of its constancy is really justified.

2. SHOULD WE TAKE A MAXIMUM DISK ?

Van Albada et al. (1985) outline the range of choices for the mass-to-light ratio of the disk component : i) if taken maximum, ("maximum disk" solution), the rotation curve in the inner parts is attributed as much as possible to the visible components, ii) if taken a bit lower, a possible halo does not have a hollow core. Any other value below these solutions is in principle allowed as well. Athanassoula et al. (1987) propose to limit the range by applying criteria derived from Toomre's swing amplifier theory (Toomre 1981). Since many spirals are two-armed, it seems reasonable to require that the disk is massive enough to allow amplification of m = 2 structures, and not massive enough to allow amplification of m = 1 structures. Models just suppressing m = 1 structures give mass-to-light ratios for the disk which are in good agreement with models where the halo is required to have a non-hollow core. Moreover, these models seem to produce "acceptable" gas mass fractions, and allow for a trend often seen in smaller late-type spirals to have m = 4 structures in their outer parts.

Kent (1986, 1987), Begeman et al. (1991) and Broeils (1992) present "best least-squares fit" models, which are more or less like maximum disk models with no hollow core. Lake and Feinswog (1989) note that even for the best studied cases the fits are not well constrained in a statistical sense, and allow for a large range in the derived parameters for the halos. Thus has sprung up the quest for galaxies for whose rotation curves extend to the largest number of disk scalelengths, with specific search programmes as in e.g. Broeils (1992).

One of the arguments for maximum disk models is that they fit very well small scale features in optical rotation curves. (cf. Kalnajs 1983, Kent 1986, Van Albada & Sancisi 1986, Freeman 1992). However, the most striking case discussed in Freeman (1992), ESO 352-G27, is a galaxy with very strong spiral arms. This has two consequences : 1) a position velocity curve along the major axis will show non-circular motions directly associated with the spiral arms, and 2) a fit of an ellipse to the isophotes of such a galaxy is approximate at best, and might show systematic distortions correlated with the spiral arms. Thus these small scale features tracing neither the rotation curve, nor the axisymmetric part of the potential, cannot constitute evidence for or against the maximum disk solution.

Van der Kruit (1992) questions the assumption of maximum disk by pointing out that the edge-on galaxies NGC 7814 and NGC 891 have similar rotation curves despite the fact that the former is bulge dominated and the latter disk dominated. This leads him to think that for both galaxies the mass distribution is strongly dominated by the dark halo, even in the inner parts.

Further evidence for non-maximum disk solutions comes from work on stellar velocity dispersions in the disks of spiral galaxies, summarized by Bottema (1993). For a simple thin disk, exponential in the radial direction and obeying a $\mathrm{sech}^2(z/z_0)$-law in the vertical direction, he finds that, for a constant mass-to-light ratio and a constant thickness parameter z_0, the maximum velocity of the disk is on average 63% of the observed maximum rotation. For the galaxy NGC 3198, he finds a maximum velocity of the disk close to the value found in a no m = 2 solution. (cf. Figure 1).

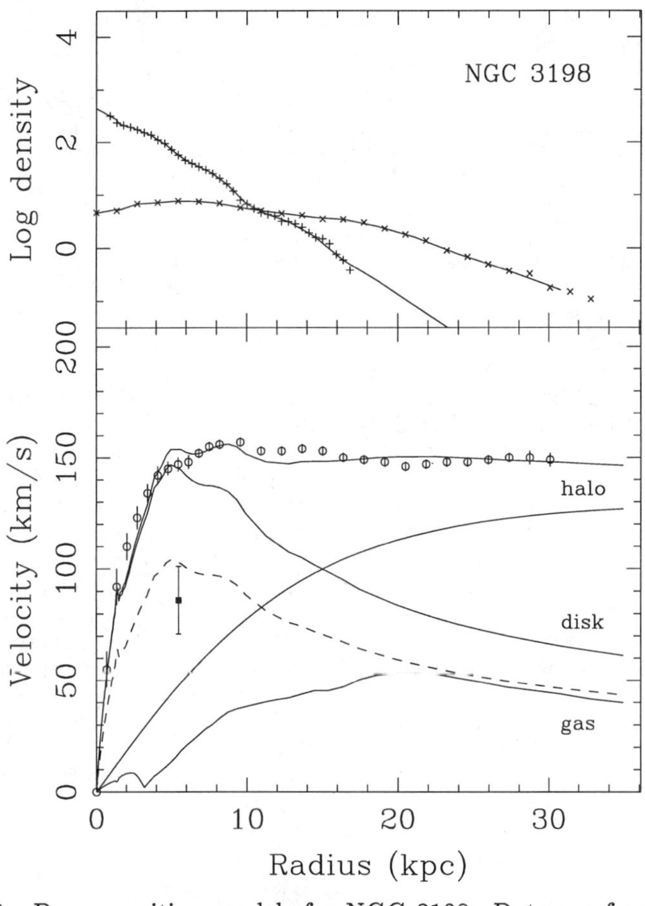

Figure 1. Decomposition models for NGC 3198. Data are from Van Albada et al. (1985) (rotation curve), and Wevers et al. (1986). The top panel gives the stellar density (IIIaF-band luminosity times the mass-to-light ratio) (plusses) and the gas density (crosses) in $M_\odot\ pc^{-2}$. The bottom panel shows the observed HI rotation curve as circles. The full drawn lines are for the maximum disk with no m = 1 solution : the labelled curves are the rotation curves for each component, and the upper curve the total circular velocity. The square with error bar denotes the maximum velocity of the disk according to Bottema (1993). The dashed curve is the disk rotation curve belonging to the no m = 2 solution.

3. IS THE DISK MASS-TO-LIGHT RATIO REALLY CONSTANT ?

The problem of whether maximum disk solutions are correct has been compounded by the opacity problem : it is not so clear anymore that spiral galaxies can be considered as (semi-)transparent, and perhaps their inner parts are more opaque than previously believed. Since the opacity is wavelength dependent, the assumption of a constant mass-to-light ratio for the disk may then break down. It is impossible here to discuss the full ramifications of the opacity problem, in fact a special meeting was dedicated to it recently (cf. Davies 1994). After the initial papers of Disney et al. (1989) and Valentijn (1990), a large number of studies have been conducted on this question, and a consensus has not yet been reached.

From the point of view of HI observations, a number of case studies of individual galaxies seen edge-on were reported in Bosma et al. (1992) and Bosma (1994b). These compare the kinematics of HI, CO, and Hα along their major axis. From these observations it can be concluded that the outer parts of large Sb and Sc galaxies, as well as small Sc galaxies, are transparent. The inner parts of Sb and larger Sc galaxies, out to roughly 2 scalelengths, could be opaque, and the material responsible for this is traced rather well by the CO molecule. The amount of absorption is moderate, as can be determined from case studies of near edge-on galaxies (e.g. Jansen et al. 1994). From a study of a backlit galaxy, the highest opacities are found in the spiral arms (White & Keel 1993).

To get around this opacity problem, it may be useful to use near-infrared observations for the derivation of the luminosity distribution. These have the advantage of a much lower extinction, and sample the underlying mass distribution of the disk better, even though the influence of red supergiants cannot be neglected. Block et al. (1994) demonstrated that it is possible to map the cold dust using blue and near-infrared images, and advanced models of dust extinction and scattering. The cold dust can also be mapped directly at (sub-)mm wavelengths using bolometer arrays (e.g. Guélin et al. 1993).

Several surveys are now underway to study the scalelengths of spiral disks in the near-infrared. Both Peletier et al. (1994) and De Jong & Van der Kruit (1994) find that the infrared scalelengths are, on average, shorter than the scalelength in the blue. As a consequence, the hypothesis that the mass-to-light ratio is constant is not justified anymore, except for the cases where the scalelengths are the same. In general, shorter scalelenghts of the disk increase the need for a halo component, imply a more concentrated halo, and a higher fraction of the gas-to-star ratio in the outer parts of the disk.

4. TRENDS IN THE HALO PARAMETERS

Since the "standard" decomposition model is far from satisfactory, proposed trends in derived halo parameters can only serve as useful guides to further work.

It has been argued that there is a close coupling between the disk and the halo, a so-called "conspiracy", which causes two structures of completely different shape and density distribution to conspire to make a flat and featureless rotation curve. Such a situation could arise in a model of galaxy formation in which gas collapses in a dark halo, and pulls the halo inwards while collapsing and forming the disk (e.g. Blumenthal et al. 1986). It is true that for galaxies with rotation speeds of order 200 km s^{-1} this situation might arise, and indeed

Athanassoula et al. (1987) report a correlation between the maximum velocity of the disk and the velocity dispersion of the halo.

However, this breaks down for dwarf galaxies, where the maximum velocity of the disk is much smaller than the velocity dispersion of the halo. Moreover, for several fast rotating galaxies there is a drop in rotational velocity just outside the optical image, sometimes of order of 20 km s^{-1} or more. In those cases, the velocity dispersion of the halo is related more directly to the rotation speed in the outer HI envelope, and not to the rotation speed in the inner disk. Such behaviour has been found in NGC 5055 (Bosma 1978, 1981), in several edge-on galaxies (e.g. Sancisi 1983), for NGC 2683 and NGC 3521 (Casertano & Van Gorkom 1991), a few cases studied by Broeils (1992), and now for several other galaxies (work in progress with Van Gorkom, Gunn, Knapp and Athanassoula). Thus for these galaxies the conspiracy never was perfect anyway. Casertano and Van Gorkom (1991) argue that such declining rotation curves are more readily associated with galaxies having short disk scalengths, but Broeils (1992) concludes differently.

Another trend often discussed is the tendency for dwarf galaxies to have higher central densities of dark matter then giant galaxies. This is particularly clear if one takes the reults of the maximum disk with no m = 1 solutions as in Athanassoula et al. (1987), and compare them with dwarf spirals and dwarf spheroidals, as done by Kormendy (1991). The presence of these compact lumps of dark matter in the dwarf spheroidals has led Lynden-Bell (1994) to suggest a model for the dark halo of our Galaxy which becomes increasingly lumpier when going outwards in radius.

A correlation seems to exist between the ratio of halo core radius to optical radius and the occurence of warped HI disks (Bosma 1991), in the sense that no warps are found for galaxies with small relative core radius. Sparke & Casertano (1988) find that for warp modes to exist in oblate halos, a large halo core radius is necessary. Dubinski & Kuijken (1994) find in numerical simulations with a live halo that the dynamical friction between a warped disk and the halo causes the two to align faster if the halo is more concentrated.

Much work has been done on the correlation of slopes of the rotation curves with other physical quantities of the parent galaxies, but few correlations are really standing out. Dwarf galaxies are more dark matter dominated then giant galaxies, cf. Carignan & Freeman (1988), Persic & Salucci (1988), Casertano and Van Gorkom (1991) and Broeils (1992). Amram et al. (1993) argue that the slopes of optical rotation curves do not seem to depend on environment, although selection effects are not well understood in their studies.

5. LOW SURFACE BRIGHTNESS GIANTS

A special class of galaxies which merits attention are those which have a disk of low surface brightness yet maintain a high rotational velocity. Representatives of these are the galaxies Malin 1, Malin 2, and some their "cousins" studied by Bothun and collaborators (Bothun et al. 1987, 1990, Van der Hulst et al. 1993, Sprayberry et al. 1993). Bosma and Freeman (1993) showed that such galaxies are relatively rare. However, some smaller "nephews" have been identified, which are nearer to us and thus more accessible to detailed studies.

Bosma et al. (1998) have observed one of those "nephews", NGC 5963, and find a nearly flat and featureless rotation curve, even though the luminosity distribution is far from featureless : it has a bright inner disk and a faint outer disk. A composite disk/halo model, when compared to the "normal" galaxy NGC 2403, shows several aspects : 1) the radial distributions of HI gas density are similar in the two cases, 2) the observed rotation curves are also similar, 3) the stellar disks are very different, and 4) the halo of NGC 5963 is more concentrated. A finer analysis of the expected radial velocity dispersion in the two cases illustrates a clue to the origin of such systems : in the case of NGC 5963, the stellar disk is so weak that the gas velocity dispersion more or less equals the critical velocity dispersion of the stellar disk. Thus the gas can remain stable, and not form stars. A similar situation exists at the edge of the disk of NGC 2403. This physical mechanism for determining disk edges, and disks of low surface brightness seems rather universal, and introduces a threshold effect in the star formation in disks. The basic idea goes back to Quirk (1972), see also Athanassoula & Bosma (1988), Kennicutt (1989), and Van der Hulst et al. (1993).

Several colloraries follow from the situation as found in NGC 5963 : 1) since the rotation curves of the gas and the observed rotation curves are similar for NGC 2403 and NGC 5963, there should be no reason to expect that anything else is different. In particular, the ratio of hidden molecular gas to atomic gas should be the same, so if the dark matter is indeed undetectable cold gas in a disk (cf. Pfenniger et al. 1994), there should be no reason why the stellar disks of NGC 2403 and NGC 5963 are so different. The fact that they are thus points against such an explanation for the dark matter. 2) as already argued in Bosma & Freeman (1993) the real difference between NGC 2403 and NGC 5963 lies in the amount of baryons normally ending up in the visible disk. The low surface brightness giants (and their family) have less of them, which suggests that the initial perturbations leading to these galaxies did not have the universal ratio of baryons to dark matter purported to exist (cf. Blumenthal et al. 1986).

Even darker disks are found around a few blue compact dwarf galaxies. Brinks & Klein (1988) find a large HI disk around IIZw40, with an axial ratio of about 8.5 to 1, and a radial velocity field showing two axes along which there is a gradient in radial velocity. Moreover, the optically visible part is not in the central part of the HI disk. Meurer (1993) finds that the blue compact dwarf NGC 2915 is surrounded by a very large and regularly rotating HI disk. There is structure in the gas, and the optical galaxy sits in the middle. The velocity field indicates that the central parts might contain an oval or bar-like distortion, and indeed the optical image is not well aligned with any of the kinematical axes. The presence of these huge gas disks with structure is surprising, since they are expected to flare : to keep the flaring down, we have to assume a rather flattened distribution of (part of) the dark material.

6. SHAPES : TRIAXIAL HALOS AND ELLIPTICAL DISKS ?

The natural shape for a dark halo is triaxial (cf. Binney 1978). Indeed, cosmological N-body simulations produce objects identifiable as galaxies which have typical axial ratios of 10 : 6-7 : 4-5 (e.g. Dubinski & Carlberg 1991). This poses the question about the shape of the disk as well ; can it really be round while sitting in a non-axisymmetric halo ? In a general discussion about HI velocity fields Bosma (1978, 1981) distinguishes several classes of deviations from axisymmetry, the most important being oval distortions and kinematical warps. For galaxies with oval distortions, the kinematical major axis is not perpendicular to the kinematical minor axis, while for kinematical warps the principal kinematical axes stay perpendicular also in the warped regions. Thus for galaxies with warps the inner disks are round, while for the galaxies with oval distortions usually a non-axisymmetric substructure in the disk can be found responsible for the kinematic signature of the deviation (bar or fat oval). This means that the dark halo, at least close to the visible disk in non-barred galaxies, cannot be strongly triaxial.

Recently, interest in this problem has been revived, partly in the context of non-axisymmetric structures in our Galaxy. Kent (1992), using CCD images, reports on axial ratios of face-on galaxies selected from a catalog of narrow single dish HI line profiles. Zaritsky & Rix (1994) study a similar sample in the near-infrared K′-band. Round face-on galaxies do seem to be present, and an upper limit to the ellipticity of disks of < 0.05 is found. A similar upper limit was arrived at by Franx & De Zeeuw (1992), from a consideration about the scatter in the Tully-Fisher relation. Franx et al. (1994) find a round halo for the galaxy IC 2006, an elliptical with a faint outer ring which is remarkably circular in shape. Dubinski (1994) discusses the "rounding out" of halos due to the dissipative formation of the visible parts.

7. SHAPES : FLATTENED OBLATE HALOS

Polar ring galaxies are relatively small S0's with a polar ring whose kinematics can give indications about the overall shape of their dark halos (and thereby perhaps about halos in general). Though early work pointed to a round halo, later work (Sackett & Sparke, 1990) finds best fitting shapes between E3 and E6, by detailed modelling of all the components. New work on NGC 4650A by Sackett et al. (1994a), which fits simultaneously the gas velocities of the ring and the stellar velocity and velocity dispersion of the disk, further constrains the models, and the preferred shapes are now about E5 - E8.

Deep CCD photometry of the edge-on Sc galaxy NGC 5907 (Morrison et al. 1994) reveals the presence of faint light centered around the bulge, but with a drop off roughly as r^{-2}. Sackett et al. (1994b) propose to consider this as tracing directly the dark matter, and present fits to the rotation curve, which are dominated by a dark component with a mass-to-light ratio of about 450, and a shape of this component about E5. However, in view of the presence of a significant warp in the optical disk, it seems preferable to model this warp first correctly, and then to see what is really left over. It is also unclear to what extent we are seeing the faint extension of the bulge which has been seen in the near-infrared data (Barnaby & Thronson, 1994). Bulges do not necessarily obey a $r^{\frac{1}{4}}$-law or a $r^{-3.5}$-law in the vertical direction (cf. Kormendy 1982)

Composite disk/halo mass models allow the study the vertical equilibrium of the gas, and the behaviour of the flaring of the HI layer as a function of radius. Bosma (1994a) discusses some of the parameters governing this problem : the shape of the halo and the velocity dispersion of the gas are major parameters. We have modelled the flaring for the edge-on galaxy NGC 100, for which we have both the necessary high resolution HI data, and the surface photometry. From the HI data we can determine the rotation curve, an estimate of the gas velocity dispersion (from the terminal radial velocities as shown in a position-velocity diagram), and a direct measure of the flaring (which might be overestimated if the galaxy is not entirely edge-on or is warped in the line of sight). The optical data provide the parameters for the stellar disk. The best fit indicates a round halo, but halos as flat as E6 cannot be entirely excluded.

8. THE LOCAL RATIO OF (DARK) MASS TO HI GAS MASS

Lequeux et al. (1993) reiterate that not much is known about the H_2 content in the outskirts of spiral galaxies, and that perhaps there is enough of it in tiny cold clouds to explain the flat rotation curves. This point has been studied by Pfenniger, Combes & Martinet (1994), who want these clouds to be in the disk, and by Gerhard & Silk (1994), who prefer them in a more spherical distribution.

Bosma (1978, 1981) found that the ratio of the mass surface density to that of the HI gas density in the outer parts of spiral galaxies is constant with radius, and that its value is a function of the mean amplitude of the rotation curve. For dwarf galaxies this ratio is of order 2 - 5, while for giant galaxies it can reach up to 100. This provides a strong argument against Pfenniger et al. (1994)'s picture of Sd's evolving to Sa's : the conversion of cold gas into stars would have to entail more production of cold gas (by accretion ?) so as to increase the local ratio of dark mass to HI gas mass, while the stellar disk grows (Sd's being smaller than Sa's). Furthermore, if the dark matter is in the disk, it is not clear why warps are preferentially found in later type spirals (cf. sec. 4), why low surface brightness giants should exist (cf. sec. 5), how disk stability is assured, and how the flaring results would square with this.

If the cold clouds are in a halo they should survive in a harsh X-ray environment such as in clusters, since otherwise every HI stripped spiral might also have lost its halo, show a falling rotation curve, and should become bar unstable unless it has a large bulge or hot inner disk. Cursory inspection of the stripped Virgo spirals (Cayatte et al. 1990) shows that this hypothesis probably cannot be ruled out there, since all but one are classified as SB or SAB.

9. HI IN ELLIPTICALS

Few elliptical galaxies have HI extended enough to unambiguously decide the presence of a dark halo from the determination of their HI rotation curve. Recent work has concentrated on IC 2006 (Franx et al. 1994) for which the dark halo is found to be round, and on systems with shells, such as NGC 5128. In the latter case, Siminowich et al. (1994) find the HI displaced with respect to the optical shells, but the velocity field is relatively regular, and indicates a flattish rotation curve, and hence a dark halo.

ACKNOWLEDGEMENTS

Thanks are due to Lia Athanassoula and Ken Freeman for frequent discussions on dark matter. Jacqueline van Gorkom provided me with an outline of what in ordinary circumstances would have been her talk.

REFERENCES

Amram, P., Sullivan III, W.T., Balkowski, C., Marcelin, M., Cayatte, V. 1993, ApJ, 403, L59

Athanassoula, E., Sellwood, J.A. 1986, MNRAS, 221, 213

Athanassoula, E., Bosma, A., & Papaioannou, S. 1987, A&A, 179, 23

Athanassoula, E., & Bosma, A. 1988, in Large Scale Sctructures in the Universe, IAU Symp. No. 130, eds. J. Audouze et al. (Dordrecht : Reidel), p. 391

Barnaby, D., & Thronson Jr, H.A. 1992, AJ, 107, 1717
Begeman, K.G. 1987, Ph.D. thesis, University of Groningen
Begeman, K.G. 1989, A&A, 223, 47
Begeman, K.G., Broeils, A.H., & Sanders, R.H. 1991, MNRAS, 249, 523
Binney, J.J. 1978, MNRAS, 183, 779
Block, D., Witt, A.N., Grosbøl, P., Stockton, A., Moneti, A. 1994, A&A, 288, 383
Blumenthal, G.R, Faber, S.M., Flores, R., & Primack, J.P. 1986, ApJ, 301, 27
Bosma, A. 1978, Ph.D. thesis, University of Groningen
Bosma, A. 1981, AJ 86, 1791 & 1825
Bosma, A. 1991, in Warped Disks and Inclined Rings around Galaxies, eds. S. Casertano et al. (Cambridge : Cambridge Univ. Press), p. 181
Bosma, A. 1994a, in ESO/OHP workshop on Dwarf Galaxies, eds. G Meylan & P. Prugniel, (ESO Garching), p. 187
Bosma, A. 1994b, in The Opacity of Spiral Disks, ed. J.I. Davies (Cambridge : Cambridge Univ. Press) (in press)
Bosma, A., & Van der Kruit, P.C. 1979, A&A, 79, 281
Bosma, A., Van der Hulst, J.M., & Athanassoula, E. 1988, A&A, 198, 100
Bosma, A., Byun, Y.I, Freeman, K.C., & Athanassoula, E. 1992, ApJ, 400, L21
Bosma, A., & Freeman, K.C. 1993, AJ, 106, 1394
Bothun, G.D., Impey, C.D., Malin, D.F., & Mould, J.R. 1987, AJ, 94, 23
Bothun, G.D., Schombert, J.M., Impey, C.D., Schneider, S.E. 1990, ApJ, 360, 427
Bottema, R. 1993, A&A, 275, 16
Brinks, E., Klein, U. 1988, MNRAS, 231, 63P
Broeils, A.H. 1992, Ph.D. thesis, University of Groningen
Carignan, C., & Freeman, K.C., 1988, ApJ 332, L33
Carignan, C., Charbonneau, P., Boulanger, F., Viallefond, F. 1990, A&A, 234, 43
Casertano, S., Van Gorkom, J.H. 1991, AJ, 101, 1231
Cayatte, V., Van Gorkom, J.H., Balkowski, C., Kotanyi, C. 1990, AJ, 100, 604
Davies, J.I. 1994, ed., The Opacity of Spiral Disks, (Cambridge : Cambridge Univ. Press) (in press)
De Jong, R.S., Van der Kruit, P.C. 1994, in Infrared Astronomy with Arrays, ed. I. Mc Lean (Dordrecht : Kluwer), p. 123
Disney, M.J., Davies, J.I., & Phillips, S. 1989, MNRAS, 239, 939
Dubinski, 1994, ApJ, 431, 617
Dubinski, J., Carlberg, R.G. 1991, ApJ, 378, 496
Dubinski, J., Kuijken, K. 1994, preprint
Franx, M., De Zeeuw, P.T., 1992, ApJ, 392, L47
Franx, M., Van Gorkom, J.H., De Zeeuw, P.T. 1994, ApJ (in press)
Freeman, K.C. 1992, in Physics of Nearby Galaxies : Nature or Nurture ?, eds. T.X. Thuan et al., (Gif-sur-Yvette : Editions Frontières), p. 201
Gerhard, O.E., Silk, J. 1994, preprint
Guélin, M., Zylka, R., Mezger, P.G., Haslam, C.G.T., Kreysa, E., Lemke, R., Sievers, A.W. 1993, A&A, 279, L37
Jansen, R.A., Knapen, J.H., Beckman, J.E., Peletier, R.F., Hes, R. 1994, MNRAS, 270, 373
Kalnajs, A.J. 1983, in Internal Kinematics and Dynamics of Galaxies, IAU Symp. No. 100, ed. E. Athanassoula, p. 87
Kennicutt, R.C. 1989, ApJ, 344, 685
Kent, S. 1986, AJ, 91, 1301

Kent, S. 1987, AJ, 93, 816
Kent, S. 1992, in Morphological and Physical Classification of Galaxies, eds. G. Longo et al. (Dordrecht : Kluwer), p. 153
Kormendy, J., 1982, in Morphology and Dynamics of Galaxies, 12th Saas Fee lecture course, eds. L. Martinet & M. Mayor, p. 113
Kormendy, J., 1991, in Evolution of the Universe of Galaxies, ed. R.G. Kron, ASP Conf. Series, vol 10, p. 33
Lake, G., Feinswog, L. 1989, AJ, 98, 166
Lequeux, J., Allen, R.J., Guilloteau, S. 1993, A&A, 280, L23
Lynden-Bell, D. 1994, in ESO/OHP workshop on Dwarf Galaxies, eds. G Meylan & P. Prugniel, (ESO Garching), p. 589
Maloney, P. 1993, ApJ, 414, 41
Meurer, G.R. 1994, in ESO/OHP Workshop on Dwarf Galaxies, eds. G. Meylan & P. Prugniel, (ESO Garching), p. 351
Morrison, H., Boroson, T.A., Harding, P., 1994, AJ, 108, 1191
Ostriker, J.P., Peebles, P.J.E. 1973, ApJ 186, 467
Peletier, R.F., Valentijn, E.A., Moorwood, A.F.M., & Freudling, W. 1994, A&AS (in press)
Persic, M., Salucci, P. 1988, MNRAS, 234, 131
Pfenniger, D., Combes, F., Martinet, L. 1994, A&A, 285, 79
Quirk, W.J. 1972, ApJ, 196, L9
Sackett, P.D, & Sparke, L.S. 1990, ApJ, 361, 408
Sackett, P.D., Rix, H.-W., Jarvis, B.J., Freeman, K.C. 1994a, ApJ (in press)
Sackett, P.D., Morrison, H.L., Harding, P., Boroson, T.A. 1994b, Nature, 370, 441
Sancisi, R. 1983, in Internal Kinematics and Dynamics of Galaxies, IAU Symp. No. 100, ed. E. Athanassoula (Dordrecht : Reidel), p. 55
Schiminovich, D., Van Gorkom, J.H., Van der Hulst, J.M., Kasow, S. 1994, ApJ, 423, L101
Sparke, L.S., Casertano, S., 1988, MNRAS, 234, 873
Sprayberry, D., Impey, C.D., Irwin, M.J., McMahon, R.G., & Bothun, G.D. 1993, ApJ, 417, 114
Toomre, A. 1981, in The Structure and Evolution of Normal Galaxies, eds. S.M. Fall & D. Lynden-Bell (Cambridge : Cambridge Univ. Press), p. 111
Valentijn, E. 1990, Nature, 346, 153
Van Albada, T.S., Bahcall, J.N., Begeman, K.G., Sancisi, R. 1985, ApJ, 295, 305
Van Albada, T.S., & Sancisi, R., 1986, Phil. Trans. R. Soc. London, A230, 447
Van der Hulst, J.M., Skillman, E.D., Smith, T.R., Bothun, G.D., McGaugh, S.S., De Blok, W.J.G. 1993, AJ 106, 548
Van der Kruit, P.C. 1992, in Morphological & Physical Classification of Galaxies, eds. G. Longo et al. Kluwer (Dordrecht), p. 39
Wevers, B.H.M.R, Van der Kruit, P.C., Allen, R.J. 1986, A&AS, 66, 505
White III, R.L., Keel, W.C. 1992, Nature, 359, 129
Zaritsky, D., & Rix, H.-W. 1994, in Infrared Astronomy with Arrays, ed. I. McLean (Dordrecht : Kluwer), p. 457

The shape of the dark matter halo of NGC 4244

Rob P. Olling, J. H. van Gorkom
Columbia University
538 West 120th street box 42, New York, NY10027
olling@tosca.phys.columbia.edu

ABSTRACT

We present a preliminary analysis of the flaring behavior of the hydrogen layer of the nearby almost edge-on ($i \approx 85^o$) Sc galaxy NGC 4244 (Fig. 1). We developed a new technique to determine the inclination and the thickness of the gas layer from many channels simultaneously. Furthermore we show that this method is not limited to edge-on galaxies only. Each side of the galaxy has been analyzed separately. Comparing the measured thickness of the hydrogen layer with model predictions we conclude that the galaxy mass model which reproduces the measured widths best is composed of a stellar disk which is about one-third as massive as maximum-disk and a DM-halo with $q = 0.5 +/- 0.3$.

1. Introduction

Little is known about the shape of dark halos : the equatorial rotation curve is a one dimensional probe only. Both spherical and flat dark matter (DM) halo models can reproduce the same observed rotation curve. The polar ring galaxy NGC 4650A has a dark halo as flattened as E6-E7 (Sackett & Sparke, 1990 and Sackett *et al.*1994a). Recently, Sackett *et al.*(1994b) reported the discovery of a flattened ($q = c/a \approx 0.5$) *luminous* halo (possibly of dynamical significance) around NGC 5907.

Like the polar rotation curve, the thickness of the gas layer depends on the shape of the DM-halo (Olling and van Gorkom 1992, OvG92; Maloney 1993, M93; Kundić, Hernquist and Gunn 1993, KHG93). In this poster we present the first results of a new method (proposed by Olling 1995, O95) to determine the flattening of DM-halos from high resolution HI measurements.

2. Some theory

Flatter DM-halos have larger midplane densities (M93, O95), so that gas layers immersed in such a flat halo will be comparatively thin. For the analysis of NGC 4244 we follow the approach taken by O95 and determine the vertical distribution of the gas from the global potential of the galaxy. The global mass distribution is composed of an exponential stellar disk and a flattened isothermal DM-halo. For this global mass distribution we calculate the global potential. If the gas is isothermal in z, then we can use the equation of hydrostatic equilibrium

$$\sigma_z(R)^2 \frac{d\ (\rho_{gas}(z))}{d\ z} = \rho_{gas}(z)\ K_z(z) \ = \ -\rho_{gas}(z)\ \frac{d\ \Phi(R,z)}{d\ z}, \qquad (1)$$

from which the (zeroth-order) density distribution of the gas follows :

$$\rho_{gas}(R,z) = \rho_{gas}(R,0) \exp\left(-\Phi(R,z)/\sigma_z(R)^2\right). \qquad (2)$$

Equation (2) can be used to calculate an approximation to the potential due to the self-gravity of the gas :

$$\Phi_{GAS}^{approx}(R,z) = -2\pi G \int_0^z dz' \Sigma_{GAS}(R,z') \,. \tag{3}$$

The next order approximation to $\rho_{gas}(R,z)$ is calculated from equation (2), where we take Φ to be $(\Phi_{GAS}^{approx} + \Phi_{STARS} + \Phi_{HALO})$. To the resulting distribution we fit a Gaussian to determine its Full Width at Half Maximum (FWHM_z). Beyond the optical disk the thickness of the gas layer increases approximately linearly (OvG92; M93; KHG93; O95) with radius :

$$\mathrm{FWHM}_z \approx 2.35 \left(\frac{\sigma_z(R)}{V_{rot,max}} \right) \text{ (kpc)}, \tag{4}$$

where $V_{rot,max}$ is the peak rotation velocity.

3. The gaseous disk of NGC 4244

High resolution (175 pc x 5 km/s) VLA (B-, C-, and D-array) HI observations were obtained for NGC 4244. The rotation curve was measured from major axis-velocity plots. Several different disk-halo decompositions are shown in Figure 2., both maximum-disk and one-third-maximum-disk models fit the observed rotation curve equally well.

3.1. Determining the thickness of the gas layer

For a perfectly edge-on galaxy the width of the gas layer at a given galactocentric radius,R, is measured in that channel which has a velocity difference w.r.t. the systemic velocity equal to the rotation speed at that radius.

If the galaxy is not exactly edge-on, the relation between the measured widths and the intrinsic gas layer thickness ,$\mathrm{FWHM}_z(R)$, is less simple. Consider for example a galaxy with a uniform surface density, a flat rotation curve, and zero gaseous velocity dispersion ($\sigma_z(R)$=0). If observed with square channels (ΔV_{chan}), any channel will be sensitive to gas in a wedge centered around $\theta = \arccos\left(V_{chan}/(V_{rot}\sin i)\right)$, while the opening angle of the wedge is given by:

$$\Delta\theta = \arccos\left\{\frac{(V_{obs} - 0.5\Delta V_{chan})}{V_{rot}(R)\ \sin(i)}\right\} - \arccos\left\{\frac{(V_{obs} + 0.5\Delta V_{chan})}{V_{rot}(R)\ \sin(i)}\right\}. \tag{5}$$

For almost edge-on systems the two wedges (the near and far side respectively) will be superimposed. At lower inclinations, the wedges bifurcate, and we observe the classical "butterfly" shape. The projected width on the sky is thus a combination of the projected gas layer thickness ($\mathrm{FWHM}_z \times \sin(i)$) and the projected "wedge" width* ($\mathrm{FWHM}_w \times \cos(i)$). As both the vertical and the planar distributions are Gaussian, the projection on the sky will also be Gaussian. The observable width (FWHM_O) is given by :

$$\mathrm{FWHM}_O^2 \approx \mathrm{FWHM}_B^2 + (\mathrm{FWHM}_z \sin i)^2 + \left(\mathrm{FWHM}_w^2 + (2.35\ P_w)^2\right)\cos^2 i\ , \tag{6}$$

where FWHM_B is the beam size. When the inclination is so low that the near and far sides are clearly separated, the P_w-term has to be dropped. As we have

* For real galaxies, the wedge-width is affected by the rotation curve, the velocity dispersion and the surface density distribution. Once these parameters are measured, the exact shape of the "wedge" can be calculated for all radii. Model calculations indicate that, at any major axis distance, these "wedges" can be well approximated by two Gaussian of width FWHM_w centered at $\pm P_w$.

measurements of FWHM_O, and are able to calculate FWHM_w and P_w, for many (≥ 10) channels, we can solve (6) for inclination and FWHM_z in a least squares sense. We applied this technique to simulated HI-observations of a galaxy like NGC 4244 at inclinations of $85^o, 72^o, 63^o$ and 55° : the input values for FWHM_z and i are recovered very well for all but the lowest inclination.

Since each spectral line channel samples a small range of azimuth only, we can determine the thickness of the gas layer for non edge-on systems as well. From calculations at different gaseous velocity dispersion we find that the width (of a typical channel), as projected on the sky, of the "wedge" can be approximated as : $\mathrm{FWHM}_w \approx 4\{MAX(\sigma_z, \Delta V_{chan})/(V_{rot}\sin(i))\}\ R\cos(i)$. As the projected width of an isothermal gas layer in the potential of a DM-halo increases linearly with radius (eqn. [4] $\times \cos(i)$) as well, we find that the ratio of the vertical and planar projected widths depends on inclination only : $\{\mathrm{FWHM}_z(R)/\mathrm{FWHM}_w(R)\} \approx 0.57\sin(i)\tan(i)$. In order to reliably determine the thickness of the gas layer its width projected on the sky needs to be larger than, the other contributors to the apparent width (beam size and FWHM_w). Thus we find a lower limit to the inclination of about 63° : similar to inclination cutoff observed in our numerical experiments.

4. Conclusions

For NGC 4244, we present the results in Fig. 3. Both sides, as well as the average are indicated. High (10") and low (38") resolution data are presented. A small warp (lower panel) and a flaring gas layer (upper panel) can be seen, notice the similarity of the two sides. For the comparison with the model predictions we fixed the inclination to 85° so that the widths beyond 10 kpc are upper limits. In Figure 4 we show these widths superimposed on model predictions. The radial coordinate is normalized by the optical scale-length (h_R). The numbers on the right side of the box indicate the corresponding halo flattenings. Inside the optical disk ($R \leq 5h_R$), the measurements clearly require a low $\mathcal{M}/\mathcal{L}$ stellar disk. For the "maximum-disk" value ($\mathcal{M}/\mathcal{L}$ =5.10) the observed widths are much larger than the model predictions. For the preliminary analysis presented in this paper the lowest $\mathcal{M}/\mathcal{L}$ -value (1.57, or about 1/3-maximum-disk) agrees better. As the measurements do not extend very far beyond the optical disk, the DM-halo flattening can not be determined with great accuracy. For q-values in the range 0.2-0.8, the predictions are consistent with the observations.

ACKNOWLEDGEMENTS

This work was supported in part through an NFS grant (AST-90-23254) to Columbia University.

REFERENCES

Kundić T., Hernquist, H., Gunn, J.E., 1993, p 592 in "Back to the Galaxy", AIP Conference proceedings # 278, Editors S.S. Holt, F. Verter (KHG93)

Maloney, P., 1993, Ap.J ,414, 41 (M93)

Olling, R.P., 1995, submitted to AJ (O95)

Olling, R., Gorkom, J.H. van, 1992, p. 374 in "The Evolution of Galaxies and their Environment", NASA conference Publication 3190, Editors D. Hollenbach, H. Thronson, J.M. Shull (OvG92)

Sackett, P.D., Rix, H.W., Jarvis, B.J., Freeman, K,C., 1994, ApJ, 436, 629

Sackett, P.D., Morrison, H.L., Harding, P., Boroson, T.A., 1994b, Nature, 370, 11 August p. 441

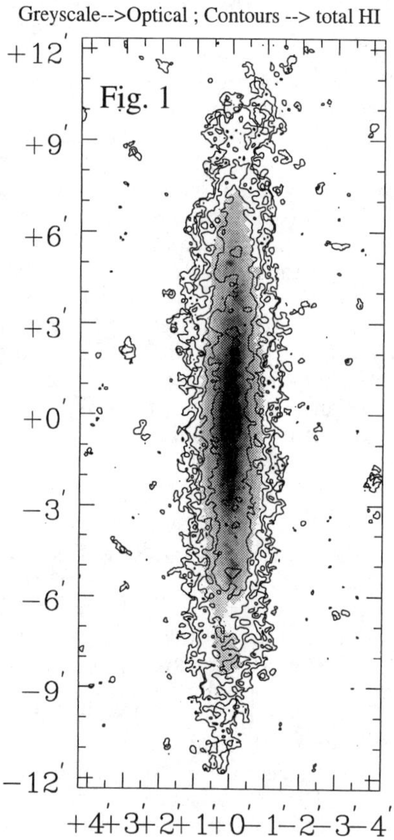
Greyscale-->Optical ; Contours --> total HI
Fig. 1
+12′
+9′
+6′
+3′
+0′
−3′
−6′
−9′
−12′
+4′ +3′ +2′ +1′ +0′ −1′ −2′ −3′ −4′

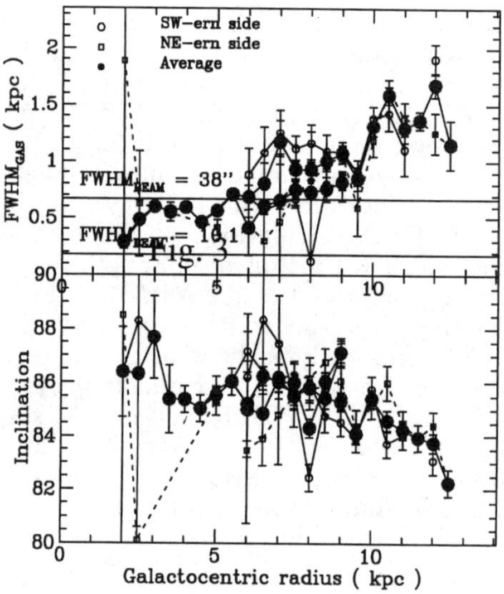
SW-ern side
NE-ern side
Average
$FWHM_{GAS}$ (kpc)
$FWHM_{BEAM}$ = 38''
$FWHM_{BEAM}$ = 10''
Fig. 3
Inclination
Galactocentric radius (kpc)

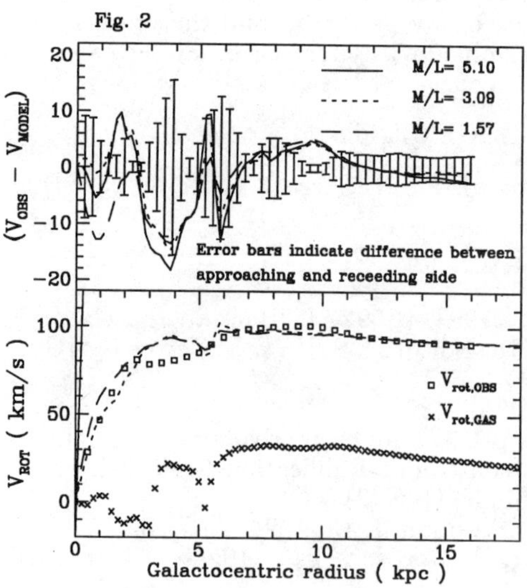
Fig. 2
M/L= 5.10
M/L= 3.09
M/L= 1.57
(V_{OBS} − V_{MODEL})
Error bars indicate difference between approaching and receeding side
V_{ROT} (km/s)
$V_{rot,OBS}$
$V_{rot,GAS}$
Galactocentric radius (kpc)

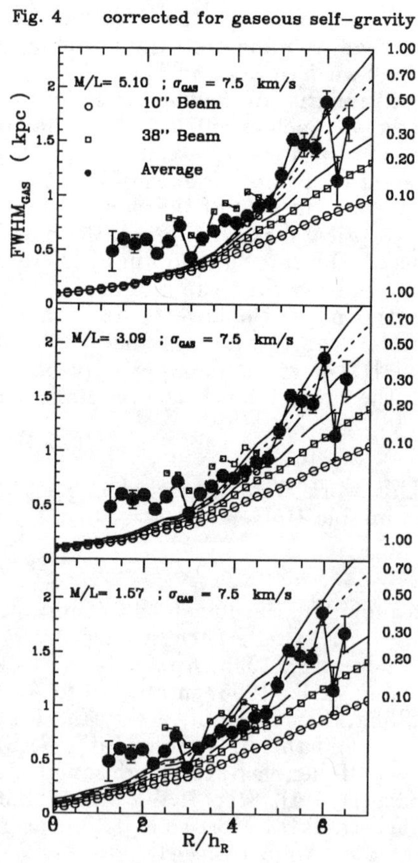
Fig. 4 corrected for gaseous self-gravity
M/L= 5.10 ; σ_{GAS} = 7.5 km/s
10'' Beam
38'' Beam
Average
M/L= 3.09 ; σ_{GAS} = 7.5 km/s
M/L= 1.57 ; σ_{GAS} = 7.5 km/s
$FWHM_{GAS}$ (kpc)
1.00
0.70
0.50
0.30
0.20
0.10
R/h_R

DARK MATTER AND THE HUBBLE SEQUENCE OF DISK GALAXIES

Adrick Broeils
Department of Astronomy, Cornell University,
524 Space Sciences Bldg, Ithaca, NY 15853-6801
broeils@astrosun.tn.cornell.edu

ABSTRACT

For a sample of 23 disk galaxies with accurate neutral hydrogen rotation curves detailed mass models were constructed. They are based on the maximum-disk hypothesis with the assumption that disk and bulge mass-to-light ratios do not vary with radius. The models provide estimates of the mass of each individual component and a number of interesting correlations between between the properties of dark and luminous matter on the one hand and morphological type on the other emerges One of the most fascinating trends is the decrease of the dark-to-luminous mass ratio towards earlier spiral types. This trend suggests that Sa galaxies contain very little dark matter relative to the visible mass and prompted us to start a project studying the internal kinematics in Sa galaxies. Sa's have been understudied as a class and were not represented in the sample of 23 galaxies. Finally I present the remarkable results on the first VLA observations in this project.

1. INTRODUCTION

Our knowledge on the properties of Dark Matter (DM) in individual galaxies is based on observations of a relatively small number of galaxies. Since the bulk of the invisible matter resides beyond the edges of optical disks extended neutral hydrogen (H I) rotation curves are needed to trace the DM halos. Furthermore, to determine the amount and distribution of the unseen halo material detailed mass modeling of individual galaxies – taking the distributions of all luminous components into account – is necessary.

Using 21-cm synthesis observations with the WSRT I derived the distributions and kinematics of H I in twelve spiral galaxies. These data were combined with data from the literature to form a sample of 23 galaxies with high precision H I rotation curves and CCD surface photometry. To establish the properties of the dark and luminous material in a consistent way, the same fitting procedure (maximum disk, see Sect. 2) was used for all galaxies.

In Sect. 3 I present examples of mass model fits and discuss the properties of dark and luminous material of the galaxies in relation to their morphological type. In the past years observers have focused a lot of attention to the H I dynamics of low-luminosity galaxies, however, the Sa's at the other end of the Hubble sequence for disk galaxies are clearly understudied and not represented in the original sample of 23 galaxies. Their H I rotation curves are more difficult to obtain, because Sa galaxies generally contain less neutral hydrogen and the H I surface densities are lower. Based on the trends described in Sect. 3 the Sa's as a class are expected to have "extreme" properties, like very small dark-to-luminous mass ratios and low central halo densities. In Sect 4 I report on results from H I observations of the Sa galaxy NGC 4138 and mass modeling of two other galaxies with data taken from the literature.

2. MAXIMUM DISK HYPOTHESIS

The basic assumption in the models is that the gravitational potential of a galaxy can be described by the sum of the potentials of three or four separate mass components: stellar disk and bulge (if present), neutral gas disk and dark halo. The rotation curves for the two disk components are calculated from the radial distribution of stellar surface brightness and gas surface density (i.e. there is no M/L scaling parameter involved for the gas component) using the formulae given by Casertano (1983) and Begeman (1987); the equations for calculating the bulge rotation curve are given by Kent (1986). The stellar mass-to-light ratios are assumed to be independent of radius. For the halo we take the density profile $\rho(R) = \rho_0 \left[1 + (R/R_C)^2\right]^{-1}$, where R_C is the halo core radius.

In order to investigate correlations between optical and dark matter properties one has to find a way to constrain either the stellar M/L ratios or the halo parameters. I used the so-called maximum disk (+bulge) method in which first the inner portion of the observed rotation curve is fitted by the luminous components and subsequently the residual is fitted by a dark halo rotation curve. In other words one assumes that the stellar component(s) dominate(s) the potential in the optical part of a galaxy. This has two advantages: 1) the models are constructed in a nearly unambiguous way, and 2) the models provide a limiting case by yielding a lower limit to the amount of dark matter inside the last point of the rotation curve. At the same time there are several indications that the maximum-disk hypothesis might be realistic:

- Features in observed (mostly optical) rotation curves can often be reproduced from features in the light distribution (Kent 1986, Buchhorn 1992, Broeils & Courteau 1994). This suggests that luminous mass must contribute dominantly in inner region.
- Non-restricted least-squares fits to observed H I rotation curves generally have solutions with M/L within 15% of the maximum-disk value (Begeman et al. 1991).
- The small scatter in Tully-Fisher relations suggests that the amount of visible matter sets the amplitude of the rotation curve.
- The disk M/L values are "reasonable", i.e. similar to the value for the solar neighborhood. The dependency of M/L on B-V colors is the same as found for population synthesis models of evolving disks (Larson & Tinsley 1978).

3. RESULTS OF THE MODELING

Examples of the maximum-disk model fits for 8 galaxies, ordered by their maximum rotation velocity and morphological type, are shown in Fig. 1. One can clearly recognize trends with maximum rotation velocity and morphological type, like the decreasing relative halo contribution towards earlier spiral type.

In Figures 2-5 a variety of properties for the dark and luminous material is plotted versus morphological type. The various correlations can be summarized as follows:

- Maximum-disk M/L decreases from Sa $\rightarrow$ Irr.

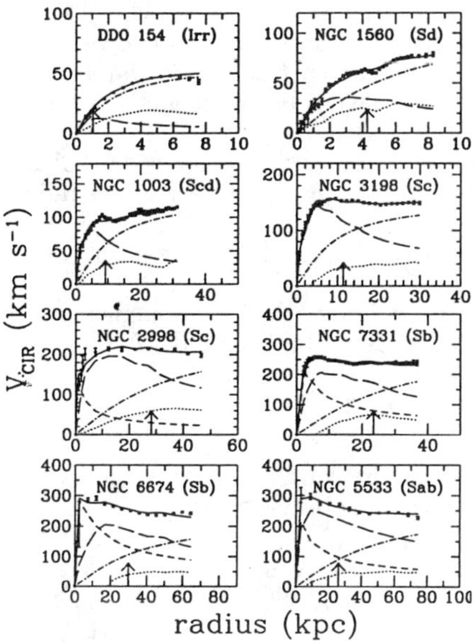
DDO 154 (Irr)
NGC 1560 (Sd)
NGC 1003 (Scd)
NGC 3198 (Sc)
NGC 2998 (Sc)
NGC 7331 (Sb)
NGC 6674 (Sb)
NGC 5533 (Sab)
V_{CIR} (km s^{-1})
radius (kpc)

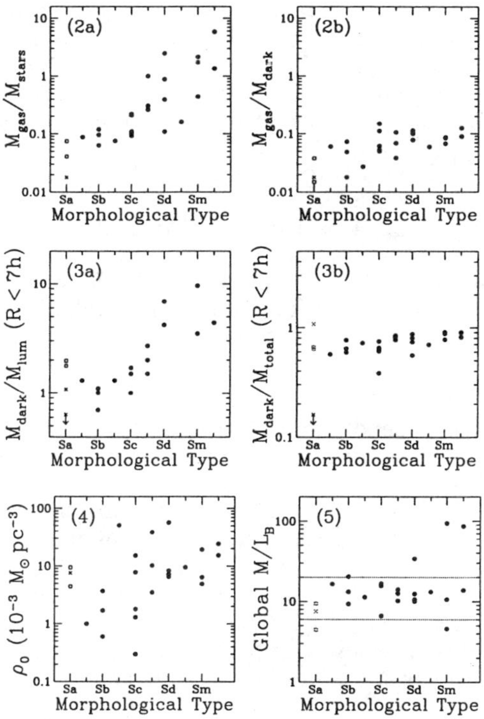
(2a)
M_{gas}/M_{stars}
(2b)
M_{gas}/M_{dark}
(3a)
M_{dark}/M_{lum} (R < 7h)
(3b)
M_{dark}/M_{total} (R < 7h)
(4)
ρ_0 (10^{-3} $M_\odot$ pc^{-3})
(5)
Global M/L_B
Sa Sb Sc Sd Sm
Morphological Type

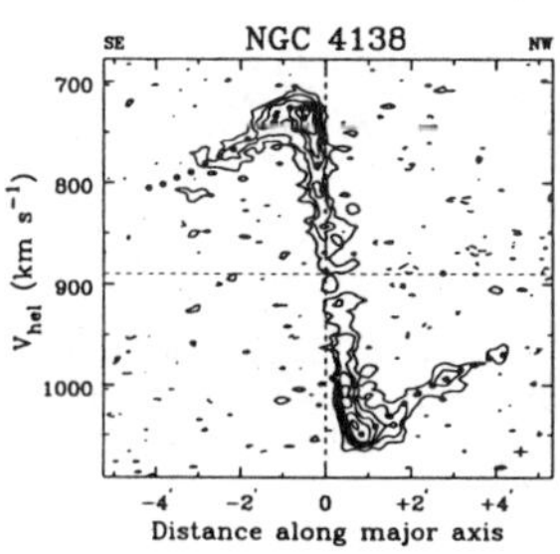
NGC 4138
SE
NW
V_{hel} (km s^{-1})
Distance along major axis

Fig. 6

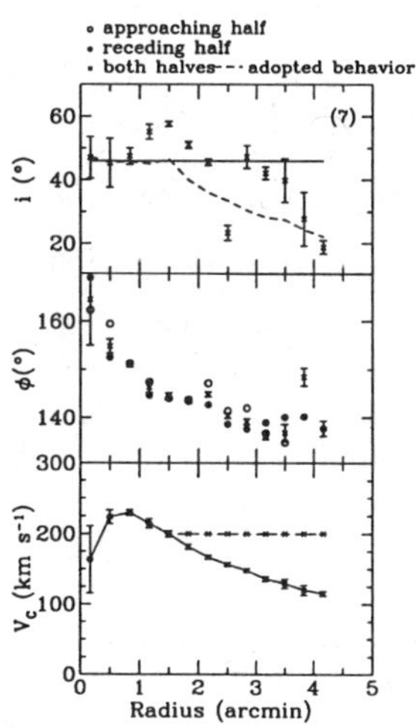
approaching half
receding half
both halves - - - adopted behavior
(7)
i (°)
ϕ(°)
V_c (km s^{-1})
Radius (arcmin)

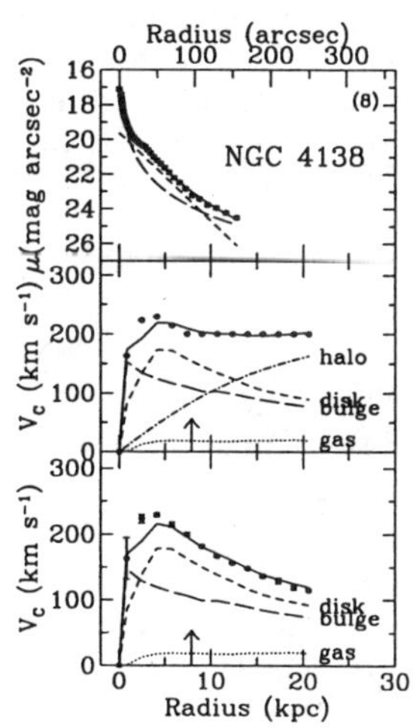
Radius (arcsec)
(8)
NGC 4138
μ(mag arcsec^{-2})
V_c (km s^{-1})
halo
disk
bulge
gas
Radius (kpc)

- (Fig. 2a:) M_{gas}/M_* increases from Sa $\rightarrow$ Irr. (Fig. 2b) M_{gas}/M_{dark} shows little variation with type. Also the distribution is much narrower than for M_{gas}/M_{dark} . Not only the gas-to-dark matter *fraction* is remarkable constant for the whole sample, also the similarity between gas and dark matter radial surface densities *distributions* is striking.
- (Fig. 3a:) M_{dark}/M_{lum} decreases from 5-10 for dwarf galaxies to 1-2 for Sab's. (Fig. 3b) M_{dark}/M_{tot} shows the same trend, but with a smaller amplitude.
- (Fig. 4:) Central halo density (ρ_0) decreases from late to earlier spiral types.
- (Fig. 5:) Global M/L ratio shows no strong trend with morphological type. Only a few (low-luminosity) galaxies have extreme high global M/L's.

4. RESULTS FOR SA GALAXIES

Here I report on the results of H I mapping with the VLA of the first galaxy in the Sa sample, NGC 4138 (Broeils et al. 1994); more observations are underway. I also constructed models for two other Sa galaxies (NGC 1169 & NGC 3898) based on less sensitive H I data taken from van Driel & van Woerden (1994).

The distribution of H I gas along the major axis of NGC 4138 (Fig. 6) shows that the radial velocities beyond one arcminute drop by more than 80 km s^{-1} on both sides. This can be the result of either a declining rotation curve (solid lines in Fig. 7) or a severe warp in the hydrogen disk beyond the optical edge (dashed lines in Fig. 7). Both models reproduce the observed velocity field and xv-diagram (dots in Fig. 6) equally well.

For NGC 4138 I made two mass models, for a declining and a flat rotation curve (Fig. 8). For the former model *no* dark halo is needed, for the other a small dark halo is required. The other two galaxies require somewhat larger halos to explain the observed rotation curves. The results for the Sa galaxies are indicated in Figs 2-5 by crosses for NGC 4138 and squares for NGC 1169 and NGC 3898. Clearly the Sa's do not always follow the trends described in Sect. 3; the results for NGC 4138 are more extreme than for the other two Sa's. The results of future H I mapping of other Sa's will tell whether NGC 4138 is the exception or the rule.

ACKNOWLEDGEMENTS

The author wishes to thank R. Sancisi, M. Haynes and K. Jore for stimulating discussions. This work was partly supported by the NSF grant AST 9023450.

REFERENCES

Begeman, K.G. 1987, Ph.D. thesis, University of Groningen
Broeils, A.H.. 1992, Ph.D. thesis, University of Groningen
Buchhorn, M, 1992, Ph.D. thesis, Australian National University
Broeils, A.H., Courteau, S. 1994, in preparation
Begeman, K.G., Broeils, A.H., Sanders, R.H. 1991, MNRAS 249, 523
Casertano, S. 1983, MNRAS 203, 735
van Driel W., van Woerden H. 1994, A&A 286, 395
Kent, S.M. 1986, AJ 91, 1301

DISKS AND DARK HALOES OF DWARF SPIRALS

V. Friese, B. Fuchs and R. Wielen
Astronomisches Rechen-Institut Heidelberg, Germany
S34@mvs.urz.uni-heidelberg.de

1. INTRODUCTION

In a number of dwarf spiral galaxies the HI-emission has been studied with sufficient resolution to derive the rotation curves of the galaxies. These show that the disks of dwarf spirals are imbedded in extended haloes of dark matter, quite similar to the disks of giant spiral galaxies. The ratios of the dark halo to disk masses which have been found in dwarf galaxies are equal to or even larger than those found in giant spirals, although they are subject to the substantial uncertainties of the decomposition of the rotation curves into disk and halo contributions (Lake, Schommer, and van Gorkom 1990).

The presence of such massive, extended haloes affects the dynamics of the disks of the galaxies. We concentrate on two aspects here: First we discuss the effects of large amounts of dark matter on the spiral structure of the disks. Unfortunately the velocity dispersion of the stars in dwarf spirals is not known, but we have to use the vertical scale heights of the disks to derive estimates of the velocity dispersions. For this purpose we have performed a statistical flattening analysis of faint dwarf spirals in the ESO-Uppsala catalogue in order to derive the intrinsic ratio of vertical to radial scale lengths for these types of galaxies. Once the velocity dispersions are known Toomre's Q stability parameter and 'X = 2' criterion (Toomre 1964, 1981, Athanassoula *et al.*1987) can be evaluated.

Second, it has been suggested that the main constituents of the dark haloes of giant spiral galaxies might be very compact objects such as black holes with masses of the order of a few 10^6 $\mathcal{M}_\odot$ (Lacey and Ostriker 1985). This scenario has been put forward in order to explain the observed increase of the velocity dispersion of the stars in the disk of the Galaxy (Wielen 1977). Such an increase would result from stochastic acceleration processes of the stars by any massive perturbers of the disk. So other physical mechanisms have been discussed as sources of the 'heating' of the disk as well, among these are interactions of stars with giant molecular clouds, transient spiral density wavelets, or infalling disrupting satellite galaxies. We have discussed the scenario of massive black holes scattering the stars, when penetrating the disk, in some detail elsewhere and found many attractive features, fitting nearly ideally the locally or globally observed kinematics of the stars in the Galaxy (Wielen *et al.*1992). Thus it is interesting to carry this conjecture over to the haloes of dwarf spiral galaxies, especially since the disks of these galaxies are much less massive and thus more fragile than in giant spirals (Rix and Lake 1993, Fuchs and Wielen 1993).

2. SPIRAL STRUCTURE

We have analyzed in detail the disks of a set of 9 dwarf spirals listed in table 1 for which rotation curves are available in the literature, mainly by Carignan and collaborators (*cf.*Côté *et al.*(1991) and references therein). To the rotation curve of each galaxy a mass model comprising an exponential disk and a dark halo component described by a pseudo isothermal density law has been fitted. Dwarf spirals have no bulge component (Freeman 1987). In the following we discuss usually a best χ^2-fit and a 'maximum disk' fit which maximises the disk contribution to the rotation curve. The resulting disk and halo masses were, with a few exceptions, in good agreement with the results of Carignan and collaborators who derived the disk contributions to the rotation curves using directly the observed optical surface brightness and radial HI density profiles of the galaxies instead of an exponential law. In the case of larger divergences we have adopted the parameters given in the original publications. In each case the decomposition of the rotation curves is by no means unique but allows for a large range of disk parameters.

It is straightforward to show that the characteristic scale length of the vertical distribution of a self-gravitating disk in hydrostatic equilibrium is

$$z_0 = \frac{< w^2 >}{\mathcal{F}\pi G\Sigma_d}, \tag{1}$$

where $< w^2 >$ is the vertical velocity dispersion, Σ_d denotes the surface density of the disk and G is the constant of gravitation. $\mathcal{F}$ is a form factor depending on the shape of the vertical distribution. For an isothermal disk with a $sech^{-2}(z/z_0)$ profile, which is often used for modelling galactic disks (van der Kruit and Searle 1981), $\mathcal{F}$ =1. Wielen and Fuchs (1983) have pointed out that galactic disks, at least the Milky Way, are non-isothermal because the disks are made up by many generations of stars with velocity dispersions varying according to their age. Correspondingly each generation of stars has a different vertical scale height and the superposed distributions resemble closely an exponential distribution. Exponential density profiles are also seen in many edge-on galaxies (Wainscoat *et al.*1989, Barteldrees and Dettmar 1994). For such a model of the vertical structure of the Milky Way disk the form factor in equation (1) is $\mathcal{F} = 2.2$ which we adopt also for the disks of dwarf spirals. Since the vertical scale heights of the galaxies in table 1 have not been measured we estimate the vertical scale heights from the radial scale lengths h. For this purpose we have performed a statistical flattening analysis of faint dwarf spirals in the ESO-Uppsala catalogue in order to derive the intrinsic ratio of vertical to radial exponential scale lengths z_0/h. From a sample of 115 galaxies with $M_B \geq$ -18 and Hubble type Sc or later we find $z_0/h = 0.2 \pm 0.05$ corresponding to an intrinsic flattening $q_0 = 0.14$. This intrinsic ratio is in good agreement with scale height ratios found by van der Kruit and Searle (1982) and Barteldrees and Dettmar (1993) for giant spirals seen edge-on or by Bottema (1993) modelling spiral galaxies with known velocity dispersions. The intrinsic flattening q_0 which we find is also in good agreement with the value of 0.12 given by Bottinelli *et al.*(1983) for Sd galaxies or 0.10 found by Guthrie (1992) for Sc galaxies. For the determination of the stability parameter we need the dispersion of the radial velocity components of the stars. These are estimated with the help of equation (1) and converting the vertical to radial velocity dispersions using the same axial ratio of the velocity ellipsoid as in the Milky Way.

In this way we have modelled individually the disk of each galaxy. There is a clear distinction between the dynamical states of galactic disks which develop ordered spiral structure or which appear irregular. NGC 300 is a typical example of the first class. The stability parameter is in the range where swing amplification is effective. The expected number of spiral arms, estimated using the 'X = 2' criterion, indicates a two-armed pattern in the inner parts of the disk which breaks up in filaments in the outer parts, exactly as observed. NGC 55 is a typical irregular galaxy. The values of the stability parameter are so large that any spiral structure is suppressed. The transition from Hubble types Sc, Sd to Sm, Im is not correlated to the absolute brightnesses of the galaxies or to the absolute masses of the dark haloes, but, as can be seen from table 1, to the disk to dark halo mass ratios. In table 1 we have listed estimates of the masses of the disks and dark haloes of the galaxies within one Holmberg radius based on the best χ^2-fit and the 'maximum disk' fit to the rotation curves, respectively. Although our dynamical models rely on rather crude estimates of the stellar velocity dispersions and ambiguous decompositions of the rotation curves our conclusions appear to be rather robust because we could interpret the Hubble type of each galaxy consistently. The only exception is NGC 5585 which is classified as Sd but does not show any well defined spiral structure and is like the other irregular galaxies in table 1 extremely gas-rich.

Table 1

	type	M_B	Q	M_d	M_h	M_d/M_h
NGC 2259	Scd	-16.5	< 3	31 (41)	30 (15)	1.03 (2.86)
NGC 247	Sd	-18.0	< 3	87 (102)	121 (79)	0.72 (1.28)
NGC 300	Sd	-17.9	< 3	35 (52)	67 (41)	0.52 (1.27)
NGC 7793	Sd	-18.3	< 3	35 (—)	106 (—)	0.33 (-)
DDO 170	Irr	-15.2	> 3	1.4 (2.7)	9.4 (7.7)	0.15 (0.35)
NGC 5585*	Sd	-17.5	> 3	7.4 (14)	108 (99)	0.07 (0.14)
DDO 154*	Im	-13.8	> 3	— (0.28)	— (2.9)	— (0.10)
NGC 55	Sm	-18.6	> 3	3.0 (14)	170 (159)	0.02 (0.09)
NGC 3109	Sm	-16.8	> 3	— (1.8)	— (56)	- (0.03)
				$10^8 M_\odot$	$10^8 M_\odot$	

*HI disk more massive than stellar disk

3. DISK HEATING

Assuming now that the stochastic heating of the disks is due to the black holes of the dark haloes we can deduce from the velocity dispersions of the stars typical black hole masses. We consider first briefly the situation in giant spiral galaxies. Wielen (1977) has determined empirically the diffusion coefficient which describes the increase of the velocity dispersion with stellar age of stars in the solar neighbourhood. Lacey and Ostriker (1985) have calculated the diffusion coefficient theoretically and shown that it is proportinal to the space density times the individual mass of the black holes. Adopting a density of 0.01 $\mathcal{M}_\odot \mathrm{pc}^{-3}$ for the mass density of the dark halo this implies a mass of $3 \cdot 10^6$ $\mathcal{M}_\odot$ of a typical black hole in the Galaxy. Quite a similar value of $2 \cdot 10^6$ $\mathcal{M}_\odot$ can be derived for NGC 3198, a giant Sc galaxy (M_B = -19.4), using the the velocity dispersion determined by Bottema (1988) and the parameters of the dark matter halo derived by van Albada *et al.*(1985). The typical black hole masses which we find for the Sd galaxies are of the same order of magnitude as in giant spirals. The velocity dispersions of the disk stars of the irregular galaxies, however, are hardly larger than the turbulent velocity dispersion of the interstellar gas, which we have assumed as 5 km/s, so that the disks of these galaxies cannot have experienced much stochastic heating. Consequently, the masses of the hypothetical black holes in these galaxies turn out to be low, of the order of 10^4 to 10^5 $\mathcal{M}_\odot$. This was noted before by Rix and Lake (1993) and Fuchs and Wielen (1993). Rix and Lake (1993) argue that such low masses contradict the supposed universal nature of the black holes and rule out the possible existence of massive black holes in the dark haloes of galaxies at all.

On the other hand the large gas content and mass to light ratios less than one of the disks of the irregular galaxies indicate that the light from the galaxies is dominated by a rather young bright population of stars and that the old disk is not seen in the integrated light. This is corroborated by the colours of the disks. From the colours of NGC 5585, NGC 55 and DDO 154 listed in the RC3 catalogue and the stellar population synthesis calculations of Larson and Tinsley (1978) we conclude that the component dominating the surface brightness of the disks of the galaxies must be younger then $2 \cdot 10^9$ yrs. Such a component has certainly not experienced much stochastic heating and might be thus not suitable to study the stochastic heating of stellar disks, i.e. to determine reliable black hole masses. So, in our view, the issue of whether there are massive black holes in the dark haloes of dwarf spirals or not remains still open.

REFERENCES

van Albada, T. S., Bahcall, J. N., Begeman, K., & Sancisi, R. 1985, ApJ, 295, 305
Athanassoula, E., Bosma, A., & Papaioaunou, S. 1987, A&A, 179, 23
Barteldrees, A., & Dettmar, R. J. 1994, A&A in press
Bottema, R. 1988, A&A, 197, 105
Bottema, R. 1993, A&A, 275, 16
Bottinelli, L., Gougenheim, L., Paturel, G., & de Vaucouleurs, G. 1983, A&A, 118, 4
Côté, S., Carignan, C., & Sancisi, R. 1991, AJ, 102, 904
Freeman, K. C. 1987, IAU Symp. No. 117, p.119
Fuchs, B., & Wielen, R. 1993, AG Abstract Ser. 8, 145
Guthrie, B. N. G. 1992, A&A Suppl., 93, 255
Lacey, C., & Ostriker, J. P. 1985, ApJ, 299, 633
Lake, G., Schommer R. A., & van Gorkom 1990, AJ, 99, 547
Larson, R. B., Tinsley, B. M. 1978, ApJ, 219, 46
Rix, H. W., & Lake, G. 1993, ApJ, 417, L1
Toomre, A. 1964, ApJ, 139, 1214
Toomre, A. 1981, in The Structure and Evolution of Normal Galaxies, eds. S. M. Fall & D. Lynden-Bell (Cambridge: Cambridge Univ. Press) p.111
van der Kruit, P. C., & Searle, L. 1981, A&A, 95, 105
van der Kruit, P. C., & Searle, L. 1982, A&A, 110, 61
Wainscoat, R. J., Freeman, K. C., & Hyland, A. R. 1989, ApJ, 337, 163
Wielen, R. 1977, A&A, 60, 263
Wielen, R., & Fuchs, B. 1983, in Kinematics, Dynamics and Structure of the Milky Way, ed W. L. H. Shuter (dordrecht: Reidel) p.81
Wielen, R., Dettbarn, C., Fuchs, B., Jahreiß, H., & Radons, G. 1992, IAU Symp. No. 149, p.81

The Nature of Dark Matter and Galactic Rotation Curves

Irshadullah Khan
9 Ericsson Street, Belmont, MA 02178.
ikhan@mitlns.mit.edu

Abstract

It is shown that if dark matter consists of particles of masses m_i, spins J_i then $m_i = 2kT/v_m^2 = 156.9E^{-1/4}$ keV (v_m is the galactic rotation velocity at large r and T is the temperature of dark matter at the galactic location, E is the emissivity of dark matter at 6450 angstroms. Relations determining the emissivity of dark matter and the flux of radiation originating in a halo of dark matter are given (eqns. (7), (8)).

1. Introduction

The rotation velocities at large distance (r) from the centers of galaxies do not show the $1/\sqrt{r}$ fall off expected if mass is assumed to be associated exclusively with stars. To explain this discrepancy it has been suggested that more than ninety per cent of the mass producing gravitational effects on luminous matter resides in the form of dark matter. The latter appears to be nearly a spherically symmetric distribution around galactic centers (as suggested by observations of the 'anonymous galaxy' AO 136-0801 which seems to have two perpendicular disks of stars around a common halo of dark matter) [Rubin, 1985, 1983] showing that dark matter did not participate in the evolution of matter that developed into a disk of stars. If dark matter interacts gravitationally with itself then the stability of the halo leads us to conclude that gravity must be balanced by forces due to pressure gradients in the dark matter (hydrostatic equilibrium). From this requirement we deduce the equation of state for dark matter in the limit of low densities.

2. Derivation of the equation of state for dark matter

In the presence of a spherical distribution of the dominant non-luminous mass such that the total mass within a sphere of radius r around the galactic center is $M(r)$, the equation of averaged motion for stars gives

$$\frac{GM(r)}{r^2} - \frac{1}{\rho_s}\frac{dP}{dr} = \frac{v(r)^2}{r} \tag{1}$$

where $M(r) = 4\pi\int_0^r r^2\rho(r)dr$, ρ_s is the average density of a star and ρ is the density of dark matter. The second term on the left hand side is negligible for $r > 10^{10}$ cm, as will be seen in the following. The condition of hydrostatic equilibrium of dark matter gives

$$-\frac{dP}{dr} = G\rho(r)\frac{M(r)}{r^2} \tag{2}$$

We shall now solve the above equations in the limit of large r under the assumption that the second term on the left hand side of eqn. (1) can be neglected. Observations of galactic rotation velocities have indicated that $v(r) \to v_m (= 50 - 250$ km/s) for large r. The solution for large r is therefore

$$M(r) = Cr \quad ; \quad \rho(r) = \frac{C}{4\pi r^2} \quad ; \quad \frac{dP}{dr} = -\frac{C^2 G}{4\pi r^3} \quad ; \quad GC = v_m^2 \qquad (3)$$

With this solution we may now verify that the above assumption (neglecting the second term in eqn. (1)) holds provided $v_m^2 << 4\pi r^2 G\rho_s$ which is valid down to distances as small as 10^{10} cm from the galactic center. The equation of state for dark matter in the limit of low densities is therefore $P = \frac{1}{2}v_m^2\rho(r) + K$, where $-(1 + \frac{1}{2}v_m^2/c^2)\Lambda = K$ and Λ is the cosmological constant. For a gas of dark matter particles i of mass m_i, spin J_i at the dark matter temperature T under the assumption of validity of the classical limit (verified below for $r > 10^{-3}$ pc), so that equipartition of energy holds, we may write

$$P = kT\sum_i n_i \quad ; \quad \rho = \sum_i m_i n_i \qquad (4)$$

Consistency with the equation of state gives $m_i = 2kT/v_m^2 = m$. Thus the squared velocity of galactic rotation (v_m^2) at large r must be proportional to the dark matter temperature T in the galactic neighborhood. No regulatory mechanism for maintaining this relation between v_m and T is proposed here. However it is clear that if this relation does not hold then in this model the luminous matter will either escape from the dark matter halo or be crushed into its core and the apparently peaceful coexistence between dark and luminous matter (except perhaps in active galaxies) will no longer hold.

Recent observations [Sackett, 1994] of the spiral galaxy NGC 5907 show that its dark matter halo has luminosity in the R-band (having a width of 1500 angstroms around 6450 angstroms) nearly equal to 2.12×10^{-7} Joule m^{-2} s^{-1} corresponding to a surface temperature nearly equal to $633E^{-1/4}$ Kelvin where E is emissivity of dark matter at 6450 angstroms. Since v_m for the galaxy is nearly 250 km/s we derive $m = 156.9E^{-1/4}$ keV. Notice that the interparticle spacing is at least $n^{-1/3}$ which is bigger than the de-Broglie wavelength $h/\sqrt{3mkT} = 7.75 \times 10^{-7}E^{1/8}$ cm provided $n < 2.15 \times 10^{18}E^{-3/8}$ cm^{-3} i.e. $\rho < 6 \times 10^{-10}E^{-5/8}$ gm cm^{-3}, $r > 3.7 \times 10^{-4}E^{5/16}$ pc. Thus the classical approximation justifying the applicability of the equipartition theorem in deriving the equation of state holds for distances from the galactic center greater than about 10^{-3} pc which is a very small fraction of the radius of the luminous disk associated with galaxies (about 15000 pc)

3. Rotation curves for galaxies

Assume that the dark matter temperature in the vicinity of a galaxy is nearly constant over the extent of the galaxy. Letting $v(r) = v_m f(r)$ in eqns. (1), (2) and using the equation of state which is valid all the way down to 10^{-3} pc we derive

$$r^2(ff'' + f'^2) + f^2(2rff' + f^2 - 1) = 0 \qquad (5)$$

The solutions of this equation regular at $r = 0$ have the behaviour $f(r) \to ar$: $r \to 0$. Denoting the solution with derivative at the origin equal to a by $f(r, a)$, we have the scaling relation $f(r, a) = f(ar, 1)$. Thus the the function $f(r, 1)$ contains complete information about all the solutions of interest. Graphs of $f(r, 1), f(r, 2), f(r, 3)$ are shown in Fig.(1).

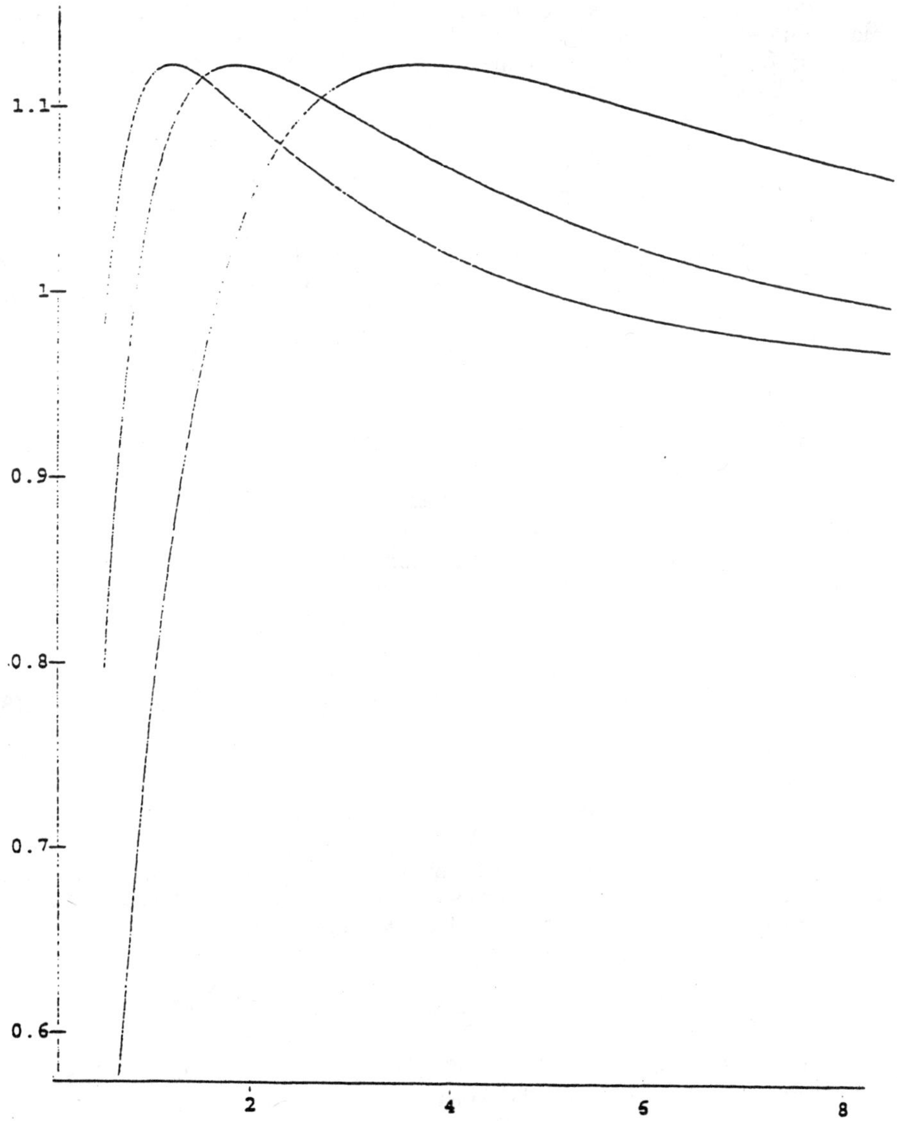

Fig. (1). Graphs of $f(r, a)$ *vs.* r : $a = 1, 2, 3$

4. Cosmic microwave background temperature

It will now be shown that the cosmic microwave background temperature [Smoot, 1992] can be understood as a consequence of the above picture of hot dark matter blobs in association with galaxies. The simplest hypothesis leading to this is the following re-interpretation of the Cosmological Principle : the emissivity $E(\nu, x)$ at frequency ν of dark matter located at distance x is equal to the emissivity at frequency $\nu/(1+Hx)$ of dark matter in our own vicinity (H is Hubble's constant). Assume that the dark matter halos associated with the galaxies have an average radius $R(T)$ and number of halos (galaxies) per unit volume is $N(T)dT$ in the dark matter temperature range $T, T+dT$. Then the flux of radiation at any location is

$$\frac{192\pi^2\zeta(5)k^5E(\nu)}{h^4Hc^2\nu^2}\int_0^\infty R(T)^2N(T)T^5dT \quad d\nu \tag{6}$$

where $E(\nu)$ is the emissivity of dark matter at our location ($x = 0$). Equating this flux to the observed flux of microwave radiation gives

$$E(\nu) = \frac{\nu^5}{F} \quad (exp(\frac{h\nu}{kT_0}) - 1)^{-1} \quad ; \quad F = \frac{96\pi\zeta(5)k^5}{h^5H}\int_0^\infty R(T)^2N(T)T^5dT \tag{7}$$

where T_0 ($= 2.7$ Kelvin) is the temperature of the cosmic microwave background.

5. Spectrum of radiation originating in the dark matter halo

Now that we have determined the emissivity of dark matter (eqn. (7)) we may determine the spectral distribution of radiation emitted by a nearby dark matter halo at temperature $T = \frac{1}{2}mv_m^2/k$. It is given by

$$\frac{2\pi h\nu^8}{Fc^2} \quad (exp(\frac{h\nu}{kT_0}) - 1)^{-1} \quad (exp(\frac{h\nu}{kT}) - 1)^{-1} \quad d\nu \tag{8}$$

Acknowledgements

I would like to thank Professor Badri Aghassi for hospitality at Department of Natural Sciences, Bentley College, Waltham, Professors Alan Guth, Jill Mason and Eric Wollman for discussions and Mr. Bilal Abdullah Khan for help. I would like to thank Dr. Todd Boroson for correspondence.

References

Rubin. V. C., Burstein, D., Ford Jr., W. K., & Thonnard, N. 1985 ApJ, 81, 289, Rubin, V. C., 1983 Scientific American (June), 31

Smoot G. et al, 1992 ApJ, 396, L1

Sackett, P. D. , Morrison, H. L., Harding, P. & Boroson, T. A. 1994 Nature, 441, 370

ROTATION CURVES OF 967 SPIRAL GALAXIES: IMPLICATIONS FOR DARK MATTER

Massimo Persic[1], Paolo Salucci & Fulvio Stel[2]

SISSA, Strada Costiera 11, 34014 Trieste, Italy
[1] and Osservatorio Astronomico, Trieste, Italy
[2] and Dipartimento di Astronomia, Università di Trieste, Italy
persic@tsmi19.sissa.it
salucci@tsmi19.sissa.it
fulvio@tsmi19.sissa.it

ABSTRACT

We present the rotation curves of 967 spiral galaxies, obtained by deprojecting and folding the raw Hα data originally published by Mathewson *et al.*(1992). Of these, 80 meet objective excellence criteria and are suitable for individual detailed mass modelling, while 820, individually less compelling mainly because of the moderate statistics and/or limited extension, are suitable for statistical studies. The remaining 67 curves suffer from severe limitations that may largely restrict their use in galaxy structure studies. A preliminary analysis of their properties confirms that RCs are a universal function of luminosity and that the dark matter fraction in spirals increases with decreasing luminosity.

1. INTRODUCTION

Rotation curves (hereafter RCs) are the prime mass tracers within spiral galaxies. Therefore, knowledge of their morphology and structural implications is essential for theories/experiments on galaxy formation (*e.g.*: Cen & Ostriker 1993; Navarro & White 1994; Evrard *et al.*1994). Such a knowledge is of course crucially increased when large samples of good-quality curves become available.

The Hα RCs of nearly one thousand spirals published by Mathewson *et al.*(1992; hereafter MFB) represent by far the biggest database on galaxy rotation available to date. In the MFB paper (and related public files), however, the Hα data are presented only as measured velocities, adequate though for MFB's purpose of studying the peculiar cosmological velocity field. In a related paper (Persic & Salucci 1995; hereafter PS95) we make available the actual RCs, after folding, deprojecting and smoothing the raw MFB data. Because of its size, homogeneity, quality, and spanned range of luminosities and asymptotic velocities, the sample of RCs so obtained will serve as a main database for studies of galaxy structure.

The plan of this contribution is as follows. In section 2 we outline the procedure used to obtain the RCs from the raw data. In section 3 we classify the 967 RCs into three quality subsets. Section 4 briefly illustrates the results of a preliminary analysis of these RCs.

2. DATA ANALYSIS

As for the observations and data acquisition and reduction, the reader is referred to MFB. Here we outline the basic procedure leading to the final RCs. For each RC it includes: 1) a selection of the individual velocity data; and 2) the identification of the kinematical center about which to fold such data. In addition, 3) we have compared the radius corresponding to the farthest measured velocity with the optical size in order to evaluate the extension of each RC.

1) The quality of each velocity point is measured by the cross-correlation coefficient ρ (given in the public MFB data files) between the observed and the template Hα line profile. Based on the rms velocity error as a function of ρ, we have found that a good correlation ($\rho > 0.35$) is essential in order to ensure a high reliability of the measure.

2) The photometric center indicating the maximum emission in the I-band does not necessarily coincide with the dynamical center of the galaxy. In order to obtain the kinematical center we have proceeded as follows: we have assumed that, once folded, each curve must be perfectly symmetric around its kinematical center, so that the curve defined by the approaching side coincides (within errors) with the one traced by the receding side. In practice, we have started from the photometric center; then, if warranted, we have estimated the kinematic center by folding the data around slightly different zero-points until the global symmetry of the velocity arms is maximized. Table 1 of PS95 contains the solution of this folding procedure, *i.e.*: the heliocentric systemic velocity, and the offset of the kinematic center from the photometric center.

3) It is important to have a reference scale for the optical size of each galaxy, in order to assess the extension of the RC vs. the visible matter. For this purpose, from the I-band CCD surface photometry of MFB we have computed the radius, R_0, encompassing 83% of the integrated light (see Table 1 of PS95; for an exponential disk this corresponds to 3.2 lengthscales, which in turn corresponds, for a Freeman disk, to the de Vaucouleurs 25 B-mag/arcsec^2 photometric radius.)

3. THE ROTATION CURVES

We have identified 900 RCs that are symmetric, with reasonably low rms internal scatter and reasonable high sampling density. These curves are shown in Fig.1 of PS95. This set of 900 safe RCs can be split into a subset of 80 *excellent* curves (Set A), and one of 820 *fair* curves (Set B).

The 80 RCs in Set A have the following properties, which ensure that they are good tracers of the gravitational potential in the region of interest and hence are well suited for a detailed mass modelling: 1) the approaching and receding sides are very symmetric; 2) the data are extended out to (at least) R_0; and 3) there are ≥ 30 data points homogeneously distributed with radius and per arm. For each galaxy, from the folded RC we produce a smooth RC as follows. We smooth the folded velocities by binning the $\leq N$ nearest data points contained within a fixed maximum bin size W. The values used for W are mostly $0.050\,R_0$ and $0.075\,R_0$, and for N are mostly 4 and 6. These values are chosen for each RC according to its sampling density: however, the profiles of the RCs do not change by other (reasonable) choices as can be seen by comparing the smoothed RCs with the original folded ones. In each bin we compute the average rotation velocity and its uncertainty. The resulting smoothed curves are shown in Fig.2 of PS95.

The 820 RCs in Set B fail, to various extents, at least one of the criteria stated above and therefore may be not suitable for accurate and direct mass modelling. However, they constitute a large database for those methods able to recover the DM properties by needing less stringent requirement on the RC quality. With this in mind, for each folded RC we estimate the velocity at each radius $R = \frac{n}{5} R_0$, with $n = 1, ..., 9$ by taking the average of the data points over a bin of size $0.2\,R_0$ and centered on the reference radius. After checking these values for stability (see PS95 for details), we have tabulated these estimated rotation velocities in Table 2 of PS95.

Finally, the 67 RCs of Set C have severe global asymmetries, large rms internal scatter, insufficient sampling and/or large-scale deviations from circular motion. These curves may be useful to study the non-axisymmetric disturbances (*e.g.*, bars, spiral arms) in spiral galaxies and are shown in Fig.3 of PS95.

4. CONCLUSIONS

By size, homogeneity, and intrinsic quality of the individual curves, the sample of 967 RCs of PS95 by far constitutes the prime sample of RCs available to date. As such, it offers a unique opportunity for investigating in considerable depth the properties of dark matter in galaxies, *e.g.*its radial distribution, total amount, scaling laws of structural parameters (see review by Ashman 1992). As a part of a preliminary study (Persic, Salucci & Stel 1995), we have co-added the 616 RCs (out of the safe 900) whose extension exceeds $0.5\,R_0$, to form 10 synthetic curves (see Rubin *et al.*1985), each relative to a different range of luminosity (spanning approx. 5 magnitudes overall) and extending out to $1.8\,R_0$ (see Fig.1). We find that RCs are well represented by a double straight line, with the change of slope occurring at about R_0 (see Fig.1). We definitely confirm the strong luminosity dependence of the RC shape claimed by Rubin *et al.*(1985) and supported by Persic & Salucci (1991) and Casertano & van Gorkom (1991). In detail, RCs depend on radius and galaxy luminosity as:

$$V(r) = V(R_0)\left[1+\left(0.30-0.27\log\frac{L_I}{10^{10}L_\odot}\right)\left(\frac{r}{R_0}-1\right)\right] \quad \ldots 0.4 \le r/R_0 \le 1.0$$

$$V(r) = V(R_0)\left[1+\left(0.12-0.16\log\frac{L_I}{10^{10}L_\odot}\right)\left(\frac{r}{R_0}-1\right)\right] \quad \ldots 1.0 \le r/R_0 \le 1.8$$

where $V(R_0) = 100\ (L_I/10^{10}L_\odot)^{0.34}$ km s^{-1}. In addition, it can be shown that the RC shape probes in a model-independent way the presence of dark matter and provide a robust estimate of its amount and distribution. As a result, we find that the dark component is ubiquitous in spirals and that the dark matter fraction increases with decreasing luminosity as L^{-x}, with $x \sim 0.5$, in agreement with previous studies (*e.g.*: Persic & Salucci 1988, 1990, 1991; Salucci & Frenk 1989; Salucci *et al.*1991; Casertano & van Gorkom 1991; Broeils 1992).

REFERENCES

Ashman, K. 1992, PASP, 104, 1109
Broeils, A.B. 1992, Ph.D. thesis, University of Groningen
Casertano, S., & van Gorkom, J.H. 1991, AJ, 101, 1231
Cen, R., & Ostriker, J.P. 1993, ApJ, 417, 415
Evrard, A.E., Summers, F.J., & Davis, M. 1994, ApJ, 422, 11
Mathewson, D.S., Ford, V.L., & Buchhorn, M. 1992, ApJS, 81, 413 (MFB)

Navarro, J.F., & White, S.D.M. 1994, MNRAS, 267, 401
Persic, M., & Salucci, P. 1988, MNRAS, 234, 131
Persic, M., & Salucci, P. 1990, MNRAS, 245, 577
Persic, M., & Salucci, P. 1991, ApJ, 360, 68
Persic, M., & Salucci, P. 1995, ApJS, in press (PS95)
Persic, M., Salucci, P., & Stel, F. 1995, submitted
Rubin, V.C., Burstein, D., Ford, Jr., W.K., & Thonnard, N. 1985, ApJ, 289, 81
Salucci, P., & Frenk, C.S. 1989, MNRAS, 237, 247
Salucci, P., Ashman, K., & Persic, M. 1991, ApJ, 379, 89

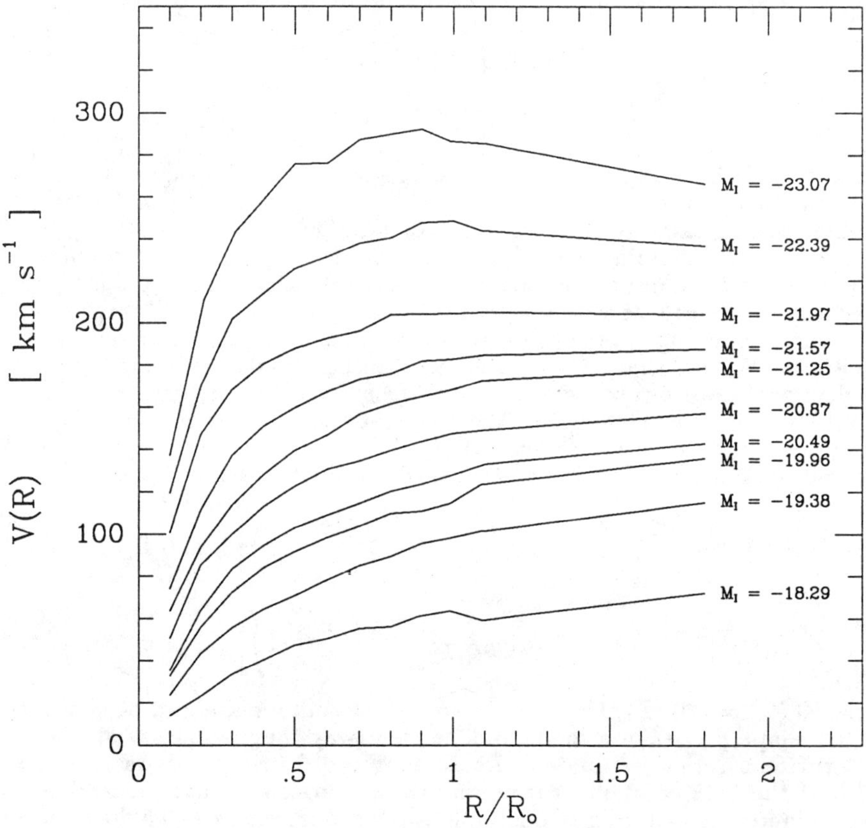

Fig. 1. The synthetic RCs resulting from the coaddition of 616 curves from PS95. Galactocentric radii are normalized to the optical disk radius. The progression of RC shape with luminosity, whereby steep (flat) curves are associated with low (high) luminosities, is clearly recognizable.

ANOTHER FLATTENED DARK HALO: POLAR RING GALAXY A0136−0801

Penny D. Sackett
Institute for Advanced Study, Princeton, NJ 08540
psackett@guinness.ias.edu

Richard W. Pogge
Ohio State University, Columbus, Ohio 43210-1106
pogge@payne.mps.ohio-state.edu

ABSTRACT

Knowledge of the shape of dark matter halos is critical to our understanding of galaxy formation, dynamics, and of the nature of dark matter itself. Polar ring galaxies (PRGs) — early-type galaxies defined by their outer rings of gas, dust and stars on orbits nearly perpendicular to those of the central host — provide a rare probe of the vertical-to-radial axis ratio ($q_\rho = c/a$) of dark halos. We present a Fabry-Perot velocity field for the Hα gas in the kinematically-confirmed PRG A0136−0801. By comparing ring orbits evolved in a generalized mass model to the observed ring velocity field and morphology of A0136−0801, we conclude that $q_\rho \sim 0.5$ and rule out a spherical geometry.

1. WHY POLAR RINGS?

Many PRG, including A0136−0801, appear to be at or near equilibrium, so that the ring gas can be assumed to follow closed orbits. The shape and velocity structure of closed orbits contain information about the shape of the underlying mass distribution: orbits are elongated along the short axis of a flattened potential and have speeds that vary with ring azimuth (slower over the poles). Given a viewing geometry, the unique relationship between orbit position and line-of-sight velocity allows one to use the 2-D velocity field of a polar ring to constrain the shape of its dark halo.

A0136−0801 is particularly suitable since its ring rotation curve is flat to more than 12 disk scale lengths above the plane of its galactic disk (Schweizer *et al.* 1983, SWR), indicating that the ring dynamics are dominated by the dark halo. We can therefore expect that the flattening we infer from our Fabry-Perot data reflects the flattening of the *dark* mass. If the mass distribution is not spherical and the polar axis does not lie in the sky plane, the kinematic major and minor axes of the ring will be twisted relative to the morphological axes (Sackett 1991). This signature is in our data (Fig. 1), and has the opposite sense as would be expected for deviations in the velocity field due to inflow. Our goal is to determine the value of the axis ratio of the density distribution, q_ρ, that produces the best fit to the geometry and velocity field of the ring.

2. THE TWO-DIMENSIONAL VELOCITY FIELD

We used the Rutgers CCD Fabry-Perot spectrograph on the CTIO 4m telescope to obtain a velocity cube of A0136−0801 in the Hα emission line. Seventeen 900-sec interferograms, sampled in 1Å intervals, were taken with the ∼2.3Å resolution (FWHM) etalon. Median seeing over the ∼4 hour run was ∼1.″5; atmospheric transparency was excellent. The processed interferograms were stacked into a data cube for calibration and analysis using the OASIS Fabry-Perot reduction package developed at OSU (Pogge 1991).

Since the polar ring is completely filled with HII regions, we are able to derive the (relative) Hα emission-line flux and radial velocity for each of the over 2000 pixels in the ring (Fig. 1). The measured Hα emission-line centroids are subject to two principal sources of uncertainty: signal-to-noise (S/N) in each spectrum pixel and an etalon drift that introduces an additional random error of about ± 7 km sec^{-1}. Adding the errors in quadrature yields a total median uncertainty of 12 km s^{-1} in the heliocentric velocities.

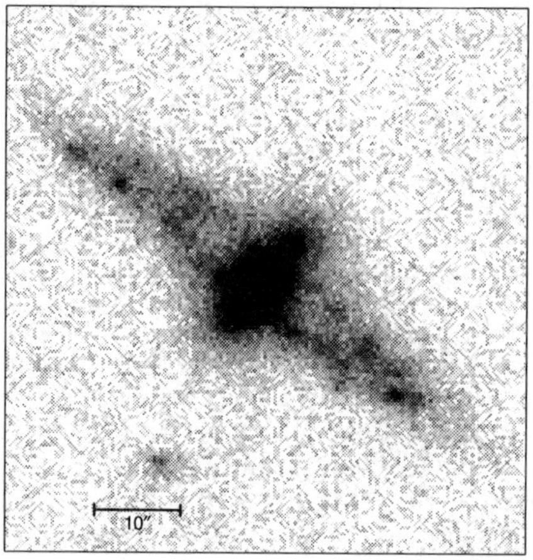

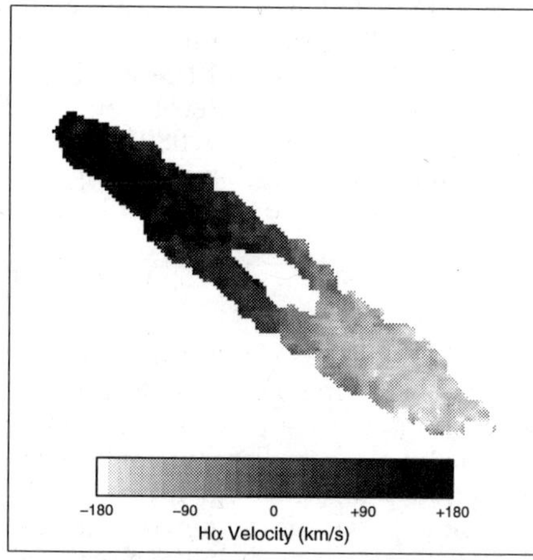

Fig. 1. *Left:* Narrow band image of A0136−0801 created by integrating the Hα data cube along the wavelength axis; image scale is $\sim 0''.4$ pixel^{-1}. *Right:* Ring velocity field; greyscale indicates the relative velocity scale.

3. THE MASS MODEL

The ring gas is assumed to follow closed, co-planar orbits in the gravitational potential of an axisymmetric, oblate, flattened mass distribution, with radial profile of the standard pseudo-isothermal form,

$$\rho(R,z) = \frac{\rho_o\, r_c^2}{r_c^2 + R^2 + (z^2/q_\rho^2)} \quad ,$$

and isodensity surfaces that are stratified on concentric, similar ellipsoids of axis ratio $q_\rho = c/a$. Such a mass distribution produces asymptotically flat rotation curves (with $v_\infty^2 = 4\pi G \rho_o\, r_c^2\, (1-q_\rho^2)^{-1/2} q \arccos q$) in the midplane of the potential, which is assumed here to be perpendicular to the polar axis.

We use a shooting algorithm to compute discrete, closed polar orbits in this mass model. After projection onto the sky plane, a cloud-in-cell algorithm is used to create a smooth velocity field with the resolution of the data. This model velocity field is then compared to the data via the χ^2 statistic; each pixel is weighted by the uncertainty in its measured velocity. The free parameters of a model are varied to minimize χ^2 and thus achieve the best *kinematic* fit to the data. With q_ρ held fixed at values in the range $0.2 \leq q_\rho \leq 1$ (corresponding to the range E8 (flat) to E0 (round) in the notation of elliptical galaxies), the free parameters are: r_c, v_∞, the systemic velocity $v_{\rm sys}$, and the projection angles, α, β, and γ, which control the pivot of the ring about its long axis, the tip of the long ring axis out of the sky plane, and the position angle of the ring on the sky.

For each halo flattening, the fitting algorithm chooses the viewing geometry that best reproduces the kinematic data, without imposing constraints from the ring shape. Only after fitting the kinematics do we examine the *morphological* consequences of varying q_ρ.

4. EVIDENCE FOR A FLATTENED HALO

The *kinematic* data alone favor models with density axis ratios of $0.5 \leq q_\rho \leq 0.9$ (Fig. 2), regardless of whether the data are weighted equally or by their statistical uncertainties, and regardless of whether or not minor axis data are masked. Only the flatter models with $q_\rho \leq 0.5$, however, are able to reproduce the observed *morphology* of the ring (Figs. 2 & 3). Spherical models provide extremely poor fits to both the kinematics and geometry of the ring. We conclude that it is likely that the mass distribution probed by the polar ring of A0136−0801 is flattened with an axis ratio $q_\rho \sim 0.5$. This result is marginally consistent with previous work for this galaxy based on long-slit data (SWR, Whitmore *et al.* 1987), and with the flattened halos derived for many other systems (for reviews, see Sackett *et al.* 1994 and Rix 1995).

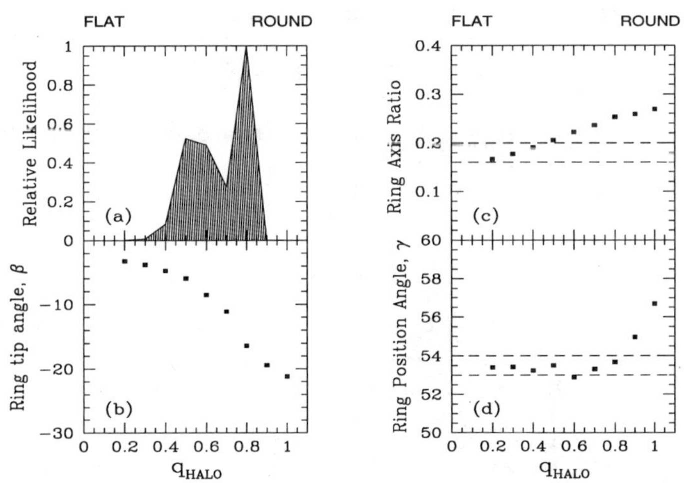

Fig. 2. The results of a typical run. Shown as a function of halo flattening: (a) relative likelihood of the *kinematic* fit, (b) ring tip angle, β, (c) resulting axis ratio of ring on sky, and (d) ring position angle. The dashed lines in (c) and (d) indicate the range of observed values.

An independent measurement of β, the angle between the polar axis and the sky, would further constrain q_ρ (Fig. 2). The visual photometry of SWR and the K-band photometry of Arnaboldi *et al.* (1994) may indicate the presence of an edge-on disk embedded in a more spheroidal central component of A0136−0801. If confirmed, this would implicate a small ring tip angle, a flatter halo, and an intrinsic flattening of the spheroidal component similar to that of the dark halo, a situation already noted for NGC 4650A (Sackett *et al.* 1994).

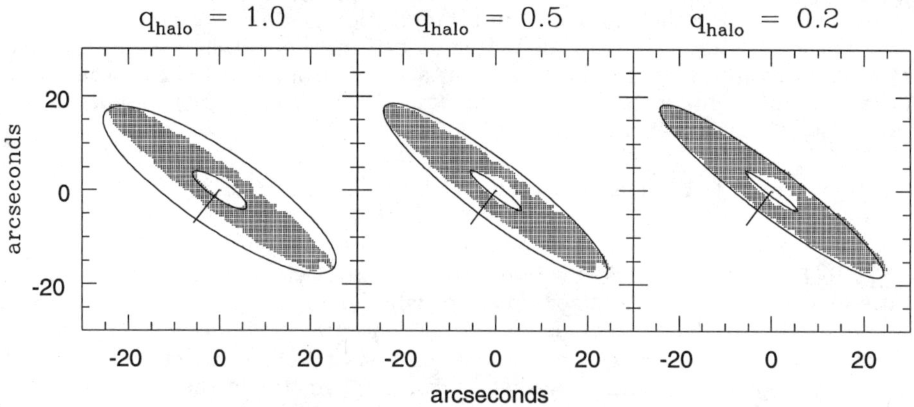

Fig. 3. Projected model orbits (solid lines) at 7″ and 30″ superposed on the masked ring data for the run shown in Fig. 2.

We will present elsewhere (Sackett & Pogge 1995, in preparation) a more complete analysis that examines the effects of warping and different choices for the parameterization of the gravitational potential.

ACKNOWLEDGEMENTS

Work by PDS was supported by the J. Seward Johnson Charitable Trust and NSF grant AST 92-15485. CTIO observing was supported in part by AURA and travel to Chile by OSU start-up funds. Support for RWP was also provided by NSF grants AST 88-22009 and AST 91-12879.

REFERENCES

Arnaboldi, M., Freeman, K.C., Sackett, P.D., Sparke, L.S., and M. Capaccioli, 1994, submitted to Planetary and Space Science

Pogge, R.W. 1991, OASIS; An Imaging Fabry-Perot Reduction Package, OSU Astronomy Internal Report 91–02

Rix, H.-W. 1995, to appear in the proceedings of *IAU Symposium 169*, ed. L. Blitz (Reidel: Dordrecht)

Sackett, P.D. 1991, in *Warped Disks and Inclined Rings around Galaxies,* eds. S. Casertano, P.D. Sackett & F.H. Briggs (CUP: Cambridge), p73

Sackett, P.D., Rix, H.-W., Jarvis, B. & Freeman, K.C. 1994, Ap.J., 436, 629

Schweizer, F., Whitmore, B.C., & Rubin, V.C. 1983, A.J., 88, 909 (SWR)

Whitmore, B.C., McElroy, D., and Schweizer, F. 1987, Ap.J., 314, 439

DARK MATTER IN ELLIPTICAL GALAXIES

Francesco Bertola & Alessandro Pizzella
Dipartimento di Astronomia Università di Padova, Italy
bertola/pizzella @astrpd.pd.astro.it

1. INTRODUCTION

Dark massive halos have been discovered in spiral galaxies due to the presence of extended HII and HI disks. In contrast the general lack of gaseous structures in elliptical galaxies has not allowed a detailed study of their mass distribution. Recently the problem of detecting dark massive halos has been attacked using three different approaches. The first one based on the use of the stellar kinematics in order to derive the potential. The second and the third ones make use, as tracer of the potential, of the ionized and HI component respectively of the gaseous disks, which sometimes are found to be present in elliptical galaxies.

2. STELLAR KINEMATICS

The kinematical study of the stellar components in elliptical galaxies has been recently the subject of an ESO Key Programme (Saglia *et al.* 1993, Bertin *et al.* 1994) The goal of the Key Programme has been to derive the trend of the mass-to-light ratio in elliptical galaxies by means of detailed dynamical modeling of the stellar kinematics. For this purpose, absorption line measurements out to the lowest surface brightness levels ($\mu_B \sim 24.4 mag/arcsec^2$) allowed by the state-of-the-art detectors have been obtained. Once the velocity and the velocity dispersion of the stellar component has been derived, two different and complementary methods have been applied in the dynamical modeling. One method consists on two-component (**2C** Bertin, Saglia & Stiavelli 1992) spherical models based on anisotropic distribution functions of maxwellian type multiplied by a weight factor of the form $(-E)^{3/2}$, with E the binding energy of the stars or the dark matter. These two-component models originate from a picture of dissipationless collapse that envisages that elliptical galaxies have isotropic pressure inside r_c, and have modest radial anisotropy in the outer regions. The second method is a technique of Quadratic Programming (**QP**, Dejonghe 1989), which approximates the distribution functions as a linear combination of simple components, for which powers of E and the total angular momentum J are chosen. This approach complements the first one, in that it makes fewer assumptions about the form of the distribution function, and hence allows a larger freedom in the radial density profile and in the anisotropy of the velocity distribution.

Extended velocity and velocity dispersion profiles have been measured for nine object reaching, in some cases, $2r_e$ and beyond. Three differents behaviours can be distinguished when looking at the stellar velocity dispersion profiles: i) decreasing profile (NGC 1400, NGC 4946, NGC 5812, NGC 6721, NGC 7507); ii) constant profile (NGC 4472, NGC 7796, IC 4296); iii) slowly increasing profile (NGC 7144). This simple classification leads to different trends for the M/L radial profile:

- The decreasing profile can be nicely reproduced by both the 2C e QP methods with M/L constant with radius and thus without involving the presence of dark matter.

• The constant profile is reproduced by both methods but while QP still can afford the high value of the velocity dispersion in the outer regions without need of dark matter, dark matter is necessary in the 2C method in order to fit the observed data.

• The case of the increasing profile is more complex. The 2C modeling gives unsatisfactory fits. While the QP method fails to construct a model with constant M/L, it gives a good fit to the data if a massive dark halo is added.

The results obtained with the 2C and the QP methods are plotted in Fig.1, except for NGC 7144.

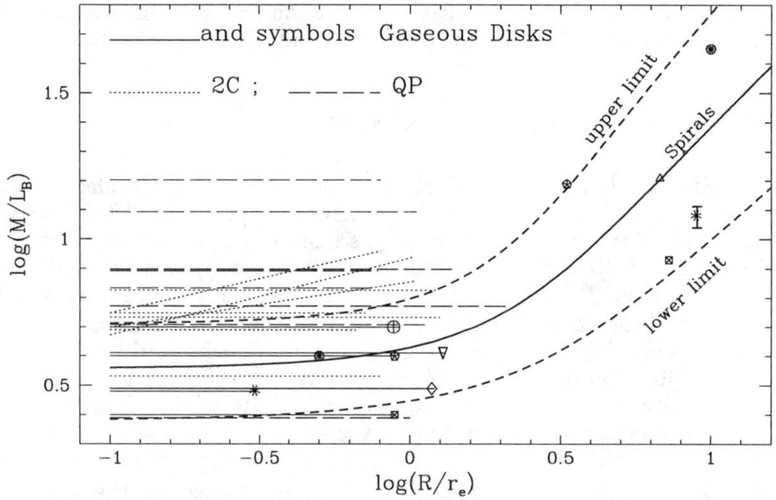

Fig.1 R/r_e versus M/L_B for a sample of elliptical galaxies. The *full line* ending with a *symbol* represents the measurement from the ionized gas disks while the isolated symbols (same symbols refer to the same galaxy) are from HI disks. *Dotted lines* are the M/L profiles obtained with the 2C method, *long dashed lines* are obtained with the QP method. The *thick lines (full and dashed)* represent the M/L profile of an L_B^* spiral galaxies with the associated uncertainties.

3. IONIZED GAS DISKS

A second approach to the problem of the determination of M/L in elliptical galaxies is based on the use of ionized gas disks, when present, in this class of galaxies (Bertola 1992). The misalignment of the gaseous and stellar isophotes and the decoupling of the angular momenta of gas and stars suggest that this gaseous material is acquired from the outside. Ionized gas disks in elliptical galaxies can be used to test the gravitational field of the host galaxy (Merritt & de Zeeuw 1983). The calculations are somewhat complex because elliptical galaxies, unlike spirals, are generally triaxial. Therefore two axis ratios and two viewing angles must be taken into account, as well as the non-circular shape of the gas orbits (de Zeeuw & Franx 1989; Bertola *et al.* 1991). In order to perform the modeling of the triaxial potential it is necessary to carry out geometrical measurements (position angles and ellipticity profiles of both the gaseous and the stellar components) as well as kinematical measurements (a complete knowledge of the whole ionized gas velocity field). This kind of analysis (Bertola *et al.* 1991; Bertola *et al.* 1993) leads to the conclusion that M/L is constant in the central region of elliptical galaxies (Fig.1), namely from the center to $0.5 - 1r_e$, which is the typical extent of the ionized gas disk. Previous results obtained analyzing the gas kinematics using the spherical approximation produced unsatisfactory M/L radial profiles (Caldwell 1984; Caldwell, Kirshner & Richstone 1986).

4. HI DISKS

The third approach to the determination of the trend M/L in elliptical galaxies makes use of the HI disks, which extend much further away than the luminous component and therefore allows determinations of M/L at large distances up to $\sim 10r_e$. At the present only few elliptical galaxies have been found to possess an HI disk suitable for a kinematical study. However the high M/L values obtained, when compared with the low values indicated by the study of the inner gas disks (Fig.1), constitute ***the strongest evidence of the presence of dark massive halos in elliptical galaxies.*** Recently Pizzella *et al.* (1994) carried out $21cm$ observation of NGC 5266, the galaxy with the largest amount of neutral hydrogen so far observed in an elliptical galaxy. A mass-to-light ratio $M/L_B = 12$ at $9r_e$ has been determined confirming the presence of a dark massive halo also in this galaxy (Fig.1).

5. COMPARISON WITH SPIRAL GALAXIES

Dark matter halos in spiral galaxies are well studied in a large number of objects. According to the systematic study of Persic and Salucci (1990, 1991) the radial trend of M/L is constant in the inner region and then it increases rapidly outwards. This relationship varies with the luminosity of the disk being studied. At a given (normalized) radius, higher luminosity disks tend to be characterized by higher M/L at $R < r_e$ (where the stellar component dominates) and by lower M/L at $R >> r_e$ (where DM dominates). Therefore the radial variation of M/L tends to be less dramatic for bright galaxies than for faint ones. In Fig.1, adapted from Bertola *et al.* (1993), the line representative of an L_B^* [where $log(L_B^*/L_{B\odot}) = 10.63$] spiral is plotted as function of R/r_e. The line has been modified according to a shift by a factor of ~ 3 of the luminous M/L in order to fit the (higher) central M/L_B found in ellipticals. Dashed lines represent the uncertainties associated with the determination of the above relationship. It is quite interesting to note that the representative points for ellipticals fall well within the dashed lines thus suggesting that the trend of M/L of ellipticals follows that of spirals, once the radial distance is expressed for both type of galaxies in terms of R/r_e. Moreover both ellipticals (van der Marel 1991) and spirals (Salucci, Ashman & Persic 1991) show the same relationship between the central value of M/L and the luminosity although obtained with independent methods based on the dynamics of the stars and gas respectively. Having established a similar trend of M/L in spirals and ellipticals it is possible to study the properties of the luminous and dark components for each class. In particular, it is found that in spirals and ellipticals of the same visible mass the radius where the density of the dark matter equals that of the visible matter occurs at the same distance in units of r_e. This suggest a similar picture of baryonic collapse into a preexisting dark matter halo for both ellipticals and spirals (Bertola *et al.* 1993). The effective radius, which is about two times larger for spirals than for E's of comparable luminosity (Kent 1985,1990), proves to be the "homology coordinate" in the unified scheme proposed. This can be understood as a consequence of the different initial conditions for the angular momentum in the two types of galaxies: the infall of baryons in spirals is halted by angular momentum at a typical radius which is about twice the radius where the baryonic infall is stopped (by violent relaxation) in ellipticals of similar visible mass.

REFERENCES

Bertin, G., Saglia, R.P., Stiavelli, M., 1992. ApJ 384, 423

Bertin, G., Bertola, F., Buson, L.M., Danziger, I.J., Dejonghe, H., Sadler, E.M., Saglia, R.P., de Zeeuw, P.T., Zeilinger, W.W., 1994, A&A in press

Bertola, F, 1992, Morphological and Physical Classification of Galaxies, ed. by G. Longo, M. Capaccioli, G. Busarello, Kluwer Ac. Publ., Dordrecht, Netherlands, p115

Bertola, F., Bettoni, D., Danziger, J., Sadler, E.M. Sparke, L., de Zeeuw, T., 1991, ApJ 373, 369

Bertola, F., Pizzella, A., Persic, M., Salucci, P., 1993, ApJ 416, L45

Caldwell, N., 1984, ApJ, 278, 96

Caldwell, N., Kirshner, R.P., Richstone, D.O., 1986, ApJ, 305, 136

Dejonghe, H., 1989, ApJ 343, 113

de Zeeuw, P.T., Franx, M. 1989, ApJ 343, 617

Kent, S.M., 1985, ApJS, 59, 115

Kent, S.M., 1990, in ASP Conference Series Vol.10: Evolution of the Universe of Galaxies, ed. R.G. Kron (Birgham Young Univ. Print Serv.), p.109

Merritt, D.R., de Zeeuw, P.T., 1983 , ApJ Letters 267, L21

Persic, M., Salucci, P., 1990, MNRAS 245, 577

Persic, M., Salucci, P., 1991, ApJ, 368, 60

Pizzella, A., Morganti, R., Sadler, M.E., Bertola, F., in proc. of IAU Symposium No. 164: Stellar population; The Hauge, Olanda 15-18 Agosto 1994 in press

Saglia, R.P, Bertin, G., Bertola, F., Danziger, I.J., Dejonghe, H., Sadler, E.M., Stiavelli, M., de Zeeuw, P.T., Zeilinger, W.W., 1993, ApJ 403, 567

Salucci, P., Ashman, K.M., Persic, M., 1991, ApJ 379, 89

Van der Marel, R.P., 1991, MNRAS 253, 710

Mapping Dark Matter around Early Type Galaxies

B. Tremblay, T. B. Williams, and D. Merritt
Department of Physics and Astronomy
Rutgers University

ABSTRACT

We have measured velocities of 67 planetary nebulae around the SB0 galaxy NGC 3384. The measurements were taken with the Rutgers Fabry-Pérot interferometer on the CTIO 4m telescope. The planetary nebulae display a well-ordered rotation field aligned with the photometric axes of the galaxy, with a peak velocity of 140 km s^{-1} at $\sim$ 2 kpc and flat thereafter. We infer that the galaxy has a mass-to-light ratio $M/L_B \gtrsim 9$. Our result implies that at least a third of the mass of the NGC 3379/3384 group – estimated by Schneider (1985) to be $\sim 6 \times 10^{11} M_\odot$ – may be accounted for in the two bright galaxies.

1. INTRODUCTION

Planetary nebulae (PNs) are bright emission line objects. Their usefulness as distance indicators has already been demonstrated (Jacoby 1989; Ciardullo *et al.* 1989), but it is only recently that Hui (1993), Ciardullo *et al.* (1993), and Arnaboldi *et al.* (1994) have used them as kinematic tracers. As part of an ongoing project to map dark matter around early-type galaxies using the kinematics of PN systems, we have measured velocities of 67 PNs in the sample obtained by Ciardullo *et al.* (1989) around the SB0 galaxy NGC 3384. Here we present the two-dimensional rotation and velocity dispersion fields of the NGC 3384 PN system and derive a lower limit on the mass of the galaxy.

2. OBSERVATIONS

We used the Rutgers Fabry-Pérot (RFP) interferometer to obtain a series of 2 Å FWHM images about the [OIII] 5007 Å emission line. The images were separated by 1.25 Å. The RFP allows for the simultaneous detection of the PNs and the measurement of their velocities. The field of view of the RFP on the 4m is 2.7 arcmin in diameter.

The data presented are from four spatial fields obtained on four nights of observation. We exposed each frame for 15 minutes and took 19 frames per field. The galaxy was offset to the east in the first series, centered in the second, offset to the west in the third, and centered again in the fourth series.

The presence of reference stars in each frame allowed the fluxes of the individual frames to be normalized using the DAOGROW prescription (Stetson 1990). We subtracted the galaxy from each frame, and then measured the PN fluxes using the point spread function routine of DAOPHOT (Stetson 1987). Each flux thus measured became part of a PN spectrum which was fit with the

RFP instrumental profile. We determined the velocity of each PN by finding the peak of the fitted spectrum (Fig. 1). More details on the observations and their reduction will be presented in a forthcoming paper (Tremblay *et al.* 1995)

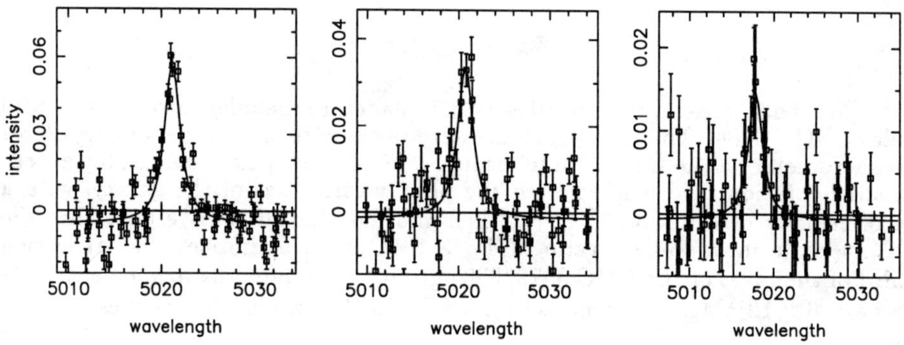

Figure 1 Spectra for three of the planetary nebulae used in this study. They are, from left to right, nos. 3 (m_{5007}= 25.66), 24 (26.07) and 77 (26.70) from the list of Ciardullo *et al.* (1989). The wavelength is measured in angstroms and the intensity is the DAOPHOT flux.

We present here velocities of 67 PNs from the sample previously identified by Ciardullo *et al.* (1989), extending to roughly two disk scale lengths from the center of NGC 3384. The velocity uncertainty is typically $\sim$ 7 km s^{-1} for PNs present in more than one field and $\sim$ 30 km s^{-1} for those found in a single field. In Figs. 2a and 2b, we show the PN mean velocity and velocity dispersion fields. The contours on both were computed using the two-dimensional thin-plate smoothing spline algorithm (Wahba 1990). The rotation of the PN system is aligned with the photometric major axis of the galaxy; the rotation velocity has a peak value of $\sim$ 140 km s^{-1} at $\sim$ 2 kpc from the center of the galaxy and remains roughly constant thereafter. The velocity dispersion field has less apparent symmetry but the dispersions are somewhat less than the peak mean velocity, implying that this system is rotationally dominated.

Assuming that the PN system is coplanar with the disk of NGC 3384, with an inclination of 67° (Nilson 1973); that the distance to NGC 3384 is 10.1 Mpc (Ciardullo *et al.* 1989); and that this galaxy is an isotropic oblate rotator, the axisymmetric Jeans equation yields a mass of $\sim 0.7 \times 10^{11}$ $M_\odot$ within 7 kpc of the galaxy center. For $m_B = 10.75$ (de Vaucouleurs *et al.* 1991), the mass to light ratio M/L_B is $\sim$ 9 in solar units.

3. CONCLUSIONS

Schneider (1985, 1991) found a total mass for the NGC 3379/3384 pair of $\sim 5.6 \times 10^{11}$ $M_\odot$ from the kinematics of the surrounding HI ring. Ciardullo *et al.* (1989) estimated the mass of NGC 3379 to be about 1.2×10^{11} $M_\odot$. Our new results for NGC 3384 imply that at least a third of the dynamical mass of the NGC 3379/3384 group may be accounted for in the two bright galaxies. If we require that half of the additional mass measured by Schneider (1985) be associated with NGC 3384 alone, and that its rotation curve remain flat, then the dark halo of this galaxy would need to extend to roughly 45 kpc. This value is within the limiting radius of 60 kpc derived from the kinematics and geometry of the HI ring (Schneider 1991). Further observations of the PN systems of both galaxies are scheduled and should reveal the extent to which the dark matter is associated with the galaxies.

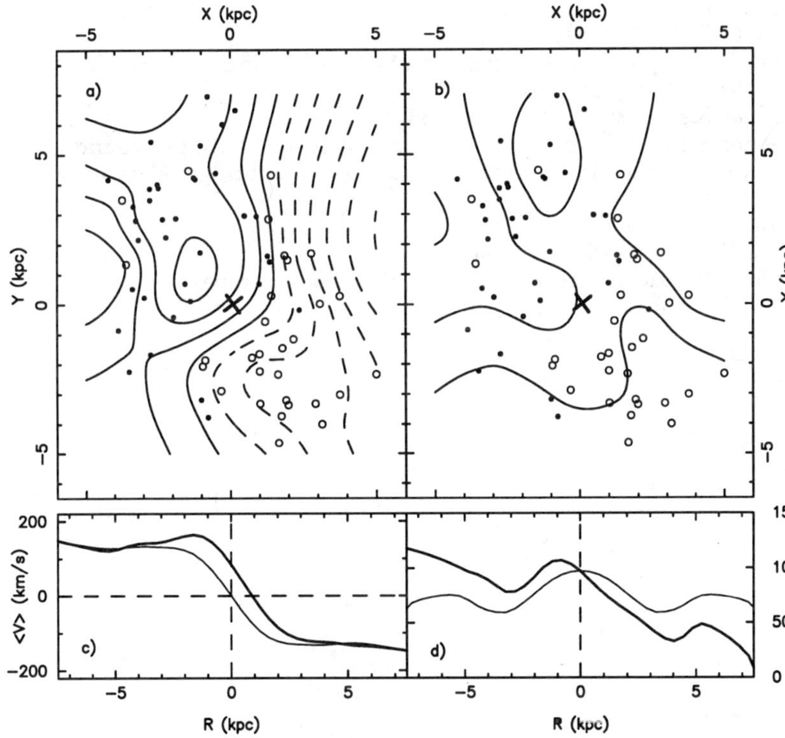

Figure 2 **(a)** Contours of constant line-of-sight, mean velocity for the NGC 3384 planetary nebula system. East is down and north is to the right. The cross indicates the position and orientation of the galaxy, with the semi-major axis extending from the upper left to the lower right. Contours are separated by 50 km s^{-1}. Dashed contours are receding; closed circles are PNs with positive velocities; open circles are PNs with negative velocities. **(b)** Contours of constant line-of-sight velocity dispersion. The contour touching the galaxy is at 100 km s^{-1}. Symbols have the same meaning as in Fig. 2a. **(c)** Heavy line: mean line-of-sight velocity along the major axis; thin line: antisymmetrized profile. **(d)** Heavy line: line-of-sight velocity dispersion along the major axis; thin line: symmetrized profile.

REFERENCES

Arnaboldi M., Freeman K., Hui X. & Ford H. 1994, Messenger, 76, 40
Ciardullo R., Jacoby G.H. & Ford H.C. 1989, ApJ, 344, 715
Ciardullo R., Jacoby G.H. & Dejonghe H.B. 1993, ApJ, 414, 154
de Vaucouleurs G., de Vaucouleurs A., Corwin H., Buta P., Patarel G.,& Fouqé P. 1991, Third Reference Catalog of Bright Galaxies,Springer-Verlag, New York
Hui X. 1993, PASP, 105, 1011
Jacoby G.H. 1989, ApJ, 339, 39
Nilson P. 1973, Uppsala General Catalog of Galaxies, Uppsala Astronomical Observatory, Uppsala
Schneider S.E. 1985, ApJL, 288, L33
Schneider S.E. 1991, in Warped Disks and Inclined Rings around Galaxies, ed. S. Casertano, P. Sackett & F. Briggs, Cambridge, Cambridge University Press
Stetson P. 1987, PASP, 99, 191
Stetson P. 1990, PASP, 102, 932
Tremblay B., Merritt D. & Williams T. 1995, in preparation
Wahba G. 1990, Spline Model for Observational Data, Philadelphia,SIAM

PROBING THE MASS DISTRIBUTION AND FORMATION HISTORY OF GALAXIES WITH GLOBULAR CLUSTERS

Stephen E. Zepf
Department of Astronomy, University of California, Berkeley, CA 94720
zepf@astron.berkeley.edu

ABSTRACT

Extragalactic globular cluster systems are valuable probes of the mass distribution and formation history of their host galaxies. Bright elliptical galaxies have globular cluster systems which are rich and spatially extended, making their globular clusters useful kinematical tracers of the mass distribution out to radii of ~ 100 kpc. Moreover, globular clusters can be used to probe the metallicity and metallicity gradient at large radii where studies in integrated light are difficult. Both the velocity and the metallicity can be determined from the spectrum of a globular cluster. Obtaining spectra of large numbers of extragalactic globular clusters is now feasible because of efficient multi-object spectrographs on large telescopes.

1. INTRODUCTION

Globular clusters have long been used as tracers of the mass distribution and the formation history of the Milky Way and other galaxies in the Local Group. Advances in multi-object spectroscopy and telescope aperture have opened up the possibility of extending this approach to more distant galaxies, encompassing the full spectrum of galaxy types. This paper will concentrate on the study of globular clusters around elliptical galaxies, with the goal of constraining the mass distribution and formation history of these galaxies.

2. MASS DISTRIBUTION

Determining whether elliptical galaxies have massive dark matter halos has proven to be more difficult than establishing their existence around spiral galaxies. For spiral galaxies, axisymmetry and the presence of HI gas in circular rotation simplify the problem greatly. Elliptical galaxies are typically triaxial, and rarely have a detectable quantity of cold gas at large radii. Spectroscopy of the integrated light of elliptical galaxies is hard to obtain beyond one effective radius.

Globular clusters have long been known to be useful tracers of the dynamics of galaxies. They are especially useful as probes of the halos of elliptical galaxies, since they are bright, numerous, and have an extended spatial distribution. Specifically, a typical bright elliptical has $\sim 5,000$ globular clusters, normally distributed in magnitude, with $< M_V > \approx -7.4$ and $\sigma \approx 1.4$, and with a more extended spatial distribution than the halo light (e.g. Harris 1991). At a distance of 1,000 - 2,000 $\mathrm{km\,s^{-1}}$, the result is that there are at least several hundred clusters in the radial range of $10 - 50h^{-1}$ kpc which are accessible to spectroscopy with 4-m class telescopes.

This is demonstrated in Figure 1 which includes plots of the radial fall-off the the globular cluster surface density for two typical cases: NGC 4472, a bright elliptical in the Virgo cluster, and NGC 3923, a bright elliptical is a loose group. Even at the bright magnitude limits plotted, the number density of globular clusters is fairly high, and significantly above the estimated background contamination. Although several hundred velocities may not be quite sufficient for a general determination of both the distribution function and the radial profile of the potential (Merrit & Tremblay 1993), such a number of velocities can play an important role in constraining various mass models. Velocities of a number of individual objects are particularly useful when combined with estimates of the mass distribution derived from X-ray studies. Loewenstein (1992) used this technique in a study of NGC 4472, and much tighter constraints are now possible with better data in both the X-ray and optical regimes.

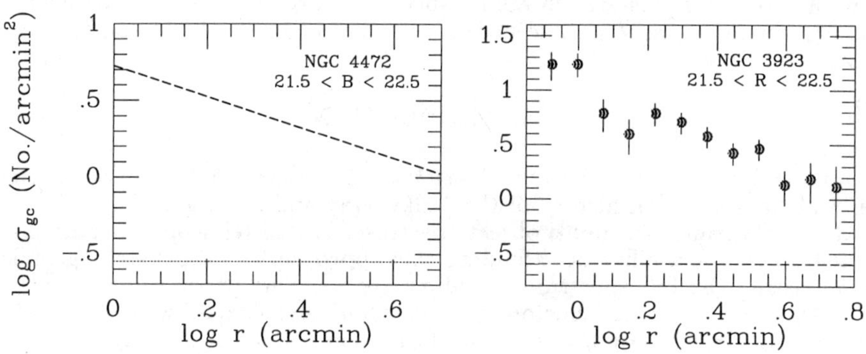

Figure 1 - The figure on the left is the surface density of unresolved objects with $21.5 < B < 22.5$ around NGC 4472 as a function of radius from Harris et al. (1991). The dotted line is the density of unresolved objects within the same magnitude range determined from two offset frames. The figure on the right is the radial profile of globular clusters with $20.4 < R < 22.1$ around NGC 3923 from Zepf, Geisler, & Ashman (1994). The dotted line is the estimated background. Like NGC 4472, the surface density of globular clusters around NGC 3923 is a factor of several greater than the estimated background out to radii of at least five arcminutes.

3. METALLICITY DISTRIBUTION

Globular clusters are invaluable probes of the formation history of galaxies. For all but a few nearby galaxies, they provide one of the few ways to study the ages and metallicities of individual objects, rather than the integrated properties. Since to first order globular clusters are bound, coeval, and chemically homogenous, they provide a distinct record of the physical conditions at the time of their formation.

Because of this property of globular clusters, they can be used to test competing models of elliptical galaxy formation (Ashman & Zepf 1992, Zepf & Ashman 1993). If elliptical galaxies form in a monolithic collapse, the metallicity distribution is expected to be smooth and single-peaked. In contrast, if elliptical galaxies form through mergers, their globular cluster systems will be a composite of at least two populations - the first associated with the halos of the progenitor spirals and the second formed in the merger itself. Because the material out of which clusters form has been significantly enriched, the metallicity distribution of the globular clusters around an elliptical formed by merging is expected to have at least two peaks.

In Figure 2, the results of color surveys of globular clusters around two elliptical galaxies, NGC 4472 and NGC 3923, are presented. In both cases, the distributions appear to be similar to the bimodal shape predicted by a merger scenario. A more quantitative analysis confirms this result (e.g. Zepf, Ashman, & Geisler 1994). Other properties of the NGC 3923 globular cluster system, such as its color gradient, are also discussed in Zepf et al. (1994).

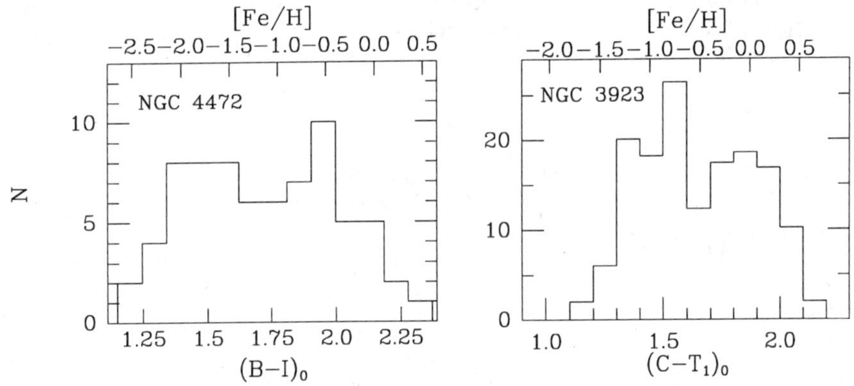

Figure 2 - Histograms of the color distribution of the globular clusters of NGC 4472 on the left and NGC 3923 on the right. The NGC 4472 data are taken from the $V \leq 22.5$ sample of Couture et al. (1991), and the NGC 3923 data are from Zepf, Ashman, & Geisler (1994). The bimodal appearance of the histograms is in agreement with the prediction of a merger scenario for elliptical galaxy formation. Specifically, a two-peaked model of these distribution is statistically preferred to a single-peak model at a 98.5% confidence level for NGC 4472, and > 99% for NGC 3923. The metallicity calibration for the color indices is given along the y-axes.

Although color traces metallicity well for old stellar systems, metallicities determined from spectroscopic absorption-line indices provide several important advantages. They are independent of reddening, allow an estimate of abundance ratios, and background contamination is removed directly. Moreover, the same spectra from which velocities are determined for kinematical studies provide metallicities for analysis of the formation history.

4. CONCLUSIONS

Globular clusters are useful kinematical tracers of the gravitational potential in elliptical galaxy halos. For typical bright elliptical galaxies, several hundred clusters within the radial range of $10 - 50h^{-1}$ kpc are accessible to spectroscopy. Radial velocities determined from these spectra will provide significant new constraints on models of the mass distribution of elliptical galaxies.

Metallicities can also be determined from spectra of globular clusters. The metallicities of globular clusters are important for distinguishing between competing theories for the formation of galaxies. The metallicity distribution inferred from broad-band colors favors an episodic formation history, as expected in a merger scenario. Metallicities determined from spectroscopic absorption-line indices are important for confirming this result and constraining metallicity gradients at large radii.

ACKNOWLEDGEMENTS

I thank my collaborators, Keith Ashman, Dave Carter, Doug Geisler, and Ray Sharples for their valuable contributions to the projects described above. I acknowledge support from NASA through grant number HF-1055.01-93A awarded by the Space Telescope Science Institute, which is operated by the Association of Universities for Research in Astronomy, Inc., for NASA under contract NAS5-26555.

REFERENCES

Ashman, K.M., & Zepf, S.E. 1992, ApJ, 384, 50
Couture, J., Harris, W.E., & Allwright, J.W.B. 1991, ApJ, 372, 97
Harris, W.E. 1991, ARAA, 29, 543
Harris, W.E., Allwright, J.W.B., Pritchet, C.J., & van den Bergh, S. 1991, ApJS, 76, 115
Loewenstein. M. 1992, ApJ, 384, 474
Merrit, D.R., & Tremblay, B. 1993, AJ, 106, 2229
Zepf, S.E., & Ashman, K.M. 1992, MNRAS, 264, 611
Zepf, S.E., Ashman, K.M., & Geisler, D. 1994b, ApJ, in press
Zepf, S.E., Geisler, D., & Ashman, K.M. 1994a, ApJL, 435, L117

A FAINT LUMINOUS HALO THAT MAY TRACE THE DARK MATTER AROUND SPIRAL GALAXY NGC 5907

Heather Morrison
Dept. of Astronomy, Case Western Reserve University, Cleveland OH 44106

Penny Sackett
Institute for Advanced Study, Princeton NJ 08540

Paul Harding
Steward Observatory, University of Arizona, Tucson AZ 85721

Todd Boroson
US Gemini Project Office, National Optical Astronomy Observatories
Tucson AZ 85726

ABSTRACT

We have detected a faint, luminous halo around the edge-on spiral galaxy NGC 5907. This halo, with a volume luminosity density that falls as $r^{-2.3\pm0.2}$, is less centrally concentrated than any previously known luminous component of a spiral galaxy. The power-law exponent suggests a possible relationship with NGC 5907's dark matter halo. In fact, the profile of light from this luminous halo is consistent with the distribution of dark mass inferred from the galaxy's rotation curve.

1. INTRODUCTION

For more than 20 years, the presence of unseen halos of dark matter around spiral galaxies has been inferred from the high rotation speeds of gas in their outer parts, beyond the optical disk. Earlier searches for possible luminous halos corresponding to the dark matter were carried out using photographic plates, aperture photometers and CCDs and IR arrays. Photographic surveys suffered from the nonlinearity of the plates and the impossibility of calibrating precisely their response. Studies using aperture photometers were unable to measure sky brightness with sufficient accuracy. CCDs have until recently been too small to study large nearby galaxies and sky-subtract from the image itself; IR arrays are still too small.

The advent of large-format CCDs on wide-field telescopes has improved the situation significantly. We present CCD surface photometry of NGC 5907 taken with a 2048^2 CCD on the KPNO 0.9m telescope, which has reached reliably some 2 magnitudes fainter than any previous work. In addition, the known noise characteristics of the CCD have made it possible to produce a reliable error model, which takes into account (a) photon noise, (b) flat fielding errors, (c) readout noise and (d) surface-brightness fluctuations. Full details of the reductions and error analysis are given in Morrison et al (1994).

2. A LUMINOUS R^{-2} HALO

For magnitudes brighter than μ_R=25, the galaxy data are well fit by an exponential disk model, with parameters close to those derived by van der Kruit and Searle (1981). At fainter magnitudes, excess light can be seen that is well fit by the addition of a flattened power-law halo of the form $\rho = \rho_o r_c^2 [r_c^2 + R^2 + (z/q)^2]^{-\alpha/2}$, with $q \equiv c/a = 0.53 \pm 0.15$, $\alpha = -2.3 \pm 0.2$, and $r_c = 2$ kpc (See Fig. 1).

The surface brightness of this faint halo is similar to the Pop II halos seen in the Milky Way and other galaxies, but it is much less centrally concentrated. Pop II halos have power-law exponents of −3.5 or steeper (Zinn 1985, Pritchet and van den Bergh 1994).

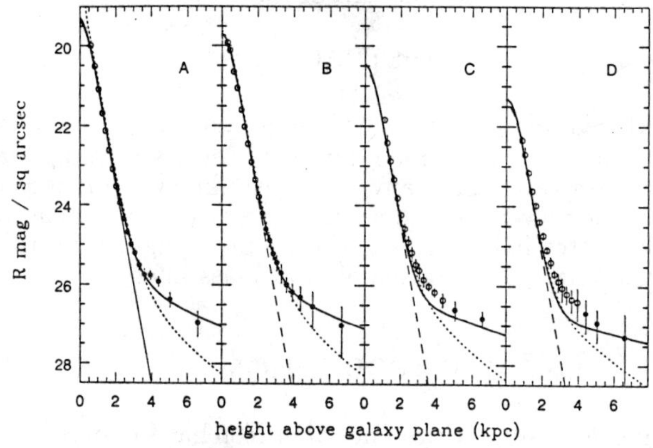

Fig. 1. The total R surface brightness is shown as a function of distance from the galactic plane for four different cuts (spaced 4.1 kpc apart) perpendicular to the plane. Light from matching positions in the four quadrants has been averaged; error bars include differences between quadrants. Open symbols denote areas underneath the mask shown in Sackett *et al.* (1994), which removes data contaminated by dust or the galaxy warp. The best fit using a thin disk only (dashed line) is shown together with the best fits for disk plus $r^{-3.5}$ halo with a 200 pc core (dotted line) and disk plus $r^{-2.26}$ halo with a 2 kpc core.

3. "ARE YOU SURE?"

Could our detection be caused by:

Sky subtraction errors? No. Because of the large field of the CCD, we could sky-subtract directly from the galaxy frame. We determined the mean sky level to within 0.01%; our signal is 1% of sky.

Flat fielding errors? We took many (15) dark sky frames to get the best possible flat field. We estimate that we characterized the response of the CCD to 0.02% of sky over small scales (10 × 10 pixel bins) and to 0.1% of sky over larger scales. This is an order of magnitude fainter than our result.

Scattered light? We checked this important point by calculating a stellar PSF to a radius of 250 pixels (3 arcmin, past our faintest halo detection) using bright stars on the galaxy frame. Deconvolution using the Lucy algorithm showed no change to the faint luminous halo.

The galaxy's warped disk? Our halo detection is L–R symmetric, which the warp is not. Also this signal is not seen in HI.

Hα emission diluted by the R passband? Although some galaxies do show diffuse Hα emission above the plane, it only extends 1–2 kpc, while we see the halo to at least 7 kpc above the plane.

4. RELATION TO NGC 5907'S DARK MATTER HALO?

Could this luminous halo be related to the only component of spirals known to have such a shallow radial distribution — the dark matter halo?

If the luminous halo traces NGC 5907's dark matter, a simple rescaling of the disk and halo light using M/L ratios would produce a mass model capable of reproducing the observed HI rotation curve. It can be seen in Fig. 2 that a remarkably good match to the rotation curve can be obtained using the shape parameters of the disk and halo from the fit to the light, implying that this faint luminous halo traces the mass distribution of the galaxy.

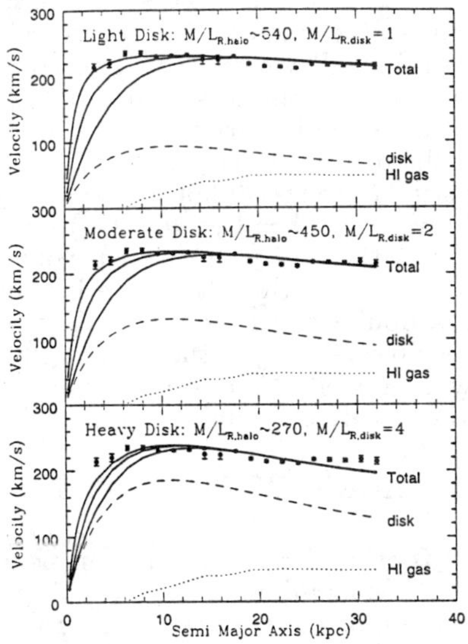

Fig. 2: The rotation curve of the galaxy as derived from HI observations is shown with model curves. The dotted line shows the contribution from the HI gas alone and the dashed line indicates the rotation curve of the thin exponential disk alone. The thick solid lines are the complete rotation curves, namely the quadrature sum of the gas curve, disk curve, and the rotation curve of halos producing similar good fits to the light: an $r^{-2.22}$ halo, with $c/a = 0.51$ and a 1 kpc core (top), an $r^{-2.26}$ halo, with $c/a = 0.53$ and a 2 kpc core (middle), and an $r^{-2.74}$ halo, with $c/a = 0.53$ and a 6 kpc core (bottom). The central dark mass densities of these models are 1.9, 0.45 and 0.08 $M_\odot/\mathrm{pc}^3$ respectively. The mass of the disk is assumed to be $5.3 \times 10^{10}\ M_\odot$, derived using an $M/L_R = 2$ (appropriate to Sc galaxies) and a luminosity based on the Tully-Fisher relation. The M/L_R of the halo component has been scaled to give a good fit to the rotation

curve: $M/L_R = 420$ for the $r^{-2.22}$ halo, $M/L_R = 450$ for the $r^{-2.26}$ halo, and $M/L_R = 455$ for the $r^{-2.74}$ halo. Plausible fits to the curve can be obtained with plausible $1 \lesssim M/L_{R,disk} \lesssim 4$ for an Sc disk, resulting in $270 \lesssim M/L_{R,halo} \lesssim 570$ for the halo, though the high $M/L_{R,disk}$ and low $M/L_{R,halo}$ give notably poorer fits. The rotation curve inward of 5 kpc is affected by light inside our mask and by the small bulge component seen in the infrared.

It is also interesting to speculate whether we could actually be observing emission from some part of NGC 5907's "dark" matter halo. A M/L_R of 450 (which provides the best fit in Fig 2) is consistent either with a population consisting entirely of late M dwarfs of moderate metal deficiency, or with a mixture of zero-metallicity late M dwarfs and brown dwarfs. Observations of NGC 5907's halo in another passband (I) will test these possibilities. Constraints on the existence of such stars in the Milky Way are discussed below.

5. A LATE M DWARF HALO IN THE MILKY WAY?

The best constraints on existence of a dynamically significant population of low mass stars in the Milky Way come from star counts. Bahcall *et al.*(1994), using deep WFPC2 images, rule out stars with V–I > 2.0: "Main sequence stars with $M_I > 10$ that are above the H-burning limit contribute < 6% of the unseen matter". (See also Boeshaar *et al.* 1994.)

Zero-metallicity late M dwarfs cannot be singled out on color alone because they are quite blue. However, Bahcall *et al.*(1994) detect very few stars of any color, which suggests that these objects (which would be expected to appear in even greater numbers because they are more luminous) also do not form a dynamically significant population in the Milky Way.

In summary, local surveys suggest that a dark halo of low mass H-burning stars in the Milky Way is unlikely. Color observations of NGC 5907 will provide more information on the nature of its faint luminous halo.

ACKNOWLEDGEMENTS

Figs 1 and 2 are reprinted with permission from Sackett *et al.*, *Nature* 370, 441, copyright 1994 Macmillan Magazines Limited, where the work on NGC 5907's faint halo is reported in more detail. PDS is a J. Seward Johnson Fellow at the Institute of Advanced Study; HLM was a Hubble Fellow at NOAO when this work was done. The National Optical Astronomy Observatories are operated by the Association of Universities for Research in Astronomy, Inc. (AURA) under cooperative agreement with the National Science Foundation.

REFERENCES

Bahcall, J.N., Flynn, C., Gould, A. and Kirhakos, S. 1994, ApJ, 435, L51.
Boeshaar, P.C., Tyson, J.A. and Bernstein, G.M. 1994, in these proceedings.
Morrison, H.L., Boroson, T.A., & Harding, P., 1994, AJ, 108, 1191.
Pritchet, C.J. & van den Bergh, S., 1994, AJ, 107, 1730.
van der Kruit, P. and Searle, L. 1981, A&A, 95, 105.
Zinn, R. 1985, ApJ, 293, 424.

COLD MOLECULAR GAS AS DARK MATTER CANDIDATE IN GALAXIES AND CLUSTERS

Daniel Pfenniger
Geneva Observatory, University of Geneva, CH-1290 Sauverny, Switzerland

Françoise Combes
DEMIRM, Observatoire de Paris, F-75014 Paris, France

ABSTRACT

Here are briefly presented the arguments given in [1-2] for the existence of a much larger amount of gas than previously thought, which could account for almost all the dark matter at the galactic and possibly cluster scales.

1. MOTIVATIONS FROM DYNAMICS AND OBSERVATIONS

Since the clear evidence in the 70's for a dark matter problem in spiral galaxies, the justifications for assuming hot spherical dark halos[3] have weakened. A considerable body of new knowledges about galaxies has been accumulated in the meanwhile, to which old and new dark matter candidates must comply. A sound strategy is to attack the dark matter problem at the galactic level, because it is better constrained than the cosmological one.

New viewpoints are emerging from galactic dynamics: bars are no longer considered as anomalies to avoid, but as common features of spirals. Bars are understood as normal engines regulating the disk angular momentum; they cause fast ($\sim 0.1 - 1\,\mathrm{Gyr}$) evolutive phase[4]. Bulges may form *after* their disks by internal instabilities related to bars and merging[5,4]. The maintenance of chaotic spirals (in Sc's) requires a constant gas supply that is not detected[6]. Episodic galaxy interactions accelerate evolution, e.g., by inducing a bar. Yet the accreted mass inside optical disks of typical spirals must remain small[7]. In short galactic disks appear as unsteady and fragile objects; they are easily subject to rapid (Gyr) phases of evolution along the spiral sequence in the sense Sd→Sa. Finally, polar rings constrain halo flattenings only weakly, in particular because ring masses are substantial[8]. Self-gravitating disks respond well to their own gravity, so they probe badly external potentials. Recent estimates about halo flattenings tend to rule out spherical halo distributions (e.g., [9]).

Another evidence for a steady growth and evolution of galaxies, which contradicts the notion of an early brief "galaxy formation epoch", is the gas consumption problem: too many galaxies seem to have nearly exhausted just now their gas reservoir by secular star formation[10,11].

Recent observational results from systematic properties of the Hubble sequence, rotation curves, and M/L ratios[12,13] show that Sd's are dark matter dominated, while Sa's require little dark matter and are well explained by stars. In fact optical disks can be made consistent without any dark matter[14] ("maximum disk hypothesis"). *These properties suggest that dark matter is transformed into stars as galaxies evolve, so dark matter in galaxies should be gas.*

Several other hints relate gas with dark matter. Not only is the HI/DM density ratio in outer HI disks nearly constant as a function of R[15] ($\Sigma_{DM} \approx 10-30\,\Sigma_{HI} \propto 1/R$), but also the global HI/DM mass ratio varies weakly along the spiral sequence[13], $M_{DM} \sim 10-30\,M_{HI}$. Furthermore, the large amounts of X-ray gas in galaxy clusters detected by ROSAT are also proportional to dark matter as a function of r outside the inner core[16,17]. Other indications that a substantial fraction of the gas mass may be invisible are:

1) The numerous Lyα absorption gas clouds in high-z quasar spectra correspond to more HI at high z than several times the total gas+star mass at low z.
2) Merging and interacting galaxies reveal more H_2 (i.e. CO) but normal HI contents as if extreme shock conditions would reveal more gas[18].
3) Apparently unrelated but in fact significant if gas may be invisible, at pc scale planetary nebulae (PN) have a "missing mass" problem too, but for once not related to the virial theorem; the expected progenitor star ($\sim 8\mathrm{M}_\odot$) has $5-9$ times more mass than the resulting system nebula+star ($\approx 0.3+0.6\,\mathrm{M}_\odot$)[19].

So several apparently unrelated problems would be essentially solved if the observed gas would be only the "tip of the iceberg" of further $10-30$ *times more gas.* This would still be consistent with the conventional Big Bang nucleosynthesis which typically requires several times more baryonic matter than observed.

2. NATURE AND MEASUREMENT OF COSMIC GAS

Interstellar gas is very often characterized by supersonic turbulence, extreme inhomogeneity ($|\delta\rho/\rho| > 10^6$), and by the coexistence of phases ($|\delta T/T| > 10^5$). Most of the detected gas mass is cold ($\lesssim 100\,\mathrm{K}$), clumpy and yet fills a small fraction of the volume, *a very unusual fluid.* There is no satisfying model of the ISM gas: usually theoretical "clouds" are imagined as round, nearly homogeneous, and stable gas blobs, with some arbitrary scale or mass, but such clouds are never observed. Recently it was realized that cold gas behaves statistically as a *fractal* (clumps are made of $N=5-10$ sub-clumps, etc.) over at least 10^4 in scale[20]. The Larson's size-line-width relation[21] $\sigma \propto r^{0.3-0.5}$ in molecular clouds indicates a fractal dimension $D \sim 1.6-2.0$ ($M(r) \propto r^D$).

Despite of all the evidences of extremely clumpy gas, the amount of HI gas is still derived assuming homogeneity below the beam, and optical thinness, which is not supported by high-resolution observations[22]. H_2 masses are calibrated from the CO emission via the virial theorem (so molecular clouds have no "missing" mass by convention), and can only be detected in regions where CO exists and is properly excited. It is time to critically question whether the detected gas masses are really exhaustive, particularly because we have not much experience about the observational biases associated to a fractal state.

3. PROPERTIES OF FRACTALS AND FINE STRUCTURES IN THE ISM

The projection of a *mathematical* fractal mass distribution in 3d-space with $D \leq 2$ onto a 2d-plane has almost always also the dimension D[23]. Then all the mass projects onto a *null* surface. In the same conditions the projection properties of a *physical* fractal, for which the scale invariance occurs over a finite scale range, depend directly on the *smallest scale* at which the fractal behavior stops. So if the fractal behavior extends over several decades, most of the mass projects onto a very small surface with high column density. In such a situation most of the mass is optically thick, while most of the volume is optically thin. Computer models allow to quantify such effects[2]. With $D < 2$ they are rapidly significant, and indeed a factor $\gtrsim 10$ of mass may be invisible.

Recently much progress has been accomplished toward resolving directly fine scale structures in the cold ISM. In particular we can mention:

1) The smallest directly resolved cold clumps in the nearest known molecular cloud MBM 12 have $d \lesssim 2000\,\mathrm{AU}$[24].
2) The nearest PN's have cold cometary globules with $d \sim 100\,\mathrm{AU}$ [25,26].
3) Varying absorption lines in pulsars or QSO's are evidences for structures in Milky Way cold clouds with $d \sim 5-100\,\mathrm{AU}$[27,28].

4. COLD GAS MODEL

Theoretically, the lightest self-gravitating isothermal clumps[29] at low $T \lesssim 5\,\mathrm{K}$ have $d \sim 60-100\,\mathrm{AU}$, $M \sim 0.001\,\mathrm{M}_\odot$, and $n \sim 10^9\,\mathrm{cm}^{-3}$, $\Sigma \sim 10^{24}\,\mathrm{cm}^{-2}$. We call these structures *"clumpuscules"*[2]. At such densities they have to be made mostly of cold H_2+He, i.e., *they are difficult to detect.* Further fragmentation and fractality at lower scale is not excluded, but then no longer due to the gravitational instability (as for terrestrial clouds).

The most conservative assumption is then to extrapolate the established fractal properties of nearby cold gas (Larson's relations) from 100 to 0.01 pc by ~ 2 dex, down to the clumpuscule scale $(10-100\,\mathrm{AU})$, and assume that such scaling laws apply also to the HI outer galactic discs[2]. With insufficient resolution the individual sky covering factor of clumpuscules is small ($\ll 1\%$). The expected velocity dispersion along the line of sight is 0.1 km/s. Most of the mass is in H_2 form. In outer disk conditions cold gas is expected to be metal poor, and self-shielded from ionizing radiation.

With $D \sim 1.6-2.0$, $N \sim 5-10$, the clumps in the fractal are *collisional* at every scale ($\tau_{\mathrm{coll}} \sim \tau_{\mathrm{dyn}}$) with a clumpuscule collision interval of $\sim 10^3\,\mathrm{yr}$[2]. So clumpuscules form and dissolve constantly, they are transient structures. Large scale clump interactions are gravitational, while clumpuscules interactions are gaseous but in near thermal equilibrium with the microwave background (MWB), though the fluctuations are expected to be large. The whole fractal medium may be little dissipative ($\tau_{\mathrm{diss}} \gtrsim 1\,\mathrm{Gyr}$). If $D = 1.67$ dissipation is scale-invariant. In such a form, most of the gas mass is invisible with today's techniques, because T is close to the MWB temperature, H_2+He have no emission lines in this regime, and absorption is not currently beam-filling.

5. CONCLUSIONS AND OPEN ISSUES

Cold and fractal molecular gas is probably the most conservative dark matter candidate in galaxies and clusters. Yet it may address many puzzles, conspiracies and problems such as 1) the disk-halo conspiracy and flat rotation curves, 2) the spiral Hubble sequence properties, 3) the gas consumption problem, 4) the ISM properties, 5) the straight line-of-nodes of HI warps, 6) the large baryonic fraction of cooling hot gas in clusters. No fundamental new physics is required, one just needs to improve the highly non-linear physics of the cold ISM. This would also help in understanding the proper conditions of star formation.

A number of interesting observational issues are raised by our model, such as the long length-scale ($\sim 15\,\mathrm{kpc}$) of energetic γ-rays ($\gtrsim 100\,\mathrm{MeV}$) in the Milky Way, and the link of the supposed 3K gas with the small-scale fluctuations of the MWB. The Lequeux et al.[30] claimed observational detection of cold molecular gas has been criticized [31,32], because it might well be relatively diffuse gas that looks cold only because of insufficient collisional excitation.

Among the theoretical and observational topics to be addressed in the future, we can mention: 1) the stability of cold outer disks: HI discs are observed to be in general very clumpy at every scale, also at the largest scale[33], so HI disks *are in fact unstable*, 2) the puzzle about the straight line of nodes of HI disk warps is naturally explained by massive rotating HI disks, basically because the angular momentum is then locally well conserved and has to diffuse progressively across the disk, 3) the fractal gas properties, and its long term persistence, and, 4) the possibility of H_2 freezing into snow[2].

REFERENCES

[1] Pfenniger D., Combes F., Martinet L., 1994, A&A 285, 79
[2] Pfenniger D., Combes F., 1994, A&A 285, 94
[3] Ostriker J.P., Peebles P.J.E., 1973, ApJ 186, 467
[4] Pfenniger D., 1993, in "Physics of Nearby Galaxies, Nature or Nurture?", T.X. Thuan, et al. (eds.), Editions Frontières, Gif-sur-Yvette, p. 519
[5] Combes F., Debbasch F., Friedli D., Pfenniger D., 1990, A&A 233, 82
[6] Toomre A., 1981, in: "The Structure and Evolution of Normal Galaxies", S.M. Fall, D. Lynden-Bell (eds.), Cambridge Univ. Press, p. 111
[7] Tóth G., Ostriker J.P., 1992, ApJ 389, 5
[8] Reshetnikov V., Combes F., 1994, A&A 291, 57
[9] Sackett P.D., Pogge R.W., 1995, these proceedings
[10] Larson R.B., Tinsley B.M., Caldwell C.N., 1980, ApJ 237, 692
[11] Kennicutt R.C., 1990, in "Evolution of the Universe of Galaxies", R.G. Kron (ed.), ASP Conf. Ser. 10, p. 141
[12] Persic M., Salucci P., 1991, ApJ 368, 60
[13] Broeils A., 1992, "Dark and visible matter in spiral galaxies", PhD Thesis, Rijksuniversiteit Groningen
[14] Kalnajs A.J., 1987, in "Dark Matter in the Universe", IAU Symp. 117, J. Kormendy, G.R. Knapp (eds.), Reidel, Dordrecht, p. 289
[15] Bosma A., 1981, AJ 86, 1971
[16] Hammer F., 1991, ApJ 383, 66
[17] Gerbal D., Durret F., Lima-Neto G., Lachièze-Rey M., 1992, A&A 253, 77
[18] Braine J., Combes F., 1993, A&A 269, 7
[19] Kwok S., 1994, PASP 106, 344
[20] Scalo J.M., 1985, in "Protostars and Planets II", D.C. Black, M.S. Matthews (eds.), University of Arizona Press, Tucson, p. 201
[21] Larson R.B., 1981, MNRAS, 194, 809
[22] Burton W.B., 1992, in "The Galactic Interstellar Medium", Saas-Fee Adv. Course 21, D. Pfenniger, P. Bartholdi (eds.), Springer-Verlag, Berlin, p. 1
[23] Falconer K.J.: 1990, Fractal Geometry, Wiley, Chichester
[24] Ingalls J.G., Bania T.M., Jackson J.M., 1994, ApJ 431, L139
[25] Meaburn J., Walsh J.R., Clegg R.E.S., et al., 1992, MNRAS 255, 177
[26] Huggins P.J., Bachiller R., Cox P., Forweille T., 1992, ApJ L43
[27] Marscher A.P., Moore E.M., Bania T.M., 1993, ApJ 419, L101
[28] Frail D.A., Weisberg J.M., Cordes J.M. Mathers C., 1994, preprint
[29] Rees M.J., 1976, MNRAS 176, 483
[30] Lequeux J., Allen R.J., Guilloteau S., 1993, A&A 280, L23
[31] Wilson T.L., Mauersberger R., 1994, A&A 282, L41
[32] Phillips J.A., 1995, in "Unsolved Problems of the Milky Way", IAU Symp. 169, in press
[33] Richter O.G., Sancisi R., 1994, A&A 290, L9

PRELIMINARY EVIDENCE FOR DUST IN GALACTIC HALOS

Dennis Zaritsky
UCO/Lick Observatory, U.C. Santa Cruz, Santa Cruz, CA, 95064
dennis@lick.ucsc.edu

ABSTRACT

I use B and I photometry of distant galaxies observed through the halos of two nearby spiral galaxies, NGC 2835 and NGC 3521, to test for the existence of dust in galactic halos out to at least a radius of 60 kpc. $B-I$ colors of background galaxies in fields centered along the nearby galaxy's major axis at a projected separation of 60 kpc are compared to the colors of background galaxies in fields centered at a projected separation of about 220 kpc. Background galaxies in the fields at smaller projected separations have average $B-I$ colors that are 0.067 ± 0.033 mag redder than those in the outer fields. Such a result will occur by chance less than 2.2% of the time and so is taken as preliminary evidence that galaxy halos contain more dust at galactic radii of 60 kpc than 220 kpc.

1. INTRODUCTION

Most of the mass in spiral galaxies is in their dark matter halos (cf. Zaritsky and White 1994), although its composition remains completely undetermined. Even if only a minute fraction of that mass is in the form of dust, there may be a significant amount of dust in galactic halos. Whether such dust exists merits attention because its presence would have serious repercussions on the study of galactic formation and evolution, dark matter, and the high redshift universe.

This paper outlines a recent search for dust in galactic halos, which is described in more detail by Zaritsky (1994). The structure of the experiment is simple: to compare the colors of background galaxies seen in fields observed through the halo of a foreground galaxy to the colors of galaxies seen in unobscured reference fields. Any evidence for a mean redenning of the inner fields relative to the outer fields may suggest that halos contain dust.

2. OBSERVATIONS

Observations of eight $21' \times 21'$ fields in B and I were made using a 2048^2 Tek CCD at the Las Campanas 40-inch telescope. Around each of two foreground galaxies, two inner fields (at about 60 kpc projected separation along the disk major axis) and two outer fields (at about 230 kpc projected separation) were observed. Each field was observed for about 14000 and 5000 seconds in B and I, respectively. Galaxies were identified using FOCAS (Jarvis and Tyson 1981) and a supplementary surface brightness criterium. Objects must be identified in both B and I to enter the catalog (note that this leads to a lower limit on reddening since highly reddened objects are excluded). The magnitude limit of the catalog is about I = 21 mag, although the I images themselves go much deeper (to about 24th mag). The I images were used for object classification.

Slight photometric zero point difference or local Galactic reddening differences could produce color offsets between fields. Therefore, the stellar color differences, as measured using the technique described below, were set to zero for all the images. Afterward, the relative colors of galaxies in the various fields can be measured. The results for the four fields around each of the two galaxies (NGC 2835 and NGC 3521) are shown in Figure 1.

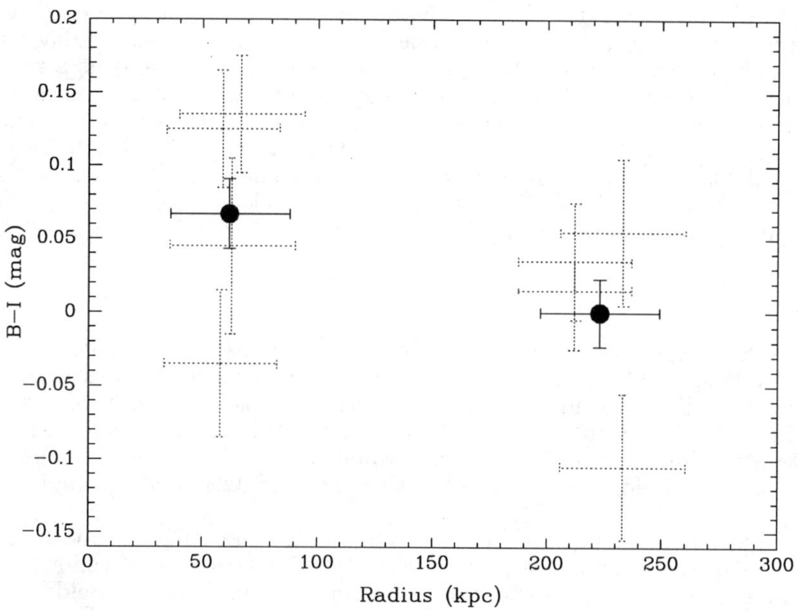

Fig. 1. Differential reddening measurements for background galaxies vs. radius for all fields. The relative B-I colors have been set to zero for the mean of the outer fields. The dotted lines represent the measurements for each field, the solid lines represent the average of the values at either the inner or outer position.

3. Discussion

A robust estimator of color difference was constructed by using one frame as a reference from which to derive the "intrinsic" color distribution of stars and galaxies. Other fields were then compared by applying a range of color shifts. The best-fit color difference between the test and reference fields was taken to be the color shift for which χ^2 was minimized, where the contribution to χ^2 from each object was taken to be the sum of Gaussians with exponents proportional to the square of the differences between its color and those of the reference objects in a small magnitude bin. This estimator proved to be more robust than either the mean or median color difference between fields.

An important test of the data and color measuring procedure is to compare the mean reddening of the fields around NGC 2835 to that of the fields around NGC 3521. Since NGC 2835 is at a much lower Galactic latitude (18.5°) than NGC 3521 (52.8°), there should be differential Galactic reddening (estimated to be 0.18 ± 0.05 mag by Burstein and Heiles (1984)). The measured value from these data, 0.11±0.03, agrees with the predicted reddening difference, providing support to the contention that relatively small color difference can be measured accurately using these data.

As evident from Figure 1, there does appear to be a tendency of the inner fields to be *on average* redder than the outer fields. Since the color differences and the observational uncertainties are of the same order, not every inner field is redder than every outer field (although three of the reddest four are inner fields). Using these results, the null hypothesis (that there is no excess reddening in the inner fields) can be discriminated against at the 98% confidence level.

The reddening difference in itself does not prove the existence of dust in galactic halos. It is possible that there are unaccounted for Galactic reddening differences or subtle variations in the intrinsic colors of background over large angular scales; either of which would increase the uncertainties and reduce the significance of the result. The first is probably not a factor since the fields were chosen to have small 100μm flux differences and because the stellar colors were zeroed. The second is unlikely because the scale at which we know environment to affect galaxy colors (the scale of galaxy clusters) corresponds to a small fraction of a single image if the typical galaxy in the survey has a luminosity of L^*. Color differences over larger areas should be much less correlated. The success of the test to recover the Galactic extinction also argues against either possibility. Nevertheless, before both of these alternatives can be completely ruled out, and before the results themselves can be unequivocally supported, additional observations over many other fields and galaxies are needed.

If confirmed, the presence of this dust has serious implications for the nature of dark matter and the distribution of baryonic dark matter, for the evolution of galaxies and the processes that result in the presence of this material at such large radii, and for our view of the high redshift universe. The optical depth of dust calculated from these observations, assuming a standard extinction law, are quite similar to those depths postulated by Heisler and Ostriker (1986) to be necessary to obscure QSO's at $z > 2$. For sufficiently effective obscuration of the high redshift universe, this dust would have to be present in high redshift galaxies, but its presence in low redshift galaxies would indirectly support their conjecture.

ACKNOWLEDGEMENTS

Financial support for this work was provided by NASA through grant HF-1027.01-91A from STScI, which is operated by AURA, Inc. under NASA contract NAS5-26555.

REFERENCES

Burstein, D., and Heiles, C. (1984), Ap. J. Supp., 54, 33.
Heisler, J., and Ostriker, J.P. (1988), Ap. J, 332, 543.
Jarvis, J.F., and Tyson, J.A. (1981), A.J., 86, 476.
Zaritsky, D. (1994) Preliminary Evidence for Dust in Galactic Halos, A.J., in press.
Zaritsky, D., and White, S.D.M. (1994) The Massive Halos of Spiral Galaxies, Ap. J., in press.

GALAXY HALOS AND THE NATURE OF DARK MATTER

Ben Moore
Department of Astronomy, University of Washington, Seattle WA 98195.
moore@hermes.astro.washington.edu

ABSTRACT

One of the few observational facts that we know about dark matter is that within about one tenth of the virial radius, the dark halos of some galaxies have density profiles which fall with radius as $\rho(r) \propto r^n$, where $n \approx 0$. Any successful dark matter candidate must be able to reproduce these observations. Cold dark matter (CDM) particles interact primarily by gravity and therefore cluster with no preferred scale, hence galaxy halos are predicted to have singular density profiles with approximately isothermal power law slopes on all scales (Filmore & Goldreich 1984, Bertschinger 1985). The theoretical predictions do not take account of the irregular merging and virialisation processes taking place during structure formation. Therefore, we must compare the observations with numerical N-body simulations which are ideally suited for this type of problem and have just begun to probe structure formation on these scales. The highest resolution simulations (Dubinski & Carlberg 1991, Warren *et al* 1992, Carlberg 1993, Crone *et al* 1994) produce halo density profiles which fall too steeply to match the data (Moore 1994, Flores & Primack 1994). Moreover, Crone *et al* (1994) claim that after correcting for force softening the density profiles match the analytic predictions. If these results are confirmed by higher resolution studies then we can reject the hypothesis that the dark matter is a cold, dissipationless particle. However, there is some disagreement between the results of the above authors and the slightly lower resolution studies of Katz & White (1993) and Navarro, Frenk & White (1994).

1. THE DENSITY PROFILES OF SPIRAL GALAXIES

The rotation curves of spiral galaxies enable the total mass distribution as a function of radius to be measured. Recovering the relative contributions to the rotation curve from the dark and luminous material is difficult, since some degree of freedom is allowed in specifying a mass-to-light ratio for the stars. What is becoming apparent from the observations, is that the rotational velocity of the stars and/or gas falls to zero towards the center. Adopting a conservative mass-to-light ratio for the stars, results in a dark halo which dominates the mass distribution beyond a couple of disk scale lengths. To avoid these problems it is informative to study dwarf galaxies or low surface brightness galaxies, which are dark matter dominated beyond a few kpc. The baryonic component of these systems exists primarily in an extended HI disk which can be accurately weighed using the observed 21cm emission.

Figure 1 shows the rotation curves of two dwarf spiral galaxies which have been observed at high resolution with the VLA (Carignan & Beaulieu 1989, Jobin & Carignan 1990). These data show a trend which is apparent in all published rotation curves of small spiral galaxies. Namely, that the rotation curve is rising slowly over many disk scale lengths into the halo, but within the core of the total mass distribution which is dominated by dark matter beyond a single optical scale length. For example, the rotation curve of the dwarf galaxy DDO 154 shown in Figure 1, is dominated by dark matter beyond about 1.5 kpc. In this system the amount of HI, $2.7 \times 10^8 M_\odot$, is 5 times the stellar component and extends over 15 optical scale lengths into the dark halo which has total mass $\gtrsim 5 \times 10^9 M_\odot$. Although dwarf spirals such as these have luminosities only a few percent of an L_* galaxy, these galaxies are the most numerous and a large fraction of the mass in the universe is associated with these systems.

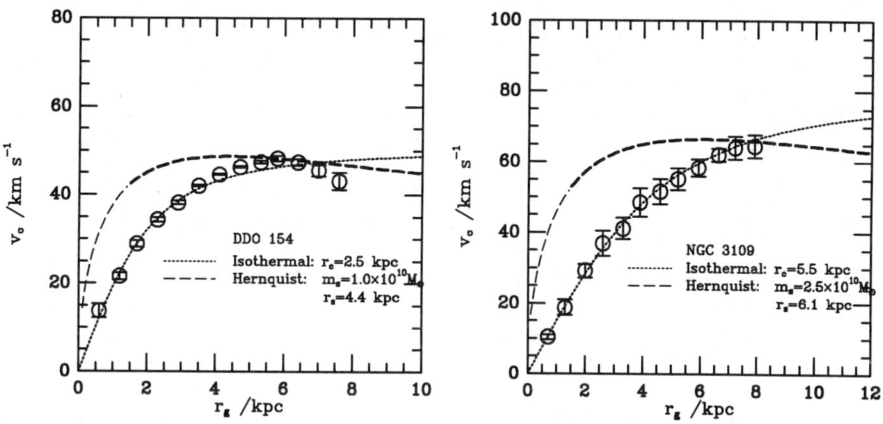

Fig 1. The observed rotation curves of two isolated dwarf spiral galaxies are plotted. The dotted lines show a fit to the observed rotation curve assuming zero contribution from the baryons and a dark halo with an isothermal density profile $\rho(r_g) \propto 1/(r_c^2 + r_g^2)$, where r_c is the core radius. These 'maximum halo' models are designed to minimise the contribution of the disk, hence provide a lower limit to r_c and they demonstrate that even if the gravity of the baryons is ignored, the dark halos must have large core radii. The dashed lines show the rotation curves obtained adopting a Hernquist density profile with effective radius scaled from the results of Warren *et al* (1992), using $r_s \propto v_{c,max(\mathrm{observed})}$ and the effective mass scaled so that the curve passes through this same point.

2. NUMERICAL RESULTS FROM CDM N-BODY SIMULATIONS

CDM dominated universes have been extensively investigated and have proven quite successful at reproducing the observed clustering pattern on scales from galaxies to clusters of galaxies (White *et al* 1987, and references therein). Structure formation within these models proceeds in a bottom up fashion, with small over-densities in the mass distribution collapsing first and subsequently merging hierarchically to form larger objects. Galaxies form from gas which has cooled within the deep potential wells provided by the dark matter halos, plus the accretion of gas rich sub-clumps which are part of the merging hierarchy. The dominant force which drives structure formation is gravity, hence numerical simulations have proven to be a useful tool with which to study the non-linear growth of structure formation over a wide range of dynamical scales.

The resulting dark halos from CDM N-body simulations can be fit quite well with a density profile $\rho(r) \propto 1/(r(r+a)^2)$, where a is a "core radius", although note that the core here does not have constant density. The natural scale length of a halo is the virial radius, r_v, at which the mean internal overdensity is 200. For an $\Omega = 1$ universe we obtain $r_v/(\text{kpc}) \approx v_c/(\text{km s}^{-1})$, where v_c is the peak circular velocity. Isolated collapse profiles, such as Dubinski & Carlberg (1991), have slightly steeper outer profiles since the higher resolution is compromised at the expense of a lower accretion rate of new material. Full cosmological simulations which resolve a larger volume of the universe and contain many individual halos, are compromised by relaxation effects due to a smaller number of particles per halo.

Reasonable fits to the observations can be obtained if $a \approx r_v/3$. The numerical simulations of Katz & White (1993) and Navarro, Frenk & White (1994) have a resolution of about 4% of the virial radius. Their halo density profiles can marginally match the data, although the fits are quite poor in the inner regions once the baryonic material is taken into account. Furthermore, if any contraction of dark halos has occurred due to adiabatic baryonic dissipation, the fits will become considerably worse. Note that when this profile is scaled to the Milky Way, the rotational velocity at 10 kpc is only $\sim$ 100 km s^{-1}. Hence, if our own dark halo is made of CDM, the central phase space density must have increased by a very large factor by a process which was not present during the formation of the dwarf spirals.

Moore (1994) used the results of Dubinski & Carlberg (1991) and Warren *et al* (1993) to compare with the observations. As White (these proceedings) points out, Dubinski & Carlberg stop their simulation at a redshift $z = 1$, hence their quoted length scale could be a factor of $\sqrt{8}$ too small. However, Carlberg (1993, private communication) evolved a similar halo with 5 times as many particles past the present day. The length resolution in that simulation was $\sim$ 1% of the virial radius and infall was still occurring at z=0. The halo density profile could be fit with $a \approx r_v/10$, which produces a rotation curve rising much more steeply than the data shown in Figure 1. This result agrees quite well with the results of Warren *et al* (1993) which were used to scale the dashed curves in Figure 1.

Recent work by Crone *et al* (1994) provides a comprehensive study of halos which form within scale free cosmological simulations. Various models were studied, including high and low Omega and different power law slopes for the initial fluctuation spectrum. The Plummer force softening of these authors gave a length resolution of $0.01r_v$ and halos from these simulations could also be fitted using $a \approx r_v/10$, yielding much steeper profiles than the observations indicate. Most interestingly, Crone *et al* attempt to correct for the softened forces by adiabatically moving mass shells to smaller radii where they would be given the correct force. The resulting density profiles became power laws with shape $\rho \propto r^{-2.2}$ on all scales smaller than the virial radius and are in direct conflict with almost all observations of spiral galaxy rotation curves.

3. DISCUSSION

Dwarf spiral galaxies are dominated by dark matter on scales larger than a kpc. The rotation curves of these galaxies can be used to determine the structure of dark halos with very little ambiguity from the baryonic material. These observations provide one of the most stringent tests for the nature of dark matter. The predictions from a CDM dominated universe cannot be "fudged" by altering the mean density parameter, the bias parameter or the cosmological constant. Furthermore, the internal structure of galaxy halos is insensitive to an additional component of hot dark matter.

Several N-body simulations of CDM universe's with enough resolution to compare with the observations have been carried out. At present there is some debate over the results since the density profiles vary significantly between the authors. These results may may not necessarily be inconsistent, but could reflect the fact that as the softening is reduced, the core radius a becomes smaller. Alternatively, the different results may be due to numerical artifacts, such as too large a time-step for a given softening or numerical relaxation from too few particles. However, most numerical artifacts such as these are expected to erase a pre-existing core. Just as likely, the halo density profiles could be sensitive to the details of the simulation such as initial conditions or the volume which is followed. It is clear that if the halos of normal spirals do consist of CDM, then the central phase space densities of these galaxies indicate that they formed in a different manner from the dwarf spirals.

The "brute force" approach to resolving this issue is to perform numerical simulations with sufficient resolution to resolve the cores of galaxy halos, using adequate numbers of time-steps and particles. To obtain the claimed accuracy of $0.01r_v$ would require at least 10,000 time-steps which is several times more than any of the above simulations. An alternative approach, is to follow the evolution of a single peak within a much larger simulation, but obtaining the correct infall rate and tidal field. In collaboration with the High Performance Computing and Communications group at the University of Washington, simulations using both of these techniques are in preparation.

ACKNOWLEDGMENTS

I would like to thank Marc Davis, Neal Katz, George Lake and Simon White for stimulating discussions on this topic. This work was funded by NASA through the LTSA and HPCC/ESS programs.

REFERENCES

Bertschinger E. 1985, ApJS, 58, 39)
Carignan C. & Beaulieu S. 1989, ApJ, 347, 760
Crone M.M., Evrard A. & Richstone D. 1994, ApJ, in press
Dubinski J. & Carlberg R. 1991, ApJ, 378, 496
Filmore J.A. & Goldreich P. 1984, ApJ, 281, 1
Flores R.A. & Primack J.R. 1994, ApJLett, 427, L1
Jobin M. & Carignan C. 1990, AJ, 100, 648
Katz N & White S.D.M. 1993, ApJ, 412, 455
Moore B. 1994, Nature, 370, 629
Navarro J., Frenk, C.S. & White S.D.M. 1994, MNRAS, in press
Warren S.W., Quinn P.J., Salmon J.K. & Zurek H.W. 1992 ,ApJ, 399, 405
White S.D.M., Frenk C.S., Davis M. & Efstathiou G.P. 1987, ApJ, 313, 505

X-RAY (ASCA) MEASUREMENTS OF ELLIPTICAL GALAXIES

Kazuo Makishima
Department of Physics, University of Tokyo, Hongo, Bunkyo-ku, Tokyo, Japan.
maxima@miranda.phys.s.u-tokyo.ac.jp

ABSTRACT

Mainly using the X-ray observatory *ASCA*, studies have been made on the gravitational potential distribution from local vicinities of luminous elliptical galaxies out to the scale of clusters in which they reside. Evidences have been obtained for clustering of X-ray emitting hot gas on two distinct spatial scales, namely on the galaxy scale and on the cluster scale. Such examples include NGC 4472 in the Virgo cluster, NGC 4696 in the Centaurus cluster, and NGC 1399 in the Fornax cluster. At least in the Fornax/NGC 1399 case, it has further been revealed that the observed hierarchical gas clustering is a direct consequence of a hierarchical structure in the gravitational potential, hence in the dark matter distribution.

1. INTRODUCTION

Most elliptical galaxies are known to contain significant amount of X-ray emitting hot interstellar matter (ISM) of temperature ~ 1 keV (*e.g.* Forman *et al.* 1985; Canizares *et al.* 1987). Since this temperature is close to virial temperature of the stellar kinetic motion in the galaxy's gravitational potential, the ISM is thought to be gravitationally bound to the galaxy in a form of hydrostatic equilibrium. This has allowed many investigators to utilize the X-ray data to measure the potential distribution around elliptical galaxies, and to get knowledge on the dark matter (DM) in these systems. Recent observations with *ROSAT* and *ASCA* (Tanaka *et al.* 1994) have significantly refined our knowledge on the DM in elliptical galaxies (*e.g.* Forman *et al.* 1993; Jones *et al.* 1994; Awaki *et al.* 1994; Loewnestein *et al.* 1994). According to these results, elliptical galaxies contain significant amount of DM, such that the mass-to-luminosity ratio (M/L) is typically in the range 20–100. Furthermore it is generally agreed that the stars are most concentrated around the galaxy, while the hot gas is most extended, and the DM may have an intermediate distribution.

The X-ray measurements of gravitational potential in elliptical galaxies are quite analogous to those in clusters of galaxies, in which X-rays are emitted from hot intra-cluster medium (ICM) instead of the ISM. (However one important difference is that the ISM mass is a minor fraction of the stellar mass in an elliptical galaxy, while the ICM mass often overwhelms the total stellar mass in a cluster.) X-ray measurements of gravitational potential have so far been performed extensively, both on galaxies and clusters, but separately in most cases. Consequently, although knowledge is quite improving on the power spectrum for the spatial distribution of stellar components, we badly lack our knowledge on the spatial power spectrum of the DM distribution. That is, we have yet to understand; how the potential changes from the galaxy scale to the cluster scale; whether the DM exhibits *hierarchical clustering* on different spatial scales or not; how the DM hierarchy compares with that in baryons; and so on.

To innovate our understanding of the DM through X-ray studies, it is now essential to treat galaxies and clusters simultaneously, paying a particular attention to the *galaxy/cluster interface.* Even if the overall cluster potential may be approximated by a smooth (*e.g.* King-type) function with a single scale parameter, there should be more or less a "dimple" in the gravitational potential around each member galaxy; otherwise galaxies would not keep their basic characteristics as self-gravitating systems. The dimples must consist of significant amount of DM; otherwise the M/L ratio would not be so large as mentioned above. Then the depth and size of such dimples would carry crucial information as to the hierarchical structure in the DM.

The potential dimples are expected to manifest themselves in high-quality X-ray data. With the superior sensitivity and the unprecedented capability of spatially-resolved spectroscopy, *ASCA* (Tanaka *et al.* 1994) for the first time allows us to challenge this problem.

2. GALAXY/CLUSTER INTERFACE

2.1 Hard component from Virgo ellipticals

From five optically-bright elliptical galaxies in the Virgo cluster (NGC 4406 = M84, NGC 4374 = M84, NGC 4472 = M49, NGC 4636, and NGC 4365), Matsushita *et al.* (1994) detected hard X-ray component in addition to the well established emission from hot ISM of temperature 0.6–1.1 keV. The hard component has a color temperature of > 2 keV, and exhibits a similar luminosity of order $10^{40-40.5}$ erg s^{-1} among the 5 galaxies. Therefore it can primarily be attributed to the assembly of discrete X-ray sources in the galaxy (Canizares *et al.* 1987). In NGC 4406 (M84) and NGC 4374 (M84), located near the cluster center, the hard component seems to be contributed additionally by foreground and background emission from the Virgo ICM that has a temperature of 2.5–3 keV (Koyama *et al.* 1991; Böhringer *et al.* 1994).

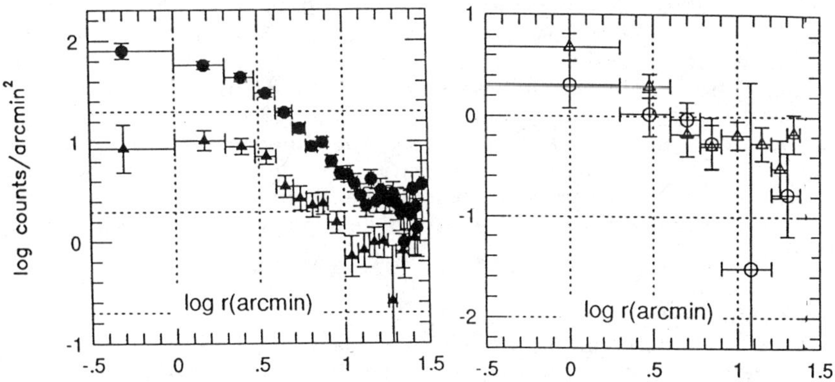

Fig. 1. X-ray count distribution around NGC 4472 in different energy bands, as observed with the *ASCA* GIS (Gas Imaging Spectrometer) instrument. Left panel: 0.5–2.0 keV (circle) and 2.0-4.0 keV (triangle). Right panel: 3–6 keV (triangle) and 4-7 keV (circle).

The most interesting case is NGC 4472. Its hard component profile as observed with the GIS (Gas Imaging Spectrometer; Kohmura *et al.* 1993) instrument on board *ASCA* appears significantly more extended than (see Fig.1), and displaced from (Matsushita *et al.* 1994), the soft component profile, indicating a significant contribution from the ICM emission. In fact, an enhancement in the ICM brightness indicative of a subcluster structure has been detected with *Ginga* from a $\sim 2°$ region centered on NGC 4472 (Takano *et al.* 1989). However the ICM brightness measured around NGC 4472 with the $\sim 1°$ beam of the *Ginga* LAC falls far short ($< 10\%$) to account for the hard component measured with *ASCA* within $\sim 20'$ of NGC 4472. Therefore we infer that the ICM emissivity around NGC 4472 is enhanced not only on the angular scale of several degrees as revealed with *Ginga*, but also on the $\sim 10'$ scale as observed with *ASCA*. This provides a possible evidence that the gravitational potential exhibits a hierarchical structure around NGC 4472, both on "subcluster" scale ($\sim$ 500 kpc) and on galaxy scale ($\sim$ 50 kpc).

2.2 Cool component around cD galaxies

With their large masses and central locations, cD galaxies provide a particularly promising case to study the galaxy/cluster interface. Actually X-ray brightness profiles have been measured in many cD clusters with *Einstein* and *ROSAT*, from a local vicinity of the cD galaxy out to a whole cluster extent. However due to limited spectral information, it has been difficult to estimate the shape of the potential from the data alone (*e.g.* Allen & Fabian 1994; Rangarajan etal 1994). Thus we have generally been unable to examine the data for hierarchical potential structure around the cD galaxies. This circumstance is best reflected in the fact that the large quantity of hot gas around M87 is considered in some cases to be the Virgo ICM (*e.g.* Binney & Cowie 1981; Böringer *et al.* 1994), while in other cases to be the ISM associated with M87 (*e.g.* Fabricant & Gorenstein 1983; Stewart *et al.* 1984).

The problem is however not easy even with *ASCA*, because the central region of a cD cluster often exhibits complex temperature structure which becomes even more complicated due to projection effects. Indeed, *ASCA* has detected cool ($kT = 1 - 2$ keV) X-ray components in central regions of almost all the observed cD clusters (Fukazawa *et al.* 1994; Ikebe *et al.*1994; Fabian *et al.* 1994). This phenomenon is closely related to the concept of "cooling flow" (CF), which is evidenced primarily by excess X-ray surface brightness around cD galaxies and augmented by many secondary indications (Fabian 1994). In terms of the CF, the excess X-ray brightness and the X-ray cool component are the same thing and can be explained as a result of radiative cooling instability. Then, is the CF model expected to give us a realistic guideline to sort out the temperature structure near the cD galaxy? The answer does not seem to be affirmative as discussed below.

2.3 Results on the Centaurus cluster

From *ASCA* observations of the Centaurus cluster, Fukazawa *et al.* (1994) have discovered cool gas with a temperature $kT_c \sim 1$ keV within $\sim 10'$ of the center, where the cD galaxy NGC 4696 is located. They have further shown that; (1) the hotter ($kT_h \sim 4$ keV) gas also coexits in the central region even removing the projection effect; (2) the values of kT_c and kT_h remain both very constant as a function of projected radius; and (3) the gas metallicity exhibits a dramatic increase within the same $\sim 10'$ region. These results are reproduced in Fig.2; note that the data for inner several arcminutes are strongly affected by the point spread function (PSF) of the X-ray telescope (XRT). These properties suggest that the apparent multi-phase nature is intrinsic to the ISM of the Cen cluster, rather than resulting from projection of a single-phase ICM with a radial temperature gradient.

The behavior of M87/Virgo appears qualitatively similar, including a strong metal concentration as first discovered with *Ginga* (Koyama *et al.* 1991) and then confirmed with *ASCA* (Matsumoto *et al.*1994). We further speculate that other galaxies with apparent central excess brightness may be under a similar condition as the Centaurus and the Virgo clusters of galaxies.

In the light of these new *ASCA* results, we suspect that the CF picture may no longer help us deepen our understanding of the X-ray phenomena seen around cD galaxies. The point of argument is twofold. One is that the observed cool component is more likely to be a manifestation of metal-rich cooler ISM associated with the cD galaxy rather than a CF, as evidenced by the increased metallicity. The ISM may coexist (*e.g.* in the form of filaments) with the hotter ICM. We must then model the temperature structure as a multi-phased medium rather than a single-phased one. The other point is that the central excess brightness may be caused partially by the potential dimple, within which the metal-rich ISM may be confined. Clearly, implication for the DM depends critically on our choice between the CF picture and this alternative view.

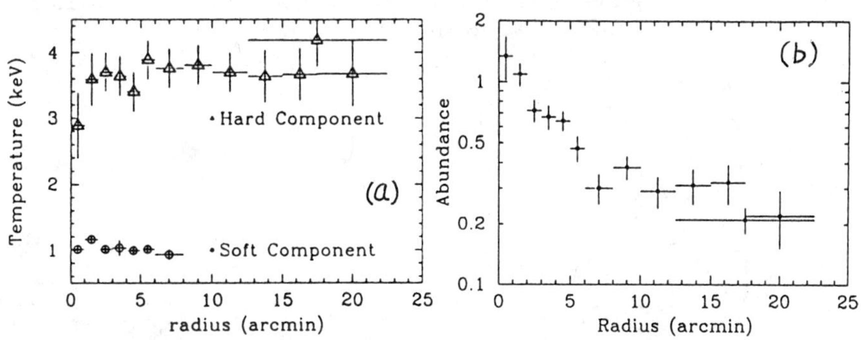

Fig. 2. *ASCA* results on the Centaurus cluster (from Fukazawa *et al.* 1994). The spectra are fit with two-temperature common-abundance plasma model. (a) is their temperature distributions, and (c) is the abundance distribution.

3. THE NGC 1399/ FORNAX CLUSTER SYSTEM

We have seen in §2 that the complex temperature structure is the main obstacle when estimating the potential distribution from viewpoint of the galaxy/cluster interface. Since a cluster usually has a higher virial temperature than its cD galaxy, in clusters with cooler ICM the overall gas would be closer to isothermality, hence easier to analyze. Furthermore in cooler clusters, the potential dimple $\Delta\phi$ around the cD galaxy would show up more clearly, because the excess gas density in response to the dimple is expected to be proportional to $\exp(\mu m_p \Delta\phi / kT)$. Here m_p is the proton mass, μ is the mean molecular weight, and T is the gas temperature. The Fornax cluster, being a near-by poor (hence cool) cluster with a convenient angular size, is ideal for this purpose.

3.1 The Fornax cluster and NGC 1399

The Fornax cluster is a southern poor cluster (Ferguson & Sandage 1990). Its recession velocity of $\sim$ 1400 km/s (Roberts *et al.* 1991) indicates a Hubble distance of $D = 19 \times (75/H_0)$ Mpc, making it one of the nearest cluster. Here we adopt $D = 20$ Mpc as a round number, which gives a plate scale of 5.8 kpc per arcmin. At the cluster center sits a cD galaxy NGC 1399 with an extended stellar envelope, and its elliptical companion NGC 1404. The velocity dispersion is $\sim$ 325 km/s (Ferguson & Sandage 1990) for the Fornax cluster and 200 – 330 km/s for NGC 1399 (Bicknell *et al.* 1989), implying that the virial temperature is relatively similar between the two hierarchical systems.

Extended thermal X-ray emission was detected with *Einstein*, centered on NGC 1399 and up to $\sim 20'$ (Killeen & Bicknell 1988). With *Ginga*, Ikebe *et al.* (1992) detected an X-ray component extending up to $\sim 50'$ from NGC 1399 with β as flat as $0.23^{+0.16}_{-0.23}$. These two facts altogether suggest that there indeed exist a galaxy-scale emission and a cluster-scale emission. The *BBXRT* observation yielded a gas temperature of 0.9–1.1 keV for NGC 1399 (Serlemitsos *et al.* 1993).

Through *ROSAT* observations (Rangarajan *et al.* 1994; Jones *et al.* 1994), the emission was found to be asymmetric around NGC 1399 in such a way that the brightness falls more slowly towards north-east. The obtained temperature is 1.0 – 1.2 keV for a radius range of $1' - 40'$ from NGC 1399. Therefore the gas can be approximated as isothermal except at the very center. These authors report that the total gravitating mass within 125 kpc is $(4.3 - 8.1) \times 10^{12} M_\odot$, and M/L increases from 25–42 at 18 kpc to 47–92 at 110 kpc.

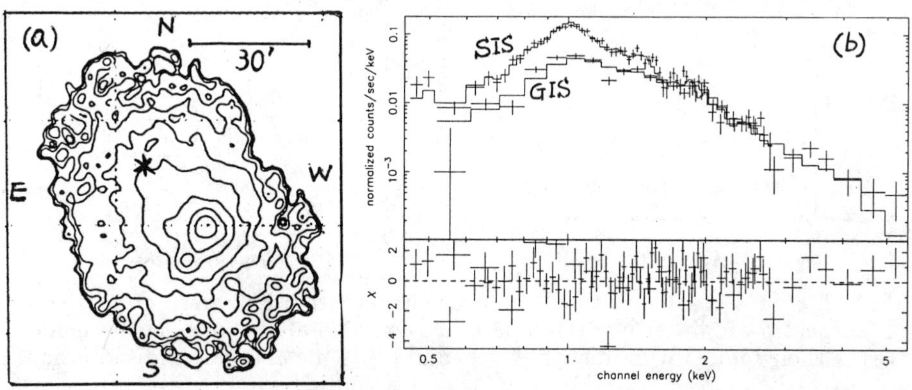

Fig. 3. The *ASCA* data on the Foranx/NGC 1399.
(a) The background-subtracted GIS image which combines three pointings. The brightest part is NGC 1399, and a smaller peak toward SE is NGC 1404.
(b) The combined SIS/GIS spectra for a circular region of 4 arcmin radius, offset from NGC 1399 by 20 arcmin toward NE (marked with * in panel a).

3.2 The *ASCA* observations

The Fornax cluster was observed with *ASCA* during the PV (performance verification) phase with two pointings, and again in the first GO (guest observation; PI=Y. Ikebe) phase with a single offset pointing. Figure 3a shows the GIS mosaic image synthesized from thee partially overlapping fields. The cosmic X-ray background and the non X-ray background have both been subtracted. The X-ray morphology in the central region is essentially the same as observed previously. In addition, diffuse X-ray emission has been detected over the entire map of $70' \times 45'$, thus confirming the *Ginga* detection (Ikebe *et al.* 1992) of a large-scale X-ray emission from the entire Fornax cluster. Particularly, the north-eastern X-ray envelope has been revealed much more clearly than before.

Figure 3b shows an example of combined SIS/GIS spectra for a region offset from NGC 1399 toward north-east. The spectra from the four detectors have been fit jointly and successfully with a common Raymond-Smith model of temperature 1.07 ± 0.06 keV and abundance in the range 0.3–0.6 solar (at 90% confidence). We have also obtained temperature in the range 0.9–1.2 keV from various other regions of Fig.3(a). This is quite consistent with the temperature measurements from *BBXRT* (Serlemitsos *et al.* 1993) and *ROSAT* (Rangarajan *et al.* 1994; Jones *et al.* 1994). In short, the emission can be considered roughly isothermal, within $\sim \pm 15\%$, over the entire image of Fig.3(a).

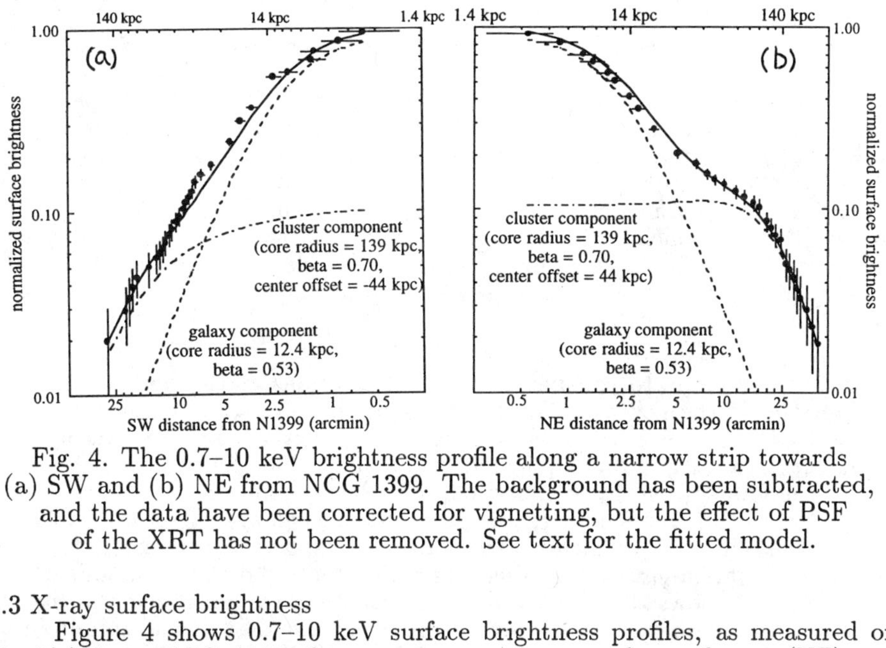

Fig. 4. The 0.7–10 keV brightness profile along a narrow strip towards (a) SW and (b) NE from NCG 1399. The background has been subtracted, and the data have been corrected for vignetting, but the effect of PSF of the XRT has not been removed. See text for the fitted model.

3.3 X-ray surface brightness

Figure 4 shows 0.7–10 keV surface brightness profiles, as measured on Fig.3(a) from NGC 1399 along a narrow strip towards north-east (NE) and south-west (SW). The overall emission size, $r = 30' - 40'$ in projected radius from NGC 1399, agrees with the *Ginga* measurement (Ikebe *et al.* 1992). Furthermore, while the SW profile is featureless, the NE profile exhibits a prominent shoulder structure (or curvature change) over $r = 5' - 20'$. Although the *ASCA* XRT has a complex PSF, this effect is unlikely to be instrumental. Furthermore, since the emission is nearly isothermal (§3.2), this means that the shoulder structure is present not only in the X-ray emissivity but also in the gas pressure, hence in the gravitational potential. We interpret that we have detected both the "galaxy" emission within $r \sim 70$ kpc and the "cluster" emission beyond. This provides one of the first clear X-ray evidences for hierarchical structure in the gravitational potential. The NE-SW asymmetry suggests that the two components have a relative offset from each other.

In order to quantify the two X-ray brightness profiles, we have introduced an empirical double-beta model consisting of "galaxy" component (with suffix

g) and "cluster" component (with suffix c), as

$$X(r) = N_g[1 + (r/a_g)^2]^{-3\beta g+0.5} + N_c[1 + ((r \pm b)/a_c)^2]^{-3\beta c+0.5}. \quad (1)$$

Here X is the surface brightness, r is the *projected* radius from NGC 1399, N is normalization, a is core radius, and b is offset between the galaxy and the cluster. We use $+b$ and $-b$ for the SW and NE profiles, respectively. This particular expression is of purely empirical purpose.

As shown in Fig.4, both profiles have been fit reasonably well with eq.(1) using a common set of parameters; $N_g/N_c = 0.12 \pm 0.03$, $a_g = 2.'1 \pm 0.'3$ (12 ± 2 kpc), $a_c = 24' \pm 2'$ (140 ± 12 kpc), $\beta_g = 0.53 \pm 0.08$, $\beta_c = 0.70 \pm 0.06$, and $b = 7.'6 \pm 1.'5$ kpc (44 ± 9 kpc). Note that N_g and a_g are strongly affected by the PSF, and severely under- and over-estimated, respectively. The obtained cluster/galaxy offset b is consistent with that reported by Jones *et al.* (1994), and the cluster core radius a_c is reasonable for the optical extent of the Fornax cluster.

The derived two beta values are both significantly larger (steeper) than those obtained with *Einstein* ($\beta = 0.41$, Killeen & Bicknell 1988) and *ROSAT* ($\beta = 0.35$, Jones *et al.* 1994) around NG 1399, and with *Ginga* ($\beta \sim 0.23$; Ikebe *et al.* 1992) over the whole cluster. Mathematically this means that a narrower/brighter beta model and a wider/fainter beta model are added together, to give a profile which resembles a single beta model with a considerably flatter slope. Physically, this means that the fainter cluster emission adds to the outskirt of the galaxy emission to make it appear flatter.

3.4 Inferred mass density profiles

We next hope to calculate spatial distributions of the hot gas and the dark matter in the Fornax/NGC 1399 system. However eq.(1) is not sufficient to specify these quantities, because spherical symmetry is broken. Accordingly we here take a rather crude approach, and assume that the two components in eq.(1) are both spherically symmetric by themselves. This ensures a cylindrical symmetry. For further simplicity, we also reset the galaxy/cluster offset parameter b to zero in eq.(1), which restores the overall spherical symmetry. Physical essence of the hierarchical clustering would not be affected by this simplification.

Figure 5 shows the derived distributions of various mass components. In deriving them we have analytically "deprojected" eq.(1) as usual to obtain an expression for the gas mass distribution $\rho(R)$, and utilized the hydrostatic equation (for an isothermal case) of $M_G(< R) = -\frac{k_B TR}{G\mu m_p}\frac{dln\rho}{dlnR}$. Here R is the three-dimensional radius and $M_G(< R)$ is the gravitating mass internal to R. We also assumed that the X-ray emitting gas is isothermal with $kT = 1.0$ keV, and the abundance is 0.5 solar, for both the galaxy and the cluster components. Dashed lines in Fig.5 show distributions when the cluster component [2nd term in eq.(1)] is neglected. Note that the effect of PSF is not removed, so that information within $R \sim 20$ kpc is not correct.

In Fig.5, we clearly see the effect of galaxy/cluster hierarchy in distributions of the gas mass and the total mass. The galaxy/cluster interface is inferred to occur at $R_{IF} \sim 70$ kpc ($12'$), which is well within the cluster core radius of 140 kpc. The total gravitating mass associated with NGC 1399 becomes $M_G = 2 \times 10^{12} M_\odot$, whereas the gas mass and the stellar mass internal to R_{IF} is $M_{gas} = 6 \times 10^{10} M_\odot$ and $M_* = 5 \times 10^{11} M_\odot$, respectively. Thus we have $M_{gas}/M_* \sim 0.1$ and $M/L \sim 20$, while the baryonic fraction becomes $f_B \equiv (M_* + M_{gas})/M_G \sim 0.28$. These agree with the general properties of elliptical galaxies (§1), including the necessity for the DM.

Beyond R_{IF}, the gas mass increases dramatically by inclusion of the cluster component, while M_G asymptotically approaches the value obtained by neglecting the cluster component because M_G is determined only by the local slope in the X-ray surface brightness. At an outer region of the cluster, say $R = 300$ kpc ($52'$), we have $M_G = 1.8\times10^{13} M_\odot$, $M_{gas} = 1.2\times10^{12} M_\odot$, and $M_* = 1\times10^{12} M_\odot$; hence $M_{gas}/M_* \sim 1$, $f_B \sim 0.12$, and $M/L \sim 180$. These values are typical of poor clusters.

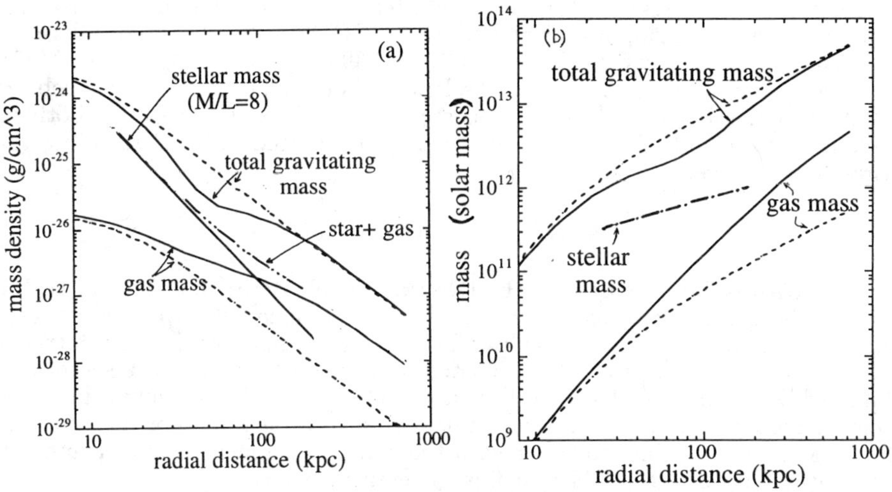

Fig. 5. The inferred differential (panel a) and integral (panel b) distributions of the stellar mass (Killeen & Bicknell 1988), the total gravitating mass, and the hot gas mass, in the Fornax/NGC 1399 system. Dashed lines show results without taking into account the "cluster" component.

3.5 Implication and discussion

Through *ASCA* observations of the Fornax/NGC 1399 system, we have obtained firm evidence for hierarchical clustering both in the hot gas distribution and in the gravitational potential. The galaxy/cluster interface occurs at $R_{IF} \sim$ 70 kpc, and the regions inside and outside R_{IF} are considered to be "galaxy" and "cluster", respectively. This also provides a clear counter-example against the general belief that the excess X-ray surface brightness seen around a cD galaxy results from the cooling flow. We are thus urged to re-examine the problem of central excess emission in other cD clusters from this new viewpoint, particularly for the Centaurus cluster and the Virgo cluster.

The galaxy/cluster interface thus established has allowed us to estimate the DM parameters for the galaxy and the cluster, separately. The baryonic fraction is ~ 0.12 for the cluster, and seems somewhat larger for the galaxy. The DM is less concentrated than the stars, but is clearly more concentrated than the hot gas.

The present results suggest that, in the Fornax/NGC 1399 system and in similar systems, the galaxy emission and the cluster emission add up together to mimic a single β-model-like distribution that can be approximated by $\beta \sim$ 0.3–0.4. For some cD galaxies this may help solving so called β-problem (Bahcall & Lubin 1994), that the value of β obtained through the surface brightness analysis tends to be less than that calculated as a ratio between the hot gas temperature and the optical velocity dispersion. We can further speculate that the β-problem might be caused in some other cases by a presence of a still fainter X-ray emission from the surrounding supercluster potential: such a large-scale faint emission would be difficult to detect, but could give a positive bias to the cluster emission to make it appear considerably flatter than it actually is.

Finally the offset between NGC 1399 and the Fornax center is intriguing. This suggests that NGC 1399 may not be a cD galaxy in a strict sense, as it does not sit at the right bottom of the cluster potential. This further suggests that ram-pressure stripping for the ISM of NGC 1399 has been considerable. The relatively low X-ray luminosity of NGC 1399 (*e.g.* as compared to M87), and a relatively high metallicity of the ICM gas of the Fornax cluster (Ikebe *et al.* 1992; Serlemitsos *et al.* 1993; Jones *et al.* 1994; see Fig.3b), may both be explained as a result of rather efficient mixing of the ISM of NGC 1399 into the intra-cluster space.

4. DISCUSSION AND SUMMARY

ASCA results suggest that the DM exhibits clustering on different scales, *e.g.* on the galaxy scale as well as on the cluster scale. We speculate that a dark halo around an elliptical galaxy in a cluster is not merely an arbitrarily defined geometrical fraction of the cluster dark halo, but instead is a segregated clump of DM with a relatively well defined scale length. We presume that each member galaxy in a cluster is surrounded by such a dark halo of its own, and the dark halo is moving, together with the stars, in a much smoother and less dense distribution of the DM that fills the entire cluster.

We confirm that, among the three mass components (stars, hot gas, and DM), the stars exhibit the strongest concentration towards smaller scales, while the hot gas is most extended. Consequently the stars form a dominant baryonic component on small scales, while on larger scales the hot gas overwhelms stars. The DM is inferred to exhibit an intermediate concentration. The M/L ratio therefore keeps increasing towards larger scales, while the baryonic fraction remains relatively constant over different angular scales.

Does f_B remain constant toward still larger angular scales, and does f_B measured with X-rays in clusters is really representing the whole universe? This is clearly a vital issue, since the value of $f_B = 0.1 - 0.2$ as inferred from X-ray data on many clusters is too large to be reconciled simultaneously with the $\Omega = 1$ belief and the value of $\Omega_b \sim 0.02$ constrained by the big-bang nucleosynthesis argument. One big concern is that the X-ray measured M_G and M_{gas} both increase almost linearly with radius, whereas we do not yet know how to define the cluster boundary. However we have obtained a novel concept as to the boundary of a cD galaxy. We therefore hope to define an X-ray boundary for clusters as well. It might be that the X-ray emission terminates at that radius, whereas the DM distribution keeps extending further out. Then f_B would currently be overestimated. This urges us to challenge a more difficult issue of finding out cluster boundaries through X-ray observations, and examine how the hot gas and the DM behave near the boundary. This will be a subject of a future challenge with *ASCA*.

ACKNOWLEDGEMENTS

The author would like to thank his colleagues at the University of Tokyo and to all the members of the *ASCA* team. His particular thanks are due to Dr. T. Ohashi, Mr. Y. Ikebe, Mr. Y. Fukazawa, Ms. K. Matsushita, Dr. M. Loewenstein, Prof. Y. Suto, and Prof. F. Takahara.

REFERENCES

Allen, S. W., & Fabian, A. C. 1994, MNRAS, 269, 409
Awaki, H., Mushotzky, R., Tsuru, T., Fabian, A.C., *et al.* 1994, PASJ, 46, L65
Bahcall, N. A., & Lubin, L. M. 1994, ApJL, 426, 513
Bicknell, G., Carter, D., Killeen, N., & Bruce, T. 1989, ApJ, 336, 639
Binney, J., & Cowie, L. L. 1981, ApJ, 247, 464
Böhringer, H., Briel, U. G., Schwarz, R. A., Voges, W., Hartner, G., & Trümper, J. 1994, Nature, 368, 828
Canizares, C. R., Fabbiano, G., & Trinchieri, G. 1987, ApJ, 312, 503
Fabian, A. C. 1994, ARA&A 32, 277
Fabian, A. C., *et al.* 1994, ApJL, in press
Fabricant, D., & Gorenstein, P. 1983, ApJ, 267, 535
Ferguson, H., & Sandage, A. 1990, AJ, 100, 1
Forman, W., Jones, C., & Tucker, W. 1985, ApJ, 293, 102
Forman, W., Jones, C., David, L., Franx., M., *et al.* 1993, ApJ, 418, L55
Fukazawa, Y., Ohashi, T., Fabian, A. C., *et al.* 1994, PASJ, 46, L55
Ikebe, Y., Ohashi, T., Makishima, K., Tsuru, T., *et al.* 1992, ApJL, 384, L5
Ikebe, Honda, H., Makishima, K., Murakami, T, Ohashi, T., Mushotzky, R. F., & Yamashita, K. 1994, in Proc New Horizon of X-ray Astronomy, eds. F. Makino & T. Ohashi (Tokyo: Universal Academy Press), p.545
Jones, C., Stern, C., Forman, W., Breen, J., David, L., Tucker, W., Franx, M., & Fabina, A. 1994, ApJ, in press
Killeen, N., & Bicknell, G. 1988, ApJ, 325, 165
Kohmura, Y., Fukazawa, Y., Ikebe, Y., Ishisaki, Y., *et al.* 1993, SPIE, 2006, 78
Koyama, K., Takano, S., & Tawara, Y. 1991, Nature, 350, 135
Loewenstein, M., Mushotzky, R. F., Awaki, H., *et al.* 1994, ApJL, in press
Matsumoto, H., Tsuru, T., Awaki, H., & Koyama, K. 1994, in Proc New Horizon of X-ray Astronomy, eds. F. Makino & T. Ohashi (Tokyo: Universal Academy Press), p.511
Matsushita, K., Makishima, K., Awaki, H., *et al.* 1994, ApJL, in press
Stewart, G., Canizares, C., Fabian, A., & Nulsen, P. 1984, ApJ, 278, 536
Rangarajan, F, Fabian, A., Forman, W., & Joes, C. 1994, MNRAS, in press
Roberts, M. S., Hogg, D. E., Bregman, J. N., Forman, W. R., & Jones, C. 1991, ApJS, 75, 751
Serlemitsos, P. J., Loewenstein, M., Mushotzky, R. F., Marshall, F. E., & Petre, R. 1993, ApJ, 413, 518
Takano, S., Awaki, H., Koyama, K., Kunieda, H., *et al.* 1989, Nature, 340, 289
Tanaka, Y., Inoue, H., & Holt, S. S. 1994, PASJ, 46, L37

X-RAY CONSTRAINTS ON THE INTRINSIC SHAPE OF THE LENTICULAR GALAXY NGC 1332

David A. Buote and Claude R. Canizares
Center for Space Research, MIT 37-241, Cambridge, MA 02139.
dbuote@space.mit.edu

ABSTRACT

We have analyzed ROSAT PSPC X-ray data of the optically elongated S0 galaxy NGC 1332 with the purpose of constraining the intrinsic shape of its underlying mass. The X-ray isophotes are elongated with ellipticity 0.10 – 0.27 (90% confidence) for semi-major axes 75″ – 90″ and have orientations consistent with the optical isophotes (ellipticity ~ 0.43). Using (and clarifying) the geometric test for dark matter introduced by Buote & Canizares (1994), we determined that the hypothesis that mass-traces-light is not consistent with the X-ray data at 68% confidence and marginally consistent at 90% confidence independent of the temperature profile of the gas; similar results are obtained considering the effects of possible gas rotation and emission from discrete sources; these same constraints apply to MOND. Detailed modeling of the underlying mass distribution yields $\epsilon_{mass} = 0.47 - 0.72(0.31 - 0.83)$ at 68% (90%) confidence for isothermal and polytropic models. Similar results are obtained when the dark matter is fit directly using the known distributions of the stars and gas. When possible rotation of the gas and emission from discrete sources are included the constraints on ϵ_{mass} are substantially weakened.

1. INTRODUCTION

The distribution of intrinsic shapes of galactic halos is of cosmological importance since it can be predicted from cosmological N-body simulations. Cold Dark Matter (CDM) simulations predict dark halos that are flatter ($\epsilon \sim 0.5$) than inferred from analysis of observed optical isophotes ($\epsilon \sim 0.3$) of ellipticals (e.g., Dubinski 1993). Since the observed stars may not trace the shape of the dark mass, this is not necessarily a discrepancy: the shape of the total gravitating matter is required to compare to the simulations. Unfortunately, reliable constraints on the shapes of dark halos exist for only a few galaxies (Sackett et al. 1994).

High quality X-ray data from ROSAT and future X-ray satellites provide the best hope for determining whether early-type galaxies, like spirals, possess large amounts of dark matter (e.g., Ashman 1992). Given the possible uncertainties involved in computing temperature gradients (e.g., Fabbiano 1989; Trinchieri et al. 1994), an X-ray method to detect dark matter that takes advantage of the high quality spatial data of current X-ray satellites but does not require detailed temperature information is of great importance. As we discuss below, the shapes of the X-ray isophotes yield robust constraints on dark matter in early-type galaxies that are independent of the temperature profile of the gas.

We have analyzed ROSAT PSPC X-ray data of the S0 galaxy NGC 1332 for the purpose of constraining the intrinsic shape of its underlying mass. Our treatment closely parallels that of Buote & Canizares (1994) who analyzed the

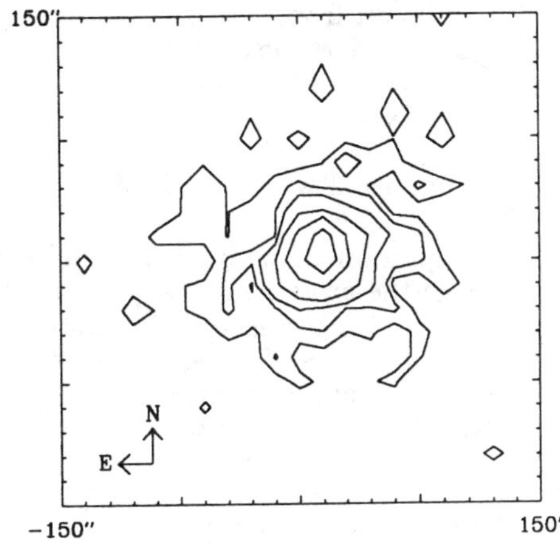

Fig. 1. PSPC Contours of NGC 1332 (0.4 - 2.4 keV)

shape of the elliptical galaxy NGC 720 using ROSAT PSPC data. NGC 1332 was selected because, along with NGC 720, it has the largest known X-ray flux from *Einstein* (Fabbiano et. al. 1992) for an early-type galaxy that is elongated in the optical ($\epsilon \sim 0.43$), relatively isolated from other large galaxies, has an angular size ($\sim 7'$) substantially larger than the PSPC PSF, and is likely to be dominated by emission from hot gas (Kim et al. 1992).

2. DATA ANALYSIS

NGC 1332 was observed for 25.6 ks on August 13-14, 1991 with the ROSAT Position Sensitive Proportional Counter (PSPC). To prepare the X-ray image for spatial analysis we (1) excluded time intervals where the background was anomalously high, (2) corrected for exposure variations and telescopic vignetting, (3) identified and removed point sources embedded in the galactic continuum emission, and (4) subtracted the background. The reduced image is shown in Figure 1.

The ellipticity of the surface brightness is computed by taking quadrupole moments of the X-ray surface brightness in elliptical apertures of different semi-major axis length. The isophote shapes are best constrained for semi-major axes $a = 75'' - 90''$ where $\epsilon_M = 0.10 - 0.27$ (90% confidence). The position angles of the X-ray isophotes for different a are consistent with each other and the optical value of 115° N-E within the estimated 95% uncertainties.

In addition to the ellipticity we also computed the azimuthally averaged radial profile of the surface brightness. Fitting the standard β-model yields core radii $a_x = 0.004'' - 3.4''$ and $\beta = 0.44 - 0.50$ both at 90% confidence. Both the radial profile and the ellipticity are used to constrain the detailed models of the intrinsic shape (see below).

The spectrum is not well constrained by the PSPC data. A single - temperature plasma with Galactic column density fits the data well, but the temperature and metallicities are not well determined; e.g., $T_{gas} = 0.40 - 0.76$ keV and abundances = 10% − 107% solar (90% confidence). Radial temperature gradients are not required by the PSPC spectrum, but large gradients $|d \ln T_{gas}/d \ln R| < 0.35$ (99% confidence) are consistent with the spectrum. Although simple models for emission from discrete sources in NGC 1332 can be ruled out as the sole source of the X-ray emission, the PSPC spectrum does not tightly constrain the relative flux of hot gas and discrete emission; generally as high as equal fluxes in the two components are allowed by the spectrum.

3. Geometric Test for Dark Matter

If the gas obeys the equation of hydrostatic equilibrium, then the gas density, temperature, and thus the gas emissivity (j_{gas}) must trace the shape of the gravitational potential (Φ). This correspondence between the shapes of j_{gas} and Φ is the basis for the geometric test for dark matter.

To ascertain whether light traces mass, we first obtain Φ from the optical surface brightness distribution Σ_L. We make the assumption that the three-dimensional luminosity density, j_L, is axisymmetric with its symmetry axis inclined by angle i with respect to the line of sight. With the additional general assumption that j_L may be represented as a finite series of spherical harmonics, Palmer (1994) has shown that j_L may be uniquely and analytically deprojected from Σ_L. For mass tracing light the mass density is just $\rho_L \propto j_L$ and the corresponding potential Φ_L follows from Poisson's equation. Setting $\Phi = \Phi_L$ we have the potential under the assumption mass traces light. By applying the same assumptions for j_L to j_{gas}, then j_{gas} may be uniquely and analytically deprojected from the X-ray surface brightness Σ_X. If Φ and j_{gas} trace the same shape , then mass tracing light is a suitable description of the galaxy. *Inconsistency of the shapes of Φ and j_{gas} signals the presence of dark matter, independent of the temperature profile of the gas and without requiring any model fitting.* This test applies to MOND as well.

The dotted lines in Figure 2. represent the X-ray isophote shapes predicted if mass traces light in NGC 1332; actually they are the mean values obtained from 1000 Monte Carlo simulations. The solid lines are the best-determined values computed from the PSPC image. The predicted ellipticity if mass follows light is not consistent with the X-ray data at the 68% confidence level, but is only marginally inconsistent at 90% confidence. Considering the maximum effects of plausible rotation of the gas and emission from stellar sources the descrepancy is only significant at the 68% level.

4. Detailed Hydrostatic Modeling

Explicitly solving the equation of hydrostatic equilibrium, we analyzed the shape and profile of the mass using the technique of Buote & Canizares (1994) derived from the work of Binney & Strimple (1978). By employing a wide class of spheroidal models we constrained the ellipticity of the underlying mass to be $\epsilon_{mass} \sim 0.5 - 0.7(0.3 - 0.8)$ at 68% (90%) confidence assuming the rotation of the gas and emission from discrete sources are negligible; a model where the mass traces the optical light is easily ruled out for an isothermal gas, but polytropic models, though inconsistent at the 68% confidence level, are marginally consistent with the data at 90% confidence. Considering the possible effects of rotation of the gas like that of the stars and the effects of emission from discrete sources, the constraints on ϵ_{mass} are weakened considerably. For all the isothermal models considered, the 90% confidence estimate of the total mass out to $a = 450''$ is $M_{tot} = (0.38 - 1.7) \times 10^{12} M_{\odot}$ assuming $D = 20$ Mpc corresponding to blue mass-to-light ratio $\Upsilon_B = (31.9 - 143)\Upsilon_{\odot}$.

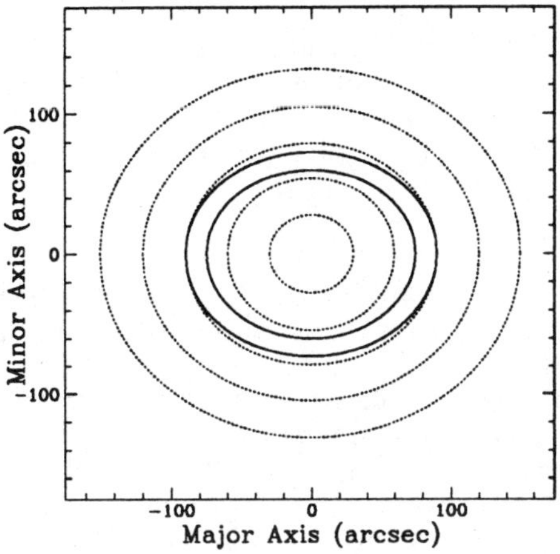

Fig. 2. Geometric Evidence for Dark Matter

We estimated the observed stellar and X-ray mass profiles from their observed surface brightness distributions and then fit the dark matter directly. For models where $M_{DM}/M_{stars} > 10$, the dark matter has the same shape as for the above composite models. For smaller values of M_{DM}/M_{stars} the lower limit on ϵ_{mass} falls by 0.03 but the upper limit is indeterminate (90% confidence). We estimate a lower limit $M_{DM}/M_{stars} > 3$ (90% confidence) assuming the dark matter extends out to $a = 450''$; these results are not sensitive to $a > 250''$.

5. CONCLUSIONS

There is marginal evidence for dark matter in NGC 1332 with a flattened, extended dark halo with $\epsilon_{DM} \sim 0.3 - 0.8$ (90% confidence). Although not as strong a case, these findings agree with our results for the E4 galaxy NGC 720 (Buote & Canizares 1994) where we found $\epsilon_{DM} \sim 0.5 - 0.7$ (90% confidence). Hence, the X-ray emission from NGC 720 and NGC 1332 may suggest that the dark halos in early-type galaxies are substantially flattened in accord with results from N-body simulations (e.g., Dubinski 1993).

REFERENCES

Ashman, K. M. 1992, PASP, 104, 1119
Binney, J., & Strimple O. 1978, MNRAS, 187, 473
Buote, D. A., & Canizares, C. R. 1994, ApJ, 427, 86
Dubinski, J. 1993, ApJ, submitted
Fabbiano, G. 1989, ARA&A, 27, 87
Fabbiano, G, Kim, D.-W., & Trinchieri, G. 1992, ApJS, 80, 531
Kim, D.-W., Fabbiano, G, & Trinchieri, G. 1992, ApJS, 80, 645
Palmer, P. L. 1994, MNRAS, 266, 697
Sackett, P., Rix, H-W, Jarvis, B. J., & Freeman, K. C. ApJL, submitted
Trinchieri, G., Kim, D.-W., Fabbiano, G., & Canizares, C., 1994, ApJ, 428, 555

CONSTRAINTS ON DARK MATTER IN CLUSTERS & GALAXIES FROM THE SHAPES OF X-RAY IMAGES

Claude R. Canizares & David A. Buote
Department of Physics & Center for Space Research
MIT Room 37-241, Cambridge MA 02139-4307
crc@space.mit.edu

ABSTRACT

We summarize an ongoing program to use the shapes of X-ray images of clusters and galaxies to gain information about the dark matter contained in those systems. Studies of five Abell clusters and one early type galaxy are briefly described (another is reported elsewhere in this volume). In each of these systems, the stellar distribution appears very flattened. For the clusters, the X-rays also appear flattened, but less than expected if light traces mass. For the galaxy, the X-ray contours are flatter at larger radii than expected if light traces mass. This is "geometrical evidence" for dark matter in both clusters and early type galaxies. Detailed modeling constrains the quantity and distribution of the dark matter. Full details can be found in several published papers.

1. INTRODUCTION & METHOD

Binney & Strimple (1978; Strimple & Binney 1979) showed that X-ray images can reveal the intrinsic shapes of the underlying mass distribution. Our first attempts used *Einstein* data on two early type galaxies (White 1987, White & Canizares 1987), but the results were inconclusive. We had greater success with Einstein cluster data (Buote & Canizares 1992, hereafter BC92), and with improved quality images of the galaxies obtained with ROSAT (Buote & Canizares 1994 [BC94], 1995 [BC95], and this

Our method exploits the fact that hot gas in hydrostatic equilibrium in a gravitational potential traces the shape of the potential *independent of the temperature or possible temperature gradients* (see BC95 for a complete discussion). Since the X-ray surface brightness traces the projected hot gas density to very good approximation, it also traces the projected gravitational potential. Gas in early-type galaxies, where sound crossing times are typically $\approx 10^7$ yr, is almost certainly in hydrostatic equilibrium. The assumption is less secure in clusters, so we have confined ourselves to the inner regions ($\leq 0.6h_{80}$ Mpc) of smooth, apparently well relaxed clusters, where the sound crossing times are $\leq 10^9$ yr. Recently Buote & Tsai (1995) show that the assumption of hydrostatic equilibrium and the method described below are valid for evolved clusters. They do so by applying the technique to the model cluster from an N-body plus hydro simulation of Katz & White (1993). They also find that assumption begins to break down for the same cluster at redshifts $z \geq 0.3$, when it is still in the process of forming from subclusters.

For a given X-ray image, we compute the radial profile and ellipticity within some radius for comparison to our models. Our primary method of determining ellipticity is to compute quadrupole moments of the distribution of X-ray photons. We then explore a wide range of possible mass models, for which we compute potentials and gas distributions. We vary any free parameters to give results throughout the 90% uncertainty ranges of our observed profile and ellipticity.

One key parameter is the gas temperature, T. Unlike conventional X-ray studies of binding mass (e.g. see Mushotzky, these proceedings), we are not sensitive to the actual value of the temperature. Rather we adjust the ratio $\Gamma = \mu\phi_0/kT$, where k is Boltzmann's constant, μ is the mean mass per gas particle, and ϕ_0 is the central value of the potential. Thus Γ is the ratio of specific potential to kinetic energy in the gas, and can be well constrained by fitting the radial profile alone. We explore isothermal models as well as those with various temperature gradients, within a range allowed by the observations. Of particular interest are models in which mass follows light. In this case, Γ is the only free parameter.

It is useful to keep in mind the general properties of the gravitational potential for ellipsoidal mass distributions, the simplest being a spheroidal shell. As shown in Figure 2.9 of Binney & Tremaine (1987; also Chandrasekar 1969), the isopotential surfaces outside such a mass shell are themselves ellipsoidal surfaces with the same foci as the the mass shell. Such confocal ellipsoids rapidly become rounder with increasing semi-major axis. The referenced figure shows a mass shell with $\epsilon = 0.7$; the potential just outside has roughly the same ellipticity, but ϵ falls to 0.24 (0.06) at 1.5 (2.6) times the semi-major axis.

2. CLUSTERS OF GALAXIES

BC92 reports our results for the central regions of five Abell clusters in which the galaxy distribution is highly flattened, with typical ellipticities of $\epsilon \approx 0.5$ (Carter & Metcalf 1980). These are A401, A426 (Perseus), A1656 (Coma), A2029 and A2199. We believe that substructure is not an important consideration for our analysis on these clusters.

Using archival images from the *Einstein* Observatory we find that the X-ray countours are indeed elongated with position angles that agree with those of the galaxy distributions to within the uncertainties, typically $\pm 15^o$. But the X-ray ellipticities are $\approx 0.1 - 0.2$, much smaller than those of the galaxies. From our qualitative description above, we would expect ellipticities closer to those of the galaxy distributions if the galaxies traced the underlying mass, because our measurements are limited to the inner regions. Therefore it appears that the underlying mass distribution must be rounder than the galaxy distribution. Our models confirm this impression. They rule out the possibility that mass follows light and constrain the mass distributions to have $\epsilon \approx 0.1 - 0.3$. (We are currently examining ROSAT images that appear to confirm these results for all but A2029, where the *Einstein* IPC point spread function seems to have affected our analysis).

We conclude that the shapes of the X-ray images require that the clusters contain dark matter. In general, we find that the galaxies trace neither the matter nor the potential in clusters. Following the statistical arguments of Plionis *et al.*(1991), it is possible that results for these apparently highly flattened systems can be generalized to most clusters.

3. EARLY TYPE GALAXIES

Our data on two early type galaxies, NGC 720 and NGC 1332, come from the ROSAT PSPC and we are acquiring higher resolution images with the ROSAT HRI (BC94, BC95, these proceedings). Both are relatively well isolated galaxies at distances of $\approx 20h_{80}Mpc$, and both have relatively slow rotation speeds. Because these images subtend only a few arc minutes on the sky, we take care to include the effects of the PSPC point spread function in our models. We also address the possible effects of rotation and constributions from discrete stellar X-ray sources. Here we concentrate on the results for NGC 720 (which is slowly rotating and reasonably well isolated); those for NGC 1332 are given by us elsewhere in these proceedings.

While the stars in NGC 720 have ellipticity $\epsilon \approx 0.4$, the X-ray image shows $\epsilon \approx 0.2 - 0.3$. But the X-rays show ellipticity extending at least 100" and possibly farther, whereas the stellar distribution is very concentrated, with core radius of 4". Therefore, a model in which mass follows light would have very nearly round X-ray isophotes in contrast to what we observe. This in itself argues for dark matter with an extended, flattened distribution. Further evidence for dark matter comes from the relative position angles of the X-ray and visible light distributions, which appear to differ by $\approx 30^o \pm 15^o$ (90%confidence). Preliminary examination of our HRI image suggests that the X-ray isophotes do align with the optical at small radii and twist to match the PSPC at larger radii.

We have explored a wide variety of models for the mass distribution of NGC 720, all of which involve extended and flattened dark matter halos. By combining the potential of the dark matter with that of the observed stellar distribution we can set a lower limit on the mass ratio of dark to visible matter from the shape alone: below this value there would be too little dark matter so the stars would dominate and give a potential that is too round to account for the X-ray image. We estimate that the halo must contain at least four times the mass in stars, with a prefered value at least twice that large. For a stellar mass to light ratio of 10, this implies a total mass to light ratio of at least 50 for the galaxy.

Our findings for early type galaxies are consistent with the recent studies a polar ring galaxy of Sackett *et al.*(1994).

4. CONCLUSIONS

Both the clusters and the early type galaxies show evidence for the presence of substantial amounts of dark matter. We stress that these results depend only on a comparison of the shapes of the X-ray isophotes and the distribution of stars, and not on any dynamical variable like X-ray temperature or stellar velocity dispersion. For that reason, we consider this to be "geometrical evidence" for dark matter in these systems. We believe that modifications of Newtonian dynamics, like those explored by Milgrom (1986 and references therein) to give flat rotation curves without dark matter, could not obviate the need for dark matter (see BC94).

Several authors have reported N-body simulations of cluster formation (*e.g.*, Katz & White 1993) and galaxy formation (*e.g.*, Dubinsky & Carlberg 1991, Dubinsky 1994). These yield flat, generally triaxial dark matter distributions that are qualitatively consistent with our results.

ACKNOWLEDGEMENTS

This work was supported in part by NASA Grant NAG5-1656, NASGW-

2681 (through subcontract SVSV2-62002 from Smithsonian Astrophysical Observatory), and NAS8-38249.

REFERENCES

Binney, J. & Strimple, O. 1978, MNRAS, 185, 227
Binney, J. & Tremaine, S. 1987, Galactic Dynamics (Princeton U. Press)
Buote, D. & Canizares, C. 1992, ApJ, 400, 385 (BC92)
Buote, D. & Canizares, C. 1994, ApJ, 427, 86 (BC94)
Buote, D. & Canizares, C. 1995, ApJ (in press; BC 95)
Buote, D. and Tsai, J. 1995, ApJ (in press)
Carter, D. & Metcalf, N. 1980, MNRAS, 191, 325
Chandrasekar, S. 1969, Ellipsoidal Figures of Equilibrium (Yale U. Press)
Dubinski, J. 1994, ApJ, 431, 617
Dubinski, J. & Carlberg R. 1991, ApJ, 378, 496
Katz, N. & White, S.D.M. 1993, ApJ, 412, 455
Milgrom, M. 1986, ApJ, 302, 617
Plionis, M., Barrow, J.D. & Frenck, C.S. 1991, MNRAS 249, 662
Sackett, P., Rix, H-W., Jarvis, B.J. & Freeman, K.C. 1994, ApJ, 436, 629
Strimple, O. & Binney, J. 1978, MNRAS, 188, 883
White, J. 1987, BS Thesis, MIT
White, J. & Canizares, C. 1987, BAAS 19, 682

DARK MATTER IN NGC 4636

M. Loewenstein and R. Mushotzky
NASA Goddard Space Flight Center, Code 666, Greenbelt, MD 20771

H. Awaki and H. Matsumoto
Department of Physics, Kyoto University, Sakyo-ku, Kyoto 606-01

K. Makishima and K. Matsushita
Department of Physics, University of Tokyo, 7-3-1 Hongo, Tokyo 113

ABSTRACT

We derive constraints on the dark matter matter distribution in the elliptical galaxy NGC 4636 using the spatially resolved temperature profile from the *ASCA* SIS. We find that NGC 4636 becomes dark matter dominated at roughly the de Vaucouleurs radius; and, the ratios of total integrated mass to light and local nonbaryonic to baryonic matter increase monotonically outward, attaining values of $\approx$150 $M_\odot/L_{B\odot}$ and $\sim$ 80, respectively, at 100 kpc.

1. INTRODUCTION

Is there dark matter in elliptical galaxies and, if so, how much? This question has important implications for our understanding of the structure and formation of these galaxies (*i.e.* if ellipticals form from mergers of spirals they should have comparable dark matter fractions). Optical astromomers have tended to answer the question in the negative since observed central velocity dispersions and surface brightnesses indicate mass-to-light ratios that are constant and consistent with elliptical galaxy stellar populations. However, if its scale height exceeds that of the stars, dark matter may manifest itself only at relatively large radii; and indeed, there is evidence for dark matter in some ellipticals with extended velocity dispersion measurements (Saglia et al. 1992), as well as in other early-type galaxies with extended HI disks, polar rings, or shells.

Analysis of X-ray observations of elliptical galaxies is, in principle, the ideal method for addressing the dark matter issue since most if not all of the brightest ellipticals have quasi-hydrostatic hot gas halos and the problems of unknown stress tensor anisotropy and limited radial coverage that arise in optical studies are obviated. The presence of dark matter was strongly suggested based on analysis of *Einstein* IPC data (Forman et al. 1985, Fabian et al. 1986) and theoretical models that showed the necessity of dark matter for retaining the hot gas in the observed amounts and distributions (Loewenstein & Mathews 1987). However, to directly and securely derive the amount of dark matter requires accurate measurements of both gas density and temperature distributions – since the latter was lacking, considerable uncertainty remained regarding the amount and even existence of dark matter in elliptical galaxies (Trinchieri et al. 1986).

Fortunately, the capability to measure the temperature of the hot gas accurately and with spatial resolution has been realized with the *ROSAT* PSPC and *ASCA* SIS. With these tools the potential to answer the dark matter question once and for all is here. We have used the temperature profile from the *ASCA* SIS to determine the dark matter distribution in the elliptical galaxy NGC 4636 as we now describe.

2. OBSERVATIONS AND SPECTRAL ANALYSIS

The *ASCA* PV phase observation of NGC 4636 was conducted with the SIS in 4-CCD mode and resulted in a net exposure of $\sim$ 40 ks. We have fit the SIS-0 and SIS-1 spectra in six concentric annuli, centered on NGC 4636, that have approximately equal X-ray fluxes with an absorbed two-temperature Raymond-Smith thermal plasma (R-S) model. The best-fit temperatures and symmetrized 90% confidence errors for the primary component are displayed in the figure below (for fits to SIS-0 spectra – best fits to SIS-1 spectra are nearly identical). We emphasize the accuracy of the temperature measurements: the temperature derived from the ratio of the H-like and He-like lines of Mg is consistent, within 4%, with that obtained by fitting the entire spectrum, and is also consistent with the temperature of the continuum beyond 2 keV. The temperatures we derive are in excellent agreement with the $Z = 0.2$ solar model of Trinchieri et al. (1994) based on PSPC analysis except that the spatial resolution of *ASCA* is too crude to detect the gradient inside $3'$ (see below).

3. DARK MATTER CONSTRAINTS – METHOD

We obtain dark matter constraints by deriving temperature (and also velocity dispersion) profiles for various dark matter models, under the assumptions of spherical symmetry and hydrostatic equilibrium, and comparing the results with the observations. The stars are modeled by a de Vaucouleurs distribution with a core, the gas by

$$\rho_g(r) = \rho_{og}\left(1 + \frac{r^2}{{a_g}^2}\right)^{-3\beta/2}, \qquad (gas)$$

and the dark matter by

$$\rho_d(r) = \rho_{od}\left(1 + \frac{r^2}{{a_{1d}}^2} + \frac{r^4}{{a_{2d}}^4}\right)^{-1}. \qquad (dm)$$

The stellar and gas parameters are fixed by published optical and X-ray observations from Bender et al. (1992), Lauer (1985), Trinchieri et al. (1994), and Trinchieri et al. (1986). We define the variable parameters of the dark matter model to be α – the ratio of dark to luminous matter interior to 100 kpc, δ – the dark matter scale-length (a_{1d}) in units of R_e, and γ – the dark matter cutoff scale (a_{2d}) in units of R_e.

4. DARK MATTER CONSTRAINTS – RESULTS

The distribution of emission-weighted kT in our favored model ($\alpha = 12.5$, $\delta = 1.0$, $\gamma = 5$), averaged over the same six projected annuli that spectra were extracted from, along with the observed SIS-0 and PSPC temperatures are shown in the figure below. The temperatures in the model for the inner two annuli are lower than those derived from the SIS; however beam dilution is a strong effect if the gas is not isothermal as these annuli are only $\sim 1.3'$ wide, which corresponds to the 68% enclosed energy radius. The PSPC kT measurements indicate that there is, in fact, a significant temperature gradient inside $3'$ that is reproduced by the model. Note that this gradient is not a result of a cooling flow, but of a transition from a stellar dominated gravitational potential to one dominated by dark matter.

Because of the spatially resolved, unambiguous kT measurements, the range of acceptable model parameters is small, and the distribution as well as the total amount of dark matter is well-constrained. Models within $3\Delta kT$ of all four outermost SIS measurements, where ΔkT is the 90% temperature uncertainty have $\delta = 0.5$-2.0 and $\alpha = 11.0$-13.5 ($M/L_B = 126$-153) and $\rho_{dark}/\rho_{lum} = 47$-$111$ at 100 kpc. Models with $\delta << 1$ yield velocity dispersion profiles that exceed the optical observations, while if $\delta >> 1$, the model temperature profiles are not sufficiently isothermal to match the SIS data.

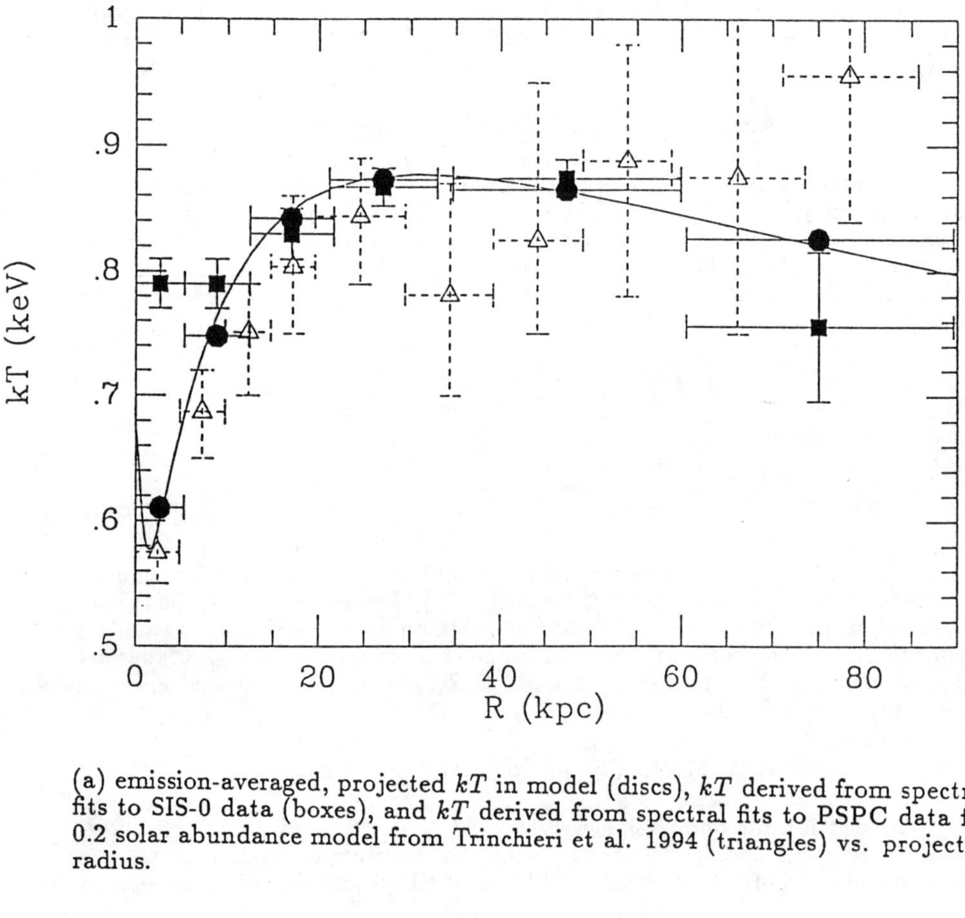

(a) emission-averaged, projected kT in model (discs), kT derived from spectral fits to SIS-0 data (boxes), and kT derived from spectral fits to PSPC data for 0.2 solar abundance model from Trinchieri et al. 1994 (triangles) vs. projected radius.

5. CONCLUSIONS

Using the temperature profile for the hot ISM derived from spatially resolved X-ray spectroscopy with the *ASCA* SIS, in conjunction with the results of analysis of higher spatial resolution PSPC data by Trinchieri et al. (1994), we infer the presence of a non-luminous halo in the elliptical galaxy NGC 4636. We find that NGC 4636 becomes dark matter dominated at roughly the de Vaucouleurs radius; and, the ratios of total integrated mass to light and local non-baryonic to baryonic matter increase monotonically outward, attaining values of $\approx$150 $M_\odot/L_{B\odot}$ and $\sim$ 80, respectively, at 100 kpc. The derived mass profile is consistent with the stellar velocity dispersion, and reproduces the observed temperature gradient without introducing a cooling flow (*i. e.* the gradient results from the transition from a stellar dominated gravitational potential to one dominated by dark matter). These numbers fit in well with the standard picture of dissipative star formation occuring within an extended nonbaryonic halo; the total baryon fraction, 0.07, is consistent with standard big bang nucleosynthesis and $\Omega = 1$.

REFERENCES

Bender, R., Burstein, D., & Faber, S. M. 1992, ApJ, 399, 462

Fabian, A. C., Thomas, P. A., Fall, S. M., & White III, R. E. 1986, MNRAS, 221, 1049

Forman, W., Jones, C., & Tucker, W. 1985, ApJ, 293, 102

Lauer, T. R. 1985, ApJ, 292, 104

Loewenstein, M., & Mathews, W. G. 1987, ApJ, 319, 614

Saglia, R. P., Bertin, G., & Stiavelli, M. 1992, ApJ, 384, 433

Trinchieri, G., Fabbiano, G., & Canizares C. R. 1986, ApJ, 310, 637

Trinchieri, G., Kim, D.-W., Fabbiano, G., & Canizares C. R. 1994, ApJ, 428, 555

THE EDGE OF DARKNESS: MEASUREMENTS OF THE CD GALAXY IN A665

Eric J. Gaidos
37-618B, Massachussetts Institute of Technology, Cambridge, MA 02139.
ejgaidos@space.mit.edu

ABSTRACT

Combined optical and X-ray imaging were used to measure the distribution of light and gravitating mass in the core of the cluster of galaxies Abell 665. Deep R-band imaging was obtained to measure the extent and shape of the luminous halo around the cD galaxy in A665. The surface brightness profile obeys a power law with index 1.4 out to a detection limit of 28 mag/sq.arc.sec. and a distance of 500 h^{-1} kpc, typical of a cD galaxy. The ellipticity was found to rise to 0.4 at 10 H^{-1} kpc and remain at this value out to the detection limit. The distibution of cluster galaxies around the cD galaxy exhibits no measurable concentration and is apparently much less relaxed. The X-ray surface brightness profile of the intracluster gas emission was found by least-squares fitting of a model to a 0.4-2 keV X-ray image obtained by the ROSAT satellite and PSPC detector. While the inner ellipticity of the X-ray emission is consistent with the shape of the optical emission, the shape of the X-ray emission is significantly rounder at larger radii, suggesting that the optical emission is tracing the mass rather than the gravitational potential in the cluster core.

1. INTRODUCTION

The giant galaxies in the center of many galaxy clusters occupy a special niche in linear scale between individual galaxies in which luminous material is plentiful and structure on larger scales, where dark matter apparently dominates. Surveys of cD galaxies by Schombert (1988) and Porter *et al* (1991) uncovered correlations between the gross characteristics of the cD galaxies and the clusters which they inhabit. The degree to which the morphology of cD galaxies is relaxed or 'smooth' indicates whether the formation mechanism is the merging of smaller objects or the initial collapse of the inner cluster. This can also be used to indirectly characterize the power spectrum of the primordial fluctuations. The extended, luminous halos of cD galaxies, up to 1 Mpc across, occupy parts of the cluster in which dark matter dominates the dynamics and for which indirect dynamical measurements are available by mapping of the X-ray emission from hot gas in hydrostatic equilibrium with the cluster potential. Thus in theory it should be possible to compare the projected optical surface brightness and the mass surface density inferred from the potential and test simple models of the mass-to-light ratio in the vicinity of the cD galaxy. Previous research based on optical line-of-sight velocity dispersions has found M/L values of $10^3 - 10^4$ in the outer halos of cD galaxies, ruling out the hypothesis that the dark matter consists of low-mass stars (Uson *et al*, 1991; Uson & Boughn, 1991). Furthermore, Schombert (1988) showed that the surface brightness profiles of cD halos were inconsistent with a constant M/L ratio if one assumes cluster velocity dispersions of $\sim$ 1000 km/s. This evidence suggests that a transition from a 'normal' galactic mix of luminous and dark matter to a cluster mix dominated by dark matter occurs in the halo.

Thus characterizating the structure of cD halos and comparing this to the cluster potential may tell us something about the formation of the galaxies and their host clusters and their subsequent dynamics. The cluster Abell 665 was selected because of its extreme richness and high X-ray luminosity.

2. OBSERVATIONS

The optical observations were performed with a Tektronix 2048^2 CCD at the 1.3 m McGraw-Hill Telescope operated by the MDM (Michigan-Dartmouth-MIT) consortium on Kitt Peak. A series of short exposures totalling 5.5 hours through a standard KPNO R filter were obtained and flattened with concurrent night sky images. The images were co-added into a single 10 arc-minute square image with a 1σ photon noise limit of 30 mags/sq.arc.sec and a 4σ limiting magnitude of $R = 24.2$. The halos around bright stars were removed using a model fit to the point spread function of a bright star image. The clean image is shown in Fig. 1(a). The cD galaxy is markedly flattened and its extended halo is readily apparent.

The X-ray imaging was performed by the Roentgen X-ray Satellite (ROSAT) with the proportional charged particle (PSPC) detector during the AO-2 phase. The effective exposure time in the central 40 arc-minute field of view is 38 ksec. The broad-band PSPC image was retrieved from the HEARSC-Legacy database maintained at NASA Goddard, and corrected using the supplied exposure map. A logarithmic X-ray countour plot is shown in Fig. 1(b). The X-ray isophotes exhibit considerable flattening, twist and centroid shift and the emission peak, centered on the cD galaxy, is offset from the rest of the cluster X-ray centroid.

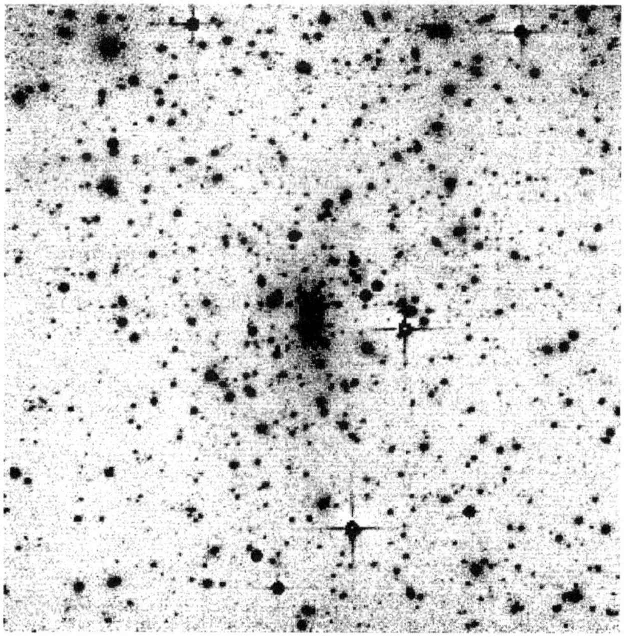

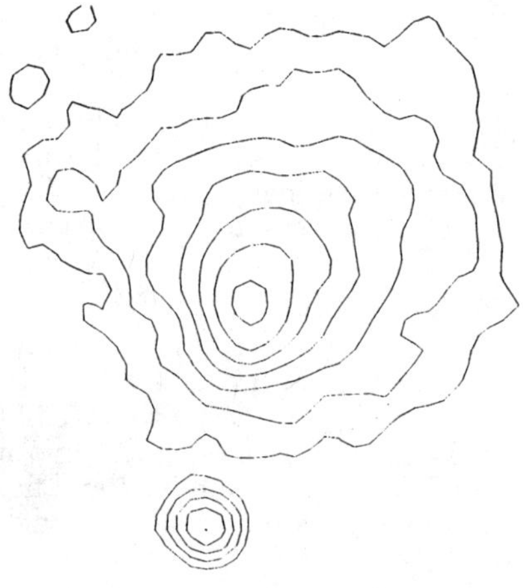

Fig. 1. (a) processed R-band image of A665; (b) X-ray contour plot.

3. ANALYSIS

I developed the routine NEEDLE to measure the radial profile and shape of the diffuse optical halo around the cD galaxy in the presence of bright stars and galaxies. The algorithms are based on previous work by Uson *et al* (1991) and bin the pixels into elliptical annuli to suppress noise, eliminate contaminating objects and find the mean background using an iterative median-finding technique. The semi-major axis of each bin is fixed, but the quadrapole moments of the background surface brightness profile are used to iteratively calculate the ellipticity and position angle. Thus the routine will iterate until the shapes of the elliptical annuli roughly reflect the isophotes. The routine was tested using Monte Carlo simulations of cD galaxy images with Poisson noise and random contaminating objects. I find that the cD halo in A665 has a smooth power law profile with index of 1.4 to the limit of detection at 500 h^{-1} kpc. The ellipticity rises to approximately 0.4 at $20h^{-1}$ kpc and stays flat to the detection limit.

The X-ray data suffers from limited resolution and signal-to-noise rather than contamination. I performed a least-squares fit of a two-dimensional model to the surface brightness profile. The model included a 'beta-model' radial profile with a central ellipticity that decays as a softened power law with radius. Centroid shift and isophotal twist are also included to achieve a good fit. Although there are twelve separate parameters in the model, only the three parameters governing shape are interpreted. A core ellipticity of 0.45, core radius of 30 h^{-1} kpc, and a power law index of 2.4 were found. This suggests that the X-ray isophotes, and by hypothesis the gravitational potential, is rounder than the outer parts of the cD halo. This is inconsistent with a picture where the light is a dynamical tracer and follows the shape of the gravitational potential, but is instead suggests a scenario where the light traces the gravitating mass (Binney & Tremaine, 1987). However, Monte Carlo simulations of the X-ray data reveal that the shape parameters are poorly determined. Furthermore, the X-ray morphology of the cluster suggests that the inner regions are not dynamically relaxed (D. Buote, private communication) and it is possible that the hydrostatic equilibrium assumption is violated.

4. CONCLUSIONS

Optical observations of the core of the galaxy cluster Abell 665 show that the cD galaxy has a smooth halo with a power law surface brightness profile $r^{-1.4}$ extending to at least $500h^{-1}$ kpc from the galaxy center. This smooth profile supports the primordial collapse theory of formation. The galaxy maintains a flat shape ($e \approx 0.4$) to the limit of detection. The X-ray isophotes are rounder at distances greater than 10^{1-} kpc, suggesting that the light is tracing the gravitating mass, rather than the potential. However, there are significant problems with accurately measuring the shape of the X-ray surface brightness and comparing it with the optical shape and these results should not be considered as definitive.

ACKNOWLEDGEMENTS

The author thanks David Buote and Claude Canizares for valuable suggestions and acknowledges the support of NASA contract NAS8-38249 and grant NAG5-1656.

REFERENCES

Binney, J., & Tremaine, S., 1987, Galactic Dynamics, Princeton Univ. Press
Porter, A. C., Schneider, D. P., & Hoessel, J. G., 1991, AJ, 101, 1561
Schmobert, J. M., 1988, ApJ, 328, 475
Uson, J.M., Boughn, S.P., & Kuhn, J.R., 1991, ApJ, 369, 46
Uson, J.M., & Boughn, S.P., 1991, ApJ, 369, 38

DARK MATTER IN CLUSTERS

CLUSTERS AND DARK MATTER

Neta A. Bahcall
Astrophysical Sciences, Princeton University
Princeton, NJ 08544-1001

ABSTRACT

Two of the fundamental questions in cosmology are what is the amount and composition of matter in the universe. Rich clusters of galaxies, the largest virialized systems known, provide a powerful tool in addressing these questions, both on cluster scales of ~ Mpc and on larger scales. We use optical, X-ray, and gravitational-lensing data of clusters to trace the mass and mass-to-light ratio in clusters. The data indicate a mass-density on cluster scale of Ω~0.2. We suggest that the dark matter in clusters may be mostly accounted for by the dark matter of large galaxy halos now stripped in the clusters. We also use cluster properties such as the mass-function and correlation-function of clusters to constrain cosmological model parameters when compared with model simulations. For the currently popular CDM and PBI models, the cluster data is best fit by a low-density universe with Ω~0.2; an Ω=1 CDM model is inconsistent with the data. If Ω~0.15-0.2, as suggested by cluster observations, then the dark-matter is mostly associated with galaxy halos and clusters, and may possibly be all baryonic. If, on the other hand, Ω=1, then the dark-matter distribution is highly biased and is mostly non-baryonic.

1. INTRODUCTION

Rich clusters of galaxies can be used as powerful tools in constraining the amount and type of dark-matter in clusters and the mass-density in the universe. I discuss in this paper some of the methods used and their results.

– Optical, X-ray, and gravitational-lensing data of clusters can be used to determine the amount of dark-matter in clusters. These data provide one of the best determination of the dynamical mass-density on ~ Mpc scales. While additional lensing data are needed in order to provide a more accurate comparison of all three methods, the available data suggest $\Omega_{dyn} \sim 0.2$.

– The mass-to-light ratio of groups and clusters of galaxies is approximately constant for scales $R \gtrsim 0.2h^{-1}$ Mpc. The flattening of M/L occurs at $M/L_B \simeq$ 200-300h, corresponding to Ω~0.2. The observations suggest that most of the dark matter may be associated with large dark galaxy halos.

– X-ray emission from clusters, which originates from the hot intracluster gas, show that a large fraction of the cluster virial mass is *baryonic* ($\gtrsim$ 30% for h = 1/2, or $\gtrsim$ 10% for h = 1, where H_o = 100h km s^{-1} Mpc^{-1} is the Hubble constant). Combined with the low baryon mass-density determined from nucleosynthesis, this suggests that the total mass-density in the universe may be low, $\Omega \sim 0.15 - 0.2$.

– Clusters of galaxies place some of the strongest constraints on cosmological models and their relevant density parameter. We find that for the popular CDM and PBI type models, the density parameter that best fits the cluster observations is, again, $\Omega \sim 0.2$.

2. CLUSTERS OF GALAXIES: OPTICAL, X-RAYS, AND LENSING RESULTS

The binding gravitational potential of clusters of galaxies has traditionally been traced by the velocity dispersion of the cluster galaxies using the virial theorem (e.g. Zwicky 1957; Peebles 1980). The cluster potential can also be estimated by the temperature of the hot intracluster gas as determined from X-ray observations (e.g., Jones and Forman 1984; Sarazin 1986) and by the weak gravitational lens distortion of background galaxies caused by the intervening cluster mass (Kaiser and Squires 1993). Sufficient data are now available in both the optical (galaxy velocity dispersion in clusters) and X-ray (the temperature of the intracluster gas) so that these two methods can be compared. In addition, some current lensing data are becoming available. We compare these three independent methods for cluster mass determination.

Optical–X-Rays. The standard hydrostatic, isothermal model for clusters assumes that both the gas and the galaxies in the cluster are isothermal and in hydrostatic equilibrium within a common potential. The relation between the galaxy velocity dispersion (σ) and the temperature (T) of the intracluster medium (ICM) is then expected to be $\sigma \propto T^{0.5}$. Lubin and Bahcall (1993) analyzed the largest available sample of clusters for which both σ and T have been measured. They find a strong correlation between σ and T, $\sigma = 332 \pm 52 (kT)^{0.6 \pm 0.1}$ km s^{-1}, where kT is in keV (Fig. 1); the observed relation is consistent with that expected from the isothermal, hydrostatic model. Previous results, obtained with smaller cluster samples (Mushotzky 1984; Edge & Stewart 1991), are generally consistent with these findings.

The observed σ (T) relation suggests that the galaxies and the intracluster gas trace the same potential. To determine the average energy per unit mass in the galaxies and the gas, we use the β parameter (Mushotzky 1984, 1988; Sarazin 1986; Edge & Stewart 1991), $\beta = \sigma^2/(kT/\mu m_p)$, where μm_p is the average particle mass, σ is the one-dimensional rms velocity dispersion of the galaxies, and kT is the temperature of the ICM. In the self-gravitating isothermal model (assuming an isotropic velocity distribution), $\beta = 1$; this implies that the galaxies and the ICM trace the same potential and that the energy per unit mass in the gas and in the galaxies are equal. Lubin and Bahcall (1993) used the observed σ(T) relation to determine the best-fit β parameter for clusters; they find $\beta = 0.94 \pm 0.08$, and a median of $\beta_{(med)} = 0.98$. These values suggest that the galaxies and gas trace each other with little or no velocity bias in the average cluster. It also suggests that both the galaxies and the gas are good tracers of the underlying cluster mass.

The comparable kinetic energies of the gas and galaxies in clusters suggest that the gas and galaxies should also have similar density profiles in the clusters. Bahcall and Lubin (1993) find that this is indeed satisfied thus resolving the long-standing "β–discrepency" problem in clusters. The average gas and galaxy distributions in clusters are therefore consistent with an approximate hydrostatic, isothermal cluster model where both the galaxy velocities and the gas temperature reflect the underlying cluster potential.

Lensing. Recent observations of weak gravitational distortions of background galaxies by intervening cluster masses (Kaiser and Squires 1993) yield positive detections, thus providing estimates of the lensing cluster mass within a given radius (see Squires *et al.* and Narayan, this proceeding). Smail *et al.* (1994) used this method to investigate two luminous X-ray clusters of galaxies, CL1455+22 at z=0.26 and CL0016+16 at z=0.55; Falham *et al.* (1994) investigated the X-ray cluster MS1224+20 at z=0.33; and Squires *et al.* (this proceedings) studied the cluster A2163 at z=0.17. Smail *et al.* find that the mass distribution derived from the lensing signal in their two clusters is strikingly similar to that traced by the galaxies and by the hot X-ray gas in the clusters. Their inferred cluster mass profile is compared with the galaxy distribution profile in Fig. 2. These findings agree well with our conclusions that the galaxies, gas, and mass approximately trace each other. They find marginal evidence for a greater central concentration of dark matter with respect to the galaxies, but the uncertainties are large. Smail *et al.* also find that the cluster masses determined from lensing are consistent with the masses determined from the hot gas using the hydrostatic assumption. For CL1455+22 they find $M_{cl}(\leq 0.45\ h_{50}^{-1}) \simeq 3.2\text{x}10^{14}M_{\odot}$ from lensing as compared with $3.6\text{x}10^{14}M_{\odot}$ from X-rays; for CL0016+16 they find ~ $10\text{x}10^{14}M_{\odot}$ from lensing and $7.5\text{x}10^{14}M_{\odot}$ from X-rays, both within $0.6h_{50}^{-1}$ (all values are for h=0.5). Falham *et al.* (1994) find a considerably larger mass (by a factor of ~ 3) for MS1224+20 from lensing as compared with the virial mass estimate. For A2163, the opposite is found: lensing yields a small or no signal compared with the virial X-ray mass. The lensing versus the optical or X-ray mass comparisons is summarized below:

– CL0016+16	z = 0.55	$r \leq 0.23h^{-1}$Mpc	$M_{lens}/M_{opt/X\text{-}ray} = 0.9 \pm (\sim 50\%)$
– CL1455+22	0.26	$r \leq 0.30$	$1.3 \pm (\sim 50\%)$
– MS1224+20	0.33	$r \leq 0.45$	$2.9 \pm (\sim 50\%)$
– A2163	0.17		<1

The ratio of the gravitational lensing mass to the optical or X-ray mass within the radius r ranges from <<1 to ~3 for the four clusters studied. What can we learn? First, it is clear that a large cluster sample is needed before reliable conclusions can be reached. Individual clusters are *expected* to vary significantly in their mass ratio estimates due to various effects including cluster elongation (elongation along the line-of-sight will yield $M_{lens}/M_{opt/X\text{-}ray} > 1$ while flattening will yield a ratio < 1), anisotropic velocity orbits, etc. Therefore, a mean of a large sample is needed. In addition, effects such as boundary corrections in the lensing mass estimate can produce a lowered lensing mass, while substructure will overestimate the lensing mass (Narayan, this proceedings). Based on the optical and X-ray results and the preliminary lensing results shown above we assume that cluster masses, *on average*, are properly traced by galaxies and gas. More accurate lensing data should clarify this assumption.

3. THE MASS, M/L, AND BARYON CONTENT OF CLUSTERS

Cluster Masses and Mass-to-Light Ratio. Detailed optical and X-ray observations of rich clusters of galaxies yield virial cluster masses that ranges from $\sim 10^{14}$ to ~ $10^{15}\ h^{-1}$ $M_{\odot}$ within 1.5 h^{-1} Mpc radius of the cluster center (§2,4). When normalized by the cluster luminosity, a median value of $M/L_B \simeq 300$ h is observed for rich clusters. This mass-to-light ratio implies a dynamical mass density of $\Omega_{dyn} \sim 0.2$. If, as desired by current theory, the universe has critical density (Ω=1), then most of the mass in the universe *can not* be concentrated in clusters, groups and galaxies; the mass distribution in this case would be strongly biased (i.e., does not follow the light).

A recent analysis of the mass-to-light ratio of galaxies, groups and clusters by Bahcall, Lubin and Dorman (1994) suggests that while the M/L ratio of galaxies increases with scale up to radii of R ~ 0.1-0.2h^{-1} Mpc, due to the large dark-halos around galaxies, this ratio appears to flatten and remain approximately constant for groups and rich clusters, to scales of ~ Mpc. The flattening occurs at $M/L_B \simeq$ 200-300h, corresponding to $\Omega \sim 0.2$. This observation may suggest that most of the dark matter may be associated with the dark halos of galaxies and that clusters do *not* contain a substantial amount of additional dark-matter, other than that associated with (or torn-off from) the galaxy halos. Unless the matter distribution is highly biased (with large amounts of dark matter in the "voids" or on very large scales), this suggests that the mass-density in the universe may be low, $\Omega \sim 0.15$-0.2.

Baryons in Clusters. Clusters of galaxies contain a large amount of baryons. Within 1.5 h^{-1} Mpc of a rich cluster, the X-ray emitting gas contributes ~ 10 $h^{-1.5}$% of the cluster virial mass (or ~ 30% for h=1/2) (Briel *et al.* 1992). Visible stars contribute only a small addition to this value. Standard Big-Bang nucleosynthesis limits the mean baryon density of the universe to $\Omega_b \sim 0.015h^{-2}$. This suggests that the baryon fraction in some rich clusters exceeds that of an Ω=1 universe by a large factor.

Detailed hydrodynamic simulations (White *et al.* 1993) suggest that baryons are not preferentially segregated into rich clusters. The above results therefore imply that either the mean density of the universe is considerably smaller than the critical value, or that the baryon density of the universe is much larger than predicted by nucleosynthesis. The observed baryonic mass fraction in rich clusters, combined with the nucleosynthesis limit suggest $\Omega \sim 0.15$-0.2; this estimate is consistent with $\Omega_{dyn} \sim 0.2$ determined from clusters.

4. CLUSTERS AND COSMOLOGICAL MODELS

Two of the fundamental observed properties of clusters of galaxies – the cluster correlation function and the cluster mass function – can be used effectively to constrain cosmological models and the density parameter Ω by comparison with model expectations. Bahcall and Cen (1992) contrasted these cluster observations with standard and nonstandard CDM models using large ($400h^{-1}$ box) N-body simulations. They find that none of the standard Ω=1 CDM models, with any bias parameter, can fit consistently both of these cluster observations. A low-density, low-bias ($\Omega \sim 0.2$-0.3, $b \sim 1$-1.3) CDM-type model, however, provides a good fit to both sets of observations (see Figs. 3 and 4). This is the first CDM-type model that is consistent with the high amplitude and large extent of the correlation function of the Abell, APM, and EDCC clusters (Bahcall and Soneira 1983, Dalton *et al.* 1992, Nichol *et al.* 1992, Bahcall and West 1992). Such low-density, low-bias models are also consistent with other observables, e.g., the angular correlation function of galaxies (Maddox *et al.* 1990), and, with $\Lambda = 1$-Ω, the recent COBE results of the microwave background fluctuations (Smoot *et al.* 1992). This result, however, constrains Ω only for the specific spectrum used. A mixed cold + hot dark matter model, for example, with Ω=1, also produces an acceptable fit to the cluster data.

5. CLUSTERS AND Ω: SUMMARY

Clusters of galaxies provide a powerful tool in constraining the mass-density in the universe. On cluster scales of ~ Mpc, the dynamical estimates of cluster masses - using optical, X-ray, and some gravitational lensing data - suggest low-densities: $\Omega \sim 0.2$. A large fraction of the cluster virial mass is baryonic: ~ 10% for h=1. This suggests that either the baryon density implied by nucleosynthesis, $\Omega_b \simeq 0.015h^{-2}$, is underestimated, or the total mass-density of the universe is low: $\Omega \sim 0.15$-0.2. This Ω estimate is consistent with the dynamical mass estimate from clusters of galaxies.

Comparing fundamental cluster properties such as the observed mass-function and correlation-function of clusters of galaxies with expectations from various cosmological models also suggests that a low-density ($\Omega h \sim 0.2$), low-bias ($b \sim 1$) model (with a CDM or PBI-type spectrum) best fits the data. This Ω constraint, however, is model dependent.

A summary of cluster properties and their implications for Ω is listed below.

- Cluster dynamics.................... $\Omega_{dyn} \sim 0.2$
- Baryon Fraction in Clusters........ $\Omega \sim 0.2$
- Cluster Mass-Function.............. $\Omega \sim 0.2$ (CDM) [but model dependent]
- Cluster Correlation-Function....... $\Omega \sim 0.2$ (CDM) [but model dependent]
- High Redshift Clusters, QSOs in Clusters......[$\Omega < 1$?]
- Substructure in Clusters............ [$\Omega (+\Lambda) = 1$?]
- $t^*_{age} > 13$ byr........................ If $H_o > 50 \Rightarrow \Omega < 1$

[*Age of the universe from globular clusters; while unrelated to galaxy clusters, it provides one of the most fundamental constraints.]

What is the nature of the dark-matter and where is it located? The answers, of course, are not yet known. However, some simple conclusions can be reached:
– If $\Omega \sim 0.15$-0.2, as suggested by most observations, then most of the dark-matter may be located in large galaxy halos. The matter may possibly be all baryonic.
– If $\Omega = 1$, most of the dark-matter must be "hidden" where light does not significantly shine (large voids, outside clusters,...). The matter distribution in this case is highly biased, and most of the matter cannot be baryonic.

The MACHO experiments, described in this proceedings, will hopefully help distinguish between these fundamentaly distinct possibilities.

This work is supported by NSF grant AST93-15368 and NASA NGT-51295.

REFERENCES

Bahcall, N.A. and Cen, R.Y., 1992, *ApJ,* **398**, L81; 1993, *ApJ*, **407**, L49.
Bahcall, N.A. and Lubin, L.M., 1994, *ApJ.*, **426**, 513
Bahcall, N.A., Lubin, M.L. and Dorman, V. 1994, *Nature*, submitted.
Bahcall, N.A. and Soneira, R.M., 1983, *ApJ,* **270**, 20.
Bahcall, N.A. and West, M.L., 1992, *ApJ,* **392**, 419.
Briel, U.G., Henry, J.P., and Boringer, H., 1992, *A&A*, **259**, L31.
Dalton, G.B., Efstathiou, G., Maddox, S., and Sutherland, W. 1992, *ApJ*, **390**, L1.
Edge, A. and Stewart, G.C., 1991, *MNRAS*, **252**, 428.
Fahlman, G.G., Kaiser, N., Squires, G., and Woods, D., 1994, *MNRAS* (in press).
Jones, C. and Forman, W., 1984, *ApJ*, **276**, 38.
Kaiser N. and Squires, G. 1993, *ApJ*, **404**, 441.
Lubin, L.M. and Bahcall, N.A., 1993, *ApJ*, **415**, L17.
Maddox, S.J., Efstathiou, G., Sutherland, W., and Loveday, J., 1990, *MNRAS*, **242**, 43.
Mushotsky, R., 1984, *Phys.Scripta.*, **T7**, L157.
Mushotzky, R., 1988, *Hot Astrophysical Plasma*, ed. R. Pallavicini, (Dordrecht: Kluwer), 114.
Nichol, R., Collins, C. Guzzo, L and Lumsden, S. 1992, *MNRAS*, **255**, 21.
Peebles, P.J.E., 1980, "The Large Scale Structure of the Universe," Princeton University Press, Princeton, NJ.
Sarazin, C.L., 1986, *Rev. Mod. Phys.*, **58**, 1.
Smail, J., Ellis, R., Fitchett, M., 1994, *MNRAS* (in press).
Smoot, G.F., *et al.*, 1992, *ApJ*, **396**, L1.
White, S., Navarro, J., Evrard, A., and Frenk, C., 1993, *Nature*, **366**, 429.
Zwicky, F., 1957, *Morphological Astronomy*, Springer-Verlag, Berlin.

FIGURE CAPTIONS

Figure 1. Cluster velocity dispersion versus intracluster gas temperature. N_z is the number of measured velocities per cluster (Lubin & Bahcall 1993). The best-fit line is $\sigma=332\pm52\,(kT)^{0.6\pm0.1}$ km s^{-1}, where kT is in keV.
Figure 2. A comparison of the surface density profile of mass (•) (from lensing) and galaxies (O) in the cluster 14455+22. From Smail *et al.* (1994). (h=1/2 in this plot).
Figure 3. The observed mass-function of clusters of galaxies (Bahcall and Cen 1993) and its comparison with various CDM model simulations (Bahcall and Cen 1992).
Figure 4. The correlation length of the cluster correlation function as a function of cluster mean separation; observations and CDM models (Bahcall & Cen 1992).

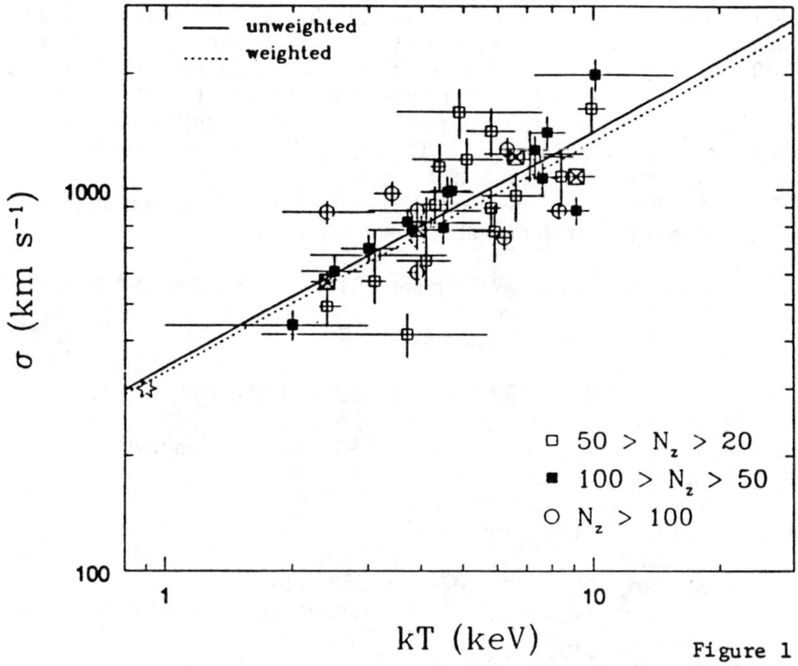

Figure 1

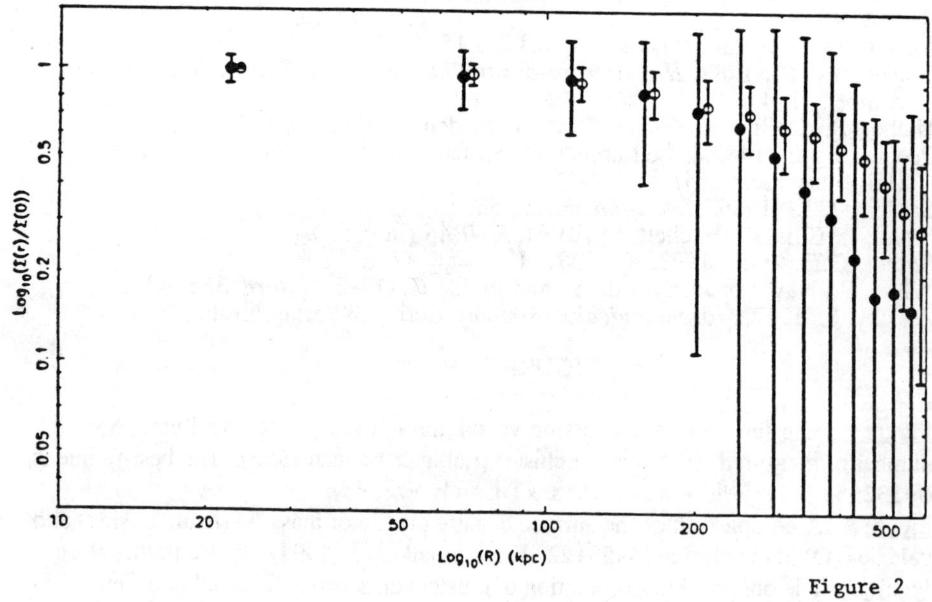

Figure 2

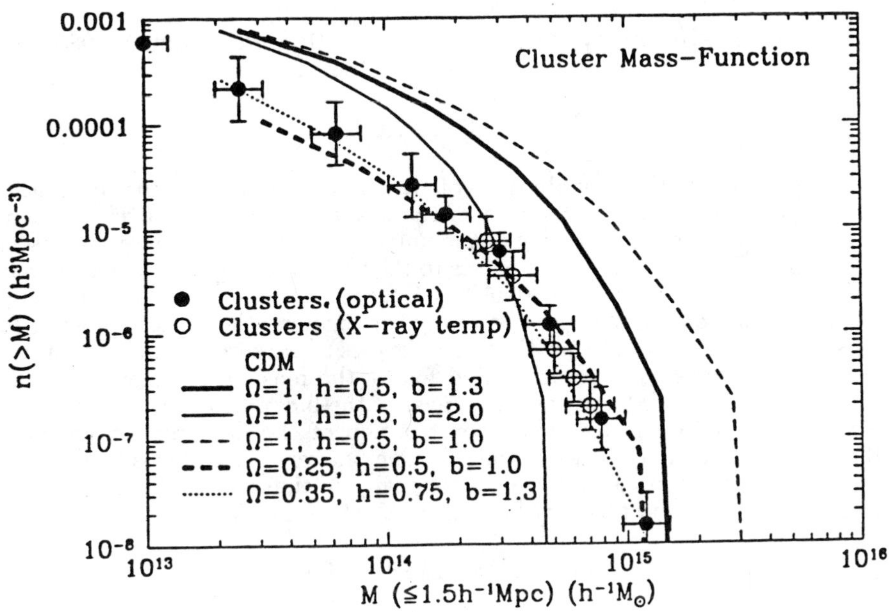

Figure 3

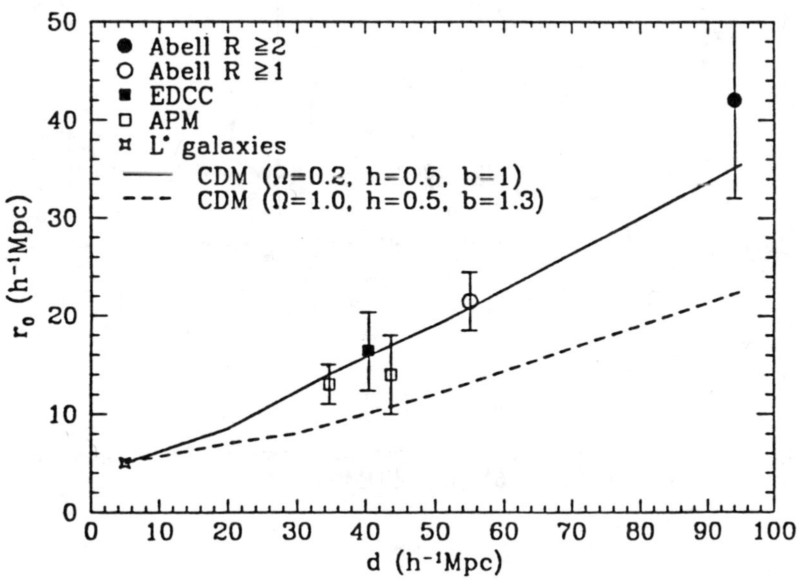

Figure 4

Gas Mass and Total Mass in Clusters of Galaxies

Lori M. Lubin
Princeton University Observatory, Peyton Hall, Princeton, NJ 08544
lml@astro.princeton.edu

ABSTRACT

We examine the contributions of gas and galaxy mass to the total mass of clusters of galaxies as an indicator of the origin of the dark matter component in clusters and the mass density of the universe. First, we compare the observed X-ray and optical relations in clusters to the results of large-scale hydrodynamic simulations and show that low-density CDM is more consistent with the observed gas mass fractions in clusters. Secondly, we examine the make-up of the total cluster mass and suggest that there is no excess of dark matter in these systems; that is, their mass can be accounted for by the mass of the member galaxies (including their dark halos) and the mass of the baryonic intracluster gas. This would imply that the fractional contribution of the dark matter component does not keep increasing with scale beyond the very large halos suggested for individual galaxies and that the mass density of the universe is of the order of the mass density observed on the scale of clusters, i.e. $\Omega \sim 0.2$.

1. GAS MASS RATIO

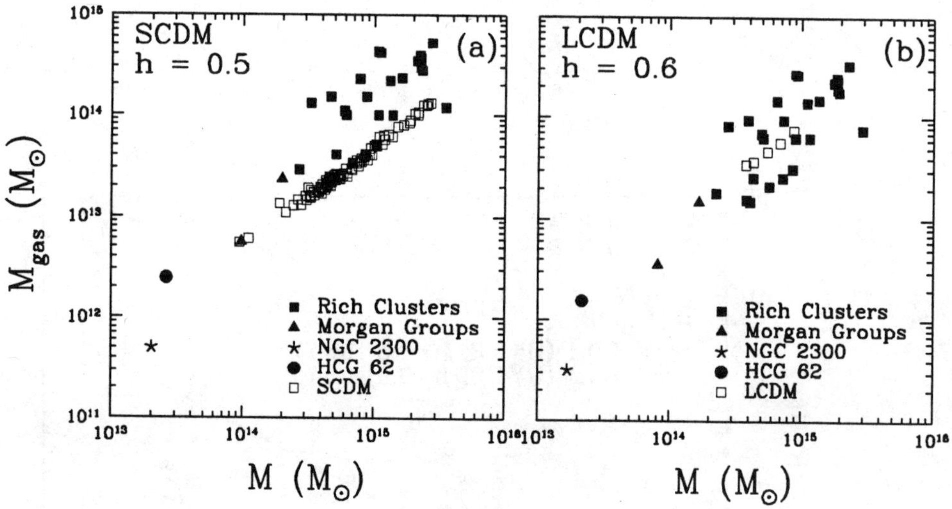

Fig. 1. Observed and simulated gas mass versus total mass for clusters and groups. (a) SCDM ($h = 0.5$); (b) LCDM ($h = 0.6$).

We show the observed relation between gas mass and total virial mass of rich clusters (within $R_A \simeq 1.5\ h^{-1}\ Mpc$) in Fig. 1. While the scatter is large, a correlation between the parameters is evident. An unweighted χ^2 fitting, including two Morgan groups (Morgan et al. 1975), the NGC 2300 group (Mulchaey et al. 1993), and HCG 62 (Ponman & Bertram 1993) yields : $M_{gas}\ h^{2.5} \simeq 2.3 \times 10^{-4}\ (M\ h)^{1.16\pm0.12}$. The best-fit linear relation satisfies $M_{gas} \simeq 0.06\ M\ h^{-1.5}$ ($\simeq 0.16\ M\ h_{50}^{-1.5}$). We compare this relation with the simulated clusters of $\Omega = 1$ standard CDM (SCDM; Fig. 1a) and $\Omega = 45, \Lambda = 0.55$ low-density CDM (LCDM; Fig. 1b). The shape of the simulated $M_{gas} - M$ relation is consistent with observations; it follows, approximately, $M_{gas} \propto M$. However, the amplitude of the relation, or equivalently the gas mass fraction in the clusters, M_{gas}/M, differs significantly in the two models. The SCDM model yields $M_{gas} \simeq 0.05\ M$ ($h = 0.5$), i.e. lower gas masses by a factor of ~ 3 for a given cluster mass than observed. LCDM, on the other hand, yields $M_{gas} \simeq 0.09\ M$ ($h = 0.6$), more consistent with the data (Lubin et al. 1994). The larger gas mass fraction in the LCDM model reflects the larger baryonic fraction of the model : $\Omega_b/\Omega \sim 0.1$ ($h = 0.6$) for LCDM, as compared with $\Omega_b/\Omega \sim 0.06$ ($h = 0.5$) for SCDM .

2. TOTAL MASS OF CLUSTERS

It is generally believed (Davis et al. 1980; Blumenthal et al. 1984) that, on average, the mass-to-light ratio increases from the bright, luminous parts of galaxies to their faint halos, with a further increase to systems of larger scale such as groups and clusters of galaxies. This would suggest that the fractional contribution of the dark matter component keeps increasing with scale. On the scale of clusters ($\sim 1\ Mpc$), the observed mass density implies $\Omega \sim 0.2$; however, if M/L keeps increasing with scale, the closure density ($\Omega = 1$) may be reached. We show that observations suggest that M/L does not increase significantly beyond the scale of the largest observed galaxy halos ($R \sim 150 - 200\ kpc$). That is, the dark matter observed in groups and clusters may be accounted for by the matter that is already believed to exist in large galaxy halos (Bahcall, Lubin & Dorman 1994a,b).

In order to determine the contribution of the member galaxies to the dark matter component in clusters, we first examine the M/L_B of individual spirals and ellipticals. Virial mass measurements within observed radii of luminous or 21–cm regions are taken from Faber & Gallagher (1979), Bosma (1981; spirals), Lauer (1985; ellipticals), and Trimble (1987). To extend the estimated galaxy mass beyond the "normally" observed optical or radio regions, we use mass estimates from motions of faint satellites around galaxies, as well as dynamical and X-ray measurements (Bahcall & Tremaine 1981; Forman et al. 1985; Kulessa & Lyndon-Bell 1992; Zaritzky et al. 1993; Mushotzky et al. 1994).

In Fig. 2, we show the M/L_B of the same galaxy on small to large scales. The lines connecting the data points indicate a *single* galaxy with measurements at different radii. As expected, the M/L_B increases with radius to $\sim 200\ h^{-1}\ kpc$, indicating the existence of the well-known dark galaxy halos. The average M/L_B ratio of elliptical galaxies appears to be larger than that of spirals by $\sim 3 - 4$ at the same radius. If the dark halos around bright galaxies extend to $\sim 150 - 200\ h^{-1}\ kpc$ (as suggested by Fig. 2), the implied total M/L_B is $\sim 100\ h\ M_\odot/L_\odot$ for spirals and $\sim 300 - 400\ h\ M_\odot/L_\odot$ for ellipticals.

Clusters of galaxies exhibit a relatively large M/L_B ratio of $\sim$ 300 h. The large value of M/L_B implied for ellipticals can thus contribute significantly to this observed increase in M/L_B in clusters since they contain up to 80% ellipticals. The extended dark halos are likely to have been stripped off in the cluster center but still reside within the cluster. Thereforc, the total mass of clusters can bc accounted for by the member galaxies (including their dark haloes) and the intracluster baryonic gas ($\sim$ 6% of the total cluster mass). For systems larger than the largest extent of individual galaxies ($\sim$ 200 h^{-1} kpc), the mass-to-light ratio remains approximately constant with values between 100 and 400 h, depending only on the spiral fraction in the system. If correct, this scenerio indicates that the mass density of the universe is low, $\Omega \sim 0.2$, and that most of the mass is located in large galaxy halos (Bahcall, Lubin & Dorman 1994a,b).

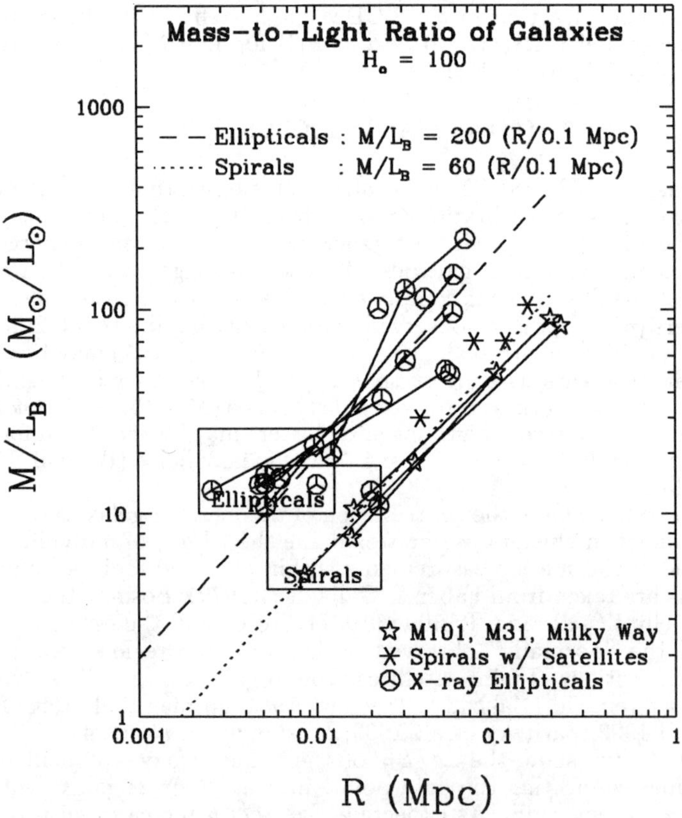

Fig. 2. The mass-to-light of spiral and elliptical galaxies as a function of scale. The large boxes indicate the typical ($\sim 1\sigma$) range of M/L_B for bright ellipticals and spirals at their luminous (Holmberg) radii. The best-fit $M/L_B \propto R$ lines for spirals and ellipticals are shown.

5. CONCLUSIONS

We use the gas mass and total mass of clusters of galaxies in order to constrain the mass density of the universe. Comparisons with large-scale hydrodynamic simulations and examinations of the mass-to-light ratio of galaxies and clusters seem to indicate that the mass density of the universe may be of the order of the mass density observed on the cluster scale, $\Omega \sim 0.2$.

REFERENCES

Bahcall, J.N., & Tremaine, S. 1981, ApJ, 244, 805
Bahcall, N.A., Lubin, L.M. & Dorman, V.L. 1994a, Nature, submitted
Bahcall, N.A., Lubin, L.M. & Dorman, V.L. 1994b, ApJ, in preparation
Blumenthal, G.R. et al. 1980, Nature, 311, 517
Bosma, A. 1981, AJ, 86, 1825
Cen, R., & Ostriker, J.P. 1994, ApJ, 429, 4
Davis, M. et al. 1984, ApJ, 238, L113
Faber, S.M., & Gallagher, J.S. 1979, ARAA, 17, 135
Forman, W., Jones, C., & Tucker, W. 1985, ApJ, 293, 102
Kulessa, A.S., & Lyndon-Bell, D. 1992, MNRAS, 255, 105
Lauer, T.R. 1985, ApJ, 292, 104
Lubin, L.M. et al. 1994, ApJ, submitted
Mulchaey, J.S. et al. 1993, ApJ, 404, L9
Mushotzky, R.F. et al. 1995, ApJ, accepted
Ponman, T.J. & Bertram, D. 1993, Nature, 363, 51
Trimble, V. 1987, ARAA, 25, 423
Zaritzky, D. et al. 1993, ApJ, 405, 464

PHOTOMETRIC PROPERTIES OF CLUSTERS OF GALAXIES

BIANCA GARILLI, DARIO MACCAGNI
IFCTR-CNR, via Bassini 15, 20133 Milano, Italy
bianca@ifctr.mi.cnr.it

LUIS CARRASCO, ELSA RECILLAS
Observatorio Astronomico Nacional, UNAM, Mexico D.F., Mexico
INAOE, Tonantzintla, Puebla, Mexico

1. INTRODUCTION

Clusters of galaxies can be selected via the galaxy density contrast over the field or through their X-ray emission.The two selection methods could complement each other to produce lists of clusters of galaxies spanning the whole range of cluster masses and morphologies, from which to derive the properties of the luminous matter and the forms of its aggregation.

Multicolor photometry of large numbers of clusters of galaxies is an important tool to study, for instance, the galaxy populations, the shape of the galaxy luminosity function and its variation from cluster to cluster and from cluster to field. It is also essential that these properties be well known for the "local" universe, so that evolutionary phenomena can be fully understood.

It is with these aims that, upon completion of the *Einstein* Extended Medium Sensitivity Survey (EMSS, Gioia et al. 1990), we set out to obtain multicolor photometric measurements of a mixed sample of clusters of galaxies. In this paper we report on the first results of such an investigation.

We have chosen a complete subsample from the EMSS, comprising all the X-ray selected clusters between 0 and +70 degrees in declination having redshift smaller than $z = 0.22$, for a total of 34 objects. As a comparative sample, we have selected a similar number of Abell clusters, with the same redshift distribution, taking care to equally sample all Bautz-Morgan morphological types.

2. OBSERVATIONS

Observations have been carried out at the 2.1m telescope of the Mexican National Astronomical Observatory at San Pedro Martir, equipped with a CCD of 384x576 23μ pixels, resulting in a field of view of $\sim$ 2x3 arcmin. Since December 1993, a new 1024x1024 22μ pixels CCD has been available, thus 12 clusters have been observed with a larger field of view ($\sim$ 4x4 arcmin). For 8 clusters we made two or more pointings, to explore a larger area. All clusters have been observed through the three Gunn filters *g,r* and *i*. The average seeing was $\sim$ 1.5 arcsec. The images generally allow to detect objects as faint as 22.5 *r* mag in all the three filters. Since the observation campaign started in 1990, we have observed 33 X-ray selected (EMSS) and 40 Abell clusters. Out of these, 57 were observed in good photometric conditions and the data have been fully reduced. For each cluster, catalogs of all objects detected in the three filters have been constructed. Magnitudes have been corrected for atmospheric extinction and galactic absorption on the basis of the measured hydrogen column density along the line of sight. As the expected number of stars brighter than 20 *r* mag is of the order of 3 in our fields (see e.g. Jarvis & Tyson, 1981), star-galaxy discrimination has been performed by carefully inspecting the images and sampling faint objects profiles. In deriving galaxy magnitudes and colors, we have considered both seeing effects and the presence of nearby objects. The number of galaxies in each cluster field for which we have multicolor information ranges from 12 to 230. Colors are therefore available for a total of 3338 galaxies.

3. DISCUSSION

The first question we want to raise is: have X-ray selected clusters similar photometric properties as Abell clusters?

In Figure 1, galaxy density is plotted vs. redshift for Abell and EMSS clusters. For each cluster, we have considered only galaxies brighter than $M_r = -18$. Abell clusters show all ranges of densities, irrespective of redshift. On the contrary, the density range for EMSS clusters is much narrower, and they never have a density higher than $1.5\ 10^{-4}$ galaxies Mpc^{-2}. There are several effects at work which must be considered in analyzing this plot. First of all, the average density should naturally decrease as the redshift increases because a larger metric area is included at higher redshifts. This is counterbalanced by a second effect: being the EMSS a flux limited sample, the X-ray luminosity of high redshift clusters is in fact higher than that of the low redshift ones, as, at the same time, is the core density. On the other hand, at $z \leq 0.1$ X-ray selected clusters naturally have lower densities, as the richer, low redshift clusters were *Einstein* targets and were not included in the EMSS. Thus, the constant galaxy density in X-ray selected clusters shown by Figure 1 is nothing more than the relation between X-ray luminosity and central galaxy density proposed by Bahcall (1980). Furthermore, in the explored redshift range, we can state that both selection methods find objects with similar luminous masses.

The mean colors of the 4 brightest galaxies of each cluster also do not depend on the selection band, and in both cases the colors of the brightest galaxies are perfectly compatible with the results of Schneider, Gunn and Hoessel (1983) and with the colors predicted from a standard galaxy model. Possible color differences of the early type population between the different clusters will be investigated through a deeper analysis of the color-magnitude diagrams and the slopes of the color-color plots in a further work. At present we can say that no difference is apparent between X-ray selected and optically selected clusters, although there appears to be a trend in the slope of the color-color diagram with redshift (possibly indicating a change in metallicity).

The color-magnitude effect is present both in X-ray and optically selected clusters (at least in those where there are enough early type galaxies to allow the color-magnitude relation to be defined). Such an effect does change from cluster to cluster, but it doesn't seem to be dependent on the selection band. In Figure 2 we show some extreme cases of color-magnitude diagrams from our sample: in the case of MS1125+4324, no color-magnitude effect is visible, and the same can be said for A175. On the other hand, in A566 and MS1050+4946 (although in this last case, galaxies are fewer), it is clearly visible. The different importance of the color-magnitude effect in the different clusters reflects the galaxy morphological mix, but in some instances might also indicate a high background contamination or the presence of more than one cluster/group of galaxies along the same line of sight.

A crucial point for any further analysis is the nature of the population of rather bright galaxies redder than the BCM and the other cluster early type galaxies which is an important feature of almost all our color-magnitude diagrams. We restricted our analysis to those clusters where the color-magnitude effect is clearly distinguishable, and limit ourselves to the $m_r = 22$, to be free from incompleteness biases. We thus have 45 cluster fields, covering an area of ~ 500 arcmin^2. In all clusters we note a few galaxies lying well above the c-m locus (see Figure 2 for an example). Most of these objects have magnitudes between 20 and 22. Their mean density accounts for half of the field galaxy counts at such magnitudes (Neuschaefer & Windhorst, 1994), and their magnitude distribution is compatible with field counts. Unfortunately, their colors pose some problems to galaxy evolution models, whether such galaxies belong to the field or to the cluster.Lacking any spectroscopic information, we can only speculate about these objects either belonging to the cluster or to the field. If they belong to the field, galaxy evolution models predict similar *r-i* colors only for high redshift (z=0.7-1.0) early type galaxies. If such were the case, their intrinsic luminosity would be greater than L*, and their number would be extremely high for objects that luminous. Moreover the *g-r* is too low compared to what predicted by the same models. On the other hand, if we look at the *g-r* colors, galaxy evolution models would suggest that these are medium redshift early type galaxies (z=0.3-0.4). In this case, the intrinsic luminosity is acceptable, but the *r-i* color remains unexplained. At present, we cannot rule out the hypothesis that at least some of these red galaxies are cluster members. Again, this is not free from theoretical problems: there is no cluster model which predicts galaxies redder than the ellipticals of the cluster itself. Galaxies with similar characteristics have been observed in other clusters by other authors: Dressler & Gunn (1992), although they do not explicitly mention it, do have galaxies with extreme red colors in their clusters, Aragon-Salamanca et al. (1993) in their infrared selected clusters, find galaxies with extreme infrared colors, compared to the optical ones.

Overall,observational evidence for a population of galaxies with anomalous red colors is piling up, posing a problem to galaxy evolution models.

ACKNOWLEDGEMENTS

We wish to thank G.P.Vettolani, for interesting discussions and suggestions, and S.Pocar for help in reducing the data.

REFERENCES

Abell, G.O., Corwin, H.G., Jr., Olowin, R.P. 1989, ApJS, 70, 1
Aragon-Salamanca et al., 1993 MNRAS 262, 764
Bahcall, N.A. 1980, ApJL, 238, L117
Dressler, A. & Gunn, J.E. 1992, ApJS, 78, 1
Gioia, I.M., Maccacaro, T., Schild, R.E., Wolter, A., Stocke, J.T., Morris,S.L., & Henry, J.P. 1990, ApJS, 72, 567
Jarvis, J.F. & Tyson, J.A. 1981, AJ, 86, 476
Neuschaefer, L.W. & Windhorst, R.A. ApJS 1995, in press
Schneider, D.P., Gunn, J.E. & Hoessel, J.G. 1983, ApJ, 268, 376

Figure 1

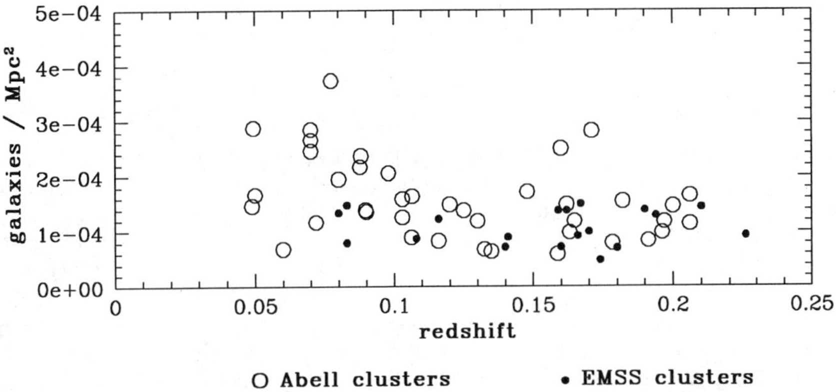

Figure 2

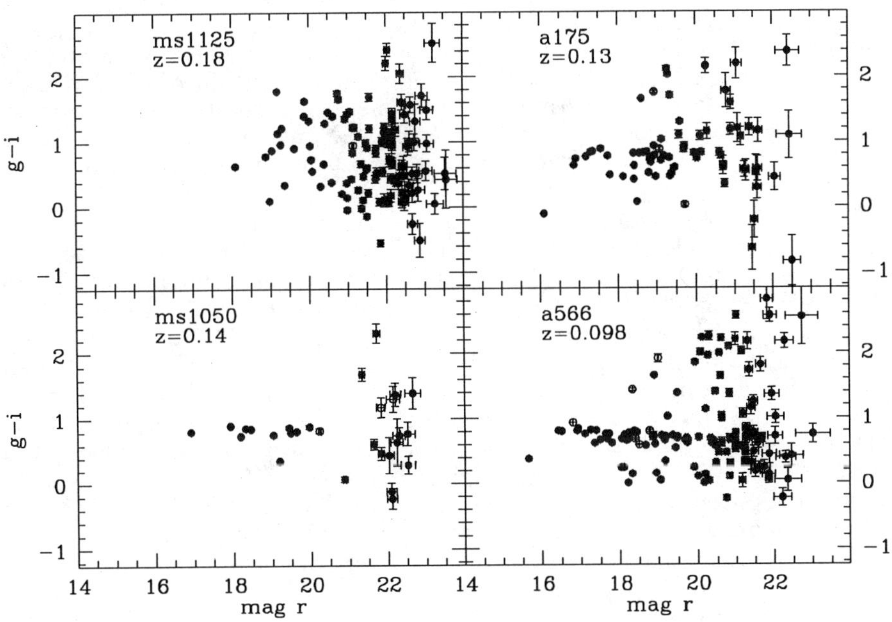

A MERGER OF SUBCLUSTERS IN THE GALAXY CLUSTER ABELL 754†

Ann I. Zabludoff and Dennis Zaritsky‡
Observatories of the Carnegie Institution of Washington
813 Santa Barbara Street, Pasadena, CA, 91101
aiz@libra.ociw.edu, dennis@lick.ucsc.edu

The complex dynamics of the gas, galaxies, and dark matter in clusters of galaxies provide clues to cluster evolution and to the underlying cosmology of the universe. In hierarchical models of the formation of large-scale structure, clusters evolve by accreting other clusters or smaller groups of galaxies from the field (Evrard 1990ab). Simulations show that the effects of such a merger can be observed for $\sim 1h^{-1}$ Gyr (where $H_0 = 100h$ km s^{-1} Mpc^{-1}). A key merger signature is the existence of hot, X-ray emitting, intracluster gas that is not in hydrostatic equilibrium with the cluster potential. If the dark matter that dominates the potential is traced by the galaxies, the peaks of the X-ray emission and the galaxy surface density will not coincide during the merger. Although the gross features of the optical and X-ray data agree in a few well-studied clusters (Fabricant *et al.* 1986, 1993), these samples lack the resolution to test the critical scales $< 0.5h^{-1}$ Mpc where the models predict differences. Today, multi-object spectroscopy and a new generation of X-ray telescopes enable us to compare the spatial distribution of hundreds of cluster galaxies with high-resolution X-ray surface brightness maps. In the first cluster we have examined, Abell 754, there is evidence for a non-hydrostatic process; neither of the two major clumps in the galaxy distribution lies on the peak of the X-ray emission. The most likely explanation is that a merger of two or more subclusters is occurring in A754.

As part of a larger study of the internal dynamics of nearby clusters, we have obtained galaxy spectroscopy and X-ray imaging data for A754, a rich cluster at a mean redshift of $\overline{z} = 0.055$ with an X-ray luminosity of $L_X = 2.2\times10^{44}h^{-2}$ erg s^{-1} (0.3-3.5 keV band; Fabricant *et al.* 1986). We used the Las Campanas multi-fiber spectrograph (Shectman *et al.* 1992) to acquire spectra of 480 galaxies with $m_B \lesssim 19$ that lie within 1° of the cluster center (Abell 1958). We determine radial velocities with a cross-correlation method (Tonry & Davis 1979) or by fitting emission lines. A comparison with published data (Dressler & Shectman 1988) indicates that our velocities have 1σ observational uncertainties of $\sim$60 km s^{-1} and a zero point offset of $\lesssim$ 60 km s^{-1}. Cluster membership is assigned to 340 galaxies in the range $0.046 \le z \le 0.063$ (limits corresponding to the cluster mean velocity $\pm$ (3 $\times$ the cluster velocity dispersion of 900 km s^{-1})). The data and detailed analyses will be presented in later work; here we compare the positions of these galaxies on the sky to the cluster X-ray surface brightness distribution.

---

† Submitted to Nature.

‡ Current address: Astronomy and Astrophysics Board, UCSC, Santa Cruz, CA, 95064

We acquired a 2322 sec exposure of A754 from the archives of the Position Sensitive Proportional Counter (PSPC) on the ROSAT satellite (Henry & Briel 1993). The X-ray emission from A754 lies within the central 40 arcmin of the 2° PSPC field, where the point response function is approximately Gaussian, and the flatfield and background levels are approximately constant. We restrict our analysis to X-ray photons in the 0.5-2.4 keV band, thereby excluding most contamination from the Galaxy.

Figure 1 shows the smoothed X-ray map superimposed on the smoothed projected distribution of cluster galaxies. Within the central $1h^{-1}$ Mpc, there is bimodal structure in the galaxy distribution (Fabricant *et al.* 1986). The southeast (SE) and the northwest (NW) galaxy clumps are separated by $\sim 0.7h^{-1}$ Mpc on the sky. The SE clump lies at $(\alpha, \delta)_{2000} \sim (09^h{:}09^m{:}30^s, -09°{:}43'{:}00'')$ and contains 47 galaxies within a projected radius of $0.3h^{-1}$ Mpc, nine galaxies comprise its bright peak. The NW clump is at $(\alpha, \delta)_{2000} \sim (09^h{:}08^m{:}30^s, -09°{:}37'{:}00'')$ and has 18 galaxies within a $0.3h^{-1}$ Mpc radius, 13 galaxies are in its bright peak.

Figure 1 is remarkable because the two major condensations of galaxies do not correspond to the most prominent feature of the highly asymmetric X-ray surface brightness map. The peak of the X-ray emission ($(\alpha, \delta)_{2000} \sim (09^h{:}09^m{:}22^s, -09°{:}40'{:}00'')$) is offset from the SE galaxy clump by a projected separation of $\sim 0.2h^{-1}$ Mpc. If the galaxies trace the collisionless dark matter, the displacement between the X-ray peak and the galaxies is strong evidence for gas that is not in hydrostatic equilibrium with the gravitational potential.

The disturbed X-ray morphology is not due to the chance superposition of an unidentified X-ray source. The X-ray luminosity of the peak is ~20-25% that of the cluster. A single X-ray point source is ruled out by the elongation of the peak. The peak shape is not reproduced by a model of two point sources at any separation. Furthermore, no point sources are revealed in a deeper PSPC exposure of this field (Briel & Henry 1994, private communication). As for contamination from extended sources, there is no galaxy with $m_B < 19$ within the brightest contour of the X-ray peak. The two closest galaxies, which are cluster members, lie near the southern and northern ends of the elongated 2nd- and 3rd-brightest contours, respectively. The northern galaxy, which has no emission lines and is classified S0/D (Dressler 1980), is visually one of the brightest in the cluster ($m_B = 16.3$; Fabricant *et al.* 1986) and has a weak associated radio source (Harris *et al.* 1980). The southern galaxy ($m_B = 17.5$; Fabricant *et al.* 1986) is typed Irr (Dressler 1980); its spectrum has [OII] $\simeq 40$Å $>$ [OIII] $\simeq 10$Å and is consistent with a starburst. It is very unlikely that both galaxies have the X-ray luminosities ($> 10^{43}h^{-2}$ erg s^{-1}) and X-ray radii ($\sim 100h^{-1}$ kpc) necessary to account for the brightness and extent of the peak (Fabbiano 1989; Stocke *et al.* 1991). Visual inspection of the POSS plates finds no additional objects near the X-ray peak, except for an 8.9 mag F star (HD 78665) which makes a negligible contribution to the X-ray flux (Fabricant *et al.* 1986). Thus we conclude that the X-ray emission of the peak originates from the intracluster gas.

What are the implications for theories of cluster evolution? The displacement between the galaxies and the X-ray emission is on a scale consistent with that seen during cluster mergers in hierarchical clustering simulations (Evrard 1990ab). During a merger of two subclusters, the models predict that the collisional X-ray gas is stripped from the free-streaming dark matter and galaxies.

The gas is not in hydrostatic equilibrium with the dark matter potential for $\sim 1h^{-1}$ Gyr afterwards. If the galaxies trace the dark matter distribution, the distributions of gas and galaxies will differ on scales $\lesssim 0.5h^{-1}$ Mpc. The models produce a second merger signature: a sharp, linear edge in the X-ray surface brightness distribution perpendicular to the merger direction. This feature, the result of the impact of the intracluster media, also relaxes in $\sim 1h^{-1}$ Gyr. In A754, the peak of the X-ray emission is elongated perpendicular to the axis joining the SE and NW optical clumps. The surprisingly good agreement of our data with the qualitative appearance and scale of the models suggests that we are viewing an active merger in A754 at a fairly large angle to the collision axis.

Quantitative comparisons require more detailed models. At present, we cannot determine whether the two clumps of galaxies are infalling for the first time or have already crossed. Because the galaxies free-stream through the collision more easily than the gas, the galaxies should fly further than the gas from the point of impact after the clumps cross. It is suggestive that the SE galaxy clump and the X-ray peak lie in this configuration. The NW galaxy clump does not have an obvious X-ray counterpart, but the east-west ellipticity of the cluster's outer X-ray isophotes implies one; deeper PSPC images (Briel & Henry 1994) may reveal this X-ray peak and its position relative to the NW clump.

Our future work will incorporate optical kinematic data, X-ray surface brightness and spectral data, and n-body/hydrodynamical models that sample a range of clump orbits and masses. Spatially-resolved PSPC and/or ASCA spectral data will show whether A754's two suspected temperature components (Henriksen 1993) are associated with the two galaxy clumps and whether the temperatures are consistent with the clump velocity dispersions. For systems like A754, models that assume both the gas and galaxies are isothermal and in hydrostatic equilibrium will underestimate the cluster binding mass by $\sim$30% (Evrard 1990b). This factor may contribute to the discrepancies sometimes found between cluster X-ray masses and the masses derived from gravitational lens models (Babul & Miralda-Escudé 1994).

The number of clusters with mergers at the current epoch is tied to the underlying cosmology — hierarchical models predict that the merger rate depends sensitively on the mean mass density parameter Ω_0 (Richstone *et al.* 1992; Evrard *et al.* 1993). The observed fraction of nearby clusters with substructure is only an estimate of the number of mergers because of uncertainties in the timescale over which substructure persists. This timescale depends on whether the subclusters are bound to the cluster and on how the dark matter is distributed. As we show in A754, it is now possible to avoid these assumptions by identifying the clusters in which mergers have occurred during the last $\sim 1h^{-1}$ Gyr. We stress that A754 is unusual only because it has been examined on scales where the effects of mergers are most visible. On-going mergers might cause the inconsistencies between the global kinematics of the gas and of the galaxies in Perseus (Gorenstein *et al.* 1978; Cowie *et al.* 1987) and in Abell 2256 (Fabricant *et al.* 1989). A detailed comparison of the optical and X-ray properties of a large, well-defined sample of clusters (Zabludoff & Smail 1994) will reveal the incidence of such mergers and will place an improved constraint on cosmological models.

ACKNOWLEDGEMENTS

We thank Marc Postman for providing digitized POSS images and Anand Sivaramakrishnan for his photography expertise. Dan Fabricant, Pat Henry, Gus Evrard, Ian Smail, Megan Donahue, and Richard Mushotzky contributed helpful information and comments.

REFERENCES

Abell, G.O. 1958, ApJS, 3, 211
Babul, A. & Miralda-Escudé, J. 1994, CITA preprint
Briel, U.G. & Henry J.P., in preparation
Cowie, L., Henriksen, M., & Mushotzky, R. 1987, ApJ, 317, 593
Dressler, A. 1980, ApJS, 42, 565
Dressler, A. & Shectman, S.A. 1988, AJ, 95, 284
Evrard, A.E., Mohr, J.J., Fabricant, D.G. & Geller, M.J. 1993, ApJL, 419, L9
Evrard, A.E. 1990a, ApJ, 363, 349
Evrard, A.E. 1990b, in Clusters of Galaxies, eds. Oegerle, W.R., Fitchett, M.J., & Danly, L. (Cambridge: Cambridge Univ. Press) p.287
Fabbiano, G. 1989, Ann Rev Astron Astrophys, 27, 87
Fabricant, D., Kurtz, M., Geller, M., Zabludoff, A., Mack, P., & Wegner, G. 1993, AJ, 105, 788
Fabricant, D., Kent, S., & Kurtz, M. 1989, ApJ, 336, 77
Fabricant, D., Beers, T.C., Geller, M.J., Gorenstein, P., Huchra, J.P., & Kurtz, M.J. 1986, ApJ, 308, 530
Gorenstein, P., Fabricant, D., Topka, K., Harden, F.R., & Tucker, W.H. 1978, ApJ, 224, 718
Harris, D.E, Constain, C.H., & Dewdney, P.E. 1984, ApJ, 280, 532
Harris, D.E., Costain, C.H., Strom, R.G., Pineda, F.J., Delvaille, J.P., & Schnopper, H.W. 1980, A&A, 90, 283
Henriksen, M.J. 1993, ApJL, 414, L5
Henry, J.P. & Briel, U.G. 1993, Adv Space Res, 13, 191
Richstone, D., Loeb, A. & Turner, E.L. 1992, ApJ, 393, 477
Shectman, S.A., Schechter, P.L., Oemler, A.A., Tucker, D., Kirshner, R.P., & Lin, H. 1992, in Clusters and Superclusters of Galaxies, ed. Fabian, A.C. (Dordrecht: Kluwer), p.351
Stocke, J.T, Morris, S.L., Gioia, I.M., Maccacaro, T., Schild, R, Wolter, A., Fleming, T.A. & Henry, J.P. 1991, ApJS, 76, 813
Tonry, J. & Davis, M. 1979, AJ, 84, 1511
Zabludoff, A.I. & Smail, I., in preparation
Zaritsky, D. & White, S.D.M. ApJ, in press

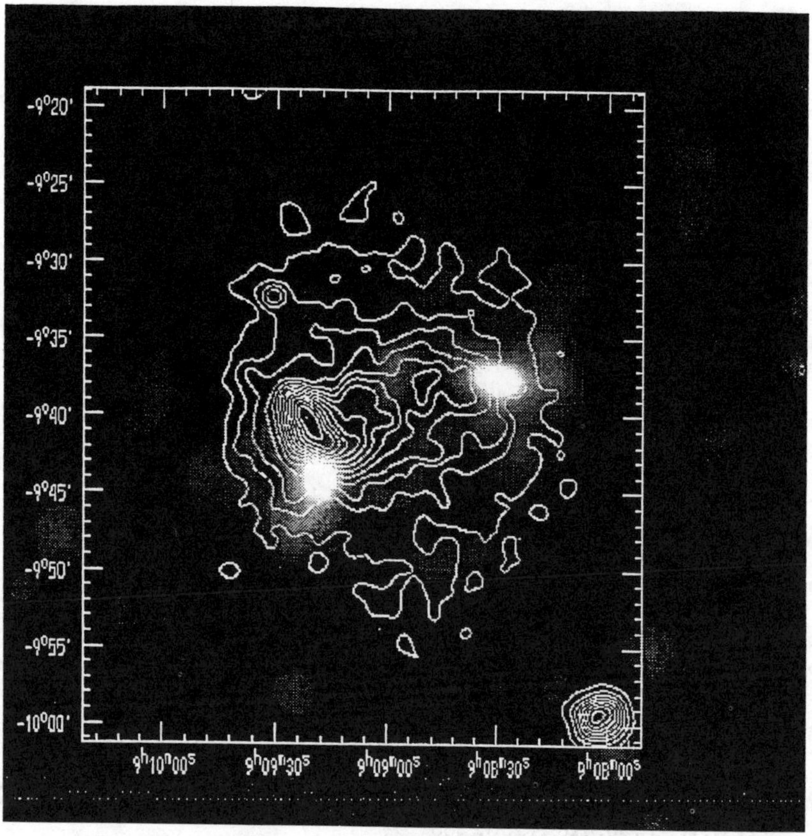

Fig. 1.
The smoothed distribution of X-ray emitting gas (white contours) superimposed on the smoothed distribution of cluster galaxies in A754. The scale of the plot is $45.6h^{-1}$ kpc arcmin^{-1}. Note that neither of the two major clumps of galaxies lies on the extended peak of the X-ray emission. The distribution of 201 galaxies is smoothed with a Gaussian of 1.23 arcmin ($\sim 60h^{-1}$ kpc) FWHM, a scale not larger than that of giant galaxy halos (Zaritsky & White 1994). The faintest greyscale level corresponds to individual galaxies. The PSPC image is smoothed with a Gaussian of 30 arcsec FWHM; the resulting map has an effective resolution of ~40 arcsec ($\sim 30h^{-1}$ kpc) when the intrinsic PSPC resolution of ~25 arcsec is added to the smoothing in quadrature. The background level is $9.1 \pm 0.5 \times 10^{-4}$ counts s^{-1} arcmin^{-2}, where 1 count s^{-1} = 1.45×10^{-11} ergs cm^{-2} s^{-1} in the 0.5-2.4 keV band after correcting for Galactic absorption (Henry & Briel 1993). The lowest contour level is three times the background. The highest level plotted is 0.04 counts s^{-1} arcmin^{-2} (44 $\times$ the background). The difference between each contour is 2.9×10^{-3} counts s^{-1} arcmin^{-2}. The bright X-ray source in the southwest is associated with a radio galaxy in a group of galaxies at the same redshift as A754 (Harris *et al.* 1984).

KINEMATICS AND DYNAMICS OF THE MORGAN POOR CLUSTERS

Christina M. Bird
Dept. of Physics and Astronomy, University of Kansas
Lawrence, KS 66045-2151
tbird@kula.phsx.ukans.edu

Timothy C. Beers and Jeffrey R. Kriessler
Dept. of Physics and Astronomy, Michigan State University
East Lansing, MI 48824
beers@msupa.pa.msu.edu; kriessler@msupa.pa.msu.edu

John P. Huchra
Harvard/Smithsonian Center for Astrophysics
60 Garden St., Cambridge, MA 02138
huchra@fang.harvard.edu

ABSTRACT

Eighteen of the 23 MKW and AWM poor clusters now have 10 or more measured redshifts within $1.5h^{-1}$ Mpc of the central galaxy; 11 clusters have at least 20 measured redshifts. Based on the 21 clusters for which we have sufficient velocity information, the median velocity scale is 336 km s^{-1}, a factor of two smaller than found for rich clusters. Several of the poor clusters exhibit complex velocity distributions due to the presence of nearby clumps of galaxies. We check on the velocity of the dominant galaxy in each poor cluster relative to the remaining cluster members. Significantly high relative velocities of the dominant galaxy are found in only 4 of 21 poor clusters, 3 of which we suspect are due to contamination of the parent velocity distribution. Several statistical tests indicate that the D/cD galaxies are at the kinematic centers of the parent poor cluster velocity distributions. Mass-to-light ratios for the 13 of the 15 poor clusters for which we have the required data are in the range $50 \leq M/L_{B(0)} \leq 200\ M_\odot/L_\odot$.

1. INTRODUCTION

The sample of 23 poor clusters of galaxies originally identified by Morgan, Kayser, & White (1975, hereafter MKW) and Albert, White, & Morgan (1978, hereafter AWM) was selected on the basis of the D- or cD-like appearance of the first-ranked galaxy. Since then, many poor clusters have been identified from the catalog of Zwicky *et al.* (1961-1968), both with and without a central dominant galaxy (Huchra & Geller 1982, White, Bhavsar, & Bornmann, unpublished, Ramella, Geller, & Huchra 1990). Observations with the EINSTEIN satellite established that many of the MKW/AWM poor clusters have associated extended X-ray emission from a hot intracluster medium (Kriss *et al.* 1980; Kriss, Cioffi, & Canizares 1983; Price *et al.* 1991). However, the nature of the X-ray emission appears to change from "intra-cluster medium dominated" to "galaxy-dominated" as one proceeds from high-temperature to low-temperature galaxy groups and poor clusters (Dell'Antonio, Geller, & Fabricant 1994). Price *et al.* (1991) have argued that the MKW/AWM poor clusters are indistinguishable (in their X-ray and optical properties) from a representative sample of other poor clusters.

Although more recent data has suggested that the extended luminous envelope typical of D/cD galaxies in richer clusters appears to be absent from many of the poor clusters (Thuan & Romanishin 1981; Morbey & Morris 1983), these clusters remain of interest for several reasons. (1) David & Blumenthal (1992) argue that the efficiency of galaxy formation is much higher in poor clusters compared to their richer cousins, and use observed properties of groups and poor clusters to place interesting limits on cold dark matter cosmologies. Alternatively, White (1991) claims that if protogalactic winds are a significant mechanism for metal enrichment of the ICM (see the poster paper by Bird & Mushotzky in this volume), then their energy contribution is greatest in the poor clusters. In this case, variations in M_{gas}/M_{virial} and M_{gas}/M_{stars} noted by David and his collaborators are due to the difference in scale length of the ICM in cool and hot clusters. (2) The low velocity dispersions of many of these clusters make it likely that additional cluster galaxies may yet be captured by the dominant central galaxy, thus we may be witnessing the state of clusters before/during the formation of a classical D/cD galaxy. Recent large-scale numerical models of poor cluster evolution indicate that a centrally located first-ranked galaxy forms via mergers of cluster galaxies within 10 Gyrs, and at later times is found at a velocity relative to the remaining cluster galaxies which is less than 50% of the cluster velocity dispersion (Bode *et al.* 1994). Thus, the presently-observed relative velocity of the D/cD galaxy might be used, with some caution, to infer the relative age sequence of different poor clusters. (3) The realization that the poor clusters form an extension to the properties of richer cluster properties, as well as the recent observations of a rather strong negative evolution of the X-ray luminosity function (Edge *et al.* 1990; Henry *et al.*1992; David *et al.* 1993) suggests that the poor clusters may be examples of the basic structures from which richer clusters eventually form. As such, they provide valuable clues to early stages of cluster formation.

The kinematical properties of the MKW/AWM poor clusters have been considered by a number of previous studies (Beers *et al.* 1984 and references therein, Malumuth & Kriss 1986; Williams & Lynch 1991; Price *et al.* 1991; Dell'Antonio *et al.* 1994). Although progress has been made, these studies have still been hampered by the small numbers of available redshifts per cluster. Over the past several years we have been working to increase the sample sizes of galaxies with measured redshifts in the region of the MKW/AWM poor clusters, concentrating on the galaxies listed in the Zwicky catalog. These observations are presented in Beers et al. (1994; hereafter B94). In this paper we concentrate on the kinematical and dynamical analysis of the poor cluster data, and we compare the properties of the poor clusters with cD-dominated Abell clusters (Bird 1994).

VELOCITY DISTRIBUTIONS AND KINEMATIC PARAMETERS

In assembling the poor cluster dataset, we have supplemented our new measurements with the extensive redshift catalog of Huchra *et al.* (1992). In general, the poor clusters have velocity dispersions which are less than 600 km

s^{-1} (larger apparent dispersions are generally due to contamination of the poor cluster velocity distribution, as argued below and demonstrated in B94). Thus, for the purpose of our detailed kinematical analysis, we identify a subset of galaxies in the region of each poor cluster which have velocities within ± 2000 km s^{-1} of the D/cD galaxy. Galaxies in each poor cluster which have velocities in this range and which are located within $1.5h^{-1}$ Mpc of the D/cD galaxy (at its redshift distance, with $H_0 = 100h^{-1}$ km s^{-1} Mpc $^{-1}$) are assigned to subsample A. Those galaxies within the velocity window which lie *outside* the $1.5h^{-1}$ Mpc radial distance but *inside* the three degree circle are assigned to subsample B.

We begin our study of the physical properties of the MKW/AWM poor clusters by quantifying the recessional velocity and the velocity dispersions for each of the clusters for which we have at least five measured redshifts. Beers, Flynn, & Gebhardt (1990) describe the application of modern robust and resistant estimators of central location (the robust analogue of the mean) and scale (the robust analogue of the dispersion). The use of estimators which perform well in the presence of possible interlopers is particularly important for the present study, as the poor clusters often have rather small velocity dispersions which might be grossly affected by the presence of a few outliers. We choose to employ the biweight estimators of central location and scale. Estimates of the confidence intervals for these estimators are obtained by application of the bias-corrected accelerated bootstrap procedures described by Beers, Flynn & Gebhardt (1990). The recessional velocities and velocity dispersions for the individual poor clusters are presented in B94.

Figure 1 is a plot of the distribution of observed velocity dispersions for the poor clusters, as quantified by the biweight estimator S_{BI}, for the sample A+B galaxies versus the same quantity for the sample A galaxies. In this figure, a number of clusters with substantial contamination from nearby systems deviate strongly from the one-to-one line.

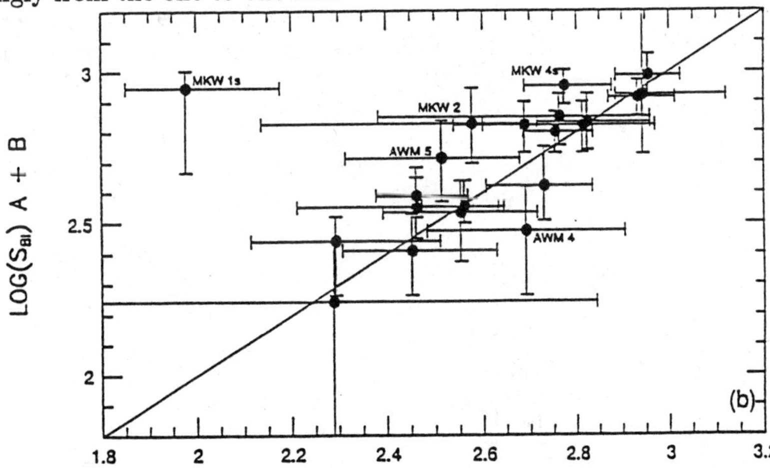

Figure 1. Observed velocity dispersions for sample A+B galaxies versus the same quantities for the sample A galaxies. The error bars indicate the 90% bootstrap confidence intervals on the respective quantities. Poor clusters which deviate from the one-to-one line are indicated.

In most cases, these deviations are caused by interactions with other massive systems in the neighborhoods of the poor clusters (see the case-by-case discussion in B94 for more details).

For 16 of the MKW/AWM clusters, our redshift information is supplemented by deep galaxy catalogs from Yamagata (1986). These catalogs are limited to galaxies within one degree of the D/cD. In addition, for poor clusters north of $\delta = 0^o$, we assemble catalogs of Zwicky galaxies within three degrees of the D/cD. These enable us to study the large-scale environment in which the poor clusters are located. Maps of the galaxy distributions, the velocity histograms, and discussions of the individual systems are given in B94. In nearly all the clusters within the Zwicky declination limit, complex structure is apparent in the extended galaxy maps, verifying our impression from Figure 1 that the poor clusters are undergoing strong interactions with their neighbors.

Even with the use of resistant estimators, the kinematic parameters (and dynamical masses) for the poor clusters in our sample can be perturbed due to the presence of multiple condensations of galaxies in their neighborhoods. Thus, for the eight poor clusters where these problems are most severe, we have clipped the apparent outliers. Though for most cases the reduction in velocity scale is modest, for both MKW 12 and MKW 2s the estimator S_{BI} dropped by roughly 700 km s^{-1}!

B94 includes a detailed discussion of the large-scale environments of each of the MKW/AWM systems, contamination of the velocity distributions and dynamical models for the clusters with massive systems in their vicinities. The majority of the poor clusters are located in complex environments and are interacting with other systems. For instance, the clusters MKW 7 and MKW 8, separated in projection by $2.6h^{-1}$ Mpc and in line-of-sight velocity by 523 km s^{-1}, are demonstrated to be gravitationally bound and likely to merge within the next 1 to 3 Gyrs. This system, with a total mass of $\sim 8 \times 10^{14} h^{-1} M_{\odot}$, may be a rich, albeit embryonic, cluster. If the MKW 7/8 system is typical of the poor clusters (as our maps of galaxy distribution suggest), then by studying the poor clusters we may be watching Abell-type clusters in the process of formation.

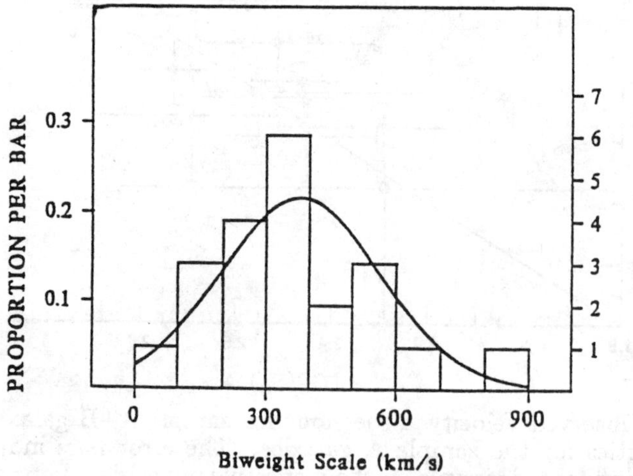

Figure 2. Distribution of velocity dispersions for the MKW/AWM clusters.

In Figure 2, we present the histogram of velocity dispersions for the poor clusters. The median velocity scale for the poor clusters in our sample is 336 km s^{-1}. The best-fit Gaussian for this distribution has a dispersion $\sigma = 183$km s^{-1}. In contrast, Zabludoff, Huchra, & Geller (1990) report that nearby rich Abell clusters have a median velocity dispersion of 744 km s^{-1}. Their sample consists of clusters with and without central dominant galaxies, and no correction for the presence of substructure is made in their determination of velocity dispersion. Bird (1994) finds a median velocity dispersion of 782 km s^{-1} for clusters which have cD galaxies and have been corrected for substructure. In any event, the median velocity dispersion of the poor clusters is less than half that of the rich systems. For the A samples, the velocity dispersions range from 95-857 km s^{-1}; this broad range suggests that rather than forming a distinct population, the MKW/AWM systems are a continuation of higher-mass clusters, as Price *et al.* (1991) noted.

MASSES AND MASS-TO-LIGHT RATIOS

We obtain dynamical mass estimates of the poor clusters using the projected mass estimator of Heisler, Tremaine & Bahcall (1985). This is

$$M_p = \frac{\xi}{GN} \sum_i v_i^2 R_i \ ,$$

where R_i is the projected distance of galaxy i from the D/cD galaxy, and v_i is the velocity of galaxy i with respect to the velocity location of the cluster, C_{BI}. The factor ξ is set equal to $\frac{32}{\pi}$, which is appropriate for a distribution of tracer particles with isotropic orbits moving in a smoothly-distributed gravitational potential. Like the classical virial mass estimator, the projected mass estimator is derived from Jeans' equation of stellar hydrodynamic equilibrium. But as Heisler, Tremaine & Bahcall (1985) demonstrate, M_p is preferred in cases where the number of tracer particles is small. The projected mass estimator for each of the poor clusters is presented in B94, Table 4, along with the virial and robust virial mass and 90% confidence intervals on all quantities.

Of course, applying any dynamical mass estimator depends critically on the assumption that the system under study is in a state of dynamical equilibrium. Given the complexity of the environments in which the MKW/AWM clusters find themselves, this assumption is questionable (as is also the case in the rich clusters). Therefore, this discussion of masses and mass-to-light ratios is uncertain. On the other hand, in the cases where X-ray and optical masses may be directly compared (Davis et al. 1994; see the contribution by Mushotzky in this volume), they generally agree. The correspondence of X-ray and optical masses suggests that either systematic errors are small, or they act in similar degrees in both the galaxies and the gas (which seems unlikely, since the physical mechanisms by which the tracer particles equilibrate are different; see discussion in Sarazin 1988).

The distribution of poor cluster masses is presented in Figure 3a. The masses range from MKW1s, with $M_p = 8 \times 10^{12} h^{-1} M_\odot$, to MKW12, with $M_p = 1.6 \times 10^{15} h^{-1} M_\odot$. (The high mass of MKW12 is probably an overestimate due to the complexity of its structure, even when clipping is performed on the velocity data.) The distribution is sharply peaked at about $10^{14} h^{-1} M_\odot$. In Figure

Figure 3. Dynamical Masses of Clusters: a) Distribution of M_p for the poor cluster sample (B94). b) Distribution of M_p for rich Abell clusters from the sample of Bird (1994). c) Total mass distribution (sum of poor and rich cluster samples) for galaxy clusters which possess D/cD galaxies. All masses are given in solar values for $H_0 = 100$ km s^{-1} Mpc^{-1}.

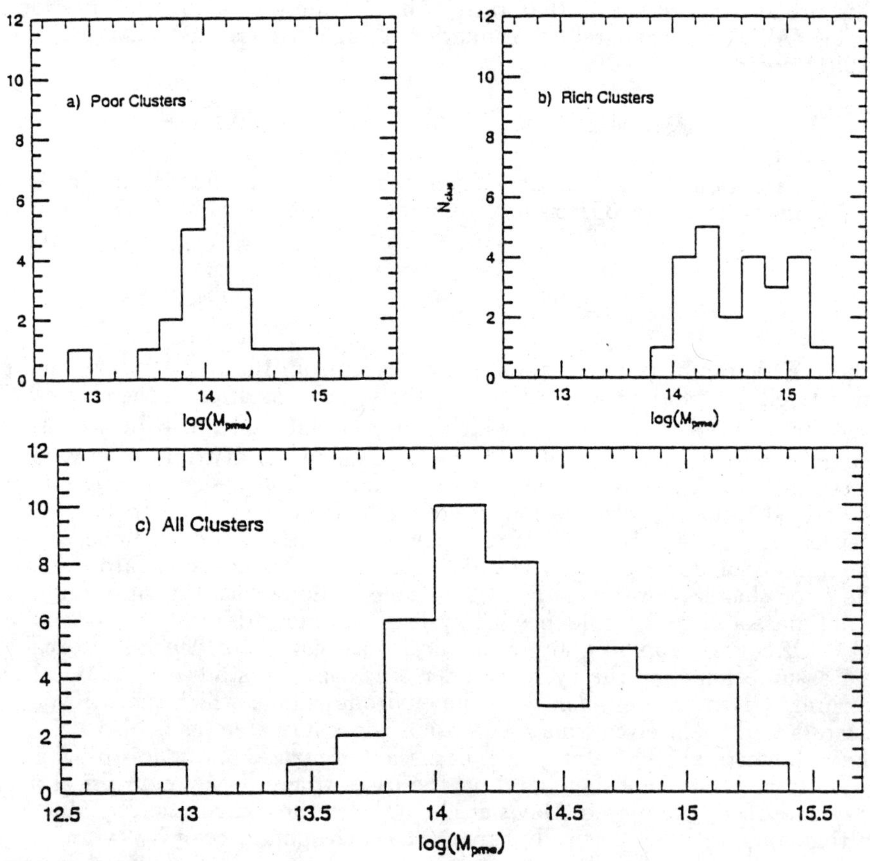

3b, the mass distribution of Abell clusters with cD galaxies is illustrated (Bird 1994, 1995). These masses have been corrected for the presence of substructure as described in Bird (1994). In contrast to the poor cluster distribution, the rich cluster mass distribution has no obvious peak. Masses vary from $\sim 10^{14}h^{-1}M_{\odot}$ for A2670 to $2.1 \times 10^{15}h^{-1}M_{\odot}$ for A399. It is possible that the broad distribution of rich cluster masses reflects their hierarchical, stochastic formation.

We calculate mass-to-light ratios for the 16 poor clusters for which deep galaxy catalogs have been obtained by Yamagata (1986). We estimate the total B(0) luminosity within $1.5h^{-1}$ Mpc (1 Abell radius) from the D/cD galaxy by assuming a $B(0)-V$ color of 0.9 for all galaxies. Because the poor clusters are all relatively close, we make no correction for K-dimming. We include all galaxies in the one-degree field without making corrections for possible interlopers. Because the Yamagata maps cover a fixed angular size, and in general, do not extend to 1.5 h^{-1} Mpc, we assume a r^{-1} projected number density profile (Beers & Tonry 1986), with a central density obtained from the 10 galaxies nearest the D/cD, and sample the present galaxies with replacement to fill in the outskirts of the cluster. To promote stability, the sampling procedure was iterated 1000 times and the average value was adopted. Note that this procedure assumes that the luminosity function obtained from the galaxies near the central portion of each cluster provides a fair estimate of the nature of the luminosity function in the outer regions of each cluster. The total luminosity for each cluster was then obtained by correcting for the luminosity contained at the faint end of the Schechter luminosity function, assuming that the knee of the luminosity function is at $M_{B(0)} = -19.2$ (Huchra 1991), and that the Yamagata (1986) samples are complete to V = 18.0. The largest such correction was no more than 20%. The individual results are included in B94, Table 4.

13 of 15 MKW/AWM poor clusters exhibit mass-to-light ratios in the range $50 \leq M/L_{B(0)} \leq 200\ h\ M_{\odot}/L_{\odot}$. The remaining two (MKW 7 and MKW 4s) have mass-to-light ratios on the order of 400 $h\ M_{\odot}/L_{\odot}$. These values are typical of results obtained for richer systems.

VELOCITY OFFSETS OF THE D/cD GALAXIES

It has long been recognized that the presence of bright central galaxies in poor clusters should provide clues to the origin and evolution of D/cD galaxies. In fact, because of their apparent simplicity, the poor clusters should be useful for assessment of the kinematic significance of their D/cD galaxies. Unlike the case in rich clusters, one does not have to worry excessively about the existence of "hidden" substructure, which might grossly disturb the observed kinematic parameters (e.g. Bird 1994).

The extensive simulations of Bode *et al.* (1994) explore the evolution of first-ranked galaxies in poor clusters, in particular the effects which result from changing the fraction of cluster mass initially tied to the galaxies rather than to the intra-cluster background. In their models, when the galaxies initially contain 50% of the cluster mass (in the form of dark halos), the galaxies experience dynamical friction and undergo mergers with each other, forming a dominant galaxy within roughly 10 Gyrs. As the fraction of mass initially tied to the galaxies is decreased, so is the rate of galaxy mergers, until a dominant galaxy is unable to form in a Hubble time. An important diagnostic of the merger process is the rate at which the "proto-FRG" reaches the kinematic center of the poor cluster. As Bode *et al.* show, once a dominant galaxy has been established (defined to be the first galaxy to achieve a total luminosity more than twice that of the brightest galaxy at the start of the simulation), its velocity with respect to the mean velocity of the other cluster members has typically decreased to less than half of the parent-cluster velocity dispersion. Because the velocity dispersions in the poor clusters are in general less than 500 km s^{-1} (in some cases as small as 200-300 km s^{-1}) , this is a particularly stringent prediction.

There have been numerous claims in the recent literature for the existence of "speeding" cD galaxies associated with rich clusters. As discussed by Beers *et al.* (1991) and Bird (1994), unrecognized substructure within the parent cluster is often the cause of the apparently high-velocity D/cD galaxy. In a number of cases the claim for a speeding cD disappears under the looking glass of statistical rigor. For the present study, we apply the Z-score method described by Gebhardt & Beers (1991):

$$Z = \frac{\text{D/cD peculiar velocity}}{\text{cluster velocity dispersion}} = \frac{v_{CD} - C_{BI}}{S_{BI}}$$

The 90% confidence intervals are determined using a bootstrap procedure as described above. The Z-score quantifies the position of the central galaxy within the overall gravitational potential of the cluster. By convention, for a central galaxy to be *significantly* displaced from the kinematic center of its host cluster, the confidence intervals must not include zero. To satisfy the Bode *et al.* (1994) requirement for the location of a dominant galaxy within a cluster, the Z-score must be less than 0.5.

The individual values of the Z-scores and the associated confidence intervals are listed in Table 3 of B94. Four of the 21 clusters (for which sufficient velocity information is available) are significantly displaced from their cluster's gravitational potential minimum: MKW2, $Z = 1.51$ (0.73, 4.02) (here we list the lower and upper 90% confidence limits within parentheses); MKW9, $Z = 1.31$ (0.36, 2.30); MKW10, $Z = 2.12$ (1.10, 3.92); and MKW12, $Z = 0.59$ (0.06, 1.08). Somewhat surprisingly, the fraction of poor clusters with displaced central galaxies is somewhat higher than a similar fraction for rich clusters ($\sim$ 10% with substructure correction; Bird 1994), but the difference is probably not significant. In general, D/cD galaxies are at rest with respect to their host systems. Of the 21 clusters, 14 have Z-scores less than 0.5 as predicted by the Bode *et al.* simulations.

SUMMARY AND CONCLUSIONS

In this paper we have considered the kinematical and dynamical properties of the poor clusters of galaxies originally identified by Morgan and collaborators. New redshift information has greatly improved determinations of velocity location and scale for these systems. Twenty-one of twenty-three poor clusters now have estimates of the kinematical parameters based on 8 or more velocities (the exceptions being MKW 3 and AWM 6). The median velocity dispersion of the MKW/AWM clusters is 336 km s^{-1}. This is slightly less than half the typical velocity dispersion of rich Abell clusters (Zabludoff, Huchra & Geller 1990; Bird 1994). The broad range in the poor cluster velocity dispersions suggests that they form a continuum with the richer systems, rather than a distinct population (consistent with the conclusions of Price *et al.* 1991).

The MKW/AWM clusters typically have masses of $\sim 10^{14}h^{-1}$ M$_\odot$, consistent with but on the low end of the mass range for rich Abell clusters (Bird 1995).

Their mass-to-light ratios are between 50 and $200h$ $M_{\odot}/L_{\odot}$, also typical of rich clusters (cluster dynamical masses provide about 20% of the density required to close the universe). As we have suggested above, a number of the MKW/AWM clusters have similar-mass clumps of galaxies in their vicinity. Two-body models (B94) suggest that they are likely to merge in the next few Gyrs, which suggests that they may be evolving into richer clusters.

A number of statistical tests have been used to verify that the D/cD galaxies in the Morgan poor clusters are, in general, at rest with respect to the potential wells of their parent cluster. A comparison with the simulation results of Bode *et al.* (1994) shows excellent agreement between the scaled velocity offsets of the D/cD galaxies and their $\beta = 50$ models with a range of dynamical ages. The four clusters with apparently significant offsets (MKW 2, MKW 9, MKW 10, and MKW 12) should be examined in more detail to verify this result.

The study of poor galaxy clusters is limited by their relatively-low number density of galaxies and their low X-ray surface brightness density. Nonetheless, observing these systems allows us to probe perhaps the most interesting regime of cluster evolution, that epoch in which both classical D/cD galaxies and their rich cluster hosts are forming. Studying poor clusters and groups of galaxies is vital to understanding not only the formation of massive galaxies, but also mechanisms for the metal and energy enrichment of the intracluster medium (White 1991) and the efficiency of galaxy formation in dense environments (David & Blumenthal 1992). Further optical and X-ray observations of these important systems, although technically challenging, will repay us with a much-improved understanding of the physical mechanisms operating within the most massive, gravitationally-bound systems in the Universe.

REFERENCES

Albert, C.E., White, R.A., & Morgan, W.W. 1978, ApJ, 211, 309 (AWM)

Beers, T.C., Flynn, K., & Gebhardt, K. 1990, AJ, 100, 32

Beers, T.C., Geller, M.J., Huchra, J.P., Latham, D.W., & Davis, R.J. 1984, ApJ, 283, 33

Beers, T.C., Kriessler, J., Bird, C.M. & Huchra, J.P. 1994, AJ, submitted (B94)

Bird, C.M. 1994, AJ, 107, 1637

Bird, C.M. 1995, in preparation

Bode, P.W., Berrington, R.C., Cohn, H.N., & Lugger, P.M. 1994, preprint

David, L.P., & Blumenthal, G.R. 1992, ApJ, 389, 510

David, L.P., Slyz, A., Jones, C., Forman, W., Vrtilek, S.D., & Arnaud, K.A. 1993, ApJ, 412, 479

Davis, D.S., Bird, C.M., Mushotzky, R.F. & Odewahn, S.C. 1994, ApJ, to appear 20 December

Dell'Antonio, I.P., Geller, M.J., & Fabricant, D.G. 1994, AJ, 107, 427

Edge, A.C., Stewart, G.C., Fabian, A.C., & Arnaud, K.A. 1990, MNRAS,245, 559

Gebhardt, K., & Beers, T.C., 1991, ApJ, 383, 72

Heisler, J., Tremaine, S., & Bahcall, J.N. 1985, ApJ, 298, 8

Henry, J.P., Gioia, I.M., Maccacaro, T., Morris, S.L., Stocke, J., & Woltjer, A. 1992, ApJ, 386, 408

Huchra, J.P., & Geller, M.J. 1982, ApJ, 257, 423

Huchra, J.P., Geller, M.J., Clemens, C., Tokarz, S., & Mickel, A. 1992, Bull. CDS, 41, 31

Kriss, G.A., Cioffi, D.F., & Canizares, C.R. 1983, ApJ, 272, 439

Kriss, G.A., Canizares, C.R., McClintock, J.E., & Feigelson, E.D. 1980, ApJ, 235, L61
Malumuth, E.M., & Kriss, G.A. 1986, ApJ, 308, 10
Morbey, C., & Morris, S. 1983, ApJ, 274, 502
Morgan, W.W., Kayser, S., & White, R.A. 1975, ApJ, 199, 545 (MKW)
Price, R., Burns, J.O., Duric, N., & Newberry, M.V. 1991, AJ, 102, 14
Ramella, M., Geller, M.J., & Huchra, J.P. 1990, ApJ, 353, 51
Sarazin, C.L. 1988, *X-ray emissions from clusters of galaxies*, (Cambridge: Cambridge University Press)
Silverman, B.W. 1986, Density Estimation for Statistics & Data Analysis, (Chapman & Hall, London)
Thuan, T.C., & Romanishin, W. 1981, ApJ, 248, 439
White, R.E. III. 1991, ApJ, 367, 69
White, Bhavsar,S. & Bornmann, unpublished
Williams, B.A., & Lynch, J.R. 1991, AJ, 101, 1969
Yamagata, T. 1986, Ann. Tokyo Astron. Obs., 2nd Ser, 21, 31
Zabludoff, A.I., Huchra, J.P., & Geller, M.J. 1990, ApJS, 74, 1
Zwicky, F., Herzog, E., Wild, P. Karpowicz, M., & Kowal, C. 1961-1968, Catalogue of Galaxies & Clusters of Galaxies, (California Institute of Technology, Pasadena)

DETERMINATION OF THE MASS DISTRIBUTION IN RICH CLUSTERS FROM IMAGING X-RAY SPECTROSCOPY

R.Mushotzky[1], M. Loewenstein[1,2], K. Arnaud[1,3] and Y. Fukazawa[4]
1 Laboratory for High Energy Astrophysics code 666, GSFC Greenbelt MD 20771
2 also USRA 3 also University of Maryland 4 Physics Department University of Tokyo

ABSTRACT

We discuss the use of x-ray imaging spectroscopy to derive the mass profiles of rich clusters of galaxies. Using the recent ASCA spectra of A496, A1795 and A2199 we find that the masses are formally constrained at R~1 Mpc to a statistical precision of a few percent and that all 3 of these systems have very similar gas to total mass ratios of ~20%. These results are in good agreement with the available Rosat results but of considerably higher precision. The derived optical velocity dispersions from our model potentials are in good agreement with the published optical data. We conclude, based on the available numerical simulations and consideration of other likely sources of pressure (magnetic fields,turbulence and cosmic rays) that the mass derived from this technique is not subject to large error.

1. INTRODUCTION

The dynamical mass of clusters of galaxies is one of the best determined parameters in the "field" of Dark Matter. The strong contrast between the amount of light detected in clusters, the "dynamically" determined mass and the mean mass to light ratio, M/L, for stellar systems is one of the best pieces of evidence for gravitationally "bound" dark matter (see Trimble 1987 for a review). However, as has been shown in detail by Binney and Strimple (1978), Merritt (1987), Richstone and Tremaine (1987) and The and White (1987) the <u>accurate</u> (e.g. factor of 2 or better) model independent, determination of the mass of a bound system from a relatively small (~300) sample of galaxy velocities <u>is not possible</u>. This is because the fundamental parameters of interest, the velocity and spatial distributions are always going to be poorly sampled and because the velocity distribution is a tensor only one of whose components can be directly measured.

However, under a wide range of assumptions about the form of the potential, the form of the velocity distribution, the ratio of the scale lengths of dark to visible matter etc, sensible bounds on the mass, inside physically interesting radii can be derived. As shown by Heisler, Tremaine and Bahcall (1985) a variety of mass estimators based on galaxy dynamics have typical errors of a factor of 2. Thus to obtain more accuracy and to derive total mass profiles other techniques are necessary.

In the last 10 years (Mushotzky 1990) the two main techniques developed to derive the mass of rich clusters have been based on either gravitational lensing (see the review by Narayan in these proceedings) or x-ray imaging spectroscopy. In this paper we will briefly review the use of x-ray imaging spectroscopy to obtain the mass profile, summarize previous work in this area and present new ASCA results. Recent studies of the correlation between x-ray temperature and cluster velocity dispersion (Lubin and Bahcall 1994, Bird and Mushotzky 1994) show that the x-ray temperature and the optical velocity dispersion are fairly well correlated and roughly consistent with $kT \sim \sigma^2$. Thus the masses derived from the "virial theorem" and the x-ray temperature measures are certainly of the same order.

2. METHODOLOGY, RELIABILITY and LIMITATIONS

As first discussed by Fabricant, Rybicki and Gorenstein (1984) and Gorenstein and Fabricant (1983), if we assume that the hot x-ray emitting gas which dominates the baryon content of clusters is in hydrostatic equilibrium with the total mass distribution of the cluster, the measurement of the temperature and density distribution of the gas determines the mass profile. The equation of hydrostatic equilibrium for a spherically symmetric system:
$GM(r)=-(r^2/\rho_g)\, d/dr(\rho_g kT_g/\mu m_H)$ where ρ_g and T_g are respectively the gas density and temperature distributions . Alternatively this may be written as
$M(<r)=(kT_g(r)/G\mu m_H)[(d\log\rho_g/d\log r)+(d\log T_g/d\log r)]r$.
All of these quantities are directly observable with x-ray imaging spectroscopy

The gas density distribution, ρ_g , is directly derivable from the x-ray surface brightness profile as measured, for example by the Einstein observatory and Rosat for many ten's of clusters. Before the advent of x-ray imaging spectroscopy one had to either make a ansatz about the form of the temperature distribution (Cowie, Henriksen and Mushotzky 1986) or combine such an ansatz with very crude imaging (Edge 1990, Hughes et al 1992) in order to obtain mass limits. Beginning in the early 1990's, the determination of the temperature profile for the massive, hot, x-ray luminous clusters, was very crude and possible only for the brightest few objects (see next section). Now, with ASCA, this measurement to be easily obtained for many clusters. The ASCA data now allow precision temperature measurements (5% errors in kT and 10% errors in abundance) in spatial resolutions elements of ~20 square arc minutes (Arnaud et al 1994).

Analysis of the effects of asphericity on the hydrostatic mass estimator (Fabricant, Rybicki and Gorenstein 1986) show that even in the case of extreme prolate or oblate configurations the errors are less than 30%. More recently there has been extensive theoretical modeling of the reliability of this technique in the context of models for the formation of large scale structure and the growth of perturbations in the universe which explicitly include the effects of the gas as well as dark matter (cf. Evrard 1994 and Navarro et al 1994 for the most recent overviews). These two papers show that for a low Ω universe the mean mass error using the equation of hydrostatic equilibrium is very low over the length scale where the density contrast with respect to average is between 1000-100 (2-10 "core radii"). For a high Ω universe the median offset is ~-20% and the 68% uncertainty is +/-10% around the mean offset. In the work of Navarro et al (1994) the estimated uncertainty in the derived mass from x-ray imaging data is about ~-13% at $R\sim0.3R_{virial}$ (~0.7 Mpc for T=4 keV); that is, the spatially resolved x-ray data give 0.87+/-0.08 of the true mass.

However in both of these papers the main "error" comes from those clusters which show strong substructure. If is not clear from these papers what the errors would be if one chose clusters which do not show, in projection at least, any evidence for strong recent mergers. While the numerical simulations gives some confidence in the following work, we know that the simulations are not yet completely rigorous in that they do not have sufficient dynamic range in the core nor do they handle cooling well. Other workers in the field (e.g.Roettiger, Burns and Loken 1993) have found considerably larger variances between the true mass and that estimated by x-ray imaging spectroscopy for very strong merger situations where large scale bulk motions are seen to persist for long timescales.

To achieve maximum reliability in the mass estimates via using of this technique one should avoid clusters that show evidence of mergers. Recent work by Bird (1994) and Davis (1994) have shown that many, if not most, rich clusters show evidence for recent mergers based either on their x-ray morphology, optical galaxy distribution or structure in the velocity field. However, at least 20% of the rich luminous clusters do not show any evidence for substructure in either the x-ray or optical data. Thus selection of relatively "unperturbed" systems is possible.

The assumption of hydrostatic equilibrium requires that the the gas be in thermal equilibrium.The electron-ion equilibration timescale is t_{eq}~$2x10^5$ $T_8{}^{3/2}$/n yrs (T in units of 10^8 degrees and n is the gas particle density). For the clusters used in the following analysis t_{eq}~ $2x10^9$ years, a typical time between larger mergers, at R~2 Mpc (if the gas is isothermal). Thus the assumption of hydrostatic equilibrium is probably not valid at larger radii. Numerical simulations of cluster formation show that the "virial" radius, at the present epoch, is ~3Mpc$T_8{}^{1/2}$, the 3 systems considered below have T_8~1/2; again extension of these results beyond 2 Mpc should be considered risky. Since the sound crossing time t_s~$6x10^8R_{Mpc}/T_8{}^{1/2}$ yrs is also longer than the "age" of the system at 2-3 Mpc the hydrostatic assumption is certainly not valid this far out. In conclusion, it seems that the use of x-ray imaging spectroscopy to derive accurate masses of rich clusters is limited to radii of ~2 Mpc. However temperature measurements at larger radii should reveal the "accretion shock" and details of cluster formation and thus are quite interesting in their own regard.

In deriving the mass of the clusters we assume that gas pressure balances gravity and that other sources of pressure (magnetic fields, cosmic rays, turbulence) are relatively negligible. These other potential sources of pressure would have to have an energy density similar to that of the gas to make a major difference. There is very little data on turbulence, B fields and cosmic ray densities at the large distances from the cluster centers where most of the mass lies. However where there is information, in the cluster centers, these effects are, at most, on the order of the gas pressure (cf Loeb and Mao 1994 for the possible effects of magnetic pressure and turbulence in the cluster core). Theory (e.g. Soker and Sarazin 1990) indicates that these effects should be much less at larger radii. The required energy density at R~1 Mpc, nkT~ $2x10^{-12}$, gives B~7uG (magnetic pressure $B^2/8\pi$) while the turbulent pressure ($1/3\rho_g v_t{}^2$)would require velocities of 500 km/sec for equal pressure.

Preliminary indications from Faraday rotation behind clusters (Kim et al 1991) indicate that the B fields are small and the lack of diffuse radio sources in most clusters indicates that the cosmic ray density is not important. The most difficult effect to measure is residual turbulence from mergers. While classical theory indicates that the turbulence should damp out on several sound crossing times (<10^9 yrs at 1 Mpc), it still might be important in some clusters.

3 PREVIOUS WORK

Because of the great promise of this field there have been numerous attempts to obtain mass profiles using spatially resolved x-ray spectroscopy (see Mushotzky 1992 for a summary of the situation prior to 1993) . The best such results were obtained by the Spartan and Spacelab groups for the Perseus and Coma clusters (Eyles et al 1991, Snyder et al 1990, and Watt et al. 1992). Recently BBXRT (Miyaji et al 1992) and Rosat results (Henry and Briel 994) have been obtained for A2256 and Rosat results for Coma (Briel et al 1992) and several other clusters (Bohringer 1994) have been presented. However the BBXRT results have suffered from having relatively poor signal to noise and low spatial sampling and the Rosat data have suffered from limited bandpass. In addition the effects of sub-structure on the mass estimators were not considered for A2256 and Perseus. Given that both of these clusters show strong non-azimuthally symmetric spatial variations in their temperature (Arnaud et al 1994, Henry and Briel 1994) it is not clear how reliable the previous mass estimations were.

The Rosat data have excellent signal to noise and good spatial resolution, but the limited bandpass gives relatively large errors in the temperature determination for rich clusters (see Mushotzky 1994 for a comparison). However for poor clusters (David et al 1994) and elliptical galaxies (Trinchieri et al 1994) the Rosat data are quite good and, if the spectral model is correct (cf. Mushotzky et al 1994) can be used to derive accurate mass profiles of these lower mass systems. A comparison for the Centarus cluster, a moderately massive system, can be seen by comparing Allen and Fabian (1994) with Fukazawa et al (1994) where both ASCA and Rosat find kT~3.8 keV at ~10' but the ASCA statistical errors in temperature are a factor of 5 smaller.

While the study of low mass systems is interesting in and of itself it is not clear that they are, in detail, cognates to the more massive systems considered here. These systems have very flat x-ray surface brightness profiles (Jones and Forman 1994) which are not consistent with numerical models of cluster formation unless additional energy sources (White 1991) are included. In at least two systems (Centarus (Allen and Fabian 1994) and NGC 5044 (David et al 1994) the mass fraction of gas is ~9%, considerably less than that found in rich clusters (see below). In addition for low mass systems the x-ray surface brightness is considerably lower than for rich clusters and thus one cannot measure their properties far away from the core regions. In fact, while for rich clusters one can often detect them out to their "virial radii" for poor systems the surface brightness is only ~10% of the sky at ~0.3 of the virial radius

A summary of the Rosat results (Bohringer 1994) shows that the mass is uncertain by roughly a factor of 2-3 (full range) at R~0.5-3 Mpc, but the mass due to the hot x-ray emitting gas is well determined. Other recent Rosat mass estimates are presented by Brunner et al , Siddiqui et al and David et al (all are in Petre and Schlegel 1994), White et al 1994, Allen et al 1993 and Allen and Fabian 1994. All of these results are qualitatively similar: the temperature profiles are roughly isothermal , the x-ray emission extends (for rich clusters) out to ~$2h_{50}^{-1}$ Mpc, the mass to light ratio is ~$100\text{-}200h_{50}^{-1}$, the hot gas accounts for ~20% $h_{50}^{-3/2}$ (10-45%) of the total mass of the system, the galaxies are a "minor" mass component and the mass derived is roughly consistent with the classical virial estimator if the effects of substructure on the velocity dispersion are considered. However, as stressed by Briel and Henry (1994) the Rosat data, because of their relatively large errors, do not allow precision measures of the mass profiles. Allen and Fabian (1994), using a combination of optical and x-ray data, were able to constrain the core radius of the potential of the Centarus cluster to 60-140 kpc; a King model fit to the x-ray surface brightness (Matilsky et al 1985) has a core radius of ~90 kpc.

4 PRESENT RESULTS

The ASCA satellite (Tanaka,Inoue and Holt 1994) has allowed a large increase in the our knowledge of cluster spectra. ASCA, launched in 1993, made a fundamental change in the field with CCD/GIS imaging detectors and foil optics. Its relatively large collecting area (~400 cm^2 at 6 keV), good spectral resolution ($E/\delta E$~ 50 at 6 keV) with true imaging spectroscopy, broad bandpass (0.4-10 keV), low detector background (diffuse sky limited at E<7 keV) and 20' diameter potential field of view allows essentially systematics limited temperature profiles to be obtained for dozens of clusters. However its limited angular resolution $\delta\theta$~1.5' and relatively large wings on the point spread function requires care in the analysis of small extended sources.

A. Source Selection

We have selected the only 3 clusters, A496, A1795 and A2199 from the ASCA performance verification (PV) observations that meet the following criteria: that they (1) be round and smooth and have no indication from either x-ray or optical observations of having significant substructure: (2) be rich, massive clusters whose temperatures are between 3-7 keV (to minimize statistical uncertainties in ASCA spectra) and to be representative systems: (3) be at redshifts between 0.025 and 0.07, such that the ASCA angular resolution gives an effective cell size of ~100-200 kpc and the field of view allows the temperature to be measured to R~1-2 Mpc: (4) have high quality Rosat images and Ginga spectra to allow detailed comparison. While our selection process did not require it, all of these are also cooling flow systems. The presence of cooling gas in the center somewhat limits our ability to derive mass profiles in the central 100-200 kpc.

B Analysis

We have sliced the clusters into 6-7 concentric rings whose average width is ~2.5' (which corresponds to the 90% power radius of the PSF). For all of these rings the cluster surface brightness was well determined and background subtraction is important only for the outer 2 rings. All 4 of the ASCA detectors were analyzed and the best fits were obtained for abundance and temperature in each ring. Multi-temperature fits were allowed, but were required only in the central 1-2 rings, consistent with the cooling radius derived from image analysis. The background was derived from deep fields at high galactic latitude.

The results presented in this section have not been corrected for the effects of the point spread function (Ohashi et al 1994). However recent analysis indicates that if the observed cluster temperature profile is roughly isothermal and the effective core radius is larger than 2' inside ~12' the temperatures are properly estimated. At larger radii the PSF effects act to increase the observed temperature over the true one significantly. Software has been developed to model these effects and the detailed results will be presented in the literature soon. The effect of neglecting the PSF is to increase the true errors over the statistical ones. Preliminary estimates show that the true errors in the derived temperature are 2-3 times larger than the statistical ones when the PSF is not taken into account. However, even with these larger uncertainties the mass distributions (see next section), are dominated by modeling assumptions rather than statistical errors.

C Mass Models for Clusters

Loewenstein (1994) has developed a numerical code to derive the best fit mass profiles consistent with all the observed data. The mass models are of the form:

$M_{total}(r)=M_{gal}(r)+M_{gas}(r)+M_{dark}(r)$;where $M_{gal}(r)$ is a "King" model $(1+(r/a_{gal})^2)^{-1.5}$; a_{gal} is derived from the literature and we use $M/L_V\sim5$ (to convert from L to M) (however our result is insensitive to this assumption since M_{gal} is the smallest component)

$M_{gas}(r)$ is a "β" model with a density law of the form

$\rho_{gas}(r)=\rho_{0\ gas}(1+(r/a_{gas})^2)^{-3\beta/2}$ with $\rho_0(gas)$, a_{gas} and β from our own fits to the public Rosat PSPC x-ray surface brightness data.

$M_{dark}(r)$ can have several forms. We have taken two generalized forms

1) a "generalized King profile" which has a density law of the form

$\rho_{dark}(r)=\rho_{0,\ dark}(1+(r/a_{dark})^2)^{-n}$ where n is varied between 1 and 1.5 (theoretical prejudice)

a_{dark} is varied between 0.1 ("scale free") and 2 a_{gas} and $\rho_{0,\ dark}$ is varied to fit the temperature profile or

2) a "Hernquist profile" (suggested by S. White's summary of the n-body simulations) where $\rho_{dark}(r)=\rho_{0,dark}(r/a_{dark})^{-1}[1+(r/a_{dark})]^{-n}$

The cut off radius of the mass distribution is set to $10a_{gas}$ and there is also the possibility of an external pressure (to mimic the existence of an accretion shock, ram pressure or gas beyond the cut off radius). The model thus has 3(+1) free parameters n, a_{dark} $\rho_{0,dark}$ as the 3 "fundamental parameters" and the external pressure (as a secondary parameter). The results are somewhat sensitive to H_0 since the gas/dark matter and galaxy/dark matter ratios depend on H_0; however since $M_{gal}(r) < M_{gas}(r) < M_{dark}(r)$, the effect is not large.

D. Results of fits

With the exception of the cooling flow regions, which are handled in the present analysis by the inclusion of a 2 temperature fit, the temperature profiles are remarkably isothermal. The statistical errors in the temperatures in most of the bins is <10%.

1) General properties of the solutions

The mass models constructed from the x-ray imaging spectra are very similar for all 3 clusters.(Table (1). All models which provide a good fit to the spatially resolved spectra (see figures 1,2,3) give essentially the same mass within a few percent. Extension of these solutions to larger radii are well bounded (the range of fitted masses is less than 15%). However in the central regions the fits can diverge due to limited angular resolution, projection effects and the presence of a cooling flow.

Table 1 General Properties of Solutions

NAME	Rmax Mpc	M(Rmax) $x10^{14}$	dark/lum	gas/total	galaxy/total	dark/total
A496	0.95	2.9-3.2	3.4-3.9	0.19-0.21	0.019-0.021	0.77-0.79
A2199	0.86	2.4-2.7	3.1-3.6	0.17-0.19	0.045-0.050	0.76-0.78
A1795	0.92	3.8-3.9	3.1-3.2	0.20-0.22	0.028-0.032	0.75-0.76

It is rather amazing that all the objects are consistent with the same ratios of luminous to dark matter .

The derived values of gas to total mass is 20%, completely consistent with the Rosat results where the same objects have been analyzed (cf Brunner et al 1994 figure d), Siddiqui et al 1994 figure 3).The derived statistical errors are very small, much smaller than the uncertainties introduced by modeling errors and the PSF correction term. The minimum core radius for the dark matter distribution is found for the low n models, and can be considerably less than the gas scale length. However if n=1.5, $a_{dark} \geq a_{gas}$.

For the 2 clusters with "good" optical photometry <M/L> ~150, consistent with the mean value from the virial estimator (Mushotzky 1992). Many (but not all solutions) tend to have the dark/luminous ratio rise in the center- (more true of the n=1 models). There is a tendency for the n=1, $a_{dark} \sim 1/2 a_{gas}$ models to fit best. It is difficult to remove the a_{dark},n degeneracy given the limited physical range of our data. Further analysis should be able to extend the temperature profiles to larger physical lengths which should help in this regard. Given either n or a_{dark} the other is "well" determined (for fixed n, a_{dark} is known to a factor of 2) . None of these 3 clusters have central critical densities large enough to become gravitational lenses, but these are "average" systems and hotter clusters/more massive systems certainly exist.

There is some sensitivity to the external pressure (which is not known). This external pressure is a "calculational device" meant to incorporate "other" physics, however the properties depend little on whether the external pressure is fixed at zero or allowed to vary. We caution the reader that a thorough search of parameter space has not yet been completed and that the results reported here are not yet complete.

2) Comparison with mass derived from optical data

If we make the very naive assumption that $R_H \sim 1$ Mpc, that the galaxies follow a King profile and that the orbits are isotropic, then $M_{(opt)} = 7.0 \times 10^{14} \sigma^2_{1000} R_{Mpc} M_0$. Table 2 compares these values

Table 2 Mass Estimators at 1 Mpc (in units of 10^{14} M)

NAME	σ(km/sec)	M(opt)	M(X)	M(virial)[1]
A496	714	3.6	3.05	3.3-4.9
A1795	920	5.9	3.9	3.6-20
A2199	794	4.4	2.55	3.0-15

1 Biviano et al 1993

Thus a (*very*) naive application of the "virial" theorem gives mass estimates which tend to be "high" by ~40%. Detailed fits to the optical data for A496 (Malumuth et al 1992, figure 1) show that the use of different mass estimators results in a large range of allowable solutions. Our mass estimates are in remarkably good agreement considering the assumptions involved in the optical analysis (e.g. orbit distribution, lack of biasing etc)

3) Comparison with optical velocity dispersions
We have calculated the "predicted" optical velocity dispersions from our models under the assumptions that the orbits are isotropic and that the observed galaxies "properly" sample the potential (Table 3)

Table 3
Comparison of Predicted Optical Velocity Dispersion

Name	σ(predicted)km/s best fit	range	σ(obs) km/s	σ(obs,substructure corrected)km/s[1]
A496	628	(527-654)	658-837	457-619
A1795	709	(669-734)	715-976	783-1107
A2199	583	(523-604)	711-953	711-953

1 From Bird (1994),

Taking the mean values, σ(predicted)= 0.79 σ(observed) and σ(predicted)= 0.84 σ(corrected). We notice that the largest deviation is seen for A2199, the most elliptical of the clusters (Buote and Canizares 1992). Thus while the "virial" estimator is not accurate, the observed optical velocity dispersions are in rough agreement with the potential as predicted by the x-ray data.There is no evidence for velocity bias; if anything the galaxies are a little "hot". If we exclude A2199 σ(predicted)= 0.93 σ(corrected) - it is clear that larger samples are required to check to see if there is any velocity bias at the levels predicted from n-body calculations.

5) Problems with Data/Models
We have assumed spherical symmetry and isotropic orbits- this probably fails at some level (cf A2199 whose image is clearly elliptical and which has the worst fits). At large radii the detector calibration is uncertain- thus the data are "too precise" for the likely systematic errors, and the effects of the point spread function have not been taken into account. The effects of subclustering, while small in these systems might be significant at the level of precision of the x-ray data. We thus conclude that the ASCA data have already reached the limits of this technique for clusters of z<0.1. At higher redshifts the ASCA spatial resolution is too poor to detect all but the strongest substructure and is inadequate to obtain more than 2-3 spectral resolution rings.

5. CONCLUSIONS

The ASCA data have demonstrated the power of x-ray imaging spectroscopy to derive the mass profiles of rich clusters of galaxies. The formal errors in the derived mass are very small. The derived masses are in rough agreement with those derived using the virial theorem. The systems have ~20% of the binding mass in gas, which confirms the "baryon catastrophe" found by White et al 1994 for the Coma cluster. The use of this technique for other systems observed by ASCA in the first 18 months of its operation should increase by an order of magnitude the number of clusters with precise mass profiles. While there is a formal disagreement by a factor of 2-3 (Loeb and Mao 1994 , Miralda-Escude and Babul 1994),between the masses in the central regions derived from x-ray imaging spectra and giant gravitational lenses, recent work by Bartelmann (these proceedings) shows that the "naive" analysis of the gravitational lens data tends to systematically overestimate the derived mass by a factor of ~1.5. Our work indicates that the use of the gas core radius as the scale length of the potential may result in an error of similar magnitude. Consideration of both of these terms may reconcile the differences.

The future of this technique is very bright. The next generation x-ray telescopes (AXAF,XMM and Astro-E), to be launched in 1998-99, will have considerably better angular resolution and thus can measure the potential for both more distant systems and on smaller scale lengths. By the end of this decade I expect that the mass profiles of many ten's of clusters over a wide range of redshifts will be well determined.

Acknowledgements : We would like to thank the organizers of this meeting for the opportunity to present the exciting new ASCA results. We would like to thank Prof. Y Tanaka and the rest of the ASCA team for a marvelous job in developing and operating this giant step forward.

References

Allen, S. W., Fabian, A. C., Johnstone, R. M., White, D. A., Daines, S. J., Edge, A. C., & Stewart, G. C. 1993, MNRAS, 262, 901

Allen, S. W., & Fabian, A. C. 1994, MNRAS, 269, 400

Arnaud, K. A., et al. 1994, ApJ, 436, L67

Bartelmann, M. 1994, these proceedings

Binney, J., & Strimple, O. 1978 MNRAS, 185, 473

Bird, C. M. 1994, AJ, 107, 1637

Bird, C. M., & Mushotzky, R. F. 1994, ApJ, submitted

Biviano, A., Girardi, M., Giuricin, G., Mardirossian, F., & Mezzetti, M. 1993, ApJ, 411, L13

Bohringer, H. 1994, in Proceedings of the XIVth Moriond Astrophysics Meeting, Clusters of Galaxies, ed. F. Durret et al. (Paris: Editions Frontieres)

Bohringer, H., Briel, U. G., Schwarz, R. A., Voges, W., Hartner, G., & Trumper, J. 1994, Nature, 368, 828

Briel, U. G., Henry, J. P., & Bohringer, H., 1992, A&A, 259, L31

Briel, U. G., & Henry, J. P. 1994, Nature, 372, 439

Brunner, H., Weimer, S., Westphal, H., & Staubert, R. 1994, in AIP Conference Proceedings 313, The Soft X-Ray Cosmos, ed. E. M. Schlegel & R. Petre (New York: AIP), 372

Buote, D. A., & Canizares, C. R. 1992, ApJ, 400, 385

Cowie, Henriksen, Mushotzky 1987, ApJ, 317, 593

David, L. P., Jones, C., Forman, W. & Daines, S. 1994, ApJ, 428, 544

David, L. P., Jones, C., Forman, W. in AIP Conference Proceedings 178, The Soft X-Ray Cosmos, ed. E. M. Schlegel & R. Petre (New York: AIP), 383

Davis, D. 1994, PhD Thesis, University of Maryland

Edge, A. C., 1989, PhD Thesis, Leicester University

Evrard, A. E. 1994, in Proceedings of the XIVth Moriond Astrophysics Meeting, Clusters of Galaxies, ed. F. Durret et al. (Paris: Editions Frontieres)

Eyles. C. J., Watt, M. P., Bertram, D., Church, M. J., Ponman, T. J., Skinner, G. K., & Willmore, A. P. 1991, ApJ, 376, 23
Fabricant, D., & Gorenstein, P. 1983, ApJ, 267, 535
Fabricant, D., Rybicki, G., & Gorenstein, P. 1984, ApJ, 286, 186
Fukazawa, Y., Ohashi, T., Fabian, A. C., Canizares, C. R., Ikebe, Y., Makashima, K., Mushotzky, R. F., & Yamashita, K. 1994, PASJ, 46, L55
Heisler, J., Tremaine, S., & Bahcall, J. N. 1985, ApJ, 298, 8
Hughes, J. P., & Tanaka, Y. 1992, ApJ, 398, 62
Hughes, P. A., Aller, H. D., & Aller, M. F. 1992, ApJ, 396, 469
Jones, C., & Forman, W. 1994, in AIP Conference Proceedings 313, The Soft X-Ray Cosmos, ed. E. M. Schlegel & R. Petre (New York: AIP), 129
Kim, K., Kronberg, P., & Tribble, P. 1991, ApJ, 379, 80
Loeb, A., & Mao, S. 1994, ApJ, 435, L109
Loewenstein, M. 1994, ApJ, 431, 91
Lubin, L. M., & Bahcall, N. A. 1993, ApJ, 415, L17
Malumuth, E. M., Kriss, G. A., Van Dyke Dixon, W., Ferguson, H. C., & Ritchie, C. 1992, AJ, 104, 495
Matilsky, T., Jones, C., & Forman, W. 1985, ApJ, 291, 621
Merritt, D. 1987, ApJ, 313, 121
Miralda-Escude, J., & Babul, A. 1994, ApJ, submitted
Miyaji, T., et al 1993 ApJ, 419, 66
Mushotzky, R. F. 1990, in AIP Conference Proceedings 222, After the First Three Minutes, ed. S. S. Holt, C. L. Bennett & V. Trimble (New York: AIP), 394
Mushotzky, R. F. 1992, in Clusters & Superclusters of Galaxies, ed. A.C. Fabian, (Dordrecht: Kluwer), 91
Mushotzky, R. F. 1994, in New Horizon of X-Ray Astronomy - First Results from ASCA, ed. F. Makino & T. Ohashi (Tokyo: Universal Academy Press), 243
Mushotzky, R. F., Loewenstein, M., Awaki, H., Makishima, K., Matsushita, K., Matsumoto, H. 1994, ApJ, 436, L79
Narayan,R. 1994 These proceedings
Navarro, J. F., et al. 1994, MNRAS, in press
Ohashi, T., et al. 1994, these proceedings
Richstone, D. O., & Tremaine, S. 1987, ApJ, 327, 82
Roettiger, K., Burns, J., & Loken, C. 1993, ApJ, 407, L53
Siddiqui, H., Stewart, G., Butcher, J., Edge, A., & Johnstone, R. 1994, in AIP Conference Proceedings 313, The Soft X-Ray Cosmos, ed. E. M. Schlegel & R. Petre (New York: AIP), 383
Snyder, W. A., Kowalski, M. P., Cruddace, R., Fritz, G., Middleditch, J. 1990, 365, 460
Soker, N., & Sarazin, C. L. 1990, ApJ, 348, 73
Tanaka, Y., Inoue, H., & Holt, S. S. 1994, PASJ, 46, L37
The, L. S., & White, S. D. M. 1987, AJ, 95, 15
Trimble, V. 1987, ARAA, 25, 425
Trinchieri, G., Kim, D.-W., Fabbiano, G., & Canizares, C. R. C. 1994, ApJ, 428, 555
Watt, M. et al. 1992, MNRAS, 258, 738
White, R. E. III 1991, ApJ, 367, 69
White, S. D. M., Navarro, J., Evrard, A., & Frenk, C. 1994, Nature, 366, 429

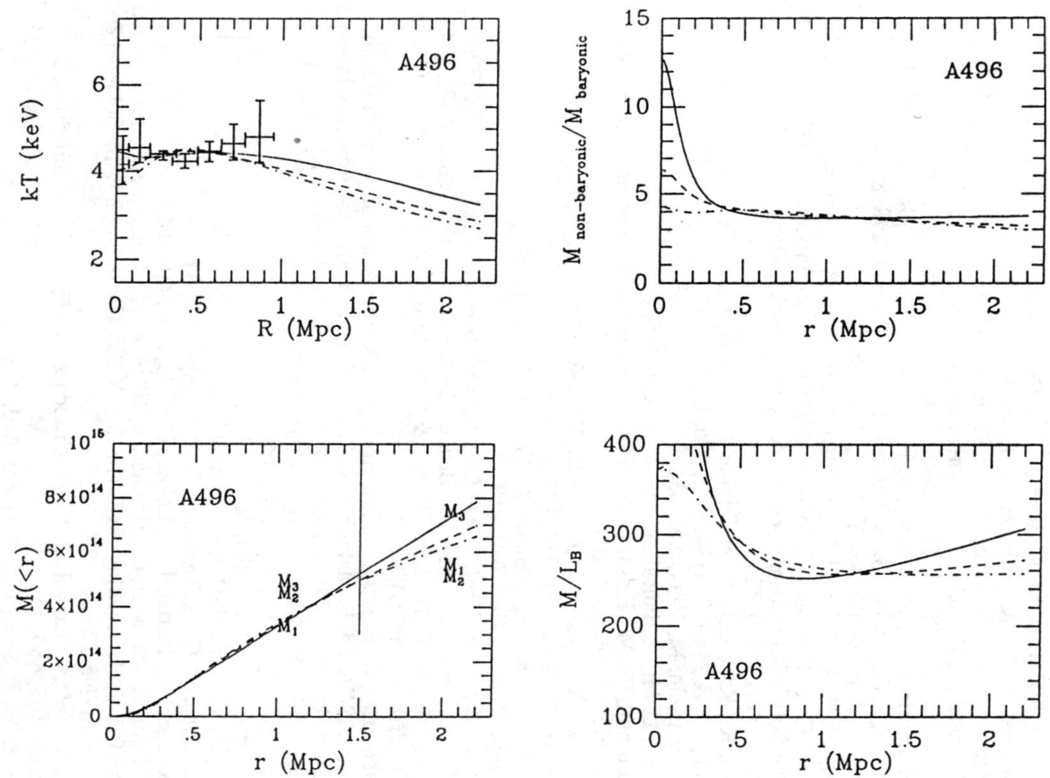

Fig 3. Same as fig 1 for Abell 496. Solid line shows the best-fit for n=1 (a_{dark}=0.5 a_{gas}), dashed line n=1.25 (a_{dark}=0.75 a_{gas}) , dash-dotted line n=1.5 (a_{dark}= a_{gas}). The vertical dashed line shows the limits onthe virial mass from Biviano et al (1993) and M_1,M_2,M_3 are dynamical mass estimates from Malmuth et al 1992.

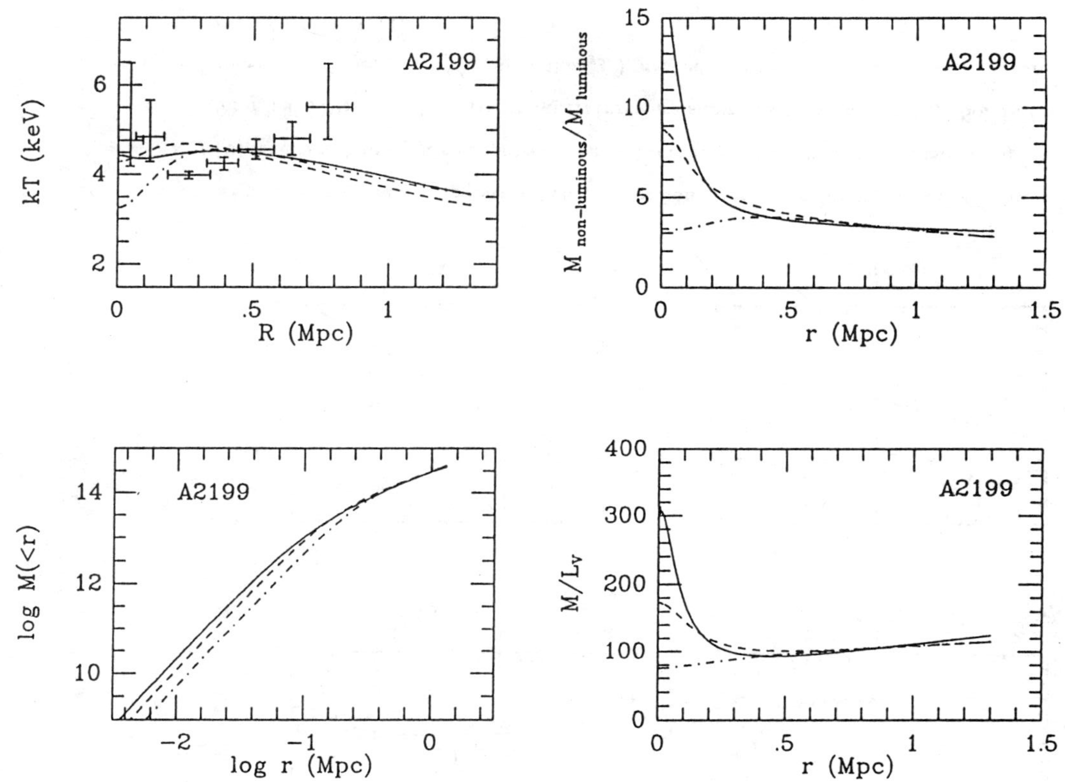

Fig 2. Same as fig 1 for Abell 2199. Solid line shows the best-fit for n=1 (a_{dark}=0.5 a_{gas}), dashed line n=1.25 (a_{dark}=a_{gas}) , dash-dotted line n=1.5 (a_{dark}=2 a_{gas}).

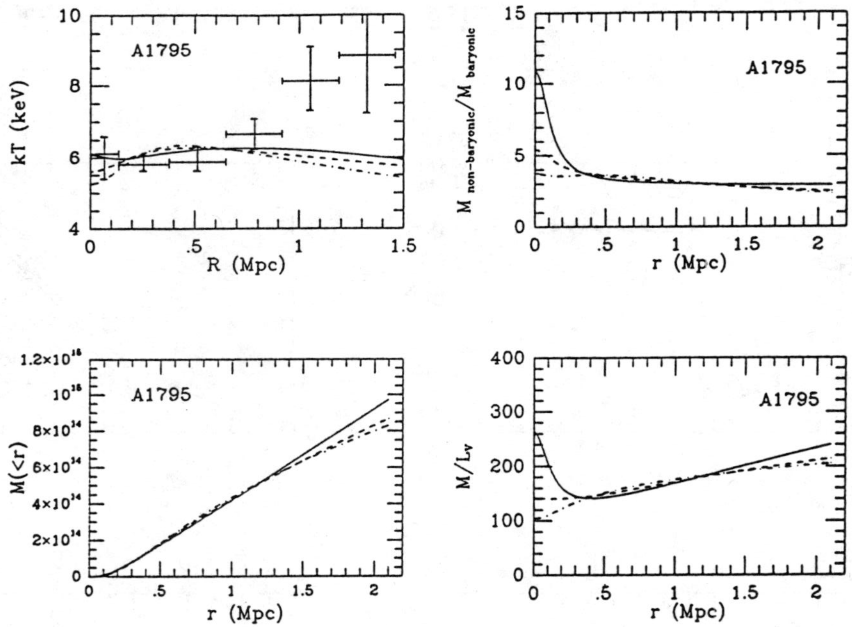

Fig. 1.– Projected temperature (a), ratio of nonluminous to luminous enclosed mass (b), enclosed mass (c), and mass-to-light ratio (d) distributions for the best-fit models for Abell 1795. ASCA-derived temperatures are shown as error bars in (a). Solid line shows best-fit for n=1.0 ($a_{dark} = 0.5a_{gas}$), dashed line n=1.25 (($a_{dark} = a_{gas}$), dash-dotted line n=1.5 ($a_{dark} = 1.5a_{gas}$).

X-RAY CONSTRAINTS ON DARK MATTER IN GROUPS OF GALAXIES

John S. Mulchaey
The Observatories of the Carnegie Institute of Washington
813 Santa Barbara St., Pasadena, CA 91101
mulchaey@pegasus.ociw.edu

ABSTRACT

A summary of recent ROSAT and ASCA observations of poor groups is presented with an emphasis on mass determinations from X-ray data. Although the number of groups with detected diffuse X-ray emission is still small, some trends are beginning to become apparent. Detectable diffuse X-ray emission appears to be largely restricted to those systems with low spiral fractions. Despite a considerable range in gas mass, the range in total group mass is very small with all of the systems having a mass around $\sim 2 \times 10^{13}$ $M_{\odot}$. The baryonic fraction of these groups is very low implying they are dominated by dark matter.

1. INTRODUCTION

The recent detection of diffuse hot gas in poor groups of galaxies allows mass estimates of these systems that are independent of the optical methods. The X-ray studies have several advantages over previous work. First, the mere presence of an intragroup medium provides strong evidence that these groups are real, physical systems and not just chance superpositions on the sky. Second, with the X-ray methods we are not limited by the number of "particles" in the potential. For example, optical mass determinations are based on application of the virial theorem, which can be unreliable when applied to a system with only a few galaxies. However, with X-ray observations we are generally limited by poor signal-to-noise ratio, so by increasing our exposure times, we can get better estimates of the mass. X-ray measurements also allow us to calculate the mass profile. Eventually it should be possible to learn about the distribution of the various components in the group, including the dark matter. This is particularly important because in some cases the X-ray gas extends well-beyond the edge of the group as determined by the galaxies and we can probe the mass on larger scales than possible with the optical data alone.

Despite the great potential for X-ray studies of groups, it was not until halfway through the lifetime of the ROSAT PSPC that it was realized that such studies were reasonable. There were hints from Einstein observations that an intra-group medium might exist in some systems, but it was not possible to unambigiously separate a diffuse component from emission from the individual galaxies (e.g. Biermann, Kronberg & Madore 1982; Bahcall, Harris & Rood 1984). The ROSAT PSPC with its low energy coverage ($\sim$ 0.1-2.4 keV), large field of view and improved spatial resolution and sensitivity was an ideal instrument to study the low surface brightness diffuse gas found in groups.

2. MASS ESTIMATES FROM X-RAY DATA

The method used to estimate the masses of groups from X-ray data is analogous to the method developed for clusters of galaxies (see, for example, Mushotzky, these proceedings). The fundamental assumption is that the hot gas is trapped in the potential well of the group and is in rough hydrostatic equilibrium. This assumption is probably a fair one given the short sound crossing times in these systems (it is assumed that there is minimal energy input from other sources such as supernovae). Making these assumptions, the total mass of the group at radius r is given by (Fabricant, Rybicki & Gorenstein 1984):

$$M_{\rm total}(<r) = \frac{kT_{\rm gas}(r)}{G\mu m_p}\left[\frac{d\log\rho}{d\log r} + \frac{d\log T}{d\log r}\right]r \qquad (1)$$

The gas density profile can be derived from the X-ray surface brightness assuming, for example, a standard β-model (e.g. Jones & Forman 1984):

$$S(r) = S_0\left[1 + \frac{r^2}{r^2_{\rm core}}\right]^{\frac{1}{2}-3\beta} \qquad (2)$$

It then follows that:

$$n(r) = n_0\left[1 + \frac{r^2}{r^2_{\rm core}}\right]^{-\frac{3}{2}\beta} \qquad (3)$$

Unfortunately, in most cases we do not have enough photons to determine the temperature profile directly, so usually it is assumed that the gas is isothermal:

$$\frac{d\log T}{d\log r} = 0 \qquad (4)$$

Making the assumption of isothermality, equation (1) reduces to:

$$M_{\rm total}(<R) \approx 1.1 \times 10^{11}\beta T_{\rm kev} R_{\rm kpc} M_\odot \qquad (5)$$

The two "baryonic" components in groups are the galaxies and the hot gas. The mass in the galaxies can be estimated from the optical luminosities assuming a standard mass-to-light ratio for each morphological type. The gas mass is calculated by integrating the appropriate β model out to the radius of interest.

3. THE NGC 2300 GROUP OF GALAXIES

The first published mass estimate for a poor group using ROSAT data was for the NGC 2300 group (Mulchaey et al. 1993). This system consists of three galaxies: NGC 2300, an elliptical, NGC 2276, a distorted spiral and IC 445, a small S0 galaxy. Figure 1 shows an optical image of the group with contours representing the X-ray emission seen by ROSAT overlayed. The X-ray emission shown is the "diffuse" component (i.e. the emission due to the individual galaxies has been removed). The diffuse gas is centered several arcminutes away from NGC 2300 and appears to extend to $\sim 25''$ ($\sim$ 300 kpc; H_o=50 km s^{-1} Mpc^{-1} is assumed throughout this paper). The ROSAT X-ray spectrum is well-described by a Raymond-Smith plasma model with temperature $\sim$ 1 keV (10^7 K) and very low metallicity ($\sim$ 10% solar).

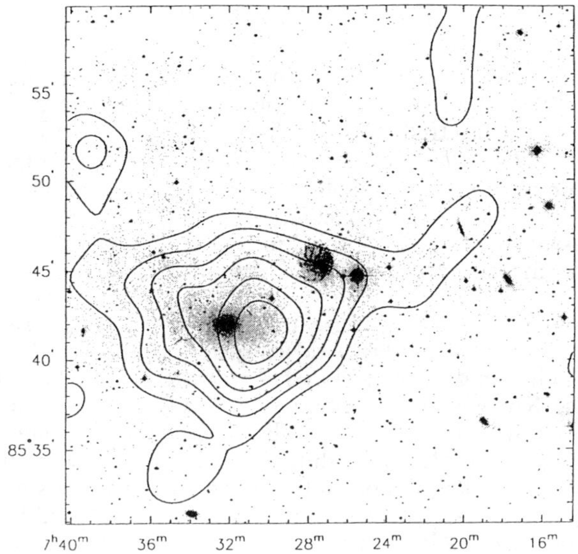

Figure 1–A contour map of the diffuse X-ray emission in the NGC 2300 group of galaxies overlayed on the Palomar Sky Survey reproduced to the same scale (Mulchaey et al. 1993).

Taking the temperature and surface brightness profile from our original analysis of the NGC 2300 group, equation (5) gives a total group mass at R=300 kpc of $M_{total} \approx 2 \times 10^{13}$ $M_\odot$. The mass of the hot gas to this radius is $\approx 6 \times 10^{11}$ $M_\odot$ and the mass of the three galaxies (assuming standard mass-to-light ratios for each morphological type) is also $\approx 6 \times 10^{11}$ $M_\odot$ (Mulchaey et al. 1993). Thus, the observed baryonic mass of the group is $M_{baryonic} \approx 1.2 \times 10^{12}$ $M_\odot$ and the observed baryonic fraction is very low ($\approx$ 6%). Subsequent re-analysis of the original ROSAT data (Pildis, Bregman & Evrard 1995) using improved PSPC flatfields yields a higher baryonic fraction ($\approx$ 14%). Regardless, the ROSAT data require a significant dark matter component.

Recently, a much deeper PSPC observation of the NGC 2300 group was obtained (Davis et al. 1995). The new data confirm the extent of the hot gas found in our initial analysis ($\sim$ 300 kpc) and suggest the gas is isothermal. NGC 2300 has also recently been observed by the Japanese-US X-ray telescope ASCA (Tawara et al. 1994, these proceedings). The ASCA data verify the temperature and low abundance implied by the ROSAT data (T $\approx$ 0.9 kev, abundance $\approx$ 9% solar). The low metal abundance of the diffuse gas is important because it suggests that the enrichment history of poor groups is very different from that of richer clusters and goes against the trend for low temperature systems to have higher abundances (e.g. David, Forman & Jones 1991). The ASCA GIS data suggest the diffuse gas may extend even further out than suggested by the ROSAT data.

4. ROSAT STUDIES OF GROUPS

Soon after the initial NGC 2300 group results were published, Ponman & Bertram (1993) reported the detection of a diffuse x-ray component in the Hickson compact group HCG 62. The X-ray properties of this system are similar to those of NGC 2300, with a temperature of $\sim$ 1 kev and a gas metallicity of $\approx$ 15% solar. The total mass of the group out to a radius of 500 kpc is $\approx 3 \times 10^{13}$ $M_{\odot}$, only 15% of which can be accounted for in observed baryonic matter. The ROSAT observation of HCG 62 contains enough photons that the temperature profile can be measured directly and the assumption of isothermality is not required.

In the two years since these initial ROSAT discoveries, an intragroup medium has been found in over a dozen other groups. Table 1 contains a compilation of all the small groups that I am aware of that have been reported to contain diffuse X-ray emission (pointed ROSAT observations only; see Section 8). While the number of systems is still relatively small, some trends are beginning to become apparent. First, diffuse hot gas is not seen in all systems, but rather appears to occur preferentially in groups with low spiral fractions. The spiral fraction of the detected systems is typically less than 0.5, while the spiral fraction of poor groups in optical catalogs is $\approx$ 0.7 (e.g. Henry et al. 1995). The trend for hot gas to be found in mostly elliptical systems is probably not a selection effect, since many spiral-rich groups have been observed by ROSAT and diffuse gas is not seen (Pildis et al. 1995; Mulchaey et al. 1995). The one possible exception to the above trend is HCG 92, Stephan's Quintet. Sulentic, Pietsch & Arp (1995) have analyzed the PSPC data on this compact system and find that the X-ray emission is extended and does not appear to peak on any of the galaxies. However, Pildis et al. (1995) claim the X-ray emission is unresolved and probably due to the individual galaxies. Even with this possible exception, it is clear that the groups with the strongest and most extended diffuse component are those with low spiral fractions.

For the most part, the X-ray properties of the newly detected groups are similar to those of the NGC 2300 and HCG 62 groups. In particular, the temperature of the gas is fairly well-determined by ROSAT and is around $\approx$ 1 keV in all cases. Unfortunately, less is known about the range of gas metallicities. Most authors have not attempted to fit the abundances directly given the poor spectral resolution of the ROSAT PSPC. When the abundances have been estimated from the ROSAT spectra, the numbers tend to be very low (10-15% solar). An exception to this is the NGC 5044 group, for which David et al. (1994) determine an abundance of $\approx$ 0.8 solar. The ASCA data for NGC 5044 suggest a much lower value of $\approx$ 25% solar (Mushotzky 1994), which is still higher than that of the NGC 2300 group. Future ASCA determinations of the gas metallicity in many more groups will provide clues into the origin and evolution of the intragroup medium.

5. THE MASS OF POOR GROUPS

From the derived properties of the X-ray emitting gas, the mass of each group can be estimated by applying equation (1). As discussed earlier, it is usually necessary to make further assumptions such as isothermality. In two of the cases listed in Table 1 (NGC 5044 and HCG 62), the ROSAT data allow the temperature profile to be determined directly. For both cases, the assumption

Table 1
Optical and X-ray Properties

Group (1)	$N_{galax.}$ (2)	f_{spiral} (3)	kT (keV) (4)	abund. (solar) (5)	$R_{ext.}$ (kpc) (6)	β (7)	ref. (8)
NGC 2300	3	0.33	0.86/0.93	0.06/-	300/210	0.66/0.32	1,2
NGC 4104	4	0.00	1.36	-	>260	~ 0.6	3
NGC 4261	8	0.25	0.85	0.02	620	0.31	4
NGC 5044	6	0.55	0.98	~ 0.8	250	0.53	5
NGC 6269	5	0.40	1.36	-	>260	~ 0.59	3
NGC 6329	3	-	1.16	-	>260	~ 0.77	3
HCG 12	5	0.20	0.72	-	190	0.92	2
HCG 42	4	0.00	0.80	0.15-0.2	200	0.7	6
HCG 62	4	0.00	0.96/1.16	0.15/-	480/360	0.54/0.51	7,2
HCG 68	5	0.20	0.98	-	130	0.61	2
HCG 92	4	0.75	0.92/-	-	120/110	0.98/-	8,2
HCG 97	5	0.40	0.97	-	270	0.48	2
RXJ 1340.6	1	0.00	0.92	0.36	500	1.0	9

Notes to Table 1

(1) Group name.

(2) Number of galaxies in the group ($L \approx L_*$).

(3) Spiral fraction.

(4) Measured gas temperature.

(5) Measured gas abundance.

(6) Radial extent of the X-ray emission.

(7) β value derived from surface brightness profile.

(8) Reference for the X-ray analysis: 1) Mulchaey et al. 1993; 2) Pildis, Bregman & Evrard 1995; 3) Thiering & Dahlem 1995; 4) Davis et al. 1995; 5) David et al. 1994; 6) Böhringer 1994; 7) Ponman & Bertram 1993; 8) Sulentic, Pietsch & Arp 1995; 9) Ponman et al. 1994

that the gas is isothermal is a fairly good one leading to errors in the total mass of approximately 10% (cf. David et al. 1994). Several of the groups have been analyzed by several different teams, often with a significant difference in quantities such as gas mass, total mass and baryonic fraction. The large range in derived properties based on the same data sets is a reminder that this field is still in its infancy and there are considerable uncertainties in the present methods.

Figure 2a gives a histogram of the estimated gas mass for the 13 systems listed in Table 1. Even with this small sample, there appears to be a range in gas mass over three orders of magnitude. There is less range in the mass of galaxies (Figure 2b). A comparison of Figure 2a and 2b reveals that in general the mass of the galaxies is greater than that of the gas mass. In several cases, the hot gas makes up less than $\approx$ 10% of the *baryonic* mass. This is in contrast to rich clusters of galaxies where the hot gas is typically the dominant baryonic component. The relative importance of the galaxies in these systems underscores the need for quality optical studies of the group membership.

Despite the relatively large range in gas mass, there is a remarkably narrow range in total derived group mass (Figure 3a) with all of the groups clustered around $\approx 2 \times 10^{13}$ $M_{\odot}$. This number agrees fairly well with the typical group mass derived from optical methods (e.g. Ramella, Geller & Huchra 1989). The baryonic fractions are in the range 0.05–0.3, with much of the scatter probably due to the uncertainties discussed earlier (Figure 3b). Even with the most conservative of assumptions, the X-ray observations require the dominance of a dark matter component in poor groups.

6. X-RAY SELECTED GROUPS OF GALAXIES

All of the groups discussed so far were previously known from optical surveys and were chosen as targets for studies with ROSAT. This heterogeneous sample is not ideal for determining the frequency of diffuse gas in groups. A more appropriate data set for this purpose is the ROSAT all-sky survey. Unfortunately, the typical integration time of the sky survey was very short and in general, not sufficient to detect most systems. However, the North Ecliptic Pole (NEP) region was scanned several times by ROSAT and data from this region can be used to estimate the number density of X-ray emitting groups. An 84.7 square-degree segment of the NEP has recently been analyzed by Henry et al. (1995), who detect eight groups with redshifts less than z=0.04. The x-ray and optical properties of these X-ray selected groups are similar to those of the groups listed in Table 1. In particular, these groups have low spiral fractions ($\approx$ 0.3). Henry et al (1995) also find that the X-ray luminosity and temperature functions of groups are a smooth extrapolation from rich clusters. From this data, X-ray selected groups of galaxies appear to contribute only about 5% of the critical density of the universe. When one corrects for the bias against the more common spiral-rich systems (assuming similar masses), groups can produce $\Omega \approx 0.25$, a number consistent with that derived from optical mass estimates.

7. CONCLUSIONS

Diffuse X-ray emission has now been detected in $\sim$ 20 small groups of galaxies with ROSAT. The X-ray properties of these groups are similar to those of the NGC 2300 group. In particular, the X-ray gas is characterized by a temperature of $\approx$ 1 keV and typically extends to a radius of several hundred

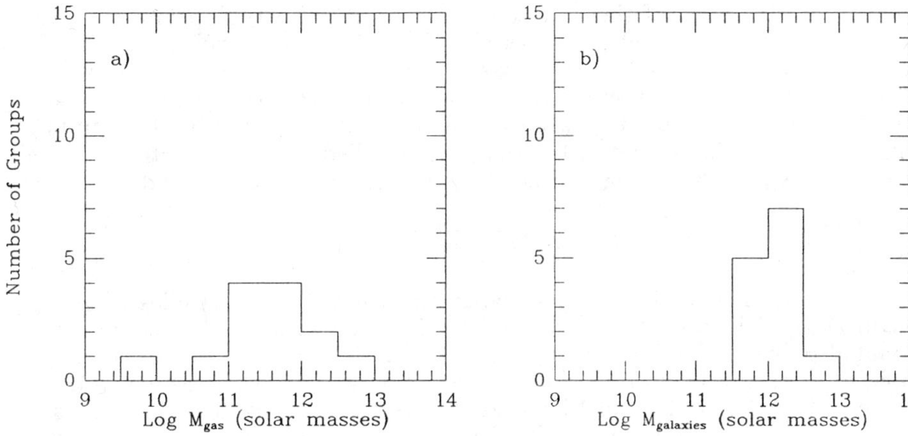

Figure 2–Histogram of a) gas mass and b) galaxies mass for the groups given in Table 1.

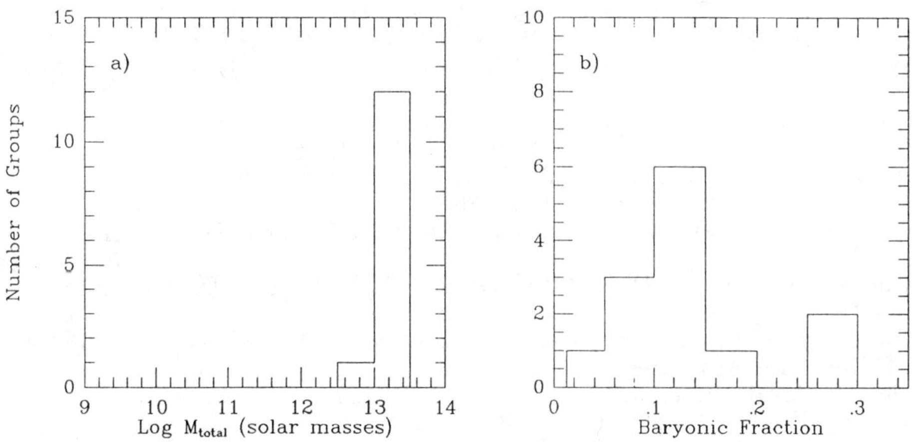

Figure 3–Histogram of a) total mass and b) baryonic fraction for the groups given in Table 1.

kiloparsecs. In many of these systems, the gas metallicity appears to be very low ($\sim$ 10–15% solar). The low gas abundance has been verified for the NGC 2300 group by ASCA. While the mass of the hot gas is comparable to the mass of the galaxies in a few cases, it is much lower in some. Despite a large range in observed gas mass, there appears to be a very small range in total group mass with a typical value of 2×10^{13} $M_{\odot}$. The implied baryonic fraction in small groups is low ($\sim$ 5-30%) suggesting such systems are dominated by dark matter.

ACKNOWLEDGEMENTS

I would like to thank my collaborators on the X-ray studies of groups, David Davis, Richard Mushotzky and David Burstein for useful discussions. I would also like to thank the many authors who shared their new results on this topic prior to publication, including Hans Böhringer, Joel Bregman, Michael Dahlem, Pat Henry, Rachel Pildis, Jack Sulentic and Inge Thiering. This research was supported in part by NASA grant NAG5-2455 and a Carnegie post-doctoral fellowship.

REFERENCES

Bahcall, N. A., Harris, D. E., & Rood, H. J. 1984, ApJ, 284, L29
Biermann, P., Kronberg, P. P., & Madore, B. F. 1982, ApJ, 256, L37
Böhringer, H. 1994, in Clusters of Galaxies, the 14th Moriond Astrophysics Meeting, in press
David, L. P., Forman, W., and Jones, C. 1991, ApJ, 380, 39
David, L. P., Jones, C., Forman, W., & Daines, S. 1994, ApJ, 428, 544
Davis, D. S., Mushotzky, R. F., Mulchaey, J. S., Worrall, D. M., Birkenshaw, M., & Burstein, D. 1995, ApJ, in press
Fabricant, D., Rybicki, G. B., & Gorenstein, P. 1984, ApJ, 286, 186
Henriksen, M. J., & Mamon, G. A. 1994, ApJ, 421, L63
Henry, J. P. et al. 1995, ApJ, submitted
Jones. C, & Forman, W. 1984, ApJ, 276, 38
Mulchaey, J. S., Davis, D. S., Mushotzky, R. F.. & Burstein, D. 1993, ApJ, 404, L9
Mulchaey, J. S., Davis, D. S., Mushotzky, R. F.. & Burstein, D. 1995, in preparation
Mushotzky, R. F. 1994, in New Horizons of X-ray Astronomy: First Results from ASCA, in press
Pildis, R. A., Bregman, J. N., & Evrard, A. E. 1995, ApJ, in press
Ponman, T. J. et al. 1994, Nature, 369, 462
Ponman, T. J., & Bertram, D. 1993, Nature, 363, 51
Ramella, M., Geller, M. J., & Huchra, J. P. 1989, ApJ, 344, 57
Sulentic, J. W., Pietsch, W., & Arp, H. 1995, AA, in press
Thiering, I., & Dahlem, M. 1995, in preparation

THE VELOCITY DISPERSION-TEMPERATURE CORRELATION FROM A LIMITED CLUSTER SAMPLE

Christina M. Bird
Dept. of Physics and Astronomy, University of Kansas
Lawrence, KS 66045-2151
tbird@kula.phsx.ukans.edu

Richard F. Mushotzky
MC 661, NASA Goddard Space Flight Center
Greenbelt, MD 20771.
richard@xray-5.gsfc.nasa.gov

ABSTRACT

Most studies of correlations between X-ray and optical properties of galaxy clusters have used the largest samples of data available, regardless of the morphological types of clusters included. Given the increasing evidence that morphology is related to a cluster's degree of dynamical evolution, we approach the study of X-ray and optical correlations differently. We evaluate the relationship between velocity dispersion and temperature for a limited set of galaxy clusters taken from Bird (1994), which all possess dominant central galaxies and which have been explicitly corrected for the presence of substructure. We find that $\sigma_r \propto T^{0.69\pm0.13}$, significantly steeper than the $\sigma_r \propto T^{0.5}$ relationship predicted by the virial theorem. The protogalactic winds model of White (1991) reproduces the observed relationship between σ_r and T.

1. INTRODUCTION

Clusters of galaxies contain two luminous components, gas and galaxies. If a cluster is sufficiently old and unperturbed, these tracer particles will have equilibrated within the cluster gravitational potential. This enables use of the equations of hydrostatic and dynamical equilibrium to explore the physical properties of these systems. For a hot gas in equilibrium with a spherical gravitational potential, the hydrostatic equation may be written

$$M_X(r) = \frac{-kTr}{G\bar{m}}\left(\frac{d\ln n_{gas}}{d\ln r} + \frac{d\ln T}{d\ln r}\right)$$

(e.g. Fabricant, Lecar & Gorenstein 1981), where M_X is the X-ray determined-virial binding mass, T is the temperature of the X-ray emitting gas, $n_{gas}$is the gas density and $\bar{m}$ is the average mass per gas particle. Similarly, the Jeans equation relates the kinetic energy of the galaxies to the binding mass of the cluster:

$$-\frac{G\, n_{gal} M_{opt}(r)}{r^2} = \frac{d(n_{gal}\sigma_r^2)}{dr} + \frac{2n_{gal}}{r\sigma_r^2}A$$

(Merritt 1987), where M_{opt} is the optically-determined binding mass, n_{gal} is the galaxy number density, σ_r and σ_t are the radial and tangential velocity dispersions respectively, and A is the anisotropy parameter describing the distribution of galaxy orbits.

If gravity is the only source of energy for the galaxies and the gas, and if both the galaxies and the gas trace the gravitational potential of the cluster, then the masses determined from the Jeans equation and the equation of hydrostatic equilibrium are equal:

$$\frac{\sigma_r^2}{\frac{kT}{\bar{m}}} = \frac{d \ln n_{gas}/d \ln r}{d \ln n_{gal}/d \ln r + 2A}.$$

(where A, the anisotropy factor, is generally taken to be zero). Therefore we predict that the velocity dispersion (as measured from galaxy velocities) and the temperature of the intracluster medium (as determined from X-ray spectra) should be correlated, with $\sigma_r \propto T^{0.5}$. The ratio of the kinetic energies is called β_{spec}. The ratio of the slopes of the density profiles is β_{fit}.

2. THE LIMITED CLUSTER SAMPLE AND THE $\sigma_v - T$ CORRELATION

The morphology of a cluster may be described by its gas and/or galaxy distribution. It has become clear, as our observations of clusters have improved, that morphology is related to the dynamical age of a cluster. Irregular clusters are dynamically young, and tend to be spiral-rich and gas-poor. They tend to have non-Gaussian velocity distributions and kinematically-distinct subconcentrations of galaxies. Regular clusters are dominated by ellipticals, have Gaussian velocity distributions and tend to be luminous X-ray emitters (cf. Sarazin 1988 and references therein; Bird 1993,1994).

Bird (1994) presents a detailed analysis of the dynamics of nearby clusters ($z < 0.1$) with central galaxies. These clusters tend to have smooth morphologies and X-ray cooling flows, and in the past it has been assumed that they represent the most relaxed, dynamically-evolved clusters in the Universe. However, Bird (1994) shows that these clusters also suffer from contamination by substructure. An objective partitioning algorithm called KMM is used to identify galaxies belonging to subsystems in the clusters, and the dynamical properties of the "cleaned" cluster datasets are presented. It is the 25 clusters in this "cD database" which will form the limited cluster sample of the present analysis. They are listed in Bird & Mushotzky (1994). X-ray temperatures are taken from ASCA and *Ginga* where available, and *EXOSAT* (Edge & Stewart 1991) and *Einstein* (David et al. 1993) otherwise.

In Figure 1, we present the velocity dispersion and temperature data for the 22 clusters in the limited sample (the clusters A1809, A1983 and A2124 do not have X-ray temperature determinations and are therefore excluded from the rest of this discussion). The velocity dispersions are corrected for substructure. The dashed lines are the correlations predicted by the virial theorem, for $\beta_{spec} = 1$ and for $\beta_{spec} = 0.67$. For these data $\langle\beta_{spec}\rangle = 0.90^{+0.12}_{-0.14}$ with an rms scatter of 0.37. The solid line is the best fit to the data, $\sigma_r = 10^{2.44\pm0.09}T^{0.69\pm0.13}$ This fit is not weighted by the errors in the measurements; as Lubin & Bahcall (1993) showed, including the uncertainty in the X-ray and optical measurements does not significantly affect the analysis. In other words, the intrinsic scatter in the data dominates the scatter due to experimental uncertainty.

Previous authors have claimed that their data was consistent with the canonical virial theorem dependence of velocity dispersion on temperature, $\sigma_r \propto$

$T^{0.5}$ (Edge & Stewart 1991, Lubin & Bahcall 1993). The primary effect of substructure correction on this correlation is to steepen the dependence of the velocity dispersion on temperature, to the point where it is *no longer consistent* with the canonical value. That is, although the scatter to high velocity dispersions has been noted by all previous studies and generally attributed to velocity substructure, correction for substructure actually makes the discrepancy *worse.*

3. DISCUSSION

Although Lubin & Bahcall (1993) found that the correlation between cluster velocity dispersion and temperature was somewhat steeper than that predicted by the virial theorem, the scatter in their dataset was too broad for them to rule out consistency with the hydrostatic isothermal model. By restricting our study to a morphologically-limited cluster sample and explicitly correcting for the presence of substructure, we show that $\sigma_r \propto T^{0.69\pm0.13}$, significantly steeper than that predicted by the virial theorem. It seems improbable that this is an artifact of the substructure correction algorithm. The mixture modelling technique used to remove substructure from the cluster datasets does not preferentially raise the velocity dispersion of high-σ_r clusters and lower that in low-σ_r systems.

Many physical mechanisms for causing clusters to violate the "virial relationship" between velocity dispersion and temperature have been proposed, including anisotropy in the galaxy orbits (see, for instance, Eyles et al. 1991); incomplete thermalization of the gas and/or galaxies (Evrard 1990); and magnetic pressure support of the cluster gas (Miralda-Escudé & Babul 1994; Loeb & Mao 1994). We discuss the effects of these processes in more detail in Bird & Mushotzky (1994), but find that none of them are adequate for predicting the variation in slope that we find using the limited cluster dataset.

Protogalactic winds provide an additional source of heating of the ICM. White (1991) discusses ram pressure stripping and protogalactic winds as mechanisms for the metal enrichment of the ICM. Using White's Equation 2, we generated a distribution of temperatures for velocity dispersions ranging from 350-1200 km sec^{-1} (taking his values for the fraction of intracluster gas coming from winds ($w = 0.5$) and the typical wind velocity in terms of the galactic velocity dispersion ($f_w = 3$)). Fitting these simulated data, we find that the protogalactic winds model predicts a correlation between the velocity dispersion and the temperature of a cluster: $\sigma_r \propto T^{0.68}$. This depends slightly on the choice of w and f_w; for $f_w = 2$ we find that $\sigma_r \propto T^{0.62}$. The protogalactic wind model reproduces nearly exactly the dependence of velocity dispersion on ICM temperature that we find in the limited cluster sample (and which is consistent with the slopes found by earlier studies).

We have explicitly included substructure and morphology in this study of X-ray and optical correlations for rich clusters of galaxies. The steepness of the $\sigma_r - T$ relationship suggests that the assumptions of isotropy of the galaxy orbits and gravity as the only source of energy for the galaxies and/or the gas are violated. The most important clarification between the possible mechanisms to explain this violation – primarily anisotropy of the galaxy orbits and the effects of protogalactic winds – will come through the measurement of more temperatures and velocity dispersions of low-temperature systems, which are currently being made available.

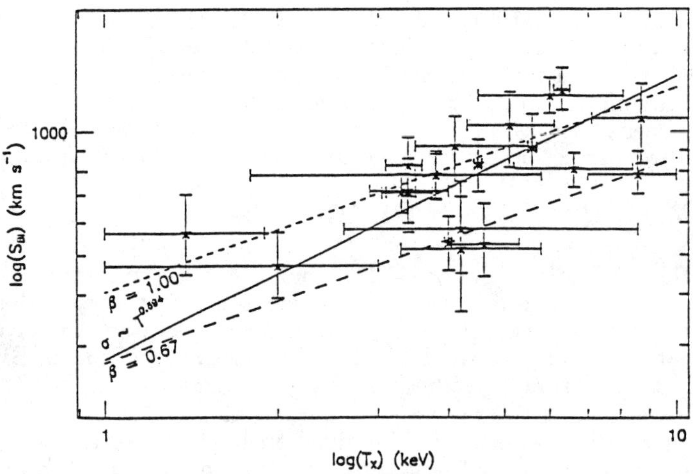

Fig. 1. The $\sigma_r - T$ correlation for the limited cluster sample. Errors in the velocity dispersions (the vertical axis) are taken from Bird (1994). Errors in the temperatures are taken from the literature. The dashed lines are the predicted correlations for the isothermal β-model, with $\beta = 1$ or $\beta = 0.67$ and $\sigma_r \propto T^{0.5}$. The solid line is the best fit to the data.

ACKNOWLEDGEMENTS

This research was supported by NSF Grant No. OSR-9255223 to the University of Kansas.

REFERENCES

Bahcall, N.A. & Lubin, L.M. 1994, preprint
Bird, C.M. 1993, Ph.D thesis, University of Minnesota and Michigan State University
Bird, C.M. 1994, AJ, 107, 1637
Bird, C.M. & Mushotzky, R.F. 1994, ApJ, submitted
David, L.P., Slyz, A., Jones, C., Forman, W., Vrtilek, S.D. & Arnaud, K. 1993, ApJ, 412, 479
Edge, A.C. & Stewart, G.C. 1991, MNRAS, 252, 428
Evrard, A.E. 1990, in *Clusters of Galaxies*, eds. Oegerle, W.R., Fitchett, M.J. & Danly, L., (New York: Cambridge University Press), 287
Eyles C.J., Watt, M.P., Bertram, D., Church, M.J., Ponman, T.J., Skinner, G.K. & Whitmore, A.P. 1991, ApJ, 376, 23
Fabricant, D.M., Lecar, M. & Gorenstein, P. 1980, ApJ, 241, 552
Loeb, A. & Mao, S. 1994, to appear ApJL
Lubin, L.M. & Bahcall, N.A. 1993, ApJL, 415, L17
Merritt, D. 1987, ApJ, 313, 121
Miralda-Escudé, J. & Babul, A. 1994, preprint
Sarazin, C.L. 1988, *X-ray emissions from clusters of galaxies*, (Cambridge: Cambridge University Press)
White, R.E. III. 1991, ApJ, 367, 69

ASCA MEASUREMENTS OF X-RAYS FROM POOR CLUSTERS

Takaya Ohashi
Department of Physics, Tokyo Metropolitan University
1-1 Minami-Ohsawa, Hachioji, Tokyo 192-03, Japan
ohashi@phys.metro-u.ac.jp

ABSTRACT

Results from ASCA on the mass, abundance, and temperature profiles in poor clusters of galaxies are described. The average iron abundance, obtained by excluding the central region where excess abundance is seen for a few clusters, so far turns out to be almost constant at 0.35 solar for $kT = 2.5 - 7$ keV. Energy-dependent point-spread function of the X-ray telescope, which creates apparent rise of temperature with radius, is carefully dealt in the analysis. The maximum extent of the cluster X-ray emission is constrained to be $r > 1.9$ Mpc for A4059, suggesting that M_{gas} is more than 20% of M_{total}. Effort in search for the possible temperature drop near the cluster edge is not yet successful due to systematic uncertainties.

1. INTRODUCTION

In this talk, we would like to report on our current results mainly on poor clusters. ASCA offers the first opportunity of the imaging spectroscopy (Tanaka et al 1994), which should bring a significant emprovement in our understanding of clusters of galaxies. In particular, GIS instrument (Ohashi et al 1991, Kohmura et al 1993) enables measurement of the spatial distribution of tempearture and metal abundance over a fairly large region of the sky (about 45 arcmin in diameter). The good efficiency for high energy X-rays up to 10 keV with very low internal background is another advantage of GIS for the study of clusters.

Before ASCA, extensive studies on the hot-gas properties of poor clusters have been made by Hatsukade (1989), Tsuru (1992) and Arnaud et al (1992) mainly from Ginga. They find that M_{Fe} in intra-cluster medium (ICM) varies in proportion to $M_{stellar}$, and the iron abundance in ICM shows negative correlation with the ICM temperature (see also Ikebe et al 1991). These features support a view that poor clusters contain less amount of the hot ICM compared with rich clusters (David et al 1990) and that observed iron in the ICM is injected from early-type galaxies (e.g. Mihara and Takahara 1994).

In the previous studies, it has been unable or very difficult to know about how much iron is concentrated in the central region of clusters. Koyama et al (1991) report significant concentration of iron around M87 in the Virgo cluster by a factor of $3 - 5$. Rather discrepant results have been found for the iron concentration in the Perseus cluster (Ponman et al 1990, Arnaud et al 1991, 1993). Systematic studies on this problem have now been carried out with ASCA. Reports on the ASCA results about the metal distribution in clusters are given by Ohashi et al (1994), as well as by Yamashita (1994) and Ikebe et al (1994). I shall summarize the current results and briefly mention on their implications.

Since emission profile of clusters is too much centrally peaked, the energy dependent behavior of the X-ray telescope causes some difficulties in deriving the true temperature profiles. In section 3, I shall describe the problem and the method of our analysis to cope with this feature.

Estimation of the gravitational mass needs both density profile and temperature profile as a function of radius. The temperature profile has been examined for several clusters along with the constraint on the maximum extent of the cluster emission (maximum radius, r_{max}). The data often agree with a constant temperature exept for the cool gas in the center, and preliminary results on the profiles for the total mass M_{total} and the gas mass M_{gas} are just coming out. Finally, I will mention about the ASCA capability of locating the cluster boundary by analyzing the temperature profile in the last section.

2. METAL ABUNDANCES

First, we analyse the cluster data by slicing them as a function of radius into several rings. This method has a problem of resulting somewhat wrong temperature profile for hot clusters as described later. However, since such an effect is not more than about 20% for clusters with $kT \lesssim 3.5$ keV, I shall present the GIS results from the ring-cut analysis here. The measured radial distributions of iron abundance are plotted in Fig. 1. One should note that the profile is not deconvolved and the true feature would be much narrower than the "raw" profiles in Fig. 1. In the spectral fits, we have fixed the relative abundances for different metal elements at the cosmic value and varied only the overall metal (or iron) abundance. A large enhancement in the abundance by more than a factor of 3 is seen in the central region of the Centaurus cluster (Fukazawa et al 1994). The central abundance which even exceeds the cosmic value is one of the highest levels observed from galaxy clusters (see e.g. Yamashita 1992). As a result of this abundance gradient, the mean abundance (such as that measured with non-imaging detectors) increases by 30% compared from the value in the outside region. A smaller but significant enhancement of the abundance is also seen from AWM7 and possibly from PKS2354-35 (A4059). Similar feature has been reported for the Virgo cluster (Matsumoto et al 1994).

So far, all the clusters showing the abundance gradients belong to poor clusters ($kT \lesssim 4$ keV) accomodating a cD galaxy. Since it has been pointed out that sedimentation of iron in the ICM is too slow to account for any metal concentration (Rephaeli 1978), it seems most reasonable to assume that the excess iron in the center is indeed produced in situ. The ram-pressure stripping is thought to be efficient in the center, however, it is unclear why metal concentration seems to require presence of a cD galaxy if pressure stripping is the only mechanism. It looks more reasonable to assume that the excess iron is produced in the cD galaxies. Since optical luminosities of cD galaxies do not differ much from cluster to cluster and lie within a factor of several, the cD may inject similar amount of excess iron into the intergalactic space. In this case, such a contribution would be larger in gas-poor clusters. However, it is not clear how cDs can produce excess iron than ordinary elliptical galaxies.

We investigate the relation between the average metal abundance and the temperature of ICM. In deriving the mean abundance, we have excluded the central region ($r \lesssim 300$ kpc) in order to avoid the effect from the metal concentration. The results obtained from our poor cluster sample are shown in Fig. 2. Even though the number of data is few, the observed iron abundance

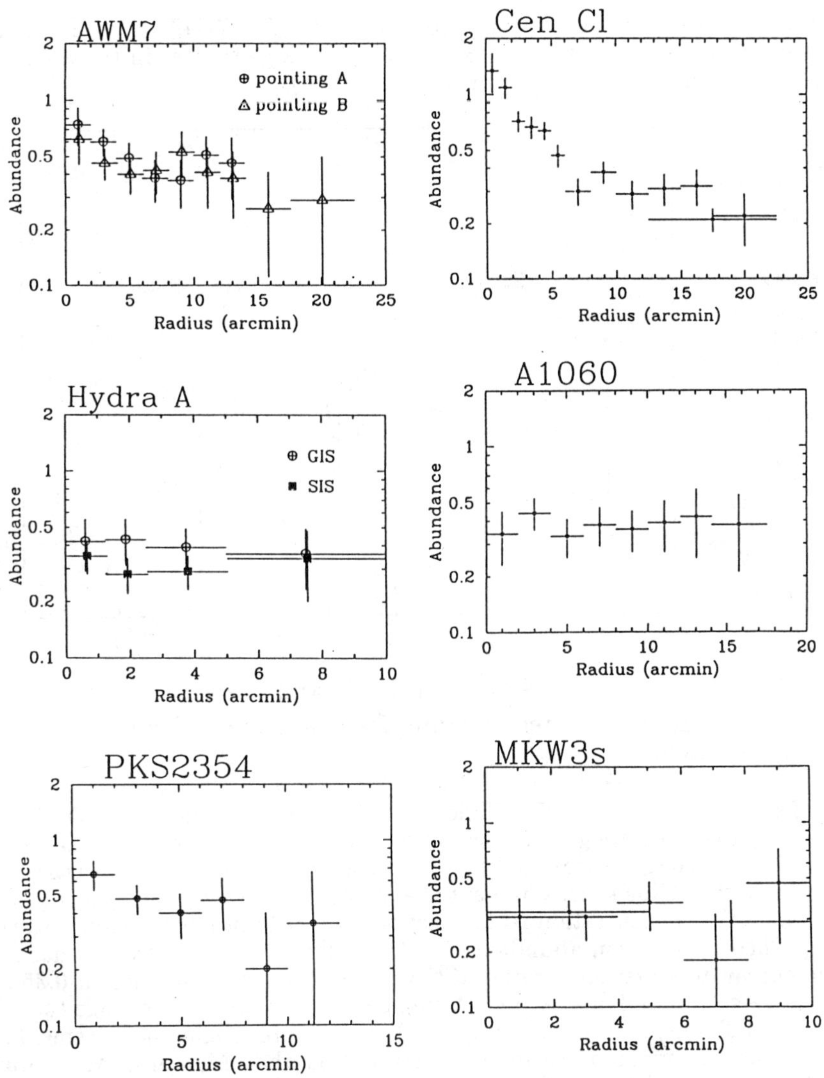

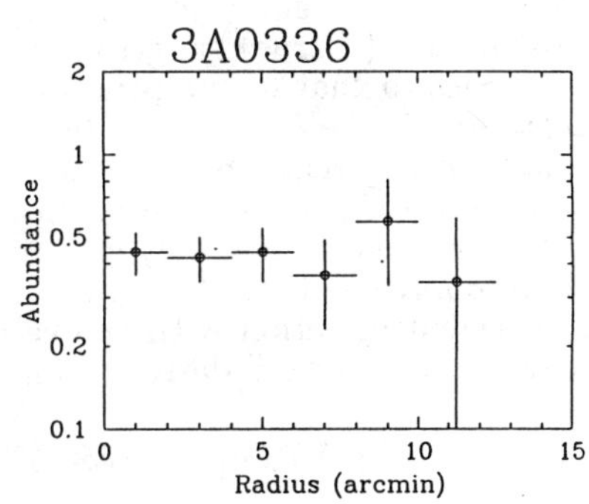

Fig. 1. Abundance distribution in poor clusters of galaxies based on ring-cut analysis.

is fairly constant for the ICM temperatures of 2.5 – 7 keV. Note, however, that very rich hot clusters ($kT > 8$ keV) show a large scatter in the abundance between 0.1 and 0.5 times solar (Yamashita 1994).

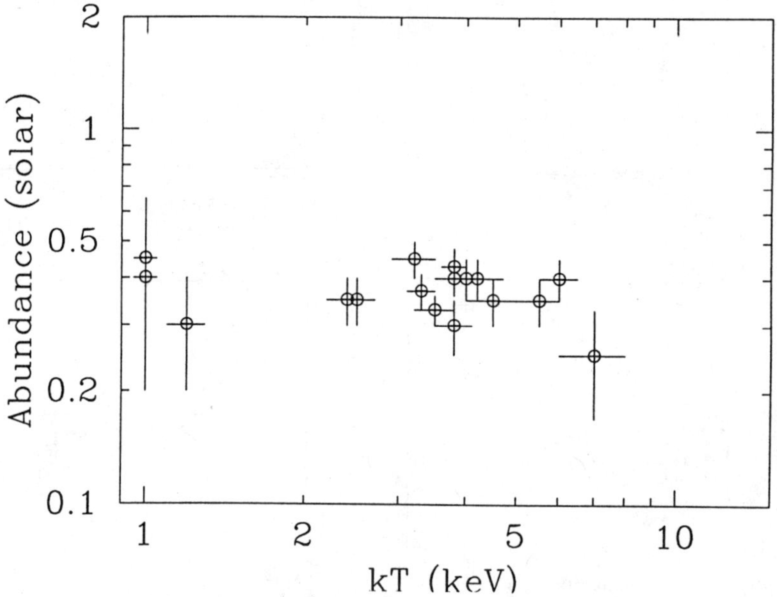

Fig. 2. Iron abundance vs. temperature. Data in the central region of clusters are excluded.

The constant metal abundance with temperature may give a different impression than the "negative correlation" first reported by Hatsukade (1989) based on the Ginga measurements. We must be cautious for both cases: It is conceivable that Ginga measurements have been affected by the iron concentration. Since the concentration so far occurs frequently in poor clusters, it would have pushed up the iron abundance for those systems. On the other hand, the sample of clusters measured with ASCA is still too small to assess if 0.35 solar is a "universal" abundance or not. A single exception will easily make the whole plot very different, and we certainly need much more ASCA measurements.

If one can make a reasonalble assumption that $M_{\rm Fe}$ in ICM is proportional to $M_{\rm stellar}$ (Tsuru 1992, Arnaud et al 1992), then the constancy of the iron abundance (if true) suggests that the ratio of the amount of gas injected from galaxies to that of the primordial gas is nearly constant for clusters with temperatures 2.5 keV to 7 keV. In this case, the efficiency of the galaxy formation may not be greatly dependent on cluster richness. This also suggests that the ratio $M_{\rm gas}/M_{\rm stellar}$ is constant in this temprature range. Therefore, it might be that poor clusters generally contain large amount of hot ICM such as several times more massive than $M_{\rm stellar}$. This possibility suggested here in fact points to a consistent direction with the value of $M_{\rm gas}$ derived in section 4 using the updated maximum radii derived from the ASCA data.

3. PROPERTIES OF X-RAY TELESCOPES

One has to be vary careful in analyzing cluster data because of the energy dependent effect of point-spread function of the X-ray telescope (XRT, Serlemitsos et al 1994). Figure 3 shows the 2-dimensional profiles of Crab Nebula taken with GIS for 3 different energy bands. One can clearly see that the distribution is more extended for higher X-ray energies. Exactly the same feature is seen in other hard point sources such as Cyg X-1. This is caused by the scattering of X-rays. We have been trying to reproduce all the properties of XRT through ray-tracing program. However, in analyzing the cluster data one needs to worry about the profile where the surface brightness goes less than 1% of the peak level. It takes some time to upgrade our ray-tracing code to meet this standard, so we carry out in the mean time a Monte-Carlo analysis by incorporating the observed point-source profiles as the trial point-spread function.

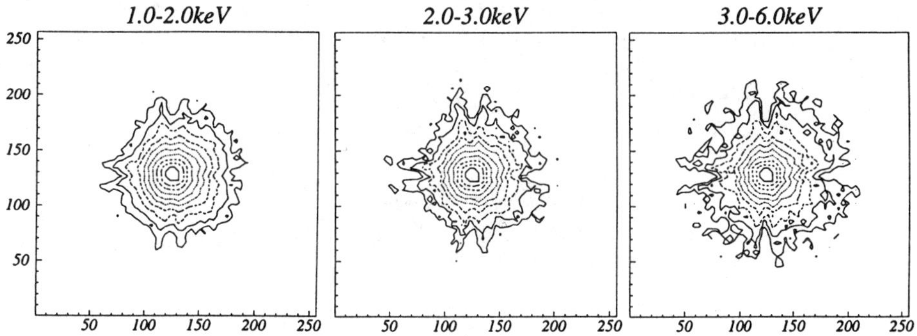

Fig. 3. GIS image of the Crab Nebula in 3 energy bands. Tails in the image are more extended for harder X-rays.

The most serious effect of the XRT is that the radial distribution of cluster pulse-height spectrum always shows a systematic rise as a function of the radius. This is illustrated in Fig. 4. The effect is drastic for hot clusters: the apparent temperature near the field edge is wrong by almost a factor of 2 when ICM temperature $kT > 7$ keV. This systematic error in temperature also affects iron abundance, since theoretical equivalent width of iron K emission lines is weaker for high temperature plasmas. Regarding these features, one should be extremely careful in deriving a science based on the traditional ring-cut analysis in particular for hot ($kT > 5$ keV) clusters.

The method of the analysis to cope with this XRT properties is briefly described here. First we assume incident radial profile of energy spectra. Practially β-model is usually assumed as a density distribution, and power-law or exponential dependence is assumed for the temperature and metal abundance as a function of radius. Then we generate some many ($> 10^5$) photons to be put into the XRT, and simulate the observed pulse-hight spectra as a function of radius in the detector coordinate. In this process, we incorporate actual observed data from Cyg X-1, Crab Nebula and other bright point sources taken at various offset positions as the point spread function. The simulated radial profiles are compared with the observed data from clusters, and χ^2 values are calculated. Then parameter values such as β or kT are incremented and then 10^5 photons are generated again. This analysis is very slow. But we can be sure about the reliability of the results since it incorporates true point spread function.

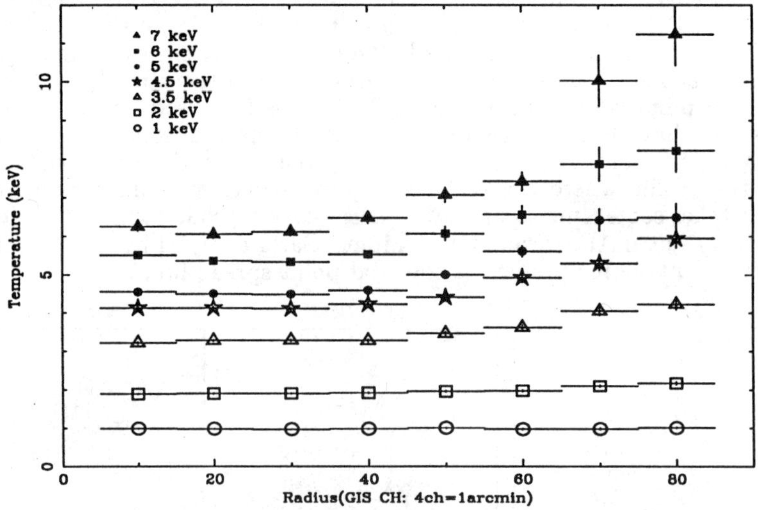

Fig. 4. Apparent temperature as a function of radius for a ring-cut analysis. This plot assumes core radius of $1'$ and $\beta = 0.55$.

4. ESTIMATION OF GAS MASS

Here, we try to constrain the total mass $M_{\rm total}$ and gas mass $M_{\rm gas}$ in ICM based on the ASCA data. The most sensitive parameter in determining these values is how far out the cluster emission is extended, namely the maximum radius $r_{\rm max}$. A suitable cluster for this analysis should be small enough (or distant enough) to be accomodated nicely in the f.o.v. of GIS and at the same time bright enough (or close enough) for statistics. PKS2354-35 (A4059) at $z = 0.048$ ($1'$ corresponds to 96 kpc) satisfies these conditions. This cluster has a cD galaxy in the center (Schwartz et al 1991), but we find that a simple β model can adequately describe the GIS data. Therefore, as suggested from the results of ring-cut analysis, we assume a single β-model with constant temperature to describe the cluster emission. We are finding that this is a good description of the cluster X-ray data in general. Even when the cluster center is peaked due to so-called cooling flows, profiles outside of a few times the core radius well agrees with a β model. Metal abundance has been assumed to be constant. Temperature kT, abundance, β, core radius r_c, central density n_0, and the outermost radius $r_{\rm max}$ have been varied as free parameters. We find that the χ^2 value shows a broad minimum with $r_{\rm max}$ greater than about 20 arcmin (1.9 Mpc in the actual distance). The χ^2 value stays nearly the same for larger values of $r_{\rm max}$ even when $r_{\rm max}$ exceeds the radius of the GIS f.o.v. (about 25 arcmin). This preliminary analysis, therefore, suggests that $r_{\rm max}$ is greater than 20 arcmin. The best-fit values are $\beta = 0.65$, $kT = 3.9$ keV, and $r_c = 160$ kpc, respectively. The lower limit of $r_{\rm max}$ is useful in deriving the lower limits for $M_{\rm gas}$ and $M_{\rm total}$.

Based on these reulsts, the total mass is calculated from the following formula.

$$M_{\rm total}(r) = -\frac{kTr}{\mu m_{\rm H} G}\left(\frac{\partial \ln \rho}{\partial \ln r} + \frac{\partial \ln T}{\partial \ln r}\right)$$

The radial profies of the gas mass and the total mass are plotted in Fig. 5. As seen commonly in other clusters, the ratio of $M_{\rm gas}$ to $M_{\rm total}$ becomes larger as a function of radius (Böhringer 1994, Stewart 1994). Applying the condition, $r_{\rm max} > 20$ arcmin, $M_{\rm gas}$ turns out be larger than about 20% of $M_{\rm total}$. Here, $H_0 = 50$ km s^{-1} Mpc^{-1} is assumed.

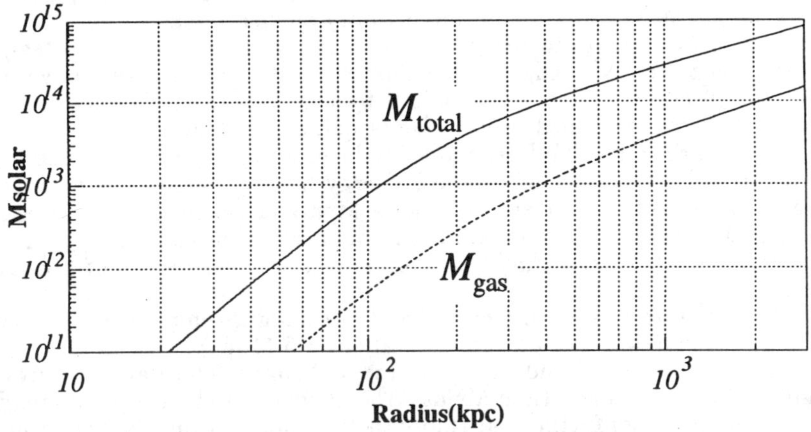

Fig. 5. Radial profiles of $M_{\rm gas}$ and $M_{\rm total}$ for A4059. X-ray emission is extended beyond a radius of 1.9 Mpc.

The present model (single β model with constant temperature) may be inadequate for the central region of the cluster, such as ignoring the cool component and absorption etc. However, since $r_{\rm max}$ constrained here from the data is much larger than the core radius, the results should not be too much sensitive to the ICM properties in the central retion. A further refinement in the model such as inclusion of a second emission component in the center is under way. The same analysis has been carried out for several other clusters whose whole emission fits in the GIS f.o.v. (Hydra-A, MKW3S, 3A0336+098), and it is likely that their emission is extented with $r_{\rm max} > 1.5 - 2$ Mpc for all of the above sample. The comprehensive results should be reported soon (Ikebe 1995).

The preliminary results coming out for A4059 and other clusters suggest that the amount of baryon may be as large as some 20% of the total mass in poor clusters. It has been discussed recently that the baryon fraction of about 30% observed in the Coma cluster is too high if the university has the critical density (White et al 1992). The big-bang nucleosynthesis predicts that the baryon can have less than 6% of the critical density (e.g. Kolb and Turner 1990), and numerical simulations show that the baryon cannot be much overabundant when averaged over a few Mpc (White et al 1992). The ASCA results suggest that the baryon content of clusters may be generally in excess of about 20% of the total mass. The value of $r_{\rm max}$ obtained here is only a lower limit, therefore the baryon fraction may be even larger.

5. SEARCH FOR CLUSTER BOUNDARY

The last topic here is on the possibility of detecting the cluster boundary from the temperature profile. This is an area where only ASCA can make a unique contribution.

Numerical simulations suggest that the density profile of clusters tend to follow a smooth function (such as β model) until it reaches near the average density of the universe (see e.g. Evrard 1990). In contrast to this, the temperature seems to show a more clear-cut boundary (Takahara et al 1976, Evrard 1990). This encourages the effor of finding the boundary based on the temperature profile. We may also emphasize that the sensitivity of ASCA up to an energy of 10 keV brings us almost the first opportunity of such a study.

The data from poor clusters have been examined to search for a temperature variation toward the edge. Even though most of the clusters have been observed for 40 ksec, it is severely limited by the statistics. To maximize the S/N ratio, data are examined in two energy bands: $1-3$ keV and $3-10$ keV. As mentioned earlier, the XRT response produces an apparent rise of the hardness ratio (count rate ratio beteen the 2 energy bands) with radius even when ICM has a constant temperature. We run a Monte-Carlo simulations based on an isothermal assumption and create the radial profiles for each energy band to be compared with the data. Any deviation would be a suggestion of a variable temperature.

The count rate profiles indicate that the constant temperature is a good description of the data out to a radius of about 1.5 Mpc for A4059 ($z = 0.48$), 3A0336+098 ($z = 0.035$) and Hydra-A ($z = 0.052$). The deviation from a constant temperature is less than about 20%. However, the data points tend to scatter if one tries to go further out (more than about $20'$ from the field center). In this region, the effective area of XRT becomes around 1/10 of the center and the data is dominated by scattered X-rays entering from outside of the field. Because of the large systematic uncertainties, we are unable to conclude at this stage about the detection of cluster boundaries. We hope that future long observations and more sophisticated analysis of the data should reveal a signature of the real boundary in clusters of galaxies.

ACKNOWLEDGEMENTS

The author is grateful to many colleagues who are carrying out very original analysis for the cluster data. Much of the results reported here come out from powerful contributions from ASCA ANL, and SimASCA software development teams. Particular thanks are due to, Yasushi Ikebe, Tadayuki Takahashi, Yasushi Fukazawa, Yoshitaka Ishisaki, Ken'ichi Kikuchi, and Noriko Yamasaki. Intriguing discussion with Kazuo Makishima, and Fumio Takahara are also gratefully acknowledged.

REFERENCES

Arnaud, K.A., et al 1991 in Frontiers of X-ray Astronomy, ed. Y. Tanaka & K. Koyama, (Tokyo: Universal Academy Press), p 481

Arnaud, K.A., et al 1993 in Proc. Ginga Memorial Symposium, ed. F. Makino & F. Nagase, ISAS

Arnaud, M., Rothenflug, R., Boulade, O., Vigroux, L., & Vangioni-Flam, E.

1992, A&A, 254, 49
Böhringer, H. 1994, in Cosmological Aspects of X-ray Clusters of Galaxies, ed. W.C. Seitter, (Kluwer Academic Publishers), p 123
David, L.P., Arnaud, K.A., Forman, W., & Jones, C. 1990, ApJ, 356, 32
Evrard, A.E. 1990, in Clusters of Galaxies, ed. W.R. Oegerle, M.J. Fitchett, & L. Danly, (Cambridge: Cambridge University Press), p 287
Fukazawa, Y., et al 1994, PASJ, 46, L55
Hatsukade, I. 1989, Ph. D. Thesis, Osaka University (ISAS RN-435)
Ikebe, Y. 1995, Ph. D. Thesis, University of Tokyo, in preparation
Ikebe, Y., et al 1992, ApJL, 384, L5
Ikebe, Y., et al 1994, in Clusters of Galaxies, ed. F. Durret (Gif sur Yvette: Editions Frontières), in press
Kohmura, Y., et al 1992, SPIE, 2006, 78
Kolb, E.W., & Turner, M.S. 1990, Early Universe, (Addison-Wesley)
Koyama, K., Takano, S., & Tawara, Y. 1991, Nature, 350, 135
Matsumoto, H., et al 1994, in New Horizon of X-ray Astronomy, ed. F. Makino & T. Ohashi (Tokyo: Universal Academy Press), p 511
Mihara, K., & Takahara, F. 1994, PASJ, 46, 447
Ohashi, T., et al 1991, SPIE, 1549, 9
Ohashi, T., Fukazawa, Y., Ikebe, Y., Ezawa, H., Tamura, T., & Makishima, K. 1994, in New Horizon of X-ray Astronomy, ed. F. Makino & T. Ohashi (Tokyo: Universal Academy Press), p 273
Rephaeli, Y. 1978, ApJ, 225, 335
Schwartz, D.A., Bradt, H.V., Remillard, R.A., & Tuohy, I.R. 1991, ApJ, 376, 424
Serlemitsos, P.J., et al 1994, PASJ, submitted
Stewart, G.C. 1994, in Cosmological Aspects of X-ray Clusters of Galaxies, ed. W.C. Seitter, (Kluwer Academic Publishers), p 139
Takahara, F., Ikeuchi, S., Shibazaki, N., & Hoshi, R. 1976, Prog. Theor. Phys., 56, 1093
Tanaka, Y., Inoue, H., & Holt, S.S. 1994, PASJ, 46, L37
Tsuru, T., 1992, Ph. D. Thesis, University of Tokyo (ISAS RN-528)
White, S.D.M., Navarro, J.F., Evrard, A.E., & Frenk, C.S. 1993, Nature, 366, 429
Yamashita, K. 1992, in Frontiers of X-ray Astronomy, ed. Y. Tanaka & K. Koyama, (Tokyo: Universal Academy Press), p 475
Yamashita, K. 1994, in Clusters of Galaxies, ed. F. Durret (Gif sur Yvette: Editions Frontières), in press

MAPPING THE PROPERTIES OF GALAXY CLUSTERS WITH ROSAT

Damian Cannon & Trevor Ponman
University of Birmingham, UK
dbc@star.sr.bham.ac.uk, tjp@star.sr.bham.ac.uk

ABSTRACT

We present results from the early phases of a project to fit models to ROSAT PSPC data for clusters and groups of galaxies. We find that in all systems studied, the distribution of hot gas is flatter than that of the dark matter and that the gas fraction is ∼15-25% within the region studied. However, profiles differ significantly in shape across the sample.

1. INTRODUCTION

It is now well established that the hot gas seen in X-ray images of galaxy clusters is the dominant visible component, with a mass several times that of the galaxies. If this gas is in hydrostatic equilibrium in the cluster potential, then its pressure gradient allows one to map out the potential.

We have begun a programme of X-ray mapping of a sample of clusters using Rosat PSPC spectral images. We fit a variety of spherically symmetric, parameterised models[1] to the data. By comparing results from different models we can compare their relative success, and ensure that our results are not dependent upon the assumption of some specific analytical form. A comparison can also be made with the results derived from spectral fits to the X-ray emission in discrete annuli.

Preliminary results for two Abell clusters (A2199 & A496), one poor cluster (AWM7) and one Hickson group (HCG62) are presented. These were selected on the basis of relatively low gas temperatures ($\leq$4.5keV) and smooth, relaxed X-ray emission. These will form part of a sample selected from the ROSAT archive. $H_0 = 50$ km s^{-1} Mpc^{-1} is used throughout our analysis.

2. SURFACE BRIGHTNESS AND CORE RADIUS

For each cluster, we generate a radial surface brightness profile of the X-ray emission (Fig 1a). The absence of self-similarity between the profiles is apparent. The presence of a cooling flow, in all of these systems, is indicated by the rise in surface brightness in the central regions. The X-ray 'core radius' has been widely used as a scale size for galaxy clusters. To determine whether it is really a well defined quantity in the presence of a cooling flow, we have investigated how well it is constrained by the data, and the extent to which the value depends on the method of derivation. If we assume a King model for the X-ray surface brightness distribution, with core radius r_c, then the radial variation is given by $S(r) = S_0[1 + (r/r_c)^2]^{-3\beta+0.5}$. To determine the best-fit core radius we compared this model to the observed brightness profile and adjusted the available parameters until a minimum in χ^2 was achieved. Similar results were obtained from analysis of two-dimensional X-ray images (spectrally integrated) and three-dimensional spectral images (see Section 3). Our results indicate that X-ray core radii are reasonably well constrained for the richer systems and consistent between different methods of analysis, when the increased central surface brightness (provided by a cooling flow) is accounted for. However in HCG62 the X-ray core radius is *not* well defined, since the cooling flow is extensive enough to hidc any core which might have been present. We intend to investigate whether a similar effect is seen in richer systems with very strong cooling flows.

3. CLUSTER MODEL FITTING

In order to extract the physical properties of the cluster gas, one needs to remove the effects of projection and instrumental blurring. We have done this by setting up spherically symmetric models of the cluster, and fitting them to the spectral imaging data. In this process emission from the models is computed using a hot plasma code, projected along the line of sight and folded through the spatial and spectral response of the PSPC for comparison with the data. The model parameters are then optimised to achieve a maximum likelihood fit. This allows us to extract the distribution of the gas and dark matter, with formal statistical errors. Outside the cooling flow, the gas density is represented by a core-index model, $\rho(r) = \rho_0[1 + (r/r_\rho)^2]^{-3\beta/2}$. As the temperature distribution is less well defined, several models can be fitted, such as a linear ramp, a power law distribution or a polytropic distribution. Alternatively, we can fit a model for the gravitational potential and derive $T(r)$ from the hydrostatic equation. For clusters containing cooling flows a separate model component can be included to account for the increased central emission.

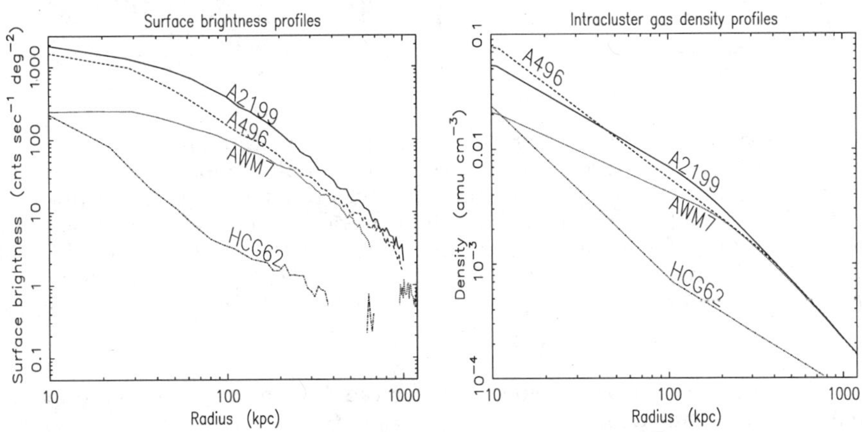

Figure 1: a) Overlay of surface brightness profiles for the clusters studied. The absence of self-similarity between the profiles and the effect of the central cooling flow are apparent. Gaps in the profiles of HCG62 and AWM7 are caused by the PSPC detector support ring. b) Radial gas density distributions derived from cluster fits. At large radii the richer systems converge towards a similar β value. Inside the cooling flow region AWM7 exhibits a flatter slope due to its lower central gravitating mass density.

4. GAS DENSITY

The spectral bandpass of ROSAT is limited to relatively low energies resulting in the detected count rates from an X-ray emitting gas with a temperature greater than ≈2 keV being relatively insensitive to the temperature. Hence the gas density profile is well determined for the clusters in this sample and doesn't vary significantly between cluster models. Gas density profiles derived with a core-index model are shown in Fig 1b. The three richer systems all have β values typical of those found[2] in rich clusters (0.52-0.62), whilst the gas profile of HCG62 is exceptionally flat by comparison, with $\beta = 0.32$. At small radii all of the clusters have cooling timescales of less than 10^9yr, consistent with the presence of cooling flows.

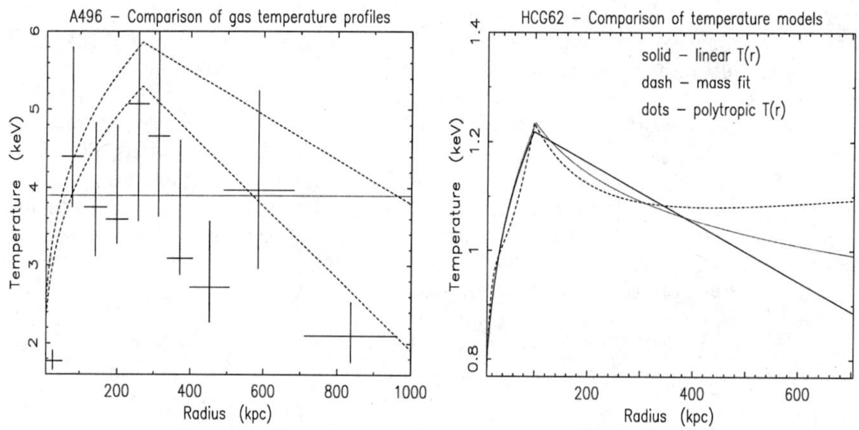

Figure 2: a) Fit of the temperature ramp model to the data for A496 . The results of annular spectroscopy are overlaid as a simple comparison, although the values are not directly comparable since the cluster fitting accounts for projection and the instrumental PSF. The Ginga best-fit value is indicated by a horizontal line. b) The fit of several temperature models to the data for HCG62. This comparison illustrates that while the models differ their general form is similar.

5. GAS TEMPERATURE

The statistical error envelope (1σ) for the 3D temperature profile fitted to the A496 data (using a linear ramp $T(r) = T_0 - \beta_T r$) is plotted in Fig 2a, together with the temperatures obtained from spectral fits to the data binned into annuli around the cluster centre. The radially averaged temperatures from these fits are comparable with those obtained by non-imaging spectrometers such as Ginga. The effect of the cooling flow is apparent at small radii while at large radii the three-dimensional fit indicates higher gas temperatures than those derived from annular spectroscopy. This difference is explained by the effects of projection and blurring. The variation in the fitted temperature profile with cluster model is shown in Fig 2b where the profiles of three different model fits to the HCG62 data are compared. The models shown are a linear temperature ramp, a polytropic temperature profile and the temperature derived from a mass fitting model, where the total enclosed mass M(r) is fitted. Although the differences between the fitted profiles can be larger than their statistical errors, they all indicate the same trend of a central cooling flow together with a decreasing gas temperature at larger radii.

6. MASS COMPONENTS

Radial profiles of the total gravitating mass are shown for each cluster in Fig 3a, whilst Figure 3b shows the radial variation in the fraction of the total mass provided by the intracluster medium. These are derived from the linear temperature ramp models fitted. The gas fractions increase with radius for all of the clusters, indicating that the gas profile is flatter than that of the dark matter, and reaches $\gtrsim$15% of the enclosed mass within the region analysed. These derived gas mass fractions are obviously lower limits to the baryonic mass fractions of the clusters since no account has been taken of the contributions from baryons in the cluster galaxies. If $\Omega = 1$ these fractions are greater than the levels suggested by cosmic nucleosynthesis models which limit the baryon mass fraction to $\approx$6%. This suggests that either $\Omega \lesssim 0.3$, that baryons are preferentially concentrated in clusters or that nucleosynthesis models are incorrect[4,5].

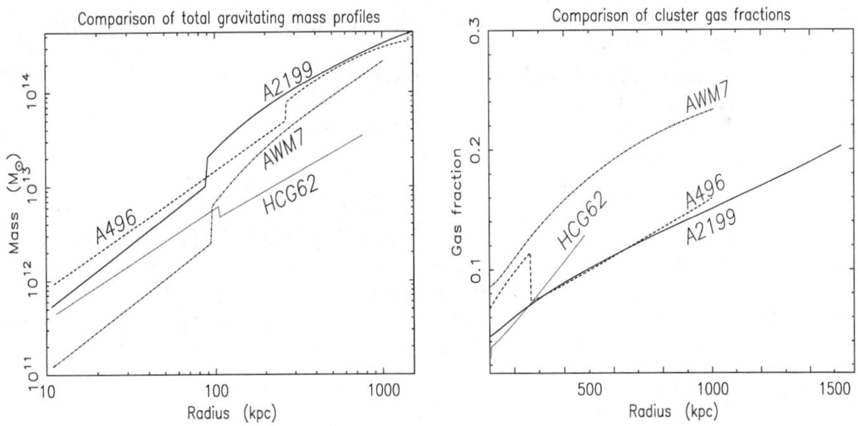

Figure 3: a) Total gravitating mass densities inferred from cluster fits. These are extended to the limits of reliable surface brightness information. The 'steps' in the profiles are caused by the temperature gradient discontinuity at the boundary of the cooling flow model. b) Radial variations in the fraction of the enclosed cluster mass provided by the intracluster medium. This is an underestimate in the baryon fraction of these clusters since the galaxy contribution (typically a few percent) is not included.

REFERENCES

1 - Watt, M. P., et al. 1992, MNRAS, 258, 738
2 - Jones, C., & Forman, W. 1984, ApJ, 276, 38
3 - White, D. A., et al. 1991, MNRAS, 252, 72
4 - White, D. A., & Fabian, A. C. 1993, preprint
5 - White, S. M., et al. 1993, Nature, 366, 429

ASCA OBSERVATION OF THE NGC 2300 GROUP OF GALAXIES

Y. Tawara, Y. Sakima, K. Yamashita
Department of Astrophysics, School of Science, Nagoya University
Furo-cho, Chikusa-ku, Nagoya 464-01, Japan

R. Mushotzky
NASA/Goddard Space Flight Center, Greenbelt MD 20701, U.S.A.

ABSTRACT

We report on NGC2300 Group of galaxies observed with ASCA in the performance verification (PV) phase. Extended diffuse X-ray emission surrounded this galaxy group, which is first reported by ROSAT observation, was confirmed with its total luminosity of 4 x 10^{43} erg s^{-1} in the 0.6 - 2.6 keV band. We also found that the temperature of this hot gas was 0.90 $\pm$ 0.14 keV and its elemental abundance was 0.09 $\pm$ 0.05 of the solar value in the 90 % confidence, if we adopt the thermal emission model derived by Raymond and Smith.

1. INTRODUCTION

The compact groups of galaxies are in the smallest end of galaxy clustering and their diffuse X-ray emission has special importance for the study of origin and evoluion of intergalactic diffuse hot gas. Among tens of groups of galaxies, the NGC 2300 group was the first one from which intergalactic diffuse X-ray emission was detected with ROSAT(Mulchaey et al. 1992). Assuming its spectrum as that from thin hot plasma, the low luminosity of 2.4 x 10^{43} erg s^{-1} and the low temperature of 0.86 keV were derived in the range expected from X-ray/optical correlation such as σ-T_X and from the L_X-T_X relation, where σ is velocity dispersion. In contrast to this trend, the elemental abundance of 0.06+0.12/-0.05 is very low relative to that in clusters, which is against the trend for low-luminosity clusters to be metal-rich. To confirm these X-ray properties and to get further information, ASCA observed this group in the PV phase and we report its results in the following sections.

2. OBSERVATIONS

A 19 ks ASCA GIS and SIS observations were obtained on 1993 May 29-30, centered on $\alpha=7^h28^m$, $\delta=85°38'$ (1950). The spiral galaxy NGC 2276 and the bright elliptical galaxy NGC 2300 are in the field of one of the SIS 4 chips, where 4 CCD mode was used. An S0 galaxy IC 455 was at the edge of SIS field and the GIS fileld covered these three galaxies. NGC 2268, the rest of three galaxy members of HG92 (Huchra and Geller 1982) was not in the fileld of both SIS and GIS. Off sky background observation was also made for about 20 ks with its filed center of $\alpha=7^h41^m$, $\delta=85°52'$ (1950). X-ray emission is clearly detected from NGC 2276 and NGC 2300, and marginally detected from IC 455 as shown in Figure 1a and 1b.

Figure 1a. GIS image of the NGC 2300 group of galaxies (E<2 keV)

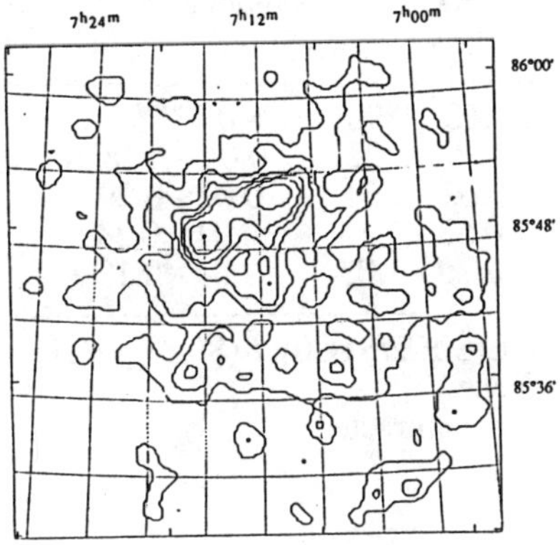

Figure 1b. GIS image of the NGC 2300 group of galaxies (E>2 keV)

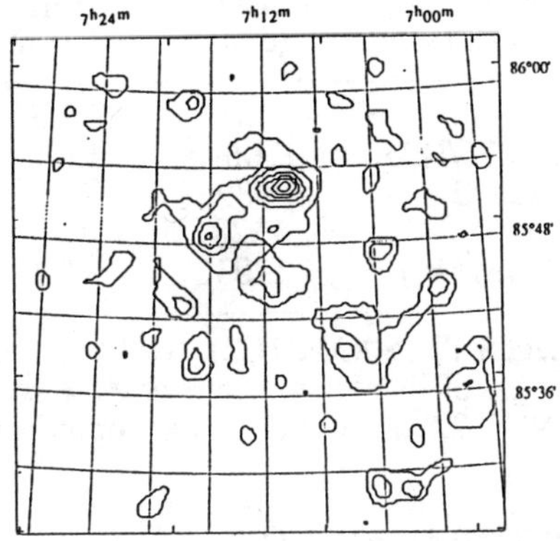

To search for a diffuse component, photon fluxes are integrated for several different regions for SIS, as shown in Table 1. The source regions B and C, including the galaxies, are taken as a circle of radius 1'.5, which corresponds to a half power radius of X-ray telescope. Though there can be contamination of galaxy components in the intragroup medium (IGM) region, the fraction is roughly estimated at most 25 % of IGM flux of 5400 photons and thus IGM component is confirmed.

Table 1. Photon Fluxes of NGC 2300 Group of Galaxies

Region	Source	Raw Photon Counts (cts for SIS0+SIS1)	Region size (arcmin^2)	Surface Brightness (x 10^{-4} cts s^{-1} arcmin^{-2})
A	all *	6540	22 x 22	9.4
B	NGC 2300	718	3.0 ϕ	70.8
C	NGC 2276	413	3.0 ϕ	40.7
A-B-C	IGM	5409	470	8.0

* SIS 4 CCD region

The spectrum of the diffuse gas was extracted by summing the counts in each SIS chip with and without a mask of the circle of radius 1'.5 centered on each galaxy and subtracting a blank sky data obtained at high galactic latitude as a background. The intragroup medium spectrum is corrected for vignetting. Results of spectral fits are summarized in Table 2.

Table 2. Results of Spectral Fitting for each SIS chip data

Region * Chip Pos.(source)	Counts (cts)	N_H (10^{20} cm^{-2})	kT ** (keV)	Abundance** (Solar)	Lx ($x10^{43}$ erg s^{-1})
W	585	9.6	0.68 - 0.90	> 0.03	1.0
N (N2276/N2300)	1908	12.8	0.76 - 1.02	0.04 - 0.18	4.9
E (N2300)	989	3.9	0.62 - 1.31	0.01 - 0.29	0.9
S	1228	15.1	0.58 - 1.13	0.01 - 0.34	0.7
N (IGM***)	1203	8.1	0.76 - 1.03	0.04 - 0.13	3.8

*Detector :W(S0C0/S1C2), N(S0C1/S1C3), E(S0C2/S1C0), S(S0C3/S1C1)
** ranges of fitted parameters at 90 % confidence
*** NGC 2276 and NGC 2300 components are subtracted

3. SPECTRAL RESULTS

The spectrum of the diffuse gas in the group is shown in Figure 2. A raymond-Smith (1977) plasma model with variable abundance and N_H was fitted to this spectrum, with results given in Table 2 and the residuals of the fit displayed in Figure 2. The X-ray absorbing H_I column densities were not well constrained but consistent with the Galactic value of 5.6 x 10^{20} cm^{-2} (Stark et al. 1992). The best fit temperature for the diffuse gas is 0.90 +0.13/-0.14 keV. The abundance of the diffuse emission ranges from 0.04 to 0.13 solar at 90 % confidence. The confidence contours for the best-fit model in the abundance-kT plane are shown in Figure 3.

On the spectra for the individual galaxies, we can see relative hardness in the Figure 1a and 1b. At least NGC 2276 has harder spectra than NGC 2300 and IC 455, but further analysis has not been completed.

Figure 2. The SIS spectrum of the diffuse gas in the NGC 2300 group of galaxies

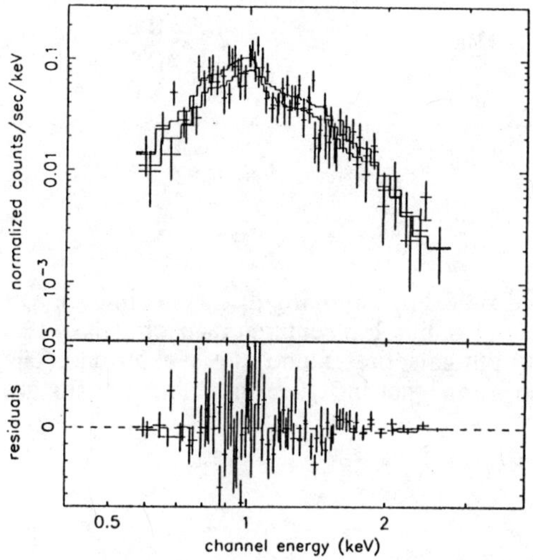

Figure 3. The confidence contours for the best-fit model in the abundance-kT plane

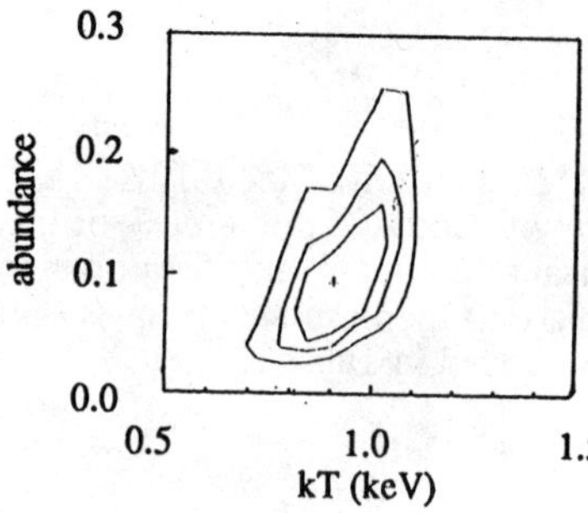

4. DISCUSSION AND CONCLUSIONS

In this observation, we confirmed the presence of a diffuse hot gas as the intragroup medium of HG92 and found that the spectral features of this component are fairly consistent with the ROSAT results (Mulchaey et al. 1992). On the abundance of the diffuse emission, we could get better constraint than ROSAT and confirmed that it is the lowest abundance intergalactic gas known. Adding to the possible interpretations of this low abundance made by Mulchaey et al. (1992) , we point out another possibility of a depletion of metals in the dust.

REFERENCE

Huchra, J.P., & Geller, M.J. 1982, ApJ, **257**, 423

Mulchaey, J. S., Davis, D. S., Mushotzky, R.F., Burstein, D., 1992, ApJ, **404**, L9

Stark, A.A., Gammie, C.F., Wilson R. W., Bally, J., Linke, R. A., Heiles, C., & Hurwitz, M., 1992, ApJS, **79**, 77

Raymond, J.C., Smith, B.W., 1977, ApJS, **35**, 419

A COMPARISON OF X–RAY AND OPTICAL WAVELET MAPS IN THE CENTRAL REGION OF COMA

D. Gerbal [1,2]
and
A. Biviano [1], F. Durret [1,2], O. Le Fevre [2],
F. Le Peintre [1], C. Lobo [1], A. Mazure [3], E. Slezak [4].

1–Institut d'Astrophysique de Paris, 2–DAEC–Observatoire de Paris–Meudon
3–Laboratoire d'Astrophysique Spatiale de Marseille, 4–Observatoire de Nice
1–authors@iap.fr, 2–lefevre@gin.obspm.fr
3–mazure@astrsp-mrs.fr, 4–slezak@rameau.obs-nice.fr

ABSTRACT

We have applied a wavelet transform analysis on ROSAT images and on an optical catalogue of the Coma cluster of galaxies. The comparison of these two sets of images allows us a detection of X–ray emiting galaxies as well as particular emiting regions.

INTRODUCTION

The wavelet transform analysis (WTA) is a powerful technique to investigate structures. It has been successfully applied to clusters of galaxies both at optical (e.g. Escalera et al 1992) and x-ray wavelengths (e.g. Slezak et al 1994) Hereafter we present a comparison of optical and x-ray data on the Coma cluster of galaxies using WTA.

THE DATA

OPTICAL SAMPLE :

We have defined a complete sample of 296 cluster member galaxies with $m_b < 20$ located in a rectangle defined by -1200"$\leq x \leq$1700" and -750"$\leq y \leq$750" with respect to the center defined by Godwin et al.(1983). The membership has been established on the basis of redshift for 205 galaxies (observed with MOS-SIS at the CFHT), and using the color-magnitude relation for 91 galaxies (see Biviano et al.1995, for the actual definition of the sample).

X–RAY DATA :

Our data were taken with the ROSAT PSPC (see White et al. 1993 for a description).

RESULTS AND COMMENTS

Four X-ray maps corresponding to four wavelet scales (15", 30", 60", 120") are displayed in fig.1 a-d, while in figs.2 a-d we display four optical maps corresponding to the same four wavelet scales.

In fig.1a, details on the scale of 1-2 pixels can be seen. Some of the most

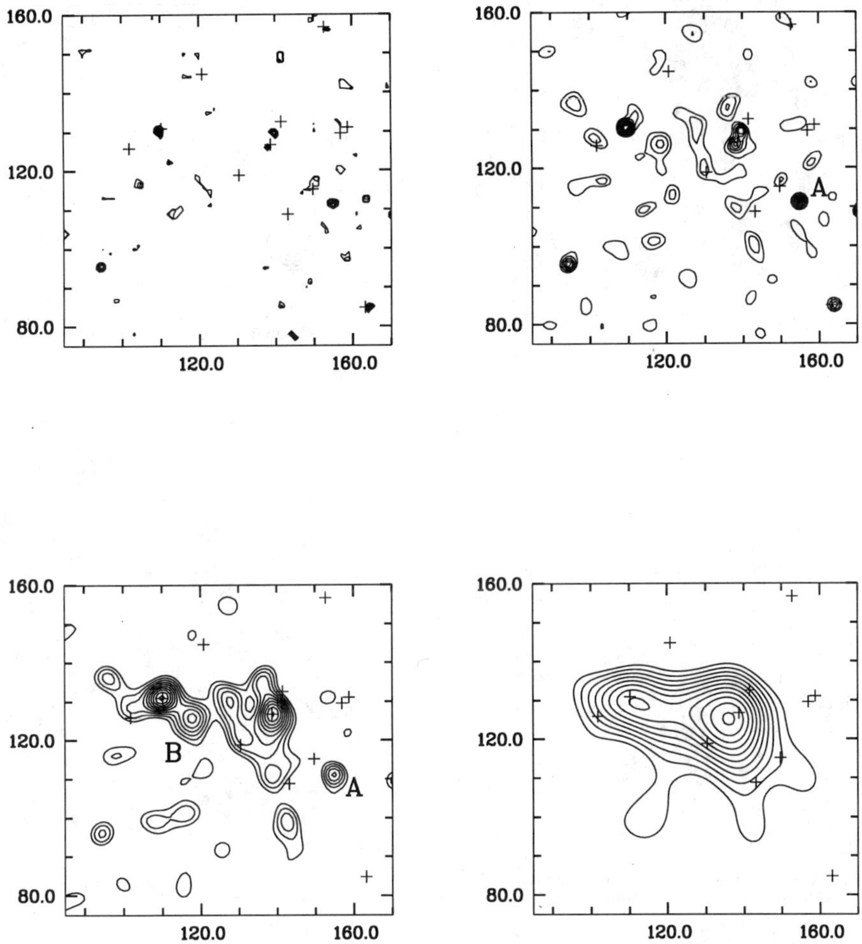

Fig.1 (a,b,c,d) 15", 30", 60" & 120"–scale wavelet X– ray image

luminous galaxies are associated with these features. This is consistent with the finding of Dow & White (1994) that some Coma galaxies are X-ray emitters (the emission of 7 out of a sample of 35 is significant at a 3 σ –level). we do not detect N4898. However this galaxy is connected with a feature in fig.1b, probably indicating a local potential well.

The two brightest galaxies (N4874 & N4889) are associated with structures seen at all scales, and their positions perfectly coincide with local maxima.

Notice an emission point (labeled A) in fig.1a becoming stronger in fig.1b not correlated with any of the 12 most luminous galaxies. In fig.2a, details on the scale of a single galaxy are seen. Significant structures are best seen on

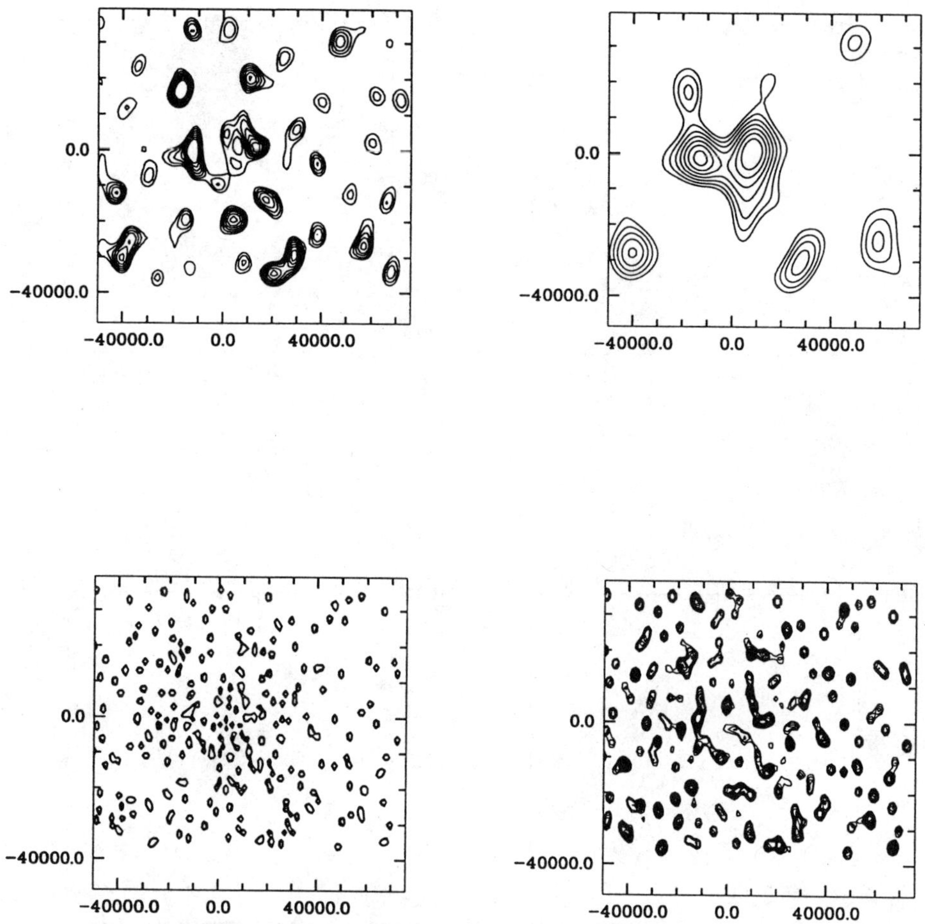

Fig.2 (a,b,c,d) 15", 30", 60" & 120"–scale wavelet optical image.

the largest scale (fig.2d). The two groups associated with the two central giant galaxies are clearly detected. Notice however that while N4874 is at the center of the west peak, this is not the case for N4889 which is located somewhat north of the east peak.

By comparing figs.1d and figs.2d, we see similarities between the optical and x-ray maps, suggesting that the galaxies and the plasma *approximately* trace the same potential. However, there are notable differences. In particular, the optical maximum in the east structure does not coincide with its equivalent in the x-ray map. Moreover, in fig.1c there is a maximum (labeled B) which is not found in the optical maps of fig.2, and not connected with luminous galaxies. It must be kept in mind that, following various hydrodynamic simulations (e.g. Evrard 1990), x-ray emission does not always follow the potential well.

Number	Name	m_b
1	N4874	12.78
2	N4889	12.62
3	N4898	14.85
4	N4867	15.44
5	N4883	15.43
6	I3959	15.27
7	I3973	15.32
8	N4869	14.97
9	N4873	15.15
10	N4864	14.70
11	N4865	14.54
12	1257.3 + 2758	

Table of galaxies quoted in Fig. 1
Galaxies are ordered following decreasing X–ray luminosity. – from Dow & White

ACKNOWLEDGEMENTS

We are very grateful to S.D.M. White for providing us with his ROSAT images of Coma.

We acknowledge financial support from the Groupe de Recherche "*Cosmologie*" of the "Centre National de la Recherche Scientifique"

REFERENCES

Biviano A., Durret F., Gerbal D., Le Févre O., Lobo C., Mazure A., Slezak E., 1995, A&A, in press

Dow & White 1994, CFA preprint N 3918

Escalera E. et al. 1992, A&A,264,379

Evrard A.E. 1990, ApJ, 363, 349

Slezak E.,Durret F., Gerbal D.1994, AJ, in press

White S.D.M. et al. 1993, MNRAS, 261, 18

A HIERARCHICAL CLUSTERING METHOD FOR THE DESCRIPTION OF STRUCTURES IN GALAXY CLUSTERS

Daniel Gerbal [1], Arturo Serna [2]
1–Institut d'Astrophysique de Paris
2–Observatoire de Paris–Meudon
1–gerbal@iap.fr, 2–serna@mesiob.obspm.fr

ABSTRACT & INTRODUCTION

The mathematical theory of cluster analysis (see, e.g., Andenberg 1973) provides different methods to distribute a set of objects into a set of clusters. We have adapted this method for the founding and the description of (sub-)structures in galaxy clusters. An overview of the method is here rapidly described as well as application upon three Abell clusters.

2. METHOD

We intend to apply this technique to clusters of galaxies. We start with a sample of N galaxies.

1) We initially consider that each galaxy constitutes a group by itself,

2) we define an affinity parameter $\{s\}$, between these galaxies or groups of galaxies (distances, inverse of forces,...) and construct an array, i.e. the value of $\{s\}$ for each couple $\{i, j\}$,

3) we then extract the two groups presenting the highest affinity following the value $s_{min} \in S_{ij}$. These two groups are then ¯merged˘ into a single group. We then have (n-1) groups,

4) the merging process is repeated until only one cluster is left.

The choice of the affinity parameter s appears crucial. Because we want to study how bound a sub-cluster is, we have made the *natural* choice of

$$E_{ij} = -G\frac{m_i m_j}{|r_i - r_j|} + \frac{1}{2}\mu_{ij}(v_i - v_j)^2 \qquad \mu_{ij} = \frac{m_i m_j}{m_i + m_j} \tag{1}$$

as the affinity parameter.

3. STABILITIES

We have tested two kinds of stabilities :

1 – We have performed a N-body simulation of the evolution of a self-gravitating system (using the code developed by Scholl & Alimi, 1992, on Connection Machine). Then, we have tested that sub-structures identified by our hierarchical method are reasonably stable during the system evolution.

2 –Each particle in a N-body simulation is characterized by its 6 coordinates¿: $\{x, v\}$. However, astronomical observations provide only 3 coordinates $\{\alpha, \delta, v_z\}$. We have tested that the previously identified sub-structures are still present when we apply our method but *just using these 3 coordinates.*

4. ASTROPHYSICS

We have applied our method to three galaxy clusters : A151, A2670, A3558. The data of ABCG151 and ABCG3558 come from the ESO-Keyprogram (EKPGM) on nearby clusters of galaxies (Mazure et al 1992), while the data of ABCG2670 are taken from Sharples et al. (1988). Notice that the data from the EKPGM concern *the field of view* of the two clusters, while the data for ABCG2670 have been cleaned from interlopers.

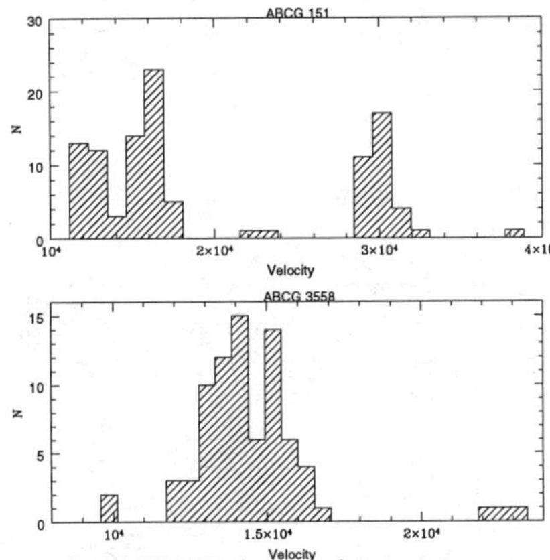

Figure 1. A151 and A3558 Velocity histograms

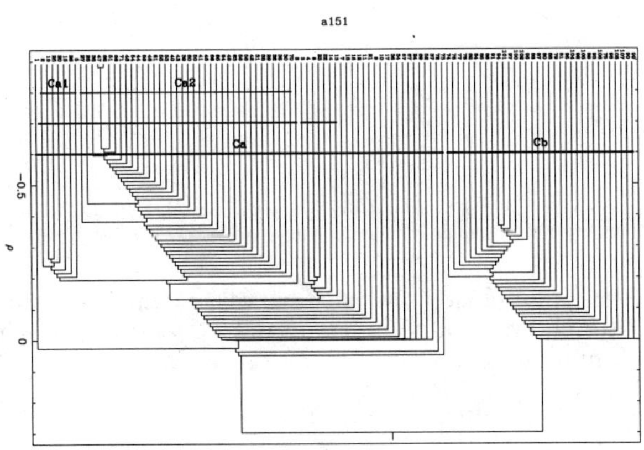

Figure 2. A151 dendogram analysis

We display the 3 resulting dendograms, on which the dynamical structure are striking visible. The analysis of these dendograms indicates that ABCG151 and ABCG3558 are very complex systems.
ABCG151 :

We can see clearly on the velocity histogram fig.1 that at least two clusters are located in the field of A151. The analysis of the A151 dendogram is instructive : The two previously noted clusters are present; C_a and C_b . C_b has no sub-structure, while C_a is again complex. Various sub-structures appear at different levels of binding (boxes in boxes). Note the two most obvious sub-structures : $C_{a,1}$ and $C_{a,2}$

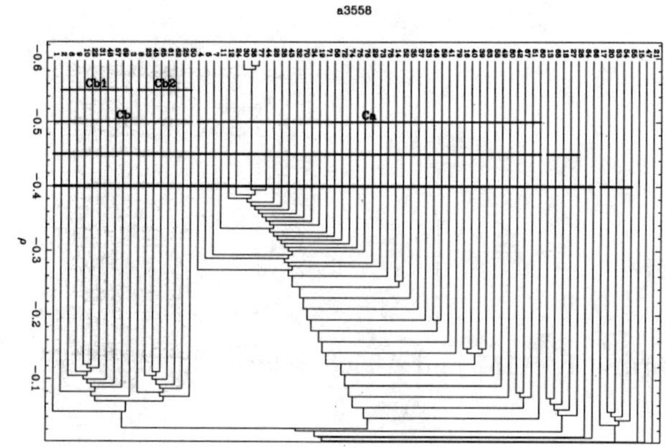

Figure 3. A3558 dendogram analysis

ABCG3558 :

we find one cluster but with small satellites (in the dynamical meaning). This cluster shows a bimodal structure C_a and C_b, one of these sub-structures is again bimodal.
ABCG2670 :

no sub - structure appears (except for some pairs - one very bounded).

5. CONCLUSIONS

We have adapted a "hierarchical clustering method" for the founding and the description of (sub-)structures in galaxy clusters. Applied on A151, 2670 & 3558, our method appears efficient : it allows us to distinguish clusters in a field of view, and sub-clusters in a cluster. More work are needed to compare with other methods.

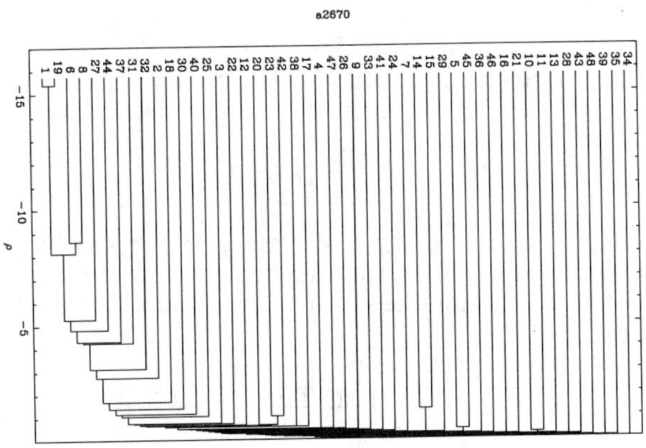

Figure 4. A2670 dendogram analysis

ACKNOWLEDGEMENTS

We aknowledge Dr. V. Gurzadyan for illuminating discussions.

REFERENCES

Alimi, J.-M. & Scholl, H. 1993, Int. J. Mod. Phys. C 4, 197
Anderberg, M. R., 1973, Cluster analysis for applications (London: Academic Press)
Mazure, A. & al 1992, in The Distribution of Matter in the Universe, eds G.A. Mamon & D. Gerbal (Paris : Observatoire de Paris) p.79
Sharples, M. C. 1988, MNRAS, 231, 479

THE EFFECT OF A PARTIALLY VIRIALIZED GAS ON THE BARYON FRACTION DERIVED FROM X-RAY MEASURMENTS

Mark J. Henriksen
Department of Physics and Astronomy, University of Alabama, Tuscaloosa, AL 35401
mark@bama.astr.ua.edu

ABSTRACT

We have quantified the uncertainty in the gas baryon fraction due to a partially virialized gas distribution by calculating the baryon fraction at various epochs in two simulations of cluster formation and evolution. The uncertainty is a factor of $\simeq$ 2 - 4.

1. INTRODUCTION

Perhaps the primary uncertainty in calculations of the cluster mass distribution is whether or not the intracluster medium is in hydrostatic equilibrium. If the intracluster medium is only partially shock heated, what would be the effect on the cluster mass distribution calculated from the X-ray emitting gas parameters? To answer this question, we fit the X-ray data in cluster simulations at the current and z=0.5 epoch. The early epoch gas parameters are for a partially virialized gas. At the current epoch the gas is virialized and the mass calculated from the X-ray parameters using hydrostatic equilibrium would be the true mass distribution. A comparison gives an indication of how uncertain the mass calculation may be due to an unvirialized cluster gas.

2. DATA ANALYSIS

The density and temperature profiles from Evrard (1990) and Thomas and Couchman (1991) for both the z=0.5 and z=0 epochs are fit with density and temperature profiles over the extent of the gas. The density profiles are fit with the function:

$$n(r) = n_c(1 + (r/a)^2)^{-\delta}$$

The temperature profiles are obtained with the polytropic density-temperature relation:

$$T(r)/T_c = (n(r)/n_c)^{\gamma-1}$$

The simulation data is given an error of 10% and fit with free parameters allowed to vary within a large range of values: γ (1 - 5/3), δ (0.5 - 1.5), and a (100 - 500 kpc). The central density and central temperature are calculated from the normalization which minimizes χ^2. A single temperature profile can not be fit to the central region in either simulation. For the data in Evrard (1990), a second component is fit to the region inside of the radius at which the temperature turns over. In the inner region, γ is allowed to vary within the range 0.1 - 1.0. The data of Thomas and Couchman were restricted to the region outside of roughly 200 kpc on the advice of the authors and only a single component is needed.

The best fit parameters derived from fitting the data at z = 0.5 is shown in Table 1 for both simulations.

TABLE 1: Best Fit Parameters: z = 0.5

	χ^2_ν	δ	γ	a (mpc)	$n_c(cm^{-3})$	T_c(keV)
TC (0.2 - 1.0 mpc)	1.5	1.5	1.32	0.4	0.013	6.9
E(0. - 1. mpc)	0.8	1.0	0.9	0.1	0.031	3.6
E(1. - 5. mpc)	15.	1.2	1.66	0.1	-	-

Table 2 contains best fit parameters for the current epoch.

TABLE 2: Best Fit Parameters: z = 0.

	χ^2_ν	δ	γ	a (mpc)	$n_c(cm^{-3})$	T_c(keV)
TC (0.2 - 1.0 mpc)	0.52	1.2	1.24	0.1	0.091	21.
E(0. - 1. mpc)	0.6	1.0	0.8	0.14	0.021	4.1
E(1. - 5. mpc)	7.9	1.4	1.4	0.42	-	-

Comparison shows that for the simulations of Thomas and Couchman, the density and temperature profile is steeper at the z=0.5 epoch; there is more cool gas in the outer region of the cluster which has not been virialized. This reduces the total mass calculated from the gas parameters since,

$M_{tot} \simeq x^{3-2(1+\delta(\gamma-1))}$, for $x >> 1$, where $x = r/a$.

The total mass is calculated from the equation of hydrostatic equilibrium using the best fit density and temperature profiles from Evrard (1990). The gas mass is calculated from the density profile and the ratio of gas to total mass is the baryon fraction in gas. The distribution of baryon fraction with radius is important since the highest quality X-ray observations typically contain only the central region of the cluster atmosphere. The difference in baryon fraction between the z=0.5 and z=0 epoch (shown in Figure 1) is less than a factor of two over the extent of the gas. The best fit temperature profiles from Thomas and Couchman (1990) are used to calculate the total mass. The baryon fraction in gas (gas mass/total mass) has a strong radial dependence at the z=0.5 epoch. The difference in baryon fraction (shown in Figure 2) increases with radius and is as large as a factor of 4 at 1 Mpc.

3. CONCLUSIONS

The available simulations offer different strengths and weaknesses and for this reason produce some gross features presently observed with modern X-ray observatories, however, they fail to produce some of the detailed features (note the difference in core radius evolution). Using Evrard's simulation, the effect of evolution on the calculated mass would be negligible, relative to the uncertainty resulting from the present capability to actually measure the temperature profile over the extent of the gas. On the other hand, to the extent that the simulations of Thomas and Couchman accurately describe the cluster atmosphere, the effect of partially virialized intracluster gas on the mass determination is potentially quite significant.

REFERENCES

Evrard, A. 1990, ApJ, 369, 349
Thomas, P.A., & Couchman, H.M.P. 1992, MNRAS, 257, 11

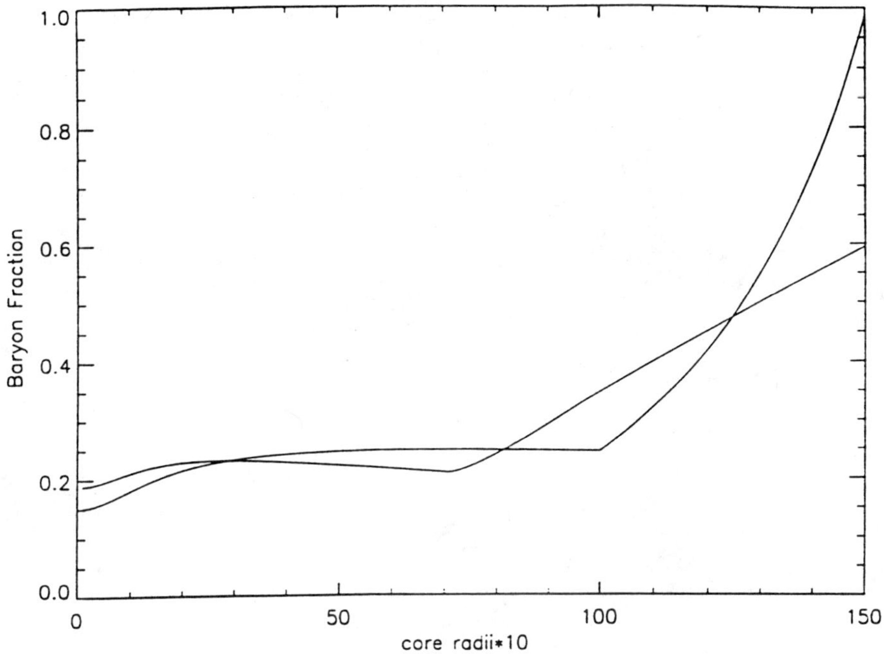

Figure 1.

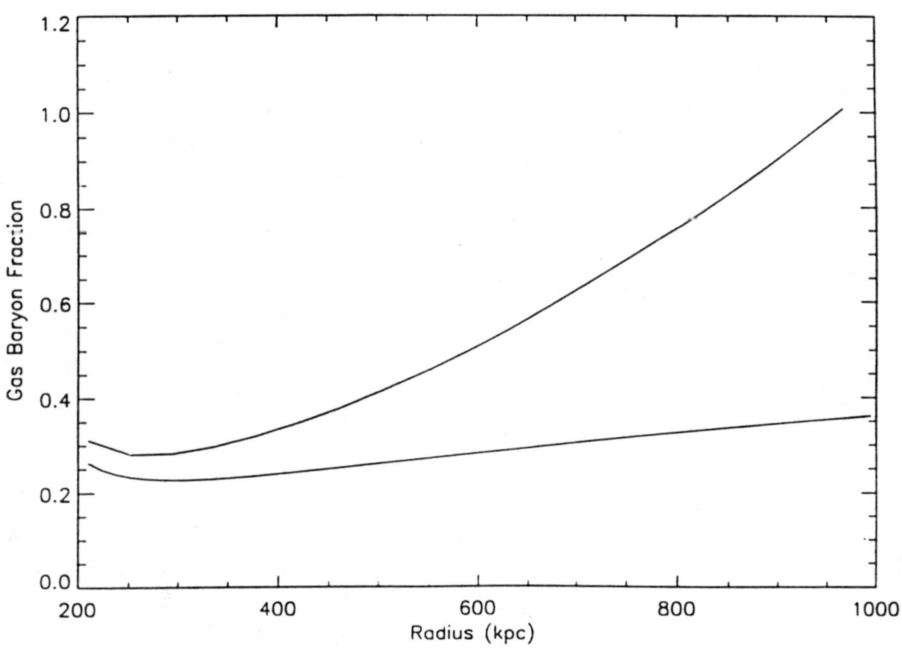

Figure 2.

GRAVITATIONAL LENSING AS A PROBE OF DARK MATTER

DARK MATTER MAPPING BY GRAVITATIONAL LENS TOMOGRAPHY

J. A. Tyson
AT&T Bell Laboratories, Murray Hill, NJ 07974
tyson@physics.att.com

ABSTRACT

The high surface density of faint blue galaxies at high redshift, selected by their low surface brightness, have become unique tools in mapping the projected mass distribution of foreground mass concentrations. The systematic gravitational lens distortions of thousands of these background galaxies in 10-arcmin fields reveal for the first time detailed mass profiles for intervening clusters of galaxies. Mass in the rich cluster A1689 follows smoothed light, outside a core region, with a rest-frame V band mass-to-light ratio of $400 \pm 60\ h\ (M/L)_{\odot}$. Out to a radius of $1\ h^{-1} Mpc$ there is evidence for a steeper than isothermal falloff.

These techniques are applied to a sample of 28 clusters of galaxies and trends among their mass, light, and X-ray properties are reviewed.

1. INTRODUCTION

At faint optical surface brightness one finds about 300,000 blue galaxies per square degree. Because of their relatively large angular sizes and redshifts they may be employed in statistical gravitational lens studies of foreground mass distributions at much lower redshift. To be useful for mass mapping by statistical gravitational lens inversion there are several source galaxy requirements: (1) redshifts must be known or large compared with the lens, (2) number density on the sky sufficient to sample the lens on relevant scales, (3) intrinsic angular diameter larger than the ratio of seeing FWHM to the magnification of the lens, and (4) other properties (blue color and very low surface brightness) permitting separation of the sources from lens and other foreground objects.

Luckily, the majority of faint blue galaxies apparently satisfy these requirements; galaxies fainter than 24 Bj mag with color bluer than $B - R = 0.5$ and mean surface brightness fainter than 26 B mag arcsec^{-2} appear to be strongly lens distorted when they are placed behind a foreground cluster mass at redshift 0.4 (CL0024+165). There is also a trend towards large angular diameters; the seeing-deconvolved half light radii of these galaxies range between 0.3 and 0.8 arcsecond at the faint end. The surface brightness cut is perhaps the most important. This may be seen in toy models of galaxy evolution (Tyson 1988, Yoshii & Peterson 1991, Im *et al.* 1994); the number of galaxies found at $z > 0.7$ is severely reduced by a surface brightness cut of 26 mag arcsec^{-2}.

2. SOURCE REDSHIFT AND SIZE DISTRIBUTION

Figure 1 summarizes redshift surveys of field galaxies, excluding AGNs and QSOs. The trend to redshift ≈ 1 at 25th B_j magnitude is clear. Because of competing volume and faint-end luminosity function effects, galaxies at 25th magnitude will have a wide range in redshifts extending from 1–3; low surface brightness sources will be at the highest redshifts.

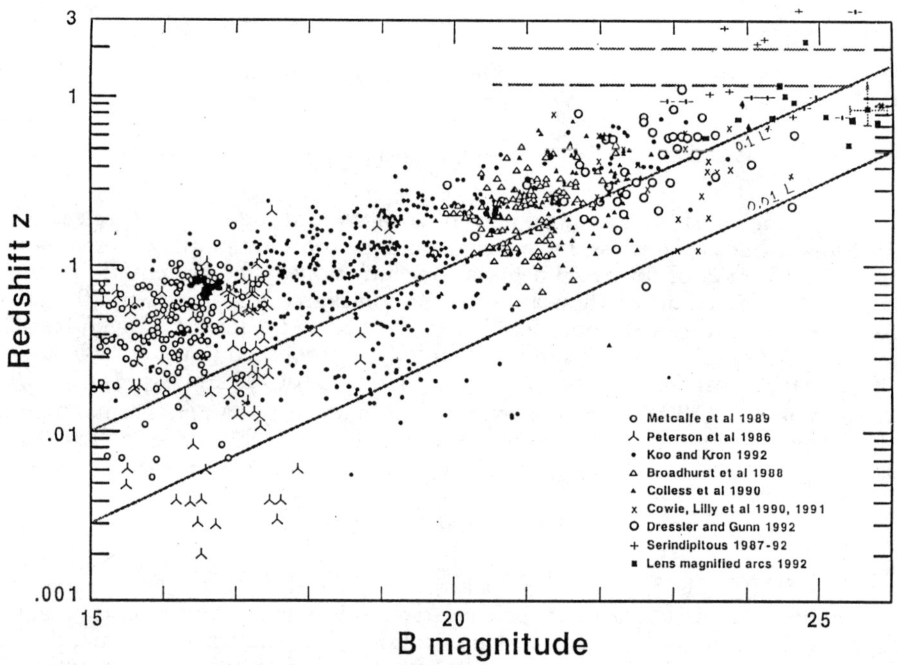

Figure 1. *A plot of field galaxy redshift vs* B_j *magnitude, from all the redshift surveys to date (adapted from Koo & Kron 1992). New survey data have been added.*

Sources for mass mapping by statistical gravitational lens distortion must have apparent angular diameters larger than the point spread divided by the lens magnification at their position. For 1 arcsec FWHM and the range of lens mass profiles typical of either CDM overdensities at z=0-0.2 or cluster mass profiles this minimum useful source diameter ranges from 0.1 to 0.5 arcsec. If the faint blue galaxies are imaged at $>$ 29 mag arcsec^{-2} surface brightness, the majority have apparent angular diameters larger than the minimum required for statistical lensing studies of foreground mass in observing conditions with 0.5-1 arcsec FWHM seeing. This must be calibrated by realistic simulations.

3. MAPPING DARK MATTER WITH FBGs

A review of gravitational lenses is given by Blandford and Narayan (1992) and the talk by Narayan at this conference. The path of a photon from a distant source is bent as it passes by the foreground mass (gravitational lens), making the source appear at an altered position and systematically distorting its image. If the background source image is resolved then this image stretching is observable. A galaxy of angular size one arcsecond may appear to be moved by an angle $\beta = 4GM(<r)/rc^2$ and distorted into an arc many arcseconds long by sufficient foreground mass $M(<r)$ interior to projected radius r. The scattering time is small compared with the period of orbiting test particles, thus avoiding orbit assumptions. This technique is independent of luminous matter in the scatterer (lens), so that it may be used to discover mass concentrations which are totally dark.

The measurement of cluster mass distribution via the gravitational lens technique is a direct measure of the projected total mass density. Since clusters of galaxies are dominated by dark matter, the detailed distribution of mass in clusters may be a useful constraint on the nature of dark matter. More generally, its nature and dynamical history will be constrained by the distance scales and time scales on which it clumps and its correlation with luminous matter. Mass estimators based on dynamics or X-ray flux maps rely on models of orbits or models of the density and temperature distribution of the hot gas.

For sufficiently deep images the number density of sources is so high that there is a high probability that one or more of these sources will lie near a critical line, resulting in a strong lensing "long arc". Even for a potential which is not perfectly spherical the equation relating the radius of a long arc to the projected mass within this radius is approximately

$$\bar{\Sigma}(r) = \Sigma_c \, . \tag{1}$$

Here, r is the angular distance from the arc to the center of the lens, and $\bar{\Sigma}(r)$ is the average surface density within r, related to the projected mass (i.e., the mass contained within a cylinder of radius r) by $M_p(r) = \bar{\Sigma}(r)\, \pi\, r^2\, D_l^2$, where D_l is the angular diameter distance to the lens. The critical surface density is given by

$$\Sigma_c = \frac{c^2}{4\pi\, G} \frac{D_s}{D_l\, D_{ls}} \, , \tag{2}$$

where D_s is the angular diameter distance to the source, and D_{ls} is the angular diameter distance from the lens to the source (Turner *et al.* 1984). Given reliable limits on the source redshifts, these long arcs then provide a self calibration of the mass scale for a given cluster.

With respect to the foreground lens centroid, the source appears at angle θ, rather than the smaller angle ϕ (the intrinsic source position in the absence of the lens): $\vec{\phi} \;=\; \vec{\theta} \;-\; \vec{\delta}(\vec{\theta})$, where $\vec{\delta}(\vec{\theta}) \;=\; (D_{ls}/D_s)\, \vec{\beta}(\vec{r})$. For a thin lens the light bending angle $\vec{\beta}$ is given in terms of the projected surface mass density Σ:

$$\vec{\beta} \;=\; (4G/c^2) \int d^2u\ \Sigma(\vec{u})\ (\vec{r}-\vec{u})\ /\ \mid \vec{r}-\vec{u}\mid^2 . \tag{3}$$

Since we do not know how the sources were distributed on the sky before the lensing, this deflection itself is not observable. But its gradient, the shear, is observable. If the lens mass is elliptical in shape or has multiple components the circular ring symmetry is broken, so that circular rings are rare. If the source angle ϕ is less than θ_E then two images of the source with separation approximately $2\theta_E$ are formed.

The virial theorem estimates the velocity dispersion and the mass interior to radius r, M/r: $\Delta v^2 \;\geq\; GM(<r)/r$. For a singular isothermal distribution of gravitationally bound masses of line-of-sight velocity dispersion σ_v, $\beta(r) \;=\; 4\;\pi\;(\sigma_v\;/\;c)^2$, and $\theta_E \;=\; 29\;(\sigma_v\;/\;10^3\;km\;sec^{-1})^2\;(D_{LS}\;/\;D_S)$ arcseconds. Lens distortions by more realistic mass distributions, such as a soft core isothermal sphere or β-models, may be calculated by substituting the corresponding density $\Sigma(r)$ (Hinshaw and Krauss 1987; Navarro *et al.* 1994) into equation 3.

If the source angle relative to the lens ϕ is larger than θ_E multiple images of that source will not be formed, but the large mass associated with rich clusters of galaxies distorts all the faint background galaxies within several arcminutes of the cluster. Foreground galaxy clusters at redshifts 0.2-0.5 with radial velocity dispersions above 700 km s^{-1} have sufficient mass density to significantly distort background galaxies of redshift greater than 0.4-1. Lensing preserves the surface brightness and spectrum of the source, so that arcs have the very faint surface brightness and blue color of the sources.

The background galaxy population is selected by its extreme blue color and low surface brightness compared with the red cluster galaxies. The high-redshift background galaxies projected within the central square arcminute of a foreground compact cluster of galaxies will be strongly lensed by the cluster mass distribution, distorting background galaxy images into faint arcs 1-30 arcsecond long. These are not the relatively rare caustic alignments: every background galaxy will be seen as stretched along a circle centered on the lens. By obtaining images which reveal the high density of faint sources, the improved statistics of these distortions permits the tomographic mapping of the projected dark matter distribution (Valdes *et al.* (1983); Tyson *et al.* 1990; Miralda-Escudé 1991a; Kaiser & Squires 1993).

4. MASS FROM IMAGE DISTORTION

For each source image this distortion is quantified using the intensity weighted second moment of the image orthogonal and along the radius relative to the lens center. A dimensionless image distortion T(x,y), related to the projected mass density clumping, is defined at each point (x,y) in the image plane:

$$T_g(x,y) = (a^2 - b^2)/(a^2 + b^2) \tag{4}$$

where a(x,y) and b(x,y) are the (r,θ) principal-axis transformed second moments of background galaxy image g. Referred to position (x,y) the radial moment is b(x,y) and the tangential moment is a(x,y): $a_g = M_{20} \sin^2\phi + M_{02}\cos^2\phi$ - 2 $M_{11}\sin\phi\cos\phi$, and $b_g = M_{20}\cos^2\phi + M_{02}\sin^2\phi + 2\,M_{11}\sin\phi\cos\phi$, where ϕ is the position angle of the vector from the point (x,y) to the background galaxy, relative to the x-axis. The intensity-weighted second moment $M_{lm,i}$ of background galaxy g is defined by $M_{lm,g} = M_{0,g}^{-1}\int(\delta x)^l\,(\delta y)^m\,W(\delta x,\delta y)\,[I_g(\delta x,\delta y) - I_0]\,dx\,dy$, with $M_{0,g} = \int W(\delta x,\delta y)\,[I_g(\delta x,\delta y) - I_0]\,dx\,dy$, where $\delta x = (x - <x>_g)$ and similarly for δy, the sky intensity near this background galaxy is given by I_0, and the optimal normalized weights $W(\delta x,\delta y,I(x,y))$ are calculated from the half luminosity radius. A randomly placed unlensed population of galaxies randomly oriented will give a net distortion $T(\vec{r})$ of zero at every point in the image plane, while a population of lensed galaxies will give a positive value at the point corresponding to the lens center. To construct a map of the gravitational lens the distortion statistic $T(\vec{r})$ is computed over a grid of positions as candidate lens centers. In the weak lensing limit, and for a circularly symmetric lens, it can be shown that the tangential alignment $T(r,\theta)$ is a measure of the mass contrast (Valdes, *et al* 1983; Miralda-Escudé 1991a):

$$T(\vec{r}) = 2\,[\bar{\Sigma}(<r) - \Sigma(r)]\,\Sigma_c^{-1}. \tag{5}$$

At any point $\vec{r}$ in the image plane we can sum over the tangential alignment of all source images about that point, creating a smoothed scalar distortion statistic $D(r,\theta)$:

$$D(\vec{r}) = \int K(\vec{u})\, T(\vec{r}-\vec{u})\, d\vec{u}, \tag{6}$$

where the apodization kernel $K(\vec{u})$ weights source images at large radius less, and is generally of the form $K(\vec{u}) = (u^2 + u_0^2)^{-1}$. Kaiser and Squires (1993) showed that K(s) must asymptotically approach the power law s^{-2} at large s in order for equation 6 to give the correct mass profile for a symmetric lens. Since light bending angles from different mass components add, the distortion D at any point is related to the mass density contrast, if the mass can be represented as a sum of cylindrically symmetric distributions.

While complicated mass distributions can be recovered from their distortion images in this way, the technique fails for mass concentrations corresponding to velocity dispersions less than 500 km sec^{-1} because the random ellipticities of the background galaxies are larger than the shear produced by the lens. The distortion image D(x,y) uniquely locates the lens mass, obtains $M(<r)/r$, and gives its morphological shape on the sky.

We test this inversion software by feeding it realistic simulated data, simulating cluster lens distortions of background galaxies by ray-tracing realistic sources as a function of cluster and dark matter parameters and comparing them with the observations. Using HST MDS imaging and seeing deconvolved ground-based data we derive the intrinsic statistical properties of the source galaxy images before lensing and before atmospheric bluring. Synthetic source galaxies are distributed in six redshift shells; background galaxy fields are synthesized to have the same properties (number-magnitude counts, ellipticity distribution, galaxy intensity profiles). The main dark matter component in the cluster is superposed on the relatively small mass cluster galaxies. The mass and core radius of the dark matter are varied in different Monte Carlo simulations. After the distortion from all lens components is calculated, the image is convolved with the atmospheric blur profile, binned down to the CCD resolution, and the shot noise of the night sky diffuse background is added. Automated detection, sub-image splitting and photometry reduce these images to a catalog of galaxy image properties including positions, total magnitudes, and second moments. Lens inversion is performed on these synthetic catalogs, and the result compared with thc input mass profile of the lens.

While equation 6 yields a useful map of the mass distribution, a higher S/N plot of the radial mass distribution of a given lens may be obtained from a variant of equation 6: it is easy to show that assuming cylindrical symmetry the average projected mass density interior to radius r is given by

$$\bar{\Sigma}(r) \;=\; 2\Sigma_c C(r) B(r) \int_r^{r_{max}} T(r)\; dlnr \;+\; \bar{\Sigma}(r, r_{max}), \tag{7}$$

where C(r) is the seeing correction found in Monte Carlo tests of the data, $B(r) = (1 - r^2/r^2_{max})^{-1}$, and $\bar{\Sigma}(r, r_{max})$ is the average density in the annulus between r and r_{max}. For a sufficiently large field $\bar{\Sigma}(r, r_{max})$ is small compared with with peak density. Outside the Einstein radius, it can be shown that equation 7 is approximately true in the more general non-spherical lens case. A radial plot of the projected surface mass density found by equation 7 is shown in Figure 2 for two simulations with different mass laws. The points are the

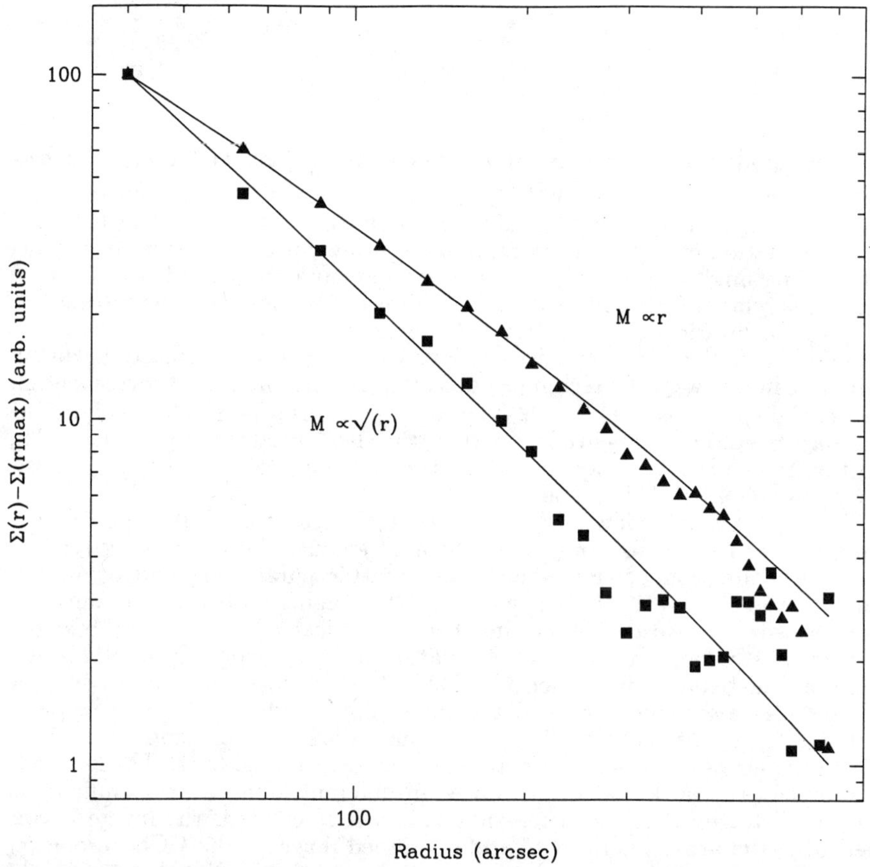

Figure 2. *A radial plot of the projected mass density obtained by inversion of the arclets in a* 29 *B mag arcsec*$^{-2}$ *noisy simulation for two mass laws in a lens at* $z = 0.3$. *Note the agreement between the reconstructed mass and the input (solid line) for each model mass profile.*

reconstructed mass surface density and the lines are the input surface densities used in these noisy simulations.

5. MASS DISTRIBUTION OF ABELL 1689

Large CCDs and mosaics of CCDs now enable tomographic studies of dark matter over a wider field surrounding clusters. Here we apply this tomographic inversion to 12,000 blue arcs and arclets in a 16 arcminute field centered on A1689, a cluster which we studied earlier with a small CCD (Tyson, Valdes, & Wenk 1990). Data on 5950 low surface brightness blue arclets found in this area are passed to the apodized moment analysis. First, an image of the mass distribution in the field is made via equation 6. Figure 3 shows this mass distribution in the field of the cluster A1689. Note that the distribution is not symmetric, having an extension towards the sub-clump of cluster red galaxies to the northeast seen in the CCD images.

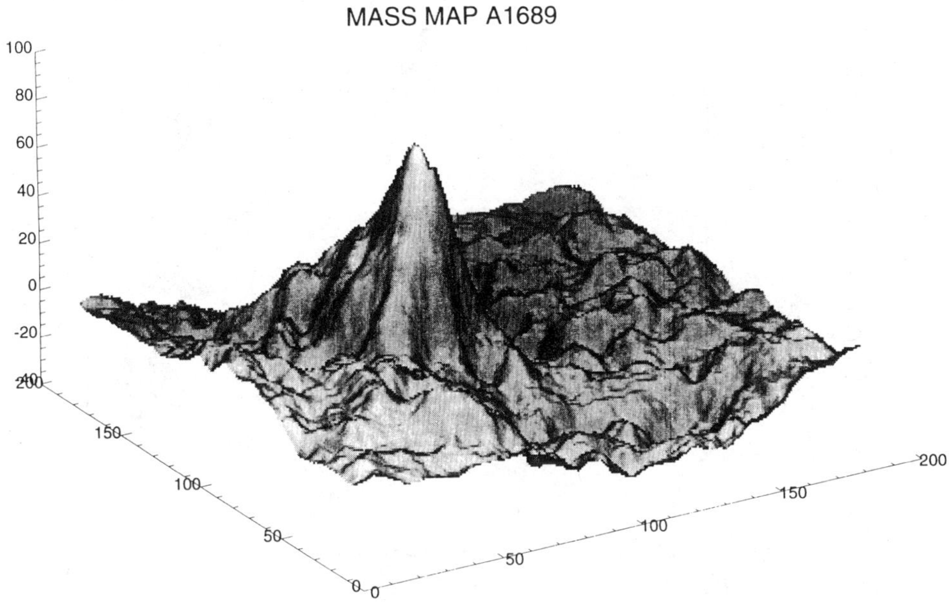

Figure 3. *A rough map of the mass distribution in the 1.8 h^{-1} Mpc field of the z = 0.18 cluster A1689.*

Having found the center of the lens, equation 7 is used to derive an accurate radial mass profile out to 1 Mpc. Mass calibration is obtained in two ways: (1) Einstein critical radius as given by several faint long arcs, and (2) simulations of the lens statistics using the observed PSF and known source galaxy properties as discussed above. These agree within 10%, method (1) giving $1.8 \pm 0.1 \times 10^{14}$ $M_{\odot}$ inside the critical radius of 52 arcsec for $z_{source} = 1-2$. A calibrated plot of the projected mass density along with the luminosity density in the rest-frame V band is given in Figure 4 as a function of radius out to 1 h^{-1} Mpc. Mass follows smoothed light, outside a core region, with a rest-frame V band mass-to-light ratio of $400 \pm 60\ h\ (M/L_V)_{\odot}$. Out to a radius of 1 $h^{-1}Mpc$ the best fit power law for mass and light is $\approx r^{-1.3}$.

6. A DEEP SURVEY OF 28 CLUSTERS

In 1988 we began a deep survey of arcs and arclets around clusters of galaxies. While poor seeing and weather have prevented completion of all 40 clusters, sufficient data now exist for preliminary conclusions regarding their total mass and light distribution. We have completed a multicolor CCD imaging survey of 28 clusters of galaxies to uniform surface brightness limits of 29 B_J mag arcsec^{-2} and 28 R mag arcsec^{-2} covering a range of optical richness and

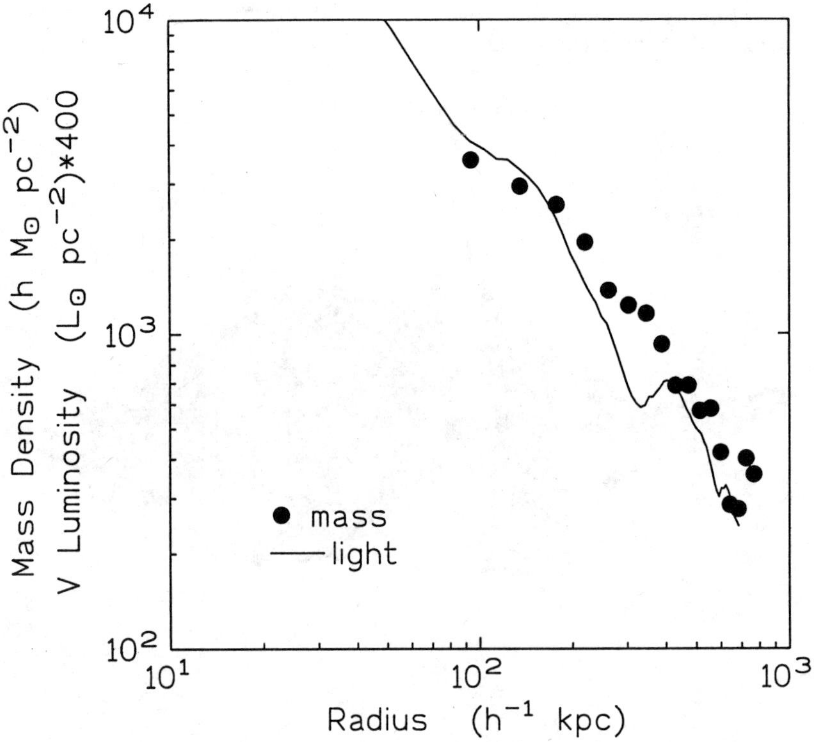

Figure 4. *The projected profiles of the mass density and the luminosity density in the rest-frame V band of A1689.*

X-ray luminosity. A null result is found in fifteen control fields devoid of any foreground cluster of galaxies. Briefly, here is a summary of the properties of the clusters in our sample.

We find mass core radii are significantly smaller than some observed X-ray core radii in nearby clusters, suggesting that the X-ray gas may be less relaxed dynamically than the dark matter. Here I define a core radius as r_{core} in a projected mass distribution on the sky given by $\Sigma(r) = \Sigma_o(r^2 + r^2_{core})^{-1/2}$. It might be more relevant to compare both the X-ray and lens -derived projected mass distributions with the Navarro *et al.* (1994) density profile obtained in N-body SPH CDM simulations, and we shall do that in future comparisons. At least for rich clusters these lens studies appear to confirm the large mass which was implied by virial calculations using velocity dispersion. The distribution of dark matter in rich galaxy clusters is found to be similar to the distribution of total red luminosity (although on a smoother scale of about 100 h^{-1} *kpc*). Many of the less compact clusters show no strong lens distortion effects and, thus, have velocity dispersions smaller than 700 km sec^{-1} and/or dark matter core radii larger than 200 kpc. The central projected surface density (or the cluster velocity dispersion), being an integral over the distortion function, is the best determined parameter.

For the optically compact clusters we find the dark matter peak densities are about 10^5 times the critical density for closure, with core radii up to 70 kpc. The scale of the sub-clumping of the dark matter within clusters of galaxies is also of interest. The dark matter cannot all be clumped on galaxy mass scales, otherwise the long arcs seen in several relaxed cluster lenses would be broken up on several arcsecond scales and arcs would form around cluster galaxies. Individual cluster galaxies appear to be no more massive than field galaxies, and the dark matter in clusters is primarily in the diffuse component.

7. COMPARISON WITH XRAY DERIVED MASSES

The mass indicated by gravitational lensing in the clusters A2218 and A1689 is larger by a factor of ~ 2.5 compared with the mass obtained from the X-ray observations, assuming that the gas is supported by its thermal pressure (Babul & Miralda-Escudé 1994; Daines *et al.* 1995). Interestingly, the cluster A2163 appears to be a counter example: it is one of the most luminous X-ray sources but exhibits a very shallow mass profile when studied by gravitational lens tomography. It may be that this cluster has a mass core larger than several arcminutes. While a large X-ray luminosity from hot gas trapped in a deep gravitational potential well is a suggestive tracer of mass overdensity at late times, there are a variety of reasons why one might not expect a detailed similarity between the mass and X-ray surface brightness or temperature.

Turbulent motions, clumping, and lack of pressure equilibrium in non-relaxed clusters would increase the required mass and help solve the discrepancy. If the central part of the cluster where the arcs are seen is not relaxed, the gas will have bulk motions which will contribute to its pressure support. Additional pressure could be provided by a magnetic field in equipartition, or a multiphase intracluster medium with an effective temperature higher than that determined from the emission-weighted X-ray spectrum. Thus, masses determined from X-ray observations may be systematically underestimated and the spatial scale overestimated for compact clusters.

8. DARK LENSES: THE ORIENTATION CORRELATION FUNCTION

If $\Omega_{DM} \approx 1$, dark matter will likely exist in places where there is no current star formation activity. Since dark matter clusters on galaxy scales, "field" clustering of dark matter on larger scales at late times will occur. There is some evidence for dark foreground lenses in the wide separation double QSO fields 0957+561 and 2345+007 (Fischer *et al.* 1994). Larger scale applications of this dark matter mapping may eventually find clumped dark matter unrelated to foreground galaxies or clusters of galaxies. If we wish only to extract the statistics and spectrum of the foreground mass overdensities, rather than map them, there is a more general statistic analogous to the scalar two-point correlation function (Valdes, *et al* 1983; Miralda-Escudé 1991b; Blandford, *et al* 1991). Mass fluctuations along any line of sight produce a slight shear in source galaxy images, coherent over the angular scales of the dark matter overdensity. An integral tensor statistic called the orientation correlation function (OCF) may be built from the lensed source galaxy orientation data. The OCF reveals any trend for background galaxies to be oriented parallel or orthogonal to one another; it is a direct measure of the power spectrum of mass fluctuations at late times, rather than luminosity or galaxy number density fluctuations.

Faint galaxy orientation correlations induced by clumped dark matter in various theoretical scenarios may be detected by deep CCD imaging over sufficiently large fields. CDM theory predicts an observable correlation at about the 2% level, so contiguous samples of 1,000 to 10,000 faint galaxies are a useful test of CDM. Field galaxies have an intrinsic ellipticity scatter of ≈ 0.3, so one must observe $\approx 10^3$ source galaxies to obtain unity S/N on a coherent shear of 1%. Mosaics of CCDs will make such a large scale search for coherent alignment in the distant faint galaxies particularly attractive, and dark matter on angular scales of degrees can be studied in this way. A representative survey of many 2-3 degree fields, in order to guage the cosmic scatter, may require a dedicated telescope; while mosaics of CCDs make effective use of the focal plane it is necessary to spend several nights of large telescope time per sub-field.

The observational challenge in OCF measurement is systematic error due to variations across the detector of the point spread function, optics distortion, trailing and atmospheric refraction. These systematics may be overcome by chopping and controlled trailing techniques and calibration in rich star fields. The OCF mass calibration may be checked on smaller fields centered on rich clusters, and also with realistic lensing simulations. Preliminary measurements of the OCF in random fields are below the CDM prediction.

ACKNOWLEDGEMENTS

My collaborators in this research are Gary Bernstein, Phil Fischer, Raja Guhathakurta, Greg Kochanski, Jordi Miralda-Escudé, Genevieve Soucail, Frank Valdes, and Rick Wenk. Data for A1689 were taken at CTIO, in three 4m runs during 1989-94; we are grateful to the CTIO staff for their help.

REFERENCES

Babul, A. & Miralda-Escudé, J. 1994, ApJ, in press
Blandford, R.D., Saust, A.B., Brainerd, T., and Villumsen, J.V. 1991, MNRAS, 251, 600
Blandford, R.D. and Narayan, R. 1992, ARA&A, 311
Daines, S., Jones, C., Forman, W., & Tyson, A. 1995, ApJ, in press
Fischer, P., Tyson, J.A., Bernstein, G.M., & Guhathakurta, P. 1994, ApJ Lett., 431, L71
Hinshaw, G. and Krauss L. 1987, ApJ, 320, 468
Im, M., Casertano, S., Griffiths, R.E., Ratnatunga, K.U., Tyson, J.A. 1995, ApJ, in press
Kaiser, N. and Squires, G. 1993, ApJ, 404, 441
Koo, D. & Kron, R. 1992, ARA&A, 30, 613
Miralda-Escudé, J. 1991a, ApJ, 370, 1
Miralda-Escudé, J. 1991b, ApJ, 380, 1
Navarro, J. F., Frenk, C. S., & White, S.D.M. 1994, MNRAS, in press
Turner, E.L., Ostriker, J.P. and Gott, J.R. 1984, ApJ, 284, 1
Tyson, J.A., 1988, AJ, 96, 1
Tyson, J.A., Valdes, F., and Wenk, R.A. 1990, ApJ, 349, L1
Valdes, F., Tyson, J.A. and Jarvis J.F. 1983, ApJ, 271, 431
Yoshii, Y. & Peterson, B.A. 1991, ApJ, 372, 8

GRAVITATIONAL LENS CONSTRAINTS ON COSMOLOGY & COSMOGONY

Christopher S. Kochanek
Harvard-Smithsonian Center for Astrophysics, Cambridge, MA 02138
kochanek@cfa.harvard.edu

ABSTRACT

Standard $\Omega_0 = 1$ cold dark matter (CDM) needs $0.27 < \sigma_8 < 0.63$ (2σ) to fit the observed number of large separation lenses. This range is strongly inconsistent with the COBE estimate of $\sigma_8 = (2.8 \pm 0.2)h$. Tilting the primordial spectrum $\propto k^n$ from $n = 1$ to $0.3 \lesssim n \lesssim 0.7$, using an effective Hubble constant of $0.15 \lesssim \Gamma = h \lesssim 0.30$, or reducing the matter density to $0.15 \lesssim \Omega_0 h \lesssim 0.3$ either with no cosmological constant ($\lambda_0 = 0$) or in a flat universe with a cosmological constant ($\Omega_0 + \lambda_0 = 1$) can bring the lensing estimate of σ_8 into agreement with the COBE estimates. The statistics of gravitational lensing by galaxies are incompatible with large cosmological constants λ and low cosmological densities Ω. The 90% confidence limits are $\Omega > 0.5$ in flat cosmologies, and $\Omega > 0.2$ in standard cosmologies. The statistics of lenses and models of individual lenses are inconsistent with constant mass-to-light ratio models of E/S0 galaxies and consistent with isothermal models of E/S0 galaxies. Even the inner regions of E/S0 galaxies must contain dark matter.

1. INTRODUCTION

The population of multiply imaged quasars and radio sources is divided into lenses smaller than 3."0 produced by galaxies, and lenses larger than 3."0 produced by groups and clusters. The galaxy scale lenses are most suited for probing the cosmological model and the structure of massive galaxies, while the cluster scale lenses are most suited for probing models of structure formation. In this review we tersely summarize the cosmological, cosmogonical, and dynamical implications of multiply imaged quasars and radio sources produced by either galaxies or groups and clusters.

2. GALAXIES: COSMOLOGICAL MODELS AND DARK MATTER

The number of gravitational lenses found in a survey for lensed quasars is a volume test of the cosmological model because the number of lenses depends on the expected number of lens galaxies between the observer and the quasar. Compared to the theoretically favored $\Omega = 1$ flat cosmological model, both flat cosmologies with a positive cosmological constant λ and low density $\Omega < 1$ cosmologies produce larger numbers of lenses (Turner et al. 1984, Turner 1990). Flat universes with $\lambda = 1$ produce ten times as many lenses, and empty $\Omega = 0$ universes produce twice as many lenses as the Einstein-DeSitter model.

The number of lenses also depends on the masses of E and S0 galaxies. Spiral galaxies, despite their larger numbers, do not represent an important contribution to the expected number of lenses because of their smaller masses (Turner et al. 1984, Kochanek 1993b, Fukugita & Turner 1991). For example, in an isothermal lens model the number of lenses depends on σ_*^4 where σ_* is the characteristic velocity dispersion of an L_* galaxy. Small uncertainties in the dynamical variables lead to large uncertainties in the expected number of lenses: a 10% uncertainty in σ_* leads to a 40% uncertainty in the expected number of lenses. This sensitivity to dynamical uncertainties severely limits the utility of lens statistics to estimate cosmological models unless we can find some other, internal means of controlling the dynamical variables.

Fortunately, each lens also has a characteristic length, the image separation. The separation depends largely on the dynamical variables, with only a 5% variation from the cosmological model (Kochanek 1993a). In the isothermal model the separation depends on σ_*^2, so by simultaneously fitting the expected number of lenses and their observed separations we can avoid the intrinsic uncertainties of galaxy dynamics. After fitting the lens data we can always go back and compare the estimates of galaxy masses from the statistical model of the gravitational lenses to the estimates from traditional, stellar dynamical methods.

We examine the optical surveys for lensed quasars (Crampton et al. 1992, Yee et al. 1993, Surdej et al. 1993, Maoz et al. 1993), which include 680 quasars and five lenses, and the image separations for the MG (Burke et al. 1992) and Jodrell Bank (Patnaik et al. 1992) radio lenses with known source redshifts. A detailed discussion of the data set is given in Kochanek (1993b). We include only E/S0 galaxies (see Kochanek (1993b) for the effects of spirals), normalized to local galaxy counts (Efstathiou et al. 1988), and galaxy type distributions (Postman & Geller 1984). We adopt the quasar number counts model of Wallington & Narayan (1993).* We use the maximum likelihood method for lenses (Kochanek 1993b) to examine models in the space of cosmology and the primary dynamical variable for both isothermal (dynamical variable σ_*) and de Vaucouleurs (dynamical variable $(M/L)_*$) models of the mass distribution in E/S0 galaxies.

Figure 1 shows contours of the likelihood for flat universes with a cosmological constant, λ. Independent of the mass normalization and of the type of potential, we find that $\lambda < 0.5$ at 90% confidence. Models with $\Omega = 0.2$ and $\lambda = 0.8$ sometimes suggested to solve problems with structure formation are ruled out by lens statistics with more than 99% confidence. The dominant systematic uncertainties are already included by looking at two potential models and allowing the mass normalization to vary. It is very unlikely that there is a systematic error large enough to allow a significant cosmological constant.

Figure 2 shows the contours of the likelihood for normal cosmological models as a function of the cosmological density, Ω. Because the differences between the $\Omega = 0$ and $\Omega = 1$ cosmological models are much smaller than when $\Omega + \lambda = 1$, the limits on Ω are much weaker. Nonetheless, $\Omega \simeq 1$ models are favored, and $\Omega > 0.2$ at 90% confidence. Thus lensing joins the evidence from large scale velocity flows and theoretical prejudices in favoring the Einstein-DeSitter model. The differences are, however, small enough that systematic and statistical errors could still shift the results by a significant amount.

Figures 1 and 2 also show the stellar dynamical normalizations of the mod-

*We have to correct a conversion error in Wallington & Narayan (1993) (also present in Kassiola & Kovner (1993)) from absolute to apparent magnitudes caused by a shift between $h = 0.5$ in Boyle et al. (1990) and $h = 1$ in the theoretical papers.

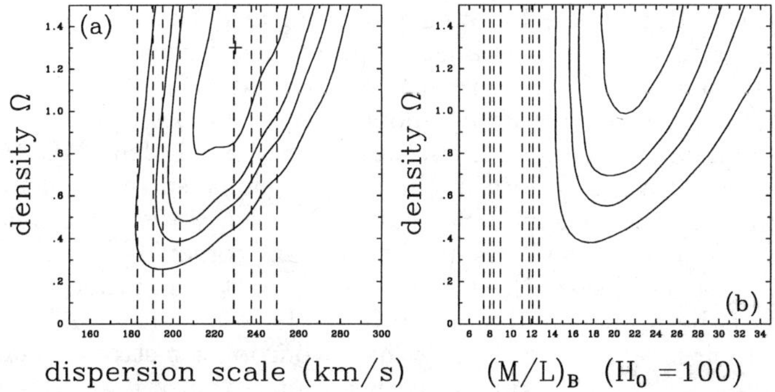

Fig. 1. Likelihood contours for the lens data (solid) and dynamical data (dashed) in flat universes with a cosmological constant using either (a) singular isothermal lenses or (b) de Vaucouleurs lenses. The likelihood contours are the 68.3%, 90%, 95.4%, and 99% confidence intervals on one parameter. The second variable is either the velocity dispersion σ_* or the blue mass to light ratio $(M/L)_*$ of an L_* galaxy. The cross marks the peak likelihood.

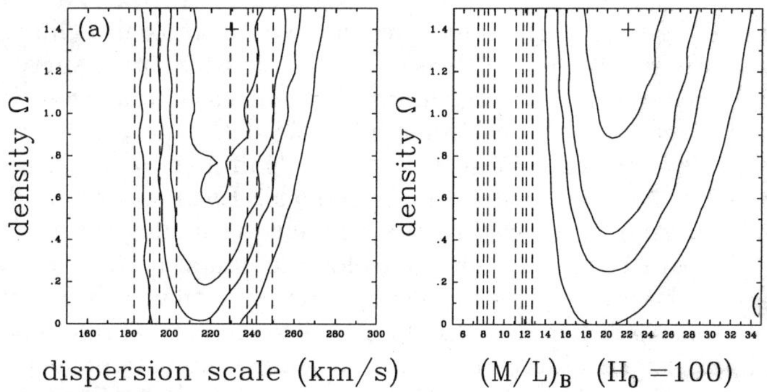

Fig. 2. Likelihood contours for standard cosmologies ($\lambda = 0$) using either (a) singular isothermal lenses or (b) de Vaucouleurs lenses. Contours and peaks have the same meaning as in Figure 1.

els. Constant M/L stellar dynamical models of a sample of 37 E/S0 galaxies (van der Marel 1991) give a blue mass to light ratio scale of $(M/L)_* = (10 \pm 2)h$ where $H_0 = 100h$ km s^{-1} Mpc^{-1}, while isothermal models of the same galaxies by Kochanek (1994a) give $\sigma_* \simeq (223 \pm 10)$ km s^{-1}. Most lens studies (Turner et al. 1984, Turner 1990, Fukugita & Turner 1991, Wallington & Narayan 1993, Maoz & Rix 1993) used a dispersion scale of $\sigma_* \simeq 270$ km s^{-1} because they assumed that the velocity dispersion of the dark matter is $(3/2)^{1/2}$ larger than the measured dispersions (Turner et al. 1984, Gott 1977). This theoretical correction uses an oversimplification of the luminosity profiles, and for real E/S0 luminosity profiles there is no such correction (Kochanek 1994a). Studies with the $(3/2)^{1/2}$ correction overestimated the number of lenses by a factor of 2.25, and overestimated the image separations by a factor of 1.5. Several studies (Kochanek 1991, 1992, 1993b) include results both with and without the correction or normalize the masses of the galaxies from the lens data. As first pointed out by Maoz & Rix (1993) the stellar dynamical normalization of $(M/L)_*$ is incompatible with the observed image separations. Figures 1 and 2 show that the de Vaucouleurs model requires $(M/L)_* = (20 \pm 3)h$ to fit the lens data. The normalization of the isothermal model is the same for both lens separations and stellar dynamics. Detailed models of the radio ring lens MG1654+134 by Kochanek (1994b) confirm both of these conclusions, so the evidence for dark matter in the inner regions of E/S0 galaxies is not a statistical artifact.

The previous models all assumed a constant comoving density of unevolving lens galaxies. Lens statistics are most sensitive to the properties of galaxies at one half of the proper motion distance to the source ($z \sim 0.3$-0.5 typically) rather than their local properties, so galaxy evolution can strongly effect the cosmological limits. The basic principle of evolution is that if high redshift galaxies cease to be effective lenses at high redshifts due to evolution, then we can reduce the expected number of lenses and allow cosmologies that overproduce lenses in the no evolution models. The models cannot, however, examine the expected number of lenses alone – evolution models that change the expected number of lenses by a factor of two also strongly modify the expected lens separations.

Three studies (Mao 1991, Mao & Kochanek 1994, Rix et al. 1994) examined how the evolution of E/S0 galaxies changes cosmological limits and is in turn limited by the statistics of lenses. After an exhaustive survey of formation and merger scenarios, we concluded that no reasonable evolution scenario could allow a large cosmological constant ($\lambda \gtrsim 0.8$). Only scenarios in which E/S0 galaxies are first formed at very low redshifts ($z \sim 0.5$) lead to significant increases in the likelihoods of these models. Even so, the models are still statistically rejectable compared to the standard cosmological models with no evolution. The very late formation times for the galaxies are also incompatible with standard galaxy formation scenarios. Merger scenarios in which the E/S0 galaxies are built up either from the locally observed spiral population or from smaller and smaller subclumps lead to little or no improvement in the models with a large cosmological constant. Most merger scenarios preserve the expected number of lenses while making their average separation smaller.

3. CLUSTERS: LIMITS ON CDM COSMOGONIES

A constant comoving density of lenses normalized by the local number density is not a realistic model of groups and clusters, so we need to examine the number of cluster lenses predicted by models of structure formation. Narayan & White (1988) pointed out that the standard cold dark matter (CDM) model predicts many more gravitational lenses with separations above 5."0 than we find. Their predictions were based on the Press-Schechter (1974, PS hereafter) model, but more recent numerical models by Cen et al. (1994) and Wambsganss et al. (1994) confirm the fundamental result. Lensing is, however, a qualitatively different test of cosmogonic models than the local density of clusters (eg. Bahcall & Cen 1992, 1993) or correlation functions (eg. Vogeley et al. 1992, Loveday et al. 1992). Any massive, collapsed, virialized halo will produce lenses, so the test is independent of the luminosity of the lenses and unaffected by problems with detecting and counting complete cluster samples locally. The probability of lensing peaks at intermediate redshifts, and it goes to zero at low redshift, so it is a test of the number density of groups and clusters at $z \sim 0.3$-0.5 rather

Table 1: Quasar Lenses and Pairs Larger Than 3.″0

Name	z_s	$\Delta\theta$	m_A	m_B	?	Why
0957+561	1.41	6.″1	17.5	17.7	Y	Q
2016+112	3.27	3.″6	22.9	23.2	Y	L
1120+019	1.46	6.″5	16.2	20.8	?	L
1429−008	2.08	5.″1	17.7	20.8	?	L
1635+267	1.96	3.″8	19.2	20.8	?	Q
2345+007	2.15	7.″3	19.5	21.0	?	Q
1145−071	1.35	4.″2	18.0	18.8	N	L
1343+264	2.03	9.″5	20.8	20.9	N	Q

Notes: The ? column entry is a Y if the pair is a lens, an N if it is not a lens, and ? if its status is uncertain. The Why column entry is a Q if the pair was found in a quasar survey, or an L if it was found in a lens survey.

than at $z = 0$.

We examine the statistics of large separation lenses in the Hewitt & Burbidge (1993, HB93) catalog (Kochanek 1994c). Table 1 summarizes all the known lensed pairs, candidate pairs, and associated quasars with separations larger than 3.″0 in the HB93 catalog. Wide separation lenses are always resolved, so given one image of the lens with magnitude m_1 brighter than the magnitude limit m_l, the selection function requires that any second lensed image with magnitude m_2 must also be brighter than the survey magnitude limit. The magnitude limit varies from survey to survey, but we would like to have a plausible model for the average dynamic range between a quasar in a heterogeneous catalog and its magnitude limit. We can do this based on two plausible assumptions: (1) all surveys have magnitude limits, and (2) surveys for fainter quasars do not substantially overlap surveys for brighter quasars. The second assumption is justified because the typical bright quasar is not part of a faint quasar survey.

We model the quasar apparent magnitude number counts by a broken power law

$$\frac{dN}{dm} = N_0 \begin{cases} 10^{\alpha(m-m_0)} & m < m_0 \\ 10^{\beta(m-m_0)} & m > m_0 \end{cases} \tag{1}$$

where $\alpha \simeq 1.12$, $\beta \simeq 0.18$, and $m_0 \simeq 19.1$ B mags (Hartwick & Schade 1990, Boyle et al. 1990, Wallington & Narayan 1993). The apparent magnitude of the break m_0 is nearly constant for the redshift range $1 < z < 3.5$. From the number counts model we calculate the probability distribution for the magnitude difference $\Delta m = m_l - m$ between a survey quasar with magnitude m and the magnitude limit. For surveys with magnitude limits brighter than the break magnitude, $m_l < m_0$, the mean dynamic range is $\langle \Delta m \rangle = (\alpha \ln 10)^{-1} = 0.34$ mag, the median is $\Delta m_{1/2} = \alpha^{-1} \log 2 = 0.24$ mag, and 90% of the quasars have $\Delta m < \alpha^{-1} = 0.79$ mag. The dynamic range between a quasar and the magnitude limit in any magnitude limited bright quasar survey is very small. When the magnitude limit becomes fainter than the break magnitude m_0, the average dynamic range increases. In the limit that $m_l \gg m_0$ the mean dynamic range is $\langle \Delta m \rangle = (\beta \ln 10)^{-1} = 1.55$ mag, the median is $\Delta m_{1/2} = \beta^{-1} \log 2 = 1.67$ mag, and 90% of the quasars have $\Delta m < \beta^{-1} = 5.5$ mag. For comparison, surveys for lensed quasars (eg. Maoz et al. 1993, Surdej et al. 1993) have

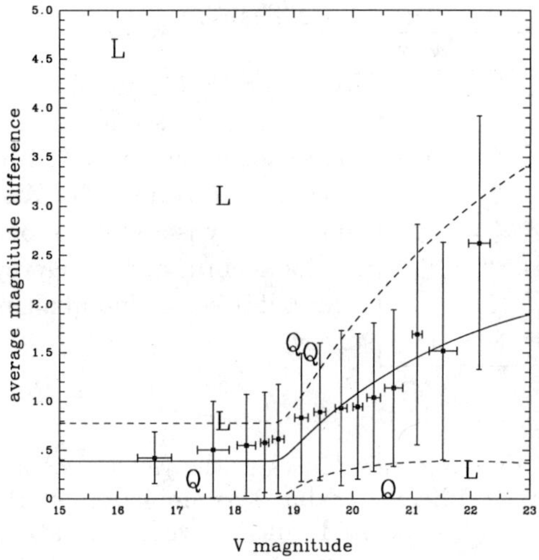

Fig. 3. Average (points) and rms scatter (vertical error bars) of nearest neighbors with redshifts $1 < z < 4$ in the HB93 catalog as a function of magnitude The horizontal error bars show the width of the magnitude bins. The lines show the predicted average (solid) and rms (dashed) width of the distribution predicted by the model selection function. The quasar pairs in Table 1 are marked by a Q if found in a quasar survey, and an L if found in a lens survey.

typical dynamic ranges of 4 to 5 magnitudes.

We test this model by examining the magnitude differences between pairs of quasars in the HB93 catalog. We took the ~ 5000 quasars with measured V magnitudes and $1 < z < 4$ in the HB93 catalog, and found the nearest neighbor for each quasar excluding known lenses and the pairs in Table 1. Figure 4 shows the mean and the dispersion of the magnitude differences of the pairs as a function of the V magnitude of the brighter quasar. Figure 4 also shows the mean and dispersion predicted for Δm by the selection function model assuming each quasar is at the median magnitude for its magnitude limit. The agreement between the data and the model for both the average magnitude difference and the spread in the magnitude difference is remarkably good. Figure 4 also shows the eight quasar pairs from Table 1, labeled by the type of survey that found the pair. As expected, the large magnitude difference pairs (Q1120+019 and Q1429−008) were found as part of a lens survey.

Following the approach of Narayan & White (1988) we use the PS model to estimate the comoving number density of clusters $dn/d\sigma$ as a function of redshift z and velocity dispersion σ. Let $\delta_c(z)$ be the critical overdensity $\delta\rho/\langle\rho\rangle$ that can collapse before redshift z, and let the velocity dispersion of the cluster formed from a perturbation of comoving scale r_0 be $\sigma \propto r_0$. Then for any cosmological model and a power spectrum normalized by the rms fluctuation σ_8 on the scale of $r_0 = 8h^{-1}$ Mpc, we can compute the expected number of large separation lenses and their properties (see Kochanek 1994c).

We use the approximate power spectrum (eg. Efstathiou et al. 1992)

$$|\delta_k|^2 = Bk^n\{1 + [ak + (bk)^{3/2} + (ck)^2]^\nu\}^{-2/\nu} \tag{2}$$

where $a = (6.4/\Gamma)h^{-1}$ Mpc, $b = (3.0/\Gamma)h^{-1}$ Mpc, $c = (1.7/\Gamma)h^{-1}$ Mpc and $\nu = 1.13$. The calculation depends on the Hubble constant h, the power spectrum normalization σ_8, the shape of the power spectrum (n and Γ), and the cosmological model. When we study cluster lenses with separations larger than 5.″0, we are examining objects with velocity dispersions between 300 km s^{-1} and

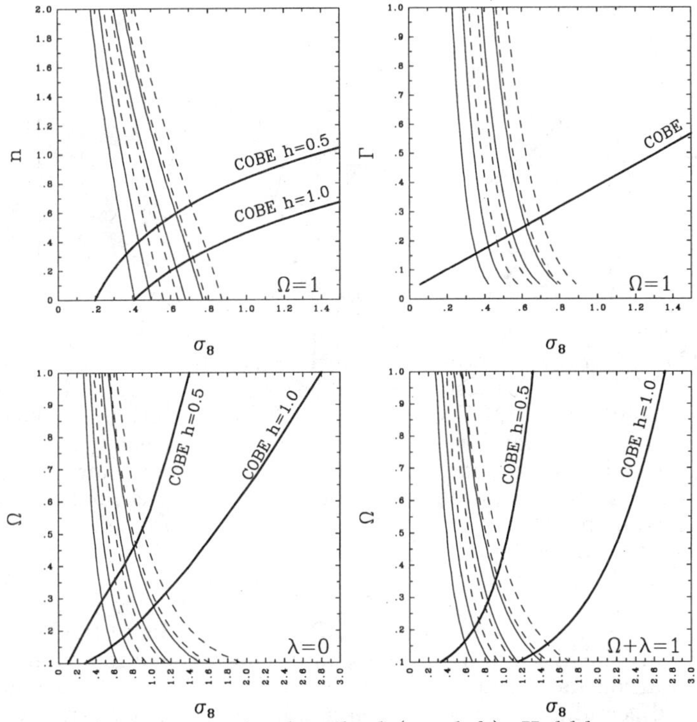

Fig. 4. Cluster limits on σ_8 in the tilted (top left), Hubble constant or fitted (top right), low density (bottom left), and flat with λ_0 (bottom right) CDM variants. The solid lines are likelihood contours for finding one lens (Q0957+561) and the dashed lines are likelihood contours for finding three lenses (Q0957+561, Q2345+007, and Q1343+264). The contours are the 1σ and 2σ confidence intervals of the likelihood ratio for one parameter. The best fit models are bounded by the contours. The heavy solid lines show the COBE normalized estimates for σ_8 when $h = 0.5$ and $h = 1.0$.

the largest objects that can collapse and virialize before the present epoch. The velocity dispersion σ for a perturbation of scale r_0 is

$$\sigma(r_0) = 950 \left[r_0/8h^{-1}\mathrm{Mpc} \right] (1+z)^{1/2} \text{ km s}^{-1} \tag{3}$$

in an $\Omega_0 = 1$ cosmology. Thus the scale used to normalize the power spectrum ($r_0 = 8h^{-1}$ Mpc) is comparable to the scale producing the lenses, and the number of large separation lenses is controlled by $dn/d\sigma \propto \sigma_8^{-3} \exp(-\delta_c^2/2\sigma_8^2)$ at the $8h^{-1}$ Mpc scale. The expected number of lenses depends exponentially on the normalization when $\sigma_8 < \delta_c$, and the number varies as a power law when $\sigma_8 > \delta_c$. Since the critical overdensity is $\sigma_c = 1.69(1+z)$ for $\Omega_0 = 1$, the number of lenses will vary exponentially with σ_8. For fixed σ_8 the number of lenses depends weakly on the slope of the power spectrum, but the slope does control the image separation distribution. The expected number of lenses depends weakly on the Hubble constant through the parameter Γ. The COBE estimates for σ_8 are sensitive to the Hubble constant (eg. Efstathiou et al. 1992, Bond 1994), so we

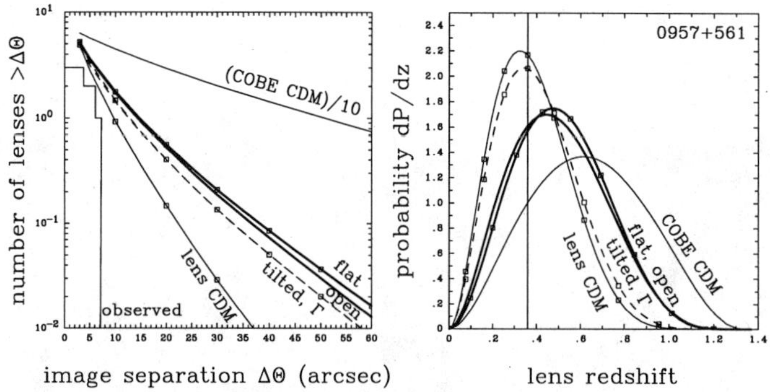

Fig. 5. The integral separation distributions (left) and the differential lens redshift probability distribution for 0957+561 (right) for several models with $h = 0.5$. The solid line (COBE CDM) shows the distributions for COBE normalized CDM ($\sigma_8 = 1.4$) and the solid line with points (lens CDM) shows the distributions for CDM normalized to produce 3 lenses larger than 5″.0 ($\sigma_8 = 0.5$). In the left panel, the number of lenses produced by the COBE normalized CDM model is divided by a factor of 10. The figure also shows a tilted model ($n = 0.6$, $\sigma_8 = 0.6$, dashed lines), a Γ model ($\Gamma = 0.25$, $\sigma_8 = 0.6$, dashed lines with points), an open model ($\Omega_0 = 0.45$, $\sigma_8 = 0.8$, heavy solid lines), and a flat model with λ_0 ($\Omega_0 = 0.35$, $\sigma_8 = 0.9$, heavy solid lines with points) that produce 3 lenses larger than 5″.0 and are consistent with COBE. The tilted and the Γ models are nearly identical. The histogram in the left panel shows the observed integral separation distribution including the known lens Q0957+561 and the Q2345+007 and Q1635+667 candidates. The vertical line in the right panel marks the lens redshift in 0957+561.

show the lens estimates for $h = 0.5$ compared to the COBE estimates for $h = 0.5$ and $h = 1.0$.

We determined the values of σ_8 required to produce the observed number of wide separation lenses in standard CDM, tilted CDM, low Ω_0 cosmologies, and flat cosmologies with λ_0. Figure 4 summarizes the range of values for σ_8 compatible with finding between one and three lenses larger than 5″.0 in the HB93 catalog in the various models, and Figure 5 shows the image separation and lens redshift distributions for standard CDM and several models normalized to fit both COBE and the lens data. The general results agree with other estimates of σ_8. For standard CDM models we find $\sigma_8 \simeq 0.45 \pm 0.2$ (2σ) for $h = 0.5$, in strong disagreement with the normalization $\sigma_8 = (2.8 \pm 0.2)h$ needed to fit the COBE observations (Smoot et al. 1992, Wright et al. 1994, Górski et al. 1994). As previously noted by Narayan & White (1988), Cen et al. (1994), and Wambsganss et al. (1994), COBE normalized CDM overpredicts the number of lenses by more than an order of magnitude. It also predicts a flatter distribution of image separations than is observed, and higher average lens redshifts (see Figure 5). Tilted models with a primordial spectrum $\propto k^n$ and $0.3 \lesssim n \lesssim 0.7$ ($h = 0.5$) are consistent with the COBE estimates for σ_8 at those values of the exponent n. The limits on the exponent n from the COBE data are $n = 1.2 \pm 0.3$ (Wright et al. 1994, Górski et al. 1994), so the overall consistency is poor. Changes in the Hubble constant have negligible effects on the lens models, with the best fit value for σ_8 decreasing by 0.1 when we increase h from 0.5 to 1.0. Treated purely as a fitting function in the power spectrum, effective Hubble constants (the Γ parameter) of $0.15 \lesssim \Gamma \lesssim 0.30$ are consistent with the COBE data and the observed number of lenses. All the variant models that produce the observed number of lenses and are consistent with COBE have steeper distributions of lens separations and lower average lens redshifts than standard CDM (see Figure 5). Since lensing determines the σ_8 of the mass distribution, the fact that it agrees with estimates of σ_8 from luminous objects suggests that there is little or no bias on cluster scales.

Models with low matter densities whether flat with λ_0 or open, are consistent with COBE and the lens observations if $0.3 \lesssim \Omega_0 \lesssim 0.5$ for $h = 0.5$. Unlike galaxy scale lenses where evolution appears to be unimportant (Mao 1991, Mao & Kochanek 1991, Rix et al. 1994), evolution matters far more than the extra volume of the low matter density universes. The same number of lenses is produced by higher values of σ_8 largely because the mass of a collapsed perturbation for a fixed comoving scale r_0 decreases rapidly when Ω_0 is reduced. This conflicts with the limits from lensing by galaxies, and (more weakly) the low value of the COBE quadrupole compared to the higher ℓ fluctuations (Sugiyama & Silk 1994). The flat models presented here neglect some corrections in the power spectrum (Kofman et al. 1993).

4. CONCLUSIONS

Gravitational lensing is a powerful new tool for examining cosmology and the structure of galaxies. Discrimination between models is restricted by the small number of lenses in the sample. Even so, models with an interesting cosmological constant are firmly ruled out, the theoretically preferred $\Omega = 1$ models are favored over low density models, and lenses require dark matter in the inner regions of E/S0 galaxies. The sample of large separations lenses is very incomplete (20% complete) but the expected number of lenses is very sensitive to the value of σ_8. Models of structure formation that correctly predict the local density of clusters generally produce the correct numbers of large separation lenses. Several new lenses are found each year, so the statistical errors will begin to decline rapidly over the next few years. The ultimate results will depend on when the problem is dominated by systematic errors (as all cosmological tests ultimately are), but a sample with a large number of lenses has many internal checks from the structure of the observed lenses to control and check the statistical model.

Acknowledgements: Research supported by the Alfred P. Sloan Foundation and NSF grant AST-9401722. Computations done on the CM-5 at NCSA.

REFERENCES

Bahcall, N., & Cen, R., 1992, ApJ, 398, L81

Bahcall, N., & Cen, R., 1993, ApJ, 407, L49

Bond, J.R., 1994, in *Relativistic Cosmology*, M. Sasaki, ed., (Academic Press)

Boyle, B.J., Fong, R., Shanks, T., & Peterson, B.A., 1990, MNRAS, 243, 1

Burke, B.F., Lehàr, J., & Connor, S.R., 1992, in *Gravitational Lenses*, R. Kayser, T. Schramm, & L. Nieser, eds., (Springer: Berlin) 237

Cen, R., Gott, J.R., Ostriker, J.P., & Turner, E.L., 1994, ApJ, 423, 1
Crampton, D., McClure, R.D., & Fletcher, J.M., 1992, ApJ, 392, 23
Efstathiou, G., Bond, J.R., & White, S.D.M., 1992, MNRAS, 258, 1P
Efstathiou, G., Ellis, R.S., & Peterson, B.A., 1988, MNRAS, 232, 431
Fukugita, M., & Turner, E.L., 1991, MNRAS, 253, 99
Górksi, K.M., Hinshaw, G., Banday, A.J., et al., 1994, ApJ, 430, L89
Gott, J.R., 1977, ARA&A, 15, 235
Hartwick, F.D.A., & Schade, D., 1990, ARA&A, 28, 437
Hewitt, A., & Burbidge, G., 1993, ApJS, 87, 2
Kassiola, A., & Kovner, I., 1993, ApJ, 417, 450
Kochanek, C.S., 1991, ApJ, 379, 517
Kochanek, C.S., 1992, ApJ, 348, 1
Kochanek, C.S., 1993a, MNRAS, 261, 453
Kochanek, C.S. 1993b, ApJ, 419, 12
Kochanek, C.S., 1994a, *(CfA 3791) ApJ, in press.*
Kochanek, C.S., 1994b, *(CfA 3933) ApJ, in press.*
Kochanek, C.S., 1994c, *CfA 3985 preprint*
Kofman, L.A., Gnedin, N.Y., & Bahcall, N.A., 1993, ApJ, 413, 1
Loveday, J., Efstathiou, G., Peterson, G., & Maddox, S., 1992, ApJ, 400, L43
Mao, S., 1991, ApJ, 380, 9
Mao, S., & Kochanek C.S., 1993, MNRAS, 268, 569
Maoz, D., Bahcall, J.N., Schneider, D.P., et al., 1993, ApJ, 409, 28
Maoz, D., & Rix, H.-W., 1993, ApJ, 416, 425
Narayan, R., & White, S.D.M., 1988, MNRAS, 231, 97P
Patnaik, A.R., Browne, I.W.A., King, L.J., et al., 1992, in *Gravitational Lenses*, R. Kayser, T. Schramm, & L. Nieser, eds., (Springer: Berlin) 140
Postman, M., & Geller, M.J., 1984, ApJ, 281, 95
Press, W.H., & Schechter, P., 1974, ApJ, 187, 425
Rix, H.-W., Maoz, D., Turner, E.L., & Fukugita, M., 1994, *IAS preprint*
Smoot, G.F., et al. 1992, ApJ, 396, L1
Sugiyama, N., & Silk, J., 1994, PhysRevLett, 73, 509
Surdej, J., Claeskens, J.F., Crampton, D., et al., 1993, AJ, 105, 2064
Turner, E.L., 1990, ApJ, 365, L43
Turner, E.L., Ostriker, J.P., & Gott, J.R., 1984, ApJ, 284, 1
van der Marel, R.P., 1991, MNRAS, 253, 710
Vogeley, M.S., Park, C., Geller, M.J., & Huchra, J., 1992, ApJ, 391, L5
Wallington, S., & Narayan, R., 1993, ApJ, 403, 517
Wambsganss, J., Cen, R., Ostriker, J., & Turner, E., 1994, *Princeton preprint*
Wright, E.L, Smoot, G.F., Kogut, A., et al., 1994, ApJ, 420, 1
Yee, H.K.C., Fillipenko, A.V., & Tang, D., 1993, AJ, 105, 7

Gravitational Lensing and the Mass Distribution of Clusters

Matthias Bartelmann[1,2] & Ramesh Narayan[1]

[1] Harvard-Smithsonian Center for Astrophysics
60 Garden Street, Cambridge, MA 02138, USA

[2] Max-Planck-Institut für Astrophysik
Karl-Schwarzschild-Strasse 1, D–85740 Garching, FRG

Introduction

The possibility that clusters of galaxies may act as efficient gravitational lenses was discussed by Noonan (1971) and Dyer & Roeder (1976) even before the first gravitational lens was discovered (Walsh, Carswell & Weymann 1979). Paczyński & Gorski (1981) interpreted a group of three high-redshift QSOs as multiple images of a single QSO split by two intervening galaxy clusters, and Narayan, Blandford & Nityananda (1984) suggested that cluster lensing may be as important as lensing by galaxies if clusters have small core radii. Webster (1985) first noticed that extended background objects may be coherently distorted by foreground galaxy clusters, and that this effect could be used to constrain cluster mass distributions. Hammer & Nottale (1986) argued that the magnification by galaxy clusters could measurably affect cosmological information obtained from the Hubble diagram of brightest cluster galaxies. Blandford, Phinney & Narayan (1987) simulated lensing by a cluster-scale mass and showed that arcs and arclets will result.

Despite these early theoretical realizations that lensing by clusters of galaxies may be common and could provide important information, the subject moved to the forefront only after the first giant blue luminous arcs were discovered in the clusters A 370 and Cl 2244−02 (Lynds & Petrosian 1986, 1989, Soucail *et al.* 1987a,b). The interpretation of these arcs in terms of gravitational lensing (Paczyński 1987) was confirmed when the first arc redshifts were measured (Soucail *et al.* 1988, Miller & Goodrich 1988, Mellier *et al.* 1991). The further development of techniques to detect smaller arclets (Tyson, Valdes & Wenk 1990) has now created a very fertile and promising field of research which is providing valuable information on the mass distribution of clusters.

The chief advantage of gravitational lenses in the study of clusters is that the lensing effect depends directly on the total mass distribution, whereas most other methods need to make assumptions about how the luminous matter is distributed relative to the mass, and about the dynamical state it is in. The giant arcs provide information about the total masses (and hence the mass-to-light ratios) of the lensing clusters, and impose constraints on the symmetry of the cluster potentials, their core radii, radial density profiles, and the frequency of substructure. Using arclets, methods have been developed recently which even allow us to reconstruct full two-dimensional mass maps of the lensing clusters. We summarize these results in the following sections.

For general information on gravitational lensing see the review by Blandford & Narayan (1992) or the monograph by Schneider, Ehlers & Falco (1992). A detailed review on arcs and arclets in galaxy clusters has been recently compiled by Fort & Mellier (1994).

Basic Gravitational Lens Optics

Following the thin-lens approximation, which is valid in basically all situations of astrophysical interest, the mass distribution of the lens can be projected onto a lens plane orthogonal to the line-of-sight from the observer to the source. Another plane, the source plane, is constructed parallel to the lens plane at the distance of the source. It is convenient to trace light rays backwards on straight lines from the observer to the lens plane, where they are deflected, and to continue them backwards on straight lines from the lens plane to the source plane.

If the surface mass density Σ of the lens somewhere becomes comparable to or exceeds the critical value, $\Sigma_{\rm crit} \approx 0.35$ g cm$^{-2}/D$(Gpc), then caustics will be formed on the source plane (*e.g.*, Blandford & Narayan 1986). Here, $D = D_{\rm d} D_{\rm ds}/D_{\rm s}$ where $D_{\rm d,s,ds}$ are angular-diameter distances from the observer to the lens, observer to the source, and from the lens to the source, respectively. Sources inside caustics are split into multiple images. If an extended source like a galaxy happens to touch a caustic, then an arc is formed (Grossman & Narayan 1988, Blandford & Kovner 1988, Narayan & Grossman 1989, Kovner 1989). A galaxy touching a fold caustic produces a moderate arc, while one which overlaps a cusp or the related beak-to-beak or lips caustic, produces a giant arc. If the lens is subcritical, or if the source lies outside the caustics, then the image of the source is only weakly distorted and is called an arclet. Illustrations of these different kinds of imaging situations are shown in the following figure.

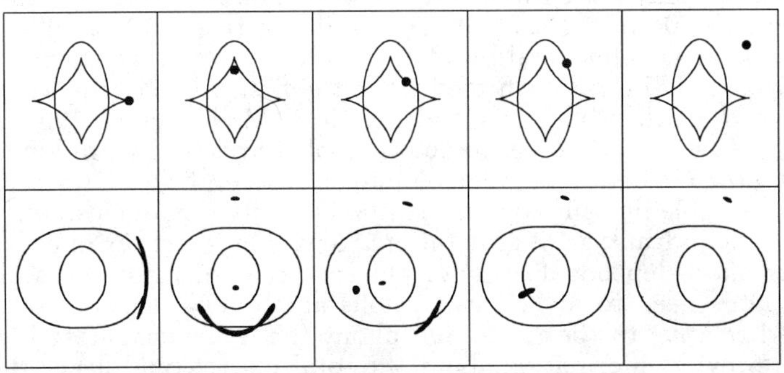

Figure 1 – Illustration of five different imaging situations for an elliptical lens with finite core. The top row shows the position of the source relative to the caustics, the bottom row shows the images of the source relative to the critcal curves. From left to right, the columns display: source on a "naked cusp" forming a large arc without counterarc; source inside a cusp forming a large arc from three merging images and counterarcs; source on a fold forming a shorter arc from two merging images; source on the radial caustic forming a short radial arc; source outside the caustic forming an arclet. The sizes of the outer and the inner critical curves approximately correspond to the Einstein radius and the core radius of the cluster, respectively. The shapes of large arcs shown in the first three panels are characteristic of elliptical lenses. Spherically symmetric lenses behave very differently (*e.g.*, Grossman & Narayan 1988).

To sufficient accuracy, the deflection of light rays by a lens is described by a two-dimensional potential,

$$\psi = \frac{2D_{\rm ds}}{D_{\rm d}D_{\rm s}} \int \Phi \, dz \,, \tag{1}$$

where Φ is the three-dimensional Newtonian potential of the lens and the integration is performed along the line-of-sight. The local properties of the lens mapping are contained in the curvature matrix or Hessian $(\psi,_{ij})$, where the comma preceding an index denotes partial differentiation with respect to the corresponding coordinate. The trace-free antisymmetric part of $(\psi,_{ij})$ measures the distortion of images and is called the shear matrix, with independent components $\gamma_1 = \gamma \cos 2\varphi$ and $\gamma_2 = \gamma \sin 2\varphi$, where φ is the orientation of the shear. The symmetric part measures the isotropic magnification or demagnification of images and is called the convergence κ. Thus

$$(\psi,_{ij}) = \begin{pmatrix} \kappa + \gamma \cos 2\varphi & \gamma \sin 2\varphi \\ \gamma \sin 2\varphi & \kappa - \gamma \cos 2\varphi \end{pmatrix} . \tag{2}$$

Note that, by Poisson's equation, $\kappa = (\psi,_{11} + \psi,_{22})/2 = \Sigma/\Sigma_{\rm crit}$, where Σ is the local surface density of the lens. The axis ratio of an arc(let) is given by

$$\frac{\text{semi-minor axis}}{\text{semi-major axis}} = \frac{1-\kappa-\gamma}{1-\kappa+\gamma} \,, \tag{3}$$

so that by measuring image distortions we can obtain information on γ.

Enclosed Mass and Mass-to-Light Ratio

If a large arc is found in a galaxy cluster, the total cluster mass inside a circle with radius equal to the distance $\theta_{\rm arc}$ of the arc from the cluster center can be roughly estimated to be

$$M \approx 1.1 \times 10^{14} \, M_\odot \left(\frac{\theta_{\rm arc}}{30''}\right)^2 \left(\frac{D}{1 \text{ Gpc}}\right) . \tag{4}$$

This result follows from assuming that $\theta_{\rm arc}$ measures the angular Einstein radius of the cluster (*cf.* Fig. 1). Equation (4) can be converted to an estimate for the line-of-sight velocity dispersion σ_v of the cluster,

$$\sigma_v \approx 10^3 \left(\frac{\theta_{\rm arc}}{28''}\right)^{1/2} \left(\frac{D_{\rm s}}{D_{\rm ds}}\right)^{1/2} \text{ km s}^{-1} . \tag{5}$$

Of course, to use these formulae, we need to know the redshifts of both the lens and the source. Eqautions (4) and (5) provide only rough estimates of M and σ_v. More reliable values are obtained with detailed lens models. Table 1 lists representative results for masses, velocity dispersions, and mass-to-blue-light ratios $M/L_{\rm B}$ of three arc clusters, obtained by means of detailed lens models. More results of this kind can be found in Fort & Mellier (1994). The accuracy

Table 1 – Examples of masses, mass-to-blue-light ratios, and velocity dispersions derived from lens models for three arc clusters.

Cluster	M ($M_\odot$)	$M/L_{\rm B}$ (solar)	σ_v (km/s)	Reference
A 370	$\approx 10^{14}$	≈ 200	≈ 1350	Grossman & Narayan (1989) Bergmann *et al.* (1990) Kneib *et al.* (1993)
A 2390	$\approx 1.5 \times 10^{14}$	≈ 120	≈ 1250	Pello *et al.* (1991)
MS 2137	$\approx 5 \times 10^{13}$	≈ 250	≈ 1100	Mellier *et al.* (1993)

of these mass estimates is hard to evaluate, but they are probably correct to a factor of two.

Asphericity of Clusters

The fact that the observed giant arcs never have counter-arcs of comparable brightness, and rarely have even small counter-arcs, implies that the lensing geometry has to be non-spherical (*e.g.*, Grossman & Narayan 1988, Kovner 1989; *cf.* Fig. 1). Therefore, cluster potentials must have substantial quadrupole and perhaps also higher multipole moments. In the case of A 370, for example, there are two cD galaxies, and the potential quadrupole estimated from their separation is consistent with the quadrupole required to model the observed giant arc (Grossman & Narayan 1989). The more detailed model of A 370 by Kneib *et al.* (1993) shows a remarkable agreement between the lensing potential and the strongly aspheric X-ray emission of the cluster.

Large deviations of the lensing potentials from spherical symmetry also help increase the probability for producing large arcs. Bergmann & Petrosian (1993) argued that the apparent abundance of large arcs relative to small arcs and arclets can be reconciled with theoretical expectations if aspheric lens models are taken into account. Bartelmann, Steinmetz & Weiss (1994) showed that the probability for large arcs can be increased by more than an order of magnitude if aspheric cluster models with significant substructure are used instead of smooth spherically symmetric models.

Core Radii

If a cluster is able to produce large arcs, its surface-mass density in the core must be approximately supercritical, $\Sigma \gtrsim \Sigma_{\rm crit}$. If applied to simple lens models, this condition requires

$$\theta_{\rm core} \lesssim 15'' \left(\frac{\sigma_v}{10^3\ {\rm km/s}}\right)^2 \left(\frac{D_{\rm ds}}{D_{\rm s}}\right) .$$

Narayan *et al.* (1984) were the first to show that cluster mass distributions need to have smaller core radii than those derived from optical and X-ray observations if they are to produce strong effects of gravitational lensing. This message has been confirmed many times by later efforts to model giant arcs. In virtually every case the core radii estimated from lensing are significantly smaller than the estimates from optical and X-ray data. Some representative results on lens-derived core radii are listed in Table 2, where the estimates correspond to $H_0 = 50\ \mathrm{km\,s^{-1}Mpc^{-1}}$.

Table 2 – Limits on cluster core radii derived from modelling large arcs.

Cluster	$r_{\rm core}$ (kpc)	Reference
A 370	$\lesssim 60$	Grossman & Narayan (1989)
	$\lesssim 100$	Kneib *et al.* (1993)
MS 2137–23	≈ 50	Mellier *et al.* (1993)
Cl 0024+1654	$\lesssim 130$	Bonnet *et al.* (1994)
MS 0440+0204	$\ll 90$	Luppino *et al.* (1993)

Statistical analyses based on spherically symmetric cluster models lead to similar conclusions. Miralda-Escudé (1992, 1993a) argued that cluster core radii can hardly be larger than the curvature radii of large arcs. Wu & Hammer (1993) claimed that clusters either had to have singular cores or density profiles much steeper than isothermal in order to reproduce the abundance of large arcs. Although this conclusion can substantially be weakened once deviations from spherical symmetry are taken into account (Bartelmann *et al.* 1994b), it remains true that $r_{\rm core} \lesssim 100$ kpc is required in all observed clusters, and cores of this size can also be reconciled with large-arc statistics.

Interestingly, there is at least one observation which indicates that cluster cores, although small, must be finite. Fort *et al.* (1992) discovered a radial arc near the center of MS 2137–23. To produce a radial arc, the core radius has to be roughly equal to the distance between the cluster center and the radial arc (*cf.* Fig. 1). Mellier, Fort & Kneib (1993) find $r_{\rm core} \gtrsim 40$ kpc in MS 2137–23. Bergmann & Petrosian (1993) presented a statistical argument in favor of finite cores by showing that lens models with singular cores produce fewer large arcs (relative to small arcs) than observed. The relative abundance increases with a small finite core.

Radial Density Profile

Many of the observed giant arcs are unresolved in the radial direction, some of them even with excellent seeing. Since the faint blue background galaxies, which provide the source population for the arcs, can be resolved (*e.g.*, Tyson, these proceedings), the giant arcs appear to be demagnified in width. It was realized by Hammer & Rigaut (1989) that spherically symmetric lenses can radially demagnify giant arcs only if their radial density profiles are steeper than isothermal; the maximum demagnification is obtained for a point mass lens, where it is a factor of two. Kovner (1989) and Hammer (1991) demonstrated that, irrespective of the mass profile and the symmetry of the lens, the demagnification is given by $\approx 2(1-\kappa)$, where κ is the convergence at the position of the arc. Thin arcs therefore require $\kappa \lesssim 0.5$. Since giant arcs have to be located close to those critical curves in the lens plane along which $1-\kappa-\gamma=0$, large thin arcs additionally require $\gamma \gtrsim 0.5$.

In principle, the radius of curvature of large arcs relative to their distance from the cluster center can be used to constrain the steepness of the radial density profile (Miralda-Escudé 1992, 1993a), but results obtained from observed arcs are not yet conclusive (Grossman & Saha 1994). One problem with this method is that substructure in clusters tends to enlarge curvature radii irrespective of the mass profile of the dominant component of the cluster (Miralda-Escudé 1993a, Bartelmann *et al.* 1994b).

Wu & Hammer (1993) argued for steep mass profiles on statistical grounds, because the observed abundance of large arcs appears to require highly centrally condensed cluster mass profiles in order to increase the central mass density of clusters while keeping their total mass constant. However, their conclusions are based on spherically symmetric lens models and are significantly altered when the symmetry assumption is dropped (Bartelmann *et al.* 1994b).

It should be kept in mind that not all arcs are thin. Some "thick" arcs are known (*e.g.*, in A 2218, Pello-Descayre *et al.* 1988, and in A 2390, Pello *et al.* 1991), and it is quite possible that thin arcs predominate just because they are more easily detected than thick ones due to observational selection effects. Also, Miralda-Escudé (1992, 1993a) has argued that intrinsic source ellipticity can increase the probability of producing thin arcs, while Bartelmann *et al.* (1994b) showed that the condition $\kappa \lesssim 0.5$ which is required for thin arcs can be more frequently fulfilled in clusters with substructure where the shear is larger than in spherically symmetric clusters.

The measurement of a coherent weak shear pattern out to a distance of $\lesssim$ 1.5 Mpc from the center of the cluster Cl 0024+1654 by Bonnet, Mellier & Fort (1994) demonstrates a promising method of constraining cluster mass profiles. Those observations show that the density decreases rapidly outward, but the data are compatible both with the isothermal profile and a steeper de Vaucouleurs profile.

Mass Sub-condensations

A 370 has two cD galaxies and is a clear example of a cluster with multiple mass centers. A two-component mass model centered on the cD galaxies (Kneib *et al.* 1993) fits the lens data as well as X-ray and deep optical images of the cluster. A 2390 is an interesting example because it contains a straight arc (Pelló *et al.* 1991, see also Mathez *et al.* 1992) which can be produced only with either a lips or a beak-to-beak caustic (Kassiola, Kovner & Blandford 1992). If the arc is modeled with a lips caustic, it requires the mass peak to be close to the location of the arc, but this is not where the cluster light is maximum. With a beak-to-beak caustic, the model requires two separate mass condensations, one of which could be at the peak of the luminosity, but then the other has to be a dark condensation.

A 370 and A 2390 are the most obvious examples of what is probably a widespread phenomenon, namely that clusters are in general not fully relaxed but have substructure as a result of ongoing evolution. If clusters are frequently substructured, this can lead to systematic effects in the statistics of arcs and in the derived cluster parameters (Bartelmann *et al.* 1994b, Bartelmann 1994b).

Arclets and Mapping the Cluster Surface Density

Giant arcs are relatively rare, but almost all clusters with velocity dispersions $\sigma_v \gtrsim 500$ km/s produce measurable distortions of background galaxy images (Webster 1985, Miralda-Escudé 1991a). Since at faint magnitudes ($B \approx 27$) there are as many as ≈ 100 faint galaxies per square arc minute (Tyson 1988), most of which probably are at fairly large redshifts $z \approx 0.7-1.4$ (Guhathakurta, Tyson & Majewski 1990, Fort 1992, Colless *et al.* 1993), this means that it is possible to obtain a huge amount of information from image distortions induced by foreground clusters. Indeed, deep images of cluster fields reveal large numbers of arclets (*e.g.*, Tyson *et al.* 1990, Smail *et al.* 1991) and one of the most exciting areas today involves the use of these arclets to derive information about the foreground mass.

The idea of using arclet data to study the mass distributions of clusters was suggested first by Webster (1985) and again by Grossman & Narayan (1989) after the discovery of the faint blue galaxies by Tyson (1988). Tyson *et al.* (1990), Kochanek (1990) and Miralda-Escudé (1991a) showed how it is possible to determine the cluster center, to constrain parametrized cluster models, and to map the cluster surface density, and Tyson *et al.* (1990) actually obtained some preliminary results on a few clusters. Later, Kaiser & Squires (1993) described a well-defined algorithm for constructing surface-density maps of clusters from coherent distortions of galaxy images in the field. This method, frequently called "cluster lens inversion", has recently been discussed and refined in great detail (Schneider & Seitz 1994, Seitz & Schneider 1994, Kaiser 1994; see also the articles by Tyson and Squires in this volume).

The reconstruction technique proposed by Kaiser & Squires is based on the fact that the convergence κ and the shear components γ_1, γ_2 of a lens are combinations of second derivatives of the same scalar potential ψ [*cf.* Eq. (2)]. Therefore they are not independent but are related through relations such as

$$\kappa(\vec{\theta}) = -\frac{1}{\pi} \int d^2\theta' \left[\mathcal{D}_1(\vec{\theta} - \vec{\theta}')\gamma_1(\vec{\theta}') + \mathcal{D}_2(\vec{\theta} - \vec{\theta}')\gamma_2(\vec{\theta}') \right] , \tag{7}$$

where the convolution kernels $\mathcal{D}_1, \mathcal{D}_2$ are given by

$$\mathcal{D}_1(\vec{\theta}) = \frac{\theta_1^2 - \theta_2^2}{|\vec{\theta}|^4} , \; \mathcal{D}_2(\vec{\theta}) = \frac{2\theta_1\theta_2}{|\vec{\theta}|^4} . \tag{8}$$

Kaiser & Squires (1993) realized that in the weak-shear limit, $|\gamma| \ll 1$, the shear can be directly obtained from the distortion of galaxy images. The intrinsic ellipticity of galaxies, however, introduces noise into the reconstruction, requiring appropriate smoothing of the data. The method appears feasible in principle, as demonstrated by Fahlman *et al.* (1994), Smail *et al.* (1994), and Bonnet *et al.* (1994). It suffers, however, from intrinsic degeneracies and systematic effects. Schneider & Seitz (1994) showed that, if the shear is not weak, the coherent ellipticities of galaxy images measure a combination of κ and γ rather than γ alone, and Kaiser (1994) and Seitz & Schneider (1994) extended the Kaiser & Squires method into the strong-lensing regime. Since the area integral in Eq. (7) formally extends over the complete plane, the finiteness of the area covered by arclet observations introduces boundary effects which are addressed by Schneider (1994). Even after these systematic effects are taken into account, the distortion pattern caused by a lens is still invariant under the transformation

$$\kappa \to (1 - \lambda) + \lambda\kappa$$

with arbitrary $\lambda \neq 0$, because adding a sheet of constant surface-mass density to the lens while simultaneously rescaling κ and γ does not change the distortion pattern (Schneider & Seitz 1994). This degeneracy can be removed if κ can be determined at least at one point in the cluster field by another method, *e.g.*, if large arcs are present or if a change in the number counts of galaxies behind the cluster can be observed (Broadhurst, Taylor & Peacock 1994).

The reconstruction method by Kaiser & Squires has been employed to estimate the mass distributions of a few clusters. For the cluster MS 1224, the estimated mass is higher than the virial mass by a factor $\approx 2-3$ (Fahlman *et al.* 1994, Carlberg *et al.* 1994), while for the cluster A 2163, the hottest and most luminous X-ray cluster in the sky, the reconstruction is surprisingly compatible with there being no mass at all in the cluster (Squires, these proceedings). The method requires careful investigation before it can reliably be used to determine absolute cluster masses rather than relative cluster mass maps.

Comparison of Lensing Results with other Determinations

Enclosed Mass

Three different methods are currently used to estimate cluster masses. Galaxy velocity dispersions yield a mass estimate from the virial theorem. The X-ray emission of rich galaxy clusters is dominated by thermal free-free emission and therefore depends on the squared density of the intracluster gas, which in turn traces the gravitational potential of the clusters. Such estimates usually assume that the cluster gas is in thermal hydrostatic equilibrium and that the potential is at least approximately spherically symmetric. Finally, large arcs in clusters provide a mass estimate through Eq. (4) or by more detailed modeling. These three mass estimates are in qualitative agreement with each other up to factors of $\approx 2-3$.

Miralda-Escudé & Babul (1994) compared X-ray and large-arc mass estimates for the clusters A 1689, A 2163 and A 2218. They took into account that these clusters deviate from spherical symmetry, so that their masses may deviate from the estimate provided by Eq. (4), and obtained lensing masses from individual lens models which reproduce the observed arcs. They arrived at the conclusion that in A 1689 and A 2218 the mass required for producing the large arcs is higher by a factor of $2-2.5$ than the mass required for the X-ray emission, and proposed a variety of reasons for such a discrepancy, among them projection effects or non-thermal pressure support. Loeb & Mao (1994) specifically suggested that strong turbulence and magnetic fields in the intracluster gas may constitute a significant non-thermal pressure component in A 2218 and thus render the X-ray mass estimate too low.

Bartelmann (1994b) showed that cluster mass estimates obtained from large arcs by straightforward application of Eq. (4) are systematically too high by a factor of ≈ 1.6 on average, and by as much as a factor of ≈ 2 in 1 out of 5 cases. This discrepancy arises because Eq. (4) assumes a smooth spherically symmetric mass distribution whereas realistic clusters are asymmetric and have substructure. Note that Daines, Jones & Forman (1994) found evidence for at least two separate mass condensations along the line of sight towards A 1689, while the arclets in A 2218 show at least two mass concentrations. It appears that cluster mass estimates from lensing require detailed lens models in order to be accurate to better than $\approx 30-50\%$. In the case of MS 1224, Fahlman *et al.* (1994) and Carlberg *et al.* (1994) have obtained masses using the Kaiser &

Squires weak-lensing cluster reconstruction method. Their mass estimates are 2–3 times higher than the cluster's virial mass. Carlberg *et al.* find evidence from velocity measurements that there is a second poor cluster in the foreground of MS 1224 which may explain the result. All of these mass discrepancies illustrate that cluster masses must still be considered uncertain to a factor of ≈ 2.

Core Radius

Lensing estimates of cluster core radii are generally much smaller than the core radii obtained from optical or X-ray data. Note, however, that the X-ray core radii depend on whether the cooling regions of clusters are included into emissivity profile fits or not, because the cooling radii are of the same order of magnitude as the core radii. If they are included, the best-fit core radii are reduced by a factor of ≈ 4 (Gerbal *et al.* 1992, Durret *et al.* 1994). The estimates from lensing are fairly robust and probably reliable. The results generally suggest that the dark matter is much more centrally condensed than the luminous matter. Many clusters with large arcs have cD galaxies, so the cD may be suspected to be the cause of the small core. However, there are also non-cD clusters with giant arcs, *e.g.*, A 1689 and Cl 1409 (Tyson 1990) and MS 0440+02 (Luppino *et al.* 1993). In fact, Tyson (1990) claims that the former two clusters have cores smaller than 100 kpc, just like the other arc clusters containing cD galaxies.

Does Mass Follow Light?

Leaving the core radius aside, does the mass follow the light? It is clear that the mass cannot be as concentrated within the galaxies as the optical light is (*e.g.*, Bergmann, Petrosian & Lynds 1990). However, if the optical light is smoothed and assumed to trace the mass, then the resulting mass distribution is probably not very different from the true mass distribution. For instance, in A 370, the elongation of the mass distribution required for the giant arc is along the line connecting the two cD galaxies in the cluster (Grossman & Narayan 1989) and in fact Kneib *et al.* (1993) are able to achieve an excellent fit of the giant arc and several arclets with two mass concentrations surrounding the two cDs. Their model potential also agrees very well with the X-ray emission of the cluster. In MS 2137, the optical halo is elongated in the direction indicated by the arcs for the overall mass asymmetry (Mellier *et al.* 1993), and in Cl 0024, the mass distribution is elongated in the same direction as the galaxy distribution (Wallington, Kochanek & Koo 1994). Smail *et al.* (1994) find that the mass maps of two clusters reconstructed from weak lensing agree fairly well with their X-ray emission. An important counterexample is the cluster A 2390, where the straight arc requires a completely dark mass concentration (Kassiola *et al.* 1992).

What Kinds of Clusters Produce Giant Arcs?

Which parameters determine whether or not a galaxy cluster is able to produce large arcs? Clearly, large velocity dispersions and small core radii favor the formation of arcs. As argued before, intrinsic asymmetries and substructures also increase the ability of clusters to produce arcs because they increase the shear and the number of cusps.

The abundance of arcs in X-ray luminous clusters appears to be higher than in optically selected clusters, and this might suggest that X-ray brightness of the cluster is the critical factor that determines how efficient a cluster will be in lensing. At least a quarter, maybe half, of the 38 X-ray bright clusters selected by Le Fèvre *et al.* (1994) contain large arcs, while Smail *et al.* (1991) find only one large arc in a sample of 19 distant optically selected clusters. However, some clusters which are prominent lenses (A 370, A 1689, A 2218) are moderate X-ray sources, while other clusters which are very luminous X-ray sources (A 2163, Cl 1455) are poor lenses. This shows that, while there probably is a correlation between X-ray brightness and the ability to produce strong lensing, the correlation is not at all tight especially if very prominent arcs are considered. In fact, substructure appears to be at least as important as X-ray brightness. For example, A 370, A 1689 and A 2218 all seem to have clumpy mass distributions.

Another possibility is that giant arcs preferentially form in clusters with cD galaxies. A 370, for instance, even has two cDs. However, non-cD clusters with giant arcs are known, *e.g.*, A 1689, Cl 1409 (Tyson et al. 1990), and MS 0440+02 (Luppino et al. 1993).

Weak Lensing by Large-Scale Structures in the Universe

Lensing by even larger scale structures than clusters has been discussed. Fugmann (1990) found that there is an excess of Lick galaxies in the vicinity of high-redshift, radio-loud QSOs, and showed that the excess extends to $\approx 10'$ around the QSOs. If the excess is real, it is most likely caused by the magnification bias due to gravitational lensing, and the scale of the lens must be very large; galaxy-sized lenses have Einstein radii of a few arc seconds and are therefore irrelevant for this kind of association.

Bartelmann & Schneider (1993a) showed that current models of large-scale structure formation can explain large-scale QSO-galaxy associations, provided a double magnification bias (Borgeest, von Linde & Refsdal 1991) is assumed, and they confirmed correlations of high-redshift, radio-loud QSOs with optical and infrared galaxies and with the diffuse X-ray emission (Bartelmann & Schneider 1993b, 1994; Bartelmann, Schneider & Hasinger 1994). The result by Bartelmann & Schneider (1993b) was recently confirmed by Seitz & Schneider (1994b), who investigated correlations between the same sample of QSOs and Zwicky clusters, following in part an earlier study by Rodrigues-Williams & Hogan (1994). Bartelmann (1994a) has shown that constraints on the density perturbation spectrum and the bias factor of galaxy formation can be obtained from the angular cross-correlation function between QSOs and galaxies.

The idea of looking for the weak lensing action of large-scale mass distributions through the distortions induced on background galaxy images was first suggested by Kristian & Sachs (1966) and Gunn (1967) and was analyzed in more detail by Babul & Lee (1991), Jaroszyński *et al.* (1990), Miralda-Escudé (1991b), Blandford *et al.* (1991), Bartelmann & Schneider (1991) and Kaiser (1992). The effect is expected to be weak (magnification and shear are typically on the order of $5-10\%$), and a huge number of galaxies has to be imaged with utmost care before a coherent shear signal can be observed. Mould *et al.* (1994) have set a limit on the observed distortion in a blank field. Their limit is consistent with standard models of structure formation. Bonnet *et al.* (1993) found weak shear in the vicinity of the twin quasar 2345+007 and confirmed the presence of a dark cluster-like mass concentration. This is a promising technique to check for the presence of suspected dark matter clumps.

Concluding Comments

Gravitational lensing has proved to be a powerful technique to measure the amount and distribution of mass in clusters of galaxies. Using lensing, it is possible to estimate the total gravitating mass, the core size, the density profile, the mass distribution, and the amount of substructure in clusters. With the advent of methods to utilize data on arclets, lensing promises to develop into an even more powerful diagnostic for cluster studies. Comparative studies of clusters using their lens effect, their X-ray emission, and measurements of the dynamics of the cluster constituents promise to yield unambiguous information about the structure and the evolution of clusters.

References

Babul, A., Lee, M.H.: 1991, MNRAS, 250, 407
Bartelmann, M., Schneider, P.: 1991, A&A, 248, 349
Bartelmann, M., Schneider, P.: 1993a, A&A, 268, 1
Bartelmann, M., Schneider, P.: 1993b, A&A, 271, 421
Bartelmann, M., Schneider, P.: 1994, A&A, 284, 1
Bartelmann, M., Schneider, P., Hasinger, G.: 1994a, A&A, 290, 399
Bartelmann, M.: 1994a, A&A, submitted
Bartelmann, M., Weiss, A.: 1994, A&A, 287, 1
Bartelmann, M., Steinmetz, M., Weiss, A.: 1994b, A&A, in press
Bartelmann, M.: 1994b, A&A, submitted
Bergmann, A.G., Petrosian, V., Lynds, R.: 1990, ApJ, 350, 23
Bergmann, A.G., Petrosian, V.: 1993, ApJ, 413, 18
Blandford, R.D., Narayan, R.: 1986, ApJ, 310, 568
Blandford, R.D., Phinney, E.S., Narayan, R.: 1987, ApJ, 313, 28
Blandford, R.D., Kovner, I.: 1988, Phys.Rev.A, 38, 4028
Blandford, R.D., Saust, A.B., Brainerd, T.G., Villumsen, J.V.: 1991, MNRAS, 251, 600
Blandford, R.D., Narayan, R.: 1992, ARAA, 30, 311
Bonnet, H., Fort, B., Kneib, J.-P., Mellier, Y., Soucail, G.: 1993, A&A, 280, L7
Bonnet, H., Mellier, Y., Fort, B.: 1994, ApJ, 427, L83
Borgeest, U., von Linde, J., Refsdal, S.: 1991, A&A, 251, L35
Broadhurst, T.J., Taylor, A.N., Peacock, J.A.: 1994, ApJ, in press
Carlberg, R.G., Yee, H.K.C., Ellingson, E., 1994, ApJ, Nov.20
Colless, M., Ellis, R.S., Broadhurst, T.J., Taylor, K., Peterson, B.A.: 1993, MNRAS, 261, 19
Daines, S., Jones, C., Forman, W., 1994, preprint
Durret, F., Gerbal, D., Lachièze-Rey, M., Lima-Nieto, G., Sadat, R.: 1994, A&A, 287, 733
Dyer, C.C., Roeder, R.C.: 1976, Nature, 260, 764
Ellis, R., Allington-Smith, J., Smail, I.: 1991, MNRAS, 249, 184
Fahlman, G.G., Kaiser, N., Squires, G., Woods, D., 1994, ApJ, in press
Fischer, P., Tyson, J.A., Bernstein, G.M., Guhathakurta, P.: 1994, ApJ, 431, L71
Fort, B., Prieur, J.L., Mathez, G., Mellier, Y., Soucail, G.: 1988, A&A, 200, L17
Fort, B.: 1992, in: *Gravitational Lenses*, ed. Kayser, R., Schramm, T., Nieser, L. (Heidelberg: Springer), p. 267

Fort, B., Le Fevre, O., Hammer, F., Cailloux, M.: 1992, ApJ, 399, L125
Fort, B., Mellier, Y.: 1994, A&AR, in press
Fugmann, W.: 1990, A&A, 240, 11
Gerbal, D., Durret, F., Lima-Nieto, G., Lachièze-Rey, M.: 1992, A&A, 253, 77
Grossman, S.A., Narayan, R.: 1988, ApJ, 324, L37
Grossman, S.A., Narayan, R.: 1989, ApJ, 344, 637
Grossman, S.A., Saha, P.: 1994, ApJ, 431, 74
Guhathakurta, P., Tyson, J.A., Majewski, S.R.: 1990, ApJ, 357, L9
Gunn, J.E.: 1967, ApJ, 150, 737
Hammer, F., Nottale, L.: 1986, A&A, 167, 1
Hammer, F., Rigaut, F.: 1989, A&A, 226, 45
Hammer, F., Rigaut, F., Le Fevre, O., Jones, J., Soucail, G.: 1989, A&A, 208, L7
Hammer, F.: 1991, ApJ, 383, 66
Jaroszyński, M., Paczyński, B., Park, C.B., Gott, J.R.: 1990, 365, 22
Kaiser, N.: 1992, ApJ, 388, 272
Kaiser, N., Squires, G.: 1993, ApJ, 404, 441
Kaiser, N.: 1994, preprint
Kassiola, A., Kovner, I., Fort, B.: 1992, ApJ, 400, 41
Kassiola, A., Kovner, I., Blandford, R.D.: 1992, ApJ, 396, 10
Kneib, J.-P., Mellier, Y., Fort, B., Mathez, G.: 1993, A&A, 273, 367
Kochanek, C.S.: 1990, MNRAS, 247, 135
Kovner, I., Paczyński, B.: 1988, ApJ, 335, L9
Kovner, I.: 1989, ApJ, 337, 621
Kristian, J., Sachs, R.K.: 1966, ApJ, 143, 379
Lavery, R.J., Henry, J.P.: 1988, ApJ, 329, L21
Le Fèvre, O., Hammer, F., Angonin-Willaime, M.-C., Gioia, I.M., Luppino, G.A.: 1994, ApJ, 422, L5
Loeb, A., Mao, S.: 1994, ApJL, in press
Luppino, G.A., Gioia, I.M.: 1992, A&A, 265, L9
Luppino, G.A., Gioia, I.M., Annis, J., Le Fèvre, O., Hammer, F.: 1993, ApJ, 416, 444
Lynds, R., Petrosian, V.: 1986, BAAS, 18, 1014
Lynds, R., Petrosian, V.: 1989, ApJ, 336, 1
Mathez, G., Fort, B., Mellier, Y., Picat, J.-P., Soucail, G.: 1992, A&A, 256, 343
Mellier, Y., Fort, B., Soucail, G., Mathez, G., Cailloux, M.: 1991, ApJ, 380, 334
Mellier, Y., Fort, B., Kneib, J.-P.: 1993, ApJ, 407, 33
Miller, J.S., Goodrich, R.W.: 1988, Nature, 331, 685
Miralda-Escudé, J.: 1991a, ApJ, 370, 1
Miralda-Escudé, J.: 1991b, ApJ, 380, 1
Miralda-Escudé, J.: 1992, ApJ, 390, L65
Miralda-Escudé, J.: 1993a, ApJ, 403, 497
Miralda-Escudé, J.: 1993b, ApJ, 403, 509
Miralda-Escudé, J., Babul, A.: 1994, ApJ, in press
Mould, J., Blandford, R.D., Villumsen, J.V., Brainerd, T.G., Smail, I., Small, T., Kells, W.: 1994, ApJ, in press
Narayan, R., Blandford, R.D., Nityananda, R.: 1984, Nature, 310, 112
Narayan, R., Grossman, S.A.: 1989, in: *Gravitational lenses*, ed. Moran, J.M., Hewitt, J.N., Lo, K.Y. (New York: Springer)
Noonan, T.W.: 1971, AJ, 76, 765
Paczyński, B.: 1987, Nature, 325, 572
Paczyński, B., Gorski, K.: 1981, ApJ, 248, L101

Pello-Descayre, R., Sanahuja, B., Soucail, G., Mathez, G., Ojero, E.: 1988, A&A, 190, L11
Pelló, R., Sanahuja, B., Le Borgne, J.-F., Soucail, G., Mellier, Y.: 1991, ApJ, 366, 405
Pelló, R., Le Borgne, J.F., Sanahuja, B., Mathez, G., Fort, B.: 1992, A&A, 266, 6
Rodrigues-Williams, L.L., Hogan, C.J.: 1994, AJ, 107, 451
Schneider, P., Ehlers, J., Falco, E.E.: 1992, Gravitational Lenses (Heidelberg: Springer)
Schneider, P.: 1994, A&A, submitted
Schneider, P., Seitz, C.: 1994, A&A, submitted
Seitz, C., Schneider, P.: 1994a, A&A, submitted
Seitz, S., Schneider, P.: 1994b, A&A, submitted
Smail, I., Ellis, R.S., Fitchett, M.J., Norgaard-Nielsen, H.U., Hansen, L., Jorgensen, H.E.: 1991, MNRAS, 252, 19
Smail, I., Ellis, R.S., Fitchett, M.J., Edge, A.C., 1994, MNRAS, in press
Soucail, G., Fort, B., Mellier, Y., Picat, J.P.: 1987a, A&A, 172, L14
Soucail, G., Mellier, Y., Fort, B., Mathez, G., Hammer, F.: 1987b, A&A, 184, L7
Soucail, G., Mellier, Y., Fort, B., Mathez, G., Cailloux, M.: 1988, A&A, 191, L19
Tyson, J.A.: 1988, AJ, 96, 1
Tyson, J.A., Valdes, F., Wenk, R.A.: 1990, ApJ, 349, 1
Wallington, S., Kochanek, C.S., Koo, D.C.: 1994, ApJ, in press
Walsh, D., Carswell, R.F., Weymann, R.J.: 1979, Nature, 279, 381
Webster, R.L.: 1985, MNRAS, 213, 871
Wu, X.-P., Hammer, F.: 1993, MNRAS, 262, 187

MASS DISTRIBUTIONS OF CLUSTERS FROM GRAVITATIONAL MAGNIFICATION

Tom Broadhurst
Physics and Astronomy, The Johns Hopkins University, Baltimore USA
tjb@skysrv.pha.jhu.edu

ABSTRACT

Lensing in the context of rich clusters is normally quantified from small image distortions, yielding a relative mass distribution in the limit of weak lensing. Here we show the magnification effect of lensing can also be mapped over a cluster, resulting in absolute mass determinations for the weak limit. Furthermore, given both magnification and distortion measurements, the mass distribution may be constrained in the strong regime. Methods for obtaining the magnification using spectroscopic and/or photometric information are discussed, for object detection within a fixed isophote or to a given flux limit. A map of the magnification around A1689 is constructed from the observed depletion of background red galaxy counts.

1. Introduction

A large scale measurement of cluster mass profiles provides information on the formation history of structure - a problem inexorably bound up with the thermal properties of the dark matter and the cosmological model. Traditionally cluster masses have been estimated from line of sight velocity dispersions and X-ray properties, indicating a density parameter of only $\Omega = 0.2$ if the M/L of cluster galaxies are representative. These methods, however, are limited in two respects. Firstly, they assume spherical symmetry and that the gas or galaxies are in hydrostatic equilibrium, yet most if not all clusters show substructure (White, Briel & Henry 1994) in accord with idea that clusters undergo continuous hierarchical merging. Secondly, at best these methods can only hope to constrain the mass interior the the radius inside which the galaxies or gas lie; they are blind to mass in an extended halo. Yet it is just such haloes that provide the most natural way to reconcile small scale measurements of Ω with the expectation of an overall $\Omega = 1$ on theoretical grounds.

A powerful new technique is to measure the gravitational distortion of background galaxies by the gravitational field of galaxy clusters. Rich clusters are found to form very strong lenses and now provide new detail on the shape of the dark matter profiles (Tyson et al 1990, Bonnet et al 1994, Fahlman et al 1994, Smail et al 1994). In principle cluster lensing also allows mass profiles to be measured out to several Mpc, if sufficiently accurate photometry is available, providing by far the cleanest way to measure the mass of clusters.

Distortions alone provide a relative mass distribution, being sensitive to gradients of the mass distribution. However, we show here how an absolute mass measurement can be obtained via the magnification of the background. We describe methods for detecting the magnification and apply one simple method to new photometry of A1689.

2. Gravitational Magnification

The importance of measuring the magnification follows from considering the effect of lensing on a small area of sky beyond the lens. An extended image is in general stretched asymetrically, increasing its size and changing its shape. Any anisotropy of lens mass distribution about a given image produces a change in its shape and a corresponding magnification. This effect or shear, γ, is non-local and can be considered separately from the local convergence, κ, which is a purely magnifying effect due to the focusing by matter local to the image. The convergence relates directly to the mass distribution and is therefore the quantity of interest, but is not directly observable, being responsible for only part of the overall magnification. However, because the observable quantities of magnification and ellipticity, depend differently on the convergence and shear it is possible to isolate κ, as follows.

An image is stretched by factors $(1-\kappa-\gamma)^{-1}$ and $(1-\kappa+\gamma)^{-1}$, along orthogonal axes defined by the local shear direction, generating an observed ellipticity, e (=b/a), of an initially circular source at angular coordinate θ:

$$e(\theta) = \frac{1-\kappa-\gamma}{1-\kappa+\gamma}. \tag{1}$$

The total magnification, $A(\theta)$, which is just the ratio of the lensed to unlensed area becomes,

$$A(\theta) = ((1-\kappa)^2 - \gamma^2)^{-1}. \tag{2}$$

(Young et al 1981), so that the convergence is related to these observables via,

$$\kappa(\theta) = 1 - \frac{1+e}{2\sqrt{Ae}}. \tag{3}$$

Strictly, $A(\theta)$, and therefore $\kappa(\theta)$ depend on redshift, asymptoting to a maximum for sources at infinity, so that the convergence relates geometrically to the surface density, $\Sigma(\theta)$, with knowledge of the lens and source distances so that,

$$\Sigma(\theta) = \kappa(\theta)\frac{c^2}{4\pi}\frac{D_s}{D_d D_{ls}}. \tag{4}$$

which simplifies when expressed in terms of the asymptotic value of the convergence κ_∞, as then the redshift dependence can be separated, becoming,

$$\Sigma(\theta) = \kappa_\infty 10^{14.44} h M_\odot Mpc^{-2} \frac{(1+z_L)^2}{\sqrt{(1+z_L)} - 1}. \tag{5}$$

(Broadhurst, Taylor and Peacock 1994). An issue we shall return to later is the dependence of the derived $\Sigma(\theta)$ on the n(z) of the background sources which because of the magnification is deeper for larger A, to a given flux limit, and if unaccounted for leads to an overestimate of $\Sigma(\theta)$.

The above is only useful for circular sources and in practice must be generalised to cover the more realistic case of images with random initial ellipticities which bias the estimate of κ. However, this exercise demonstrates the basic behaviour in terms of observables and allows us to explore limiting cases of interest.

2.1 Weak Limit

In the general weak limit the convergence can be determined from the magnification alone (eqn 3):

$$\kappa(z) = (A(z) - 1)/2 \qquad (\kappa << 1, \gamma << 1). \tag{6}$$

This is interesting as it shows how, in principle, information on the mass distribution can be obtained independently of the image distortions, hence improved S/N results from combining this information with the detections obtained by the 'classic' Kaiser & Squires (1992) technique for weak distortions.

Another exciting possibility is the extension of mass determinations into the strong regime, along the lines of eqn 3, since κ is a function of e and A. So far this has not been possible using only image distortions, as the derived surface density is subject to an unknown transform (Schneider & Seitz 1994, Kaiser 1994).

3. Measuring Magnification

Magnification as an observable effect has recently been discussed in the context of rich clusters, as an alternative way of measuring cluster masses (Broadhurst, Taylor and Peacock 1990, BTP). We describe here methods developed by BTP and other ideas, paying attention to the distinction between isophotal and total measures of galaxy magnitudes. This distinction is important as lensing preserves surface brightness, so that in principle if galaxies can be measured to a fixed isophote, then the magnification is just the change in area (of individual galaxies or in the surface density of images) relative to the unlensed sky. In practice defining such an isophote requires very high spatial resolution and this is not yet possible even with HST, given the small sizes of faint galaxies, as described below. However, with good estimates of total magnitudes the magnification can be obtained reliably by its effect on the background surface density of galaxies.

3.1 Areal Sizes

The ideal selection of images for measuring magnification would be purely isophotal because of surface brightness conservation. In principle then, the ratio of images areas measured within a given isophote in lensed sky and unlensed comparison sky is the magnification. The difficulty of making this simple test is obvious from examining deep HST images of random sky areas. These images show clearly the intrinsic distribution of image sizes and surface brightness is very broad, extending to very small sizes so that many objects are only partially resolved even with 0.1″sampling (see Kaiser,Squires and Broadhurst 1994, KSB hereafter). This has interesting implications for galaxy formation but a detrimental effect on the ability to set a common isophote for measuring the magnification and certainly means that ground based measures of small lens induced changes in shape are more strongly circularised by the seeing than suspected, (KSB).

3.2 Galaxy Counts

A more practical approach is to measure the effect of magnification from the change in surface density of galaxies on the sky. For the ideal case of galaxy selection to a fixed isophotal threshold, the ratio of field and cluster background counts is equal to the magnification. Again in practice we don't yet have such data and must consider the effects of seeing on intrinsically small faint galaxies which make up a good fraction of the background. In the case of ground based photometry it is best to attempt a total magnitude to some reliable limiting flux, which is approximately achieved using a large enough aperture on high S/N images. In this case the counts are depressed equal to the magnification but countered by objects brightened up above the flux limit, so that the relation between lensed an unlensed counts is:

$$N(m)' = N(m)_o A(z)^{2.5s-1}. \tag{7}$$

where $s = dlogN(m)/dm$ is the intrinsic slope of the number counts. The blue selected galaxy counts are no use for this test as s=0.4, as lensing leaves the counts unaltered (eqn 7). However the reddest galaxies do have a very flat slope $s \approx 0.15$ (see section 4), due to a combination of k-correction and evolution, so that a significant count *depletion* is expected from the areal expansion, A^{-1}, which dominates. In practice the galaxy selection will fall between purely isophotal and total magnitudes requiring the curve of growth distribution of the images to be accounted for in an accurate estimate of the magnification but at bright enough magnitudes this is a negligible uncertainty (KSB).

Of course, the counts are subject to fluctuations from the small scale projected clustering of background galaxies and pure shot noise, translating into a limiting 'resolution' for detection of a desired amplification in a given patch of sky behind the lens. For example, at B=26 the clustering fluctuations in the counts are equal to the poisson noise in 1′areas, where the mean count is 10^2 producing 20% fluctuations in surface density. Going fainter with HST or large ground based telescopes can reduce the shot noise considerably but a residual $\approx$ 5% clustering or minimum magnification of 15% is the best that can be hoped for on this scale. For radial measurement of the mass, clustering can be smoothed over on large angles since the counts increase at r^2, so that the S/N of the magnification measurement is roughly independent of radius (for an isothermal profile). This technique is applied below to photometry of A1689.

3.3 Fluctuation Independent Galaxy Counts via Redshifts

Considering the behaviour of the number counts as a function of redshift N(m,z) leads to a clustering independent measure of the amplification but at the cost of requiring redshift information for samples of a few hundred redshifts (BTP). The idea is to make use of the break away from the powerlaw of the luminosity function at the bright end, so that the shape of the distribution of N(m,z) is altered compared to a similarly selected control field sample, and *independent* of density fluctuations (BTP). The probability distribution is constructed:

$$P(m,z) = \frac{N(m + 2.5logA(z), z)}{\int N(m + 2.5logA(z), z)dm}. \tag{8}$$

which is normalised in each redshift bin to the mean field value to remove the uncertain density dependence. BTP have shown that the mass of a cluster can be constrained out to r $\approx$ 1Mpc for the massive clusters (average $\kappa_\infty = 0.05$ over this radius) with only ≈ 300 redshifts. The gross redshift distribution is altered by the magnification since at every redshift one sees further down the luminosity function for a flux limited sample so that the relationship between the lensed and unlensed N(z) becomes,

$$N(z)' = N(z)_o A^{\beta(z)-1}. \quad (9)$$

Where $\beta = -dlog\Phi(L_{lim}(z), z)/dlogL(z)$, the effective index of the luminosity function at the lower limiting luminosity L_{lim} corresponding to the flux limit. This index increases with redshift in a flux limited sample because at high redshift where the flux limit falls on the steep part of the luminosity function the observed number is increased relative to an unlensed sample, but at lower redshift, just behind the lens, N(z) is suppressed by $\approx A^{-1}$, since the flux limit falls on the flat part of the luminosity function. This effect leads to a stretching of the N(z) to higher redshift and hence a downward correction to the measure of κ based only on projected data (BTP).

In general this method can be thought of as an extension to the current distortions approach by the addition of redshift information for the faint images. In the near future we may expect redshifts for many thousands of faint background galaxies (e.g. the HET project now underway, Hill et al 1994) and hence we may realistically hope to establish the form of $A(z)$ over the surface of a cluster.

3.4 Standard Candles

Given internal velocity measurements, every galaxy is a rough standard candle via the L-ΔV relations for discs and early type galaxies. Since the velocities are unaffected by lensing, then at fixed ΔV any shift in L is just the magnification. Unfortunately, internal velocity measurements are not yet practical at cosmological distances and in any case the intrinsic dispersion in the relation limits the 'resolution' of the magnification measure. 10% precision in A requires $3\Delta L/\sqrt{n} \approx 0.1$ or a minimum of 100 accurate internal velocity measurements, so that in order to achieve 1′resolution in the magnification measurement, given the surface density of galaxies on the sky translates into 1A spectroscopy down to B=26, which is not going to be easy.

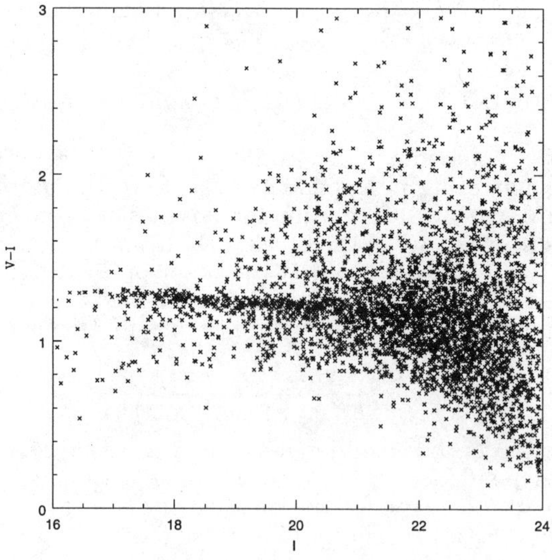

Fig. 1. Colour-Magnitude diagram for A1689 showing the cluster E/S0

sequence as a near horizontal line at V-I=1.3 in this plot. Many redder galaxies are evident and these are taken to be background to the cluster since higher redshifts are required to generate colours redder than the cluster E/S0's.

4 Number Counts Behind A1689

Good photometry for A1689 (z=0.18) now exists with which to examine both the magnification and the distortion field. The data covers 70 square arcmins on the cluster and an offset control field in seeing 0.7″, from the ESO NTT 3.6m telescope (Broadhurst et al 1994). The passbands and exposures were chosen such that the cluster E/SO galaxies would be bluer than a good fraction of the background, requiring much deeper imaging in the bluer passband for detection. For red objects we can be confident they are in the background since redshifts greater than the cluster are required to generate colours redder than an early type at the cluster redshift. The red counts have the further important advantage that they are relatively flat so that the magnification is dominated by the areal expansion (eqn 7).

The photometry was carried out by Nick Kaiser using a carefully optimised scheme (see KSB for details of procedure and calibration), obtaining near total magnitudes within 3 times the Petrosian radius and accurate image shapes. Figure 1 is the colour magnitude distribution, showing a well defined E/SO sequence for this cluster, and also the spread of redder background galaxies. The distortion map obtained from the KS technique is discussed by Kaiser et al (1994) showing a 9σ detection of the cluster and a smooth radial mass profile out to the limit of the data, making it the best detection of a cluster to date by the KS technique.

Here we explore the magnification via the number counts of the background galaxies. The far field red galaxy number counts are observed to have an intrinsically flat slope for $I > 20$, as predicted, $s = 0.15$, and is well fitted by extrapolating the local early type luminosity function to higher redshift (solid curve). Figure 2 shows the far field red counts and the no-evolution model (upper panel) scaled to the area of the field for V-I>1.5, which is 0.2 mag redder than the upper limit of the E/SO sequence. In contrast the blue counts, V-I<1, have the steep $s = 0.4$ slope which can be reproduced with density evolution at a rate $(1 + z)^3$, as shown in Figure 2 (lower panel). This relative evolution in V-I is well known from evolutionary studies and is most clearly demonstrated by the luminosity functions of Lilly et al (1994) in these passbands.

In measuring the surface density of background objects the coverage of the background by the cluster galaxies is determined. This is easy to quantify by tracing the radial distribution of areal coverage of all the galaxy images and finding the level at which it flattens off to the background level. The difference is then the coverage which is found to be important only in the central 30″of this cluster due to a large central concentration of cD's. In any case this central region is within the Einstein radius as defined by the prominant giant arc surrounding the center and therefore cleared of background images apart from counter images (see figure 3). The number counts in the central region, r<3′, are shown in Figure 2, for the red and blue galaxies and compared with isothermal model lensed predictions by modifying the count model (eqn 9) for total magnitudes (dotted histograms and curves respectively). Examination of the effects of seeing on this assumption are being investigated by Kaiser & Broadhurst from 2 colour HST data of the 'Medium Deep Survey'). For the red counts a strong effect is observed, well matched by an isothermal mass distribution of 1600km/s and showing the predicted depletion from the areal expansion of the sky by the lens.

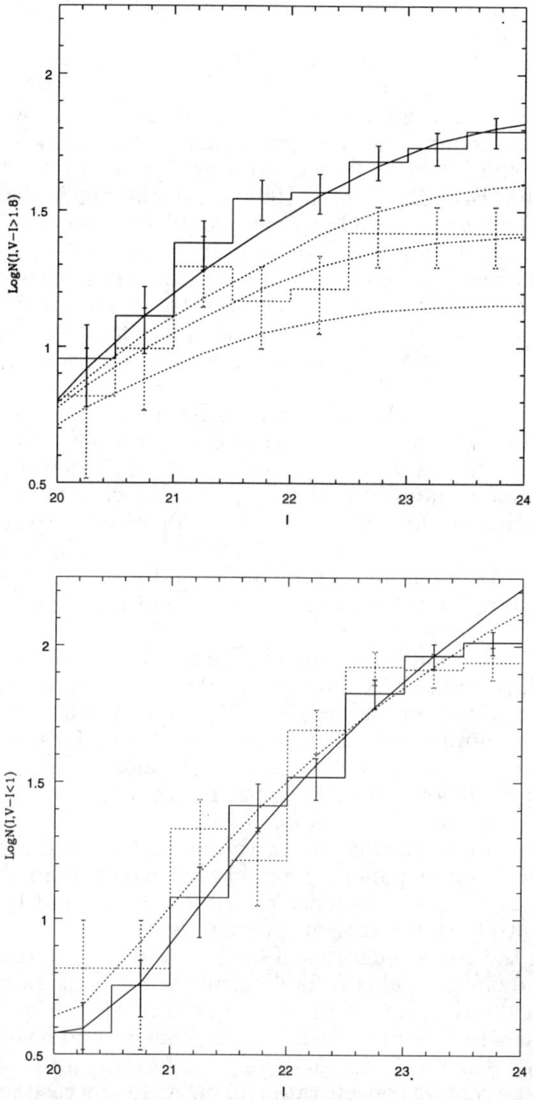

Fig. 2. Comparison of the far field number counts of red galaxies (upper panel solid histogram) and blue galaxies (lower panel solid histogram) with the number-count model predictions for the far field and evolution as found by Lilly et al (solid curves) in V and I. The red and blue counts within a radius of 3′are also shown (corrected for coverage as described in the text) as dotted histograms and compared with the lensed counts with the lensing modified count predictions for isothermal profiles with finite core radius as described in the text. It is clear that the red counts are very sensitive to lensing as expected given their intrinsically flat slope, matching best v_{11} = 1600km/s. This contrasts with the

insensitivity of the blue counts which have the invariant dlogN/dm=0.4 slope (lower panel).

For the blue counts no useful constraint is possible as expected, given the invariant s=0.4 count slope, and demonstrated by the small difference between the lensed and unlensed expectation shown in Figure 2 (lower panel), in this case the 1600km/s isothermal model is plotted - dotted curve. The radial depndence of the red count depletion is shown in Figure 3 versus the three isothermal models of figure 2a with a core radius of $0.3 \times \theta_{crit}$ (normalised to the far field). The observed profile is shallower than isothermal, with some evidence of a break in the profile at 4′, although further checks are required. The model profiles show the critical radius as a minimum in the surface density. The increase in density interior to this radius comes from the multiple images and is very sensitive to the core radius. Unfortunately the bright galaxies in the core effectly blot out this interesting region of the sky, in the optical at least. Figure 3 also shows agreement between the minima of the data and the v_{11}=1600km/s model at $\approx$ 50″. This is interesting as it predicts well the observed critical radius. Several giant arcs are seen around the center in our data, ranging in radius from 45-65″(Broadhurst et al 1995). Note, the dispersion in radius of these giant arcs is a combination of split images straddling the critical curve and also the range in redshift of the background sources. This agreement is an encouraging check on the consistency of the method.

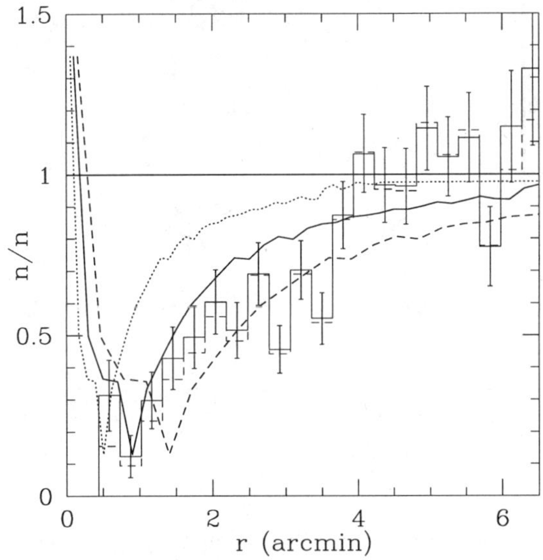

Fig. 3. Radial counts (unsmoothed) normalised by area and corrected upward for coverage by the cluster galaxies. Dashed histogram shows the counts without the correction for by the coverage mask, demonstrating the small relative coverage by the cluster. The curves are the expected form of the radial counts for the isothermal cases, with $v_{11} = 1200, 1600, 2000$km/s (dotted, solid and dashed curves respectively) showing the turn up interior to the critical radius from split images (core radius=$0.3\theta_{crit}$). The observed profile is shallower than isothermal and may show a break at 4′, requiring confirmation.

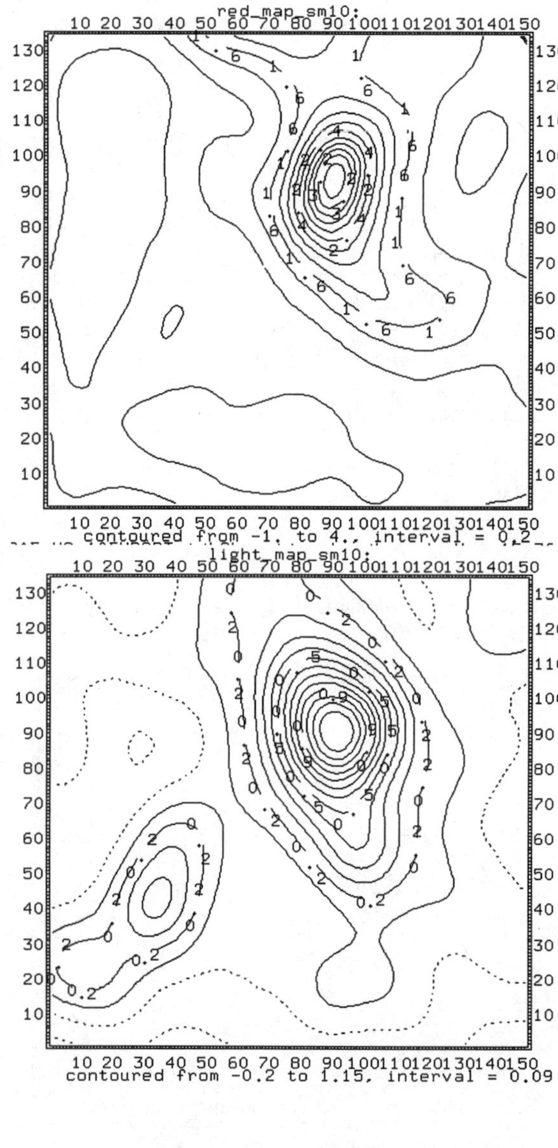

Fig. 4. Comparison of the amplification map over the full 8′of the data (upper panel) for comparison with the cluster light map (lower panel) at the same resolution, (30″gaussian smoothing). The agreement in position of the peaks of light and amplification is very good (Small light peak in lower left is due to the halos of 2 bright stars not fully dealt with as yet). Conversion of this to a mass map requires incorporating the image distortions via eqn 3 in the strong limit.

A simple direct estimate of the mass over the area covered by the data (0.8Mpc/h on a side) using the the simple weak limit relation derived above $\kappa = (A(z) - 1)/2$ applied to $A(z)$ measured via eqn 9, yields a mean surface density of $3 \approx 10^{15} h M_{\odot} Mpc^{-2}$. However, more work is required to obtain sensible constraints in the inner strongly lensed regions as the shear is strong here generating magnification in addition to that of the local convergence (eqn 3). requiring use of the ellipticity information contained in the image distortion map of this cluster.

Interesting detail is seen the map of the amplification distribution compared to that of the light distributions as shown in Figure 4. Both maps are at the same 30″resolution (scale of the maps is 8′on a side). Data on this cluster is also good enough to produce a detailed distortion map (Kaiser et al 1994), and hence we will be able to reconstruct fully the mass distribution over the surface of this cluster by combining measurements of ellipticity and magnification.

CONCLUSIONS

In conclusion, it would appear that the magnification effect by the cluster A1689 is clearly detected, depressing the surface density of the background red galaxies. The detection is strong enough to allow a crude map of the magnification. Further work is required to reconstruct the mass distribution by combining this information with the image distortion field of the cluster. At that point we can expect to define an accurate mass distribution and also a map of the mass-to-light ratio, in both the strong and weak limits.

ACKNOWLEDGEMENTS

Many thanks to Steph Côté, Alex Szalay, Ed Turner and especially Nick Kaiser for helpful discussions.

REFERENCES

Bonnet, H., & Mellier, Y., Fort B., 1994 ApJ Lett 427, L83
Broadhurst T., Taylor A., Peacock J., 1994 ApJ in press
Broadhurst T., Szalay A., Couch W., Moller P., Kaiser N., in prep
Kaiser N., Squires G., 1993 ApJ, 404,441
Kaiser N., ApJ Lett 1994 in Press
Kaiser N., Squires G., Broadhurst T., 1994 ApJ submitted
Kaiser N., Squires G., Fahlman G., Woods D., Broadhurst T., 1994 "Wide Field Spectroscopy of the Distant Universe" eds Haddox S. et al, in press
Fahlman G., Kaiser N., Squires G., Woods D., 1994 ApJ lett in press
Hill G. et al 1994 "Wide field Spectroscopy and the Distant Universe" eds Maddox S. et al, in press
Lilly S., Crampton D., Hammer F., LeFerve O., Tresse L., 1994 "Wide field Spectroscopy and the Distant Universe" eds Maddox et al, in press
Schneider, P., & Seitz, C., N. I., 1994, A&A, in Press
Smail, I., Ellis R., Fitchett, M, Edge, A., MNRAS 1994 in press
Tyson, T., Valdes F., & Wenk, R. 1990, ApJ, 349 L19
White, S.D.M., Briel, U.G., Henry, J.P., 1993, MNRAS 261, L8.
Young, P. 1981, ApJ, 244, 756.

Evidence From Gravitational Lensing For A Non-Thermal Pressure Support In The Cluster of Galaxies A2218

Abraham Loeb and Shude Mao

Harvard-Smithsonian Center for Astrophysics

60 Garden Street, Cambridge, MA 02138

ABSTRACT

The central mass distribution of clusters of galaxies can be inferred from gravitationally lensed arcs with known redshifts. For the Abell cluster 2218, this method yields a core mass which is larger by a factor of 2.5 ± 0.5 than the value deduced from X-ray observations, under the assumptions that the gas is supported by thermal pressure and that the cluster is spherical. We examine quantitatively various potential causes for this discrepancy, and show that a non-thermal pressure support is a plausible explanation. Such a pressure can be naturally provided by strong turbulence and equipartition magnetic fields ($\sim 50\mu$G) that are tangled on small scales in the cluster core. The turbulent and magnetic pressures do not affect the measured Sunyaev-Zel'dovich (SZ) effect for this cluster. Intracluster magnetic fields with a comparable magnitude ($\sim 10^{1-2}\mu$G) have already been detected by Faraday rotation in other clusters. If generic, a small-scale equipartition magnetic field should affect the structure of cooling flows and must be included in X-ray determinations of cluster masses.

[For a full account of this work, see: ApJL, 435, L109 (1994)]

GRAVITATIONAL LENSING OF VARIABILITY SELECTED QSOs BY GALAXY CLUSTERS

L. L. Rodrigues-Williams
Astronomy Dept, FM-20, Univ. of Washington, Seattle WA 98195, USA
llrw@astro.washington.edu

M. R. S. Hawkins
The Royal Observatory, Blackford Hill, Edinburgh EH9, UK
mrsh@starlink.roe.ac.uk

ABSTRACT

Distribution and properties of QSOs behind galaxy clusters in the UKJ287 field were studied. QSOs were selected using variability criteria and are confined to $z \geq 0.4$ and $m_B \leq 19.5$. No reddening or obscuration of background QSOs due to dust in clusters is detected.

A statistically significant positive angular correlation between clusters ($\langle z \rangle \simeq 0.15$) and QSOs ($0.4 \leq z \leq 2.2$) on scales of several arcminutes is found. This number overdensity of QSOs behind clusters is ascribed to gravitational lensing by foreground galaxy clusters. The overdensity of QSOs increases as their limiting magnitude gets brighter in a way that would be expected from a simple lensing model derived using Boyle, Shanks & Peterson (1988) QSO number counts and an average amplification of $A = 2$ due to clusters. The implied amplification due to clusters is substantially larger than would be expected from isothermal sphere clusters with a velocity dispersion $\sigma_v \sim 1000$ km/s.

1. QSO NUMBER COUNTS

Since UKJ287 QSOs were selected using variability criteria (see Hawkins & Veron 1993, HV for details) as opposed to a more conventional color criteria it will be instructive to compare the HV and BSP (Boyle, Shanks & Peterson, 1988) QSO number-magnitude counts. The latter were UVX selected.

The HV QSO number counts are shallower than BSP at the faint end, around $m \simeq 19$. Assuming magnitude calibration is the same in both samples QSO number density at $m = 19.5$ is 5.6 and $8.4/\Box^\circ$ for HV and BSP respectively. However using the Kolmogorov-Smirnov test we cannot rule out the hypothesis that HV and BSP QSOs were drawn the same distribution. For the total numbers of HV and BSP QSOs down to $m_{lim} = 19.5$—106 and about 97 QSOs respectively—the KS test confidence level is only 67%.

2. QSO-CLUSTER CORRELATION

Galaxy cluster positions were cross-correlated with those of background QSOs ($z \geq 0.4$). In order to take into account different angular sizes of clusters angular separations between QSOs and cluster centers were calculated in terms of individual cluster radii, then QSO-cluster separations were multiplied by the average cluster radius (0.06°) to yield degrees. Fig.1 shows two-point correlation functions for four different QSO flux cutoffs, m_{lim}=18.0, 18.5, 19.0, and 19.5. Each point in Fig.1 represents a separation of one average cluster radius.

Clusters and (at least the bright) QSOs are seen to be positively correlated. How significant are these correlations and can they be attributed to gravitational lensing? We will explore these two questions together. We make no attempt in this paper to investigate alternative explanations of foreground cluster/background QSO associations (but see Rodrigues-Williams & Hogan 1994).

A useful way of evaluating statistical significance of such observations is with binomial statistics (see Seitz & Schneider 1994). We calculated binomial probability of having the observed number of QSOs within 0.5, 1.0, 1.5, 2.0, and 2.5 cluster radii with different QSO flux limits. The most statistically significant associations occur for QSOs within 0.5 cluster radii and with m_{lim} = 18.0, 18.1, and 18.2 (99.3%, 99.3%, and 98.6% c.l. respectively), and within 2.0 cluster radii and m_{lim} = 18.8 (98.6% c.l.). While these probabilities may not be overwhelmingly convincing by themselves there is other evidence that QSO-cluster associations are not chance associations but are due to weak lensing by clusters: (a) the most statistically significant results occurs for the brightest QSOs found closest in projection to cluster centers; (b) QSO number excess behind clusters increases closer to cluster centers, see Fig.1; (c) QSO number excess increases with brighter QSO flux limit, see Fig.1 and 2.

Thus we conclude that there is a statistically significant association between QSOs and clusters in our sample and the association is due to gravitational lensing.

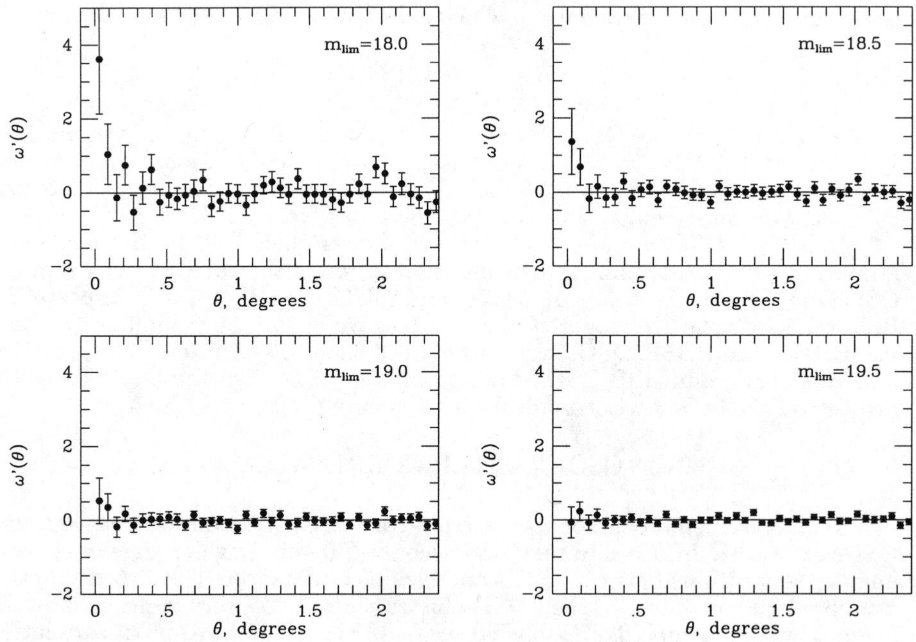

Fig. 1. QSO-cluster angular correlation functions for different QSO flux limits. QSO-cluster separations are calculated in terms of individual cluster radii. Each point (bin) in the figure represents one cluster radius.

3. GRAVITATIONAL LENSING

With our limited number of QSOs we cannot afford to investigate QSO overdensity as a function of distance form cluster centers. Instead, let us "fix" cluster radii at twice their cataloged values, and thus divide QSOs into "association" QSOs, *i.e.* those found behind extended galaxy clusters, and "field", *i.e.* the rest of the QSOs. With this division association area is 17.4% of the total UKJ287 field area. Let us define overdensity, q, as the ratio of the actual number of association QSOs to the number that would have been found had the field QSO number density applied to the area behind clusters.

Filled dots in Fig.2 are the observed overdensities of QSOs behind extended clusters. Error bars represent r.m.s. dispersion between clusters. We will now compare these observations with what might be expected from a simple model. Assume the clusters are uniform circular disks with constant amplification, A, at an average redshift of 0.15, while QSOs are all located at large redshifts and obey BSP magnitude-number counts. This model is represented by three solid lines in Fig.2a, each for a different value of A: 1.14, 2.0, and 3.0. A=2 model seems to fit the observations reasonably well.

The dashed lines in Fig.2a were obtained using the modified BSP counts: dashed lines to the left of the solid lines are the same as BSP model, *except* the break in the counts has been moved brightward by 0.5 mag. to m_{break}=18.65. Similarly, the dashed lines on the right have m_{break}=19.65, or 0.5 mag. fainter that in the BSP counts. The q vs. B_{lim} plot is sensitive to the m_{break} location, in particular, modified BSP counts with m_{break}=19.65 are not consistent with the observations regardless of the value of A.

Solid lines in Fig.2b use the number counts fit to the observed HV counts. Again, three amplifications are represented by three solid lines: A=1.14, 2.0, and 6.0. A=6 clusters could reproduce the observations, however as we will see below such large amplifications are not characteristic of galaxy clusters.

What kind of clusters would produce observed overdensities? A back-of-an-envelope calculation shows that most clusters should not be able to generate the observed QSO overdensities. With $\langle z_{cluster}\rangle \simeq 0.15$ and $\langle \theta_{cluster}\rangle \simeq 0.12°$ average extended cluster radius is $R_{cluster} \simeq 0.75h^{-1}Mpc$. $A = 2$ implies average surface mass density of clusters, $\sigma \simeq 0.3$ in terms of critical surface mass density for lensing. The latter is about 1 gm/cm^2 for lenses at $\langle z_{cluster}\rangle \simeq 0.15$ and sources at $\langle z_{QSO}\rangle \simeq 1.5$. Assuming a singular isothermal sphere model for clusters, their velocity dispersion is calculated to be $\sigma_v \simeq 2130km/s$. This value is too high compared to observations of cluster velocity dispersions, hence clusters appear to be producing larger overdensities than expected.

What amplification would reasonable clusters be expected to produce? Using the same model as above, clusters with $\sigma_v \sim 1000km/s$ will have A=1.14. The overdensity curves for this value of A are shown in Fig.2a,b. In both cases these are not consistent with the observations.

4. DUST IN THE UKJ287 CLUSTERS

Dust in clusters will affect background QSOs in two ways: (1) obscuration will diminish the number density of QSO behind clusters, and (2) QSOs behind clusters would be reddened. In a sample of variability selected QSOs one would not expect to "lose" QSOs to reddening. Since $A_B \approx 4 \times E(U-B)$, obscuration is expected to be the more important of the two dust effects.

(1) Obscuration: An overdensity of QSO behind clusters is observed. Underdensity is not observed at any QSO limiting magnitudes (see Fig.1), hence there is very little, if any obscuration due to dust in clusters.

(2) Reddening: All QSOs with $z \geq 0.4$ were divided into two samples: those behind extended clusters and the rest. The $(U - B)$ color distribution of association and field QSOs is not consistent with dust reddening by clusters. In fact, the average color of association QSOs is somewhat bluer than field QSOs.

We conclude that there is little or no smoothly distributed dust in UKJ287 galaxy clusters.

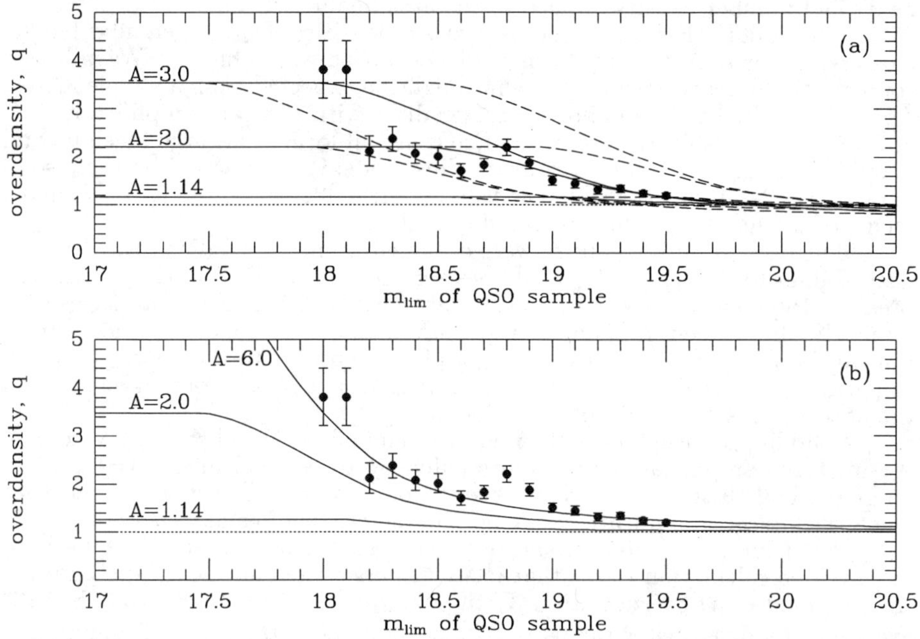

Fig. 2. QSO number overdensity behind extended clusters. See text for details.

5. CONCLUSIONS

Statistically significant association between high redshift variability selected QSOs and foreground galaxy clusters were detected. The amplitude of associations displays trends expected if weak gravitational lensing by clusters were responsible: (a) most statistically significant associations are found for brightest QSOs found very close in projection to clusters centers; (b) QSO number excess behind clusters increases closer to cluster centers (Fig.1); and (c) QSO number excess increases with brighter QSO flux limit (Fig.1,2).

REFERENCES

Boyle, B. J., Shanks, T., & Peterson, B. A. 1988, MNRAS, 235, 935, BSP
Hawkins, M. R. S., & Veron, P. 1993, MNRAS 260, 202, HV
Rodrigues-Williams, L. L., & Hogan, C. J. 1994 AJ, 107, 451
Seitz, S., & Schneider, P. 1994, Preprint

Measuring the Bias Factor With the Weak Lensing Effect

Matthias Bartelmann[1,2]
[1] Harvard-Smithsonian CfA, 60 Garden St., Cambridge, MA 02138
[2] MPI für Astrophysik, Postfach 1523, D–85740 Garching, FRG

Abstract

The bias parameter b of galaxy formation relates the relative fluctuations in the galaxy number density to the relative fluctuations in the dark-matter density distribution. It lies at the heart of all interpretations of large-scale structure models, because they are compared to observations in terms of the spatial distribution of luminous matter they predict, and this can be inferred from the distribution of dark matter only by the assumption of some galaxy formation prescription, of which the biasing hypothesis is one example (Kaiser 1984, Dekel & Rees 1987). I propose here a method based on weak lensing which allows a measurement of the bias factor. This method requires galaxy counts down to $21^{\rm st}$ to $22^{\rm nd}$ magnitude, preferably in the red or near infrared, in fields of $20' \ldots 30'$ radius centered on $50 \ldots 100$ high-redshift, bright QSOs. The result, a cross-correlation function between QSOs and galaxies, will be sensitive to the shape of the dark-matter perturbation spectrum on scales between 1 and 10 Mpc/h, where $100\,h$ km/s/Mpc is the Hubble constant. The dependence of the so-derived b on the density parameter Ω_0 is different from the dependence $\propto \Omega_0^{0.6}$, hence the measurement proposed here can break the Ω_0-b degeneracy in dynamical measurements of b.

Introduction

Weak lensing by large-scale structures in the Universe can cause angular correlations between background (high-z) QSOs and foreground (low-z) galaxies on angular scales of several $10'$ (Bartelmann & Schneider 1993a). The dark matter distribution in the Universe is inhomogeneous to some model-dependent degree. Such inhomogeneities act as lenses, because they differentially deflect light and thereby magnify or demagnify background sources relative to the case of homogeneously distributed matter. In a flux-limited source sample, such sources will preferentially be included which are magnified, provided the number counts of the background sources are sufficiently steep, because the magnification can "lift" sources above the flux threshold which would otherwise be too faint. This effect is called magnification bias (*e.g.*, Schneider, Ehlers & Falco 1992, Narayan & Wallington 1993). Therefore, the sources of a flux-limited sample will lie preferentially in the background of matter overdensities. By the biasing hypothesis of galaxy formation, matter overdensities coincide with "light" overdensities, *i.e.*, galaxies or other forms of light sources are suspected to be overdense where the underlying matter distribution is. Taken together, the magnification bias due to lensing and the biasing hypothesis of galaxy formation predict a lensing-induced cross correlation between a flux-limited sample of background sources and foreground light sources. Bartelmann & Schneider (1993a) have shown that the dark-matter overdensities predicted by current models of structure formation are sufficient to produce such correlations on angular scales of $\lesssim 30'$, provided the number counts of the background sources are a steep function of flux.

Such correlations have been detected on a very high significance level ($\lesssim$

99.8%) between the 1-Jansky sample of radio-loud, optically identified QSOs (Kühr *et al.* 1981, Stickel & Fried 1989, Stickel, Kühr & Fried 1993, Stickel & Kühr 1993a,b) and optical galaxies (Fugmann 1990, Bartelmann & Schneider 1993b), infrared galaxies (Bartelmann & Schneider 1994), and diffuse X-ray emission (Bartelmann, Schneider & Hasinger 1994). The correlations detected are most easily interpreted in terms of gravitational lensing by structures on scales between 1 and 10 Mpc/h.

Angular cross-correlation function between QSOs and galaxies

The statistical test employed in the previously cited papers (Spearman's rank-order correlation test, *e.g.* Kendall & Stuart 1973) was designed to detect the expected weak correlations with a high significance level. Its result is a single number, the correlation coefficient, which indicates the significance of correlations. Once the existence of these large-scale correlations is established, methods can be applied which allow to extract more detailed information. An example for such a measure is the cross-correlation function.

The angular two-point cross-correlation function between QSOs and galaxies, $\xi_{\rm QG}(\phi)$, can be analytically derived in the limit of weak lensing (Bartelmann 1994). There, it was shown that $\xi_{\rm QG}(\phi)$ is related to the correlation function between magnification μ and density contrast δ, $\xi_{\mu\delta}(\phi)$, by

$$\xi_{\rm QG}(\phi) = (-a-1)\, b\, \xi_{\mu\delta}(\phi)\;, \tag{1}$$

where a is the slope of the QSO number counts $n_{\rm Q}$ effective for the magnification bias,

$$-a-1 = \frac{d\ln n_{\rm Q}}{d\ln\mu}\;. \tag{2}$$

The cross-correlation function $\xi_{\mu\delta}(\phi)$ gives the excess probability above random to have a density contrast δ along a line-of-sight which encloses an angle ϕ with a line-of-sight along which the magnification is μ. $\xi_{\mu\delta}(\phi)$ is worked out in Bartelmann (1994).

For model spectra resembling the CDM and the HDM power spectra, series expansions of $\xi_{\mu\delta}(\phi)$ in ϕ can be analytically derived. They read

$$\begin{aligned} \xi_{\mu\delta}^{\rm CDM}(\phi) &= C_0 - C_1\phi + C_2\phi^2 + \mathcal{O}(\phi^3)\;, \\ \xi_{\mu\delta}^{\rm HDM}(\phi) &= D_0 - D_2\phi^2 + \mathcal{O}(\phi^4)\;, \end{aligned} \tag{3}$$

where the coefficients C_i and D_j are determined by the scale length in the power spectrum, its amplitude, and weighted integrals over the redshift distribution of galaxies and QSOs. Eq. (3) describes the difference in shape of $\xi_{\mu\delta}(\phi)$ for CDM and HDM power spectra. Figure 1 illustrates $\xi_{\mu\delta}(\phi)$ for CDM and HDM. For the figure, the QSOs were assumed to be at redshifts $z_{\rm Q} \geq 1.5$, and the different curves in each panel were obtained for different limiting redshifts of the galaxy distribution, as indicated.

Numerical tests

Given the analytic cross-correlation function between QSOs and galaxies, numerical tests are interesting to see whether a measurement of $\xi_{\rm QG}(\phi)$ appears

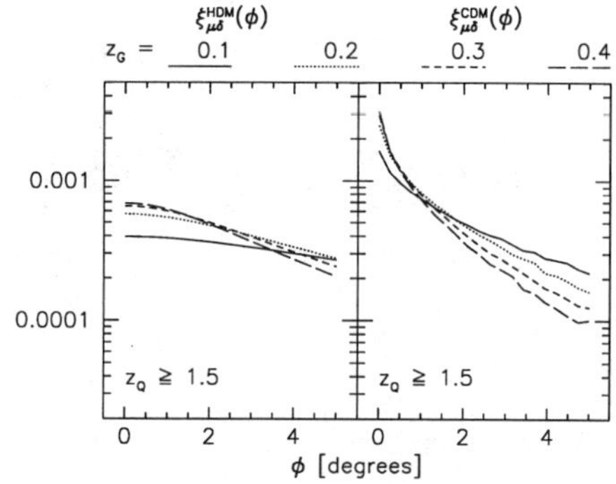

Fig. 1 – Cross-correlation function $\xi_{\mu\delta}(\phi)$ for a HDM (left panel) and a CDM (right panel) power spectrum of density fluctuations. The background QSOs were assumed at redshifts $z_Q \geq 1.5$ for the figure, and the limiting redshift of the galaxy distribution z_G was chosen as indicated. Note that ϕ is given in degrees.

feasible. Fig. 2 compares the QSO-galaxy correlation function derived from a numerical large-scale structure simulation with the analytic form of this function. For a detailed description of the large-scale structure model, see Bartelmann & Schneider (1992).

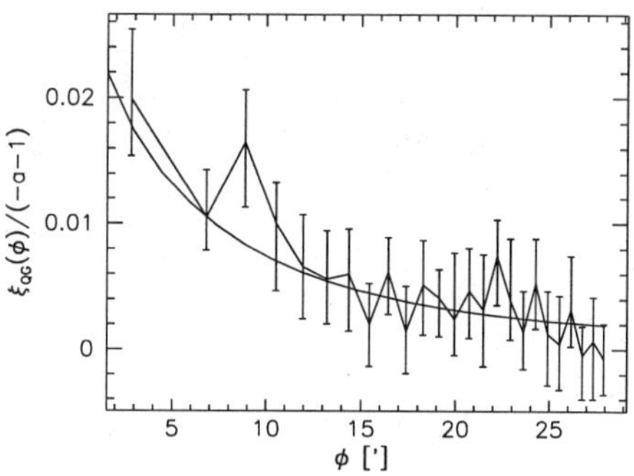

Fig. 2 – Comparison of the QSO-galaxy correlation function derived from numerical large-scale structure simulations (curve with error bars) and the analytic form of this function, adapted to the parameters of the simulation (smooth curve). The analytic curve fits the numerically determined function very well.

To estimate statistical errors, a large number of synthetic QSO samples, each containing 100 sources, was correlated with the foreground galaxies in the model. The error bars in the figure give 80% bootstrap errors. Galaxies down to a limiting magnitude of 22 were included, the bias factor was adapted to the (known) value of the model, and the number-count slope was taken to be $a = -5$. Fig. 2 shows that, within the error bars, the analytic function fits the numerically determined function very well.

The QSO sample should contain at least about 50 QSOs to ensure the QSO-galaxy correlation function can be distinguished from zero. This can be estimated from Fig. 3, which shows $\xi_{QG}(\phi)$ similar to Fig. 2, but for decreasing QSO sample size, as indicated.

It can be estimated from a similar numerical experiment that the galaxies in the QSO fields should be observed down to a limiting magnitude of 21 to 22. Shallower galaxy observations are only sensitive to lensing matter overdensities at lower redshifts, which produce a weaker result and therefore increase statistical uncertainties. To suppress the influence of reddening, galaxy observations

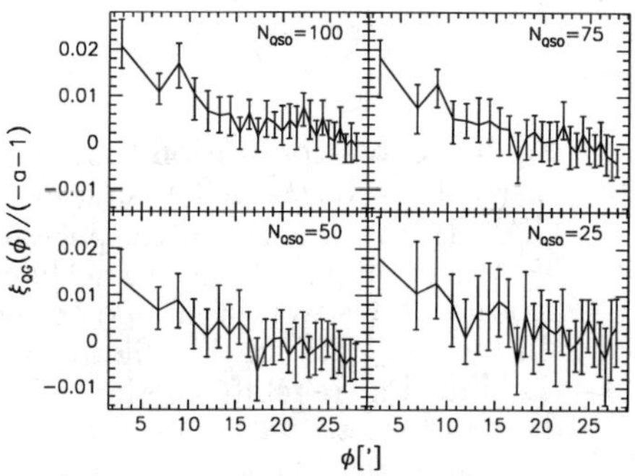

Fig. 3 – Numerically determined QSO-galaxy cross-correlation function similar to the one shown in Fig. 2, but for decreasing QSO sample sizes, as indicated. The figure shows that the sample should contain $\gtrsim 50$ QSOs in order to sufficiently reduce the statistical uncertainty of the result.

in the red or near infrared are favorable.

Summary

To summarize, the QSO-galaxy angular two-point cross-correlation function induced by weak lensing due to large-scale structures provides information about the spectrum of density perturbations in the Universe and on the value of the bias parameter of galaxy formation. It can be measured from sufficiently deep galaxy counts in fields of $20'$ to $30'$ radius around 50 to 100 high-redshift QSOs. The sensitivity of the result to the density parameter Ω_0 is different than the $\Omega_0^{0.6}$-dependence of dynamical measurements of the bias factor, so that the measurement proposed here can possibly break the Ω_0-b degeneracy. A combination of both types of measurement on the same galaxy population can give access to both b and Ω_0.

References

Bartelmann, M., Schneider, P., 1992, A&A, 259, 413
Bartelmann, M., Schneider, P., 1993a, A&A, 268, 1
Bartelmann, M., Schneider, P., 1993b, A&A, 271, 421
Bartelmann, M., Schneider, P., 1994, A&A, 284, 1
Bartelmann, M., Schneider, P., Hasinger, G., 1994, A&A, 290, 399
Dekel, A., Rees, M.J., 1987, Nature, 326, 455
Fugmann, W., 1990, A&A, 240, 11
Kaiser, N., 1984, ApJ, 284, L9
Kendall, M.G., Stuart, A., 1973, The Advanced Theory of Statistics, Vol.2 (London: C. Griffin & Co.)
Kühr, H., Witzel, A., Pauliny-Toth, I.I.K, Nauber, U., 1981, A&AS, 45, 367
Narayan, R., Wallington, S., 1993, in: Gravitational Lenses in the Universe, Proc. 31st Liège Astroph. Coll., eds. J. Surdej *et al.*
Schneider, P., Ehlers, J., Falco, E.E., 1992, Gravitational Lenses (Heidelberg: Springer)
Stickel, M., Fried, J.W., Kühr, H., 1989, A&AS, 80, 103
Stickel, M., Kühr, H., Fried, J.W., 1993, A&AS, 97, 483
Stickel, M., Kühr, H., 1993a, A&AS, 100, 395
Stickel, M., Kühr, H., 1993b, A&AS, 101, 521

DETECTION OF LENS CANDIDATES FOR THE DOUBLE QSO Q2345+007

Philippe Fischer and J. Anthony Tyson
AT&T Bell Laboratories, 600 Mountain Ave., Murray Hill, NJ 07974

Gary M. Bernstein
Steward Observatory, University of Arizona, Tucson, AZ 85721

Puragra Guhathakurta
Space Telescope Science Institute, 3700 San Martin Dr., Baltimore, MD 21218

ABSTRACT

Luminous objects associated with the 7.06″ separation double quasar Q2345+007 have, until now, escaped detection. In this letter we present the results of the deepest known imaging of the region surrounding the quasar. The total exposure times were 47600 seconds in B_j (101 frames) and 32160 seconds in R (82 frames). The frames came from CFHT, CTIO, and KPNO. We detect a $B_j = 25.0$ mag galaxy ($B_j - R = 0.5$ mag) in close proximity to the fainter QSO image, possibly at $z = 1.49$ given by several absorption features in the QSO spectra. Furthermore, there is a 2.9σ enhancement in the number density of faint galaxies ($24 \leq B_j \leq 28$, $B_j - R \leq 1.5$) near the quasar and another 3.1σ enhancement further away. These discoveries support the hypothesis that Q2345+007 is being lensed by one or more distant mass concentrations and may imply that compact "clusters" already exist at early epochs. We discuss several lens models for the system and the cosmological implications.

1. INTRODUCTION

The evolution of cluster-size concentrations and their expected numbers and peak densities at moderate redshifts are related to the nature of the dark matter and Ω. The existence of compact mass concentrations of $10^{13} - 10^{14}$ $M_\odot$ at large redshift would have significant implications for cosmology and may favor a low value of Ω (Cen, Ostriker & Peebles, 1993).

In this context, Q2345+007AB has attracted considerable attention. More than a decade after its first discovery (Weedman *et al.* 1982) there is still controversy regarding its true nature. Is this a lensed system or simply two quasars with remarkably similar spectra and redshifts (z = 2.15, $\Delta V = 15 \pm 20$ km s^{-1}; Steidel & Sargent 1991)? Despite having the largest separation of any known double quasar, no good lens candidate has been found in deep broadband images (Tyson *et al.* 1986).

Both the QSO images have metal absorption lines with redshift z = 1.491 (Foltz et al 1984, Steidel & Sargent 1991) and the B image has a second C IV doublet ($z = 1.483$) with $\Delta V = 1060 \pm 60$ km s^{-1}. Is this evidence for lensing by a large cluster at $z = 1.49$? Absorption lines at other redshifts in the range $z = 0.75 - 1.98$ are also seen although the largest number (15) are from the two systems at $z = 1.483$ and $z = 1.489$. Further evidence comes from Bonnet et al. (1994) who claim the detection of a possible lens from weak lensing arclet candidates near the QSO.

2. IS Q2345+007 A LENSED QUASAR?

We have constructed very deep B_j and R images from the nine different epochs [47600 s in B_j (101 frames) and 32160 s in R (82 frames)]. Since the seeing and signal-to-noise of the individual frames varied we used an optimal weighting algorithm to combine them (Fischer & Kochanski 1994). The RMS scatter in the sky backgrounds are approximately 31.2 mag arcsecond^{-2} (B_j) and 29.9 mag arcsecond^{-2} (R).

Since there are strong metal absorption lines in the QSO spectra, it is likely that there are galaxies projected close to the QSO images. Ideally, one would like to model the stellar point spread function (PSF) and subtract the QSO images. The complete overlap region of all the frames is approximately $80'' \times 80''$ and contains no bright stars; it is impossible to obtain a useful PSF independent of the QSO. We, instead, subtracted a scaled version of A from B, and B from A. The scaling and centering were determined from iterative weighted least-squares fitting. The A/B flux ratio is approximately 3.4 in B_j and 3.1 in R. Lensing should be achromatic; differences can arise due to: variability of the QSO and the net timing delay between A and B, or differential reddening. Alternatively, this may be evidence against the lensing hypothesis. This issue could be settled with long-term photometric monitoring. The mean brightnesses for the 9 individual epochs of image A are $B_j = 19.33 \pm 0.08$ mag, $R = 18.78 \pm 0.05$ mag, while image B has $B_j = 20.64 \pm 0.06$ mag and $R = 19.96 \pm 0.03$ mag.

In both the subtracted B_j and R images there is a source (G1) located 1.0 - 1.2$''$ north and 0.8$''$ east of image B (0.4 - 0.6$''$ northwest of the line joining A to B) (Fischer et al 1994). This detection is highly significant; the peak intensity of G1 is greater than one third the subtracted B_j flux at that position, and one sixth the R flux. In both cases, the peaks are greater than twelve sky standard deviations. The centers agree at the 0.2$''$ level between the B_j and R frames. The brightness of G1 is $B_j = 25.0 \pm 0.5$ and $R = 24.5 \pm 0.5$. These convert to: B(Johnson) = 25.0, V(Johnson) = 24.6, R_c(Cousins) = 24.4 (Gullixson *et al.* 1994).

We have run FOCAS on QSO-subtracted versions of the B_j and R images. There is a density enhancement located 17$''$ north and 0.7$''$ west of G1 (Fischer et al 1994). There are two other large maxima located further to the west. The detection of the large density maximum near the QSO images is very robust to changes in how the galaxies were culled from the catalogs (e.g. relatively independent of color and brightness cut-offs), however, the position is somewhat less robust; the uncertainties are several arcseconds. If we are missing faint galaxies projected near the QSO images, this would tend to bias the center away from the quasar. The detection nearest the QSO is significant at around the 2.9σ level; a $40'' \times 40''$ box centered on the density enhancement contains 49 galaxies while 32.4 are expected. The enhancement just to the west is significant at the 3.1σ level (50 galaxies). These density enhancements may be one or more clusters at large redshift.

4. CONCLUSIONS

In this paper we have endeavored to find lens candidates for the double QSO 2345+007. We constructed R and B_j images from exposures taken at CFHT, CTIO, KPNO; the total exposure times were 47600s in B_j (101 frames) and 32160 in R (82 frames).

The conclusions are: 1) there is a faint galaxy (B_j = 25.0 mag, B_j – R = 0.5 mag) located 1.5″ from the B image. 2) There is a significant excess of galaxies with $24 \leq B_j \leq 28$ and $B_j - R \leq 1.5$ near the QSO, evidence for clustering at large redshift. 3) There is no evidence for subcomponents at the 0.1″ level, contradicting two previous claims. 4) It is possible to reproduce the image positions and flux ratios using simple lens models although the problem is currently underconstrained. Multiple lenses with a foreground mass screen are favored (see Fischer et al. 1994). 5) The evidence we have uncovered in this letter supports the lensing hypothesis. To conclusively decide the issue will require long-term photometric monitoring of the QSO images to search for a timing delay.

Gravitational lensing directly measures M($\leq$R)/R. If the lensing is occurring at z = 1.5, then a mass of at least M(R$\leq$ 20kpc)$\sim 10^{13}$ $M_\odot$ is required. There is evidence for a large lens at $z = 1.49$: the previously detected absorption line systems separated by 1000 km s^{-1} and the overdensity of faint galaxies reported in this letter. If future redshift determinations confirm the cluster at $z = 1.49$, then our result has direct cosmological implications. The existence of massive compact structures as early as z=1.5 is less likely in cold, hot or mixed dark matter models (Ω = 1) than in an open (Ω = 0.2) primordial baryonic isocurvature model (see Cen, Ostriker & Peebles 1993, Fig. 13) which can form structure earlier.

5. FUTURE WORK

Based on thirteen epochs of imaging data, it appears that bot QSO images are varying (see Figure 1). Unfortunately the data are too sparse to measure a time-delay. We have begun monthly queue observing of the QSO at KPNO with the goal of detecting a time-delay for the system. Detection of a time-delay would be definitive proof of the lensing hypothesis.

REFERENCES

Bonnet, H., Fort, B. Kneib, J. P., Mellier, Y., & Soucail, G. 1994, Preprint.
Gullixson, C. A., Boeshaar, P. C., Tyson, J. A., & Seitzer, P. 1994, submitted to ApJ Supp.
Cen, R., Ostriker, J. P. & Peebles, P. J. E. 1993, ApJ, 415, 423
Fischer, P., & Kochanski, G. P. 1994, AJ, 107, 802.
Fischer, P., Tyson, J. A., Bernstein, R., Guhathakurta, P. 1994, ApJL, 431, L71.
Foltz, C. B., Weymann, R. J., Roser, H. J., & Chaffee, F. H. 1984, ApJ, 281, L1.
Steidel, C. C. & Sargent, W. L. W. 1991, AJ, 102, 1610.
Tyson, J. A., Seitzer, P., Weymann, R. J., & Foltz, C. 1986, AJ, 91, 1284.
Weedman, D. W., Weymann, R. J., Green, R. F., & Heckman, T. M. 1982, ApJ, 255, L5

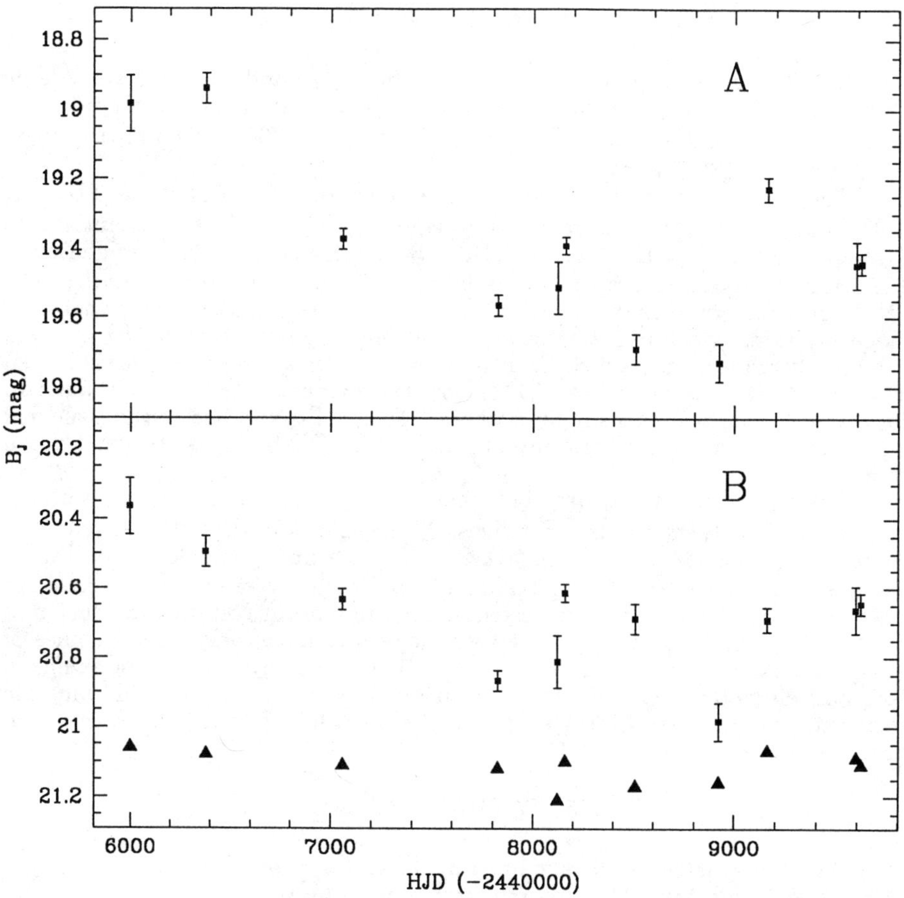

Fig. 1 – Variability of the QSO images in the J band; A is on top and B on bottom. The triangles indicate the brightness of a star in the field. The χ^2 is 278 for A and 117 for B for 10 degrees of freedom. Therefore, the variability is highly significant. Similar variability is seen in the R band. Measurement of a time-delay is neccesary to definitively verify the lensing hypothesis.

GRAVITATIONAL LENSING OF QUASAR 0957+561 AND THE DETERMINATION OF H_0 *

George Rhee[1], Gary Bernstein[2], Tony Tyson[3] and Phil Fischer[3]
[1] Department of Physics University of Nevada Las Vegas,
[2] Department of Astronomy, University of Michigan,
[3] AT & T Bell Laboratories

ABSTRACT

We present results of a deep imaging study of the cluster lensing the double quasar Q0957+561. Using data obtained at KPNO and CFHT we have detected distortions in the lensed blue background galaxy population. From these arclet distortions we have constructed a map of the mass density in the field of the double QSO. We use this map to provide constraints on the mass concentrations responsible for the lensing. Using these constraints in conjunction with mass models and the known time delay between the A and B images we place limits on the value of the Hubble constant.

1. INTRODUCTION

The double quasar 0957+561 was the first discovered instance of multiple imaging via gravitational lensing and has been observed in the radio, optical, infrared and X-ray. It is the only gravitational lens system for which the magnification matrix and the time delay between the two images are well constrained. This is at present the only system for which one can follow the suggestion of Refsdal (1966) to derive the Hubble constant from the time delay between multiple images. Unfortunately the lens geometry cannot be explained using only the galaxy G1 found between the quasar images. The galaxy cluster surrounding G1 is an important deflector as well and this complication has so far precluded construction of a unique, satisfactory model of the lens, reducing the accuracy of the derived Hubble constant. To address this problem have obtained deep images of the system at CFHT. The cluster is sufficiently massive to cause distortions on distant background galaxy images. We have used a mass map derived from lensing distortions to improve the accuracy of the cluster center location and place new limits on H_0.

2. THE TIME DELAY

To derive the Hubble constant using the double quasar one must measure the difference in the light travel time along the paths from the source to the observer. This has been done in the radio and optical by monitoring the two quasar images. Vanderriest *et al.*(1989) obtain a value of 415 ± 20 days based on optical data. Schild (1991) obtains a value of 404 ± 10 days also based on optical monitoring. Press *et al.*(1992) using a global χ^2 statistic obtain a value 536 ± 12 days using

* Based on observations collected at the Canada-France-Hawaii Telescope at Mauna Kea, Hawaii, USA

the same optical data. They emphasize that smooth interpolation followed by cross correlation is an unreliable technique. From radio monitoring studies Lehar *et al.*(1991) conclude that the time delay is 513 ± 40 days. We conclude that the true value of the time delay lies between 415 and 540 days.

3. THE LENS PROPERTIES

The galaxy principally responsible for the lensing (G1) lies between the two quasars about 1″ north of the B image. It has a redshift of 0.36. Bernstein *et al.* (1993) have published photometry of this galaxy. They find that the surface brightness profile is well fit by a power law with index $n = -1.94 \pm 0.06$. The profile is also well fit by a de Vaucouleurs law with $R_e = 4.5 \pm 0.65''$. The central surface brightness is $I_e = 23.7 \pm 0.1$ R mag arcsec $^{-2}$. The position angle of the galaxy may twist slightly about a mean value of 55°. The I_e and R_e values quoted above are typical for brightest cluster member ellipticals or cD's. Applying the Faber-Jackson relation to these data one infers a velocity dispersion for G1 of ~ 400 km s^{-1} . From spectroscopic observations Rhee (1991) derives a velocity dispersion 300 ± 50 km s^{-1} for G1. This is somewhat less than predicted from the Faber Jackson relation but it is known that brightest cluster ellipticals tend to have lower velocity dispersions than the Faber-Jackson law would suggest.

The cluster in which the galaxy G1 is found has a velocity dispersion of $500 \sim 700$ km s^{-1} . From the galaxy distribution the cluster appears to be located 20-30″ west of the B image, it has an estimated core radius of 30″ and ellipticity of 0.55 .

4. MODELS OF THE MASS DISTRIBUTION

Based on the observations described above we have derived models for the galaxy and cluster lensing the double quasar. We model the galaxy mass to have the profile $\Sigma_0(r) = 0.033(r/3)^n$ in the range $0.3'' \leq r \leq 30''$, the power law index is allowed the following range $-1.9 \leq n \leq -0.5$. There are 4 free parameters for the galaxy model: the velocity dispersion, the ellipticity, the galaxy position angle and the power law index. The cluster is modeled to first order in the ellipticity as an expansion of an elliptical pseudo-isothermal sphere.

$$\phi_c = b_c(s_c^2 + r^2)^{\frac{1}{2}} + \frac{\gamma}{3} b_c r^2 (s_c^2 + r^2)^{-\frac{1}{2}} \cos 2(\theta - \theta_\gamma)$$

The 6 free parameters are: the cluster velocity dispersion, the cluster core radius, the cluster ellipticity, the cluster position angle the distance between the cluster center and the galaxy center and the position angle of the line joining the cluster center to the galaxy center. The cluster maps show that the cluster potential is not perfectly smooth. This can be taken into account by adding a shear component to the two components listed above. The shear component has two parameters the magnitude of the shear and its position angle. The effect of the shear may be taken as some indication of the changes in the models that would result from the addition of lumps to the mass distribution.

5. THE OBSERVATIONS

The observations were made with FOCAM and the SAIC CCD at CFHT in January 1994. We obtained a 3 hour integration in B band and a 4 hour integration in I-band. Sky flats of high precision were secured from dithered exposures on a control field. The B image is shown in figure 1. Figure 2 shows a map of the surface mass density of the matter lensing Q0957+561. This map is obtained by inverting the estimated shear field inferred from the observed shapes and orientations of background galaxies.

6. CONSTRAINTS ON THE MASS MODELS

From the map of the mass distribution we can locate the center of the cluster relative to the lensing galaxy. This is a key parameter for the determination of H_0. We measure the cluster center position to be at a distance (r_c) of 9 arcseconds from the lensing galaxy in the direction $\theta_c = 193°$. In a recent study Dahle et al. (1994) using deep images obtained with the Nordic Optical Telescope find r_c= 6 arcseconds and $\theta_c = 123°$. The uncertainties associated with these numbers are ~3 arcseconds in the distance and ~ 30° for the position angle.

7. CONCLUSION

From the constraints on two of the parameters of our model of the cluster and galaxy lensing Q0957+561 we can infer limits on the Hubble constant. Using the distortion of background galaxies we have derived a map of the mass distribution lensing the QSO. We have used this map to locate the position of the cluster lensing the QSO. The two parameters are the position angle of the vector joining the first ranked galaxy (G1) to cluster center, and the length of this vector. By limiting the range in values that these two parameters can have we eliminate all models except those having H_0 less than 70 km/s/Mpc.

ACKNOWLEDGEMENTS

George Rhee acknowledges support from a UNLV Barrick travel grant and the UNLV Physics department Bigelow fund.

REFERENCES

Bernstein, G.M., Tyson, J.A., & Kochanek, C.S. 1993, AJ, 105, 816

Dahle, H., Maddox, S.J., and Lilje, P.B. 1994, ApJL, in press

Lehar, J., Hewitt, J.N., & Roberts, D.H., 1989, in Gravitational Lenses, ed. J.M. Moran, J.N. Hewitt, & K.Y. lo (Berlin: Springer)

Refsdal, S. 1966, MNRAS, 132, 101

Rhee, G. 1991, Nature, 350, 211

Press, W.H., Rybicki, G.B., & Hewitt, J.N. 1992, ApJ, 385, 416

Schild, R., AJ, 1990, 100, 1771

Vanderriest, C., Schneider, J., Herpe, G., Chevreton, M., Moles, M., and Wlerick, G. 1989, A&A, 215, 1

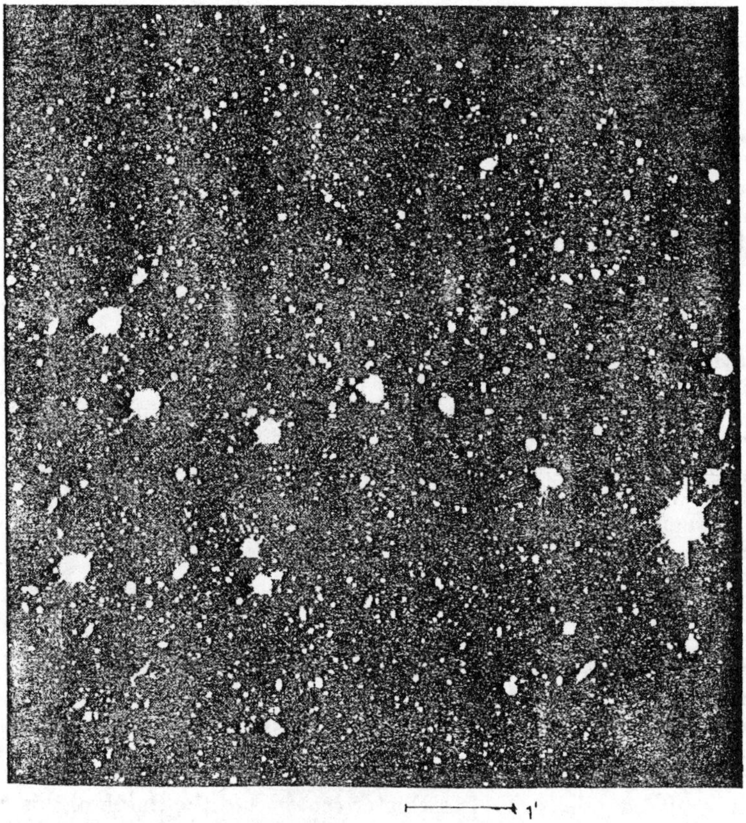

Fig. 1. B band exposure of the 0957 field. CFHT FOCAM.

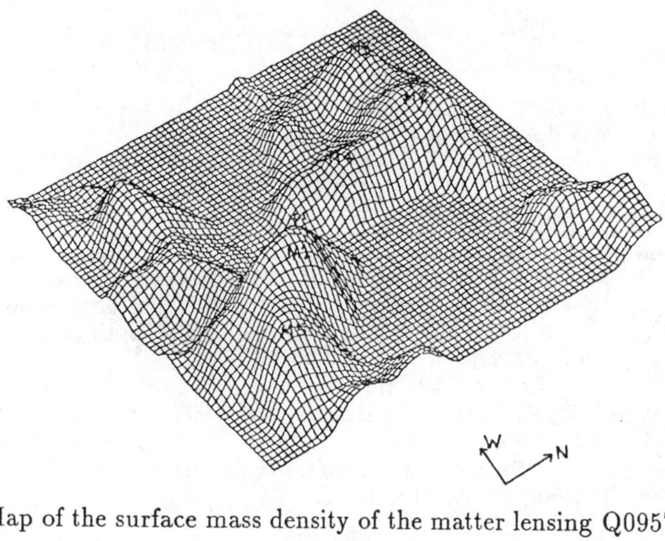

Fig. 2. Map of the surface mass density of the matter lensing Q0957+561

TESTING COSMOGONIC MODELS WITH GRAVITATIONAL LENSING

Joachim Wambsganss
Astrophysikalisches Institut Potsdam, An der Sternwarte 16, 14482 Potsdam, Germany
jwambsganss@aip.de

Renyue Cen, Jeremiah P. Ostriker, and Edwin L. Turner
Princeton University, Princeton, NJ 08540, USA
cen, jpo, elt@astro.princeton.edu

ABSTRACT

Gravitational lensing provides a strong test for cosmogonic models. We attempt to determine the lensing properties of different models, e.g. the frequency of multiply imaged quasars, the distribution of separations and the lens redshifts. This is done by detailed numerical propagation of light rays through a universe having a distribution of inhomogeneities derived from particular cosmogonic models. Here we present results for matter distributions obtained with the standard CDM (cold dark matter) scenario. This model predicts far more widely split quasar images than have been found. This means the COBE normalized CDM model with $\Omega = 1$ and $H_0 = 50$km/s/Mpc is inconsistent with current observations. It might be possible to construct variants of this model which could pass the stringent tests provided by strong gravitational lensing.

1. INTRODUCTION

Conventional tools for comparing cosmogonic theories with observations rely on either galaxy density or velocity information, both of which unavoidably suffer from the uncertainties with regard to density or velocity bias of galaxies over the underlying mass distribution, hampering our attempts to understand the more "fundamental" questions concerning the mass evolution and distribution. In contrast, gravitational lensing *directly* measures fluctuations in the gravitational potential along lines of sight to distant objects. Thus, gravitational lensing provides a powerful independent test of cosmogonic models. Narayan and White (1988) were the first to study this using the Press-Schechter formalism, Cen *et al.* (1994, subsequently paper I) used surface mass density distributions from realistic n-body simulations.

Each model for the development of cosmogonic structure (e.g. the HDM [hot dark matter] or CDM scenario) has at least one free parameter, the amplitude of the density (or potential) power spectrum. But now in the light of COBE observations (Smooth *et al.* 1992), that parameter is fixed by the ($\pm 15\%$) determination on the $5^o - 10^o$ scale in the linear regime. With its amplitude fixed, a secure determination of the potential fluctuation on any scale provides a test; any single conflict between the theory and reality can falsify the former. The most leverage is obtained for tests made on scales as far as possible from the COBE measurements. The reason is that all models have an assumed power spectrum that passes through the COBE normalization point at the very large comoving scales ($\lambda \approx 1000$Mpc) fixed by that measurement. Since the slope of the power spectrum is a primary model dependent feature, the maximum variations amongst models occur typically at the smallest scales. Thus one looks for tests at scales as small as possible, but they should not be so small as to be greatly influenced by the difficulty in modelling the physics of the gaseous, baryonic componcnts ($\lesssim$ 10kpc). Thus critical tests are best made on scales 0.01Mpc $< r <$ 1Mpc. Here we will present first results from using gravitational lensing from matter distributions on these scales to test the standard CDM scenario.

2. METHOD

The model used is the "standard" CDM scenario with $\Omega = 1$, $\lambda = 0$ and $H_0 = 100h = 50$km/s/Mpc. The normalization was done according to the COBE first year results (Efstathiou *et al.* 1992), it corresponds to $\sigma_8 = 1.05$. In order to have detailed small scale information (which is crucial for effects of "strong lensing") we ran a total of 10 independent simulations with $L = 5h^{-1}$Mpc, having $500^3 = 10^{8.1}$ cells and $250^3 = 10^{7.2}$ particles. Because this box size does not account for power on large scales, we also ran an $L = 400h^{-1}$Mpc size box with $500^3 = 10^{8.1}$ cells and $250^3 = 10^{7.2}$ particles.

Knowing the distribution of overdensities on the $5h^{-1}$Mpc scale from the large simulation box, we statistically convolve the small and large scale runs to produce simulated sheets or screens of matter spaced $5h^{-1}$Mpc apart between the observer at $z = 0$ and a putative galaxy or quasar in the source plane at $z = z_S$. A large number of independent runs (ten runs were done) is required so that identical structures do not repeat along a line of sight. The details of the convolution method and tests of it using a high resolution P^{3}M simulation provided by Bertschinger and Gelb (1994) will be presented in a subsequent paper. The method is statistically reliable for describing structures in the range 30kpc $< \Delta L h <$ 1.2Mpc, which corresponds roughly to splitting angles $5'' < \theta < 200''$. On these scales we expect that dark matter dominates over baryons so that a dark matter only simulation is approximately valid.

A very preliminary attack on this problem has been presented in Paper I. In that work no ray tracing was done. Rather it was simply checked whether or not surface mass densities were greater than the critical surface mass density (Turner *et al.* 1994) at which multiple imaging will occur. In addition to the much better method of treating the lensing aspect used here, the convolution algorithm has been modified and improved over the one adopted in Paper I.

We follow light rays in a cone of fixed angular size (about 6 arcmin) through realistic matter distributions in many lens planes, determine their deflection angles in each plane and collect them at source planes of different redshifts. We use the multi-plane lens equation for the light rays, as described in Schneider, Ehlers & Falco (1992). The deflection angels are calculated for 500^2 rays in each plane, according to all the matter inside this plane. Subsequently we study the lensing properties of the matter in these lens planes.

3. RESULTS AND DISCUSSION

We performed 100 independent lensing runs of a square field with about 6 arcmin at a side. In Figure 1 we present the probability of multiple imaging as a function of quasar redshift. We considered only multiple images with separations greater than $5''$ and magnitude difference less than 1.5 mag. In fact "amplification bias" (TOG) will increase the probabilities over those shown in Figure 1 by a significant amount. From this figure it can be seen that splittings larger than 5" should be common (when several thousand quasars have been examined), if this cosmogonic model were correct.

A very revealing statistic is the distribution of image separations expected for multiple images as shown in Figure 2 for quasar redshifts of $z_S = 1$, 2 and 3. Large and very large splittings should be the rule. Also revealing is the distribution of expected lens redshifts as shown in Figure 3. Most lenses in the standard CDM scenario should be relatively nearby, close enough to be seen in almost all known gravitational lens cases. On this issue the recent observation (Fischer *et al.* 1994) of a lens candidate for the double quasar QSO2345+007 is quite relevant. The separation of the two images is 7".06, the quasar redshift is $z_S = 2.15$ and the putative lens is at $z_L = 1.49$. We see from Figure 3 that, although 7 arcsecs separation can be produced in the CDM model, the probability that the lens is as far away as $z = 1.49$ is very small (2%) due to the relatively late formation of structure in this model. In open models structure formation occurs earlier.

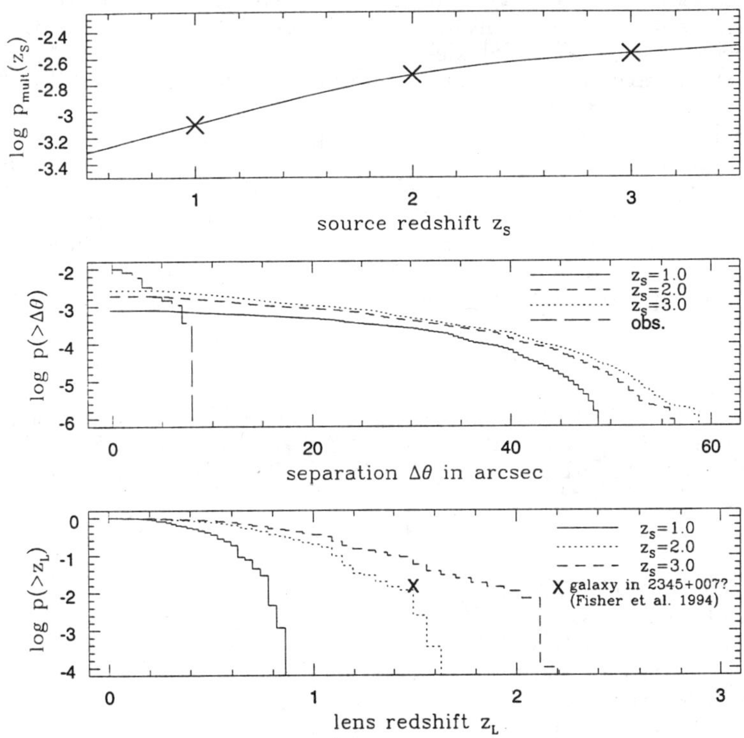

Fig. 1 (top) Probability of multiple lensing ($\Delta\theta \geq 5''$, $\Delta m \leq 1.5$mag) as a function of z_S;
Fig. 2 (middle) Probability distribution of splitting angles $\Delta\theta$;
Fig. 3 (bottom) Distribution of lens redshifts z_L.

It appears that all three of these results (the total frequency of large separation multiple quasars, the distribution with splitting angle, and the distribution of lens redshifts) do not agree with the existing observations, the first two are in serious conflict. Surveys and occasional serendipitous discoveries have revealed 27 confirmed or possible multiply-imaged QSO's according to a recent compilation (Surdej & Soucail 1993). Detailed analysis of the most statistically useful of these surveys yields a lensing rate in the vicinity of a few tenth to one percent (Kochanek 1993, Maoz & Rix 1993), consistent with the CDM predictions quoted above making allowance for plausible magnification biases. However, as shown in Figure 3, all observed QSO lens systems have image splittings of less than 10", and the large majority, less than 5".

This sharply contradicts and thus falsifies the model. Since the large splitting, modest brightness ratio systems predicted by the model would be typically much easier to detect and recognize than those 27 which have actually been found, no escape by appeal to observational selection seems possible.

This failing of the model is not presented as an entirely new result, but only as a new and more robust manifestation of a previously recognized problem, namely the excessively deep potential wells produced by the dark matter component in COBE normalized standard CDM. These excessively deep potential wells lead both to excess galaxy pair-wise velocity dispersions and to the predicted excessive rate of large splitting lensing events. The virtue of the lensing test is that it is independent of other tests and is not subject to the same caveats concerning "bias" of galaxies with respect to dark matter.

Are there variant models that would not fail these tests? A lower Ω is clearly useful, but it would be premature to argue that the present results by themselves indicate $\Omega < 1$, as many other properties of the scenario ("temperature" of the dark matter, shape of the power spectrum etc) contribute to lensing properties. However the *directness* of gravitational lensing as a test for the growth of inhomogeneities, coupled with the rapidly increasing power of computers and numerical algorithms, makes one optimistic that calculations of the type reported on here should become a major tool for testing and discriminating among competing cosmological scenarios.

ACKNOWLEDGEMENTS

We thank J.R. Gott, C.S. Kochanek and P. Schneider for useful discussions, NCSA for use of the Convex-3880, and NASA grants NAGW-2448 and NAGW-2173, NSF grants AST91-08103 and HPCC ASC-9318185 for financial support.

REFERENCES

Bertschinger, E., & Gelb, J. 1994, ApJ, in press
Cen, R., Gott, J.R., Ostriker, J.P., & Turner, E.L. 1994, ApJ, 423, 1
Efstathiou, G., Bond, J.R, & White, S.D.M 1992, MNRAS, 258, 1p
Fischer, P., Tyson, J.A., Bernstein, G., & Guhathakurta, P. 1994, ApJL, 431,L71
Kochanek, C.S. 1993, ApJ, 417, 438; 419, 12
Maoz, D., & Rix, H.-W. 1993, ApJ, 416, 425
Narayan, R. & White, S.D.M. 1988, MNRAS, 231, 97p
Schneider, P., Ehlers, J., & Falco, E.E., *Gravitational Lensing* (Springer Verlag, Berlin, 1992)
Smoot, G.F., *et al.* 1992, ApJ(Letters), 396, L1
Surdej, J., & Soucail, G. 1993, in "Gravitational Lenses in the Universe", Proceedings of the 31st Liege International Astrophysical Colloquium, p205
Turner, E.L., Ostriker, J.P., & Gott, J.R. III 1984, ApJ, 284, 1 (TOG)

Reconstructing Cluster Lensing Mass Distributions from Gravitational Arclets

Gillian Wilson, Shaun Cole and Carlos Frenk
University of Durham, South Road, Durham, DH1 3LE, England
gillian.wilson@durham.ac.uk

ABSTRACT

A weak lensing mass reconstruction technique proposed by Kaiser and Squires is analysed in practice.

INTRODUCTION

High mass concentrations present in rich galaxy clusters distort images of background galaxies through gravitational lensing. Lensed galaxies tend both to be displaced radially outwards from the cluster centre and to be elongated in the tangential direction. Although background galaxies are occasionally lensed serendipitiously to form spectacular giant arcs, they are far more likely to undergo only small distortions. These more common but less distorted galaxies are too faint for spectroscopy and are useless individually as measures of the lensing mass. However, statistical analysis of these weakly but coherently lensed arclets can lead to recovery of the cluster mass distribution.

The mass reconstruction technique proposed by Kaiser and Squires was unique in that it made no a priori assumptions about the amount of mass or its distribution. The technique produces a "map" of the cluster's surface density.

The inferred values of surface density are complicated by the influence of inherent observational effects such as noise, seeing, crowding and pixelation (intensity being averaged over each pixel area and sampled discretely). In order to evaluate the usefulness of the technique in a practical scenario we incorporate these effects in simulated CCD frames of galaxies which have undergone lensing. We use an artificial galaxy cluster from N-body simulations as the lens. We measured galaxy shapes using the same techniques of faint galaxy data reduction (namely FOCAS software) as would commonly be applied to real data.

RESULTS AND CONCLUSIONS

The first two figures overleaf illustrate the power of the technique. Figure 1 shows the original simulated cluster shown in projection. Figure 2 shows a theoretical Kaiser-Squires reconstruction, that is to say the galaxy distortions are calculated directly rather than measured from a CCD frame. Note that the reconstruction technique involves inherent smoothing.

Figures 3 and 4 show the reconstructions obtained after generating artificial CCD frames. As the figures indicate the technique only works well in good seeing conditions. This is because high seeing tends to circularise images and suppress the signal.

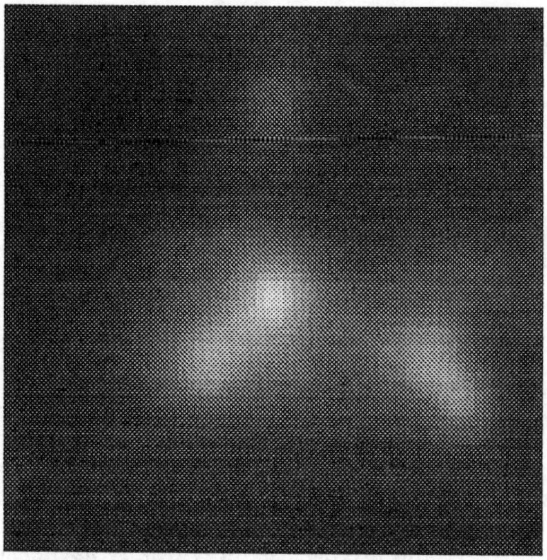

Fig. 1. Original lensing cluster shown in projection (cluster is from N-body simulations).

Fig. 2. Theoretical Kaiser-Squires reconstruction (Note that the technique involves inherent smoothing).

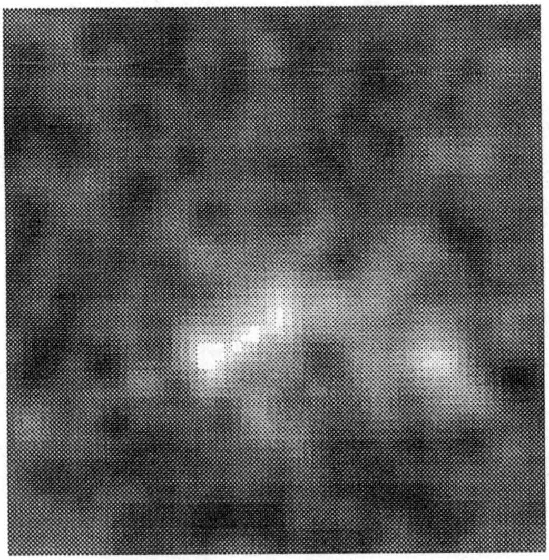

Fig. 3. Reconstruction in 1 arcsec seeing.

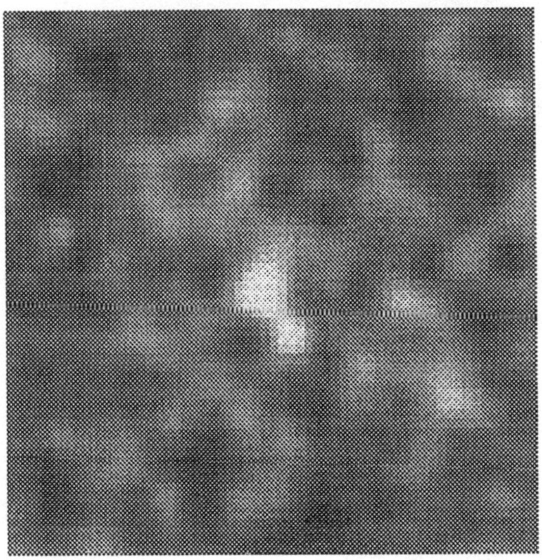

Fig. 4. Reconstruction in 2 arcsec seeing.

REFERENCES

Bartelmann,M. & Weiss,A. 1994, A&A, 287, 1
Bartelmann,M., Steinmetz,M. & Weiss,A. 1994, preprint
Bonnet,H., Mellier,Y., & Fort,B. 1993, A&A , 280, L7
Broadhurst, T.J., Taylor, A.N., & Peacock, J.A. 1994, ApJ, in press
Fahlman,G., Kaiser,N., Squires,G. & Woods,D. 1994 ApJ, in press
Kaiser,N. 1992, ApJ, 388, 272
Kaiser,N. & Squires,G. 1993, ApJ, 404, 441
Kochanek,C. 1990, MNRAS, 247, 135
Miralda-Escude,J. 1991, ApJ, 370, 1
Schneider,P., Ehlers, J., & Falco, E.E, 1992, Gravitational Lenses (Heidelberg:Springer)
Schneider,P., & Seitz, C. 1994, A&A, submitted
Seitz, C., & Schneider,P. 1994, A&A, submitted
Smail,I., Ellis,R. ,Fitchett,M. & Edge,A. 1994, M.N.R.A.S., in press
Tyson,J.A., Valdes,F., & Wenk,R. 1990, ApJ 349, L1
Tyson,J.A. 1994, in: Cosmology and Large Scale Structure, Proceedings Les Houches Summer School, August 1993, ed R Schaeffer (Elsevier)

Arc Statistics With Realistic Cluster Potentials

Matthias Bartelmann[1,2], Matthias Steinmetz[2] and Achim Weiss[2]
[1] Harvard-Smithsonian CfA, 60 Garden St., Cambridge, MA 02138
[2] MPI für Astrophysik, Postfach 1523, 85740 Garching, FRG

Abstract

We construct a sample of numerical models for clusters of galaxies and employ these to investigate their capability of imaging background sources into long arcs. The clusters are simulated within the CDM cosmogonic scheme in an Einstein-de Sitter universe. Emphasis is laid on the statistics of the arcs formed, and optical depths for arc formation are determined. We also compare the results to predictions based on simplified, radially symmetric cluster models. We find that the capability of the numerically modeled clusters to produce long arcs is increased by a factor of $\lesssim 50$ compared to a sample of softened isothermal spheres with the same observable parameters (core radii and velocity dispersions), and that they are comparably efficient as singular isothermal spheres with the same velocity dispersion. This largely enhanced capability to produce large arcs of the numerical cluster models can be understood in terms of substructure and intrinsic asymmetry, which enhance the tidal field (shear) of the clusters compared to the radially symmetric cases. We also find that the intrinsic ellipticity of the background sources has a noticible influence on arc statistics; the optical depth for arcs with a length-to-width ratio $\gtrsim 10$ is significantly larger for elliptical than for circular background sources.

Introduction

The statistics of long arcs in clusters of galaxies can provide important information about the distribution of the probably dominant dark-matter component in clusters which is otherwise hard to extract. Unfortunately, the interpretation of arc statistics is by no means straightforward since it critically depends on the lens models chosen to parametrize the lensing cluster population. Several studies have been performed on the basis of analytical lens models (see, e.g., Grossman & Narayan 1988, Miralda-Escudé 1993a,b, Wu & Hammer 1993, and references therein). Specifically, Wu & Hammer arrive at the conclusion that either cluster core radii have to be much smaller than observed in both the galaxy distribution and the X-ray emission of clusters, or cluster mass profiles must be much steeper than isothermal in order to concentrate a larger fraction of the cluster mass towards the cluster center; otherwise, the total number of long arcs in clusters of galaxies would be far too low. This conclusion raises the question of whether it may be an artefact of the high intrinsic symmetry of the lens models employed.

In order to investigate the statistics of arcs in a more realistic way, we have constructed a sample of numerical galaxy-cluster models forming in a hierarchical structure-formation scenario like *e.g.*the Cold Dark Matter (CDM) model. Great care was taken to include the tidal effects of the matter distribution surrounding the collapsing objects. The lensing properties of these cluster models were then analyzed by means of an adaptive-grid ray-shooting method.

Simulations

Our simulation strategy is as follows. First, a coarsely grained, 36^3-particle, COBE-normalized (Efstathiou, Bond & White 1992) CDM N-body simulation of a box with $150h^{-1}$ Mpc side length is performed. The cosmological background model was of Einstein-de Sitter type, with a Hubble constant of $H_0 = 100\,h$ km/s/Mpc and $h = 0.5$. The box size was chosen such as to contain a resonable number of massive clusters. At $z = 0$, we identified regions of radius $5h^{-1}$ Mpc containing more than 100 particles, *i.e.*, having an average overdensity in excess of ≈ 15. We found 13 such regions in our simulation. Each of these regions was then re-calculated in a high-resolution treecode run (75000 particles of $\approx 4\,10^{11}\,M_\odot$ each), using the same large-scale density modes, but adding smaller-scale modes with up to the Nyquist frequency of the refined grid. The mass distribution outside the refined region is approximated by ≈ 8000 macro-particles of radially increasing mass. The gravitational softening within the high-resolution region is $\varepsilon = 25h^{-1}$ kpc. This resolution is marginally sufficient for resolving the central 100 kpc of a cluster, the region critical to strong lensing, by a few hundred particles. For details of the simulation technique we refer to (Bartelmann, Steinmetz & Weiss 1994).

At $z = 0$, the 13 clusters typically contain a dominant, but not entirely relaxed clump. The radial density stratification of these main objects is well described by a power law with index $n \simeq -2.3$ and a core radius of about 50 kpc/h. Superposed on this density profile, there exist smaller subclumps with a typical mass of about $10^{13}\,M_\odot$. The typical line-of-sight velocity dispersion is ≈ 1200 km/sec. In recent gasdynamical simulations (Haehnelt & Steinmetz, in preparation) we found a typical X–ray luminosity of 10^{45} erg. In most of our cluster models, we find $2\ldots4$ accompanying secondary clumps with a mass ranging between $5\ldots30\%$ of the main object's mass. However, the mass of these subclumps turns out to be too small to become critical to strong lensing. None of these clusters has a dominant progenitor at $z \gtrsim 0.3$, but is typically formed out of a dozen low-mass progenitors, which is typical for clusters formed in a high-Ω hierarchical universe. We tested with a bootstrap analysis that none of the cluster potentials is significantly influenced by individual particles.

Arc Statistics

To determine the cross sections for the formation of large arcs, we investigate the imaging of extended sources on the source plane, assumed at redshift $z_s = 1$ throughout, using the deflection angles determined from the numerical cluster model. For details of the procedure briefly sketched below, we refer to Bartelmann & Weiss (1994). The source are positioned on a grid covering the source plane, whose resolution is refined towards caustics, and the the images of each source are assigned statistical weights in proportion to the inverse of the squared local grid resolution. This is to avoid waste of CPU time by imaging sources in regoins of the source plane which are far from caustics and thus uninteresting for our purposes. The images are then classified using the algorithm described by Bartelmann & Weiss (1994). For each image, we determine length, width, and curvature radius. This imaging procedure is done for each time step and each of the three independent projection directions for each cluster model, if the cluster model is critical, *i.e.*, if it produces critical curves, and if the cluster model has a of redshift $z_d < z_s$. The sources are chosen either circular or elliptical, where the circular sources have a diameter of $1''$. The axis ratio (semi-minor divided by semi-major axis) for elliptical sources is drawn randomly from the interval $[0.5, 1]$, and their area is chosen equal to that of the circular sources.

A total of 73395 images were classified, which should provide a reasonably large sample to investigate statistically. From this sample of images, we determined cross sections for the formation of large arcs with a variety of properties. Having determined the cross sections, we computed the total optical depth for the formation of large arcs, the redshift distributions of probable cluster lenses, and the distribution of arc curvature radii and arc widths within samples of large arcs. We also investigated the influence of intrinsic source ellipticity vs. intrinsically circular sources.

The cross sections and optical depths found from the numerical cluster models were compared to analogous results derived from two spherically symmetric analytic approximations to the numerical cluster models, the singular and the non-singular (smoothed) isothermal sphere. We adapt the analytic models to the numerical models by choosing the velocity dispersion and the core radius obtained from the numerical models. See Fig. 1 for various optical depths.

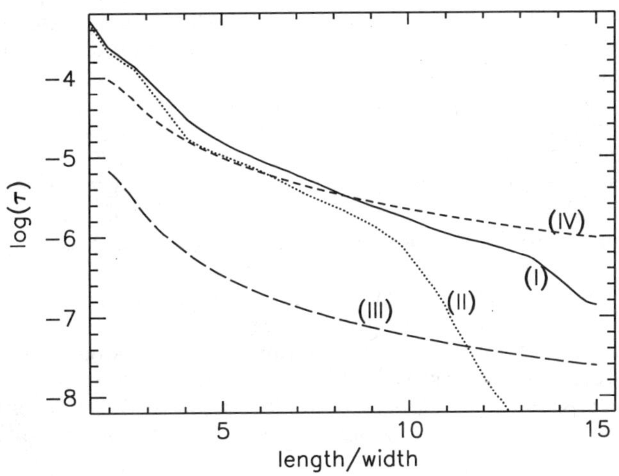

Figure 1. – Optical depths for large arcs as a function of length-to-width ratio. (I, solid line): numerical cluster models with elliptical sources; (II, dotted line): numerical cluster models with circular sources; (III, long-dashed line): non-singular isothermal spheres; (IV, short-dashed line): singular isothermal spheres. The final decline in curves (I) and (II) for high (L/W) ratio is influenced by the finite source size and the finite grid resolution.

The solid and dotted lines in Fig. 1 display the optical depths for the arc length-to-width ratio (L/W) obtained from numerically determined cross sections for elliptical and circular sources, (curves I and II, respectively). Curves

(III) and (IV) (long- and short-dashed, respectively) show optical depths obtained from the analytic approximations to the cluster models, curve (III) from the non-singular and curve (IV) from the singular case. For all cases except case (III), the optical depth for long and thin arcs ranges between 10^{-5} and 10^{-6}, while the optical depth for the non-singular approximation is lower by roughly a factor of $\lesssim 50$. The optical depths from the numerical cluster models for either elliptical or circular sources (curves I and II, respectively) are similar for $(L/W) \lesssim 10$, while for larger length-to-width ratios the optical depth for elliptical sources is substantially larger. This indicates that the source ellipticity is crucial for arcs with a large length-to-width ratio. For arcs of moderate (L/W), the numerically determined optical depth for elliptical sources is about as large as for the singular isothermal cluster models, but it falls below for arcs with $(L/W) \gtrsim 13$. These results show that the numerical cluster models are significantly more efficient in producing large arcs than spherically symmetric analytic models when the latter are adapted to the observable properties of the numerical models, and that the numerical models are roughly as efficient as singular isothermal models despite their extended cores.

Conclusions

For the sake of brevity, we have focussed in this contribution on the optical depth for large arcs. The findings of our study (Bartelmann, Steinmetz & Weiss 1994) can be summarized as follows:

1. Cross sections for the formation of giant arcs can have several local minima and maxima as functions of the cluster redshift. (Not discussed here; see Bartelmann, Steinmetz & Weiss 1994.)
2. The cross sections for the formation of arcs with a length-to-width ratio $\gtrsim 10$ derived from the numerical cluster models are of order 10^{-3} $(\mathrm{Mpc}/h)^2$ for intermediate cluster redshifts ($0.2 \lesssim z \lesssim 0.6$) and the source redshift $z_s = 1$.
3. Due to substructure and an enhanced shear, the numerically modeled clusters produce cross sections for large arcs which are larger by a factor of roughly $\gtrsim 50$ than corresponding cross sections derived from softened isothermal spheres with the same velocity dispersions and the core radii of the numerical models. They are comparably efficient as singular isothermal spheres with the same velocity dispersions despite their finite core.
4. Intrinsic source ellipticity is important in mainly two respects: first, the optical depth for arcs with $(L/W) \gtrsim 10$ is significantly larger for elliptical than for circular sources, and second, the large arcs formed from elliptical sources are narrower than those formed from circular sources.

References

Bartelmann, M., Weiss, A. 1994, A&A, 287, 1
Bartelmann, M., Steinmetz, M., Weiss, A. 1994, A&A, in press
Efstathiou, G., Bond, J.R., White, S.D.M. 1992, MNRAS, 258, 1P
Grossman, S.A., Narayan, R., 1988, ApJ, 324, L37
Miralda-Escudé, J. 1993a, ApJ, 403, 497
Miralda-Escudé, J. 1993b, ApJ, 403, 509
Wu, X.-P., Hammer, F. 1993, MNRAS, 262, 187

THE LARGE SCALE DISTRIBUTION AND MOTIONS OF GALAXIES

INDICATIONS OF DARK MATTER DERIVED FROM LARGE SCALE FLOWS

Marc Davis[1] and Adi Nusser[2]
[1] Depts. of Astronomy and Physics, U.C. Berkeley
marc@coma.berkeley.edu
[2] Institute of Astronomy, Cambridge
adi@mail.ast.cam.ac.uk

ABSTRACT

Recent progress in the measurement of relative distances to galaxies has been quite substantial, and catalogs of 3000 galaxies with distances are soon to become available. The peculiar *velocity* field (deviations from Hubble flow) derivable from these catalogs, when compared to the peculiar *gravity* field derived from all sky redshift surveys of galaxies such as the 1.2Jy IRAS survey, leads to a unique and extremely powerful test of the density parameter $\beta \equiv \Omega^{0.6}/b_I$, where b_I is the possible linear bias of the IRAS selected galaxies relative to the mass fluctuations. We review the status of these large scale flow measurements and present a new methodology to describe the two fields by means of an expansion in a set of orthogonalized functions describing a general potential flow to any chosen resolution. The parameters of the flow can be estimated by minimization of the χ^2 describing the scatter of observed versus predicted linewidths from an inverse Tully-Fisher relation. By this method one can intercompare the gravity and velocity fields coefficient by coefficient, deriving a precise fit for the density parameter and an assessment of the degree of coherence between the fields. The present situation is transitory– different analyses of the same data are not yielding consistent results. Until this embarassment is untangled, estimates of β should be taken with a large grain of salt.

1. INTRODUCTION

The study of large scale flows permits an estimate of the density parameter on a characteristic scale of $10 - 30h^{-1}$ Mpc, which is thirty times larger than the central regions of rich clusters, where mass estimates can be derived on the basis of hydrostatic equilibrium or virial equilibrium. As is well known, mass estimates from cluster analysis imply a cosmological density of order $\Omega \approx 0.1 - 0.2$, assuming that galaxies in and out of clusters have similar mass to light ratios, or that the baryon fraction in the clusters is typical of the Universe as a whole (White *et al.* 1993). The large scale flow analysis would be expected to yield similar estimates of the density parameter if the dark matter fully congregates with the galaxies on cluster scales, but could well yield a different value if galaxy formation is more efficient in the denser regions or if the dark matter is too hot to cluster on any but the very largest scales.

Dekel (1994) and Strauss and Willick (1995) have recently reviewed this subject in detail, and we shall confine our remarks here to a few comments on the discrepancies of the β derived from different analyses. A decade ago the quantitative analysis of large scale flows consisted largely of the question of Virgocentric flow (Davis and Peebles 1983), but now the focus has broadened to analysis of the details of the flow on a scale four times as large, which has led to new insights about large scale structure in the Universe. The observational situation is improving rapidly, and will continue to improve in the next few years, particularly if high precision distance estimators, such as SN-Ia (Riess *et al.* 1994), or surface-brightness fluctuations (Tonry *et al.* 1994) in early type galaxies, lead to useful catalogs. Giovanelli *et al.* (1994) are preparing a full sky sample of some 2400 galaxies with Tully-Fisher distances, while Willick *et al.* (1995) have compiled a sample of 2900 galaxies over the full sky (Faber *et al.* 1994). These large samples probe the velocity field to a redshift of approximately 8000 km/s, and the distance accuracy of the Tully-Fisher indicator is estimated to have a precision of 15-20%.

The cosmological interpretation of large scale flows is dependent on late time linear perturbation theory, which extensive testing has shown to be valid on scales $> 10h^{-1}$ Mpc. This is convenient, because linear theory is easy to work with and is the fundamental prediction of alternative models of large scale structure. If one confines the analysis to simple moments of the velocity field, such as a bulk flow or the alignment of the flow on a given scale with the dipole of the cosmic microwave background, the cosmological test is limited to questions of the likelihood that a given model would produce sufficient power to generate such a flow.

To sharpen the cosmological probe, it is instructive not simply to analyse the *power* in the velocity flow, but to compare the *alignment* of the velocity field with the inferred *gravitational* field derived from the observed galaxy distribution. Toward this end, complete redshift surveys of whole-sky galaxy catalogs are an essential ingredient. The IRAS selected 1.2Jy survey (Fisher *et al.* 1995) of 5300 galaxies covering 88% of the sky is the largest available catalog at present, but it should soon be superceded by at least two other catalogs (Santiago *et al.* 1994) (Saunders *et al.* 1995). The IRAS catalog is very suitable for this type of analysis because the 60 micron flux measured by IRAS is unaffected by galactic extinction and it is possible to select galaxies down to galactic latitude $|b| = 5°$ with only modest confusion. However, the IRAS sample is quite dilute, particularly in cluster centers, and the inferred gravitational field has considerable noise, especially for structures with redshift $cz > 6000$ km/s.

If one had full knowledge of the mass fluctuation field $\delta_\rho(\mathbf{r})$ over all space, one could trivially write the gravitational fluctuation field $\mathbf{g}(\mathbf{r})$ as

$$\mathbf{g}(\mathbf{r}) = G\bar{\rho} \int d^3\mathbf{r}' \delta_\rho(\mathbf{r}') \frac{\mathbf{r}' - \mathbf{r}}{|\mathbf{r}' - \mathbf{r}|^3} , \tag{1}$$

where $\bar{\rho}$ is the mean mass density of the Universe. The great simplification of late time linear theory is the intuitive relation between the gravity field $\mathbf{g}$ and the velocity field $\mathbf{v_L}$, namely the equating of the velocity as the acceleration times the time, where the only possible time is the Hubble time. The correct expression (Peebles 1980) is

$$\mathbf{v_L}(\mathbf{r}) = \frac{2}{3H_0\Omega^{0.4}} \mathbf{g}(\mathbf{r}) . \tag{2}$$

Assuming the galaxy distribution at least approximately traces the mass on large scale, with linear bias b between the galaxy fluctuations δ_G and the mass fluctuations, ($b = \delta_G/\delta_\rho$), and replacing the integral over space with a sum over the galaxies in a catalog, with radial selection function $\phi(r)$, we have

$$\mathbf{v_L}(\mathbf{r}) = \frac{H_0}{4\pi\bar{n}} \sum_i \frac{1}{\phi(r_i)} \frac{\mathbf{r}_i - \mathbf{r}}{|\mathbf{r}_i - \mathbf{r}|^3} , \tag{3}$$

where $\bar{n}$ is the mean galaxy density in the sample. Note that the result is insensitive to the value of H_0, as the right hand side has units of velocity. Thus in this game it is traditional to quote distances in units of km/s.

Because the effective weight per galaxy $1/\phi(r)$ diverges at large distance, one must cutoff the calculation at some limiting redshift, which for the 1.2Jy IRAS sample is usually chosen to be 20,000 km/s. To limit the direct influence of one galaxy on another and to filter out nonlinear effects, it is necessary to smooth the small scale interaction, typically with a smoothing scale of $r_s \geq 500$ km/s. The method assumes that the associated mass per galaxy is constant, but in fact it is straightforward to piece together different redshift catalogs with different weights and selection functions.

Two different schemes have been used to compute $\mathbf{v_L}$ from redshift catalogs. The original scheme of Yahil *et al.* (1991) solved Equation (3) by iteration, adiabatically turning on the gravity field to advance the position of the galaxies from redshift space to real space. Recent refinements on this method by computation on a grid (Strauss and Yahil 1994) will allow nonlinear corrections to be applied to the predicted velocity field. An alternative formulation that can be applied directly in redshift space has recently been suggested by Nusser and Davis (1994a). This method solves a modified form of the Poisson equation using spherical harmonic decomposition and can readily deal with variable smoothing. To lowest order the two methods agree, but differ on how they treat the triple valued zones, regions around cluster centers in which three separate distances can have the same redshift. The Nusser and Davis algorithm intrinsically assumes a monotonic relation between distance and redshift, thereby eliminating multivalued zones. The older, iterative algorithm allows the existence of multivalued zones but has no idea of how to position an object that finds itself in such a zone; several ad hoc algorithms have been adopted over the years. There is no good solution for this problem which is one reason to filter out nonlinear effects or to compare alternative algorithms. Fortunately, most of the local density field appears to be single valued when smoothed on scales of 500 km/s or larger; only around cluster centers is the reconstruction of questionable validity. Nonlinear reconstruction algorithms are of marginal utility because they do not eliminate the problems of the multi-valued zones, yet the nonlinear effects are only important in regions which are nearly identical to the multivalued zones. Furthermore these are the regions where the assumption of linear bias is most questionable.

2. $\delta - \delta$ COMPARISON

The POTENT algorithm (Dekel *et al.* 1990) is a scheme to convert the observed radial velocity field into a density field, and depends on the reasonable assumptions that the large scale velocity field is a potential flow in the linear limit. The assumption of potential flow allows one to construct a three dimensional velocity field from the observed radial flow, and the divergence of this field yields an estimate of the fluctuating *mass* field driving the flows. Comparison of this mass field with the observed galaxy distribution leads to an estimate of β. Details are given by Dekel (1994); the method has been applied to the Mark-2 catalog (Faber and Burstein 1988) by Bertschinger *et al.* (1990) and then compared to the galaxy density field derived from the 1.9Jy IRAS redshift catalog (Strauss *et al.* 1992), resulting in a very high density estimate, $\beta = 1.3 \pm 0.3$ (Dekel *et al.* 1993).

In order to define the velocity potential, it is necessary to heavily smooth the velocity data; Dekel *et al.* (1993) use a guassian smoothing of $\sigma = 1200$ km/s, which is sufficiently large to filter out the infall of the local group toward the Virgo supercluster! In its present implementation the POTENT method depends on a forward Tully-Fisher approach, in which a calibrated Tully-Fisher relation is used to infer the absolute magnitude of a galaxy given an observed line width. As is well known, this procedure suffers from homogeneous, and more particularly, inhomogeneous Malmquist bias (Schechter 1980). If the scatter in the distance estimator is accurately known, the Malmquist biases can be removed. If the incorrect scatter is used, of if no correction is applied, an overdensity of points in redshift space will appear to have the signature of a massive cluster, even if the cluster has no mass at all. Dekel *et al.* (1993) undertook an extensive series of Monte-Carlo tests to calibrate these biases, and their result is consistent with a number of earlier comparisons of IRAS predictions to peculiar velocity data (e.g. Kaiser *et al.* 1991, Straus 1988, Strauss and Davis 1988).

3. v-v COMPARISONS

Because this result is the only observational evidence for a high density Universe, it is certainly worth pursuing alternative methods that could provide confirming or refuting evidence. In the interval since Dekel *et al.* (1993), the available database has improved considerably. The IRAS sample has doubled with the availability of the 1.2Jy survey (Fisher 1992), but more importantly, the Mark-2 sample of 496 galaxies is being superceded by the Mark-3 sample of 2900 galaxies (Faber *et al.* 1994, Willick etal 1995a). Several new analyses are underway and results will be reported in due course. Here we give a progress report.

Willick *et al.* (1995b) compare the observed distribution of the magnitudes, redshifts, and linewidths for the Mark-3 data, and have cast the problem in terms of a likelihood of measuring a set of observed magnitudes, given the positions of the galaxies in redshift space and observed linewidths. The method accounts for the inhomogeneous space distribution by means of the observed low resolution density field measured from the IRAS sample. They have devised an algorithm that explicitly attempts to neutalize the effects of the triple valued zones in a statistical sense. Distortions from real to redshift space are provided by flow models derived from the IRAS gravity field, with β the only adjustable parameter. The procedure then asks which value of β is most likely to fit the observed distribution. The test favors a tentative value of $\beta = 0.55 \pm 0.13$. Details are given by Strauss and Willick (1995). While the method provides a powerful measure of which IRAS flow model is the best fit to the data, it does not in itself provide an independent picture of the flow field, which POTENT does provide, and which gives POTENT such a visual appeal.

Two alternative schemes of intercomparing the velocity and gravity fields have recently been presented by Nusser and Davis (1994a,b). The first is limited to a consideration of the dipole component of the radial peculiar velocity field and depends on a simple, but little used fact of potential theory. Consider a decomposition of the radial field on a given redshift shell into spherical harmonics. Let the shell have radius r and position a charge (mass) at some location $\mathbf{R}$. For a potential field Φ satisfying the Laplace equation, the lth multipole Φ_l will scale as r^l or $r^{-(l+1)}$, depending on whether $R > r$ or $r < R$, and the gradient of the potential ($\mathbf{v}$ or $\mathbf{g}$) will scale as r^{l-1} or as $r^{-(l+2)}$. We are used to the iron sphere theorem of Newton: for $l = 0$, a particle feels no force from the mass external to it. Note that for $l = 1$, $d\Phi_1/dr$ is constant for $r < R$. That is, by transforming to the local group frame, the dipole amplitude of the velocity on all shells with $r < R$ is zero, and the influence of an external charge is measurable only for $l \geq 2$. In the local frame, the measured dipole on a given shell must be due entirely to the distribution of charge (mass) internal to that shell.

Such a situation overcomes a serious objection to the use of Equation (3)- that the gravity field is nonlocal. Since the estimates derived from IRAS do not integrate the mass density over all space, they might be missing some large scale components. This problem is explicitly avoided by limiting the consideration to the behavior of the dipole amplitude. Nusser and Davis (1994a) extracted the dipole component of the velocity field from an early version of the Mark-3 POTENT map and compared it to the dipole field of the IRAS gravity maps, for various values of β (Figure 1). They find that the velocity and dipole vector directions are aligned within 11°, and furthermore that wiggles in the amplitude of the dipole versus redshift z are remarkably consistent. Note that the POTENT dipole does not go to zero at the origin because of the large smoothing. $\beta = 0.6 \pm 0.2$ seems to be the preferred fit, in agreement with Willick *et al.* (1995). One very important point from Figure 1 is that 450 km/s of shear is clearly generated within a distance of 5000 km/s, a distance to which the IRAS density field should be reliable. Only the 250 km/s difference of this measured shear amplitude and the 620 km/s amplitude of the CMBR dipole can be induced by larger scale bulk flows. This must be kept in mind in the interpretation of claims of the detection of a bulk flow on a much larger scale (e.g. Lauer & Postman 1994).

4. INVERSE TULLY-FISHER RECONSTRUCTION OF A SMOOTHED FIELD

To generalize this approach, one might consider directly comparing the velocity field derived from POTENT with the IRAS gravity field for higher multipoles. This approach is not advisable, because the distribution of points comprising the Mark-3 sample are not uniformly distributed on the sky or in redshift, so that the multipoles are not orthogonal. Furthermore POTENT is constructed on a cubic grid, which is not suitable for higher order multipole comparison.

Consider an alternative approach that starts afresh with the measured magnitudes and linewidths. Nusser and Davis (1994b) describe an inverse Tully-Fisher algorithm based on a flow model which is a low order, general description of the velocity field in redshift space. Assume that the Tully Fisher relationship is linear, i.e.

$$\eta_i \equiv ln(\Delta v_i) = \eta_0 + sM_i \pm \epsilon \ , \tag{4}$$

where η_0 and s are the unknown zero point and slope of the regression, ϵ is the error in the relation with $\langle \epsilon \rangle = 0$, and

$$M_i = m_i - 5log(z_i - u_i) - 15 \ . \tag{5}$$

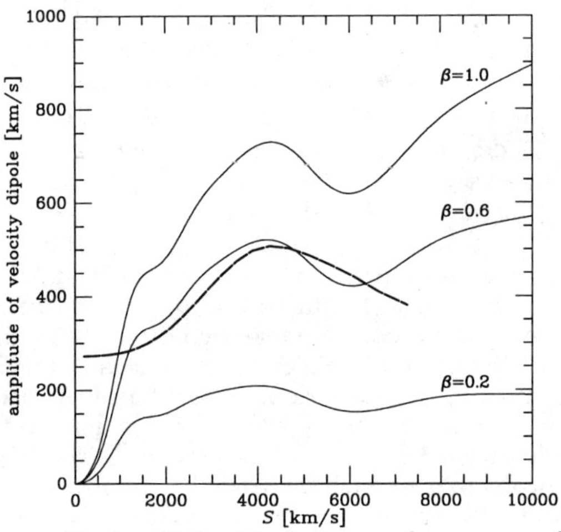

Fig. 1: The dipole amplitude of the Mark-3 sample compared to the IRAS predicted dipole amplitude for various β values.

Here m_i is the apparent luminosity, z_i is the redshift, and u_i is the peculiar velocity of the ith object in the LG frame. Let $M_i = M_{0i} + P_i$ with $M_{i0} = m_i - 5log(z_i) - 15$ and $P_i = -5log(1 - u_i/z_i)$. Then, if the TF scatter of galaxies in the η direction is σ_η, one can compute a χ^2 statistic that minimizes the η scatter as a function of the parameters that describe the peculiar velocity field P. Such an *inverse* fit to the Tully-Fisher relationship does not suffer the usual Malmquist biases, but is biased in the presence of a "hot" velocity field or if there is a selection in the sample based on η. The typical bias in the inferred velocity is $\Delta u_{bias} \approx \sigma_u^2/z$. Given that the rms scatter of distances at a given redshift σ_u is thought to be less than 300 km/s in most regions of redshift space, the bias of this technique is quite modest.

The problem with inverse methods is that, in contrast to a forward method such as POTENT, one does not generate pictures, but can only fit parameters to a model. However, with large datasets such as Mark-3, which have nearly full sky coverage to substantial depths, this is not a serious limitation. Consider for example an expansion of P in terms of a set of functions that are orthogonal over a given dataset,

$$P_i = \sum_j \alpha_j \tilde{F}_i^j$$

where the α_j are the coefficients and $\tilde{F}_i^j$ is the value of the jth orthogonal function for object i, and

$$\sum_i \tilde{F}_i^j \tilde{F}_i^{j'} = \delta_K^{jj'} ,$$

with δ_K the Kronecker delta function. Orthogonal functions are a convenient choice, as the χ^2 minimization is linear in all the coefficients α_j as well as η_0 and s, and each term of the orthogonal expansion is decoupled from the others so that terms can be added one by one.

A natural choice of functions for a general expansion of the radial peculiar velocity field is again motivated by potential theory. Because we are working in the local group frame, $P(0) = 0$. One can guarantee the correct asymptotic behavior near the origin by expanding the field P in terms of the derivatives of spherical Bessel functions times spherical harmonic functions $Y_{lm}(\theta, \phi)$,

$$P(z, \theta, \phi) = \sum_{nlm} a_{nlm} F_i^{nlm} = \sum_{nlm} a_{nlm} \frac{j_l'(k_n z)}{z} Y_{lm}(\theta, \phi) \ . \qquad (6)$$

See ND2 for details. One can choose k_n by the desired boundary conditions at some $cz_{max} = 10000$ km/s, such as requiring $F(z_{max}) = 0$, but the solutions are insensitive to the details of this choice since the Mark-3 sample sets little constraint on the flow at this redshift. Fisher *et al.* (1994) have used similar sets of wavefunctions to describe the density field of the IRAS 1.2 Jy catalog. The choice of radial wave function can be flexible, as the method is only meant to be a fitting formula, and the irregular distribution of galaxies in redshift space prevents the original basis functions from being orthonormal. A convenient choice that leads to lower χ^2 with fewer degrees of freedom is to use radial wavefunctions $j'(k_n y)/z$ where $y = (\ln(1 + (z/2000)^2)^{1/2}$. This has the effect of stretching the oscillations of the radial wavefunctions at larger redshift, allowing more radial resolution in the foreground than in the background, so as to match the resolution gradient of the gravity field and the absolute accuracy gradient of the observed peculiar velocities. By working to order $n_{max} = 5$, $l_{max} = 4$, we have 125 independent modes. Five additional modes describing an external quadrupole field ($F =$ constant $\times Y_{2m}(\theta, \phi)$) are also included for good measure. We transform the functions F_i^j into an orthogonal basis set $\tilde{F}_i^j$ by means of an SVD inversion (Press *et al.* 1992).

To confirm that the method works, we constructed 2600-point mock Mark-3 catalogs derived from nbody simulations, including inhomogeneous sky selection and scatter in the TF relation of $\sigma_\eta = 0.03$. Of the 130 modes, only approximately 80 are determined with significance greater than one sigma, but these 80 coefficients contain a full description of the velocity field to the chosen resolution. The scatter in η about the fit solution is spatially random in the simulations, indicating that fits to this resolution are adequate for the size and precision of the catalog. More detailed fits can be performed with larger set of Tully-Fisher data, or with more precise distance indicators.

To test whether the estimated mode amplitudes are correct, we have evaluated each mode amplitude using the true peculiar velocity of each simulation point. Figure 2a shows the comparison of the true versus ITF inferred mode amplitude for the 130 coefficients derived from the mock catalog. No bias in the mode amplitudes are detected, and the scatter of the amplitudes is completely consistent with the expected noise. Furthermore, most of the power of the large scale flow is contained within a few basis functions. For the true sky, we of course do not know a priori the true peculiar velocity of each point, and we can only substitute the linear-theory predicted peculiar velocity derived from the IRAS gravity maps. Figure 2b shows this test with the same mock Mark3 catalog, now substituting an IRAS derived estimate of the peculiar velocity of

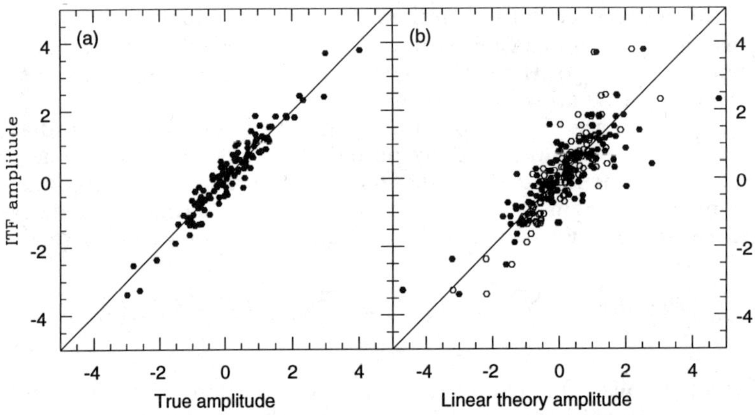

Fig. 2: (a) The ITF inferred amplitude of the orthogonal modes of a mock Mark-3 catalog are plotted versus the true mode amplitude. (b) The solid symbols show the ITF inferred amplitudes versus the amplitude measured from the linear theory predictions for $\beta = 1$ (the correct answer in this simulation). The scatter is larger than Fig. 2a, but the slope remains unbiased. The open symbols are for a gravity field with $\beta = 0.6$; note that the diagonal line is now a poor fit.

each point in the evaluation of the mode amplitudes. A flux limited mock IRAS catalog with the same selection function as the 1.2Jy survey was extracted from the simulation data and used to estimate the gravity field, so this test is a complete simulation of the real situation. Now a comparison of ITF inferred versus IRAS inferred mode amplitude shows more scatter, but still a complete absence of bias. The simulation is evaluated for $\beta = 1$, the true answer in this case. Evaluating the gravity field for other β values changes the slope of the scatter diagram, so estimation of β is completely straightforward. A gravity field evaluated using $\beta = 0.6$ is shown as the open symbols, and is clearly a poor fit to the diagonal line. With a suitable catalog of distance measurements, this method should lead to a definitive measure of β.

The great advantage of this scheme is that it allows the comparison of the velocity and gravity field mode by mode, with identical resolution to each field. The Monte-Carlo simulations with realistic mock catalogs confirm that the bias in the extracted velocity field is negligible. The method allows for the gradient in radial resolution of the two fields, so it permits the extraction of the maximum useful information for each field. One can rank order the modes by their signficance, or contribution to χ^2, keeping only the most significant modes, so as to provide a very compact description of the field. This is an example of a Karhunen-Lo‘eve transformation (Therrien 1992). Recent applications of this technique to astrophysics include the CMBR fluctuations of the COBE data (Gorski 1994, Bunn 1994, Bond 1994), the power spectral analysis of fluctuations in redshift surveys (Vogeley 1994), and the spectral classification of galaxies (Connolly *et al.* 1994).

However, at the present time, this method is not yielding sensible results when applied to the Mark-3 catalog. There is strong coherence in the residuals between the ITF velocity field and the gravity field which is unphysical and suggestive of a calibration error. Work is in progress to iron out the problem and results will hopefully be available soon.

5. WHY DON'T THE ALTERNATIVE METHODS YIELD A CONSISTENT β ?

The POTENT-IRAS density comparison resulted in an estimate of $\beta > 1$, while the v-v comparisons discussed above are giving a best value half as large. What could account for such a discrepancy?

It is important to note that, while the different methods might start with the same basic data, their complex processing is rather different. In the $\delta - \delta$ analyses, the weighting is dominated by infall to cluster centers and outflow from voids. Larger scale components of the velocity field are less important than the local divergence. The direct velocity-velocity comparisons, on the other hand, are dominated by the signature of the dipole reflex of the motion of the local group. Figure 1 shows that 2/3 of the motion of the LG is reflected within the galaxy distribution to redshift 6000 km/s. Giovanelli *et al.* (1994) are claiming the reflex dipole is larger than shown in Figure 1, which would increase the inferred β toward the POTENT value. The coherent residuals between the ITF inferred field and the IRAS predicted field for the Mark-3 sample are certainly cause for concern. Until we understand better the reason for this discrepancy, we must beware of possible systematic errors in β values so derived.

The subject of large scale flows is rapidly evolving. The large datasets from Giovanelli *et al.* (1994) and Mark-3 are likely to be released soon, and other, more accurate catalogs of peculiar velocities (e.g. using surface-brightness fluctuations) are under construction. The 1.2Jy IRAS catalog will be soon superceded by the IRAS catalog of 15,000 galaxies with flux greater than 0.6Jy (Saunders *et al.* 1994) and by the ORS/IRAS catalog (Santiago *et al.* 1994), both of which will offer denser sampling and therefore improved statistical precision to the gravity field. The large scale flow analyses should yield a β value accurate to 10%, and the various estimates should certainly be consistent. Since progress continues at a rapid pace, it is perhaps premature to expect consistency on this important parameter at the present moment.

ACKNOWLEDGEMENTS

This research was supported by grants from the NSF and NASA.

REFERENCES

Bertschinger, E., Dekel, A., Faber, S., Dressler, A., & Burstein, D 1990, ApJ 364, 370
Bond, J. R. 1994, Les Houches Lectures, preprint
Burstein, D. 1990, Rep. Prog. Phys., 53, 421
Bunn, T., 1995, Ph.D thesis, UC Berkeley
Connolly, A. J., Szalay, A. S., & Bershady, M. A. 1995, AJ, in press
Davis, M. & Peebles, P. J. E. 1983, Ann. Rev. of A&A, 21, 109
Dekel, A., Bertschinger, E., & Faber, S. 1990, 364, 349
Dekel, A. 1994, Ann. Rev of A&A, 32, 371
Dekel, A., Bertschinger, E., Strauss, M., Yahil, A., Davis, M., & Huchra, J. 1993, ApJ, 412, 1

Faber, S.M., & Burstein, D. 1988, in *Large Scale Motions in the Universe: A Vatican Study Week*, eds. V. C. Rubin & G. V. Coyne, (Princeton University Press), 116
Faber, S.M., Courteau, S, Dekel, A., Dressler, A., Kolatt, T., Willick, J. & Yahil, A 1994, in *Cosmic Velocity Fields*, eds. F. Bouchet & M. Lachi'eze-Rey, (Gif-sur-Yvette Cedex; Editions Frontieres), 15
Fisher, K. 1992, Ph.D thesis, UC Berkeley
Fisher, K., Scharf, C., & Lahav, O. 1994, MNRAS, 266, 219
Fisher, K., Huchra, J., Strauss, M., Davis, M., Yahil, A., & Schlegel, D. 1995, ApJ Supp, submitted
Giovanelli, R., Haynes, M., Salzer, J., Wegner, G., da Costa, L., & Freudling, W. 1994, AJ, 107, 2036
Gorski, K. 1994, ApJL. , 430, L85
Kaiser, N., Efstathiou, G., Ellis, R., Frenk, C., Lawrence, A., Rowan-Robinson, M., & Saunders, W. 1991, MNRAS, 252, 1
Lauer, T., & Postman, M. 1994, ApJ, 425, 418
Nusser, A., & Davis, M. 1994a, ApJ Lett, 421, L1
Nusser, A., & Davis, M. 1994b, MNRAS, submitted
Press, W., Teukolsky, S., Flannery, B., & Vetterling, W. 1992, *Numerical Recipes, The Art of Scientific Computing*, 2nd edition, Cambridge University Press
Riess, A. G., Press, W., & Kirshner, R. 1994, ApJ Lett., submitted
Santiago, B. X., Strauss, M., Lahav, O., Davis, M., Huchra, J., & Dressler, A., 1995, ApJ, in press
Saunders, W. *et al.* 1994, in preparation
Schechter, P. 1980, ApJ, 85, 801
Strauss, M. 1989, Ph.D thesis, UC Berkeley
Strauss, M., & Davis, M. 1988, in *Large-Scale Motions in the Universe*, ed. V. Rubin & G Coyne (Princeton University Press), p.255
Strauss, M., Huchra, J., Davis, M., Yahil, A., Fisher, K., & Tonry, J. 1992, ApJS, 83, 29
Strauss, M., & Yahil, A. 1994, unpublished
Strauss, M., & Willick, J., 1995, Physics Reports, in press
Therrien, C. W. 1992, *Discrete Random Signal and Statistical Signal Processing*, (New Jersey: Prentice-Hall)
Tonry, J. *et al.* 1994, in preparation
Vogeley, M. 1994, preprint
White, S., Navarro, J., Evrard, A., & Frenk, C., 1993, Nature, 366, 429
Willick, J. *et al.* 1994a, ApJ, to be submitted
Willick, J. *et al.* 1994b, ApJ, to be submitted
Yahil, A., Strauss, M., Davis, M., & Huchra, J. 1991, ApJ, 372, 380

Using Galaxy Clusters to Trace the Large Scale Velocity Field

Marc Postman
Space Telescope Science Institute
3700 San Martin Drive, Baltimore, MD 21208

ABSTRACT

The large scale velocity field is a key indicator of the cosmic density and of the power spectrum of the density field. I discuss recent efforts to measure peculiar motions on scales of $\sim 100h^{-1}$ Mpc with particular focus on the Abell cluster survey of Lauer and Postman. The Lauer–Postman data are compared with a variety of cosmological models and are consistent with the detection of a bulk flow of unprecedented amplitude. Several new peculiar velocity surveys are underway which will begin to probe scales comparable with the Abell cluster survey. The combined datasets should reveal important insight into the large scale distribution of matter, both light and dark.

1. INTRODUCTION

Constraints on the evolution of the large scale distribution of mass in the universe are becoming available from a number of different observational programs. All viable theories of structure formation must, therefore, explain this extensive and complementary array of data. The three most significant observational constraints on the large scale mass distribution are 1) the significant fluctuations ($\sim \pm 50\%$ about the mean density) in the present epoch space density of galaxies on scales of $\sim 100h^{-1}$ Mpc seen in large area redshift surveys (de Lapparent *et al.* 1986; da Costa *et al.* 1994; Shectman *et al.* 1992), 2) the detection of bulk flows on scales of $\sim 100h^{-1}$ Mpc, and 3) the detection of significant fluctuations on 1 degree scales in the cosmic microwave background (CMB) ($\Delta T/T = 32^{+14}_{-12}\mu$K, Clapp *et al.* 1994). Sub degree scales at the decoupling epoch ($z \approx 1100$) correspond to present epoch scales $\lesssim 300h^{-1}$ Mpc.

The large scale fluctuations in the space density of galaxies is demonstrated in Figure 4 from da Costa *et al.* and is repeated here in Figure 1. These fluctuations are due to large sheet-like structures (*e.g.*, "The Great Wall") and large voids that are seen in the combined CFA/SSRS survey discussed in da Costa *et al.*The fluctuations have large coherence lengths, comparable with the depth of the survey ($\sim 100h^{-1}$ Mpc). The combined CFA/SSRS survey spans 7900 square degrees (20% of the whole sky) and contains 9325 galaxies (with $m_B \leq 15.5$). The fluctuations are, thus, quite real (note the small error bars in Figure 1).

One consequence of such large scale, small amplitude fluctuations (assuming the gravitational instability model is accurate) is the generation of coherent streaming motions on scales comparable to the size of the largest structures. The scale and amplitude of peculiar velocities constrains the value of $\Omega_o^{0.6}/b$ (where b is the bias factor relating the mass fluctuations to the observed fluctuations in the galaxy distribution) and the initial density fluctuation power spectrum. Hence, the measurement of the large scale velocity field is an important component to understanding the nature and amount of dark matter in the universe. The amplitude of mass fluctuations is expected to decrease with increasing spatial scale, so if we average over a big enough volume of space to obtain a fair sample of the large scale structure of the universe, then the average motion of the galaxies contained within the volume will be at rest with respect to the CMB. In 1989, Lauer and I set out out to test for such "convergence" by

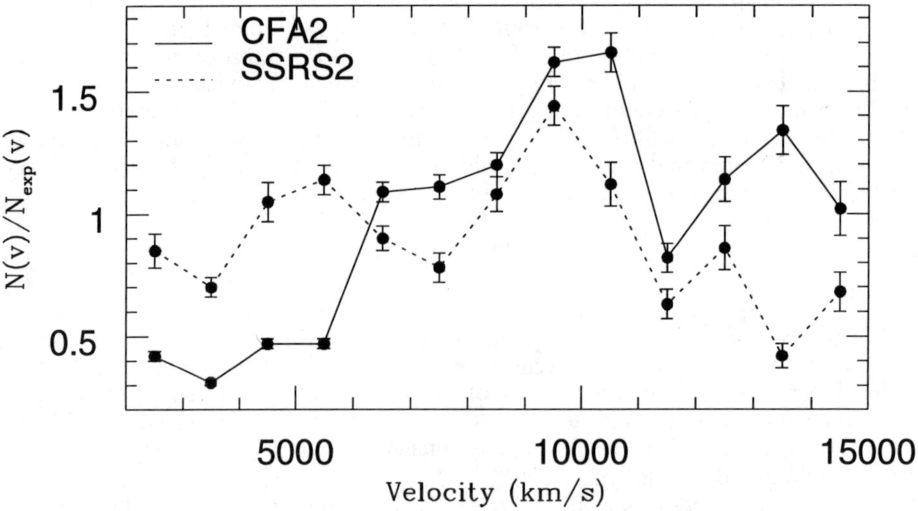

FIG 1.— Variation of the normalized luminosity function amplitude. The data are normalized to N_{exp}, the number of galaxies expected for a uniform distribution with the identical survey parameters. Error bars are 1 σ Poisson errors. Taken from da Costa *et al.* 1994.

constructing an all-sky sample of 119 brightest cluster galaxies (BCGs) out to a limiting redshift of 15,000 km s^{-1}. The use of BCGs as standard candles began with the work of Sandage (1972) and a modern re-evaluation of their status as distance indicators is presented in Postman & Lauer 1995. This was the first volume-limited peculiar velocity survey and the first to probe such large scales. The Lauer-Postman survey (1994; hereafter LP) was exclusively designed to measure the reflex motion of the Local Group with respect to the Abell cluster reference frame. With the exception of the LP survey, peculiar velocity surveys have been limited to depths of about 7000 km s^{-1}. It is fair to say that "convergence" has not been convincingly demonstrated by any survey to date. By convergence I mean that the motion of the Local Group with respect to the frame defined by the objects in a survey matches the motion implied by assuming that the CMB dipole is kinematic (i.e., that there is NO significant bulk motion seen if one switches to the CMB frame).

The "Great Attractor" model of Lynden-Bell *et al.* (1988) attempts to explain the motion of the Local Group and the apparent flow of the galaxies within the Local Supercluster as a consequence of a large mass concentration centered at 4500 km s^{-1}. Recently, Courteau *et al.* (1993) and Mould *et al.* (1993) argued for the existence of coherent bulk-flows out to at least 6000 km s^{-1} scales —

indeed Plionis *et al.* 1991 and Scaramella *et al.* 1991, who estimate the peculiar *acceleration* of the Local Group from the distribution of Abell clusters, argue that convergence may not occur until 18,000 km s^{-1}. The LP dipole measurement is inconsistent with the Local Group velocity inferred from the CMB dipole anisotropy. If the LP dipole measurement is due to a bulk flow, one possible cause could be a fluctuation with $0.2 \lesssim \delta\rho/\rho \lesssim 0.6$ on scales of $100h^{-1}$ Mpc (the lower limit is for $\Omega_o = 1$, the upper limit for $\Omega_o = 0.2$). However, this calculation assumes a single, spherical overdensity. A long wavelength density gradient may induce significantly smaller velocities. It remains to be seen if anisotropic galaxy distributions exist on large scales and over the remaining regions of the sky not yet mapped by the redshift surveys. This would further imply that the peculiar motion of the Local Group is significantly effected by mass concentrations at or beyond scales of $100h^{-1}$ Mpc. Alternate interpretations of the LP dipole are discussed later on. Evidence for significant contributions to the Local Group motion from scales $\gtrsim 100h^{-1}$ Mpc is also presented by Plionis *et al.* (1993) and Hudson (1993) based upon peculiar acceleration analyses of the IRAS/QDOT redshift survey and of a large optical redshift survey, respectively. Most recently, Mathewson & Ford (1994) find good evidence for a bulk flow extending to at least 8000 km s^{-1} based on Tully-Fisher (TF) measurements for 2273 galaxies although the extent of the flow appears to depend on angular direction.

2. CLUSTERS AS TRACERS

The use of clusters as tracers of the large scale velocity field has several advantages. First, clusters do appear to be unbiased tracers of the underlying velocity field — this conclusion is based on large scale N-body simulations recently conducted by Bahcall, Cen, & Gramann (1994) and by Strauss *et al.* (1994). In these simulations, a known velocity field was input and then "measured" by sampling it both at cluster positions (regions of high density in the simulations) and at random positions. The two resulting fields agree remarkably well. Given that clusters may form in special locations where large scale matter streams intersect, it was by no means obvious that clusters should be unbiased tracers of the velocity field. That they are is promising and validates the use of clusters to map large scale velocity fields. Second, because the mean intercluster separation is $\sim 40h^{-1}$ Mpc the clusters trace the linear density field and, thus, should provide constraints on the initial conditions of the density field. Third, many distance estimation methods work exclusively in cluster environs (metric luminosity of the BCG, knee of the cluster luminosity function) and/or the errors in distance can be greatly reduced because multiple, independent observations are possible. The large galaxy overdensity means that SNIa search efficiency is greatly enhanced and these type of supernovae appear to have great potential as distance indicators with typical distance errors of $\sim 5\%$. The $D_n - \sigma$ technique works exclusively on bulge-dominated systems and more than 70% of all elliptical galaxies reside in clusters.

While the distance error associated with using the metric luminosity of BCGs cannot be reduced by multiple observations (i.e., there is, alas, only ONE brightest galaxy per cluster!), we have chosen to use it because BCGs are remarkably good standard candles which can be easily identified and observed out to redshifts well in excess of 15,000 km s^{-1}. The details of the BCG distance indicator and the results of the LP survey are discussed in Lauer & Postman (1994) and Postman & Lauer (1995). Briefly, we reinvestigated the Hoessel

(1980) relationship between the metric luminosity, L_m, within the central $10h^{-1}$ kpc of the BCG and the logarithmic slope of the surface brightness profile, α. The $L_m - \alpha$ relationship reduces the cosmic scatter in L_m from 0.327 mag to 0.244 mag, yielding a typical distance accuracy of 17% per BCG. Residuals about the $L_m - \alpha$ relationship are independent of BCG luminosity, BCG $B - R_c$ color, BCG location within the host cluster, and richness of the host cluster. The metric luminosity is independent of cluster richness even before correcting for its dependence on α, which provides further evidence for the unique nature of the BCG luminosity function. Within the central $10h^{-1}$ kpc the colors are remarkably constant — the mean $B - R_c$ color is 1.51 with a dispersion of only 0.06 mag. The narrow photometric and color distributions of the BCG, the lack of "second parameter" effects, as well the unique rich cluster environment of BCGs, argue that BCGs are perhaps the most homogeneous distance indicator presently available for large scale structure research. We solve for the best fit dipole and find that the velocity of the Local Group relative to the Abell cluster frame ($\vec{L}$) is 561 ± 284 km s^{-1} towards $l = 220°$, $b = -28°$ ($\pm 27°$) when optimal z^{-2} weighting is used. This vector is inconsistent with the Local Group absolute space velocity inferred from the CMB dipole anisotropy at the 99.99% confidence level. If the CMB dipole is kinematic in origin, then this result implies that the Abell cluster frame itself is moving at 689 ± 178 km s^{-1} towards $l = 343°$, $b = +52°$, ($\pm 23°$). Both the Local Group and Frame velocity amplitudes have been corrected for error biasing.

Measuring the Local Group motion relative to distant objects is, like most astronomical observations, a process of extracting a small signal from a noisy background. Therefore, our foremost concern has been to understand and quantify all random and systematic errors and biases. We calculate the random errors in the $\vec{L}$ components through 10,000 Monte Carlo experiments based on the observed scatter in the $L - \alpha$ relationship. We have verified the Monte Carlo-derived errors both by analytic calculation and by the distribution of $\vec{L}$ measured from random cluster subsamples. Our sensitivity to detecting motion of the Local Group with respect to the cluster frame varies with direction, thus our random errors are anisotropic and are given in the form of a tri-axial ellipsoid.

We find no large biases or systematic effects that can generate a spurious motion of the cluster reference frame of the observed magnitude. Indeed, because this is a full sky sample selected by redshift, all variations in sample properties with redshift integrate out to first order. It is also worth emphasizing that in addition to being the first all-sky, volume limited peculiar velocity survey, our method is completely self-calibrating — the $L_m - \alpha$ relation and the Local Group velocity are derived from the same homogeneous dataset.

3. DO WE ALL SEE THE SAME THING?

Figure 2 shows a compilation of recent bulk flow measurements. The three Galactic cartesian velocity components derived from each experiment are plotted as a function of the approximate limiting redshift of the survey. All velocities are in the CMB frame and, thus, an experiment yielding perfect convergence would produce a null vector. Note, the effective depth of the LP sample is between 8000 and 11,000 km s^{-1}. However, since effective depths for the other surveys shown were not readily accessible I have opted to plot the survey limits instead.

The results in Figure 2 are derived from surveys with quite different depths,

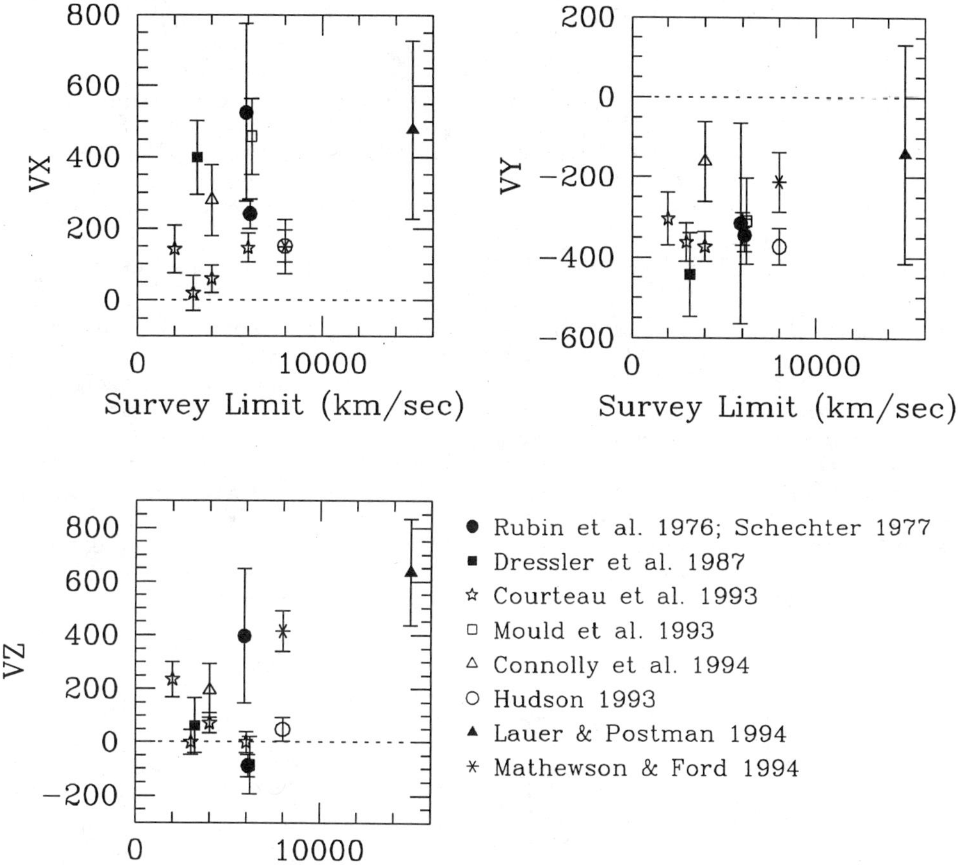

FIG 2.— The bulk flow velocity vector components for different surveys as a function of the survey redshift limit. Error bars are those quoted by the authors (usually 1 σ). All velocities are with respect to the CMB frame.

angular coverages, and methods for extracting peculiar velocity information (TF , $D_n - \sigma$, BCGs, and peculiar acceleration inferred from a redshift survey). Hence, the natural questions to ask in attempting to make sense of Figure 2 are "do we all see the same bulk flow?" and "should we expect to?" Note, even at redshifts less than 5000 km s^{-1}, there are some significant discrepancies between the observations.

In addition to variety of systematic biases discussed by LP and others, there are three effects which, in some cases, may suggest a negative answer to the question "should all surveys expect to see the same flow?" The first is simply one of scale – the LP survey is volume limited to a radius of $150h^{-1}$ Mpc. It is the only survey, to date, with reasonable sampling to such depths. Consequently, the LP survey may be sampling a component of the velocity field hitherto undetected. The other two effects are nicely discussed by Feldman & Watkins (1994). They are biases introduced by sparse sampling and non-uniform distributions. Feldman & Watkins show that these biases are small for the LP sample (introducing a bulk flow amplitude over-estimation of $\sim 10-15\%$ at most). For smaller surveys (with fewer than 20 objects), however, the effects of sparse sampling can be significant causing bulk flows to be over-estimated by as much as 50% to 100% (depending on the survey depth). This is true *even if one were to use noiseless distance indicators.* The importance of such effects must therefore be determined by analytic or numerical computations before conclusions can be made about the "observed" peculiar velocity.

4. CONFRONTING MODELS WITH THE OBSERVATIONS

The LP result is clearly the most difficult one to reconcile with popular models for large scale structure formation. In order to quantify just how seriously the data clash with theoretical expectations, Strauss, Cen, and Ostriker have collaborated with Lauer and myself to construct N-body simulations for six different models of the initial power spectrum (Standard, Tilted, and $\Omega_o = 0.3$ Cold Dark Matter (CDM), Hot Dark Matter (HDM), and two variants of the Primordial Baryon Isocurvature (PBI) model). We use these simulations to create Monte-Carlo realizations of the LP data (complete with identical selection effects) under the six models and compute the probability of detecting bulk flows as large and as statistically significant as that derived from the LP dataset. Our results are described briefly here and in full detail in Strauss *et al.* (1995).

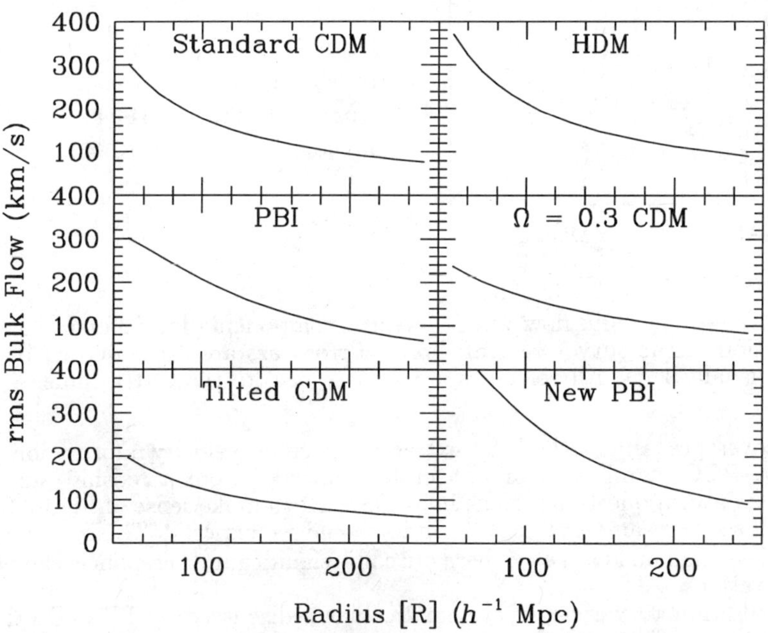

FIG 3.— The rms expected bulk flows in spheres of radius R for the six models described in the text. Taken from Strauss *et al.* 1995.

In this analysis, we assume that structure is formed by gravitational instability from initially small-amplitude perturbations with random phases. Thus, only the power spectrum of the initial density field and the present amplitude of the fluctuations need to be specified in order to determine the statistical properties of the velocity field. All models (with the exception of HDM) are normalized to the observations of CMB anisotropy at 10° scales (Smoot *et al.* 1992). The rms expected bulk flow as a function of scale is shown in Figure 3 for our six models. The parameters of the power spectra are summarized in Table 1 of Strauss *et al.* 1995. All models are spatially flat so that $\Lambda_0 = 1 - \Omega_0$.

Our Particle Mesh N-body simulations are performed using 1.56×10^7 particles in a box $800h^{-1}$ Mpc on a side. We have a single N-body simulation available for each model (these are rather CPU intensive to create!). Because the simulation volume is only about 36 times larger than that of the LP cluster sample we are subject to the finite number of modes sampled on the largest scales and will tend to underestimate the variance of the bulk flows. We can correct for this by making random realizations of the large scale power, both on scales larger than $800h^{-1}$ Mpc (because these scales are not included in the simulation) and on scales comparable to the simulation (because these scales are sampled by only a few waves and do not properly represent the cosmic variance we wish to model). Hence, our simulation procedure is as follows:

1) Subtract off the contribution to the velocity field in each N-body simulation from waves with wavelengths between 200 and $800h^{-1}$ Mpc. Replace these with a random realization of the contribution to the velocity field on scales larger than $200h^{-1}$ Mpc, given the power spectrum and linear theory.
2) Choose a "Local Group" position by selecting an N-body point with peculiar velocity (smoothed on a $1h^{-1}$ Mpc scale) between 520 and 720 km s^{-1}, and a local density (smoothed on a $5h^{-1}$ Mpc scale) between 0.8 and 2.0.
3) Identify clusters that lie within a "heliocentric" redshift of 15,000 km s^{-1}and define a coordinate system in which the Local Group peculiar velocity vector is pointed towards (277,30), the direction of the CMB dipole. Introduce Galactic extinction to model the decline in cluster surface density with latitude. Reject those clusters which fall below the LP latitude limit ($|b| = 15°$).
4) Select the most massive 50% of the clusters that remain, which yields a number of 114 ± 11. Mimic the presence of the Shapley supercluster by not sparse sampling clusters which lie within $300 < l < 360°, 20 < b < 40°$. This does not significantly effect our results but, in conjunction with careful modeling of extinction effects, properly duplicates the shape of the error ellipsoid seen in the real data.
5) Assign BCG surface brightness profiles by randomly drawing from the real BCG profile distribution. Assign an absolute magnitude to the BCG based on the observed $L_m - \alpha$ relation and add a Gaussian-distributed error with standard deviation of 0.244 mag.
6) Duplicate the iterative LP dipole fitting procedure exactly.

Steps 1-6 are repeated 500 times for each N-body simulation and statistics on the estimated bulk flow in the CMB frame are accumulated. The results are shown in Table 1.

TABLE 1

BULK FLOWS IN VARIOUS MODELS

Model	True Bulk Flow (km s^{-1})	Eq. 1 (km s^{-1})	Measured Bulk Flow (km s^{-1})	% exceeding Bulk Flow	% exceeding χ^2
Standard CDM	126 ± 67	123	486 ± 244	11.2 %	2.6 %
HDM	132 ± 58	147	539 ± 257	17.8 %	5.0 %
PBI	101 ± 45	137	445 ± 208	6.2 %	2.2 %
$\Omega_0 = 0.3$ CDM	72 ± 31	126	461 ± 209	9.4 %	1.6 %
Tilted CDM	98 ± 41	92	482 ± 232	11.0 %	4.2 %
New PBI	163 ± 63	184	472 ± 217	10.0 %	3.4 %
Poisson	—	—	519 ± 228	13.0 %	1.0 %
Standard CDM[a]	220 ± 180	176	548 ± 272	18.6 %	6.6 %
Standard CDM[b]	181 ± 86	123	441 ± 211	5.8 %	2.4 %

[a] Cut at $cz = 10,000$ km s^{-1}. [b] Sampled with the exact LP geometry.

The last two columns in Table 1 give the key results. A bulk flow at least as large as the LP result is measured between 6% and 19% of the time in the various scenarios. Hence, a value of around 700 km s^{-1} in a standard CDM model is unusual but not impossible. One must realize that this is not because the bulk flow on these scales is really so large; rather it is due to the appreciable scatter in the peculiar velocity estimation. However, since the LP sample is anisotropic (*e.g.*, there are no data at low Galactic latitudes) the sensitivity to detecting flows is a function of direction and is highest in the direction of the Galactic poles. This, of course, is true for almost any peculiar velocity survey. The CMB dipole is rejected at high significance by LP because the CMB apex lies largely perpendicular to the major axis of the error ellipsoid. Therefore, the more appropriate question to ask is what percentage of the simulations yield bulk flows that are detected with at least the same significance as the inferred LP flow. These results are given in the last column of Table 1. The χ^2 of the detected bulk flow is much more difficult to match in the simulations then its raw amplitude. The models examined are ruled out at confidence levels between 94% and 98%. These results are in very good agreement with the analytic computations of Feldman & Watkins (1994) who use linear perturbation theory and a variety of power spectra to show inconsistency between the models and the LP result at the 95–97% confidence level.

5. THE PRESENT SITUATION

If we accept the simplest interpretation of the LP dipole – that the $z \leq 0.05$ Abell cluster frame exhibits a 689 km s^{-1} bulk flow, then the present situation is that the most popular models for structure formation are "stressed but not fractured!" We also are led to conclude that the contribution to Local Group motion from matter on scales $\gtrsim 100h^{-1}$ Mpc may be as high as 50%. Alternative interpretations of the LP result are 1) it is a statistical fluke (the nearby BCG sample is a "bad" draw or there are large scale gradients in the properties of BCGs), 2) the CMB dipole in an intrinsic property of the CMB radiation field, or 3) there is some problem with the LP data. We feel quite confident in rejecting this last possibility as we have done extensive overlap observations as well as external checks of our photometry (Burstein 1994) and spectra. We have also externally verified our software and have determined that all plausible systematic errors are negligible. Non-kinematic origins for the CMB dipole have been postulated (Gunn 1988, Paczyński & Piran 1990, Turner 1991) but I believe the present observational situation does not justify adopting these scenarios just yet. As for the first alternative, there does not appear to be strong evidence for large scale gradients in BCG properties (*e.g.*, the LP dipole shows no significant dependence on limiting redshift. See also Postman & Lauer 1995). However, if there were such gradients they would presumably effect other galaxian distance indicators as well.

How will all this be resolved? There are several surveys now in progress which all reach to depths of at least $100h^{-1}$ Mpc and which should be completed within the next 2 or 3 years. These include 1) an extension of the BCG survey out to $z = 0.08$ (~ 550 clusters in a nearly volume limited survey) being performed by myself, Lauer, Strauss, and Huchra, 2) a survey by Fruchter, Moore, & Steidel to test the use of cluster luminosity functions as distance indicators, 3) a variety of TF-based surveys extending to $z \sim 0.06$ (Willick; Giovanelli *et al.*), and 4) surveys of type Ia supernovae out to $z \sim 0.1$ by Hamuy *et al.*All are being done independently and use independent distance estimations methods. The results should allow us to better understand the nature of the large scale velocity field and, ultimately, the nature of dark matter in the universe.

ACKNOWLEDGEMENTS

Tod Lauer and Michael Strauss deserve equal billing with me for the results presented here. It is a pleasure to have them as colleagues in this exciting work.

REFERENCES

Bahcall, N. A., Cen, R., & Gramann, M. 1994, 430, L13
Burstein, D. 1994, private communication
Clapp *et al.* 1994, ApJ, 433, L57
Connolly, A., Joseph, B., Collins, C., & James, P. 1994, BAAS, 1473
Courteau, S., Faber, S. M., Dressler, A., & Willick, J. A., 1993, ApJ, 412, L51
da Costa, L. N. *et al.* 1994, ApJ, 424, L1
de Lapparent, V., Geller, M. J., & Huchra, J. P. 1986, ApJ, 302, L1
Dressler, A., Faber, S. M., Burstein, D., Davies, R. L., Lynden-Bell, D., Terlevich, R. J., & Wegner, G. 1987, ApJ, 313, 42
Feldman, H. A., & Watkins, R. 1994, ApJ, 430, L17
Gunn, J. E. 1988, in *The Extragalactic Distance Scale,* eds. S. van den Bergh and C. J. Pritchet, (ASP Conf. Series), 344
Hoessel, J. G. 1980, ApJ, 241, 493
Hudson, M. 1994, MNRAS, 266, 475
Lauer, T. R., & Postman, M. 1994, ApJ, 425, 418
Lynden-Bell, D., Faber, S. M., Burstein, D., Davies, R. L., Dressler, A., Terlevich, R. J., & Wegner, G. 1988, ApJ, 326, 19
Mathewson, D. S., & Ford, V. L. 1994, ApJ, 434, L39
Mould, J. R., Akeson, R. L., Bothun, G. D., Han, M., Huchra, J. P., Roth, J., & Schommer R. A. 1993, ApJ, 409, 14
Paczyński, B. & Piran, T. 1990, ApJ, 364, 341

Plionis, M., & Valdarnini, R. 1991, MNRAS, 249, 46
Plionis, M., Coles, P., & Catelan, P. 1993, MNRAS, 262, 465
Postman, M., & Lauer, T. R. 1995, ApJ, in press
Rubin, V. C., Ford, W. C., Thonnard, N., Roberts, M. S., & Graham, J. A. 1976, AJ, 81, 687
Sandage, A. 1972, ApJ, 178, 1
Scaramella, R., Vettolani, G., & Zamorani, G. 1991, ApJ, 376, L1
Schechter, P. L. 1977, AJ, 82, 569
Shectman, S. A., Schechter, P. L., Oemler, A. A., Tucker, D., Kirshner, R. P., & Lin, H. 1992, in Clusters & Superclusters of Galaxies, ed. A. C. Fabian (Dordrecht: Kluwer)
Smoot, G.F. *et al.*1992, ApJ, 396, L1
Strauss, M. A., Cen, R., Ostriker, J. P., Lauer, T. R., & Postman, M. 1994, ApJ, in press
Turner, M. S. 1991, Phys. Rev. D, 44, 3737

A SPHERICAL HARMONIC APPROACH TO REDSHIFT DISTORTIONS

A.N. Taylor and A.F. Heavens
Institute for Astronomy, University of Edinburgh,
Blackford Hill, Edinburgh, EH9 3HJ, UK
ant@roe.ac.uk, afh@roe.ac.uk

ABSTRACT

In a recent study, we have developed new techniques for analysing galaxy redshift surveys. The radial distortion introduced by peculiar velocities and the spherical geometry of the survey motivates a Fourier-like expansion of the density field in spherical Bessel functions and spherical harmonics. The power of this technique is that the distortion can be included exactly (to linear order), and the spectral nature of the expansion means that one can isolate and analyse linear modes. For these reasons we have a formalism for treating redshift surveys which avoids the approximations and inaccuracies which have beset previous treatments of this problem, and maximizes the measureable wavelength range.

Exploiting the formalism statistically we can use the observed galaxy distribution to estimate the distortion parameter of redshift space, $\beta \equiv \Omega_0^{0.6}/b$, and the amplitude of fluctuations, quantified by σ_8. We have applied the method to the 1.2-Jy redshift survey, and find that $\beta = 1.1 \pm 0.3$, and $\sigma_{8,\mathrm{IRAS}} = 0.68 \pm 0.05$ in *real space.*

These methods have been developed further to measure the real space power spectrum, free from distortion, and for application to 2-D 'slices' in redshift space.

1. INTRODUCTION

A natural coordinate system for describing all–sky galaxy redshift surveys is that of spherical harmonics. The density field is decomposed by the Bessel–Fourier expansion;

$$\begin{aligned} \rho(\mathbf{x}) &= \sqrt{\frac{2}{\pi}} \sum_{lmn} c_{ln} \rho_{lmn} j_l(k_{ln} x) Y_l^m(\theta, \phi), \\ \rho_{lmn} &= \frac{c_{ln}}{V} \int \rho(\mathbf{x}) j_l(k_{ln} x) Y_l^{m*}(\theta, \phi) d^3x \ . \end{aligned} \tag{1}$$

where c_{ln} are normalization coefficients setting the gradient of the field to zero at the survey boundary, $r_{\max}$.

The measured redshift to a galaxy is the sum of the average recession velocity and the local peculiar velocity, $\mathbf{v}(\mathbf{x})$, projected along the line of sight,

$$\mathbf{x} \longrightarrow \mathbf{s} = H_0 \mathbf{x} + \hat{\mathbf{x}}(\hat{\mathbf{x}}.\mathbf{v}(\mathbf{x})) \ . \tag{2}$$

where H_0 is the Hubble parameter. The coordinates $(s(\mathbf{x}), \theta, \phi)$ are called redshift space.

2. METHODS

If $D(\mathbf{s})$ is the observable perturbed galaxy density field from a flux–limited redshift survey, then the harmonic coefficients measured in redshift space are

$$D_{lmn} = \frac{c_{ln}}{V} \int \Phi(x) W(\Omega) \delta(\mathbf{s}) j_l(k_{ln} s) Y_l^{m*}(\theta, \phi) d^3 s \ , \tag{3}$$

where $\Phi(x)$ is a radial galaxy selection function and $W(\Omega)$ the radial selection for the survey. From continuity $d^3 s \ \rho(\mathbf{s}) = d^3 x \ \rho(\mathbf{x})$, leaving behind s in the radial Bessel function. Expanding the radial velocity field in harmonics we find to first order

$$D_{lmn} = \sum_{l'm'n'} \sum_{l''m''n''} S_{ll''}^{nn''} W_{l''l'}^{mm'} (\Phi_{l''l'}^{n''n'} + \beta V_{l''l'}^{n''n'}) \delta_{l'm'n'} \ , \tag{4}$$

where perturbations in the Hubble flow are related to perturbations in the real space galaxy density by the galaxy continuity equation;

$$\frac{\delta v}{v} = \beta \frac{\delta \rho}{\rho} \quad \text{and} \quad \beta \equiv \frac{\Omega_0^{3/5}}{b} \ . \tag{5}$$

The parameter β is then both the distortion parameter and a measure of the cosmological density parameter, Ω_0. b is the ubiquitous linear galaxy bias parameter, where $\delta_{\text{galaxies}} = b_{\text{galaxies}} \delta_{\text{mass}}$

The transition matrices take into account the effect of observing galaxies in a finite redshift space survey. Φ is the effect of radial galaxy selection in a flux limited survey; V is the linear distortion of the density field due to peculiar velocities mixed with radial selection effects; W takes into account the coupling of modes if the survey has an angular selection function and S corrects the observed density field for distortion by small–scale nonlinear effects (the "fingers of God") and redshift errors. Details of their definitions can be found in Heavens & Taylor 1994. These equations are exact to first order.

3. LIKELIHOOD ANALYSIS

A natural condition for density perturbations formed in the Early Universe to satisfy is that of random phases. Assuming this we can form a likelihood functional, $\mathcal{L}$, for the harmonic modes of the observed redshift–space density field;

$$\begin{aligned} \mathcal{L}[D|\beta, P(k)] &= \frac{\exp(-S)}{\sqrt{\text{Det} C}} \\ S &= \frac{1}{2} \sum D_{lmn} [C_{lmn}^{l'm'n'}]^{-1} D_{l'm'n'} \ . \end{aligned} \tag{6}$$

where $C_{lmn}^{l'm'n'}[P(k), \beta]$ is the covariance matrix of observable density perturbations. With this form we can ask what is the most likely value of β and form of power spectrum, $P(k_{ln})$, that will lead to the distorted pattern of galaxies observed.

This approach has a number of clear advantages – spherical harmonics provide a natural coordinate system for all–sky surveys – radial and angular selection functions are separable; the redshift distortion is radial and can be included exactly in linear theory; small scale, nonlinear distortions and redshift errors can be treated exactly; β is measured without reference to Tully-Fisher or $D_n-\sigma$ based estimates of peculiar velocities which may be susceptible to distance errors and sampling limitations ; the likelihood analysis includes the effects of mode coupling for a finite survey and redshift distortion; phase information is included in the analysis; this is essentially a harmonic analysis, which allows us to split the field into linear and nonlinear scales, and only analyse the linear perturbations; edge effects can be correctly dealt with by imposing boundary conditions; the likelihood analysis provides *independent*, unbiased estimates of β and $P(k)$; and a set of optimal weighting functions for the data can derived from the likelihood functional.

4. RESULTS

We have applied this analysis to the IRAS 1.2-Jy all–sky redshift survey. Taking a two-power law fit to the observed power spectrum, the likelihood functional only depends on the degree of distortion of the galaxy survey, β, and the amplitude of density fluctuations, here parameterized by $\sigma_{8,\mathrm{IRAS}}$. To keep the analysis in the linear regime we restrict wavenumbers to the range $0.001 \leq k/(h\mathrm{Mpc}^{-1}) \leq 0.1$, and applied an optimal weighting scheme to the data.

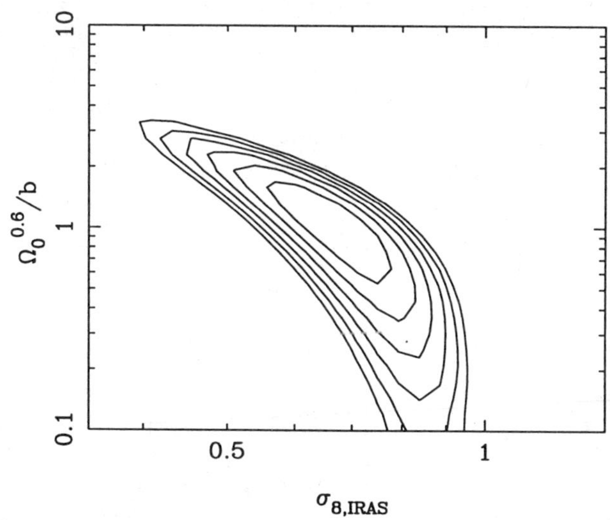

Fig. 1. Likelihood contours of the distortion parameter, $\beta = \Omega_0^{3/5}/b$ and power spectrum amplitude, parameterized by $\sigma_{8,\mathrm{IRAS}}$, measured from the IRAS 1.2-Jy redshift survey. The parametric power spectrum used is an empirical 2 power–law best fit described by Peacock (1991, MNRAS, 256, 1p). Contours are spaced in units of $\Delta \ln(\mathcal{L}) = -1/2$. The inner contour is the 1-σ uncertainty. The degeneracy between distortion and amplitude is broken here due to the inclusion of phase information, and corresponds to an anticorrelation with coefficient -0.7. We used modes in the range $2 \leq l \leq 30$ and $0.001 \leq k/(h\mathrm{Mpc}^{-1}) \leq 0.1$ to ensure linearity.

Figure 1 shows the likelihood contours of parameter space. The amplitude of fluctuations and the β factor are anticorrelated due to their similar effects in redshift space. Neglecting phase information would render these parameters degenerate. The maximum likelihood solution is

$$\begin{aligned} \beta &= 1.1 \pm 0.3 \\ \sigma_{8,\mathrm{IRAS}} &= 0.68 \pm 0.05 \ , \end{aligned} \tag{7}$$

where we quote formal 1σ errors. The correlation coefficient between these parameters is approximately -0.7.

5. FUTURE DEVELOPMENTS

To date, we have exploited the formalism statistically to estimate the density parameter of the Universe (via β, from the anisotropy of the distortion), and the normalization of the real-space power spectrum using the IRAS 1.2Jy galaxy survey (Figure 1). This is now being generalized to measure the *real-space* power spectrum, $P(k)$, free of distortions and for application to surveys which are not all-sky.

There are natural applications for the latter aspect: the IRAS QCCOD survey; the forthcoming proposed AAT 2-degree field survey and the Sloan digital sky survey. It is not obvious that an essentially two-dimensional survey (such as QCCOD) can be analysed to extract large-scale information. The usual difficulty is that three-dimensional power from short wavelengths gets aliased into the two-dimensional power, which makes the interpretation very difficult – the short-wavelength power may need to be known very precisely. This is illustrated in Figure 2(a) for Fourier modes. The problem can be circumvented by choosing a different set of modes (Figure 2(b)). As can be seen, the window function, shown here for a deep survey of characteristic depth $z = 0.1$ is sharply peaked, and the power in this mode is very nearly a genuine measure of three-dimensional power on large scales.

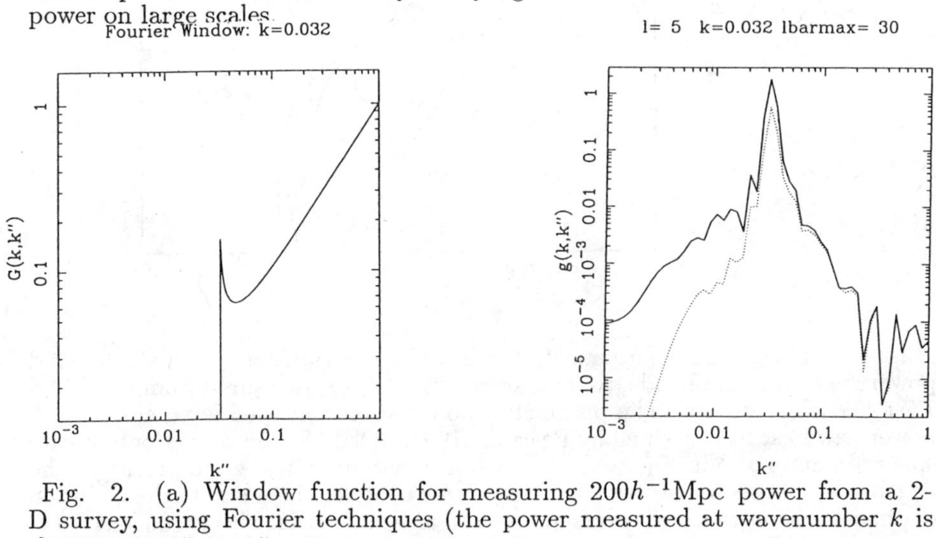

Fig. 2. (a) Window function for measuring $200h^{-1}$Mpc power from a 2-D survey, using Fourier techniques (the power measured at wavenumber k is $\int P(k')G(k,k')d\ln k'$). Notice how the 2-D power does not really measure 3-D power, but is determined by short-wavelength power in 3-D. (b) Window

function for our new method of projection and mode choice. Notice how one is effectively measuring directly large–scale 3-D power, even with a 2-D survey. Dotted line is $\beta = 0$, full line includes redshift distortions for $\beta = 1$. The method is also exact to first order, unlike the Fourier method.

Finally, combining the results for the distortion parameter, $\beta = \Omega_0^{0.6}/b_{\rm IRAS}$, from the 1.2-Jy IRAS survey, the amplitude of temperature fluctuations from the COBE observations of the microwave background and the degree of nonlinearity in the universe, we have calculated the joint distribution function for the IRAS galaxy bias parameter and mean density of the universe. These results are all independent and strongly favor a high mean mass density for the universe, $\Omega_0 > 0.2$. At present this method provides the only practical way of breaking the degeneracy in Ω_0 and $b_{\rm IRAS}$ that have plagued interpretations of large scale structure (see the accompanying paper).

ACKNOWLEDGEMENTS

We are grateful to Marc Davis and the rest of the 1.2 Jy IRAS team for providing the survey prior to publication. ANT is supported by a SERC research assistantship.

REFERENCES

Heavens, A. & Taylor, A.N., 1994, submitted to MNRAS.

GALAXY BIAS AND THE MEAN DENSITY OF THE UNIVERSE

A.N. Taylor and A.F. Heavens
Institute for Astronomy, University of Edinburgh,
Blackford Hill, Edinburgh, EH9 3HJ, UK
ant@roe.ac.uk, afh@roe.ac.uk

ABSTRACT

We have show that the degeneracy in the cosmological density parameter, Ω_0, and the IRAS galaxy bias parameter, $b_{\rm IRAS}$, which has plagued dynamical studies of large scale structure, can be broken. By combining the constraints set by the distortion pattern in the 1.2-Jy redshift survey, the amplitude of fluctuations observed by the COBE satellite, and a constraint set by nonlinear evolution of the density field, we have derived an analytic formula for the likelihood function of Ω_0 and $b_{\rm IRAS}$. While the quality of the data still precludes a tight constraint on the absolute value of these parameters, the 99% confidence contour strongly excludes a low Ω_0 universe. Preliminary results indicate that $\Omega_0 = 1.0 \pm 0.4$ and $b_{\rm IRAS} = 0.9 \pm 0.13$. At the present this method is the only practical way of separating these parameters.

1. INTRODUCTION

Measurement of the mean mass density of the universe by dynamical means has proven to be one of the main unresolved problems for modern cosmology. Arguments exist both for a high mass density, close to the critical Einstein-de Sitter value (separating open and closed universes, with vanishing cosmological constant) and low density model (either spatially curved, or flat and with a cosmological constant). In the low density camp lie small scale virialized clusters, suggesting on scales $\sim 1h^{-1}$Mpc that $\Omega_0 \equiv \rho_0/\rho_{\rm crit} = 8\pi G\rho_0/3H_0^2 \sim 0.1$; bulk flows for $H_0\lambda < 4000\,{\rm kms}^{-1}$, the expansion timescale for $H_0 > 70\,{\rm kms}^{-1}{\rm Mpc}^{-1}$; and – the weak link in the large scale dynamical armour – galaxy bias (see *e.g.* Coles & Ellis 1994)

The evidence for a large mean density comes from large scale observations, where we expect linear theory to apply, but results are perhaps more susceptible to systematic observational selection effects. Leading the high density front is the correlation between IRAS galaxies and the POTENT analysis of radial peculiar velocities. These aim to measure the ratio of density–to–velocity fluctuations, which is approximately equal to $\Omega_0^{5/3}$. But, as we only observe galaxy densities rather than mass, a new degree of freedom is introduced to relate the two. In the IRAS-POTENT case it is usual invoke a strong biasing condition, $\delta_{\rm galaxies} = b_{\rm galaxies}\delta_{\rm mass}$. These dynamical test then actually measure $\beta \equiv \Omega_0^{5/3}/b_{\rm galaxies}$. In one form or another it is this degeneracy that has plagued dynamical studies of large scale structure as a discriminator of cosmological models.

One way round this is to use bias–free constraints. Two examples are the void analysis of Dekel and Rees (1994), and skewness of $\nabla.{\bf v}$ suggested by Nusser and Dekel (1993). However, all of these discriminators rely on the interpretation of radial distance estimators and the POTENT analysis. Both of these methods suffer severely as a consequence of distance errors and sampling limitations.

An alternative is to use velocity field information implied by the distortion of the galaxy pattern viewed in redshift space, where β is now the distortion parameter.However, this still leaves us with the degeneracy between Ω_0 and b. Here we shall combine a number of recent analyses of large scale structure in the framework of gravitational instability where similar degeneracies arise. We show that combining these results leads us to a new likelihood function for Ω_0 and $b_{\rm galaxy}$, where the degeneracy *is* broken.

1. THE CONSTRAINTS

2.1 Redshift Space Distortions

In a recent paper (Heavens & Taylor 1994, and the accompanying paper in these proceedings), we have shown that the redshift space distortion of galaxy surveys is most readily described in the language of spherical harmonics. All the effects of survey geometry, galaxy selection and radial distortion are calculated. Applying a likelihood methodology to the individual modes of the galaxy density field, we arrive at the likelihood function $\mathcal{L}(\beta, \sigma_8)$, which we have measured from the IRAS 1.2-Jy reshift survey (see figure 2 in accompanying paper). The main assumptions here are the *weak* biasing condition, $\sigma_{\rm galaxies} = b_{\rm galaxies}\sigma_{\rm mass}$, and the shape of the power spectrum of density fluctuations, $P(k)$, in the range $0.001 < k/(h{\rm Mpc}^{-1}) < 0.1$. This distribution can be modelled by a bivariate Gaussian with $\beta = 1.1 \pm 0.3$, $\sigma_{8,\rm IRAS} = 0.68 \pm 0.05$, and correlation coefficient $\rho = \langle(\beta - \langle\beta\rangle)(\sigma_{8,\rm IRAS} - \langle\sigma_{8,\rm IRAS}\rangle\rangle/\sqrt{{\rm Var}(\beta){\rm Var}(\sigma_{8,\rm IRAS})} = -0.7$.

2.2 Inflation and the Microwave Background

Assuming an initial power–law spectrum, $P(k) \sim k^n$, and a spatially flat cosmology ($\Omega_0 + \Omega_{\rm vac} = 1; \Omega_{\rm vac} = \Lambda/3H_0^2$), predicted by generic inflationary models, we can calculate the observed rms quadrupole anisotropy, $Q_{\rm rms-PS}$, including the effects of gravitational wave contribution for $n \neq 1$. From this, and the empirical power spectrum from Peacock (1991), we find the relation

$$\sigma_8 \Omega_0^{0.77} = 0.7 \ , \tag{1}$$

for the mass σ_8. As this is derived from a large number of multipoles, the error distribution on this relation is nearly Gaussian. We shall restrict our analysis here to the scale invariant $n = 1$ spectrum favoured by COBE, and use the value of $Q_{\rm rms-PS}$ found by Gorski *et al.* (1994).

2.3 Nonlinear Structure in the Universe

Our final constraint on the amplitude of fluctuation and the mean mass density can be calculated from the degree of nonlinearity in the observed power spectrum. In a low density Universe, fluctuations are larger, and so become nonlinear earlier than in a high density Universe. Assuming that galaxies today trace the nonlinear power spectrum of density perturbations, and that the power spectrum of clusters is linear, we can use the Ω_0-dependent mapping of Peacock & Dodds (1994) to estimate the effect on the linear mass σ_8. Peacock & Dodds find

$$\sigma_8 \Omega_0^{0.15} = 0.75 , \tag{2}$$

with a formal rms uncertainty of 13%. To this we add an extra error of 10% for nonlinear evolution of σ_8 (in fact the observed galaxy power spectra suggests that very little nonlinear evolution has taken place).

3. THE LIKELIHOOD FUNCTION OF Ω_0 AND GALAXY BIAS

If we assume that the errors on the rms quadrupole and cluster abundance results are independently Gaussian distributed, then the likelihood function $\mathcal{L}(\Omega_0, b_{\rm IRAS})$ can be calculated by integrating over all values of σ_8. (This requires integrating over the unphysical regime of $\sigma_8 < 0$, but the distributions have fallen almost to zero here, and add little to the final result). We find that

$$\mathcal{L}(\Omega_0, b) = \frac{1}{(2\pi)^{3/2}\Delta^{1/2}} \exp\left(-\frac{1}{2}\Sigma\right), \tag{3}$$

where

$$\begin{aligned} \Delta &= \nu_1^2\nu_2^2 + (\nu_1^2 + \nu_2^2)(1-\rho^2), \\ \Sigma &= \frac{\psi^2}{1-\rho^2} + (\phi_1^2 + \phi_2^2) - \frac{[(\phi_1\nu_2 + \phi_2\nu_1)(1-\rho^2) - \nu_1\nu_2\psi\rho]^2}{\Delta(1-\rho^2)} , \end{aligned} \tag{4}$$

and

$$\begin{aligned} \nu_i &= \mathrm{Var}(\gamma_i\sigma)/\gamma_i^2\mathrm{Var}(\sigma), \\ \phi_i &= (\gamma_i\langle\sigma\rangle - \langle\gamma_i\sigma\rangle)/\sqrt{\mathrm{Var}(\gamma_i\sigma)}, \\ \psi &= (\beta - \langle\beta\rangle)/\sqrt{\mathrm{Var}(\beta)} , \end{aligned} \tag{5}$$

where $i = 1, 2$ and $\gamma_1 = \Omega_0^{0.56}/b$, $\gamma_2 = \Omega_0^{\alpha/2}/b$.

Figure 1 shows the results for IRAS, COBE and the nonlinear mapping constraints. It is clear that low values of Ω_0 and $b_{\rm IRAS}$ are strongly excluded. The 1σ formal errors are $\Omega_0 = 1.0 \pm 0.4$ and $b_{8,\rm IRAS} = 0.90 \pm 0.13$. The 99% confidence limits show $\Omega_0 > 0.2$, and $b_{\rm IRAS} > 0.5$.

4. CONCLUSIONS

We have shown that the likelihood distribution $\mathcal{L}(\beta, \sigma_{8,\rm IRAS})$, measured from the redshift space distortion of the IRAS 1.2-Jy redshift survey, the constraints set by the cosmic microwave background anisotropy and the amplitude of nonlinear distortion of the observed power spectrum, suggests that the mean mass density of the universe is high, with a preferred value of Ω_0 near unity. This result is a consequence of the large peculiar velocities inferred by redshift space anisotropy, the depth of the potential wells generating the Sachs–Wolfe effect and small degree of nonlinear density evolution, and is independent of assumptions about the dark matter species. This rules out the spatially flat,

$\Omega_0 = 0.2$ CDM model required to fit the APM correlation function and COBE (Efstathiou *et al.* 1991).

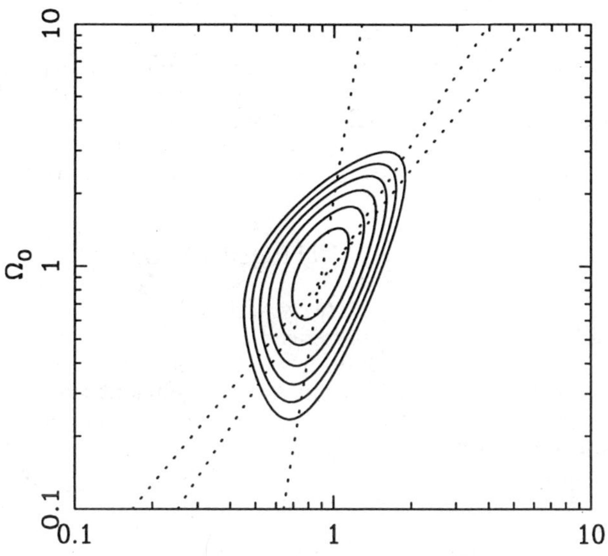

Fig. 1. The likelihood function $\mathcal{L}(\Omega_0, b_{\rm IRAS})$, given the constraints of redshift distortion in the IRAS 1.2-Jy galaxy survey, the amplitude of fluctuations in the microwave background measured by COBE, and the degree of nonlinearity of large–scale structure. Dotted lines show each degenerate solution for $\sigma_8 = 0.68$. Contours are spaced in units of $\Delta \ln \mathcal{L} = -1/2$

ACKNOWLEDGEMENTS

ANT is supported by a SERC research assistantship.

REFERENCES

Coles P. & Ellis G., 1994, Nature, 370,609
Dekel, A. & Rees, M., 1994, ApJ Lett, 422, L1
Efstathiou, G.P., Sutherland, W.J. & Maddox, S.J., 1990, Nature, 348.705
Gorski *et al.*, 1994, ApJ Lett, 430, L89
Heavens, A. & Taylor, A.N., 1994, submitted to MNRAS
Nusser, A. & Dekel, A., 1993, ApJ, 405, 437
Peacock J.A., 1991, MNRAS, 253, 1p
Peacock, J.A. & Dodds, S.J., 1994, MNRAS, 267, 1020

Spatial Clustering of Radio Sources
in the 1987 Green Banks Survey

Hugues Sicotte and P.J.E. Peebles
Jadwin Hall, Princeton University, Princeton, NJ 08544
sicotte@pupgg.princeton.edu

ABSTRACT

We analyzed a large sample of 54579 radio sources from the 1987 Green Banks survey by Condon *et al.*.(87GB) The sources in that flux limited sample have a mean redshift $z \sim 0.7$, allowing us to probe clustering on very large scales. After computing the angular auto-correlation function of these radio sources , a likelihood analysis constrains the parameters of the spatial correlation function for radio sources.

1. INTRODUCTION

Clustering on Scales $>> 40Mpc$ would indicate that the density Fluctuations that we see in the CMBR experienced a very large growth. If such very Large Scale Clustering were to be found it would therefore argue for a low density Universe, in order that the fluctuations have a sufficient time to grow significantly.

We measured the angular autocorrelation of the radio sources in the flux density limited sample(S >25mJ) of Gregory *et al.*. There is no distance or redshift information for these sources. Nevertheless, we may extract the spatial auto-correlation function from the angular correlation. We use a redshift dependant power law model for the spatial autocorrelation function, $\xi(r,z) = Ar^{\gamma}(1+z)^{\delta}$ which we can integrate using the mean redshift distribution of radio sources, dN/dz, to yield the predicted angular correlation function. We then use a Maximum likelihood criteria to set limits on the Clustering Scale and the parameters of the model.

2. THEORY

The 87GB catalog, probes distance of up to $z \sim 1$, with dN/dz peaking at around $z \sim 0.7$. There are 54579 sources over 6 sterad , therefore about 3 sources per square degree *i.e.* 2 neighbours within 0.5 degrees for each sources. This tells us we can measure an angular auto correlation function as small as $\omega(0.5) \sim \sqrt{2 * 54579} \sim 3 \times 10^{-3}$. This work had studied the 87GB sample in much detail, taking great pains to study and remove any systematic effect. We will fit to A and γ directly using maximum likelihood analysis.

Let $\omega(i)$ be the correlation function measured for some angle and $\omega_m(i)$ be the model's prediction for the correlation function. Then the deviation from the model can be denoted by $\epsilon_i = \omega(i) - \omega_m(i)$. We will assume that the deviations from the model are normally distributed, this allows us to use a multivariate Normal distribution for the likelihood function

$$\mathcal{P}(A,\gamma,\delta|data) = \frac{1}{(2\pi)^{\frac{k}{2}}|\boldsymbol{\Sigma}|^{\frac{k}{2}}} \exp\frac{-\vec{\epsilon}^T\boldsymbol{\Sigma}^{-1}\vec{\epsilon}}{2} \quad (1)$$

This is the likelyhood function for the parameters of the model given the data. Σ is the covariance matrix of the model. i.e.

$$\boldsymbol{\Sigma}_{ij} = \langle(w(i)- < w(i) >)(w(j)- < w(i) >)\rangle \quad (2)$$

with the expectation taken over the possible experiments that could yield one data sample.

To reduce our sensitivity to systematic effects, such as bias of a single measurement, we will evaluate the angular correlation in various flux intervals. So $\omega(\theta, S_{min}, S_{max})$ is the angular auto-correlation function for the data measured between the flux density S_{min} and S_{max}. In integrating $\omega(\theta, S_{min}, S_{max})$ we use the mean differential counts per unit solid angle ,dN/dz, appropriate to each flux interval (S_{min}, S_{max})

Our model for the spatial correlation function is $\xi(r,z) = Ar^{\gamma}(1+z)^{\delta}$. A represent the amplitude of the correlation function, while γ is the spectral index of the power law. For optical galaxies, $\gamma = 1.8$. The factor $(1+z)^{\delta}$ represents an evolution factor to take into account the fact that the clustering was evolving in the past. We have tried a range of γ, and 3 evolution cases ($\delta = 0, -3, \gamma$). $\delta = 0$ is the no evolution case, $\delta = \gamma$ is the plausible case that the correlation was fixed in comoving coordinates, and $\delta = -3$ is the case of constant clustering per unit volume.

To calculate the theoretical angular auto-correlation, we integrate the probability of finding one source at z_1 times the probability of finding a source at a redshift z_1 in the ring of solid angle between $\theta - \frac{\Delta\theta}{2}$ and $\theta + \frac{\Delta\theta}{2}$.

$$\omega(\theta|S_{min}, S_{max}) = \frac{\int_{0.01}^{2} dz_1 \int_{0.01}^{2} dz_2 \int_{\theta-\frac{\Delta\theta}{2}}^{\theta+\frac{\Delta\theta}{2}} d\tilde{\theta} \frac{dN}{dz_1}(S_{min}, S_{max}) \sin\tilde{\theta} \frac{dN}{dz_2}(S_{min}, S_{max}) \xi(r(z_1, z_2, \theta), z_{mid} = \frac{z_1+z_2}{2})}{n(S_{min}, S_{max})\tilde{n}(S_{min}, S_{max}, \theta, \Delta\theta)} \tag{3}$$

With

$$n(S_{min}, S_{max}) = \int_{0.01}^{2} dz_1 \frac{dN}{dz_1}(S_{min}, S_{max}) \tag{4}$$

and

$$\tilde{n}(S_{min}, S_{max}, \theta, \Delta\theta) = \int_{0.01}^{2} dz_1 \int_{\theta-\frac{\Delta\theta}{2}}^{\theta+\frac{\Delta\theta}{2}} d\tilde{\theta} \sin\tilde{\theta} \frac{dN}{dz_1}(S_{min}, S_{max}) \tag{5}$$

$dN/dz(S_{min}, S_{max})$ is an experimentally measured quantity and is independent of the cosmological model adopted. We rely on a model of the luminosity function which reproduces all known number count data. This was graciously provided to us by J. Condon, in the form of a computer program.

3. COMPUTATION OF $\omega(\theta)$

The 87GB survey covers all right ascensions from $\delta = 0$ to $\delta = 75$, it is quoted to be complete down to 25mJ. The beamsize was 3.3arcmin , which means that any extragalactic source is unresolved. We subselected a sample of the data that we believe is bias free. We have found that the 87GB survey is not complete to 25mJ over the entire survey area. For $S > 30mJ$ we have to limit ourselves to declinations $\delta > 15°$. For $S > 45mJ$, the limit is $\delta > 5°$.

In addition ,we removed a patch due to interference from the sun, we cut out the outer edges $0 < \delta < 3°$ due to incompleteness and excluded source within $b^{II} < 10°$ of the galactic plane. We removed a small area around all sources for which $S > 900mJ$, due to sidelobes contamination, and a bigger area around sources with $S > 9J$ (*i.e.*Cygnus, Cas A, *ect.*.) because of their large sidelobes. We excluded extended sources because they are most likely galactic or very local.

The angular auto-correlation function was computed using a slightly modified version of the Landy-Szalay indicator because it has small variance. Let $DD(\theta)$ be the number of pairs of data points separated by the angular distance $\theta, \theta + d\theta$. Similarly RR counts the pairs in random catalogs and DR the number of pairs between the data catalog and the random catalog. In that notation, the Landy-Szalay can be written as

$$1 + \omega(\theta) = 2 + \frac{2DD(\theta) - 2DR(\theta)}{RR(\theta)}.$$

The Landy-Szalay indicator requires that we generate a random catalog. The purpose of this random catalog is to compare the pair counts of the data to pair counts expected in the absence of any intrinsic correlation. Randomness of the "random" catalog is therefore not sufficient, it needs to reproduce all observational biases that would make it appear that there is some correlation. We generate uncorrelated Monte-Carlo simulations that reproduce all the biases that we know, including the smoothing by the beams. This proves to be necessary to achieve the level of accuracy necessary.

4. ANGULAR AUTOCORRELATION FUNCTION

The results for the angular auto-correlation function will be presented elsewhere. Our biggest problem is that due to small gradients, we have to renormalize the correlation function. We see an excess of counts at small angles, but to find a bias free normalization is difficult. We take here the conservative approach that $\int_{\theta=3^\circ}^{\theta=5^\circ} (w(\theta) d\theta = 0$. This does tend to underestimate the correlation function and will make our upper limit very conservative.

The samples we are using are $30mJ < S < 45mJ$, $45mJ < S < 100mJ$,$45mJ < S < 85mJ$, $85mJ < S < 899mJ$, $30mJ < S < 899mJ$. They were chosen because this distribution has similar number of sources in each dataset.

For this work, we only included points from $0.7 < \theta < 1.7$ degrees. The signal in the two other bins $0.3 < \theta < 0.5$ and $0.5 < \theta < 0.7$ indicate even more positive correlation function, but we are not yet confident that they are bias free. We will thus give us a very strong upper limit because we ignore most of the signal which is at small scales.

5. RESULTS

For this work, we have assumed that the covariance matrix is diagonal. This is a good approximation within 1 flux range dataset, since our error bars are so large compared to the signal that any off-diagonal contribution should be small. On the other hand, since we have overlapping datasets, there are some off-diagonal terms between any 2 flux ranges that can be important.

We show contours of likelihood in fig 1. for the physically plausible case $\delta = -\gamma$ for a low density , $\Omega = 0.1$, model. Changing the density alone does not affect the result significantly. The result is mostly sensitive to δ and H_o. We set a 90% C.L. upper limit of 85Mpc to the clustering scale of radio sources ,with $h_{100} = 0.75$, or 113Mpch^{-1}. The $\delta = -3$ likelihood (not shown) yields even tighter constraints of $50h^{-1}$ Mpc.

Currently we are working on a better normalisation of the angular correlation function, testing the small angular scale bins ,and an analytical estimate of the covariance matrix.

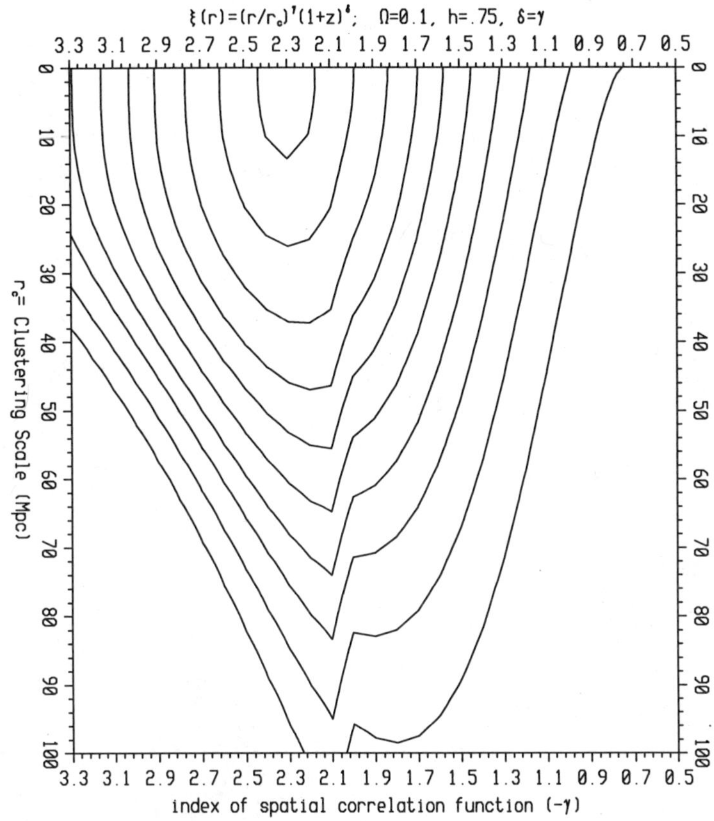

(Fig.1) Likely contours for the parameters of spatial correlation function model $\xi(r) = (\frac{r}{r_o})^{\gamma}(1+z)^{\delta}$. On the horizontal axis is the power law index of the clustering, while on the vertical axis is the clustering scale in Mpc. The redshift evolution parameter, $\delta = -\gamma$, is set so that the spatial correlation function is constant in comoving coordinates. The top contour is a 10% C.L. and each contour is separated by 10%, except for the last one which is a 95% C.L.

ACKOWLEDGEMENTS

We wish to thank J.J. Condon for his help. This work was supported in part by NSF.

REFERENCES

Condon,J.J., Broderick,J.J., & Seielstad, G.A., 1989,AJ,97,4

Peebles, P.J.E., 1980, The Large Scale Structure of the Universe,Princeton University Press

Landy,S.D., & Szalay,A. 1994, ApJ, in press

Gregory,P.C., Condon,J.J.,1991,ApJS,75,1011

COLD AND HOT DARK MATTER

THE SECOND COMING OF COLD DARK MATTER?

W.H. Zurek, B.C. Bromley, M.S. Warren
Theoretical Astrophysics, Los Alamos National Laboratory

ABSTRACT

While the standard cold dark matter (CDM) model has received numerous marks against it on the basis of a variety of observational data, the prediction of high σ_v, the pairwise-velocity dispersion between galaxies, on small (megaparsec) scales was reported as perhaps one of its greatest failings. Here, we reexamine the case of CDM and σ_v with high-resolution numerical simulations. The σ_v statistic was measured in simulations, in artificial galaxy catalogs, and in the CfA North Forty redshift survey. In our reanalysis of the CfA data, we found that $\sigma_v \sim 500$ km/s, significantly higher than the original estimates of Davis & Peebles (1983). This new value, along with our estimate of the velocity bias, leads to a virial measure of the cosmological density parameter, $\Omega \sim 0.9$, which is more than four times the traditionally cited CfA result. Analysis of the simulations indicates that while the method of Davis and Peebles works well, the σ_v statistic itself is not generally robust: there is large scatter from catalog to catalog and evidence that the recovered σ_v value depends strongly on a small fraction of galaxies in the central regions of rare, massive clusters. Thus, at present the CDM model cannot be deemed incompatible with observed small-scale peculiar velocities.

1. INTRODUCTION

The biased CDM cosmogony reigned as the premier model of structure formation in the Universe from its inception in the early 1980's through the end of that decade. Its fall from favor began with evidence from angular galaxy-galaxy correlations in the APM survey (Maddox *et al.* 1990) that the CDM model could not fit the amplitude of clustering over the full range of observed scales. At that time, there was still freedom to choose the normalization of the primordial CDM power spectrum to match galaxy clustering data on either large or small scales (a length of 8,000 km/s or $8h^{-1}$ Mpc delineates these two regimes), but not both.

Then came the Cosmic Background Explorer (COBE) data. The detection of the cosmic microwave background anisotropy constrained the amplitude of of the initial CDM power spectrum to within 20%. This result gave the CDM model little trouble on large scales when compared to redshift data, although other problems persist on these scales, notably with angular correlations of galaxies and cluster-cluster correlation functions. The real difficulty for CDM seemed to lie at small scales, the issue we address here.

There are two key indicators of small-scale structure. One is σ_8, the rms fluctuation in mass averaged within spherical tophats of radius $8h^{-1}$ Mpc. Observations indicate that for galaxies this quantity is near unity. If the galaxies are biased tracers of matter then σ_8 for the total mass may have a different value. Conventionally, the total mass fluctuations are expressed by the so-called bias parameter b, where $1/b$ is σ_8 for total mass in units of σ_8 for galaxies.

The second indicator of small-scale structure is σ_v, the radial pairwise peculiar velocity dispersion of galaxies. In the seminal work of Davis & Peebles (1983), the observational estimate of σ_v on megaparsec scales was first determined to be 340 ± 40 km/s. By contrast, early simulations of CDM indicated $\sigma_v \geq 1200$ km/s. Prior to the COBE results, reparations could be made for this discrepancy by reducing the mass normalization of CDM at the expense of increasing the bias required to keep the galaxy distribution consistent with $\sigma_8 = 1$. The bias required to lower σ_v below 500 km/s appeared to be $b \approx 2.5$

The COBE data curtailed this freedom to "tune" the CDM model with the bias factor b. These data suggest that $\sigma_8 \approx 1.1$ for total mass in a CDM model, with an initial uncertainty level around 20%. In this case galaxies are likely to be slightly antibiased relative to the CDM distribution.

Taken at face value the COBE result and the small-scale constraints condemn the CDM cosmogony when compared with numerical simulations. However, we must be wary that historically the problems of CDM on small scales were mitigated somewhat as the resolution of CDM simulations was increased. A case in point is the pairwise velocity dispersion which fell from over 1200 km/s in the 1985 work of Davis *et al.* to a value around 500 km/s, as reported by Carlberg & Couchman (1989) and Gelb (1992). Furthermore, as we shall discuss, the opposite trend has occurred for the observed value of σ_v.

There is another issue that must be considered before the CDM model is dismissed: In an inflationary scenario it is possible that a fraction of the anisotropy signal detected by COBE resulted not from density perturbations but from gravity waves—relics of the transition from the inflationary epoch to the early radiation-dominated phase. In this event, the COBE constraint will admit lower-amplitude mass fluctuations so that a biased model may still be viable.

In view of these developments we have decided to simulate formation of structure in a CDM universe with a high-resolution N-body code. Our numerical methods and some of the related astrophysical results are discussed elsewhere (Zurek *et al.* 1994; Warren & Salmon 1993; Warren & Salmon 1992). Here we shall concentrate on pairwise radial peculiar velocities, their measurement in the simulations and in observations.

2. SIMULATIONS

We consider results from three simulations with the initial conditions appropriate for CDM cosmogony with mass fluctuations of $\sigma_8 \simeq 0.7, 1.05$ and 1.6; hereafter, we refer to these runs as undernormalized, COBE-normalized, and overnormalized, respectively. These simulations were realizations of the CDM density and velocity fields within a 250 Mpc diameter sphere.

To enable a comparison with redshift catalogs, we identified halos of CDM with galaxies. As our simulations had either $N \simeq 8.8 \times 10^6$ or $N \simeq 1.76 \times 10^7$ particles each, halos with hundreds of virialized particles formed abundantly in all of them; by $z = 0$ there were typically 10^4 such halos per simulation (Fig. 1 illustrates the halo distribution in one of the runs). High spatial resolution (20 kpc Plummer smoothing) of our models appears to have prevented massive overmerging (reported by some of the earlier simulations [White *et al.* 1987; Gelb *et al.* 1993]).

We still may be subject to some overmerging: Halo-halo correlation functions (not shown here, but reproduced in Zurek *et al.* 1993) follow the observed

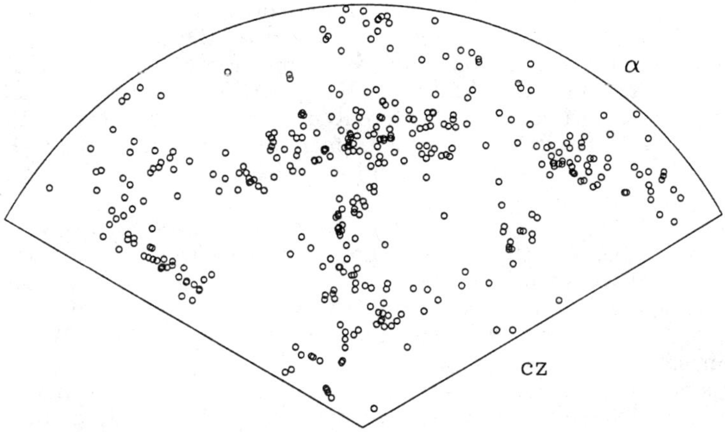

Fig. 1. A 20° redshift slice of halos from the COBE-normalized CDM simulation. The slice spans 120° in right ascension and extends to 8000 km/s.

power law shape on scales of a few megaparsecs, but are a bit more shallow on smaller sub-megaparsec scales. We regard this as evidence for a possible deficiency of objects with small ($\sim$ 100 kpc) separations, which is most logically attributed to merging.

Correlation functions tend to be equal to unity on scales $\sim 4h^{-1}$ Mpc for σ_8 of 0.7 and 1.06 (i.e., somewhat less than the observed $5.4h^{-1}$ Mpc). Moreover, on submegaparsec scales halos are significantly less correlated than mass. This slight antibias of the halo distribution is probably also due to mergers, as the correlation between the average masses of the halos and the density of their environment indicates.

In all of the simulations the least massive halos resolved by the N-body code are most abundant. The peak in the halo mass function, which corresponds to a rotation velocity of $v_{rot} \sim 150$ km/s, is an artificial cutoff resulting from our finite resolution. On the other hand, there are very few halos with "excessive" rotation velocities, especially in the undernormalized ($\sigma_8 \simeq 0.7$) run which, as we will remark in section 6, may be the best estimate of σ_8 on small scales given (i) COBE normalization on the horizon scale and (ii) "generic" inflationary scenarios for the origin of perturbations.

Before proceeding to the subject of peculiar velocities, we note that the sizes of galactic halos, their spatial distribution, and their rotation curves are all reasonably consistent with observations.

3. SMALL-SCALE PECULIAR VELOCITIES IN THE SIMULATIONS

Observations of the spatial distribution of matter have been reasonably well-matched by the CDM cosmogony, as realized in our simulations. The excess of massive galaxy halos reported in previous numerical work seems to have been caused by poor resolution, as it has been greatly alleviated by the high resolution of our N-body code. The discrepancies which remain between the CDM model and the Universe could be caused by the serious uncertainties in processes such as star formation which determine how galaxies become luminous and how they are related to the underlying dark matter. The situation with the velocity distribution is quite different: it is commonly believed that small-scale peculiar velocity statistics are extremely problematic for the CDM model. We now consider on small-scale peculiar velocities in our simulations and in observational data sets.

The pairwise peculiar radial velocity dispersion σ_v is measured in simulations on the basis of the full 6D phase space; its values for halos on megaparsec scales are approximately 600, 800, and 1600 km/s for the simulations with $\sigma_8 \simeq 0.7, 1.06$, and 1.6, respectively. In the absence of error limits these values do not compare favorably with the commonly accepted observational estimate of 340 km/s.

To some degree, the disrepancy between the observed estimate of σ_v and previous numerical results has been decreased in our simulations by a velocity bias in the halo velocity dispersion, as detected by Carlberg & Couchman (1989). It is nearly a factor of two on kiloparsec scales, becoming somewhat smaller ($b_v \sim 0.7$) at megaparsec scales. There is preliminary evidence (*e.g.* Carlberg 1994; also see Bromley, Warren & Zurek, this volume) that a significant contribution to this bias is differences in the statistical weighting of halos compared to that of the mass particle. Nonetheless, this effect is still too small to bring our results into agreement with the Davis & Peebles estimate.

The velocity dispersion can be extracted from the data by computing the distribution of number of pairs as a function of their radial peculiar velocity differences. The reader should be cautioned that a naive estimate of σ_v from a simple average over pairwise peculiar velocities is not optimal because it spuriously mixes in the first moment of the distribution. In the models investigated here, the errors introduced by following the naive prescription for σ_v are of the order of 10%.

At Mpc separations the velocity distribution is well-approximated by an exponential,

$$f(\Delta v) \sim \exp(-|\Delta v|/\sqrt{2}\sigma_v) \; ; \tag{2}$$

when the mean infall velocity is not included in the pairwise peculiar velocity Δv (see Zurek *et al.* 1994 for a more complete discussion). Thus, the slopes of an isosceles triangle, generated by plotting the logarithm of f, can be used to obtain σ_v. We shall shortly see that this is in fact the basis of the method for extracting σ_v from the observations.

The situation becomes considerably more complicated at larger distances (5–10 Mpc). There the triangular profiles of $\log f$ no longer isosceles, hence the symmetric exponential form of the distribution equation (2) is no longer valid. This is an important conclusion for the observational estimates of σ_v, since the form of the velocity distribution function must be assumed in order to extract the value of σ_v.

4. OBSERVATIONS—THE DAVIS & PEEBLES METHOD

In contrast to numerical experiments, which allow one to compute σ_v directly from complete position and velocity data, redshift surveys contain only the angular separations and proper recession velocities of galaxies. In the absence of peculiar motions, redshifts translate directly to the line-of-sight distances and the galaxy-galaxy correlations as measured in "redshift space" would be equivalent to their real-space counterparts. The effect of peculiar velocities is to smear out redshift-space correlations along the line of sight. Hence, the extent to which the line-of-sight correlations are distorted is a measure of peculiar motions.

The quantitative measure of second-order correlations in redshift space is $\xi(r_p, \pi)$, which gives the excess two-point clustering above a pure random distribution as a function of separations parallel (π) and perpendicular (r_p) to the line of sight:

$$\pi = v_1 - v_2, \quad r_p = \frac{1}{H_0}(v_1 + v_2)\tan(\theta_{12}/2) , \tag{3}$$

where v_1 and v_2 are recession velocities of two points in redshift space and θ_{12} is the sky angle between them. One estimator of $\xi(r_p, \pi)$ is

$$\frac{n_R DD(r_p, \pi)}{n_D DR(r_p, \pi)} - 1, \tag{4}$$

where DD is the number of redshift data pairs which are separated by (r_p, π); DR is the number of pairs which consist of one data object and one object from a catalog of Poisson distributed objects, subjected to the same selection criteria as the data catalog. The mean densities n_D and n_R correspond to the data and random catalogs respectively. Figure 2 shows $\xi(r_p, \pi)$ estimated in this way from the three simulations and the CfA North Forty survey.

The method of Davis & Peebles (DP) is to model the distribution of pairwise peculiar velocities, with σ_v as a parameter, and to fit the observed anisotropy in the redshift-space correlation function. However, there are assumptions in this procedure which which cannot be validated from redshift data alone; they must be tested independently through other means such as numerical experiments.

The first fundamental assumption is that the pairwise velocity distribution function of the galaxies is exponential and symmetric about the mean, as in equation (2). In our simulations this assumption is found to be quite accurate at small separations. The use of the method at 5–10 Mpc separations, however, would require modification of the assumed velocity distribution.

The second fundamental assumption is a model for the mean streaming velocity relative to the Hubble flow, since the goal is to extract the velocity dispersion about the mean pairwise peculiar infall. Davis & Peebles use the similarity solution derived in their 1977 work,

$$h(r) = F\left[1 + (r/r_0)^2\right]^{-1} . \tag{5}$$

The agreement between this model and the simulations is reasonable, but there are a discrepancies of 50–100 km/s at scales between 1 and 3 Mpc, where the model predicts less infall than is observed in the N-body data.

These two assumptions are sufficient to model the redshift-space correlations as the result of convolution of the real-space correlations with the pairwise velocity distribution along the line-of-sight direction (DP eq. [22]). In the case of power-law correlations in real-space, an exponential velocity distribution results in an approximately exponential profile in $\xi(r_p, \pi)$ along the π axis for values of r_p near or below megaparsec scales.

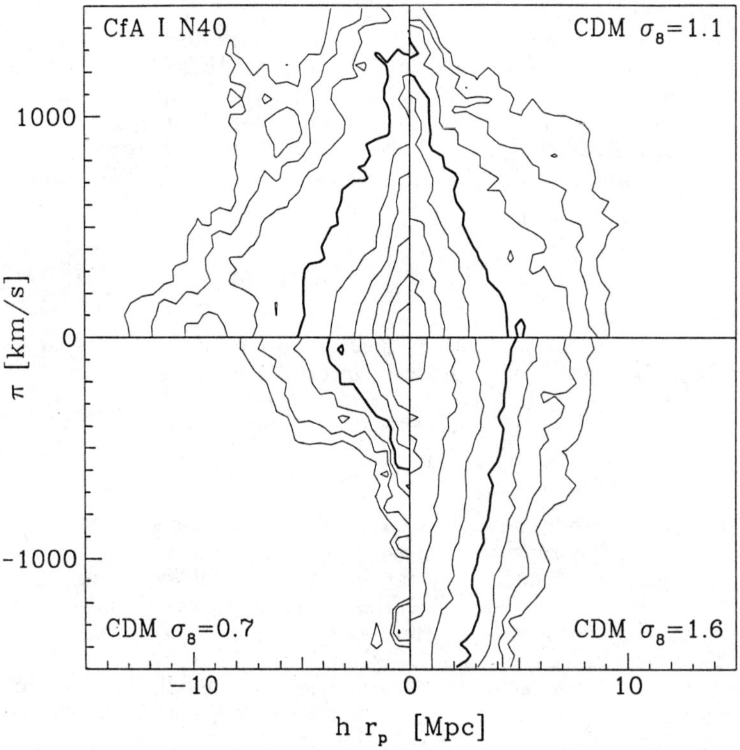

Fig. 2. Contours of $\xi(r_p, \pi)$ for the CfA North Forty data and CDM halo surveys of equivalent geometry and magnitude selection. The contour spacing is logarithmic (0.2 dex); the heavy line indicates the $\xi = 1$ contour.

To test our implementation of this method of analysis, we have applied it to a redshift catalog of all halos from the undernormalized model. For this test case, unlike the analysis of the CfA data and artificial catalogs below, we use a majority of the halos and include no flux limits. The DP method derives a value for σ_v of 628 km/s as compared with 610 km/s obtained from the full phase-space analysis. Thus, the DP method works well in this optimal case.

The key observational result of σ_v = 340 km/s from the DP analysis comes from the North Forty slice of the CfA I survey. The survey boundaries are: sky mask, $\delta > 0°$, $b > 40°$; depth, $100 < cz < 10,000$ km/s; magnitude limit, $m_B \leq 14.5$; and an absolute magnitude cut, $M_B < -18.5$ for a Hubble parameter of 100 km/s/Mpc, to semi-volume limit the sample. To assess the statistical integrity of these data, we ran the DP analysis on equivalent surveys from our halo catalogs. Due to the limited size of the computational volume, the origin for these slices is placed near the edge of the computational sphere. The center of the volume tends to be oversampled, and data sets with different view-point origins are not completely independent. Thus, the formal uncertainty limits are underestimates as they neglect the "cosmic" variations expected from truly independent samples.

Pairwise velocity dispersions obtained from the full phase-space analysis of these CfA-like slices vary between 600 and 800 km/s for 1 Mpc separations in the $\sigma_8 \simeq 0.7$ model samples. An even greater variation of 450 to 850 km/s occurs when the DP algorithm is used to infer σ_v from sky and redshift coordinates. Moreover, both of these estimates are bound to depend critically on the treatment of the regions near rich clusters, which have velocity dispersions near 1500 km/s in our simulations.

Differences of σ_v among the artificial surveys are much larger than the statistical estimates of the error inferred from any one of them. For example, the standard deviation computed for σ_v estimates from ten undernormalized slices and eight COBE-normalized slices are 115 km/s and 197 km/s respectively. We emphasize that because the slices overlap, these limits are bound to underestimate the variations in independent slices of similar size. The uncertainties estimated from a single slice on the basis of a least squares fit, presumably the source of the limits quoted by Davis & Peebles, are typically half of these values.

5. PAIRWISE VELOCITIES FROM THE CfA SURVEY

We have analyzed the publicly available CfA North Forty redshift survey using the method described in the previous section. The raw redshift data were corrected for solar motion then subjected to the sky mask, redshift limits and apparent magnitude cuts as described above; they were also corrected for an infall of 440 km/s toward the Virgo cluster to determine their distance prior to application of absolute magnitude cuts. With these selection criteria we obtain a number of galaxies, 1259, similar that of Davis & Peebles. If the Virgo infall correction for the absolute magnitudes is not performed, the number of galaxies in the sample is only 1032.

We estimated $\xi(r_p, \pi)$ from these data (Fig. 2) to obtain σ_v. The value we recovered was 517 ± 60 km/s. A lower value, 457 km/s, obtains if we exclude all galaxies beyond 8000 km/s, reducing the total galaxy count to ~ 1130.

These values are still in excess of the original DP estimate. We can alter the recovered σ_v by either neglecting the Virgo infall correction or cutting out galaxies from within the core of Virgo. For example, we made another CfA sample by excluding $\sim$ 50 galaxies within 10° of the core, from which σ_v is estimated to be approximately 365 km/s. This is surprising from the point of view that the measurement appears to be quite sensitive to this set of galaxies which constitutes less than 5% of the sample. However, it is less surprising when one notes that the method weights heavily pairs which are separated by about 1 Mpc: The galaxies which were eliminated from this latter sample had megaparsec separations and represented a good portion of the population with high peculiar velocity. This assessment is consistent with the recent reanalysis of the CfA data by Mo *et al.* (1993).

6. DISCUSSION

The principal lesson from our analysis of small-scale velocity dispersions is that the σ_v statistic is highly sensitive to the presence of rare, massive clusters. As we have found, there is a large variance in estimates of σ_v from sample to sample of the same cosmological model that stems from how clusters appear given a set of selection criteria. Indeed, with uncertainty levels at a minimum of 200 km/s for the COBE-normalized simulation and of the order 100 km/s for the CfA survey, the CDM model cannot be excluded with reasonable confidence. The undernormalized model fares even better, however, the same cannot be said of the overnormalized case for which σ_v is estimated at ~ 1600 km/s. This result, in conjunction with other recent studies (*e.g.* Brainerd & Villumsen 1994), suggests that the predictions of CDM at small scales are generally compatible with observations.

Our conclusion about the viability of the CDM model on the basis of small scale data depends not just on our reanalysis of the CfA data but on a velocity bias that appears in the simulations: The averaged pairwise velocities of halos on megaparsec scales are significantly smaller than the relative velocities of particles. This velocity bias decreases the discrepancy between the observational estimates of velocity dispersion and our numerical estimates for the CDM model. The bias we obtain is close to the one reported by Carlberg (1991) in a dissipationless simulation with a significantly lower resolution. A larger velocity bias $b_v = 0.37$ was given by Couchman & Carlberg (1992) for the "old" population of collapsed objects which had well-defined seeds at $z = 1$–3. On the other hand, our velocity bias is slightly more pronounced than the one reported by Cen & Ostriker (1992) who estimate $b_v \approx 0.8$ on megaparsec scales. This small discrepancy, which is of low statistical significance, could be certainly attributed to the difference in the resolution between the two codes. Overall, there seems to be a consensus between these estimates of b_v obtained in several rather different simulations, using different numerical schemes and somewhat different physical input.

The increase in σ_v from the reanalyzed CfA data and the detection of a velocity bias both have an impact on estimates of the cosmological density parameter Ω from the cosmic virial theorem. For clustering parameters appropriate to the CfA galaxies,

$$\Omega \approx 1.73 \times 10^{-6} b_1 \sigma_v^2 / b_v^2 \ ,$$

where b_1 is the bias in galaxy correlations at megaparsec scales. Whereas Davis & Peebles obtained $\Omega = 0.2$ with $\sigma_v = 340$ km/s and no biasing, we now derive $\Omega = 0.9$ with $\sigma_v = 500$ km/s, $b_1 = 1$, and $b_v = 0.7$. However, as we discuss next, the statistical significance of σ_v must be carefully considered when interpreting this result.

The degree of sensitivity that σ_v exhibits to the presence of galaxies within rich clusters is striking. We mentioned that the selective removal of a small fraction (5%) of galaxies in the CfA data induced a significant ($\gtrsim 40\%$) change in the estimated value of σ_v. Clearly σ_v is heavily dependent on the number density and velocities of galaxies in the centers of clusters. Biasing in the spatial distribution of these galaxies, as well as the velocity bias, can greatly affect the pairwise velocity dispersion. For example, an underabundance of galaxies in a cluster core would make these high-velocity objects relatively sparse, causing an underestimate in σ_v. Unfortunately, the biases in the phase-space distribution of galaxies in rich clusters depend on uncertain physics of galaxy formation and are presently not well determined, much less well understood. We thus find σ_v to be an unreliable measure of cosmic structure on small scales.

The dependence of σ_v on galaxies in the most massive clusters in a survey may be quite common for second-order statistical indicators of cosmological structure at megaparsec scales and below. For example, number counts of galaxies in spherical cells are known to have distribution functions with very broad tails, falling off in certain cosmological systems as a shallow power law. In this event, the variance in the cell counts can depend on the number of galaxies within the richest cluster in the survey (Bromley 1994). Although the velocity distribution itself falls off exponentially, the steepness may well depend on the dominant contributions of galaxies within the single most massive cluster.

These remarks concerning the reliability of σ_v should be kept in mind when analyzing other data sets. For example, Fisher *et al.* (1994) found $\sigma_v \approx 320$ km/s from the IRAS survey; (we find a value of about 330 km/s in the same data set; the two numbers are both well within each other's uncertainty bounds). It is expected that the dispersion derived from the IRAS survey is lower than in the CfA case, because the infrared selection criteria tend to exclude galaxies in the center of clusters. Tests of our simulations against the IRAS data at small scales will critically hinge on how we identify an equivalent halo population.

7. CONCLUSION

In an effort to examine small-scale cosmological indicators in CDM simulations, we have found that the σ_v statistic is inadequate, especially when derived from samples which are not representative of the Universe at large. Consequently, claims that COBE-normalized CDM are inconsistent with observed velocity dispersion on small scales carry little weight at this point in time.

With the caveat that the velocity statistic is unreliable, we propose that the undernormalized model, with its low σ_v prediction, may be realistic. It can, for example, match the observed abundance of Abell clusters (see White *et al.* 1993). This biased model ($b = 1.4$) is also consistent with large-scale flow measurements, for which observations indicate $\Omega/b^{1.7} \sim 0.4$–0.6 (Fisher *et al.* 1993; Nusser & Davis 1994).

The apparent discrepancy between the undernormalized model and the COBE data can be remedied by speculation about an inflationary scenario (Davis *et al.* 1992): Inflationary cosmology yields an exact Harrison-Zel'dovich power spectrum of fluctuations, $P(k) \sim k^n$, $n = 1$, when inflation is eternal. The roll-over from de Sitter expansion to the standard radiation-dominated phase at some early epoch must leave the spectral index $n < 1$. This "tilt" in the spectrum would lower the amplitude of CDM power from megaparsec scales to the Hubble radius. The tilt necessarily implies that a fraction of the fluctuations detected by COBE is the result of gravitational tensor modes which do not serve as seeds for structure formation. In two popular models for the inflaton field with quadratic and quartic potentials, the predicted small-scale fluctuation amplitude σ_8 is 0.85 and 0.75, respectively, values which are between the undernormalized and COBE normalized models discussed here.

The CDM cosmogony certainly still has hurdles to overcome in the large-scale regime. We have already mentioned a problem with angular correlations in the APM survey on scales corresponding to $\sim$ 100 Mpc. The CDM model predicts too little power compared with the optical dataset. Interestingly, there is no such problem of missing power on those same scales when compared to redshift data (*e.g.* Efstathiou 1993) although these data are sparse when compared to the two million APM galaxies. Even with a sample of a large number of galaxies, one must worry about selection effects. For example, the Great Wall, the 100 Mpc sheetlike structure found in the optically selected CfA II survey, is not similarly discernible in the IRAS data.

Other concerns include cluster-cluster correlations. The standard CDM model reportedly predicts low amplitude correlations between clusters of galaxies on scales greater than 10 Mpc (*e.g.* Dalton *et al.* 1992). Here again, selection effects are worrisome until we achieve a greater understanding of how galaxies light up within clusters.

We cannot sweep all of the outstanding problems of CDM under the carpet of uncertainty in the physics of galaxy formation. Indeed, the model's assumption that the density parameter $\Omega = 1$ would crumble if reliable measurements of the Hubble parameter lead to conflicts with direct age estimates from globular clusters. However, we do emphasize the importance of high-resolution numerical work in drawing conclusions about the viability of CDM and other theories of structure formation.

ACKNOWLEDGEMENTS

We are grateful to T.G. Brainerd and R. Laflamme for valuable discussions. This research is supported by NASA through the HPCC program.

REFERENCES

Bertschinger, E., & Gelb, J. 1991, Comp. in Phys., 5, 164
Brainerd, T.G., & Villumsen, J.V. 1994, ApJ, 435, 403
Bromley, B.C. 1994, ApJ, in press
Carlberg, R.G. 1991, ApJ, 367, 385
Carlberg, R.G., 1994, ApJ, 433, 468
Carlberg, R.G., & Couchman, H.M.P. 1989, ApJ, 340, 47
Cen, R., & Ostriker, J.P. 1992, ApJ, 399, L113
Couchman, H.M.P.,& Carlberg, R.G. 1992, ApJ, 389, 453
Dalton, G.B., Efstathiou, G., Maddox, S.J., & Sutherland, W.J. 1992, ApJ, 390, L1
Davis, M., Efstathiou, G., Frenk, C.S., & White, S.D.M. 1985, ApJ, 292, 371
Davis, M., & Peebles, P.J.E. 1977, ApJS, 34, 425
Davis, M., & Peebles, P.J.E. 1983, ApJ, 267, 465
Davis, R.L., Hodges, H.M., Smoot, G.F., Steinhart, P.J., & Turner, M.S. 1992, PRL, 69, 1856
Efstathiou, G. 1993, Proc. NAS, 90, 4859
Fisher, K.B., Davis, M., Strauss, M.A., Yahil, A., & Huchra, J.P. 1993, ApJ, 402, 42
Fisher, K.B., Davis, M., Strauss, M.A., Yahil, A., & Huchra, J.P. 1994, MNRAS, 266, 50
Gelb, J. 1991, PhD Thesis, MIT
Gelb, J.M., Gradwohl, B., & Frieman, J.A. 1993, ApJ, 403, L5
Maddox, S.J., Sutherland, W.J., Efstathiou, G., & Loveday, J. 1990, MNRAS, 243, 692
Mo, H.J., Jing, Y.P., & Börner, G. 1993, MNRAS, 264, 825
Nusser, A. & Davis, M. 1994, ApJ, 421, L1
Warren, M.S., & Salmon, J.K. 1992, in Supercomputing '92, Los Alamitos IEEE Comp. Soc.
Warren, M.S., & Salmon, J.K. 1993, in Supercomputing '93, Los Alamitos IEEE Comp. Soc.
White, S.D.M., Davis, M., Efstathiou, G., & Frenk, C.S. 1987, Nature, 330, 451
White, S.D.M., Efstathiou, G., & Frenk, C.S. 1993, MNRAS, 262, 1023
Zurek, W.H., Quinn, P.J., Salmon, J.K., Warren, W.S. 1994, ApJ, 431, 559

THE POWER SPECTRUM OF DARK MATTER HALOS IN A CDM UNIVERSE

Tereasa G. Brainerd
Los Alamos National Laboratory, MS B288, Los Alamos, NM 87545
tgb@t6-serv.lanl.gov

Jens V. Villumsen
Max-Planck-Institut für Astrophysik, Karl-Schwarzschild-Strasse 1,
D-8046 Garching bei München

ABSTRACT

It has been suggested that a COBE-normalized "standard" cold dark matter (CDM) universe has too much power on small scales compared to the power spectrum of IRAS galaxies. We present the results of an N-body simulation of such a universe and compare the redshift space $P(k)$ for dark matter halos to that of the IRAS and CfA galaxies. Over the scales that can be investigated in the simulation, $P(k)$ for halos with overdensities $\delta \sim 250$ is in excellent agreement with $P(k)$ for the IRAS galaxies. $P(k)$ for halos with overdensities $\delta \sim 2000$ (a more "biased" tracer of the mass than the $\delta \sim 250$ halos) is in moderately good agreement with $P(k)$ for the CfA galaxies. Assuming one luminous galaxy per dark matter halo, we do not find an excess small-scale power problem for galaxies in a COBE-normalized standard CDM universe.

1. INTRODUCTION

Recent results of Fisher et al. (1993) suggest that, when normalized to the microwave background fluctuations observed by COBE ($\sigma_8 \simeq 1$), a standard CDM universe yields a power spectrum for galaxies that is far in excess of $P(k)$ for IRAS galaxies. This conclusion is based on a comparison of the observed $P(k)$ of IRAS galaxies and $P(k)$ obtained for "galaxy tracer" particles in an N-body simulation of a standard CDM universe. In their simulation, Fisher et al. could not resolve individual "dark matter halos" that might be fairly associated with the locations of luminous galaxies and, so, assumed a linear bias between "galaxies" and the particles in the simulation. In addition, prior to determining $P(k)$ in redshift space, Fisher et al. smoothed the velocity field of the particles in the simulation in order to force their pairwise velocity dispersion to be ~ 300 km/s. Smoothing the velocity field effectively reduces the "Fingers of God" and results in a measured redshift space $P(k)$ that is *larger* on small scales than would have been obtained if the velocity field had not been smoothed. That is, less of the small-scale power will have been washed out by large Fingers of God.

Here we report results of $P(k)$ obtained for resolvable dark matter halos in an N-body simulation of a standard CDM universe. $P(k)$ for two halo populations is determined directly in redshift space and compared to the observed $P(k)$ of both the IRAS (Fisher et al. 1994) and CfA galaxies (Vogeley et al. 1992). The velocity fields of the halos and particles are *not* smoothed prior to the determination of $P(k)$. As excessive pairwise velocity dispersions (σ_v) can *artificially* suppress $P(k)$ in the simulation to the point where it would, serendipitously, coincide with $P(k)$ observed for galaxies, we also state σ_v of the halos and particles.

2. THE SIMULATION

A particle-mesh simulation of a "standard" CDM universe ($H_0 = 50$ km/s/Mpc, $\Omega_0 = 1$, $\Lambda = 0$, Harrison-Zeldovich primordial $P(k)$) was run. It began at a redshift, z, of 29 and ended at $z = 0$, where $z = 0$ corresponds to $\sigma_8 \simeq 1$. The comoving length of the simulation box was $80h^{-1}$ Mpc, and 128^3 particles and 256^3 grid cells were used. "Halos" with overdensities of $\delta \sim 250$ and $\delta \sim 2000$ were selected from the final timestep using a friends-of-friends algorithm with linking parameters, l, of 0.2 and 0.1, respectively.

$P(k)$ for the halos and particles was calculated using standard grid-based techniques (including white-noise subtraction). To determine $P(k)$ in redshift space, the real space locations of halos (or particles in the case of the mass $P(k)$) were distorted according to their peculiar velocities, as seen by a *moving* observer located at the center of mass of a randomly chosen halo. 100 such observers were chosen at random and the mean and variance of $P(k)$ in redshift space computed. While the choice of an observer at rest with respect to the CMBR does not affect the mean $P(k)$ in redshift space, it dramatically reduces the variance observed in $P(k)$.

The scales over which $P(k)$ can be computed accurately are determined by the size of the simulation and (approximately) the mean interparticle (or interhalo) spacing.

3. RESULTS

From the real space $P(k)$, we observe a non-linear anti-bias between the particles and $l = 0.2$ halos and an approximately linear bias between the particles and $l = 0.1$ halos (figure not shown).

The redshift space $P(k)$ for the particles agrees well with the shape of $P(k)$ for galaxies (figure not shown). Its amplitude falls between that of the IRAS and CfA galaxies on the largest resolvable scales, and agrees well with the IRAS $P(k)$ on the smallest resolvable scales. However, due to the large σ_v of the individual particles (~ 1200 km/s), the agreement between the N-body and galaxy power spectra could be entirely coincidental.

The redshift space $P(k)$ for $l = 0.2$ halos with 5 or more particles agrees well with that of the mass, and with the shape of the galaxy $P(k)$ (Fig. 1). The mean σ_v of these halos on scales $r \lesssim 137$ km/s is 524 km/s, in excess of σ_v of IRAS galaxies (Fisher et al. 1994) and, so, the agreement between the simulation and observations may be coincidental.

The redshift space $P(k)$ for large, well-resolved $l = 0.2$ halos (70 or more particles) agrees well with the IRAS $P(k)$ (Fig. 2). The sharp turndown in the halo $P(k)$ is artificial and due to the fact that we have reached the small-scale limit below which $P(k)$ cannot be computed accurately. These halos have a mean σ_v on scales $r \lesssim 137$ km/s of 319 km/s, in good agreement with σ_v for IRAS galaxies. Therefore, the agreement between the simulation and observations cannot be attributed to excessive pairwise velocity dispersions wiping out small-scale power. The mean 1-d internal velocity dispersion of these halos is 197 km/s, in agreement with the depth of potential well one would expect for luminous galaxies. So, on average, one might associate 1 luminous galaxy with each of

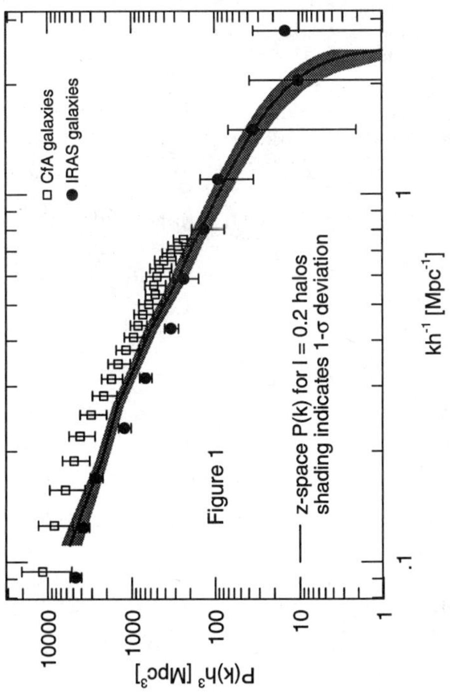

Figure 1
□ CfA galaxies
● IRAS galaxies
— z-space P(k) for l = 0.2 halos
shading indicates 1-σ deviation
P(k)h³ [Mpc³]
kh⁻¹ [Mpc⁻¹]
10000
1000
100
10
1
.1
1

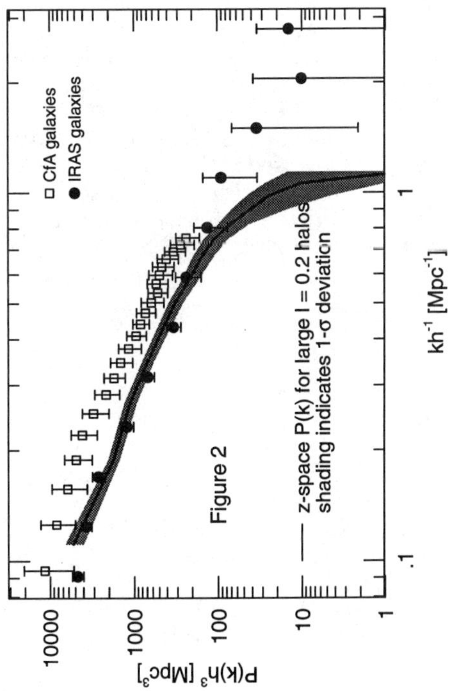

Figure 2
□ CfA galaxies
● IRAS galaxies
— z-space P(k) for large l = 0.2 halos
shading indicates 1-σ deviation
P(k)h³ [Mpc³]
kh⁻¹ [Mpc⁻¹]
10000
1000
100
10
1
.1
1

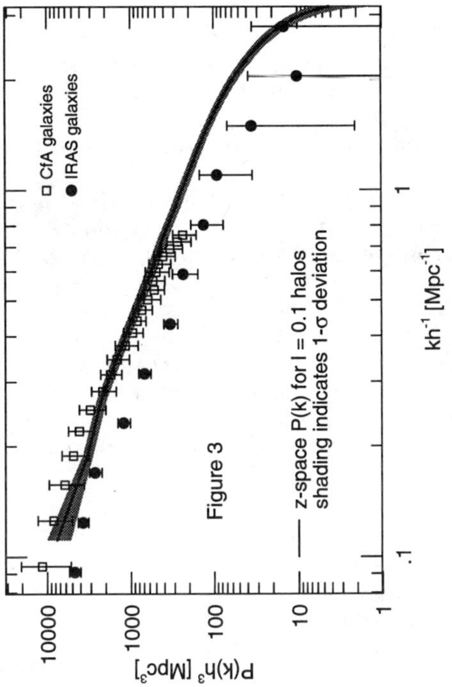

Figure 3
□ CfA galaxies
● IRAS galaxies
— z-space P(k) for l = 0.1 halos
shading indicates 1-σ deviation
P(k)h³ [Mpc³]
kh⁻¹ [Mpc⁻¹]
10000
1000
100
10
1
.1
1

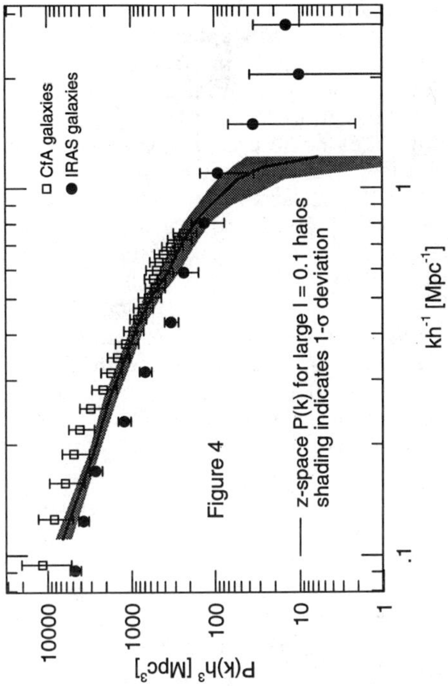

Figure 4
□ CfA galaxies
● IRAS galaxies
— z-space P(k) for large l = 0.1 halos
shading indicates 1-σ deviation
P(k)h³ [Mpc³]
kh⁻¹ [Mpc⁻¹]
10000
1000
100
10
1
.1
1

these halos.

The redshift space $P(k)$ for $l = 0.1$ halos with 5 or more particles has an amplitude that is in rough agreement with that of the CfA galaxies, but the slope is too steep (Fig. 3). From Brainerd & Villumsen (1992) we know $l = 0.1$ halos are more biased to regions of high mass density than $l = 0.2$ halos and, thus, it is not surprising that these halos have a $P(k)$ that is in excess of that of a less biased tracer of the mass.

The redshift space $P(k)$ for large, well-resolved $l = 0.1$ halos (70 or more particles) agrees moderately well with $P(k)$ for the CfA galaxies, having the right slope and falling within the 95% confidence limits (Fig. 4). Again, the sharp drop in $P(k)$ for the halos is due to the resolution limit of the simulation. The σ_v of these halos on scales $r \lesssim 137$ km/s is 483 km/s, in excess of the Davis & Peebles (1983) estimate for the CfA galaxies, but in agreement with the reanalysis of σ_v done recently by Zurek et al. (1994). However, the mean 1-d internal velocity dispersion of these halos is 279 km/s, in excess of that expected for the potential wells associated with individual galaxies, and, so, it is not clear that one can associate a single luminous galaxy with each of these halos.

4. SUMMARY & CONCLUSIONS

Using an N-body simulation of a standard CDM universe normalized by the observed CMBR fluctuations, we have shown that the redshift space power spectra of the mass particles and two populations of dark matter halos are in reasonable agreement with the observed power spectra of galaxies.

Large pairwise velocity dispersions will erase small-scale power in redshift space and may cause $P(k)$ to be artificially suppressed to the point where it would, serendipitously, coincide with $P(k)$ observed for galaxies. This is a particular concern for $P(k)$ for the mass particles and small, poorly resolved halos.

Over the scales measurable in the simulation, $P(k)$ for large, well-resolved halos of overdensities $\delta \sim 250$ and $\delta \sim 2000$ is in good agreement with $P(k)$ for the IRAS and CfA galaxies, respectively. While the pairwise velocity dispersions of these halos are in acceptable ranges (i.e. it is unlikely that $P(k)$ has been overly suppressed), the large $\delta \sim 2000$ halos have potential wells deeper than expected for individual galaxies.

Provided it is fair to associate 1 luminous galaxy with 1 dark matter halo, we find a standard CDM universe normalized by the COBE observations does not predict a galaxy power spectrum in excess of that observed for the IRAS and CfA galaxies.

REFERENCES

Brainerd, T. G. & Villumsen, J. V. 1992, ApJ, 394, 409
Davis, M. & Peebles, P. J. E. 1983, ApJ, 267, 465
Fisher, K. B., Davis, M., Strauss, M. A., Yahil, A., & Huchra, J. P. 1993, ApJ, 402, 42
Fisher, K. B., Davis, M., Strauss, M. A., Yahil, A., & Huchra, J. P. 1994, MNRAS, 266, 50
Vogeley, M., Park, C., Geller, M., & Huchra, J. 1992, ApJ, 391, L5
Zurek, W., Warren, M., Quinn, P., & Salmon, J. 1994, in Proceedings of the 9th IAP Meeting, Paris, France, July 1993

Cold + Hot Dark Matter Cosmology with $m(\nu_\mu) \approx m(\nu_\tau) \approx 2.4$ eV

Anatoly Klypin
Department of Astronomy, New Mexico State University, Las Cruces, NM 88001

Joel R. Primack
Santa Cruz Institute for Particle Physics,
University of California, Santa Cruz, CA 95064

Jon Holtzman
Lowell Observatory, Mars Hill Road, Flagstaff, AZ 86100

David O. Caldwell
Department of Physics, University of California, Santa Barbara, CA 93106-9530

ABSTRACT

Recent Kamiokande data supports the $\nu_\mu \rightarrow \nu_\tau$ neutrino oscillation explanation of the deficit of cosmic ray muon neutrinos, which requires that the muon and tau neutrinos have almost identical masses; and preliminary evidence from the current Los Alamos neutrino oscillation experiment suggests that the mass of the muon neutrino $m(\nu_\mu)$ is $\gtrsim 2.4$ eV. Two neutrinos of mass 2.4 eV each would constitute "hot" dark matter accounting for a fraction $\Omega_\nu = 0.2(0.5/h)^2$ of critical density. Here we consider the consequences of such neutrino masses for cosmological models for the formation of galaxies and large scale structure in the universe, which are spatially flat and in which most of the dark matter is "cold." The linear calculations and N-body simulations that we report here indicate that an $\Omega = 1$ Cold + Hot Dark Matter (CHDM) cosmological model with two neutrinos each of mass ≈ 2.4 eV (we will call this model Cν^2DM) agrees remarkably well with all available observations. However, we find that this is true only if the Hubble parameter $h \approx 0.5$. We also consider Cold Dark Matter (CDM) models with a cosmological constant Λ and show that evidence for hot dark matter raises serious difficulties for low-Ω ΛCDM models.

1. INTRODUCTION

Predictions of a CHDM cosmological model with a single massive neutrino and $\Omega_\nu = 0.3$ (corresponding to $m_\nu \approx 7$ eV) have been shown (Davis, Summers, & Schlegel 1992; Klypin et al. 1993 (KHPR); Schlegel et al. 1994; Jing et al. 1994; Nolthenius, Klypin, & Primack 1994) to agree well with observations, with the possible exception that galaxies may form too late to account for the observations of quasars and damped Lyα systems (Lanzetta, Wolfe, & Turnshek 1993) at high redshifts $z \gtrsim 3$ (Haehnelt 1993; Mo & Miralda-Escude 1994; Kauffmann & Charlot 1994; Ma & Bertschinger, 1994). The latter observations can be accommodated (Klypin et al. 1994) if the assumed neutrino mass in CHDM is lowered from ~ 7 eV to ~ 5 eV ($h = 0.5$).

Current experimental hints regarding neutrino masses suggest that the net neutrino mass is shared among two species of neutrinos. In particular, if the deficit of ν_μ relative to ν_e produced in the atmosphere is due to $\nu_\mu \to \nu_\tau$ oscillations, then the hot component must involve more than one species of neutrino, since because of the long baseline the ν_μ–ν_τ mass-squared difference must then be rather small, $\sim 10^{-2}$ eV^2. We show here that $\Omega_\nu = 0.2$ CHDM with the mass evenly shared between two neutrino species – Cν^2DM – agrees better with observations than the one-neutrino version, better indeed than any other variant of CDM that we have considered.

2. EXPERIMENTAL DATA ON NEUTRINO MASSES

It has been pointed out (Caldwell 1993; Caldwell & Mohapatra 1993) that if the solar ν_e and atmospheric ν_μ deficits arise from the existence of neutrino masses, there are only two viable patterns for those masses: (A) the three active neutrinos are approximately degenerate; or (B) the nearly degenerate ν_μ and ν_τ constitute the hot dark matter, and the ν_e and a sterile neutrino ν_s are lighter and are also nearly degenerate (Peltoniemi &Valle 1993; Shi, Schramm, & Fields 1993). If one also takes into account the need for about 5 eV of neutrino mass for CHDM cosmological models to be viable, pattern (A) corresponds to $m_{\nu_e} \approx m_{\nu_\mu} \approx m_{\nu_\tau} \approx 1.6$ eV, while pattern (B) requires $m_{\nu_\mu} \approx m_{\nu_\tau} \approx 2.4$ eV.

Evidence for a neutrino mass explanation of the solar ν_e deficit is now fairly convincing, since at least two of the three types of experiments have to be wrong to be compatible with some non-standard solar models. If the solar ν_e deficit is due to MSW $\nu_e \to \nu_\mu$ or $\nu_e \to \nu_s$ neutrino oscillations in the sun, the mass squared difference between either pair of particles is $\Delta m^2_{ei} \equiv |m(\nu_e)^2 - m(\nu_i)^2| \approx 10^{-5}$ eV^2. Similarly, evidence for a neutrino mass explanation of the deficit of ν_μ's relative to ν_e's in atmospheric secondary cosmic rays is also increasing, with compatible results from three experiments (Hirata et al. 1992; Becker-Szendy et al. 1992; Litchfield 1993) and especially new information from Kamiokande (Fukuda et al. 1994). The latter includes accelerator confirmation of the ability to separate ν_e and ν_μ events, as well as an independent higher energy data set giving not only a ν_μ/ν_e ratio agreeing with the lower energy data, but also a zenith-angle (hence source-to-detector) dependence compatible with $\nu_\mu \to \nu_e$ or $\nu_\mu \to \nu_\tau$ oscillations with $\Delta m^2_{\mu i} \approx 10^{-2}$ eV^2. However, almost the entire region of $\Delta m^2_{\mu e} - \sin^2 2\theta_{\mu e}$ allowed by the Kamiokande data is excluded by data from the Bugey and Krasnoyarsk reactor neutrino oscillation experiments; thus $\nu_\mu \to \nu_\tau$ oscillations are favored as an explanation of the atmospheric ν_μ deficit.

That the neutrino mass pattern (B) might be correct is indicated by early results of the LSND experiment (Louis 1994), the initial run of which showed an excess of about 8 beam-on events of a type which could be interpreted as $\bar\nu_\mu \to \bar\nu_e$, whereas a background of $\lesssim 1$ event mimicing a $\bar\nu_e$ was expected. The LSND collaboration at this time is not claiming to have observed neutrino oscillations, preferring to await results of their current run, which will provide considerably more data. Nevertheless, if the LSND result is interpreted as $\bar\nu_\mu \to \bar\nu_e$, it appears (Caldwell 1994) to give $\Delta m^2_{\mu e} \sim 6$ eV^2. This is not in conflict with the KARMEN or BNL E776 experiments, which are least sensitive at that Δm^2 where LSND is most sensitive. Furthermore, the LSND positron energy distribution (if assumed to be from $\bar{\nu}_\mu \to \bar{\nu}_e$, $\bar{\nu}_e + p \to e^+ + n$) is also consistent with this same Δm^2. This Δm^2 is inconsistent with ν mass pattern (A). If the ν_e mass is relatively small ($\lesssim 1$ eV, as indicated for Majorana neutrino mass from neutrinoless double beta decay experiments), then the ν_μ mass is ~ 2.4 eV, and the data is consistent with ν mass pattern (B). It is this scenario for the hot dark matter in a CHDM cosmology which we will show below gives predictions that are in remarkable agreement with astronomical observations.

3. COMPARISON OF COSMOLOGICAL MODELS WITH OBSERVATIONS

COBE observations of fluctuations in the microwave background radiation provide an upper limit (since they include possible tensor gravity wave as well as scalar density wave contributions) on the normalization of the spectrum of fluctuations in models of structure formation in the universe. When this normalization is used for the standard CDM model (Blumenthal et al. 1984; Davis et al. 1985; Ostriker 1993) in a $\Omega = 1$ universe with a Zel'dovich primordial power spectrum ($P(k) = Ak^{n_p}$ with $n_p = 1$) as predicted by simple inflationary models, this fits large-scale data but produces too much structure on smaller scales; for example, there are far too many clusters of galaxies (Efstathiou et al. 1992) and small-scale pairwise velocities of galaxies are too large. CDM is attractive because of its simplicity and the existence of well-motivated particle candidates (lightest superpartner particle and axion (Primack et al. 1988)) for the cold dark matter and it came remarkably close to predicting the COBE signal. So several variations have been tried to patch up the CDM model. Lowering the normalization (introducing "bias") or "tilting" the primordial spectrum (assuming $n_p \approx 0.7$) improves agreement somewhat with data on intermediate (~ 10 Mpc, e.g. cluster) scales and small ($\sim 1\ h^{-1}$ Mpc, e.g. galaxy pairwise velocity) scales, but leads to serious disagreement with larger-scale ($30 - 100\ h^{-1}$ Mpc) measurements of galaxy bulk velocities and power spectra, and galaxy and cluster correlations. From the viewpoint of agreeing with observations, the best variants of CDM that have been discussed (Holtzman & Primack 1993) add either a cosmological constant (ΛCDM) or a little hot (neutrino) dark matter (CHDM). The former assumes $\Omega \approx 0.3$ and adds a cosmological constant Λ such that $\Omega_\Lambda \equiv \Lambda/(3H_0^2) = 1 - \Omega$ to preserve flatness (predicted by inflation) as well as improve agreement with data. ΛCDM works best for a larger Hubble parameter $h \approx 0.7$ favored by many observers; however, it produces bulk velocities somewhat too low, may overproduce clusters, and predicts correlation function, which is too steep ($\gamma \approx -3$). It predicts relatively early galaxy formation since at late times structure formation ceases as the universe goes into inflation again.

The first version of CHDM to be studied in detail assumed 60% cold, 30% hot (corresponding to a neutrino of mass $94h^2\Omega_\nu \approx 7$ eV), and 10% baryonic matter, with $\Omega = 1$ and $h = 0.5$. This version of CHDM fits galaxy and larger scale structures in the present-epoch universe remarkably well, but it forms galaxies too late since the large fraction of free-streaming neutrinos washes out small- scale density fluctuations too effectively. Lowering the hot fraction to about 20% solves this problem (Klypin et al. 1994). Heretofore these CHDM models have placed the needed neutrino mass in one flavor of neutrino, assumed to be the ν_τ, whereas if the asmospheric ν_μ deficit is due to $\nu_\mu \to \nu_\tau$ oscillations this cannot be correct. If we take neutrino mass pattern (B) *or* the LSND indications seriously, then that mass ought to be shared about equally between the ν_μ and ν_τ. Having two neutrinos of 2.4 eV each, which we call the Cν^2DM model, produces an interesting effect: the ratio of the power spectrum for Cν^2DM compared to that for CHDM with the same total neutrino mass in one species is essentially unity at large and small scales, but it has a dip of about 30% centered at $\sim 10\ h^{-1}$ Mpc. The larger neutrino free-streaming length, resulting from a neutrino mass of 2.4 eV instead of twice that, lowers the abundance of clusters and gives better agreement with observations. This model appears to agree with all available data, as we will discuss further below.

We report (quasi-)linear estimates for this Cν^2DM scenario and some variants of it in the Table. All models in the Table are normalized to COBE except for the CDM model labelled "biased". Actually, our COBE normalization ($Q = 17\,\mu$K) is perhaps 10% lower than the latest analyses would suggest. We have chosen this normalization to allow for a little gravity wave contribution and tilt (consistent with the expectations from simple models of cosmic inflation), and we regard it as being both more realistic and more conservative than a higher normalization – more conservative, since the greatest problem for the CHDM models is enough early structure formation, a problem that worsens as the normalization of the spectrum is decreased.

The first two lines of numbers give our estimates of a variety of observational quantities and the uncertainties in them, from large to small scales. The bulk velocity at $r = 50\ h^{-1}$ Mpc is derived from the latest POTENT analysis (Dekel 1994); the error includes the error from the analysis but not cosmic variance. However, similar constraints come from other data on large scales such as power spectra that may be less affected by cosmic variance since they probe a larger volume of the universe. We have estimated the current number density of clusters ($N_{\rm clust}$) from comparison of data on the cluster temperature function from X-ray observations with hydrodynamic simulations (Bryan et al. 1994) as well as from number counts of clusters (Bahcall & Cen 1993 (masses were scaled up by 1.5); Biviano et al. 1993; White, Efstathiou, & Frenk 1993). All recent estimates of the cluster correlation function give fairly large values at 30 h^{-1} Mpc (e.g., Olivier et al. 1993; Klypin & Rhee 1994); this also suggests that the zero crossing of the correlation function must exceed $\sim 40\ h^{-1}$ Mpc. The linear estimate of pairwise velocities (σ_v) is not an observed value, since pairwise velocities are strongly influenced by nonlinear evolution. However, from previous experience with N-body simulations for various models, it was found that the results from simulations are about a factor of three larger than the linear estimate. The limit we quote here is our estimate of the maximum allowed linear value to match numbers derived from redshift surveys, although the statistics on velocities derived from these surveys may not be very robust (Mo, Jing, & Börner 1994; Zurek et al. 1994) since they are heavily influenced by the presence of (relatively rare) clusters. To get a better estimate of pairwise velocities in our preferred Cν^2DM model, we have performed an N-body simulation, as discussed below. The final column gives the observed density in cold hydrogen and helium gas at $z = 3.0 - 3.5$ from observations of damped Lyman α systems.

TABLE: COMPARISON OF MODELS ($Q_2 = 17\mu K$)

Model	$\Omega_{\rm bar}$ %	Ω_ν %	N_ν^a	m_ν^a	σ_8^b	V^c 50Mpc	$N_{\rm clust}^d$	ξ_{cc}^e 30Mpc	r^f	σ_v^g 1Mpc	$\Omega_{\rm gas}^h$
OBSERVED					0.7	335	4.0e-7	0.30	> 40	< 200	6.e-3
errors$\pm$					0.1	80	2.0e-7	0.15			2.e-3
CDM, h =0.5											
COBE	7.5	0	0	0.00	1.08	356	4.8e-06	0.10	36	405	3.0e-2
biased	7.5	0	0	0.00	0.70	230	1.0e-07	0.07	36	260	1.5e-2
CHDM, h =0.5											
Cν^2DM	7.5	20	2	2.35	0.67	347	2.4e-07	0.35	70	144	4.6e-3
KHPR	10.	30	1	7.04	0.66	359	3.6e-07	0.37	51	98	4.0e-4
1ν	7.5	20	1	4.69	0.75	357	7.7e-07	0.30	52	156	5.2e-3
CDM/CHDM, h =0.7											
CDM$_{0.7}$	5.0	0	0	0.00	1.56	393	1.8e-05	-0.01	28	714	3.1e-2
Cν^2DM$_{0.7}$	4.0	10	2	2.30	1.24	389	9.3e-06	0.09	38	432	2.0e-2
Cν^2DM$_{n0.8}$	4.0	10	2	2.30	0.71	271	2.1e-07	0.14	49	199	4.4e-3
CDM/CHDM+Λ models, h =0.7 and Ω_Λ =0.70											
ΛCDM	2.6	0	0	0.00	0.86	277	2.3e-08	0.20	125	113	1.2e-2
ΛCHDM	2.6	5	1	2.30	0.54	263	5.3e-11	0.35	136	48	4.1e-4
ΛCν^2DM	2.6	10	2	2.30	0.33	247	1.8e-16	0.53	144	19	6.e-10

[a] N_ν is the number of neutrino species with mass. If $N_\nu \geq 1$, each species has the same neutrino mass, given by m_ν.
[b] $(\Delta M/M)_{\rm rms}$ for top-hat filter with $R_{\rm top-hat} = 8h^{-1}$ Mpc
[c] Bulk velocity in sphere of radius $50h^{-1}$ Mpc, top-hat window function.
[d] Number density of clusters $N(> M)$ in units of $h^3{\rm Mpc}^{-3}$ above the mass limit $M = 10^{15}h^{-1}M_\odot$. The Press-Schechter approximation with gaussian filter and $\delta_c = 1.50$ was used to make the prediction.
[e] The cluster-cluster correlation function amplitude at $30h^{-1}$ Mpc; computed using linear theory and assuming a unit bias factor for dynamical contribution.
[f] Zero crossing ($\xi(r) = 0$) of the correlation function in units of h^{-1} Mpc.
[g] Linear estimate of pairwise velocity at 1 h^{-1} Mpc scale: $\sigma_v^2 = 2H_0^2 \int dkP(k)(1 - \sin kr)/kr$.
[h] Mean density of collapsed baryons at z=3–3.5, using $\Omega_{gas}= (\Omega_b/\Omega_c){\rm erfc}(\delta_c/\sqrt{2}\sigma)$, with $\delta_c = 1.4$ and σ computed for $5 \times 10^{10}h^{-1}M_\odot$ mass using gaussian smoothing and assuming all gas is neutral.

The next two lines present predictions from the CDM model, and illustrate its problems. The cluster correlation function at 30 h^{-1} Mpc is smaller than observations indicate regardless of CDM normalization, reflecting the fact that the matter correlation function becomes negative beyond ~ 40 h^{-1} Mpc. In addition, CDM normalized to COBE produces far too many clusters and excessive small-scale pairwise velocities. If CDM is normalized to $\sigma_8 = 0.7$ (or equivalently to linear bias $b \equiv \sigma_8^{-1} = 1.43$), the cluster density problem is avoided, but small-scale velocities are still too large (KHPR; Nolthenius, Klypin, & Primack, 1994), and bulk velocities on a scale of 50 h^{-1} Mpc are probably too low. Even biased CDM is able to account for observations of damped Lyα systems, judging from Ω_{gas}, the amount of gas in collapsed halos at redshift $z = 3 - 3.5$.

The problem with CDM is that it has too much power on small scales relative to power at large scales. Since the presence of massive neutrino(s) reduces small scale power (because neutrino free streaming causes perturbations to damp on smaller scales), including a neutrino component improves the agreement of model predictions with observations. The model studied by KHPR with a neutrino mass of 7 eV, corresponding to $\Omega_\nu = 0.3$ for $h = 0.5$, is a much better match to observations than CDM, but it has Ω_{gas} too small. The small-scale velocities in this model are small enough (KHPR) to agree with the old result $\sigma(1\ h^{-1}\ \text{Mpc}) =$ 340 km s^{-1} from the CfA1 survey (Davis & Peebles 1983). However, this result is now known to be in error because of the accidental omission of the Virgo cluster (Somerville, Davis, & Primack, 1994); as mentioned above, this is not a very robust statistic. A direct comparison of galaxy groups in "observed" CDM and CHDM simulations with CfA1 groups shows that CDM velocities are much too high, even with biasing, while the velocities in the $\Omega_\nu = 0.3$ CHDM model are perhaps a little too small (Nolthenius, Klypin, & Primack 1994, 1995).

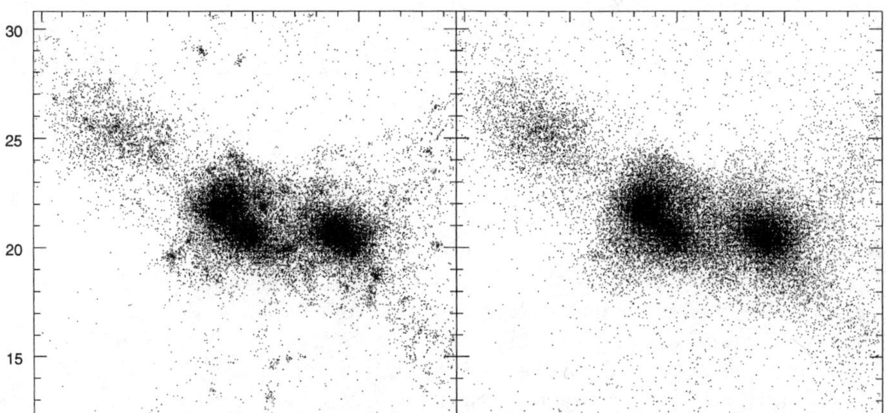

Figure 1: Rich cluster in Cν^2DM N-body simulation. This figure represents cold (left panel) and hot (right panel) particles in a $20 \times 20 \times 2$ Mpc ($h = 0.5$) slice of a high resolution Particle Mesh simulation [(1 cold + 2 hot)$\times 256^3$ particles, 800^3 force mesh].

Lowering Ω_ν to 0.20 (1ν) raises Ω_{gas} enough for early object formation, but this model may overproduce clusters and have slightly too large small-scale velocities. In order to avoid these problems, it should probably to be normalized lower – which we might imagine could reflect some gravity wave contribution to COBE – but this would result in too little early structure formation (because $\Omega_{\rm gas}$ is exponentially sensitive to the power spectrum).

Now consider $C\nu^2DM$. All quantities are in good agreement with the astronomical data if the same $\Omega_\nu = 0.2$ is divided between two neutrino species, as suggested by the neutrino data. The small-scale power in the two models is nearly identical, so $C\nu^2DM$ also produces enough $\Omega_{\rm gas}$. We have begun to analyze results from our first N-body simulation of $C\nu^2DM$ (high resolution 800^3 PM mesh in a 50 h^{-1} Mpc box with 256^3 cold and 2×256^3 hot particles). In Fig. 1 we show the distribution of cold and hot particles around a cluster from this simulation. The hot particles are much more broadly distributed than the cold ones, but few galaxy halos form in this extended cluster hot dark matter halo. This may help to explain recent observations suggesting extended mass distibutions around some clusters from studies of weak gravitational lensing and of the velocity distribution (Kaiser et al. 1994; Carlberg et al. 1994). The fact that hot particles are distributed more smoothly than galaxies implies that the usual growth rates should be lowered when velocities are estimated at $z = 0$. Projected pairwise velocities can be estimated by placing an "observer" in the box and measuring relative velicities along the line of sight for given projected separation. The dark matter particle pairwise velocity calculated in this way is $\sigma_v(\rm projected, darkmatter) = 560$ km s^{-1} at 1 h^{-1} Mpc separation. If we conservatively estimate that the velocity bias (the ratio of the rms velocity of the dark matter halos to that of the dark matter particles (Carlberg & Couchman 1989) is 0.8, this corresponds to 450 km s^{-1} for galaxies, consistent with current observations.

It is remarkable that only cosmological models with $h \approx 0.5$ match observations with the neutrino masses as suggested by the experiments. Returning to the Table, note that for $h = 0.7$ – favored by many observers – CDM ($CDM_{0.7}$) is an even worse fit to the data than for $h = 0.5$ because the larger h makes matter-dominance ($\propto \Omega h^2$) occur earlier and thus moves the bend in the CDM spectrum to smaller scales, giving more intermediate and small scale power for a given large scale normalization. Adding two 2.4 eV neutrinos ($C\nu^2DM_{0.7}$) only slightly improves the situation, because this only gives $\Omega_\nu \sim 0.1$ for $h = 0.7$, so the spectrum is not modified very much. The $h = 0.7$ can match models with a tilted spectrum of primordial fluctuations with $n_p = 0.81$ ($C\nu^2DM_{n0.8}$) – but only without gravity waves. Typical cosmic inflation models with this much tilt would suppress scalar power by a factor of $[1 + 6(1 - n_p)]^{-1} = 0.47$, which would lead to very serious underproduction of clusters and of $\Omega_{\rm gas}$. Of course, with large h, $\Omega = 1$ models also lead to too short a time since the big bang: $t_0 = 2/(3H_0) = 6.52\,{\rm Gy}/h = 9.3$ Gy for $h = 0.7$.

A larger age is obtained for an open universe; in order to be consistent with inflation (and COBE), the usual cure is to assume a positive cosmological constant. The maximum value of Λ allowed by the COBE data is $\Omega_\Lambda \equiv \Lambda/(3H_0^2) \approx 0.78$ (Bunn & Sugiyama 1994), and the maximum allowed by quasar lensing statistics is $\Omega_\Lambda \approx 0.7$ (Kochanek 1994). For a flat ($k = 0$) universe with $\Omega_\Lambda = 0.7$ and $\Omega = 0.3$, $h = 0.7$ corresponds to $t_0 = 13.5$ Gy. ΛCDM with these parameters is a good fit (Efstathiou, Sutherland, & Maddox 1990; Cen, Gnedin, & Ostriker 1993; Kofman, Gnedin, & Bahcall 1993) to the data, although our linear calculations suggest that not enough clusters are produced and bulk velocities may be too low. However, this model becomes much worse if even one neutrino of 2.4 eV is added, seriously underproducing clusters and Ω_{gas} because of the excessive fraction of hot dark matter suppressing small-scale structure. Consequently, open models have serious problems if any neutrinos have significant mass.

A similar situation occurs for $\Omega = 1$ Cν^2DM with $h = 0.4$, for which $\Omega_\nu = 0.32$ with two 2.4 eV neutrinos. (Recall that for given $m(\nu)$, Ω_ν scales as h^{-2} since critical density is $\propto h^2$.) Because the bend in the CDM spectrum moves to larger scales as h decreases, there is less intermediate and small scale power for given large scale normalization; adding hot dark matter further decreases small scale power. We find that even with only one 2.4 eV neutrino, there is just not enough power to generate the observed number of clusters or high-redshift objects.

4. CONCLUSIONS

Ever since the early 1980s there have been hints (Dekel 1981, 1984) that features on small and large scales may require a hybrid scenario in which there are two different kinds of dark matter. Preliminary studies of the CHDM scenario were carried out in 1984 (see references in KHPR) and it was first worked out in detail only in the last two years with one massive neutrino. We have shown here that the Cν^2DM model, with Hubble parameter $h = 0.5$ and both neutrinos having a mass of 2.4 eV as suggested by ongoing experiments, gives a remarkably good account of all presently available astronomical data. New data on CMB, large scale structure, and structure formation will severely test this highly predictive model. Results expected soon from new solar ν and ν-oscillation experiments will clarify whether indeed $m(\nu_\mu) \approx m(\nu_\tau) \approx 2.4$ eV. The Table shows the implications of such neutrino masses for a variety of popular CDM-type cosmological models. If even just the ν_μ has a mass of 2.4 eV, as suggested by preliminary results from the LSND experiment, low-Ω models are disfavored.

ACKNOWLEDGEMENTS

We have benefited from conversations with S. Bonometto, S. Borgani, A. Dekel, S. Faber, R. Nolthenius, and D. Seckel. This research was supported by NSF grants at UCSC and NMSU. The simulation was done on the Convex C-3880 at NCSA.

REFERENCES

Bahcall, N., & Cen, R. 1993, ApJ, 408 , L77

Becker-Szendy R., et al. 1992, Phys.Rev.D46, 3720

Biviano, A., et al., 1993, ApJ, 411, L13

Blumenthal, G.R., Faber, S.M., Primack, J.R., & Rees, M. 1984, Nature, 311, 517

Bryan, G. et al. 1994, ApJ, in press, preprint astro-ph/9409091

Bunn, E., & Sugiyama, N. 1994, preprint astro-ph/9407069

Caldwell, D.O. 1993, in Perspectives in Neutrinos, Atomic Physics, & Gravitation, ed. J. Tran Thanh Van et al. (Ed. Frontieres, Gif-sur-Yvette, France)

Caldwell, D.O., & Mohapatra, R. 1993, Phys.Rev.D, 48, 3259

Caldwell, D.O. 1994, in XVI Conf. on Neutrino Phys. & Astrophys., Eilat, Israel
Carlberg, R.G., & Couchman, H.M.P. 1989, ApJ, 340 , 47
Carlberg, R.G., et al. 1994, Toronto preprint
Cen, R., Gnedin, N.Y., & Ostriker, J.P. 1993, ApJ, 417, 387
Davis, M., Peebles, P.J.E. 1983, ApJ, 267, 465
Davis, M., Efstathiou, G., Frenk, C.S., & White, S.D.M. 1985, ApJ, 292, 371
Davis, M., Summers, F., & Schlegel, D. 1992, Nature, 359, 393
Dekel, A. 1981, A&A, 101 , 79
Dekel, A. 1984, ApJ, 284, 445
Dekel, A. 1994, ARA&A in press; privite communications
Efstathiou, G., Sutherland, W.J., & Maddox, S.J. 1990, Nature, 348, 705
Efstathiou, G., Bond, J.R., White, S.D.M. 1992, MNRAS, 258, P1
Fukuda, Y. et al. 1994, Phys.Lett.B335, 237
Haehnelt, M.G. 1993, MNRAS, 265, 727
Hirata, K.S., et al., 1992, Phys.Lett.B280, 146
Holtzman, J., & Primack, J.R. 1993, ApJ, 405, 428
Jing,Y.P., Mo, H.J., Börner, G., & L.Z. Fang 1994, A&A, 284, 703
Kaiser, N., et al. 1994, Toronto preprint astro-ph/9407004
Kauffmann, G., & Charlot, S. 1994, ApJ, 430, L97
Klypin, A., Holtzman, J., Primack, J.R., & Regős, E. 1993, ApJ, 416, 1 (KHPR)
Klypin, A., & Rhee, G. 1994, ApJ, 428, 399
Klypin, A., Borgani, S., Holtzman, J., Primack, J.R. 1994, ApJ submitted, Preprint astro-ph/9410022
Kochanek, C. 1994, preprint & private communication
Kofman, L., Gnedin, N., Bahcall, N. 1993, ApJ, 413, 1
Lanzetta, K.M., Wolfe, A., & Turnshek, D.A. 1993, Preprint
Litchfield, P.J. 1993, in Proc. Int. Europhys. Conf. on High Energy Phys., eds. J. Carr & M. Perrottel (Ed. Frontières, Gif-sur-Yvette, France)
Louis, W.C. 1994, in XVI Conf. on Neutrino Phys. & Astrophys., Eilat, Israel
Ma, C.P., & Bertschinger, E. 1994, ApJ, in press
Mo, H.J., Jing, Y.P., & Börner,G. 1994, MNRAS, 264, 825
Mo, H.J., & Miralda-Escude, J. 1994 ApJ, 430, L25
Nolthenius, R., Klypin, A., & Primack, J.R. 1994, ApJ, 422, L45
Nolthenius, R., Klypin, A., & Primack, J.R. 1995 ApJ, submitted
Olivier, S., Primack, J., Blumenthal, G.R., & Dekel, A. 1993, ApJ, 408, 17
Ostriker, J.P. 1993, ARA&A, 31, 689
Peltoniemi, J.T., & Valle, J.W.F. 1993, Nucl.Phys.B, 406, 409
Primack, J.R., Seckel, D., & Sadoulet, B. 1988, Ann.Rev.Nucl.Part.Sci., 38, 751
Shi, Z., Schramm, D.N., Fields, B.D. 1993, Phys.Rev.D, 48, 2563
Schlegel, D., Davis, M., Summers, F., & Holtzman, J.A. 1994, ApJ, 427, 512
Somerville, R., Davis, M. & Primack, J.R. 1994, in prep.
White, S.D.M., Efstathiou, G., & Frenk, C.S. 1993,MNRAS, 262, 1023
Zurek, W.H., Quinn, P.J., Salmon, J.K., & Warren, M.S. 1994, ApJ, 431, 559

MASSIVE NEUTRINOS AND GALAXY FORMATION

Chung-Pei Ma
Theoretical Astrophysics 130-33
California Institute of Technology, Pasadena, CA 91125
cpma@tapir.caltech.edu

ABSTRACT

We report the most recent results from high-resolution numerical simulations of structure formation in two flat cold+hot dark matter models with neutrino mass densities $\Omega_\nu = 0.2$ and 0.3. We find that structure forms too late in all CDM+HDM models with $\Omega_\nu > 0.2$ to account for the amount of dense neutral gas in high-redshift damped Lyman-α systems. The $\Omega_\nu = 0.2$ model at $z \approx 0$ provides a better match to observations than the pure CDM model.

1. N-BODY SIMULATIONS

Cosmological models with cold and hot dark matter (CDM+HDM) are parameterized by the neutrino mass density Ω_ν. We chose to study structure formation in two $\Omega = 1$ CDM+HDM models: (1) $\Omega_\nu = 0.3, \Omega_{\rm cdm} = 0.65, \Omega_{\rm baryon} = 0.05$; (2) $\Omega_\nu = 0.2, \Omega_{\rm cdm} = 0.75, \Omega_{\rm baryon} = 0.05$. The corresponding neutrino masses (assumed to be in one species for simplicity) are 7 and 4.7 eV, respectively. The Hubble constant is taken to be 50 km s^{-1} Mpc^{-1} (or $h = 0.5$). The non-linear growth of structure in the $\Omega_\nu = 0.3$ model has been shown (Davis et al. 1992; Klypin et al. 1993; Jing et al. 1994) to provide excellent fits to observations at $z = 0$. However, the worry that galactic systems form too late in this model prompted us to investigate model (2), which assumes smaller Ω_ν, causing less free-streaming suppression and earlier formation of galaxies.

We performed two high-resolution particle-particle particle-mesh (P^3M) N-body simulations to study the non-linear gravitational collapse in models (1) and (2). The same initial phases were used for the two simulations to give a controlled study of the effect of neutrino masses on galaxy formation. The primordial density perturbations were assumed Gaussian and scale-invariant. The simulation box was a cube with 100 Mpc comoving sides; the Plummer force softening was 50 kpc comoving (for H_0=50 km s^{-1} Mpc^{-1}). A total of 23 million particles were used: 2.1 million for the cold component, and 10 times more for the hot component to sample the neutrino momenta at a given position in phase space. The particle masses in the $\Omega_\nu = 0.2$ model were $2.6 \times 10^{10} M_\odot$ for the cold and $6.6 \times 10^8 M_\odot$ for the hot component. The dynamic range and the number of particles make these the largest cosmological simulations ever published.

Potential galactic sites were identified with the Denmax algorithm described in Gelb & Bertschinger (1994) and Ma & Bertschinger (1994). In addition, an overdensity constraint of $\delta\rho/\rho \geq 200$ was imposed to eliminate the diffuse, nonvirialized dark halos.

The effect of force resolution in N-body simulations has been studied extensively by Gelb & Bertschinger (1994), who also demonstrated that the effect of mass resolution on the mass functions at redshifts $z \lesssim 1$ is small. Since high-redshift objects are important for constraining our models, we carried out an additional test on the effect of resolution at high redshifts. Figure 1 shows the cumulative number densities of dark halos at $z = 3$ in two $b = 2.5$ (or $\sigma_8 = 0.4$) CDM simulations with different particle masses: $2.3 \times 10^{10} M_\odot$ vs. $2.9 \times 10^9 M_\odot$. Both simulations were performed in 100^3 Mpc^3 comoving boxes with H_0=50 km s^{-1} Mpc^{-1}. The lower-resolution one (CDM16 in Gelb & Bertschinger 1994) has 144^3 particles and a Plummer force softening of 65 kpc (for $h = 0.5$); the higher-resolution one (Jain & Bertschinger 1994) has a force softening of 20 kpc and 8 times more particles (and thus samples 8 times more wavenumbers initially). This low-amplitude CDM model produces a comparable number of halos as the $\Omega_\nu = 0.3$ CDM+HDM model at $z \sim 3$. Figure 1 shows only $< 25\%$ increase in the number densities when the mass resolution is increased by a factor of 8 and the force resolution increased by a factor ~ 3. Although we have not proven convergence, this test lends support to the validity of our mass resolution and high-redshift results discussed below considering that the uncertainties in cosmological observations (*e.g.* the normalization of the power spectrum) are generally larger than 25% and that we are ruling out models only based on order-of-magnitude discrepancies.

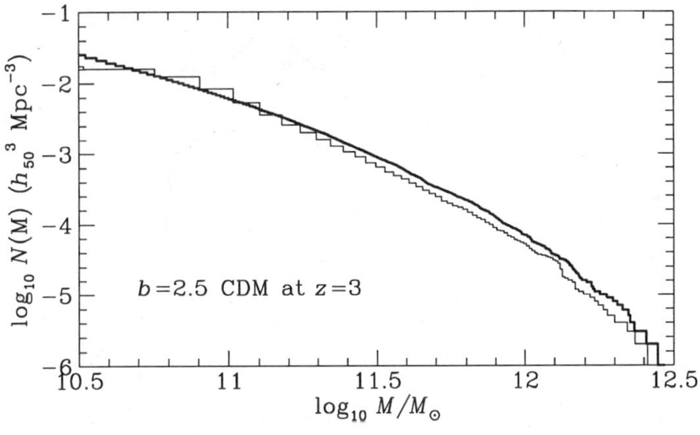

Fig. 1. The effect of resolution on the halo number densities at high redshift. The Denmax algorithm was used to identify halos; no overdensity constraint was applied. The output at $z = 3$ from two $b = 2.5$ CDM simulations is shown. The thin curve is for 144^3 particles with particle mass $2.3 \times 10^{10} M_\odot$ and the thick curve is for 288^3 particles with mass $2.9 \times 10^9 M_\odot$. An increase in mass resolution by a factor of 8 and force resolution by a factor of 3 makes only small difference in the number densities of halos.

2. HIGH-REDSHIFT CONSTRAINTS

Any hierarchical model of structure formation must account for the amount of collapsed objects observed at high redshifts. The quasar space density at redshifts up to ~ 5 has been used to constrain the CDM model (Efstathiou & Rees 1988; Turner 1991; Katz et al. 1994). Although the analytic predictions by

Turner (1991) challenge hierarchical structure formation models, recent numerical studies (Katz et al. 1994) indicate that even the low-amplitude $\sigma_8 = 0.5$ CDM model is compatible with the known distribution of high-redshift quasars.

Since structure forms later in CDM+HDM models, a better constraint is provided by the amount of gas associated with the damped Lyα systems at $z \sim 3$ (Ma & Bertschinger 1994 and references therein). Lanzetta (1993) and Lanzetta, Wolfe, & Turnshek (1994) estimated that at $3 < z < 3.5$, a fraction $(2.9 \pm 0.6) \times 10^{-3} h^{-1}$ of the critical mass density is in the form of dense neutral gas in these systems. Figure 2 shows their measured value and the predictions from our simulations for the two CDM+HDM models. The normalization of the power spectra corresponds to $Q_{\rm rms-PS} = 17.6\,\mu K$ (Bennett et al. 1994). For comparison, we also plot the predicted Ω_g in the $\Omega_\nu = 0.2$ model for the higher normalization $Q_{\rm rms-PS} = 21\,\mu K$ obtained by Gorski et al. (1994). The increase of a factor ~ 1.7 at the low mass end is too small to save the $\Omega_\nu = 0.3$ model.

We conclude from Figure 2 that only the CDM+HDM models with $\Omega_\nu \lesssim 0.2$ can produce gravitational collapses early enough to account for the Ω_g inferred from damped Lyα systems at $z \sim 3$.

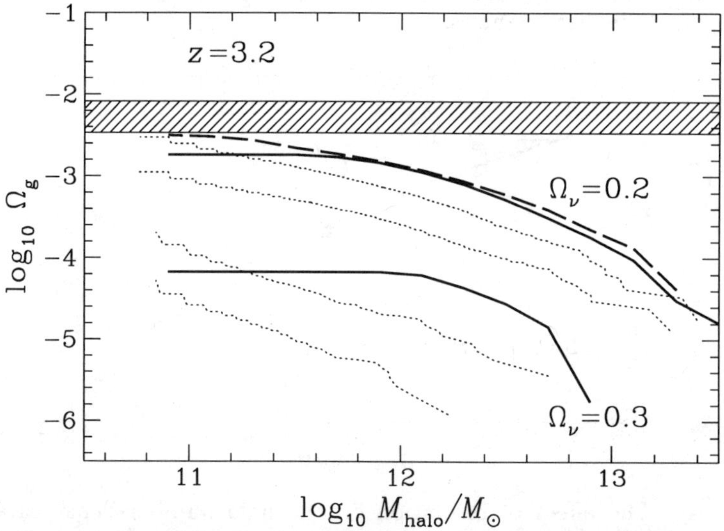

Fig. 2. The fraction of the critical density in baryons (hydrogen and helium) associated with damped Lyα systems at $z = 3.2$. The thick curves assume the hydrogen column density threshold of Lanzetta et al. (1993), where all halos (with no overdensity constraint) were used. The thick solid curves are for Ω_ν=0.2 (top) and 0.3 (bottom) with $Q_{\rm rms-PS} = 17.6\,\mu K$ (bottom), and the thick dashed curve is for Ω_ν=0.2 with $Q_{\rm rms-PS} = 21\,\mu K$ for comparison. The top two dotted curves show Ω_g predicted by the Ω_ν=0.2 model, assuming *all* hydrogen in Denmax halos with $\delta\rho/\rho \geq 200$ (bottom) and ≥ 50 (top) contributes to Lyα absorption; the bottom two dotted curves are for Ω_ν=0.3. (All dotted curves assume $Q_{\rm rms-PS} = 17.6\,\mu K$.) The shaded band shows the observed Ω_g from Lanzetta et al.

3. LOW-REDSHIFT RESULTS

While the high-redshift objects discussed above provide an upper bound on the neutrino fraction Ω_ν in CDM+HDM models, Ω_ν is bounded from below by various $z \approx 0$ observations. After all, $\Omega_\nu = 0$ leads us back to the troubled pure CDM model. (All low-redshift results discussed below are for the $\Omega_\nu = 0.2$ model. We did not continue the simulation past $z = 1.5$ for the failed $\Omega_\nu = 0.3$ model.)

Figure 3 shows the two-point correlation function and the pairwise velocity dispersions at $z = 0.25$ (for $Q_{\rm rms-PS} = 17.6\,\mu K$). These are the two statistical quantities commonly used for comparison of models and observations. These should be compared with, *e.g.*, Figures 8 and 12 for the pure CDM model in Gelb & Bertschinger (CDMII 1994) before their halo break-up procedures were applied. In general, the particle velocities have been reduced from 1000-1200 km s^{-1} for the CDM model to 500-800 km s^{-1} for the $\Omega_\nu = 0.2$ CDM+HDM, and the halo velocities have dropped from 500-600 km s^{-1} to 200-400 km s^{-1}, in much better agreement with observations (Davis & Peebles 1983; see also Zurek et al. 1994). The correlation function for the $M \geq 2.5\times10^{12} M_\odot$ halos match the observed values fairly closely, although both the amplitude and the correlation length are too small for the $M \geq 2.5 \times 10^{11} M_\odot$ halos. We are continuing the simulation to $z = 0$. It remains to be seen if there will be enough evolution between $z = 0.25$ and $z = 0$ to bring $\xi(r)$ into agreement while keeping σ reasonably low.

In Figure 3, the smaller correlation amplitude and pairwise velocity for the hot particles compared to the cold ones (dashed curves) at $r \lesssim 0.5h^{-1}$ Mpc reflect the free-streaming effect of the neutrinos. On larger scales, the two components track each other as expected.

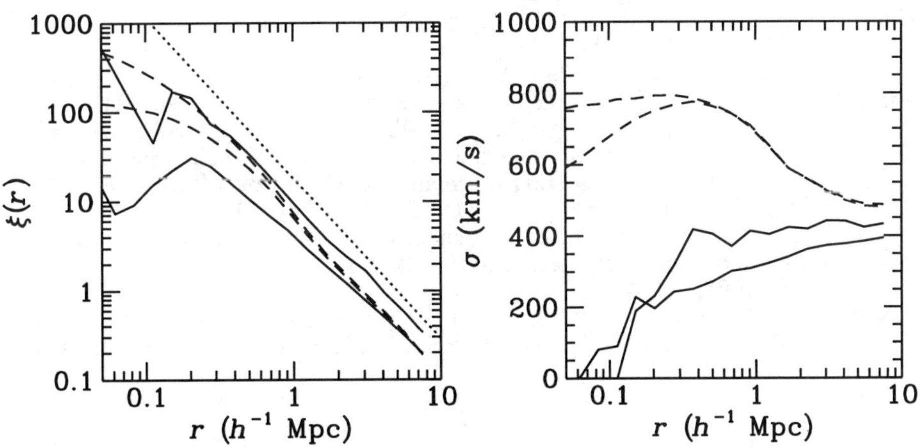

Fig. 3. The two-point correlation $\xi(r)$ and the pairwise velocity dispersion σ at $z = 0.25$ for the $\Omega_\nu = 0.2$ CDM+HDM model. The dashed curves in both panels are for the CDM (top) and HDM (bottom) simulation particles; the solid curves are for dark halos with masses above $2.5 \times 10^{12} M_\odot$ (top) and $2.5 \times 10^{11} M_\odot$ (bottom). The dotted line indicates the observed $\xi(r) = (r/5h^{-1}{\rm Mpc})^{-1.8}$ at $z \approx 0$.

4. CONCLUSIONS

Between the much-studied pure CDM and HDM models span a family of mixed dark matter models which can be parameterized by $0 < \Omega_\nu < 1$, or equivalently, $0 < m_\nu < 93h^2$ eV. We are able to rulc out all scale-invariant flat CDM+HDM models with $\Omega_\nu > 0.2$ and $H_0 = 50$ km s^{-1} Mpc^{-1} by comparing the abundance of collapsed objects at high redshifts in observations and in high-resolution N-body simulations. This leaves only a narrow parameter space open since very small Ω_ν would recover the troubled CDM model. Is $\Omega_\nu \approx 0.2$ large enough to alleviate the problems encountered in the pure CDM model at low redshifts? Our simulations show that the $\Omega_\nu = 0.2$ model can better match the observed pairwise velocity dispersions and the two-point correlation function than the pure CDM model. We are analyzing other statistical quantities and completing the simulation to $z = 0$ to further scrutinize this model.

ACKNOWLEDGEMENTS

The author acknowledges support from a PMA Division Fellowship at Caltech. The simulations were performed in collaboration with Edmund Bertschinger. Supercomputer time was provided by the National Center for Supercomputing Applications at Illinois.

REFERENCES

Bennett, C. L. et al. 1994, COBE 94-01 preprint; astro-ph/9401012
Davis, M., & Peebles, P. J. E. 1983, ApJ, 267, 465 Davis, M., Summers, F. J., & Schlegel, D. 1992, Nature, 359, 393
Efstathiou, G., & Rees, M. J. 1988, MNRAS, 230,5p
Gelb, J. M., & Bertschinger, E. 1994, ApJ, 436, in press
Gorski, K. M. et al. 1994, ApJ, 430, L89
Jain, B., & Bertschinger, E. 1994, ApJ, 431, 495
Jing, Y. P., Mo, H. J., Börner, G., & Fang, L. Z. 1994, A&A, 284, 703
Katz, N., Quinn, T., Bertschinger, E., & Gelb, J. M. 1994, MNRAS, 270, L71
Klypin, A., Holtzman, J. A., Primack, J. R., & Regos, E. 1993, ApJ, 416, 1
Lanzetta, K. M. 1993, in The Environment and Evolution of Galaxies, ed. J. M. Shull & H. A. Thronson, Jr. (Dordrecht: Kluwer), 237
Lanzetta, K. M., Wolfe, A. M., & Turnshek, D. A. 1994, ApJ, submitted
Ma, C.-P., & Bertschinger, E. 1994, ApJ, 434, L5
Turner, E. L. 1991, AJ, 101, 5
Zurek, W. H., Warren, M. S., Quinn, P. J., & Salmon, J. K. 1994, in Proc. Ninth IAP Meeting on Cosmic Velocity Fields, ed. F. Bouchet & M. Lachieze-Rey (Paris)

The influence of CDM on the dynamical estimate of Ω in the Local Neighborhood

A.M. DUNN
Institute of Astronomy, University of Cambridge
Madingley Road, Cambridge,CB3 0HA, UK

and

R. LAFLAMME
Theoretical Astrophysics, T-6, MSB288
Los Alamos National Laboratory Los Alamos, NM87545, USA

ABSTRACT

Peebles has suggested an interesting method to trace back in time positions of galaxies called the least action method. This method applied on the Local Group galaxies seems to indicate that we live in an $\Omega \approx 0.1$ Universe. We have studied a CDM N-body simulation with $\Omega = 0.2$ and $H = 50kms^{-1}/Mpc$ and compare trajectories traced back from the Least Action Principle and the center of mass of the particle forming CDM halos. We have shown that the agreement between these set of trajectories is at best qualitative. We have also shown that the line of sight peculiar velocities are underestimated. This discrepancy is due to orphans, CDM particles which do not end up in halos. By varying the density parameter Ω in the least action principle we show that using this method we would underestimate the density of the Universe by a factor of 4-5.

1. INTRODUCTION

The density parameter Ω is one the most important parameter characterising our Universe. There are many methods to measure Ω, but a factor of 10 uncertainty in its value remains. Recently Peebles suggested that it may be possible to estimate its value in the Local Neighborhood, by tracing Local Group galaxies back in time.

Peebles(1989,1990,1994) used the principle of least action to find complete trajectories for Local Group galaxies. The idea is to assume that galaxies growing out of small density perturbations in the early universe will have negligible peculiar velocities with respect to the Hubble flow. This is a reasonable assumption as we know that the microwave background has very small anisotropies. Using zero initial peculiar velocities as one boundary condition and the present positions of the galaxies as the other, trial orbits are iteratively varied so as to minimize the action. The method has been criticised since the galaxies are treated as point particles throughout their history, even though the size of the galaxies must be comparable to their separation at early times. However, the least action principle leaves the final velocities of the galaxies unconstrained, and its ability to reproduce the observed radial velocities remains a powerful test of the validity of the trajectories. For the Local Group galaxies, Peebles has obtained remarkable agreement between the observed radial velocities and those calculated from the least action principle. Obtaining reliable trajectories for nearby galaxies might shed light into the origin of their angular momentum (Dunn & Laflamme 1993).

Although the Least Action Method provides a powerful tool for investigating galaxy orbits, its predictions are only as good as the assumptions they stand on. These are, that a) galaxies initially have negligible peculiar velocity with respect to the Hubble flow, b) galaxies can be represented as point particles throughout their history, c) mergers have little effect on a galaxies motion, and d) light traces mass. One way to test the validity of some of these assumptions is to apply the Least Action Method to a numerical simulation of the universe. In the simulation we have complete information about the particle trajectories, which we compare with the predictions made by the Least Action Method. In this contribution we use a cold dark matter (CDM) simulation. Although CDM may not be able to reproduce all the observable features of our universe, it does at least represent a *possible* universe in which all the particles are governed by Hamiltonian dynamics.

2. Peebles' Least Action Method

Peebles' use of the least action method (LAM) selects a set of classical trajectories for a group of galaxies (point masses) which are interacting through gravity, against the background of an expanding universe model. This method differs from the usual application of the least action principle in that boundary conditions are applied to the beginning and end of each trajectory. The trajectories are constrained such that

$$\delta \mathbf{x}_i = 0 \text{ at } t = t_0, \qquad a^2 d\mathbf{x}_i/dt \to 0 \text{ at } a \to 0 \tag{2.1}$$

where a is the scale factor of the universe and $\mathbf{x}_i(a)$ is the trajectory of the ith galaxy in comoving coordinates. That is, the galaxies are fixed at their present positions at the present epoch, and their peculiar velocities vanish as we approach the Big Bang. Trial trajectories for the set of galaxies are adjusted in order to find a stationary point in the action (for details see Peebles 91,92).

3. CDM simulation

In order to understand the limits of the Least Action Method, it is important to compare it with some other method. We used a CDM, N-body simulation of Kauffmann and White (1992). It is a PPPM simulation with 262 144 particle, representing an $\Omega = 0.2$ universe. Scaled to $H = 50 km\ s^{-1}\ Mpc^{-1}$, it encompasses a size of 100 Mpc with particles of mass $5.2 \times 10^{10}\ M_{\odot}$.

We have studied a few groups containing 10 or so galaxy halos. The halos are determined by a friend of a friends algorithm. They were chosen in order to match the conditions in the Local Group; two dominant galaxies with peculiar velocities towards each other, a mass ratio of roughly 4:3 and somewhat isolated from high density mass concentrations. Due to the limitations of dynamic range in the simulation, we could not find any such halos with a separation of 0.7 Mpc but had to go to approximately 2 Mpc. These halos also had masses approximately 5-10 times greater than M31 and the MW. In addition to the two central galaxies, the galaxies around them up to a distance of 20 Mpc were selected to form a group. We thus intend to study the effect of the spatial distribution of the halo, the influence of the nearby galaxies on their dynamics and also the effect of particles not linked to any halo (orphans).

We have selected 9 groups which had 2 halos of the order of 200 particles within 2 Mpc. We investigated roughly 10 galaxies around these two center halos. All the groups had similar behavior and we therefore for brevity present the results of only one of them here.

4. Comparison

Once we have identified the galactic halos we can use the least action method described in section 2 to trace them back in time. We can also trace back the particles making the halo in the final step of the CDM simulation. There is only a very rough agreement between the LAM trajectories and the CDM ones.

Assuming that these are in rough agreement we can compare the line of sight velocities of these two trajectories. It is seen from figure 1 (see also Dunn & Laflamme 1994) that with the same parameter Ω and H, the LAM would overestimate this velocity. By dividing Ω by a factor of 4 we we would obtain reasonable line of sight velocity. It is this problem that we address now.

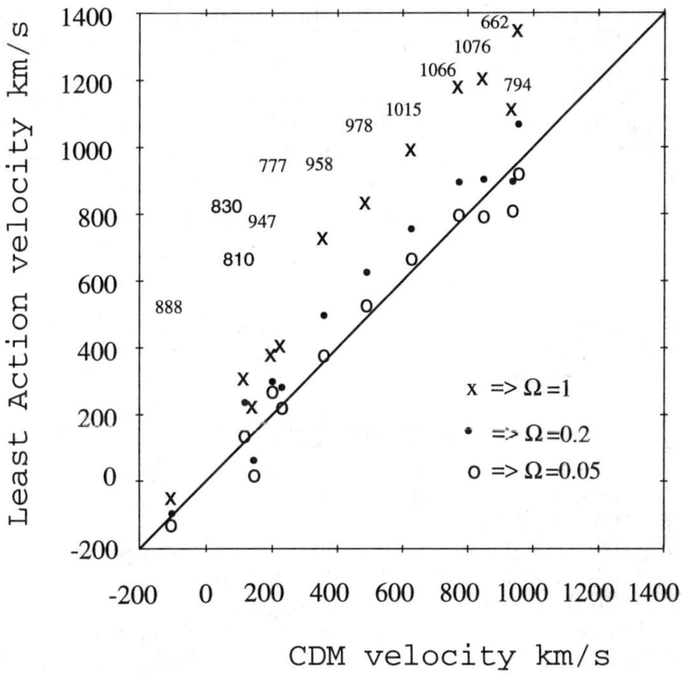

Figure 1. Least Action Method line of sight velocities (with respect to one target galaxy of the chosen group) as a function of the CDM line of sight velocity. The best fit corresponds to an adjusted density parameter $\Omega \approx 0.05$ a factor of 4-5 higher than the CDM simulation parameter.

First let's see the effect of modifying Ω in the LAM. By varying Ω in the LAM we can change the line of sight velocities. Here there are two factors to take into account. Changing Ω will change the time elapsed since the Big Bang, increasing Ω decrease the elapsed time and thus increased the velocity. The second factor is that with a larger Ω the radius R_0 a sphere which would enclose a homogeneous distribution of the total mass, is smaller. Thus we do not have to go as far to gather the mass to make the halos, thus decreasing the velocity. These two factors conspire against each other but a simple calculation for a 2 body system show that the first one wins. Thus as Ω decreases the velocity decreases.

We must now answer the question of why, for the same Ω and H we have the LAM line of sight velocities being larger than the CDM one. The answer lies in the fact that in the CDM model there are orphans, i.e. CDM particles which are not linked to halos. In the early times the background of orphans is roughly homogeneously distributed and cancel the force due to the particles which will eventually make the halos. From the LAP point of view, this will in fact reduce the force or the effective mass of the halos and thus, in order to end up at their known final positions, they will have to start at a closer distance than the LAM has given us. This has the effect of reducing the velocities, as the CDM simulation shows.

Neglecting orphans is equivalent to failure of one of the key assumptions of the Least Action Method, that is, that light traces mass (at least at kiloparsec to megaparsec scales). Of course, it is quite possible that this assumption *does* in fact hold for the universe we live in. However, since there is little observational or theoretical reason to suppose this, we must call in to doubt previously published results based on the LAM. We conclude that the dynamics of the Local Group and a careful examination of its line of sight velocities do not exclude a closed Universe.

ACKNOWLEDGEMENTS

We would like to thank B. Bromley, D. Lynden-Bell, S. White, M. Warren and W.H. Zurek for useful comments. We would also like to thank the NASA HPCC program for support.

REFERENCES

Peebles, P.J.E., 1989, ApJ, 344,L53
Peebles, P.J.E., 1990, ApJ, 362, 1
Peebles, P.J.E., 1994, Orbits of nearby galaxies, ApJ, to appear
Dunn, A.M. & Laflamme, R., 1993, MNRAS, 264, 865
Kauffman, G & White, S.D.M., 1992, MNRAS, 258, 511
Dunn, A.M. & Laflamme, R., 1994, The Least action Method, CDM and Ω, Los Alamos National Laboratory report LA-UR-94-2231, astro-ph/9411002

DARK MATTER IN VOIDS

Richard Fong,[1] Andrei G. Doroshkevich[2,3] & Victor I. Turchaninov[2]
[1]Department of Physics, University of Durham, Durham,DH1 3LE, England
[2]Keldysh Institute of Applied Mathematics, 125047 Moscow, Russia
[3]Teoretical Astrophysics Centrum, Blegsdamsvej 17, Copenhagen DK 2100, Denmark

ABSTRACT

The theory of the formation of large-scale structure in the universe through the action of gravitational instability imply the existence of substantial amounts of baryonic dark matter, of the order of 50% of the total baryon content in the universe, in the 'voids' or under-dense regions seen in the large-scale distribution of galaxies. We discuss also the large-scale structure of dark matter expected in voids and the present and future possibilities for the observation of this baryonic dark matter in 'voids'.

1. INTRODUCTION

The observational and theoretical investigations of structure on scales of $\sim$ 10–100 h^{-1} Mpc have made impressive advances in the past decade. In particular, the CfA (de Lapparent, Geller & Huchra 1988) and the SRSS (da Costa et al. 1988) redshift surveys began to show that the large-scale structure depicted by galaxies is a complex random network structure of filaments or chains of galaxies, with many embedded in 'walls' or 'sheets' of galaxies, which comprise the surfaces of a random cellular structure with an average cell size of $\sim$ 30–50 h^{-1} Mpc. This picture is now being confirmed with the new deeper, high sampling, redshift surveys, such as those of Broadbent et al. (1994), Vettolani et al. (1994) and the Las Campanas survey (Shectman et al. 1995). A most striking aspect of these new surveys to fainter magnitude limits is how structure becomes even better defined and how very rarely they pick up intrinsically faint galaxies in 'voids', finding them instead to be more or less associated with structures already found in surveys to brighter magnitude limits. Indeed, papers are now appearing showing that, although intrinsically faint galaxies are less strongly clustered, they nevertheless trace the same structures as do bright L^* galaxies (see, e.g., Mo, McGaugh & Bothun 1994 and Hopp et al. 1995).

The observations seem then to be an overwhelming confirmation of the simplest physical theory for the formation of structure, that it has formed simply through gravitational instability. The general properties of this physical process have been studied by Zel'dovich and collaborators ever since his pioneering paper in 1970 (for an excellent review, the reader is referred to Shandarin & Zel'dovich 1989). The results of this theory are discussed further in Buryak, Doroshkevich & Fong (1994, hereafter BDF), who re-emphasize that it is the *velocity field* which leads to the formation of structure in the universe, creating filaments of overdensities of non-baryonic and baryonic matter. The *gravitational field,* on the other hand, is much smoother, and *modulates* this formation process on large scales. The first structures to form are those lying in gravitational wells, whereas structures in the positive regions of the gravitational potential may normally have formed later, but are also inhibited by secondary ionization of the inter-galactic medium by the UV radiation from the first objects to form (Doroshkevich et al. 1995). This modulation by the gravitational potential produces cellular-like structure with a mean cell size of $\sim$ 50–100 h^{-1} Mpc.

In Doroshkevich et al. (1995) they put forward a new theory, an extension of the Zel'dovich approximation, to describe some general physical properties of the actual formation and evolution of large-scale structure in the Universe. It is seen that, in the usual models with Gaussian initial conditions, filaments form very rapidly around a 'creation' epoch, z_{cr}, the value of which depends on the model. This is so rapid as to merit being called the 'phase transition' of a structureless universe to one with structure. After this comes a very quiet phase, in which, although filaments are still forming, the process is dominated by the merging of filaments, so that the mean distance between filaments begins gradually to increase with time, reaching about 10–15 h^{-1} Mpc today (BDF). Also, during this phase, galaxies flow along filaments to the branch points of this random network structure of filaments, where already rich structures have formed. It is interesting then that one is now beginning to observe the extension of the X-ray gas around rich clusters of galaxies into the filaments that emanate from them (Böhringer 1995).

2. DARK MATTER IN VOIDS

Whatever non-baryonic matter there may be, the baryonic matter rapidly catches up with it once recombination is over. In the linear regime, they then simply evolve together under the influence of gravity. It is only when over-densities become non-linear and the baryonic gas can become dense enough to undergo highly dissipative physical processes that there is some biasing between baryonic matter and non-baryonic matter inside highly over-dense regions. However, it is enlightening that even in such dense environments as X-ray clusters it is only inside the central Mpc or so where there seems to be any significant difference between the galaxy distribution with the X-ray gas and dynamical mass distribution (Böhringer 1995). However, as far as this paper is concerned, clearly the baryonic and non-baryonic matter trace each other inside 'voids', i.e. regions under-dense in galaxies.

Doroshkevich, Fong & Makarova (1995) have analysed a CDM simulation (Eke et al. 1994) using the method of BDF. They found that, whereas the 'galaxy' structure is predominantly filament-like, the dark matter structure is sheet-like, i.e., they are simply Zel'dovich pancakes. Thus, in these simulations 'galaxies' trace the 'ridges' of rich dark matter pancakes, with weak pancakes containing very few 'galaxies'. However, a very crucial process has, as yet, to be incorporated into dynamical N-body simulations. This is the process of secondary ionization, occurring at z_{cr}, of the inter-galactic medium by objects in the first structures to form. As we mentioned in the introduction, this will clearly inhibit the formation of galaxies in weak pancakes. Thus, this will greatly enhance the modulation of large-scale structure by the gravitational potential, producing very clear voids in the galaxy distribution, which, indeed, is what we seem to see in the observations. But, most importantly of all, it will lead to an environment dependent bias, that is, biasing that is also a function of position.

Initially, the regions of positive and negative gravitational potential (with the mean set to zero, of course) occupy equal volumes. Then, since the 'filaments' and 'walls' of galaxies occupy so little volume - they are only $\sim$ 2–4 h^{-1} Mpc thick, small compared to the size of voids - the regions of negative potential have therefore expanded to fill most of space. We expect then that voids contain as much as $\sim 50\%$ of all matter. Now, simulations show that the homogeneously distributed background density of matter is only $\sim 15\%\ \bar{\rho}$, leaving as much as $\sim 35\%\ \bar{\rho}$ as actually collapsed matter in weak, but nonetheless over-dense, dark matter pancakes (Doroshkevich et al. 1995). We would like to emphasize that this is a most important consequence of the theory.

3. OBSERVATIONS

The most notable present method to detect and measure neutral hydrogen in the intergalactic medium is to try to measure the Gunn-Peterson effect in distant QSOs. It has proven very difficult to obtain any detection. Indeed, this is taken as a signal of the high degree of photoionization of the inter-galactic medium. As we discussed above, this occurs around z_{cr}, which, by virtue of the presence of structure to $z \geq 4$, must be $\gtrsim 5$ (Doroshkevich et al. 1995). Perhaps, the most exciting result has been the recent detection of the Gunn-Peterson effect for the He II 304 Å line of an HST observation of a $z = 3.286$ QSO (Jakobsen et al. 1994). This observation would seem to put an upper limit on the inter-galactic plasma of $\Omega_{pl} \approx 10^{-2}$, which would be consistent with the homogeneous background expected from N-body simulations and with a low value of $\Omega_b \sim 0.1$ for the baryon density in the universe. However, a model is required to estimate the amount of ionized hydrogen in the inter-galactic medium. Clearly, it is a measurement with important consequences for theories for the formation and evolution of structure in the universe.

Another interesting observation has been the detection of Ly-α clouds at $z \approx 0.065$ in a void in the line of sight of 3C273, again an HST observation (Morris et al. 1993). One of these clouds is as much as $6.6\, h^{-1}$ Mpc away from the nearest galaxy they could find in the field. Unfortunately, at this depth they could only detect galaxies brighter than about -18 mag in B. However, as mentioned in the introduction, fainter galaxies may well have a low correlation length, and yet trace exactly the same structures as L^* galaxies. Thus, it seems unlikely then that these three Ly-α clouds are associated with galaxy haloes, where, in any case, they would have difficulty in surviving. Instead, we have proposed that they are naturally associated with the haloes of galaxy-less dark matter pancakes, which occupy voids. Unfortunately, to estimate the mainly ionized medium in these filaments, a model is again needed.

A direct means of observing this ionized medium in voids is through the use of the dispersion measure, which has been so successfully used with pulsars to estimate ionized matter in the Galaxy. It was first proposed by Haddock & Sciama (1965), but they overestimated the effect (see, also, Weinberg 1972). Unfortunately, this requires cosmological objects that vary in the radio on short time scales. However, we are beginning to search for very distant supernovae in the hope of using them to estimate the Hubble constant. One might then hope that, perhaps, in the distant future the dispersion measure may become a viable technique for the direct measurement of the column density of ionized hydrogen to cosmological distances.

Although there are highly compelling arguments for 'exotic' dark matter, there is a need to bear in mind that the simplest universe supported by observations seems to be an open one with known particles (Coles & Ellis 1994). However, whatever the true situation may be, there is expected to be of $\sim 50\%$

of baryonic dark matter in 'voids' and its detection is, clearly, also of great physical importance.

ACKNOWLEDGEMENTS

We are very grateful to Shaun Cole for the data from their N-body dynamical numerical simulations. We would also like to thank Simon Morris for invaluable discussions. AGD and VIT wish to acknowledge support from the Center of Cosmo-Particle Physics, Moscow, and from the European Community's INTAS project.

REFERENCES

Broadbent, A., Shanks, T., Watson, F.G., Parker, Q.A., Fong, R., Oates, A.P. and Collins, C.A., 1994, in Astronomy from Wide-field Imaging; IAU Symp. No. 161, eds H.T. MacGillivray et al. (Dordrecht: Kluwer) p. 693

Böhringer, H., 1995, in Large Scale Structure in the Universe; Proc. XI Cosmology Workshop, Potsdam, eds S. Gottlöber, J.P. Mücket and V. Müller (Singapore: World Scientific) in press

Buryak, O.E., Doroshkevich, A.G. & Fong, R., 1994, ApJ, 434, 24 (BDF)

Coles, P. & Ellis, G., 1994, Nature, 370, 609

da Costa, L.N., Pellegrini, P.S., Sargent, W.L.W., Tonry, J., Davis, M., Meiksin, A., Latham, D.W., Menzies, J.W. & Coulson, I.A., 1988, ApJ, 327, 544

de Lapparent, V., Geller, M.J. & Huchra, J.P., 1988,ApJ, 332, 44

Doroshkevich, A.G., Fong, R., Gottlöber, S., Mücket, J.P. & Müller, V., 1995, in preparation

Doroshkevich, A.G., Fong, R. & Makarova, O.V., 1995, in preparation

Eke, V.R., Cole, S., Frenk, C.S. & Navarro, J.F., 1994, in preparation

Haddock, F.T. & Sciama, D.W., 1965, PRL, 14, 1007

Jakobsen, P., Boksenberg, A., Deharveng, J.M., Greenfield, P., Jedrzejewski, R. & Paresce, F., 1994, Nature, 370, 35

Mo, H.J., McGaugh, S.S. & Bothun, G.D., 1994, MNRAS, 267, 129

Morris, S.L., Weymann, R.J., Dressler, A., McCarthy, P.J., Smith, B.A., Terrile, R.J., Giovanelli, R. & Irwin, M., 1993, ApJ, 419, 524

Shandarin, S.F. & Zeldovich, Ya.B. 1989, RMP, 61, 185

Shectman, S.A., Landy, S.D., Oemler, A.A., Tucker, D., Kirshner, R.P., Lin, H., Schechter, P.L., 1995, in Wide Field Spectroscopy and the Distant Universe; Proc. 35th Herstmonceaux Conference, in press

Vettolani, G. et al., 1994, Astronomy from Wide-field Imaging; IAU Symp. No. 161, eds H.T. MacGillivray, et al. (Dordrecht: Kluwer) p. 687

Weinberg, S. 1972, Gravitation and Cosmology (New York: Wiley)

Zeldovich, Ya.B., 1970, A&A, 5, 20

RICH CLUSTER SIMULATION: DYNAMICS AND MASS ESTIMATES

B.C. Bromley, M.S. Warren, W.H. Zurek,
Theoretical Astrophysics, Los Alamos National Laboratory

P.J. Quinn
Mount Stromlo Observatory

ABSTRACT

We present an analysis of a simulated rich cluster dominated by cold dark matter. Here, we focus on the dynamics of collapsed dark matter halos which may be associated with galaxies. We use projected positions and velocities of both dark matter particles and of halos to assess several conventional measures of cluster mass. The results indicate that the measures give reasonable results for dark matter particles but cause overestimates of cluster mass when applied to halos. In searching for the source of the discrepancies, we find that tidal disruption and dynamical friction do not induce a significant bias in the projected velocity dispersion of halos; however, an antibias in the spatial distribution of halos relative to that of the total mass appears to be most problematic for cluster mass estimates. We conclude that over and above the usual concerns about the applicability of virial equilibrium to clusters, any bias can adversely affect cluster mass estimates and one must question their reliability as applied to observed clusters.

1. INTRODUCTION

Galaxy clusters have played a key role in the effort to determine the nature of matter in the Universe and the value of the cosmological density parameter Ω. The dynamical mass of clusters has particularly important cosmological implications, as first recognized when Zwicky (1933) deduced presence of dark matter in the Coma cluster. Several estimators designed to yield cluster mass from projected galaxy velocities have been introduced by Bahcall & Tremaine (1981) and Heisler, Tremaine & Bahcall (1985). These authors assessed such measures in numerical N-body simulations with 20 or less point-like galaxies. Subsequently, West & Richstone (1988) performed tests with simulations of $N = 10^4$ particles, and Carlberg (1994) examined virial mass estimates in a cluster simulation of 10^6 mass particles. The conclusions from these more recent studies was that virial measures underestimate cluster masses.

Ideally, mass estimates should be tested with simulations that include the large-scale gas dynamics of the baryons as well as high mass resolution to track the interaction of extended galactic halos with the rest of the cluster. However, work on gas dynamics within clusters (e.g., Evrard 1990) indicates that cooling processes may be neglected in the large-scale intracluster gas—the baryonic component may be effectively dissipationless. Thus as an approximation, dissipative effects may be neglected when one proceeds to higher and higher mass resolution.

Here we present the results of a million-particle high-resolution simulation of a cluster with dissipationless cold dark matter. The cluster is dominated by a single massive core interspersed with smaller collapsed halos which may be associated with galaxies. We track the dynamical evolution of the halos, as well as that of their constituent dark matter particles. Our principal aim is to test conventional measures of cluster mass based on projected positions and velocities, as might be available to a distant observer. We first use the mass particles to test the assumptions, such as virialization, upon which these measures rest. Then, we focus upon the application of the mass measures to the halos. At issue are the effects of tidal disruption and dynamical friction, and the extent to which the halo phase-space distribution is biased with respect to that of the total mass.

2. NUMERICAL METHOD

The simulation of the cluster began with selection of a primordial over-dense structure, the seed for a present-epoch cluster, from a palette of random realizations of the cosmic density field. Each realization was generated as a Gaussian field variate with a COBE-normalized CDM power spectrum in an $\Omega = 1$ universe. The selected initial field was a 1-in-500 event; it contained a 2.5-σ over-density in the middle 1/9th of the simulation volume. Force calculation for this 1.1×10^6 particle system was performed using a tree code with spatial resolution of 5 kpc (see Warren 1994). The final output, obtained after 3000 computational timesteps, had a radius of 15 Mpc and a Hubble parameter of 50 km/s/Mpc. The evolved cluster had a mass of approximately 5×10^{14} $M_\odot$within the Abell radius.

In the simulation, aggregates of dark matter particles were identified as galaxy halos with the algorithm described by Warren et al. (1994). The constituent particles of 40 halos were tagged and tracked from a redshift of $z = 0.85$ to the present epoch ($z = 0$).

3. CLUSTER MASS MEASURES

We examine three conventional cluster-mass measures (Bahcall & Tremaine 1981; Heisler et al. 1985; Binney & Tremaine 1987) for a cluster of N galaxies and a radius R:

$$\text{Isothermal Mass}: \ 2R\sigma_1^2/G;$$

$$\text{Virial Mass}: \ 3\pi N\sigma_1^2/2GR_H \ ;$$

$$\text{Projected Mass}: \ 32\alpha < v_\parallel^2 r_\perp > /\pi G \ .$$

In these expressions, σ_1 is the dispersion in the line-of-sight velocities $v_\parallel$; $r_\perp$ and R_H are radial distance from the cluster center and the mean harmonic radius, respectively, both measured in the sky plane of a distant observer. The factor of α in the projected-mass measure ranges from 1 for isotropic orbits to 2 for radial orbits.

These mass estimates are applied to the CDM particles in the simulation and are shown in Figure 1a as a function of plane-of-sky radius $r_\perp$. Evidently, the virial and projected mass estimates are in fair agreement with the true mass within the Abell radius. The isothermal mass estimate works well only within the central region of the cluster.

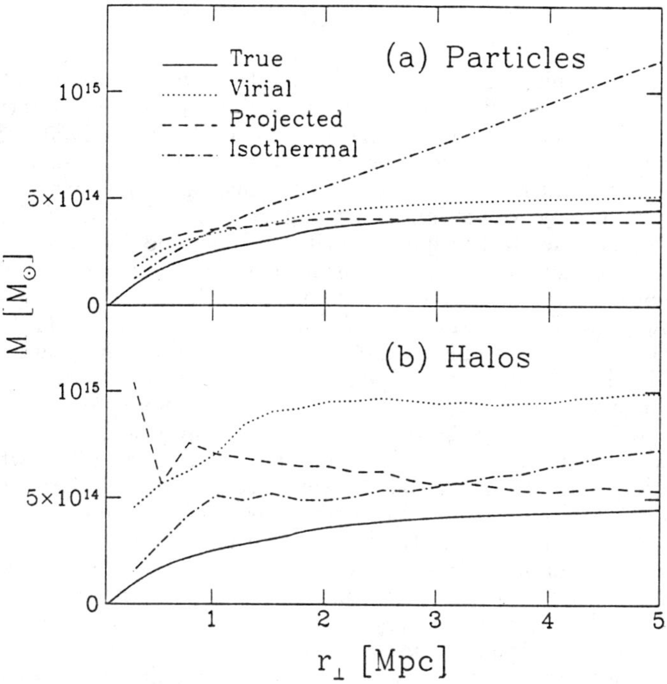

Fig. 1. Cluster mass measures from (a) total mass and (b) halos as a function of plane-of-sky radius $r_\perp$. The cluster was defined to include all objects within 7.5 Mpc of the center of mass.

When applied to the halo population, the mass estimates fare considerably worse (Fig. 1b). In general, the mass is overestimated by 40% to 250% for 1 Mpc $< R <$ 3 Mpc. We now examine possible sources of the discrepancies.

The mass and force resolution of the simulation are sufficiently high that we may expect to see the effects of tidal forces and dynamical friction on the halo trajectories. We found that tidal effects were not significant even in those orbits that passed very near the cluster core, as only 5% of the halos were disrupted.

The more subtle phenomenon of dynamical friction, which may result when halos have a large number of internal degrees of freedom, may produce a velocity bias in the halo trajectories (Carlberg 1994). Here, we do not find evidence for a significant bias in the one-point halo velocity distribution. For example, the 1D halo velocity dispersion as a function of plane-of-sky radius from the cluster core traces the dispersion profile for particles over a broad range of $r_\perp$. Thus, a velocity bias is not the source of error in the mass measures discussed above.

However a spatial bias is detected. Evidence appears in R_H, the mean harmonic radius in the sky plane. The profile of R_H for halos lies significantly above that for particles, indicating that halos are more dilute in their contribution to the cluster potential than expected if they were a random subset of the total mass. An increase in R_H at fixed velocity dispersion strengthens the magnitude of the inferred gravitational potential and drives the virial and projected mass measures upward from the true value. We conclude that spatial bias appears to be the principal source of error in these measures.

4. CONCLUSION

The small smoothing scale of 5 kpc for the force calculation and the large number of particles distinguish this experiment from previous numerical simulations of dissipationless cluster dynamics. We find that conventional mass measures based on projected velocities and positions are in reasonable agreement with the actual mass when simulation particles are used. However, these same measures generally overestimate the cluster mass when applied to halos. The overestimate is linked to the spatial antibias of the halo distribution. A halo velocity bias, such as might be expected from if tidal disruption and dynamical friction, was not detected. These conclusions compare favorably, in qualitative terms, with analyses of x-ray and optical data: Lubin & Bahcall (1993) argue for an absence of a velocity bias, while Eyles et al. (1991) find that galaxies are antibiased in clusters and dynamical masses are overestimated. We strongly caution that the spatial antibias in our data may be artificially enhanced by the halo finding algorithm and by the dissipationless assumption.

Our results are at odds with the findings of Carlberg (1994) who ran a similar simulation but with nearly 50% lower force resolution. Carlberg cites a significant velocity bias that causes underestimates in virial mass measures. The discrepancy may stem in part from the force resolution and in part because halos are tagged at an early epoch ($z = 2$)—such objects may not be representative of the halo population at $z = 0$. This discrepancy serves as a reminder that cluster mass measures are highly sensitive to the uncertain details of biasing in dense regions and that mass estimates must be interpreted accordingly.

In closing, we note that the spatial antibias of galaxies in clusters has ramifications for other cosmological measures. An example is the pairwise radial velocity dispersion, σ_{12}. Even if galaxies are exact tracers of the cosmic velocity field, σ_{12} inherits a bias simply because the high-velocity galaxies in the dense clusters are under-represented compared with the slower objects in the field. This effect can be of considerable importance for cosmological information derived from σ_{12}, as discussed at this conference (see Zurek, Bromley & Warren) and elsewhere (Zurek *et al.* 1994).

We are grateful to the Aspen Center for Physics where part of this work was completed. This research was supported by the NASA HPCC program.

REFERENCES

Bahcall, J., & Tremaine, S. 1981, ApJ, 244, 805

Binney, J., & Tremaine, S. 1987, Galactic Dynamics (Princeton: Princeton University Press)

Carlberg, R.G., 1994, ApJ, 433, 468

Evrard, A.E., 1990, ApJ, 363, 349

Eyles, C.J, Watt, M.P., Bertram, D., Church, M.J., Ponman, T.J., Skinner, G.K., & Willmore, A.P., 1991, ApJ, 376, 23

Heisler, J., Tremaine, S., & Bahcall, J. 1985, ApJ, 298, 8

Lubin, L.M, & Bahcall, N.A. 1993, ApJ, 415, L17

Warren, M.S., 1994, Ph.D. Thesis, UCSB

Warren, M.S., Quinn, P.J., Salmon, J.K., & Zurek, W.H. 1992, ApJ, 399, 405

West, M.J., & Richstone, D.O. 1988, ApJ, 335, 532

Zurek, W.H., Quinn, P.J., Salmon, J.K., Warren, W.S. 1994, ApJ, 431, 559

Zwicky, F. 1933, Helv. Phys. Acta, 6 110

HYDRODYNAMIC SIMULATIONS OF GALAXY FORMATION

Anne A. Thoul
Institute for Advanced Study, Princeton, NJ 08540.
thoul@guinness.ias.edu

ABSTRACT

We have developed an accurate, one-dimensional, spherically symmetric, Lagrangian hydrodynamics/gravity code, designed to study the effects of radiative cooling and photo-ionization on the formation of protogalaxies. We examine the ability of collapsing perturbations to cool within the age of the universe. In contrast to some studies based on order-of-magnitude estimates, we find that cooling arguments alone cannot explain the sharp upper cutoff observed in the galaxy luminosity function. We also look at the effect of a photoionizing background on the formation of low-mass galaxies.

1. INTRODUCTION

There have been two quite different approaches to the study of the effects of gas dynamics in galaxy formation. One uses simple analytic estimates (Binney 1977; Silk 1977; and Rees & Ostriker 1977): typically, one computes the characteristic density and virial temperature of a dark halo assuming a spherical collapse model, then asks whether gas at this density and temperature can cool within a dynamical time, or within a Hubble time. Combined with extended versions of the Press-Schechter (1974) formalism, these methods can yield detailed predictions for properties and evolution of the galaxy population (White & Rees 1978; White & Frenk 1991; Kauffman, White & Guiderdoni 1993; Cole et al. 1994). The second approach is to incorporate gas dynamics directly into three-dimensional numerical simulations (e.g. Katz & Gunn 1991; Cen & Ostriker 1992, 1993; Katz, Hernquist & Weinberg 1992; Evrard, Summers & Davis 1994; Steinmetz & Muller 1994).

In the work reported here (see Thoul & Weinberg 1995a,b for full versions), we take an intermediate path, modeling the collapse of individual perturbations with a *one-dimensional*, Lagrangian, gravity/hydro code. The code evolves a mixture of gas and collisionless dark matter, elements of which are represented by concentric, spherical shells. The gas responds to gravity and pressure forces; it can be heated by adiabatic compression, by shocks, and by energy input from a photo-ionizing background, and it can cool by a variety of atomic radiative processes. The collisionless dark matter responds only to gravitational forces. We adopt initial conditions appropriate to Gaussian random fluctuations, as might be produced by inflation in the early universe.

The geometry in our calculations is idealized, and one must therefore take care to keep their limitations in mind. Nonetheless, they provide a valuable complement to their more elaborate, 3-dimensional cousins because of their high resolution, their speed, and their relative simplicity. Three-dimensional hydrodynamic simulations of galaxy formation suffer from limited spatial resolution and mass resolution, making it difficult to separate genuine physical results from numerical artifacts.

Here we apply the code to two of the basic questions of galaxy formation: what causes the abrupt cutoff at the upper end of the galaxy luminosity function, and does photoionization by the UV background prevent the formation of dwarf galaxies?

2. MODEL

The gaseous component is described by the fluid equations for a perfect gas. The collisionless component is described by the equation of motion $dv_d/dt = -M(r_d)/r_d^2$, where r_d and v_d are the radii and velocities of the collisionless mass shells, and $M(r_d)$ is the total (baryonic and dark matter) mass inside radius r_d. We use the standard, second-order accurate, Lagrangian finite-difference scheme (Bowers & Wilson 1991).

The radiative cooling is calculated for a gas of primordial composition, 76% hydrogen and 24% helium by mass. We compute the abundances of ionic species as a function of density and temperature by assuming that the gas is in ionization equilibrium with a spatially uniform background of UV radiation. In other words, we choose the abundances so that the rate at which each species is depopulated by photo-ionization, collisional ionization, or recombination to a less ionized state is equal to the rate at which it is populated by recombination from a more ionized state or by photo-ionization or collisional ionization of a less ionized state. The intensity and spectrum of the UV background are specified as a function of time based on theoretical models or observational constraints. Here, we adopt a uniform and constant UV background given by $J_\gamma = J_0(\nu_L/\nu)^\alpha$, where ν_L is the Lyman limit and $J_0 = J_{21} \times 10^{-21}\,\mathrm{ergs\,s^{-1}\,Hz^{-1}\,cm^{-2}\,sr^{-1}}$. For the physical problems that we study here, the timescales for reaching ionization equilibrium are much shorter than the other timescales of interest, so our equilibrium assumption should be an excellent approximation. Photo-ionization can affect the behavior of lower mass perturbations because it eliminates the neutral hydrogen and singly ionized helium that dominate cooling at low temperatures, and because the residual energy of the photo-electrons heats low-density gas to $T \sim 10^4$K.

As initial conditions, we adopt the average density profile around a 2σ peak in a Gaussian random density field (Bardeen et al. 1986). These initial conditions have two free parameters, which we choose to be the filter mass M_f, i.e., the mass contained within a sphere of radius $2R_f$ where R_f is the filter radius, and the collapse redshift z_c, defined to be the redshift at which the $r = 2R_f$ shell would collapse to $r = 0$ in the absence of pressure.

3. RESULTS

We have performed a series of collapse calculations in which we vary the value of the filter mass M_f while keeping the collapse redshift fixed at $z_c = 2$. In these calculations, we use $\Omega_d = 0.9$ and $\Omega_b = 0.1$. The solid points in Figure 1 show the results for $J_{21} = 0$ (i.e., no UV background). We plot the ratio of the mass M_c of gas that cools by $z = 0$ to the mass $M_c(p = 0)$ of gas that would cool and collapse to the center by $z = 0$ in the case of perfectly efficient radiative cooling (zero pressure). We characterize the perturbations by the circular velocity $v_c = (GM_f/0.5r_{ta})^{1/2}$ that would be associated with the shell of interior mass M_f if it collapsed and virialized at half its turnaround radius r_{ta}. In the absence of a UV background, the radiative cooling cuts off very sharply at $T \approx 10^4$ K. The solid points in Figure 1 show three distinct regimes: for very low circular velocities, $v_c \lesssim 7$ km/s, the post-shock gas remains below 10^4K, and there is no cooling at all; at intermediate circular velocities, $7\,\mathrm{km/s} \lesssim v_c \lesssim 30\,\mathrm{km/s}$, all of the shocked gas cools; and at high circular velocities, $v_c \gtrsim 30$ km/s, only a fraction of the shocked gas cools. However, the transition from the intermediate-velocity regime to the high-velocity regime is not a sharp one. Indeed, in the latter case, the outermost shocked shells remain pressure supported all the way to $z = 0$ and do not cool, but the inner shells of these perturbations collapse earlier, at higher density and lower virial temperature, and these shells are able to cool. Therefore, even though the ratio $M_c/M_c(p = 0)$ decreases with increasing v_c in this regime, the decrease is quite slow. In particular, $M_c/M_c(p = 0)$ falls much less rapidly than $1/M_c(p = 0)$, so the actual amount of cooled gas *increases steadily* with v_c. We see that the requirement that gas be able to cool within a Hubble time cannot by itself explain the sharp upper cutoff in the luminous mass of observed galaxies.

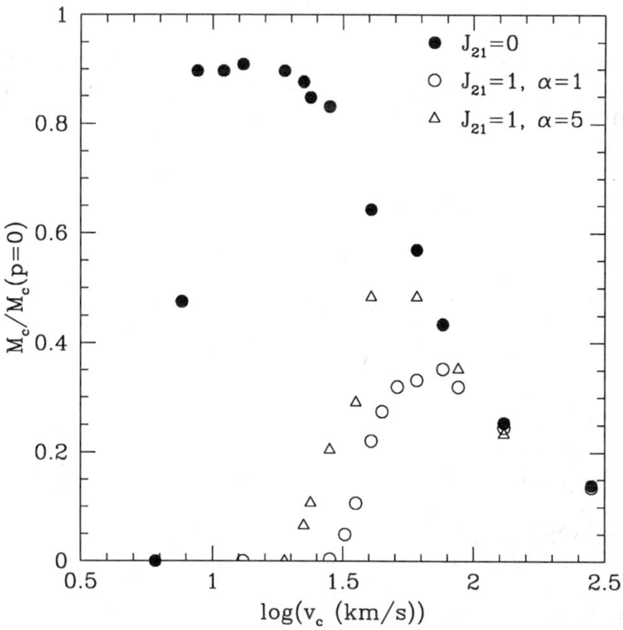

Fig. 1. Ratio of the mass M_c of gas that cools by $z = 0$ to the mass $M_c(p = 0)$ of gas that would cool and collapse by $z = 0$ in the absence of pressure, as a function of the circular velocity v_c.

It has been speculated that the presence of a background of UV radiation could inhibit the formation of dwarf galaxies (Efstathiou 1992). We have also performed collapse calculations in the presence of a uniform UV background of amplitude $J_{21} = 1$. Here again, we have kept a fixed $z_c = 2$, $\Omega_d = 0.9$ and $\Omega_b = 0.1$, and we varied the value of the filter mass M_f. The open circles in Figure 1 show the results for $\alpha = 1$, a spectrum slope we might expect if the UV radiation is mostly produced by quasars. In this case, we see that the UV background completely suppresses the formation of galaxies with circular velocities $v_c \lesssim 30\,\mathrm{km/s}$. The open triangles show the results for $\alpha = 5$, a softer spectrum that might arise if the UV radiation is mostly produced by star-forming galaxies. Here the UV background suppresses the formation of galaxies with circular velocities $v_c \lesssim 20\,\mathrm{km/s}$. The presence of a photoionizing background considerably affects the amount of gas allowed to cool and form galaxies up to circular velocities of the order of $80\,\mathrm{km/s}$ for $\alpha = 1$ and $50\,\mathrm{km/s}$ for $\alpha = 5$. For larger circular velocities, however, the virial temperature is high, the gas is collisionally ionized, and the cooling is dominated by free-free transitions. Including a photo-ionizing background does not, therefore, alter the results at high circular velocities.

Thus, even though cooling arguments alone cannot explain the sharp upper cutoff in the observed galaxy luminosity function, the presence of the UV background significantly affects the results at the low-mass end. We are presently examining the role of the spectrum and history of the UV background in more detail.

ACKNOWLEDGEMENTS

This research was supported by the Ambrose Monell Foundation (AT), the W.M. Keck Foundation (DW) and by NSF grant PHY92-45317. We enjoyed the hospitality and stimulating atmosphere of the Aspen Center for Physics during part of this work.

REFERENCES

Bardeen, J. M., Bond, J. R., Kaiser, N., & Szalay, A. S. 1986, ApJ, 304, 15
Binney, J. J. 1977, ApJ, 215, 483
Black, J. H. 1981, MNRAS, 197, 553
Bowers, R. L. & Wilson, J. R. 1991, Numerical Modeling in Applied Physics and Astrophysics (Boston: Jones and Bartlett Publishers)
Cen, R. 1992, ApJS, 78, 341
Cen, R. & Ostriker, J.P. 1992, ApJ 393, 22
Cen, R. & Ostriker, J.P. 1993, ApJ 417, 404
Cole, S., Aragon-Salamanca, A., Frenk, C. S., Navarro, J. F. & Zepf, S. E. 1994, MNRAS, in press
Efstathiou, G. 1992, MNRAS, 256, 43
Evrard, A.E., Summers, F.J & Davis, M. 1994, ApJ 422, 11
Katz, N. & Gunn, J. E. 1991, ApJ, 377, 365
Katz, N, Hernquist, L. & Weinberg, D. H. 1992, ApJ, 399, 109
Kauffmann, G., White, S. D. M. & Guiderdoni, B. 1993, MNRAS, 264, 201
Press, W. H. & Schechter, P. 1974, ApJ, 187, 425
Rees, M. J. & Ostriker, J. P. 1977, MNRAS, 179, 541
Silk, J. I. 1977, ApJ, 211, 638
Steinmetz, M. & Muller, E. 1994, A&A, 281, L97
Thoul, A.A., & Weinberg, D.H. 1995a, ApJ, in press
Thoul, A.A., & Weinberg, D.H. 1995b, ApJ, in preparation
White, S. D. M. & Rees, M. J. 1978, MNRAS, 183, 341
White, S. D. M. & Frenk, C. S. 1991, ApJ, 379, 52

LARGE-SCALE STRUCTURE FORMATION AND THE SEGREGATION OF DARK AND BARYONIC MATTER

Hugo Martel, Paul R. Shapiro, and Azita Valinia
Department of Astronomy, University of Texas, Austin, TX 78712.

Jens V. Villumsen
Max-Planck-Institut fur Astrophysik
Karl-Schwarzschild-Str. 1, Postfach 15 23, 85740 Garching, Germany

ABSTRACT

We have developed an improved version of SPH – Adaptive SPH (ASPH) – for cosmological gas dynamical simulations. We have applied our ASPH method to three models – Hot, Cold, and Mixed Hot-and-Cold Dark Matter (HDM, CDM, and CHDM) – to simulate large-scale structure formation in 2D. We have analyzed these simulations to determine how the ratio of gas to total density in each differs from that of their universal mean values – the density bias – as a function of the degree of overdensity of the matter. All three models show a tendency for the gas density to be anti-biased (i.e. gas mass fraction less than its mean value) in regions of highest overdensity, while positively biased in regions of moderate overdensity. This anti-bias trend may cause models like these with $\Omega = 1$ and with Ω_b limited by standard Big-Bang nucleosynthesis to be inconsistent with previous X-ray and optical determinations of the baryonic mass fraction in rich clusters. To test this, we focus on CDM, study the bias inside simulation clusters and pick one to simulate the standard procedure for X-ray mass determinations. Fitting X-ray brightness profiles of clusters to measure total cluster mass, assuming isothermal sphere and hydrostatic equilibrium, can give approximately correct results, while underestimating the gas fraction somewhat.

1. 2D, COSMOLOGICAL HYDRO SIMULATIONS USING ASPH/PM

We apply our ASPH method in 2D to simulate the growth of large-scale structure in a box of size $L_{box} = 200\,\mathrm{Mpc}$ in a COBE-normalized HDM, CDM, or CHDM universe. The 2D power spectrum is the same as in 3D, but multiplied by $2k$, where k is the wavenumber — i.e. equal rms $\delta\rho/\rho$ in 2D and 3D for equal fluctuation scale length. We couple our ASPH method to a standard Particle Mesh (PM) gravity solver in 2D (Shapiro et al. 1994; 1995). We take $h = 0.5$ and $\Omega_{tot} = 1$. For HDM and CDM, $\Omega_{DM} = 0.9375$, $\Omega_b = 0.0625$. For CHDM, $\Omega_{CDM} + \Omega_b = 0.7$, $\Omega_{HDM} = 0.3$, and $\Omega_b = 0.058333$. For all three models, the gas particle number is $N_{gas} = 128 \times 128$ and the PM grid has 1024×1024 cells. For HDM and CDM, the total particle number is 512×512. For CHDM, $N_{CDM} + N_{gas} = 196608$ of which 1/12 are gas, and $N_{HDM} = 512 \times 512$. For CHDM, there are 4 HDM particles assigned to each initial position, to represent random thermal velocities. The initial density fluctuations are represented as the superposition of plane wave perturbations with dimensionless wavenumber

$k = L_{box}/\lambda$ in the range $1 \leq k \leq 32$. Radiative cooling is neglected.

2. RESULTS

We define the gas density bias parameter $B \equiv (\rho_{IGM}/\rho_{tot})/\ (\langle\rho_{IGM}\rangle /\langle\rho_{tot}\rangle) = \Omega_b^{-1}\rho_{IGM}/\rho_{tot}$, where $\langle\ \rangle$ refers to the universal mean value, and $\rho_{tot} = \rho_{IGM} + \rho_{DM}$. The value of B averaged over all particles of the same total overdensity $\Delta_{tot} \equiv \rho_{tot}/\langle\rho_{tot}\rangle$ is plotted against overdensity in Figure 1a for each model. Our results indicate that the gas density in all three models is "anti-biased" (i.e. $B < 1$) at the highest total overdensities ($\Delta_{tot} > 10^2$). At moderate overdensities ($1 < \Delta_{tot} \leq 10$), there is positive bias (i.e. $B > 1$), on average, although at every value of Δ_{tot} there are mass points with a broad range of B, encompassing both bias and anti-bias. Our bias results for the CDM model agree qualitatively with those of recent simulations using the Eulerian, TVD method in 3D, of somewhat smaller spatial resolving power than ours in the densest regions (Kang et al. 1994). We are encouraged to believe that bias results for the same model in 2D versus 3D should be qualitatively similar, therefore.

X-ray and optical observations of rich clusters indicate that $\rho_{IGM}/\rho_{tot} \sim 10h^{-1.5}\%$ (Briel, Henry, & Boringer 1992). If $B = 1$, such empirical estimates imply that if $\Omega_b \leq \Omega_{b,max}^{BB} = 0.015h^{-2}$ as required by standard Big Bang nucleosynthesis (BBN) constraints, then the universe is open ($\Omega_{tot} < 1$), or else $\Omega_b > \Omega_{b,max}^{BB}$ – the so-called "baryon catastrophe" (White et al. 1993). If our antibias results for the highest overdensity regions suggest that $B < 1$ in rich clusters for all three models, HDM, CHDM, and CDM, this would strengthen the inequalities in the conclusion above, a result generic to *all* such models involving adiabatic, Gaussian-random primordial density fluctuations in a DM-dominated universe. In this case, either: (1) the models are incorrect, (2) the universe is open, (3) the baryonic content of the universe substantially exceeds the current BBN limit, or (4) the assumptions made in estimating the total mass and gas fraction in rich clusters from X-ray and optical data are flawed.

In order to check that clusters are, in fact, anti-biased and explore possibility (4), above, further we have focused on the CDM model (see Fig. 1b) and made a simulation cluster catalogue, using a friends-of-friends algorithm. Our 2D calculations can be used to calculate equivalent 3D cluster masses M_{2D} and radii R with the property that the Keplerian orbital velocity and acceleration of test particles around a uniform circular clump at radius R in 2D is identical to that for a test particle at distance R from a point-mass in 3D (Kates, Kotok, & Klypin 1991). The cluster shown in Fig. 1b has $M_{2D} = 3 \times 10^{14} M_{\odot}$ and $R = 2.6$ Mpc. The average bias B within each cluster is plotted against Δ_{tot} and M_{2D} in Figs. 1c and 1d. Apparently the anti-bias trend is strongest for small clusters, while the most massive clusters tend to be positively biased.

The cluster density biases described above suggest that estimates of cluster mass based upon X-ray surface brightness profiles and the assumption of isothermal gas spheres in hydrostatic equilibrium may yield erroneous results. To test this, we have calculated the X-ray mass profile estimate $[M(< R)]_{X-\mathrm{ray}}$ which would result if the X-ray surface brightness profile of the cluster shown in Fig. 1b were observed in the 2–10 keV band. The resulting thermal bremsstrahlung X-ray intensity profile $I_X(r)$ is plotted against position in Fig. 2, along with the emissivity-weighted gas temperature profile $T_X(r)$. The X-ray profile is

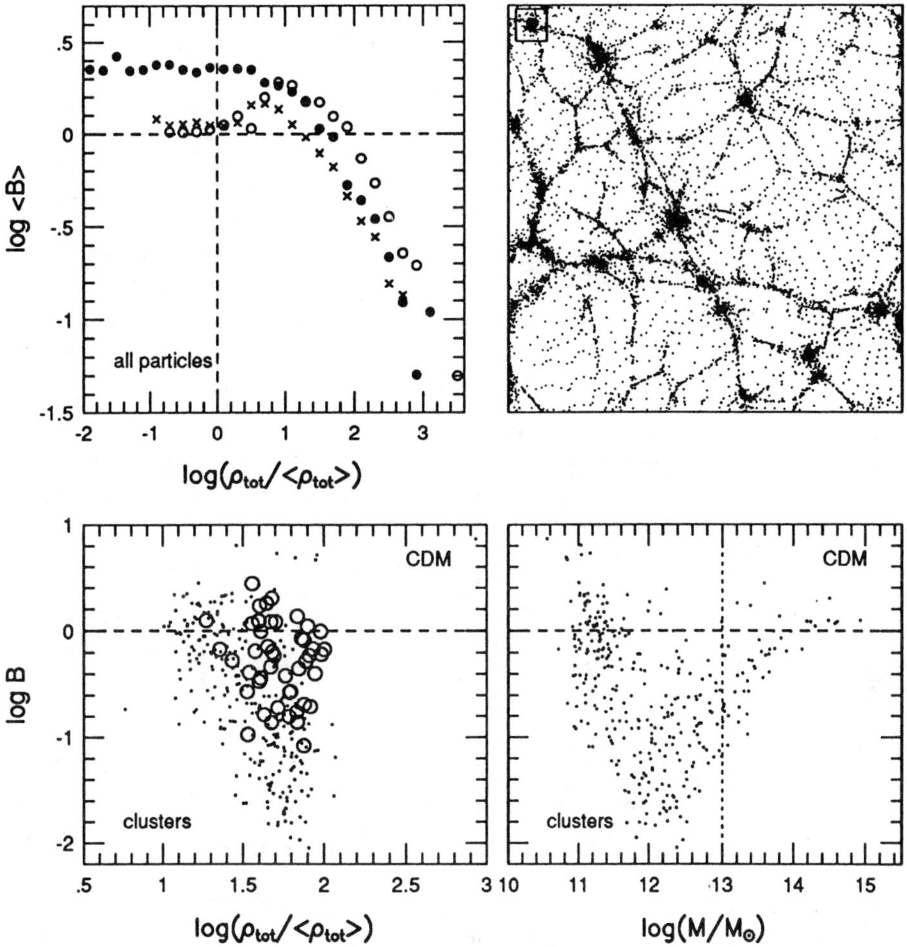

Fig. 1. (a) (top left) Average Density Bias vs. Total Overdensity for all particles in 200 Mpc × 200 Mpc simulation boxes at present for CDM (filled circles), CHDM (crosses), and HDM (open circles); (b) (top right) Gas Particle positions for CDM ($z = 0$) (small inset box indicates cluster featured in Fig. 2); (c) (bottom left) Average Bias vs. Average Total Overdensity within each cluster in CDM (open circles indicate $M_{2D} \geq 10^{13} M_{\odot}$); (d) (bottom right) same as (c), but vs. cluster mass M_{2D} (open circles in (c) are to the right of vertical dotted line).

smooth and symmetrical and more compact than either the density or temperature profiles. The inner profile is well fit by a form $I(r) \propto (1+r^2/r_c^2)^{1/2-3\beta}$, with $r_c \cong 1$ Mpc and $\beta \cong 1.1$. For an isothermal, cylindrically-symmetric cluster – the 2D analog of the isothermal sphere – the corresponding gas mass density

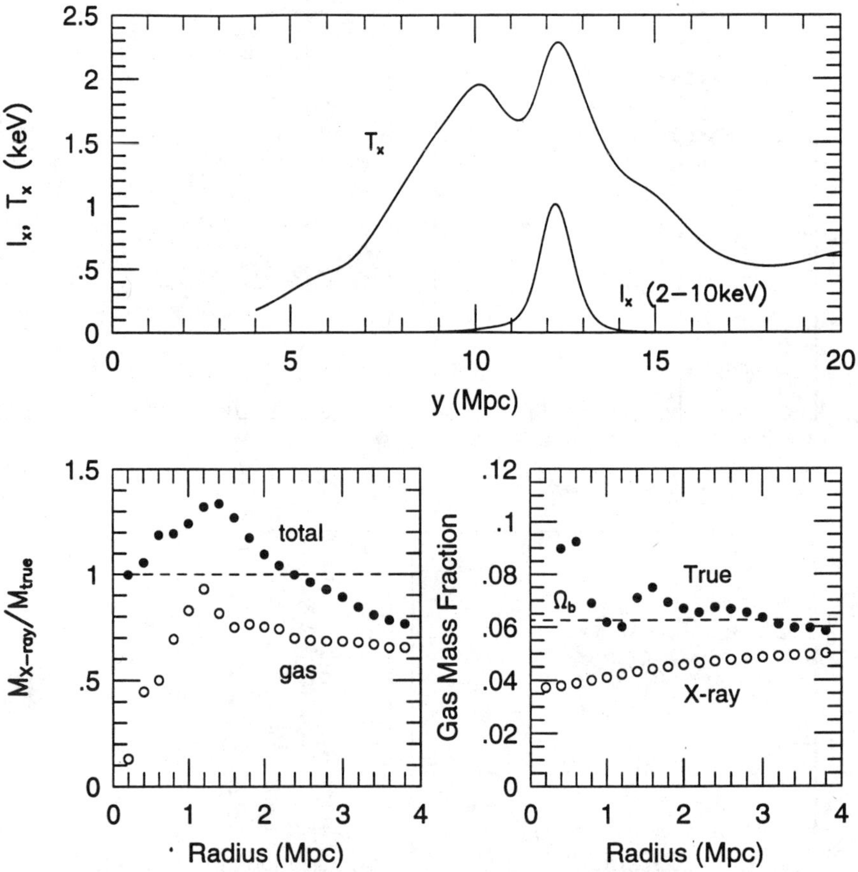

Fig. 2. Simulated X-ray Cluster in CDM: (top) 2 – 10 keV X-ray Brightness Profile I_X vs. y-position for cluster viewed along the -**x**–direction (as labelled) and emissivity-weighted line-of-sight gas temperature T_X (keV); (bottom left) Total and gas $M_{X-ray}(\leq R)/M_{true}(\leq R)$ vs. R; (bottom right) X-ray and true $M_{gas}(\leq R)/M_{tot}(\leq R)$.

profile is $\rho_g \propto (1 + r^2/r_c^2)^{-3\beta/2}$. The assumption of hydrostatic equilibrium then yields the X-ray-profile-fit estimate of the 2D total and gas masses interior to R. The results are compared with the true $[M_{tot}(< R)]$ and $[M_{gas}(< R)]$, calculated by summing the simulation particle masses within a circle of radius R centered on the X-ray brightness peak directly, in Fig. 2. Apparently, the traditional X-ray-profile-fitting model used to measure cluster total mass is weakly biased (overestimate) at small radii, $R <$ 2Mpc, and antibiased (underestimate) at larger radii. The X-ray gas mass is underestimated at all R. The X-ray-estimated gas mass fraction, thereby, underestimates the true gas fraction, which exacerbates the "baryon catastrophe."

ACKNOWLEDGEMENTS

We are pleased to acknowledge the support of NASA Grants NAGW-2399 and NAG5-2785, Robert A. Welch Grant F-1115, the University of Texas High Performance Computing Facility, the NSF NCSA, and Cray Research.

REFERENCES

Briel, U. G., Henry, J. P. & Boringer, H., 1992, A&A, 259, L31
Kang, H., Cen, R., Ostriker, J. P. & Ryu, D., 1994, Ap.J., 428, 1
Kates, R. E., Kotok, E. V., & Klypin, A. A. 1991, A&A, 243, 295
Shapiro, P. R., Martel, H., Villumsen, J. V., & Kang, H. 1994, in Numerical Simulations in Astrophysics, eds. J. Franco, S. Lizano, L. Aguilar, E. Daltabuit (Cambridge, UK: Cambridge Univ. Press), 45
Shapiro, P. R., Martel, H., Villumsen, J. V., & Owen, J. M. 1995, submitted to Ap.J.
White, S. D. M., Navarro, J. F., Evrard, A. E., and Frenk, C. S. 1993, Nature, 366, 429

THE EFFECT OF REIONIZATION OF THE INTERGALACTIC MEDIUM ON GALAXY FORMATION

Paul R. Shapiro and Hugo Martel
Department of Astronomy, University of Texas, Austin, TX 78712.

ABSTRACT

The reheating which accompanies reionization exerts a negative feedback effect on small-scale structure formation. To illustrate this, we simulate the formation of structure on galactic and sub-galactic scales in the mixed Cold + Hot Dark Matter (CHDM) model, utilizing our new Adaptive SPH method. We compare the rates of structure formation in a universe which is reionized and reheated at some epoch $z > 5$ (as required to satisfy the H I Gunn-Peterson (GP) constraint on the IGM at $z < 5$) to the rate of structure formation in a universe without such reheating. These results have strong implications for the comparison of the CHDM model with the observed abundance of damped Ly alpha clouds and the total H I content of the universe at $z \gtrsim 3$.

1. INTRODUCTION

Recently, it has been shown that the existence of the GP limit and the implied reheating of the IGM by $z \sim 5$ required to satisfy it impose a strong boundary condition on models like the CDM model, resulting in feedback on the rate of collapse of gas out of the background IGM (Shapiro, Giroux, & Babul 1994). The reheating of the IGM, which can be either to $T \gtrsim 10^4$K by photoionization or $T \gtrsim 10^6$K by bulk hydrodynamical processes (e.g. shocks from galactic superwinds and supernovae), raises the Jeans mass in the background, uncollapsed IGM and causes subgalactic and even galactic scale density fluctuations to be suppressed. The result is that the growth of the collapsed fraction of baryons with time is slowed. These calculations utilized the linear growth equations for density fluctuations and calculated the thermal and ionization state of the diffuse IGM in detail. A similar trend was found recently in gas dynamical simulations of galaxy formation in the CDM model (Cen & Ostriker 1993). This effect should be even more dramatic for a model like CHDM, in which fluctuation growth on galactic and subgalactic scales is delayed compared to CDM.

2. THE CALCULATIONS

We have performed high resolution, 2D, gas dynamical simulations of the evolution of the IGM and structure formation in the COBE-normalized CHDM model. We consider a 2D, periodic universe of comoving size $L_{box} = 1$ or 5 Mpc (present units). The 2D power spectrum is the same as in 3D, but multiplied by $2k$, where k is the wavenumber — i.e. equal rms $\delta\rho/\rho$ in 2D and 3D for equal fluctuation scale length. We couple our ASPH method (Shapiro et al. 1994; 1995) to a standard Particle Mesh (PM) gravity solver in 2D. We take $h = 0.5$, $\Omega_{tot} = 1$, $\Omega_{CDM} + \Omega_B = 0.7$, $\Omega_{HDM} = 0.3$, and $\Omega_B = 0.058$. We take $N_{gas} = 128 \times 128$ gas particles, $N_{CDM}/N_{gas} = 11$, and $N_{HDM} = 512 \times 512$, on a uniform square lattice with 1024×1024 PM cells, at initial redshift $z_i = 24$. There are 4 HDM particles assigned to each initial position to represent the random thermal velocity of the HDM. The initial density fluctuations are represented as the superposition of plane wave perturbations with dimensionless wavenumber $k = L_{box}/\lambda$ in the range $1 \leq k \leq k_{\max}$, $k_{\max} = 64$. Radiative cooling is neglected. The simulations are terminated at $z = 3$ before the rms density excess on the scale of L_{box} has reached nonlinear amplitude.

We compare 2 cases for each box size and reheating model, one with and one without reheating. We represent the effect of photoionization by setting $T = T_h = 10^4 K$ everywhere at $z \leq z_h$, the reheat epoch. For the other case, we idealize the complicated, inhomogeneous effects of reheating by widespread shocks associated with energy-release by the collapsed fraction as follows. For all $z < z_h$, we reset the gas temperature everywhere to

$$T_{new} = \begin{cases} \max[T_h, T_{old}]\,, & \rho_g/\bar{\rho}_g \leq 1\,; \\ \max\left[T_h(\rho_g/\bar{\rho}_g)^{-1}, T_{old}\right]\,, & \rho_g/\bar{\rho}_g > 1\,. \end{cases} \tag{1}$$

For $\rho_g/\bar{\rho}_g > 1$, this takes approximate account of the effect of strong shocks which travel through the diffuse, uncollapsed (i.e. mean density) IGM at some speed v_s, sometimes encountering overdensities $\rho_g/\bar{\rho}_g$. Such shocks would travel into the overdense region at a reduced speed such that $\rho_g v_s^2 \sim$ constant, thus heating the denser gas to $T_h \sim v_s^2 \sim T_h(\rho_g = \bar{\rho}_g) \times (\rho_g/\bar{\rho}_g)^{-1}$. Denser gas which is already too hot to be shock-heated by such shocks is assumed not to be reheated. For this paper, we take $T_h = 10^6$K in equation (1).

3. THE RESULTS

The importance of reheating in suppression of density fluctuation growth is anticipated by Fig. 1a, based upon the linear growth equations for density fluctuations. In Fig. 1a, we plot the comoving Jeans wavelength λ_J for the IGM (present units) versus redshift such that fluctuations at wavelength $\lambda < \lambda_J(T_h)$ are suppressed, unless the rms overdensity in spheres of this radius λ has already reached nonlinear amplitude ($\sigma > 1$) as indicated by the curve for λ_{NL}, the wavelength for which density fluctuations are just going nonlinear at $z = z_{NL}$. In addition, we have plotted λ_v, the wavelength such that the rms peculiar velocity dispersion averaged over spheres of radius λ_v is equal to the sound speed of the gas at $T = T_h$. We expect that linear fluctuations with $\lambda < \lambda_v$ will be suppressed. The results in Fig. 1a indicate $T_h = 10^4$K or 10^6K at $z_h = 10$ will strongly affect the rate of baryon collapse in the CHDM model.

Our simulation results for gas particle positions in Fig. 1b allow us to compare the structure formation in the photoionized CHDM model at $z = 3$ with the case without reheating. The effect of reheating on the distribution of gas is dramatic.

We calculate the collapsed fraction f_c of the baryons versus z for the reheated and un-reheated CHDM models in Fig. 2. The collapsed fraction f_c is taken to be all those gas particles for which the overdensity $\rho_g/\bar{\rho}_g > (1 + \delta_{th})$, for two thresholds, $(1 + \delta_{th}) = 9\pi^2/16 = 5.55$, the value at turn-around for the top-hat model of nonlinear density fluctuation growth, or 50. As Fig. 2 shows, reheating of the IGM either to 10^4K or 10^6K as early as $z = 10$ can dramatically reduce the collapsed baryon fraction at $z < 10$.

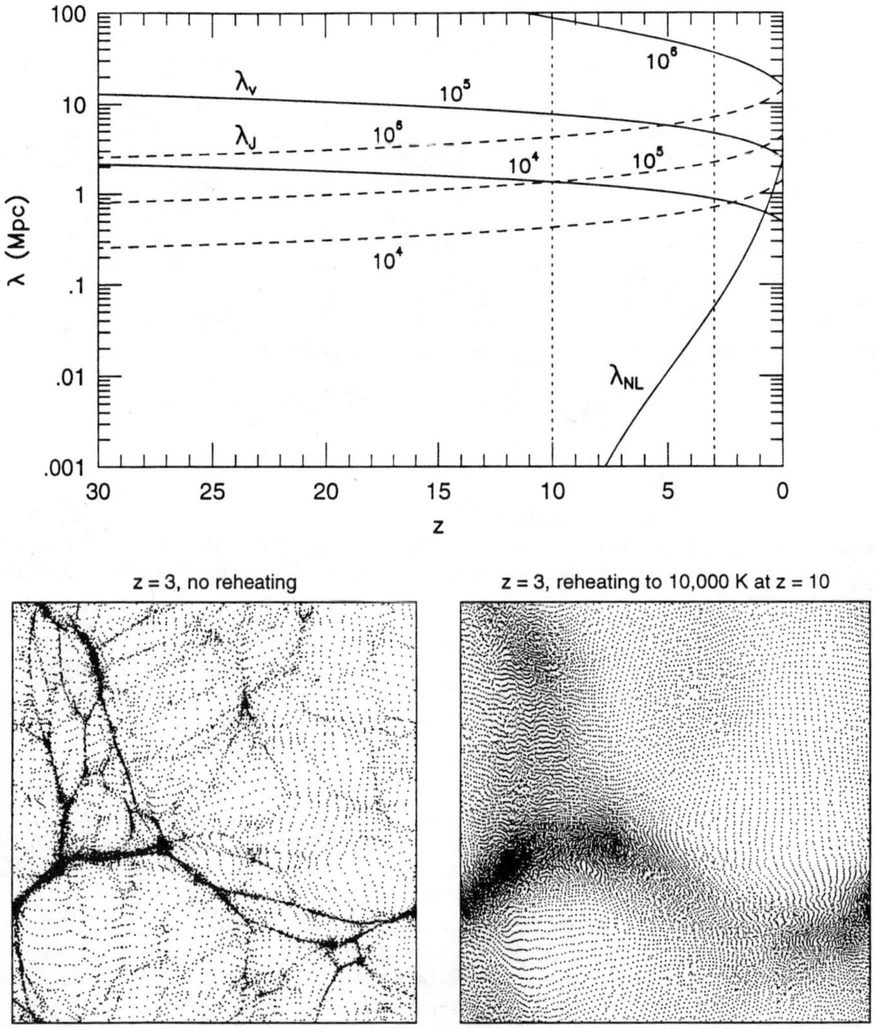

Fig. 1. (a) (top) Linear Estimate of Density Fluctuation Growth Vs. Suppression. Curves labelled λ_{NL}, λ_J, and λ_v are the wavelengths (present comoving units) of density fluctuations for which rms $\delta\rho/\bar{\rho} = 1$ at that redshift z, the Jeans length of IGM at the labelled value of T_h(K), and the radius of the sphere within which the velocity dispersion equals the sound speed at the labelled value of T_h(K). Vertical dashed lines indicate reheat epoch $z_h = 10$ and final time-slice at $z = 3$. (b) (bottom) Gas particle positions at $z = 3$ in CHDM simulation ($L_{box} = 1$ Mpc) without reheating (left) and with $T_h = 10^4$K for $z \leq z_h = 10$ (right).

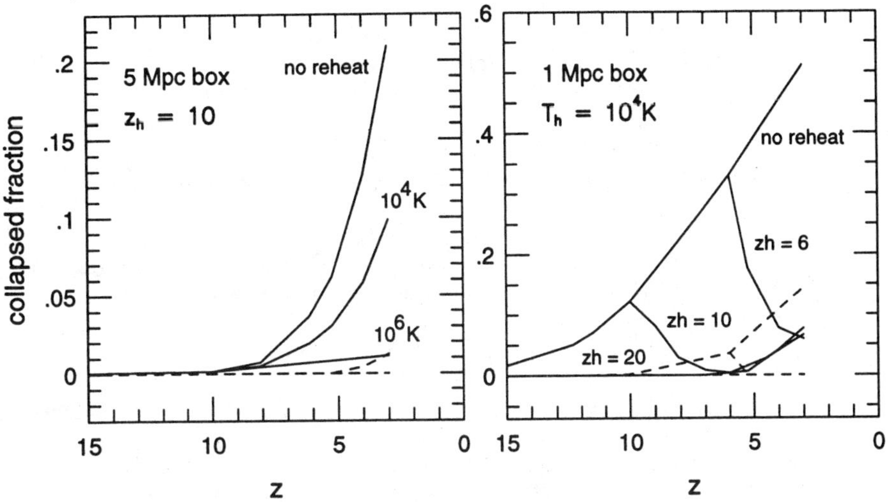

Fig 2. Fraction f_c of gas mass at gas overdensity $\rho_g/\bar{\rho}_g \geq (1+\delta_{th})$, for $(1+\delta_{th}) =$ 5.55 (solid) and 50 (dashed), without reheating or reheated at all $z \leq z_h$ to $T = T_h$(K), as labelled. For $1 + \delta_{th} = 50$, $f_c \cong 0$ for $z_h = 10$, 20 (1 Mpc, 10^4K), and $z_h = 10$ (5 Mpc, 10^4K, 10^6K).

These results reinforce the conclusions reached by Shapiro et al (1994) and Cen & Ostriker (1993) for the CDM model. The feedback effect of reionization on structure formation is even more pronounced, however, in the CHDM model. Our results imply that the difficulty which has been noted for CHDM, in supplying enough collapsed baryon fraction to explain the number of damped Lyman alpha quasar absorption line clouds at $z > 3$ and the total mean H I density of the universe at $z > 3$ (e.g. Mo & Miralda-Escudé 1994) will be significantly exacerbated by reheating. If, as seems likely, such reheating is inevitable if such models are to satisfy the GP constraint, then models like CHDM will fail to explain the observed H I content of the universe at $z > 3$.

ACKNOWLEDGEMENTS

We are pleased to acknowledge the support of NASA Grants NAGW-2399 and NAG5-2785, Robert A. Welch Grant F-1115, the University of Texas High Performance Computing Facility, the NSF NCSA, and Cray Research.

REFERENCES

Cen, R., & Ostriker, J. P. 1993, ApJ, 417, 404

Mo, H. J., & Miralda-Escudé, J. 1994, preprint (IASSNS-AST94/7)

Shapiro, P. R., Giroux, M. L., & Babul, A. 1994, ApJ, 427, 25

Shapiro, P. R., Martel, H., & Villumsen 1994, Numerical Simulations in Astrophysics, eds. J. J. Franco (Cambridge, UK: Cambridge Univ. Press) 45.

Shapiro, P. R., Martel, H., Villumsen. J. V., & Owen, J. M. 1995, submitted.

THE LITTLE BANG: PECULIARITIES IN THE LOCAL GROUP AND SOME DYNAMICAL EXPLANATIONS

M.J. Valtonen, M.L. McCall, K.A. Innanen
York University, Department of Physics and Astronomy, 4700 Keele St.
North York, Ontario M3J 1P3, Canada

J.Q. Zheng
University of Turku, Tuorla Observatory, 21500 Piikkiö, Finland

G.G. Byrd
University of Alabama, Department of Physics and Astronomy
P.O.Box 1921, Tuscaloosa, Alabama 35487, USA

ABSTRACT

There has been increasing evidence that Maffei 1/2 and IC 342 are close to the Local Group of galaxies, so close that they have to be included in the discussion of the history of group. We find that the dynamical evolution in the local area may have been quite complicated in comparison with the traditional Kahn-Woltjer model. For example, the Andromeda galaxy may be a product of a merger of two major galaxies, and Maffei 1/2 and IC 342 may have been thrown out of the Local Group about 8 billion years ago.

1. INTRODUCTION

Recently the distances from our Galaxy of Maffei 1, IC 342 and associated galaxies have been revised downwards to the weighted mean distance of about 3 Mpc (Buta and McCall 1983, McCall 1989, Luppino and Tonry 1993, Lee and Madore 1993). The first two galaxies are the dominant members of group 14–11 of Tully (1988), which includes the newly discovered Dw 1 (Kraan-Korteweg et al. 1994) and the small spiral Maffei 2 (Bottinelli et al. 1971; henceforth we will refer to this as the Maffei 1 group). Assuming constant radial motion at the observed radial velocities we find that the Maffei 1 group was in the middle of the Local Group only about 8 billion yrs ago (McCall 1987, Zheng et al. 1991). This implies an encounter event less than eight billion years ago involving Maffei 1 and M31 and their respective companions. Figure 1 illustrates the orbits of some of the galaxies involved. The dynamical event leading to the recession of the Maffei 1 group from the Local Group will be referred to as the Little Bang.

2. THE LITTLE BANG

The interaction eight billion years ago would cause a redshift asymmetry in the radial motions of the distant companions of M31, much like the observed asymmetry. As a recoil effect of the encounter with the massive Maffei 1 galaxy, the Andromeda galaxy obtains a peculiar velocity towards us which is not fully shared by its distant companions.

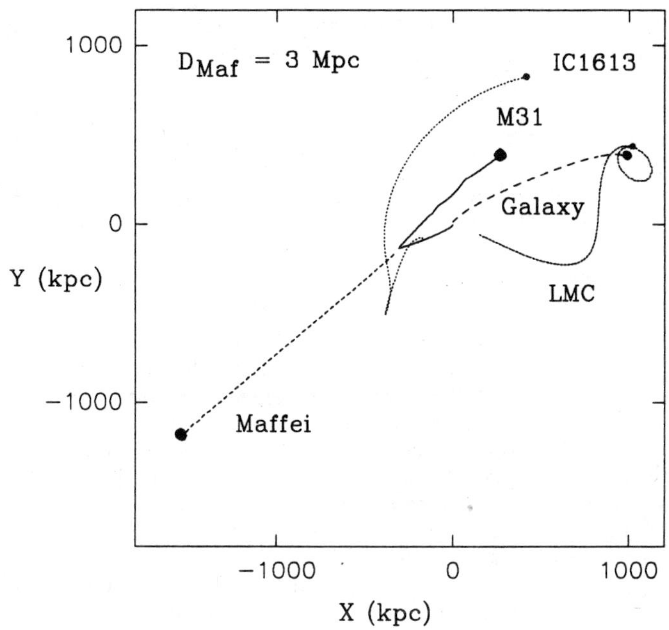

Fig. 1. An illustration of orbits of some galaxies and the Maffei 1 group since the Little Bang 7.4 billion years ago and since the Big Bang 16 billion years ago (in case of the Galaxy, LMC and the original M31 group of galaxies).

The simulations of Zheng et al. (1991) gave a 7:2 asymmetry in Δz in favour of redshifts. It is not significantly different from the 5:1 observed asymmetry (Arp 1985). Typically, some companions may be thrown toward the Milky Way galaxy. Astronomers would today not consider these to be companions of M31 but instead to belong neither to M31 nor to our Galaxy.

Shuter (1992) has suggested that the Magellanic stream can be best modeled dynamically if the LMC arrived in the neighborhood of our galaxy for the first time from the direction of the Andromeda galaxy about six billion years ago. The LMC's arrival at that time would be consistent with the ejection of small satellite galaxies at the relative parabolic encounter speed or greater of M31 in the Little Bang. It is possible that the LMC would have been captured into its present orbit around the Milky Way galaxy by gravitational drag. In fact both the Milky Way and the LMC appear to have undergone bursts of star formation six billion years ago (Barry 1988, Noh and Scalo 1990, van den Bergh 1992), possibly triggered by the close approach of the LMC (Byrd et al. 1988).

Leo I may be another intergalactic traveller (Byrd et al. 1994). It shows a younger mean age than most dwarf spheroidals (Lee et al. 1993, van den Bergh 1994). Its radial motion is also peculiarly high for an ordinary satellite of the Milky Way (Zaritsky et al. 1989). The mass of the Milky Way derived from the orbit of Leo I would be consistent with the value derived from other satellites if Leo I came from the Little Bang.

In addition to the Virgocentric streaming, Richter et al. (1987) find that the Local Group has a peculiar motion of 80 $\pm 17 km\ s^{-1}$ approximately along the supergalactic plane towards the longitude SGL $\cong$ 240°. In the model of Zheng et al. (1991) the Local group (dominated by the Andromeda galaxy) obtains a recoil velocity of 66 km s^{-1} towards SGL = 190°. The fact that the observed peculiar motion is approximately opposite to the direction of the Maffei 1 group in the sky is by itself a hint that the two peculiarities (peculiar radial velocity of the Maffei 1 group and the motion of the Local group) may have a common explanation.

5. M31 AS A MERGER

Valtonen et al. (1993) discuss various ways in which the Maffei group and the Local Group could have obtained their high relative velocity about 8 billion years ago, long after the Big Bang. They conclude that the most plausible way is a dynamical breakup of a system which contained several tight binaries. One of the binaries would have collapsed to make the present-day M31. Alternatively, several successive mergers may have taken place.

Evidence for a past merger event comes from the observation of a bump in the metallicity distribution of the bulge of M31, at $[z/z_{\odot}] = -0.5$, which could be interpreted as coming from a blend of metal-poor and metal-rich populations (Bica et al. 1990). Also, recently M31 was discovered to have a double nucleus (Lauer et al. 1993).

In M31, the surface number density of globular clusters has a peculiar drop near the center with respect to the DeVaucouleurs' Law (Harris and Racine 1979). The Milky Way does not have such a drop. Nor is it seen in other old populations of M31, such as planetary nebulae (Ciardullo et al. 1989). McCall et al. (1995) demonstrate that the deficiency can arise in a merger of two galaxies. The influence of the two merging centers cause a tidal break up of many of the inner clusters

Another interesting feature about the globular cluster system in M31 is the dependence of metallicity on radial distance. In the Milky Way, clusters with high metallicity tend to be near the center, but in M31 they are distributed over a wide range of radius (Harris and Racine 1979, Huchra et al. 1991). Again, the difference can be explained if M31 is the outcome of a merger, as dynamical perturbations would have altered cluster orbits to such an extent that the original distribution, as displayed by the Milky Way, would have been destroyed.

7. DISCUSSION

Many peculiarities of the Local Group could have a common origin in a "Little Bang" about 8 billion years ago. The Little Bang was initially inferred from the fact that the major galaxies Maffei 1 and IC 342 must have left the Local Group around this time, if their distances from us are only about 3 Mpc. Now, as described above, there is additional observational evidence to support the case for an interaction in the relatively recent past.

Since the Local Group can no longer be regarded as an isolated system, determinations of the mass of the Local Group and the age of the Universe derived from conventional timing (Kahn and Woltjer 1959) are invalidated. A revised method of timing has been proposed by Valtonen et al. (1993). They estimated the age of the Universe to be around 19 billion years, but the inclusion of the LMC and several other smaller galaxies in the simulation has subsequently brought down the age estimate to 16 ± 2 billion years. The uncertainty is largely due to the uncertain distance to the Maffei 1 group of galaxies. The mass of the traditional Local Group (M31 and the Milky Way with their companions) in this model is 2/3 of the Kahn-Woltjer value for the same age of the Universe. The mass-to-light ratio of the Local Group is about 50 solar units.

We find that the dynamical history of Local Group is probably quite a bit more complicated than the simplest Big Bang scenarios (Peebles 1989, 1994). Even though our scenario of a tight group of galaxies in the Andromeda region and its dissolution in the Little Bang may seem complicated, it provides explanations for many seemingly unrelated peculiarities in the Local Group of galaxies.

REFERENCES

Arp, H. 1985, A & A, 156, 207

Barry, D.C. 1988, ApJ, 334, 436

Bica, E., Alloin, D. & Schmidt, A.A. 1990, A & A 228, 23

Bottinelli, L., Chamaraux, P., Gérard, E., Gouguenheim, L., Heidmann, J., Kaz 264

Buta, R.J. & McCall, M.L. 1983, MNRAS, 205, 131

Byrd, G., Valtonen, M., McCall, M.K., Innanen, K. 1994, AJ, 107, 2055

Byrd, G.G., Sundelius, B., Thomasson, M. & Valtonen, M. 1988, in Active Galactic Nuclei, eds. H.R. Miller & P.J. Wiita, Springer: Berlin, p. 143

Ciardullo, R., Jacoby, G.H., Ford, H.C. & Neill, J.D. 1989, ApJ, 339, 53

Harris, W.E. & Racine, R. 1979, Ann.Rev. A & A, 17, 241

Huchra, J.P., Brodie, J.P. & Kent, S.M. 1991, ApJ, 370, 495

Kahn, F.D. & Woltjer, L. 1959, ApJ, 130, 705

Kraan-Korteweg, R.C., Loan, A.J., Burton, W.B., Lahav, O., Ferguson, H.C., Henning, P.A. & Lynden-Bell, D. 1994, Nature (in press)

Lauer, T.R., et al. 1993, AJ, 106, 1436

Lee, M.G., Freedman, W., Mateo, M. & Thompson, I. 1993, AJ, 106, 1420

Lee, M.G. & Madore, B.F. 1993, AJ, 106, 66

Luppino, J.A. & Tonry, J.L. 1993, ApJ, 410, 81

McCall, M.L. 1987, BAAS, 19, 689

McCall, M.L. 1989, AJ, 97, 1341

McCall, M.L., Valtonen, M.J. & Innanen, K.A. 1995, in preparation

Noh, H. & Scalo, J. 1990, ApJ, 352, 605

Peebles, P.J.E. 1989, ApJ, 344, L53

Peebles, P.J.E. 1994, ApJ, 429, 43

Richter, O.-G., Tammann, G.A. & Huchtmeier, W.K. 1987, A & A, 171, 33

Shuter, W.L.H. 1992, ApJ, 386, 101

Tully, R.B. 1988, Nearby Galaxies Catalog, Cambridge, Cambridge Univ. Press

Valtonen, M.J., Byrd, G.G., McCall, M.L. & Innanen, K.A. 1993, AJ 105, 886

van den Bergh, S. 1992, Sky & Telescope, 83, 508

van den Bergh, S. 1994, ApJ, 428, 617

Zaritsky, D., Olszewski, E.W., Schommer, R.A., Peterson, R.C. and Aaronson, M. 1989, ApJ, 345, 759

Zheng, J.Q., Valtonen, M.J. & Byrd, G.G. 1991, A & A, 247, 20

CMB ANISOTROPY AND DARK MATTER

WHAT CAN WE LEARN FROM FUTURE CMBR EXPERIMENTS?

David N. Spergel
Dept. of Astrophysics, Princeton University NJ 08544
dns@kocab.princeton.edu

ABSTRACT

The detection of CMBR fluctuations by the COBE experiment on large angular scales marks the beginning of a new epoch in cosmology. Future experiments should be able to determine the origin of the primordial fluctuations and constrain basic cosmological parameters such as Ω_0 and h as well as probe the ionization history of the intergalactic medium. I review the physical processes that produce CMBR fluctuations in various scenarios for the origin of large scale structure. I also describe numerical simulations that estimate the potential power of future CMBR experiments.

1. RECIPES FOR STRUCTURE FORMATION

The big bang theory is incomplete. While its simplest form, a theory in which matter and radiation is uniformly distributed through the universe and general relativity is assumed to be universally valid, can account for the isotropy of the cosmic background radiation (CBR) and the origin of the light elements, it does not explain the origin of galaxies and large-scale structure.

The big bang theory must be supplemented by a theory for the formation of structure. Almost all models under active consideration assume that initially low amplitude fluctuations in density grew gravitationally to form the galaxies, clusters and super-clusters. While hydrodynamics and electromagnetic forces are believed to play an important role in star formation and on the scale of individual galaxies, gravity is thought to be the dominant force that assembles the large-scale distribution of galaxies.

Models of structure formation under active study are composed of several ingredients: (a) a mechanism for generating structure (e.g., inflation, strings, textures); (b) assumptions about the nature of the dark matter (e.g., hot dark matter, cold dark matter, baryonic dark matter); and (c) assumptions about the background cosmology (the value of Ω_0, the ratio of the mean density of the universe to the closure density, H_0, the Hubble constant, and Λ_0, the cosmology constant). By choosing one from each column, cosmologists can design a menu of recipes for structure formation: a model with $\Omega = 1$, $H_0 = 50$ km/s/Mpc, cold dark matter and inflation (the "standard" model of the 1980's) or a string-seeded model with hot dark matter and $\Omega = 1$ and $H_0 = 100$ km/s/Mpc. Future CMBR observations can potentially place significant constraints on (a) and (c). As these observations will be complemented by optical, X-ray and radio observations of the evolution of the large-scale structure of galaxies, cosmologists will have less and less freedom in determining these parameters.

Mechanisms for generating the initial low amplitude fluctuations can be divided into two classes of models. In inflationary models, quantum fluctuations in the scalar field that drive inflation are amplified during the inflationary phase that preceded the radiation and matter dominated phase to produce superhorizon *curvature* fluctuations. Curvature fluctuations are large-scale variations in the density of the universe, hence in the local space curvature. In alternative models, the universe began the radiation dominated phase without any variations in spatial curvature. In these *isocurvature* models, spatial variations in either the equation of state generates density fluctuations in an initially smooth universe. The equation of state fluctuations can be driven by either spatial variations in the photon-to-baryon ratio or by topological defects (e.g., gauged or global strings, global monopoles, or global textures).

While dark matter searches may potentially directly determine the nature of dark matter, observations of large-scale structure can provide insight into the nature of dark matter. Cold dark matter (CDM) clusters much more easily on small scales than hot dark matter (HDM). Thus, a model with cold dark matter predicts the existence of more small scale structure in the galaxy distribution than a model that assumes that the universe is dominated by hot dark matter. In inflationary models, structure formation precedes differently in the two models: in a CDM model, dwarf galaxies merge to form larger galaxies that are then assembled by gravity into superclusters; while in an HDM model, the supercluster "pancake" fragments to form clusters, groups and galaxies.

2. MICROWAVE BACKGROUND FLUCTUATIONS

Observations of the microwave background contain a tremendous amount of information about both the process that produced the gravitational fluctuations and the fundamental properties of the universe: Ω_0, Λ_0, and the baryon-to-photon ratio.

On the large angular scales probed by COBE, the dominant source of microwave background fluctuations are Sachs-Wolfe fluctuations due to variations in the gravitational potential:

$$\frac{\delta T}{T} = \left(\frac{\Phi}{3} + \frac{V}{c} + \frac{\Delta_\gamma}{4}\right)_{t_i}^{t_0} + 2\int_{t_i}^{t_0} dt \left(\frac{\partial \Phi}{\partial t}\right)_{x(t)}$$

where the first term describes potential fluctuations at surface of last scatter, the second term describes variation in the velocity at the surface of last scatter, the third term describes fluctuations in the photon density and the final term describes fluctuations produced when the photon moves through time-varying potentials as it propagates from the surface of last-scatter to the observer. There are other additional possible sources of fluctuations: the motion of a photon through the time-varying metric due to a gravitational wave will also alter its energy.

In inflationary models, large-scale fluctuations in the gravitational potential are produced in the very early universe (Bardeen et al. 1984). These fluctuations are Gaussian, random phase variations in the energy density of the universe and are adiabatic (there are no variations in the photon to baryon ratio). If the universe is matter-dominated, then the gravitational potential is constant with time. Thus, in these models, the last term in equation (3) does not contribute CMBR fluctuations. On the large angular scales probed by COBE, the first term dominates and these models predict a scale-invariant spectrum of CMBR fluctuations.

On smaller angular scales, variations in the photon density and the photon velocity term become more important. Inflationary models predict that CMBR fluctuations will peak on the angular scale subtended by the horizon at the surface of last-scatter, $\theta \sim 1^o$. This "Doppler peak" or more, properly, "Acoustic peak" is an important prediction of these models (Bond & Efstathiou 1984, Hu & Sugiyama 1994).

Changing parameters in the inflationary scenario will alter the details of the shape of the predicted CMBR spectrum (Bond et al. 1994). If the Higgs potential causes the inflaton field to accelerate as the scales that we probe with CMBR experiments crossed the horizon, then spectrum of fluctuations will be "tilted". As this implies a higher level of fluctuations on larger physical scales, this alteration predicts an increased quadrupole and lowered acoustic peak. In most inflationary models, "tilting" the spectrum also generates gravity waves. Gravity waves produce fluctuations on large angular scales without contributing to the acoustic peak. Thus, for a model normalized to COBE's detection, they also have the effect of increasing the amplitude of the quadrupole and lower the height of the acoustic peak. Both of these alterations predict a lower level of fluctuations on galaxy scales.

The CMBR spectrum is also sensitive to assumption about the background universe today. If the universe today is dominated by a cosmological constant, then potential fluctuations on large angular scales can decay. This has the effect of increasing the quadrupole. In a flat universe, the presence of a cosmological constant term implies that matter-radiation equality occured at a later epoch. This shifts the peak in the galaxy fluctuation spectrum to larger physical scales. If the universe is open, then the angular scale subtended by the same fixed physical scale is smaller. This geometric factor implies that the acoustic peak in these models is shifted towards larger l.

Several groups have detected fluctuations near the one degree scale. These are difficult experiments and their detected fluctuation levels are not consistent. Further work is needed to clarify the experimental situation on these angular scales.

3. CBR OBSERVATIONS AFTER COBE

While these balloon observations and ground-based observations of fluctuations on small angular scales are useful, I believe that we will need a follow-up satellite to COBE to measure the CBR fluctuations accurately on the 1° scale. Such an experiment should be able to determine the nature of the primordial fluctuations.

If the fluctuations are indeed adiabatic curvature fluctuations, as predicted in inflationary models, then this should confirmed by observations of the shape of the acoustic peak. If the primordial fluctuations are entropy fluctuations or are produced by defects, then both growing and decaying modes should exist at recombination and the structure of the acoustic peak will be altered. In isocurvature models, the peak is much broader than in models with curvature fluctuations. Thus, such an experiment will allow us to directly probe fundamental physics.

If the acoustic peak is carefully mapped, then its shape will also allow the determination of fundamental cosmological parameters. The location of the Doppler peak provides a measure of the geometry of the universe (Kamionkowski et al.1994). Observations of its width will determine the baryon/photon ratio at recombination. Measurements of the height of the peak and its shape below the peak will probe $\Omega_0 h^2$. Observations of the secondary peaks will further constrain these parameters.

I have run a series of Monte-Carlo simulations to estimate how effectively future CMBR experiments could constrain cosmological parameters. These simulations are complementary to the recent work of Kogut et al. (1994) that more accurately modelled the determination of cosmological parameters from maps, but used fewer parameters in their cosmological model.

Using the Hu and Sugiyama (1994a,b) approximations to compute the expected multipole moments, I generated synthetic measurements of the CMBR spectrum. These synthetic measurements included the effects of noise and cosmic variance. The noise level was set to 15 μK per pixel for most of the realizations For each realization of the c_l's, I found the set of model parameters that best fit the multipole spectrum. For the first set of runs, the model parameters were the density of the universe in non-relativistic particles, Ω_{nr}, the Hubble Constant, h, the slope of the primordial power spectrum, n, the tensor mode contribution, r, the epoch of reionization, z_{reion}, the baryon density as a fraction of the critical density, Ω_b and the overall normalization of the spectrum. For the second set of runs, I added a cosmological constant term, Ω_Λ. I used the simulated annealing routine AMEBSA (Press et al. 1992) to probe the 8 dimensional parameter space and find the best fit parameters for each Monte-Carlo realization. To assure that a large region of parameter space was searched, the starting values for the "ameoba" spanned a factor of 2 around the "true" values. Different starting points were used for different realizations to avoid any tendency for the "amoeba" to collapse to special best fit values.

The simulations were run for four potential experiments: an all sky map with half-degree resolution, a map that cover 900 square degrees with half-degree resolution, an all sky map with quarter-degree resolution and an all sky map with half-degree resolution and a noise level of only 1.5 μK per pixel. As the uncertainties in the c_l's at large l is dominated by noise rather than cosmic variance, it is possible to achieve the constraining power of the quarter-degree resolution map by combining a half-degree resolution all-sky map with a higher resolution map of a small fraction of the sky.

Figures 1 - 3 show the results of the first set of runs. In the upper panels, the figures show the spread in Ω_0, Ω_b and h. These parameters can be accurately constrained by the all-sky maps. The partial sky maps should be able to measure Ω_0 by determining the location of the Doppler peak. The slope of the spectrum is also well constrained and its overall normalization is definitively fixed. However, the partial sky maps do not appear able to constrain the other parameters accurately as cosmic variance swamps the differences between models. The lower panels of the figures show the constraints on z_{reion} and r. These parameters are not as well constrainted by the CMBR observations.

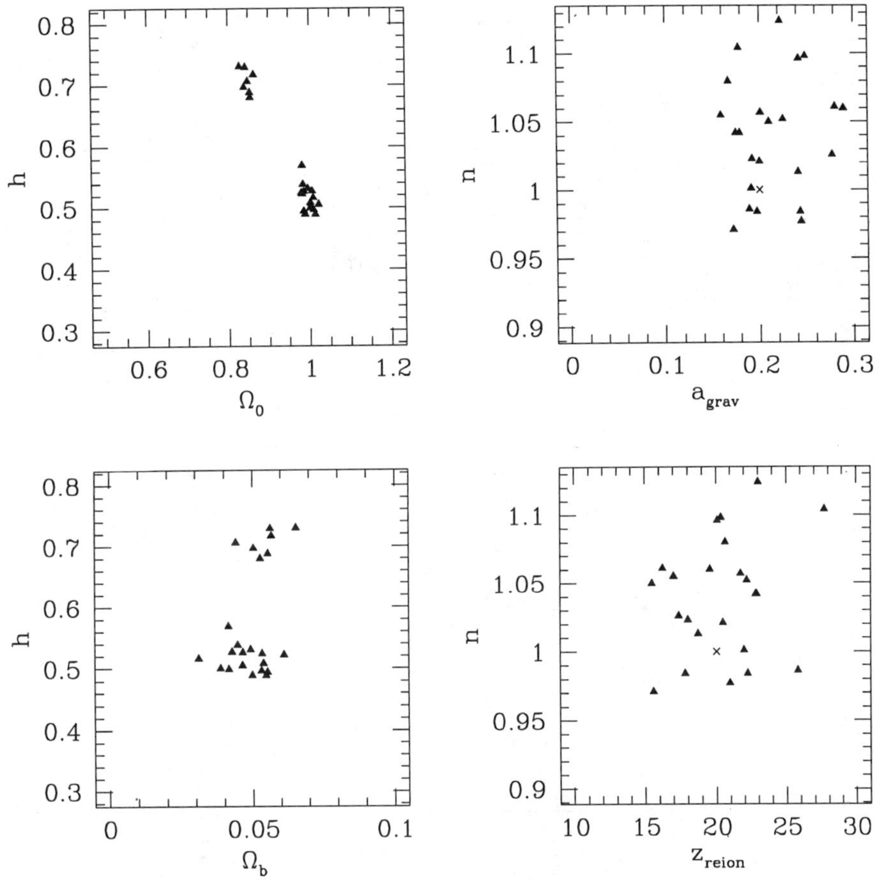

Figure 1. This figure shows the results of the Monte Carlo simulations of an all sky experiment with 0.5 degree resolution. The simulations include cosmic variance and pixel noise at the 15 μ K per pixel level on a 400,000 pixel map. The simulations were generated for a sky with "true values" of $\Omega_0 = 1, \Omega_b = 0.05, h = 0.5$ and $n = 1$ (denoted by x's in the figures). Complete reionization was assumed to occur at $z = 20$ and tensor modes were assumed to account for 20% of the quadrupole. There contribution to higher l modes were scaled accordingly. The figures show the best fit values in each realization for $\Omega_0, \Omega_b, h, n, z_{reion}$, and r, the ratio of tensor to scalar modes.

If Λ is added as an additional parameter to the theory, there is a significant increase in "cosmic confusion" (Bond et al. 1994). This can be seen in the next set of figures. While $\Omega_0 + \Omega_\Lambda, \Omega_b$ and n are still pretty accurately determined by a half-degree experiment, the Hubble constant and the value of Ω_Λ are less well constrained. This degeneracy can be significantly reduced by observations on smaller angular scales that probe the secondary acoustic peaks. These high

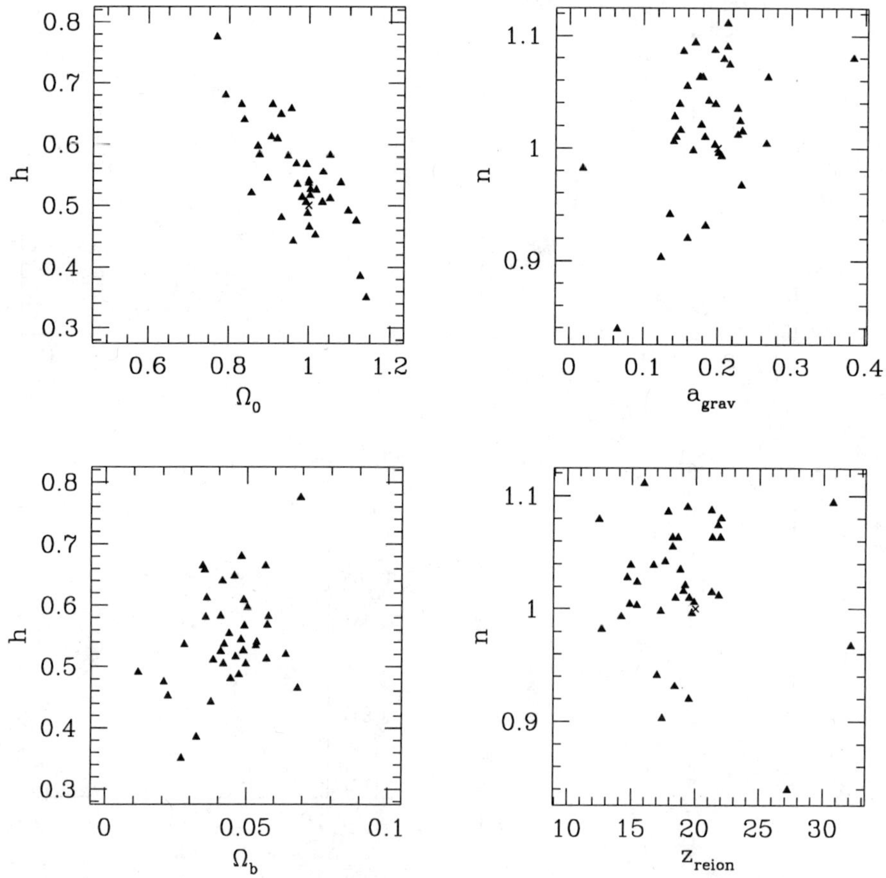

Figure 2. Same as figure 1 for a map with only partial sky coverage (900 square degrees). Here cosmic variance leads to a significant uncertainty in parameters.

resolution experiments need not cover the whole sky as there are many pixels in a 0.1^o resolution map.

Figures 4 - 6 show that even with the addition of Λ as a parameter in the models CMBR experiments place very important constraints on the basic cosmological parameters. By probing the beginning of the second acoustic peak at $l \sim 500$, there is a significant reduction in degeneracy. Figures 6 and 7 show that with higher resolution or with lower noise level, it is possible to infer information at these high l's. The height and location of the second Doppler peak is a sensitive probe of $\Omega_b h^2$ and h.

Isocurvature models predict a very different spectrum of CMBR fluctuations at these small angular scales. Preliminary calculations (Spergel 1995) suggest that the acoustic peaks are shallower and broader and that they occur

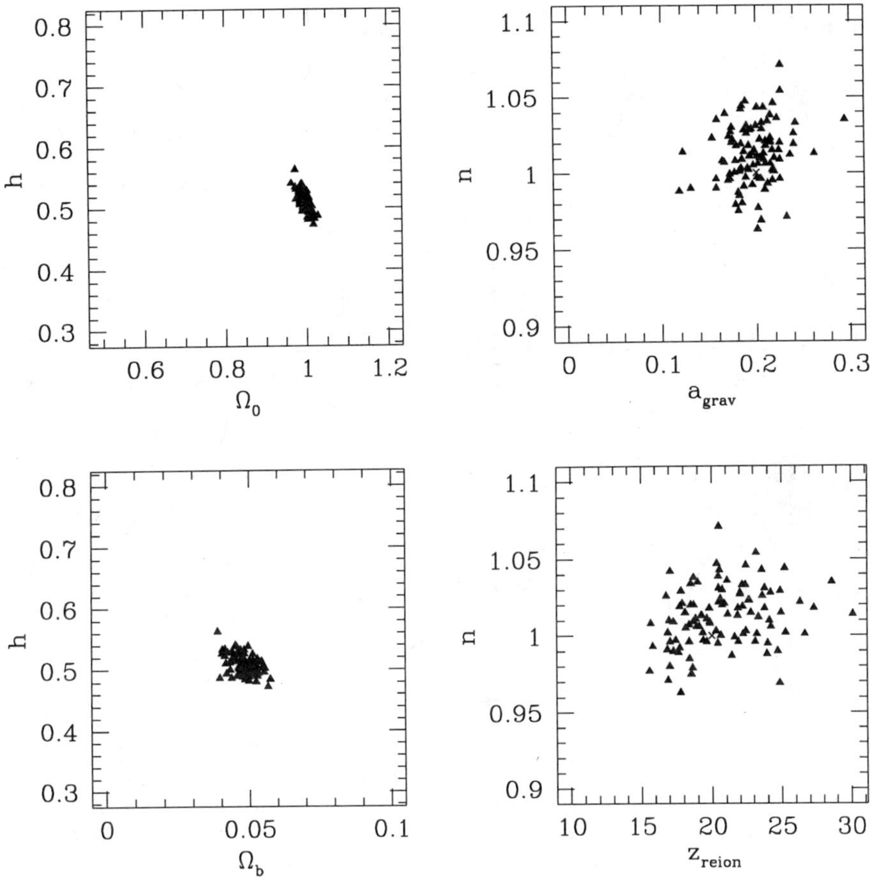

Figure 3. Same as figure 1 for a map with full sky coverage and 0.25 degree Gaussian beam.

at smaller angles. In curvature models, the acoustic waves on a given scale start oscillating as soon as they cross their causal horizon. In defect models, there are fluctuations in the photon density due to two sources: (1) variations in the energy density in defects are "compensated" on large scales by variations in the energy density in radiation, baryons and dark matter so that the mean density of the universe is constant; and (2) the motion of the defects accelerate the baryons, dark matter and photons and excite acoustic waves. As these waves are excited over a period of time, there are not "in phase" in the same way that the the acoustic waves in curvature models are in phase. Thus, there are not narrow acoustic peaks in the CMBR spectrum in these models.

4. CONCLUSIONS

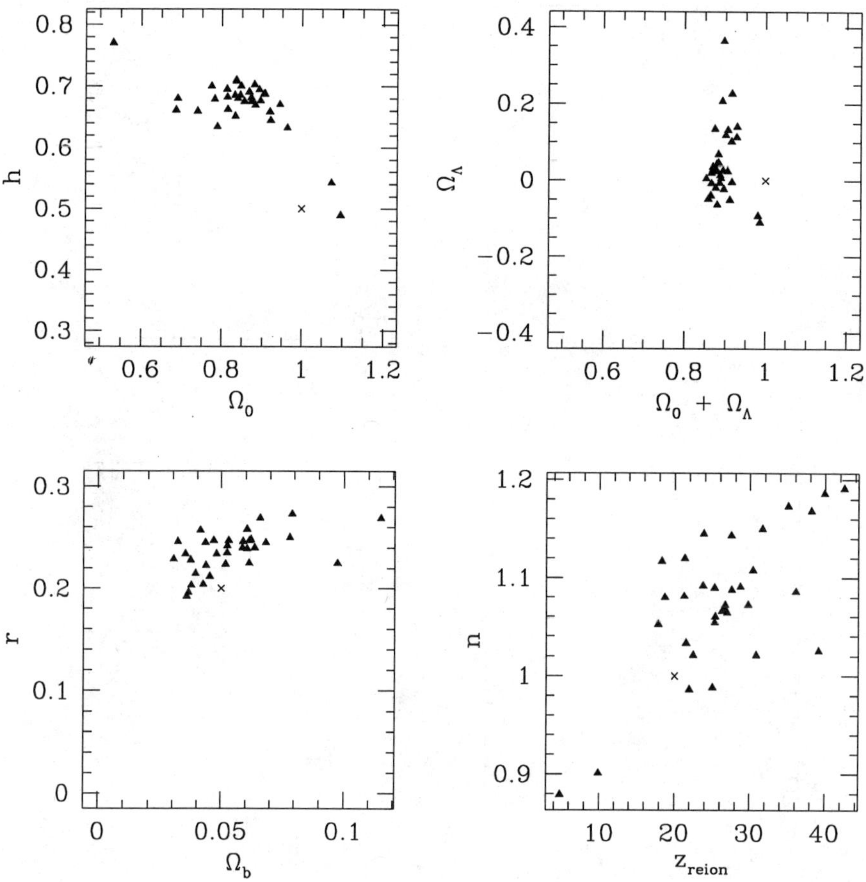

Figure 4. In this set of simulations, Λ has been added as an additional parameter to the theoretical model used to fit the simulations. This figure shows the best fit values for an all sky map with 0.5^o resolution.

COBE detection of CMBR fluctuations marks the beginning of a new era in cosmology. Future experiments will enable us to study the physical processes that produced large scale structure and probe the geometry, age and composition of the universe.

ACKNOWLEDGEMENTS

I would like to thank the members of the MAP collaboration for helpful discussions. I would also like to acknowledge NSF and NASA for support.

REFERENCES

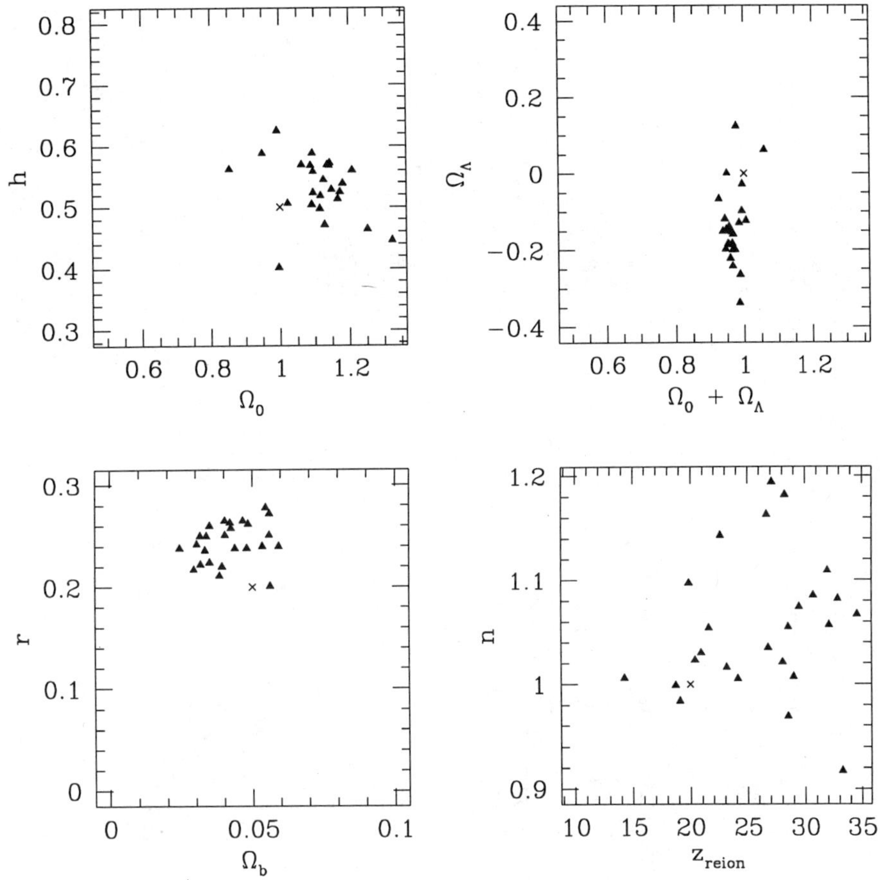

Figure 5. Same has figure 4 but with 0.25^o Gaussian beam.

Bardeen, J.M., Steinhardt, P.A. & Turner, M.S. 1983, Phys. Rev. D 28, 679
Bond, J.R. & Efstathiou, G. 1984, ApJ, 285, L45
Bond, J.R. et al. 1994 PRL, 72, 13
Hu, W. & Sugiyama, N., CfPA preprint TH-94-34, ApJ, submitted
Kamionkowski, M., Spergel, D.N. & Sugiyama, N. 1994, ApJ, 426, L57
Kogut, A., Hinshaw, G. & Bennett, C.L. 1994, ApJ, submitted
Press, W.H., Teukolsky, S.A., Vetterling, W.T., & Flannery, B.P., Numerical Recipes in Fortran: The Art of Scientific Computing, Second Edition, Cambridge University Press, Cambridge 1992

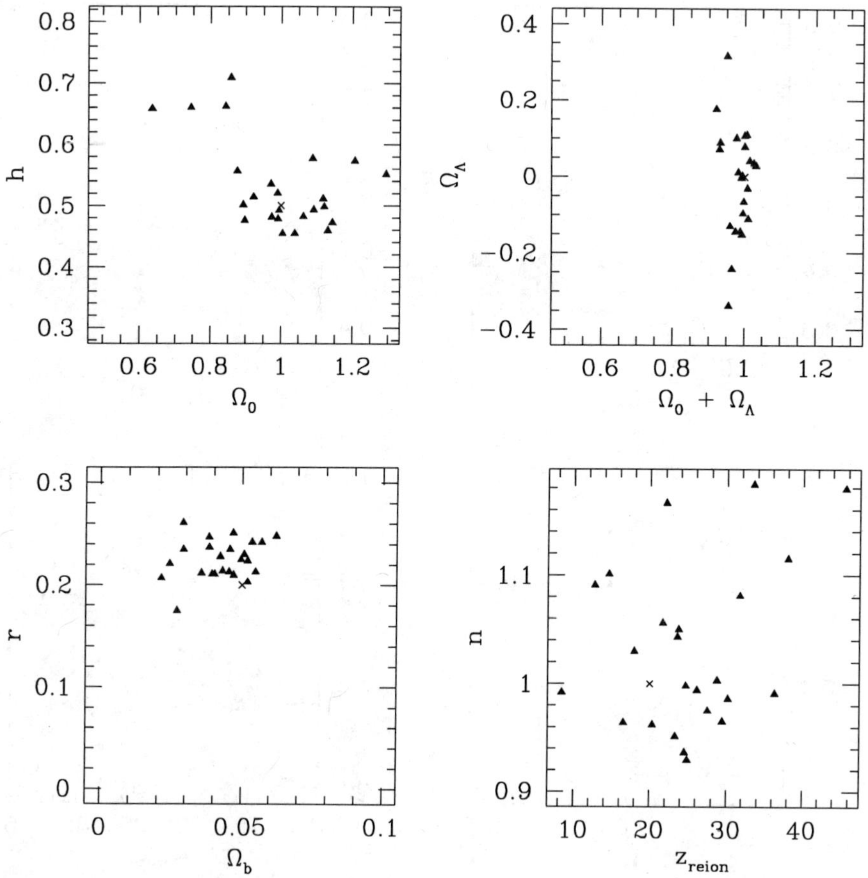

Figure 6. Same has figure 4 but with only 1.5 μK noise per pixel. beam.

Dark Matter — Cosmic Microwave Background Connection

Krzysztof M. Górski[1]
Universities Space Research Association, NASA/GSFC, Code 685, Greenbelt MD 20771
gorski@stars.gsfc.nasa.gov

ABSTRACT

A method for inference of the primordial power spectrum from the *COBE*[2] DMR sky maps is discussed. This approach involves a Fourier decomposition of the sky maps in a basis of orthonormal functions on the incompletely sampled sky, a detailed mathematical model of all noise properties, and a likelihood analysis of the data based on an exact probabilistic model of the measurable quantities, valid in the context of Gaussian theories of structure formation. Results from the two year *COBE*-DMR data analysis and the implications for inflationary (flat geometry) models dominated by either dark matter or a cosmological constant are presented. The cross power spectrum of the 53 and 90 GHz DMR sky maps is also discussed.

1. INTRODUCTION

The development of inflationary ideas during the 1980s (Blau & Guth 1987) induced a decade-long adherence to the cosmological paradigm which posits that the universe is spatially flat. Such a picture requires that the present energy density of the universe is dominated by non-baryonic dark matter or alternatively by a non-zero vacuum energy contribution (a cosmological constant term, Λ). The minimal version of the model, which invokes cold dark matter (CDM) as the major constituent of the universe, is presently in direct confrontation with astronomical observations. An extension of the model, which in addition to CDM postulates an admixture of hot dark matter (HDM), enjoys considerable popularity in contemporary cosmological research. Vigorous discussion ensues in the literature as to the plausibility of the mixed dark matter (MDM) model as a viable cosmology, (see e.g. Schaefer, Schafi & Stecker 1989; Davis, Summers, & Schlegel, 1992; Klypin et al. 1993; Pogosyan & Starobinskii 1993; Ma & Bertschinger 1994; Primack et al. 1994; Ma, and Klypin et al. in this volume). Cosmological constant dominated, spatially flat, cold dark matter models (CDM-Λ) can be considered a natural extension of the CDM scenario. By adjusting the combination of a non-zero cosmological constant and the total matter density parameter, Ω_0 (which remains poorly constrained by observations), we can maintain the spatial flatness of the universe required by the inflationary paradigm, whilst striving to improve the agreement of the theoretical model predictions to the observational data (Peebles 1984; Efstathiou, Maddox, & Sutherland 1991; Kofman, Gnedin,

---

[1]On leave from Warsaw University Observatory, Aleje Ujazdowskie 4, 00-478 Warszawa, Poland

[2]The National Aeronautics and Space Administration/Goddard Space Flight Center (NASA/GSFC) is responsible for the design, development, and operation of the Cosmic Background Explorer (COBE). Scientific guidance is provided by the COBE Science Working Group. GSFC is also responsible for the development of the analysis software and for the production of the mission data sets.

Bahcall 1993; Carrol, Press, Turner 1992). The empirical determination of the power spectrum of primordial inhomogeneities and assessment of their consistency with the predictions of these inflation-related models of the universe are critical issues in contemporary cosmology.

The implications of recent observational advancements in extragalactic astronomy, including measures of the galaxy distribution and bulk flow motions for the above models of the universe are discussed elsewhere in this volume. In this contribution I focus on some issues related to the current status of the cosmic microwave background (CMB) anisotropies. White, Scott, & Silk (1994) have recently reviewed both the theoretical and experimental status of the field — the interested reader is invited to refer to this paper. In recent years there has been rapid growth in the experimental area of medium and small angular scale CMB anisotropy, but the field is presently in a transitory stage. Therefore I shall concentrate on the *COBE* DMR results and their relation to Dark Matter in the universe.

The *COBE* DMR discovery of CMB anisotropy (Smoot et al. 1992, Bennett et al. 1992, Wright et al. 1992) has affected cosmology in both ontological and practical ways, but its predominant quantitative influence has been to provide the means for the accurate normalisation of theories of large scale structure formation. In the following I will discuss the issues of power spectrum estimation from the *COBE* DMR sky maps, the related issue of model normalisation and implications for the inflationary models of structure formation.

2. CMB ANISOTROPY POWER SPECTRUM ESTIMATION

2.1 *COBE* DMR Data

All the results discussed were derived by analysis of the *COBE* DMR two year sky maps. A summary of two years of observations has been presented in Bennett *et al.* (1994). The *COBE* DMR instrument has observed the microwave sky at three frequencies (31.5, 53, and 90 GHz) with pairs of $\sim 7°$ (FWHM) antennas separated by 60° on the sky. The temperature difference measurements were used to construct whole sky anisotropy maps (Janssen & Gulkis 1992) binned into 6144 equal area ($\sim 2.^\circ6 \times 2.^\circ6$) pixels. The 31 GHz maps were not used in the analysis, since they are relatively noisy and contain a more significant galactic signal contribution. The A and B channels of the sky maps are coadded using inverse noise variance weighting. The sky maps have been generated in both galactic and ecliptic coordinate frames, involving separate binning of the underlying time-ordered series of sky temperature measurements into 6144 sky map pixels (of equal area). In this analysis, both the galactic and ecliptic coordinate frame data sets are used as a check on the extent to which the coordinate dependent noise binning can affect the inferred normalisation. Strong galactic plane emission was excised from the maps by removing all pixels within $|b| = 20°$ of the galactic equator. Faint, high latitude galactic emission was not removed from the data (since no fully supported models or definitive measurements thereof exist). As a simple measure of the extent to which such foreground emission could affect the inferred normalisation, the power spectrum analysis is conducted with the lowest order ($\ell = 2$) anisotropy mode, the quadrupole, either included or excluded. No smoothing or filtering of the sky maps is applied prior to the implementation of the power spectrum inference.

2.2 Model Power Spectra of CMB Anisotropy

The cosmological models are specified to within an arbitrary amplitude of the perturbations as follows:
1) the global geometry is flat with $\Omega_0 + \Omega_b + \Omega_\Lambda = 1$, with the bulk mass density provided by either CDM or MDM; for the MDM models (with only $\Lambda = 0$ considered) the hot dark matter is introduced in the form of either one or two (equal mass) families of massive neutrinos, with the contributed fraction of critical density taken as $\Omega_\nu = 0.15$, 0.2, 0.25 and 0.3 for one massive flavour, and $\Omega_\nu = 0.2$, 0.3, otherwise;
2) the value of the Hubble constant, $H_0 = 100\ h$ km s^{-1} Mpc^{-1}, is sampled at $h = 0.5$, and 0.8, and, in agreement with the Big-Bang nucleosynthesis arguments, the baryon abundance obeys the relation $\Omega_b = 0.013\, h^{-2}$ (Reeves 1994);
3) random-phase, Gaussian, scalar primordial curvature perturbations (no gravity waves) are assumed with the inflationary Harrison-Zel'dovich spectrum corresponding to an adiabatic density perturbation distribution, $P(k) \propto k$.

The angular distribution of CMB temperature anisotropy induced by such curvature perturbations (Sachs & Wolfe 1967, SW) is easily described in Fourier language, using the spherical harmonic decomposition of the random field of temperature fluctuations — $\delta T/T(\theta, \phi) = \sum_{\ell m} a_{\ell m} Y_{\ell m}(\theta, \phi)$ (e.g. Peebles 1981). Individual spherical harmonic coefficients $a_{\ell m}$ are Gaussian-distributed in the theoretical ensemble of initial conditions for structure formation. The variances of the probability distributions of individual modes, $\langle |a_{\ell m}|^2 \rangle$, are uniquely expressed as integrals over the power spectrum, and depend only on ℓ due to the statistical isotropy of the CMB temperature field. The CMB anisotropy multipole coefficients and the matter perturbation transfer functions for all models were evaluated using the Boltzman equation code of Stompor (1994) by solving the propagation equations up until the redshift $z = 0$.

Over the low-ℓ range of CMB multipoles probed by *COBE*-DMR the theoretical spectra are indistinguishable for the CDM and MDM models with equivalent h and Ω_b. Thus, the power spectrum amplitude derived from the data applies equally to both CDM and MDM models. The low-ℓ shape of the CDM/MDM spectrum is a little steeper than the underlying, inflationary $n = 1$ spectrum (Bond 1993; Bunn, Scott, & White 1994; Górski, Stompor, & Banday 1995). The steepening is caused by three effects: 1) the anisotropy of the photon distribution function at the moment of decoupling, 2) the high-redshift integrated SW effect generated by the change in the growth rate of matter perturbations at the post-recombination epoch due to the non-negligible contribution of radiation (i.e. photons and massless neutrinos) to the total energy density, 3) the contribution of the first adiabatic/Doppler peak to the low ℓ multipoles. These contributions, though small compared to the overall amplitude of the usual curvature perturbation driven SW effect, should be accounted for in an accurate power spectrum normalisation based on the *COBE*-DMR anisotropy data. The CDM-Λ radiation power spectra at low-ℓ ($\lesssim 10$), with a characteristic enhancement of the lowest order modes, are determined by both the usual SW and the cosmological constant induced, integrated SW (ISW) effects (Kofman & Starobinskii 1985, Górski, Silk, & Vittorio 1992), and depend very weakly on the Hubble constant.

We express the power spectrum amplitude in terms of Q_{rms-PS}, the value of the exactly computed quadrupole, a_2, multiplied by $(5/4\pi)^{1/2}$, which depends non-trivially on cosmological parameters. This is a generalisation of the normalisation introduced in Smoot et al. (1992) for pure power law model spectra. For

$\Omega_0 \lesssim 0.5$, the ISW contributions boost the quadrupole over its pure SW value. For $\Omega_0 \gtrsim 0.5$ — models with a negligible cosmological constant induced ISW effect — the resulting exact quadrupole is lower than its pure SW counterpart. The steepest spectra arise in models near critical matter density ($\Omega_0 \simeq 0.8$).

2.3 Power Spectrum Parameter Inference

The power spectrum estimation method described in Górski (1994) was applied. Orthogonal basis functions are constructed from spherical harmonics with the exact inclusion of both pixelisation effects and galactic plane excision, which is coordinate system specific, and leaves 4016 and 4038 pixels in the galactic and ecliptic sky maps, respectively. [Note that unlike the Karhunen-Loeve, or signal-to-noise eigenmode approach (Bond 1993, Bunn and Sugiyama 1994) this approach utilizes mathematical constructs which are *entirely* independent of the properties of the theory which one wishes to test with the data.] The resulting finite Fourier series decomposition of the signal over the spectral range $\ell \in [2, 30]$ is then performed in this basis, and the Fourier mode amplitudes of the CMB anisotropy are used in the construction of an exact Gaussian likelihood function for a given theory. As explained in Górski (1994), this method is unique in allowing the algebraic elimination of the non-cosmological, $\ell = 0,\ 1$, modes (without affecting higher-ℓ modes at all) from the power spectrum inference procedure.

The correlation matrices for both theoretical signal and map noise are evaluated using the orthogonal mode expansion, i.e. explicitly reflecting the mode-coupling due to the galaxy cut. The noise is considered to be sufficiently described by a pixel-to-pixel uncorrelated random process (the 60° correlations have a negligible effect on the inference from the DMR data, see Lineweaver et al. 1994).

Monte Carlo testing has demonstrated that this method of power spectrum inference is statistically unbiased (Fig. 1).

2.4 Results of Q_{rms-PS} Fitting and Large Scale Structure Statistics

The actual two parameter contour plot of the likelihood derived in the pure power law power spectrum fit to the two year data is shown in Górski et al. (1994), and in Bennett et al. in this volume. A detailed presentation of the results for CDM, MDM, and CDM-Λ model fitting to the *COBE* DMR data can be found in Górski, Stompor, & Banday (1995), and Stompor, Górski, & Banday (1995). For comparison one may refer to Bunn, Scott, & White(1994), and Bunn & Sugiyama (1994). A typical one parameter likelihood fit of a flat, CDM/MDM model to the two year *COBE*-DMR data yields a $\sim 13\sigma$ significant determination of $Q_{rms-PS} \sim 20\,\mu$K. For the CDM-Λ models the best fit amplitude grows with decreasing Ω_0 up to $\sim 26\,\mu$K. Systematic shifts in the central value of the fit are observed due to: 1) differences in the noise pixelisation in the galactic and ecliptic coordinate frames, which result in a $\sim 0.8\,\mu$K difference between the inferred normalisation amplitudes, with higher values obtained from the ecliptic maps; 2) exclusion of the quadrupole, which produces a $\sim +0.4\,\mu$K variation in the fitted amplitude; 3) the uncertainty in the values of h and Ω_b (reflected by differences in spectral shape over the ℓ-range accessible to DMR) causes an additional small spread, $\pm 0.25\,\mu$K, of the fitted amplitudes.

A convenient summation of the proposed overall normalisation for the CDM and MDM models is $Q_{rms-PS} = (20 \pm 1.52)(\pm 0.4\ \pm 0.2\ \pm 0.25)\mu$K, and for the CDM-$\Lambda$ models $Q_{rms-PS} \simeq \{[20 + 9.1h^{0.6}\exp(-\Omega_0^2 h^{0.3}/0.057)] \times (1 \pm 0.077) \pm 0.4$ $\pm 0.2 \pm 0.25\}\mu$K. The error ranges represent the statistical error and uncertainties

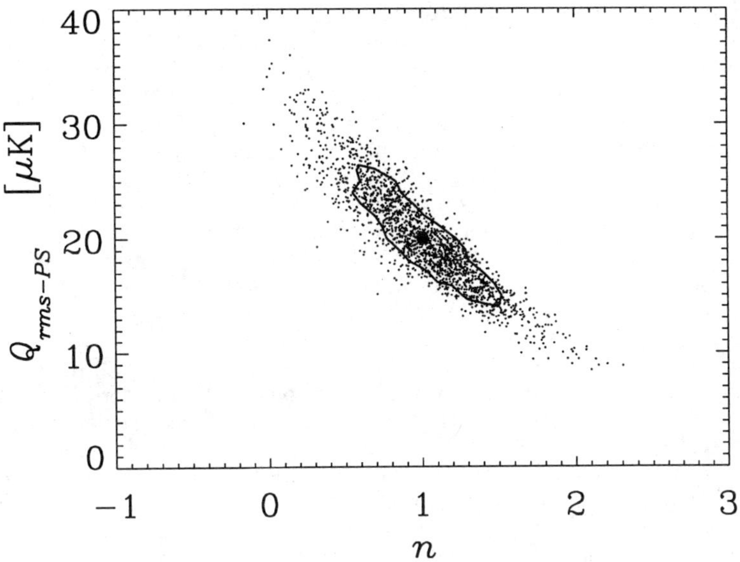

Figure 1: Results from Monte Carlo simulations of the power spectrum inference procedure. Each of the 2000 points shows the locus of the maximum likelihood parameters derived from a simulated CMB sky map for an input model with $n = 1$, and $Q_{rms-PS} = 20\,\mu$K and noise appropriate to the DMR 53 GHz receivers. The filled circle represents the average of the output parameters and demonstrates that the inference method is statistically unbiased. The solid line is the approximate 68% probability density contour derived from the simulated distribution.

associated with effects 1 through 3 above, respectively. It will be noted that the statistical error on the inferred normalisation is considerably larger than the other uncertainties.

Table 1 contains a representative selection of models and the rms values of several large-scale structure statistics computed from the matter perturbation spectra according to the usual prescriptions given in the footnotes.

The cold dark matter theory with a standard choice of cosmological parameters requires a very high normalisation in order to fit the CMB anisotropy distribution. Analysis of the first year of *COBE*-DMR data had already suggested that $(\sigma_8)_{mass} \sim 1$ (Wright et al. 1992, Efstathiou, Bond, & White 1992), and this value increases to ~ 1.4 with two years of data and an improved analysis technique. Although this normalisation allows the theory to predict POTENT-scale velocities (Dekel 1994) with a comfortably high amplitude, it is still not high enough to explain the Lauer & Postman (1994) result, and it also results in a significant overproduction of density perturbations on scales of $\lesssim 20\,h^{-1}$Mpc. CDM has often been criticised for its poor match to both galaxy and cluster distributions. Mixed dark matter models manage to circumvent, to a certain degree, these same problems by construction — massive neutrinos partially damp the perturbations at those length-scales where CDM looks problematic. Among the MDM models those with two species of massive neutrino seem to meet the observational constraints more comfortably (see also Primack et al. 1994). The larger free-streaming radius allows for the suppression of the perturbation amplitude on larger scales than in models with one massive flavour. This is reflected in the decrease of the predicted $(\sigma_8)_{mass}$ values. Nevertheless, the proponents of Dark Matter cosmology will have to address the viability of the models viz. the contradictory requirements that there *should be* a bias between the galaxy and mass distribution, as suggested by the galaxy cluster abundance argument (White, Efstathiou, & Frenk 1993) and the galaxy pair-wise velocities, and there *should not be* one, as suggested by the *COBE* DMR normalisation.

The low-Ω_0 CDM-Λ models can not be rejected solely on the basis of the *COBE*-DMR data ($\Omega_0 \geq 0.15$, 95% confidence), although the most likely Ω_0 value is strikingly close to unity. Conversely, observations of the matter distribution do require low values of the total density, with $\Omega_0 h^2 \sim 0.1$. being the favoured value. On larger scales, these models predict bulk flows in reasonable agreement with POTENT, but dramatically smaller than determined by the Lauer & Postman analysis. Therefore, if the latter observation is confirmed all CDM-Λ models (together with critical density CDM or MDM models) will be found wanting.

At present, although the position of the CDM-Λ models is more comfortable than that of critical density models, it is not free from potentially fatal flaws. Unfortunately, CMB anisotropies on a one degree scale do not offer any serious prospects for distinguishing between cosmological constant models and other viable scenarios (Stompor & Górski, 1994; Bond et al., 1994). Hence, definite conclusions will have to await more reliable observational data, particularly large-scale, deep galaxy surveys capable of unraveling the shape of the galaxy power spectrum down to $k \sim 0.01\,h\mathrm{Mpc}^{-1}$, or a comoving length-scales up to $\sim 600\ h^{-1}\mathrm{Mpc}$.

3. CROSS POWER SPECTRUM

Visualising the faint CMB structure present in the rather noisy DMR maps is a difficult task. A straightforward and relatively simple procedure is the computation of a 'cross' power spectrum between two sky maps. Such a construction can take full advantage of the multiple channel data acquisition by the DMR instrument. Since the noise properties of the different channels are not correlated, there is no noise bias in the amplitude of the quadratic power spectrum coefficients.

The cross power spectrum presented here is derived by the direct product of the individual Fourier coefficients from the 53 and 90 GHz sky maps. Monte Carlo simulations are performed to obtain comparison statistical distributions for the CDM (MDM) model, since analytic solutions are no longer easily computed.

Figure 2 displays both the sum (A+B) and difference (A-B) map cross power spectra, and clearly demonstrates a significant detection of CMB anisotropy by the *COBE*-DMR instrument. The signal-to-noise ratio in the cross-combined 53 and 90 GHz two year data exceeds unity for $\ell \lesssim 15$. The cross spectra derived in galactic and ecliptic coordinates appear to be in excellent agreement. This is not a trivial fact — the basis functions with respect to which the Fourier decomposition was performed are coordinate frame dependent due to the different geometry of the applied Galactic cut. Thus Fig. 2 is the best demonstration to date of the

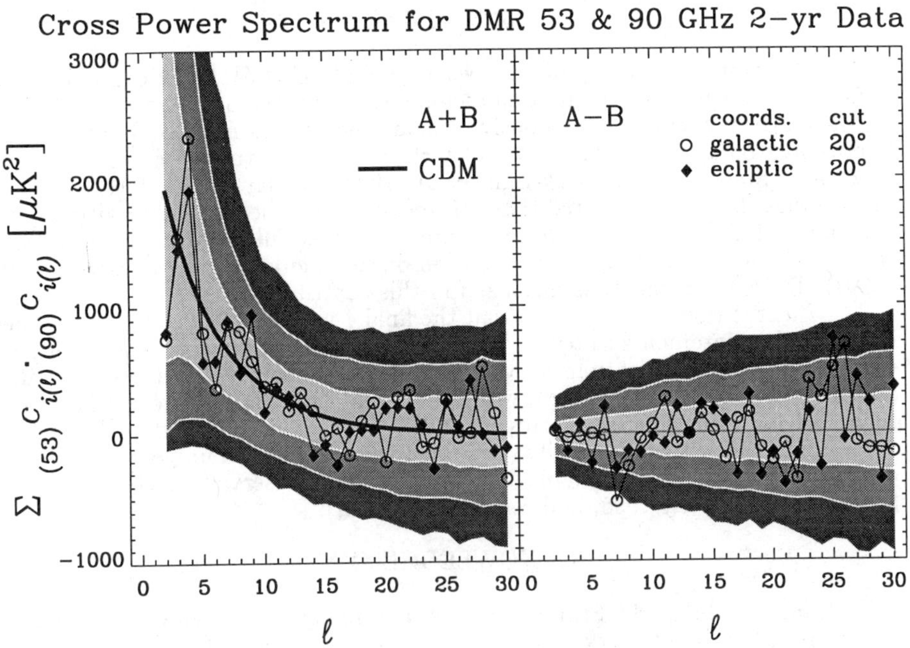

Figure 2: The cross power spectra for the DMR 53 and 90 GHz two year sum and difference sky maps derived in both galactic and ecliptic coordinate frames. The results of 40,000 Monte Carlo simulations of CDM skies normalised to $Q_{rms-PS} = 19.5\,\mu$K and with noise appropriate to the two year sky maps are shown. The heavy, solid line shows the mode of the power distribution at each ℓ. Note that this is *not* a fit to the plotted cross-power spectrum. The grey bands represent the confidence regions for the distribution of power at each value of ℓ as follows: light grey — 68% , medium — 95%, and dark grey — 99.8% .

rotational invariance of the angular spectrum of *COBE* DMR measured CMB anisotropy.

The overplotted CDM power spectrum distribution (identical to MDM over this range of ℓ) appears to be a reasonable statistical descriptor for the measured CMB anisotropy pattern. The low *COBE* DMR two year quadrupole amplitude (Bennett et al. 1994) is clearly evident, despite the fact that we have not attempted to model and correct for residual high latitude Galactic emission. Although such modelling would result in a lower estimated cosmological CMB quadrupole amplitude, it would still fall within the 95% probability zone under the distribution derived from the *COBE* DMR normalised CDM/MDM model. An analogous plot for CDM-Λ models would only show differences at low ℓ for low values of Ω_0; the average power in this range would tend to be higher than in Fig. 2. Coupled with the relatively low measured value of the quadrupole, the relative likelihood of high-Λ models is lower than that for the matter dominated models.

5. SUMMARY

The improved quality of the two year *COBE*-DMR data combined with reliable power spectrum estimation techniques allows the accurate normalisation of cosmological theories. Previously, the variance of *COBE*-DMR temperature fluctuations on a 10° angular scale was utilised for the normalisation of the power spectrum. Subsequent work (Wright et al. 1994, Banday et al. 1994, and these proceedings) has demonstrated that this technique can be unreliable without considerable attention. More appropriate methods take full advantage of the measured CMB anisotropy power distribution on all angular scales accessible to the *COBE*-DMR instrument, as discussed in this contribution.

Whilst it would be safer to await the final 4-year COBE results before offering definitive statements as to the viability of theoretical models, one should note that the CDM normalisation derived from the two year *COBE*-DMR data does appear to be irreconcilably high, while the MDM and CDM-Λ models have little room left for adjustments. Interestingly, theoretically favoured globally flat universe models notwithstanding, it is possible to construct open universe inflationary dark matter models in reasonable agreement with the observations (Górski, Ratra, Sugiyama, & Banday 1995, and references therein).

ACKNOWLEDGEMENTS

I am grateful to R. Stompor and A.J. Banday for allowing me to use the results from our joint work before publication. I am grateful to A.J. Banday, C. Bennett, G. Hinshaw and A. Kogut for numerous discussions and help in improving the manuscript. The efforts of those contributing to the *COBE*-DMR are acknowledged.

REFERENCES

Banday, A. J., et al. 1994, ApJ, 436, L99
Bennett, C. L., et al. 1992, ApJ, 396, L7
Bennett, C. L., et al. 1994, ApJ, 436, 423
Blau, S. K., & Guth, A. H. 1987, in "300 Hundred Years of Gravitation", eds. S. W. Hawking and W. Israel (Cambridge: Cambridge University Press), p.524
Bond, J. R. 1993, in Proc. of the IUCAA Dedication Ceremonies, ed. T. Padmanabhan, New York:John Wiley & Sons, in press
Bond, J.R., Crittenden, R., Davies, R.L., Efstathiou, G., & Steinhardt, P.J. 1994, Phys.Rev.Lett., 72, 13
Bunn, E., & Sugiyama, N. 1994, Berkeley preprint CfPA-TH-94-33
Bunn, E., Scott, D., & White, M. 1994, Berkeley preprint CfPA-94-TH-42, astro-ph-9409003
Carrol, S.M., Press, W.H., & Turner, E.L. 1992, ARA&A, 30, 499
Davis, M., Summers, F.J., Schlegel, D. 1992, Nature, 359, 393
Dekel, A. 1994, ARA&A, 32, 371
Efstathiou, G.P., Bond, J.R., & White, S.D.M. 1992, MNRAS, 258, 1P
Efstathiou, G., Maddox, S.J., & Sutherland, W.J. 1990, Nature, 348, 705
Górski, K. M. 1994, ApJ, 430, L85
Górski, K. M., et al. 1994, ApJ, 430, L89
Górski, K. M., Silk, J., & Vittorio, N. 1992, Phys.Rev.Lett., 68, 733

Górski, K.M., Stompor, R., Banday, A.J. 1995, ApJ submitted
Górski, K.M., Ratra, B., Sugiyama, N., & Banday, A.J. 1995, ApJ submitted
Janssen, M.A., & Gulkis, S., 1992, in "The Infrared and Submillimeter Sky after *COBE*", eds. M. Signore & C. Dupraz (Dordrecht:Kluwer)
Klypin, A., Holtzman, J.A., Primack, J.R. & Regös, E. 1993, ApJ, 416, 1
Kofman, L.A., Gnedin, N.Y., & Bahcall, N.A. 1993, ApJ, 413, 1
Kofman, L.A., & Starobinskii, A.A. 1985, Sov.Astron., 11, L5
Lauer, T. R., & Postman, M. 1994, ApJ, 425, 418
Lineweaver, C., et al. 1994, ApJ, 436, 452
Ma, C.-P., & Bertschinger, E. 1994, ApJ, 434, L5
Peebles, P.J.E. 1981, ApJ, 243, L119
Peebles, P.J.E. 1984, ApJ, 284, 439
Pogosyan, D.Yu., & Starobinsky, A.A. 1993, MNRAS, 265, 507
Primack, J.R., et al. 1994, preprint SCIPP 94/28
Reeves, H. 1994, Rev. Mod. Phys., 66, 193
Sachs, R.K., & Wolfe, A.M. 1967, ApJ, 147, 73
Schaefer, R.K., Schafi, Q., & Stecker, F.W. 1989, ApJ, 347, 575
Smoot, G.F., et al. 1992, ApJ, 396, L1
Stompor, R. 1994, A&A, 287(III), 693
Stompor, R., & Górski, K.M. 1994, ApJ, 422, L41
Stompor, R., & Górski, K.M., Banday, A.J. 1995, ApJ submitted
White, M., Scott, D., Silk, J. 1994, ARA&A, 32, 319
White, S. D. M., Efstathiou, G., & Frenk, C. S. 1993, MNRAS, 262, 1023
Wright, E.L., et al. 1992, ApJ, 396, L13
Wright, E. L., et al. 1994, ApJ, 420, 1

Table 1: Inferred cosmological statistics for spatially flat CDM, MDM and CDMΛ models (with $\Omega_0 + \Omega_\Lambda + \Omega_b = 1$, where $\Omega_0 = \Omega_{CDM} + \Omega_\nu$ and $\Omega_b = 0.013h^{-2}$), with a *COBE*-DMR normalisation expressed in terms of Q_{rms-PS}. N_ν is the number of massive neutrino species, and m_ν the neutrino mass in eV. The errors, including both statistical (1σ) and systematic deviations, are of the order of 11%.

Ω_0	h	Ω_ν	N_ν	m_ν	$Q^{(a)}_{rms-PS}$	$(\sigma_8)^{(b)}_{mass}$	$J_3(20)^{(c)}$	$v^{(d)}_{40}$	$v^{(d)}_{60}$	$v^{(d)}_{100}$
1.0	0.5	0.00	–	0.0	20.04	1.36	968	444	355	248
..	..	0.15	1	3.7	..	0.97	706	441	357	251
..	..	0.20	1	4.9	..	0.92	694	442	358	252
..	..	0.25	1	6.1	..	0.88	691	444	359	252
..	..	0.30	1	7.3	..	0.85	695	445	360	252
..	..	0.20	2	2.4	..	0.82	567	435	356	252
..	..	0.30	2	3.7	..	0.71	516	439	359	254
0.1	0.8	0.0	–	0.0	26.11	0.53	335	208	192	167
..	0.5	..	..	..	25.27	0.17	41	139	134	123
0.2	0.8	..	..	..	23.61	1.03	985	330	290	230
..	0.5	..	..	..	23.24	0.44	222	235	216	183
0.3	0.8	..	..	..	22.04	1.39	1496	403	342	258
..	0.5	..	..	..	21.79	0.66	428	301	267	214
0.4	0.8	..	..	..	21.10	1.66	1825	449	372	272
..	0.5	..	..	..	20.93	0.84	607	349	301	233

$^{(a)}$ in μK

$^{(b)}$ $(\sigma_{hR})^2_{mass} = \frac{1}{2\pi^2} \int_0^\infty w^2_{TH}(kR)\, P(k) k^2\, dk$

$^{(c)}$ $J_3(hR) = \frac{R^3}{2\pi^2} \int_0^\infty w^2_{TH}(kR)\, P(k)\, k^2\, dk$, in $(h^{-1}\mathrm{Mpc})^3$, and for $R = 20h^{-1}\mathrm{Mpc}$

$^{(d)}$ $v^2_{hR} = \frac{H_0^2}{2\pi^2} \int_0^\infty w^2_{TH}(kR)\, e^{-k^2 r_s^2}\, P(k)\, dk$, $hr_s = 12\mathrm{Mpc}$, in km s^{-1}

$w_{TH}(x) = 3j_1(x)/x$

COBE[1]-DMR Two-Year Large Scale Anisotropy Results

C.L. Bennett, A. Banday, K. Górski, G. Hinshaw, A. Kogut
Code 685, Laboratory for Astronomy and Solar Physics
NASA/Goddard Space Flight Center, Greenbelt MD 20771
bennett@stars.gsfc.nasa.gov

& E.L. Wright
Department of Astronomy
University of California, Los Angeles, CA 90024

ABSTRACT

We summarize the results of the Cosmic Microwave Background (CMB) anisotropy measurements of the COBE-DMR using the first two years of data. The inference of angular power spectrum parameters from the DMR data is complicated by the need to omit Galaxy contaminated regions of the sky, preventing the spherical harmonics from forming an orthogonal basis set over the remaining sky. The COBE-DMR team has implemented three methods for deducing the power spectrum parameters. These include: a two-point correlation function analysis (Bennett et al. 1994); an analysis in terms of orthogonal modes constructed on the cut sphere (Górski et al. 1994); and an analysis of a modified Hauser-Peebles power spectrum (Wright et al. 1994). These methods for deducing the power spectrum are discussed and the results are summarized and compared.

1. INTRODUCTION

The large angular scale cosmic temperature fluctuations are believed to reflect the gravitational potential fluctuations at the epoch of recombination. The amplitude, Q_{rms-PS}, and power law spectral index, n, of the primordial anisotropy are important cosmological parameters, and serve as initial conditions for for N-body and hydrodynamic codes that test dark matter models. We have applied three methods for determining the amplitude and slope of the primordial power spectrum to the two-year COBE-DMR data. The results from each method are presented in detail in Table 1. The three methods yield maximum likelihood values for n between 1.02 and 1.42 with uncertainties between 0.32 and 0.45. In no case is a determination inconsistent with the scale-invariant value $n = 1$, and the steeper determinations tend to be driven by the smallness of the quadrupole. All methods return a similar ridge of degeneracy between Q_{rms-PS} and n, though the slope of the ridge does vary slightly from method to method. The maximum likelihood values for the normalization of an $n = 1$ spectrum range from 18.2 to 20.4 μK, with uncertainties between 1.6 and 2.0 μK. The n values returned by each method vary by about 0.5 σ, while the normalizations for a scale-invariant spectrum vary by about 1 σ.

The two-point correlation analysis and the modified Hauser-Peebles power spectrum analysis employ maximum likelihood fits to quadratic statistics constructed from products of maps: a combination of the 53 and 90 GHz data. The

[1]The National Aeronautics and Space Administration/Goddard Space Flight Center (NASA/GSFC) is responsible for the design, development, and operation of the Cosmic Background Explorer (COBE). Scientific guidance is provided by the COBE Science Working Group. GSFC is also responsible for the development of the analysis software and for the production of the mission data sets.

Table 1: Power Spectrum Analysis of Combined 53 and 90 GHz 2-yr DMR Data

Parameter	2-pt function Bennett et al.	$\psi_{\ell m}$ modes Górski et al.	$G_{\ell m}$ modes Wright et al.
	including quadrupole		
n	$1.42^{+0.49}_{-0.55}$	$1.22^{+0.43}_{-0.52}$	---
n (marginal[a])	1.42 ± 0.37	1.10 ± 0.32	$1.39^{+0.34}_{-0.39}$
Q_{rms-PS} (μK)	$14.3^{+5.2}_{-3.3}$	$17.0^{+7.6}_{-4.8}$	---
$Q_{rms-PS\|n=1}$ (μK)	18.2 ± 1.5	19.9 ± 1.6	---
	excluding quadrupole		
n	$1.11^{+0.60}_{-0.55}$	$1.02^{+0.53}_{-0.59}$	---
n (marginal)	1.11 ± 0.40	0.87 ± 0.36	$1.25^{+0.40}_{-0.45}$
Q_{rms-PS} (μK)	$17.4^{+7.5}_{-5.2}$	$20.0^{+10.5}_{-6.5}$	---
$Q_{rms-PS\|n=1}$ (μK)	18.6 ± 1.6	20.4 ± 1.7	19.8 ± 2.0
a_ℓ at pivot (μK)[b]	$a_7 = 9.5 \pm 1.0$	$a_9 = 8.3 \pm 0.7$	---

[a]Bennett et al. and Wright et al. define the marginal likelihood on n to be $L(n) = \max L(Q|n)$, while Górski et al. define it to be $L(n) = \int_0^\infty L(Q,n)dQ$, where $Q \equiv Q_{rms-PS}$.

[b]An alternate power spectrum normalization, expressed in terms of the multipole moment that is independent of n.

techniques make extensive use of Monte Carlo simulations to construct and calibrate the derived likelihood functions. The orthogonal mode analysis employs an exact maximum likelihood fit to the linear coefficients c_i from the 53 and 90 GHz data, which, in effect, places the signal and noise on a similar footing. Monte Carlo simulations are only used to verify that the technique is statistically unbiased. Note that the three methods will tend to respond differently to unmodeled effects in the data, such as residual Galactic emission, or systematic artifacts, though we believe that any such effects are small. While the three methods return generally consistent results, small discrepancies do exist and the reasons for this are not presently understood. Studies of our methodologies are continuing as we prepare to analyze the full four-year sky maps.

2. TWO-POINT CORRELATION FUNCTION

The two-point cross correlation function at angular separation α, $C(\alpha)$, is the average product of all pixel temperatures with a fixed angular separation: $C(\alpha) = \sum_{i,j} w_i w_j \; T_{A,i} T_{B.j} \; / \; \sum_{i,j} w_i w_j$, where the sum is restricted to pixel pairs (i,j) for which the pixel angular separation resides in the bin defined by α, $T_{A,i}$ and $T_{B,i}$ are the observed temperature in maps "A" and "B" respectively, after monopole and dipole (and quadrupole) subtraction, and w_i is the statistical weight of pixel i. The analysis discussed below is based on the 53 $\times$ 90 GHz cross correlation function.

We determine the most likely quadrupole normalized amplitude, Q_{rms-PS}, and spectral index, n, by evaluating the Gaussian approximation to the likelihood

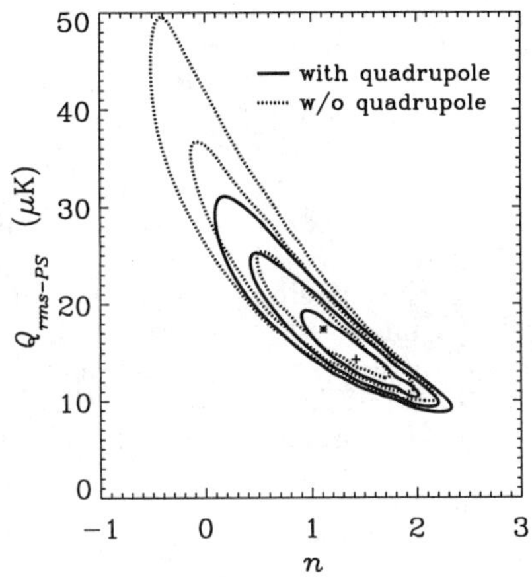

Figure 1: Likelihood contours as a function of n and Q_{rms-PS} for the 53(A+B)/2 × 90(A+B)/2 cross correlation data. The contours correspond to 68%, 95%, and 99.7% confidence regions, corrected for a bias in n of +0.11, and in Q_{rms-PS} of −1.5 μK. The solid curves are contours with the quadrupole included in the analysis; the peak of the likelihood is indicated by +. The dashed curves are contours with the quadrupole excluded from the analysis; the peak of the likelihood is indicated by $*$. See Table 1 for a quantitative summary of these likelihoods.

function

$$L(Q_{rms-PS}, n) = \frac{\exp[-\frac{1}{2}(\mathbf{\Delta C})^T \mathbf{M}^{-1} (\mathbf{\Delta C})]}{[(2\pi)^m \det(\mathbf{M})]^{1/2}}.$$

Here $(\mathbf{\Delta C})^T$ and $(\mathbf{\Delta C})$ are row and column vectors with entries $(\Delta C)_i = C(\alpha_i) - \langle C(\alpha_i) \rangle$, and $\mathbf{M} = \langle (\mathbf{\Delta C})(\mathbf{\Delta C})^T \rangle$ is the covariance matrix of the correlation function. The angled brackets denote averages over receiver noise and cosmic variance. We estimate the mean correlation function and covariance matrix, as a function of Q_{rms-PS} and n, by means of Monte Carlo simulations that account for all important aspects of our data processing. The final results are calibrated with additional Monte Carlo simulations to account for any biases introduced by the Gaussian approximation or other effects. Figure 1 (after Figure 4 of Bennett et al.) shows the resulting likelihood contours as a function of Q_{rms-PS} and n for the analyses both including and excluding the quadrupole moment. The contours correspond to 68%, 95%, and 99.7% confidence regions.

3. ORTHOGONAL MODE DECOMPOSITION

Górski et al. explicitly construct linear combinations of spherical harmonics to form orthonormal functions on the cut sphere, following a procedure developed by Górski (1994). The 53 and 90 GHz data are decomposed with respect to these modes: $T(p) = \sum_i c_i \psi_i(p)$, where $T(p)$ is the temperature in map pixel p, and c_i is the coefficient of the ith mode ψ_i. For Gaussian power-law models of anisotropy, the c_i coefficients have a multivariate Gaussian distribution, thus an exact likelihood function for the parameters Q_{rms-PS} and n may be constructed

$$L(Q_{rms-PS}, n) = \frac{\exp[-\frac{1}{2}\hat{\mathbf{c}}^T \cdot (\mathbf{C_{CMB}} + \mathbf{C_N})^{-1} \cdot \hat{\mathbf{c}}]}{[(2\pi)^{\mathbf{m}} \det(\mathbf{C_{CMB}} + \mathbf{C_N})]^{1/2}}$$

where $\hat{\mathbf{c}}$ is the vector of c_i coefficients obtained from decomposing the 53 and 90 GHz data, and $(\mathbf{C_{CMB}} + \mathbf{C_N})$ is the covariance matrix of the spectra $\hat{\mathbf{c}}$, which

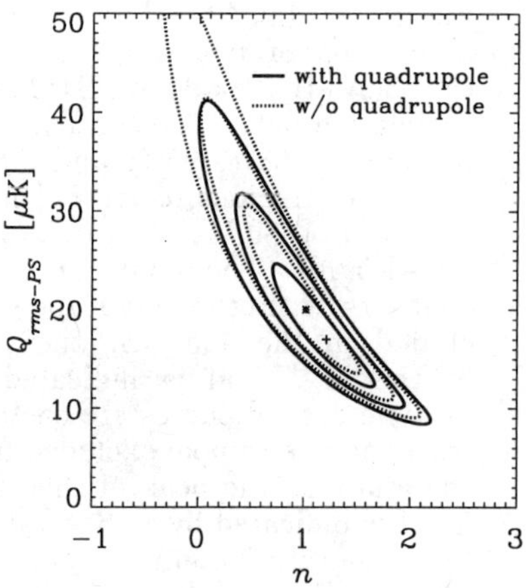

Figure 2: Likelihood contours as a function of n and Q_{rms-PS} for the 53 and 90 GHz orthogonal mode decomposition. The contours correspond to 68%, 95%, and 99.7% confidence regions. The solid curves are contours with the quadrupole included in the analysis; the peak of the likelihood is indicated by +. The dashed curves are contours with the quadrupole excluded from the analysis; the peak of the likelihood is indicated by $*$. See Table 1 for a quantitative summary of these likelihoods.

depends on Q_{rms-PS} and n. See Górski and Górski et al. for details. Monte Carlo simulations with controlled inputs indicate that the parameters inferred with the above likelihood function are not statistically biased with respect to the input. Figure 2 (after Figure 2 of Górski et al.) shows the resulting likelihood contours as a function of Q_{rms-PS} and n for the analyses both including and excluding the quadrupole moment. The contours correspond to 68%, 95%, and 99.7% confidence regions.

4. MODIFIED HAUSER-PEEBLES POWER SPECTRUM

Wright et al. solve for the angular power spectrum of the DMR two year data by modifying and applying the technique described by Peebles (1973) and Hauser & Peebles (1973) for data on a cut sphere. The input data for this analysis is a "cross" power spectrum which involves the evaluation of terms of the form $\langle G_{\ell m} T_A \rangle \cdot \langle G_{\ell m} T_B \rangle$ where T_A and T_B are two independent DMR maps, $G_{\ell m}$ is a modified spherical harmonic which, by construction, is orthogonal to the monopole and dipole (and, optionally, the quadrupole) on the cut sky, and the angle brackets denote an inner product on the cut sky. See Wright et al. for more details. They have evaluated a Gaussian likelihood, similar to eq. 1, over a variety of spectral ranges: $3 \leq \ell \leq 19$, $3 \leq \ell \leq 30$, and $2 \leq \ell \leq 30$, and for a variety of data combinations. After calibrating their results with Monte Carlo simulations they quote most likely spectral indices in the range 1.02 to 1.41 depending on the choice of data, the spectral range included, and the type of modes used. Their results, together with the other two methods are summarized in Table 1.

REFERENCES

Bennett, C.L., et al. 1994, ApJ, 436, 423
Górski, K.M. 1994, ApJ, 430, L85
Górski, K.M., et al. 1994, ApJ, 430, L89
Hauser, M.G., & Peebles, P.J.E. 1973, ApJ, 185, 757
Peebles, P.J.E. 1973, ApJ, 185, 413
Wright, E.L., et al. 1994, ApJ, 436, 443

ON THE RMS ANISOTROPY AT 7° AND 10° OBSERVED IN THE *COBE*-DMR TWO YEAR SKY MAPS

A.J. Banday, K.M. Górski, A. Kogut, G. Hinshaw, and C.L. Bennett
Code 685, NASA Goddard Space Flight Center, Greenbelt, MD 20771
C.H. Lineweaver, G.F. Smoot and L. Tenorio
LBL, SSL & CfPA, Bldg 50 - 351, University of California, Berkeley CA 94720

ABSTRACT

We summarize the recent results on the observed *COBE*-DMR two year sky rms temperature fluctuations. A 'cross-RMS' statistic is used to infer the Q_{rms-PS} normalization for a scale-invariant ($n = 1$) spectral model. The method is extended to the normalization of other cosmological power spectra.

1. INTRODUCTION

In principle, the observed sky rms on a given angular scale is a convenient number to use for the normalization of a particular cosmological model. Wright et al. (1994a) used the sky rms temperature fluctuations from the first year *COBE*-DMR sky maps smoothed to an approximate FWHM of 10° to determine the effective normalization Q_{rms-PS} for the scale-invariant Harrison-Zel'dovich power spectrum, $P(k) \propto k^n$, where $n = 1$ and k is the comoving wavenumber. We have updated and extended the analysis of Wright et al. (1994a) using the two year *COBE*-DMR data to infer the normalization for an $n = 1$ power law model, together with a number of other cosmological power spectra.

2. ANALYSIS

Banday et al. (1994) have defined a cross-RMS, $\otimes_{RMS}$, between two maps a and b as

$$(\otimes_{RMS})^2 \equiv \sum_i T_i^a T_i^b \, w_i^a \, w_i^b \; / \; \sum_i w_i^a \, w_i^b, \tag{1}$$

where the sums are over all pixels i surviving the galactic cut ($|b| > 20°$), and w_i is the weight assigned to that pixel. Unit weighting is adopted here. The best fit monopole, dipole and, where appropriate, quadrupole are removed from the sky maps prior to the computation of the $\otimes_{RMS}$.

A Monte Carlo approach was adopted to generate the $\otimes_{RMS}$ distributions for a grid of Q_{rms-PS} values (with 2500 simulations used for each value of Q_{rms-PS}). Each simulation combines a realization of sky anisotropy filtered through the *COBE*-DMR beam (Wright et al. 1994a) with noise realizations based on the specific *COBE*-DMR channels. The Q_{rms-PS}-dependent statistical means, variances and covariances of the 7° and 10° $\otimes_{RMS}$ are derived from these Monte Carlo simulations, and used to construct the gaussian approximation to the probability distribution of the $\otimes_{RMS}$. This, together with the measured $\otimes_{RMS}$ values, defines the likelihood function $\mathcal{L}(Q_{rms-PS})$. That the $\otimes_{RMS}$ provides a statistically unbiased estimator of the true cosmological Q_{rms-PS} amplitude has been verifed with additional simulations used as test input data.

We refer the reader to Banday et al. (1994) for full details. Summarizing the main results therein:

- The observed $\otimes_{RMS}$ is frequency independent and thus consistent with a cosmic origin for the temperature fluctuations.
- The $\otimes_{RMS}$ values derived from different combinations of the 53 and 90 GHz data are generally in excellent statistical agreement (Table 1), thus the inferred Q_{rms-PS} values are reasonably independent of data selection.
- Q_{rms-PS} derived from the $\otimes_{RMS}$ at 7° and 10° *including* the quadrupole is biased **low** by the small observed *COBE*-DMR sky quadrupole amplitude. This can be corrected for, after which the estimates for Q_{rms-PS} both including and excluding the quadrupole are in excellent agreement.
- the amplitude of primordial inhomogeneity in the context of a Harrison-Zel'dovich $n = 1$ power law spectral model is $Q_{rms-PS} \sim 19$ μK (in good agreement with values derived from other techniques — see Górski et al. 1994, Bennett et al. 1994, Wright et al. 1994b).

Table 1: Observed $\otimes_{RMS}$ values derived from possible 53 and 90 GHz combinations.

$\otimes_{RMS}$ combination	including Quadrupole		excluding Quadrupole	
	7° (μK)	10° (μK)	7° (μK)	10° (μK)
53A⊗53	$44.5^{+4.8}_{-4.7}$	$32.4^{+1.8}_{-1.8}$	$43.1^{+5.4}_{-5.2}$	$30.6^{+1.9}_{-1.9}$
90A⊗90B	$0.0^{+12.0}_{-10.9}$	$25.7^{+3.4}_{-3.4}$	$0.0^{+13.7}_{-13.4}$	$24.7^{+3.5}_{-3.5}$
53A⊗90A	$28.4^{+8.1}_{-7.7}$	$30.9^{+2.7}_{-2.9}$	$27.7^{+9.2}_{-8.6}$	$30.3^{+2.8}_{-2.8}$
53B⊗90B	$32.3^{+7.0}_{-6.7}$	$29.9^{+2.3}_{-2.4}$	$30.2^{+8.2}_{-7.6}$	$27.7^{+2.4}_{-2.5}$
53A⊗90B	$45.2^{+6.1}_{-5.9}$	$32.4^{+2.2}_{-2.2}$	$44.0^{+6.9}_{-6.6}$	$30.9^{+2.2}_{-2.3}$
53B⊗90A	$34.6^{+9.4}_{-8.7}$	$31.7^{+2.9}_{-3.0}$	$33.7^{+10.8}_{-10.0}$	$30.8^{+3.0}_{-3.0}$
53(A+B)⊗90A	$31.6^{+6.3}_{-6.0}$	$31.3^{+2.5}_{-2.6}$	$30.8^{+7.1}_{-6.6}$	$30.6^{+2.5}_{-2.5}$
53(A+B)⊗90B	$39.3^{+4.8}_{-4.7}$	$31.2^{+1.9}_{-2.0}$	$37.7^{+5.3}_{-5.1}$	$29.3^{+2.0}_{-2.0}$
53A⊗90(A+B)	$37.7^{+5.1}_{-4.9}$	$31.7^{+1.9}_{-1.9}$	$36.8^{+5.7}_{-5.5}$	$30.6^{+2.0}_{-2.0}$
53B⊗90(A+B)	$33.5^{+5.8}_{-5.7}$	$30.8^{+2.1}_{-2.1}$	$31.9^{+6.7}_{-6.3}$	$29.3^{+2.1}_{-2.1}$
53(A+B)⊗90(A+B)	$35.7^{+4.0}_{-3.9}$	$31.2^{+1.6}_{-1.7}$	$34.4^{+4.4}_{-4.3}$	$29.9^{+1.7}_{-1.7}$
(53+90)A⊗(53+90)B	$35.6^{+4.3}_{-4.2}$	$30.7^{+1.7}_{-1.7}$	$34.4^{+4.9}_{-4.6}$	$29.4^{+1.7}_{-1.7}$
(53A+90B)⊗(53B+90A)	$30.3^{+4.3}_{-4.3}$	$29.9^{+1.7}_{-1.7}$	$28.8^{+4.8}_{-4.6}$	$28.5^{+1.7}_{-1.7}$

MODEL NORMALIZATIONS

The normalization of other models of cosmological anisotropy has proceeded by a detailed reworking of the above. We report the inferred Q_{rms-PS} values for a number of flat geometry, critical density, scale-invariant CDM models.

An exact calculation for the power spectrum of CMB anisotropy in this context renders an effective spectral slope (in the sense of a power law approximation, $P(k) \propto k^{n_{eff}}$, used to generate the multipole coefficients a_ℓ^2 solely through the Sachs-Wolfe effect) slightly steeper than $n = 1$ over the angular scales probed by the *COBE*-DMR instrument ($\ell \lesssim 15$). A value in the range 1.1 - 1.15 would be most appropriate. The steepening is caused by three effects:
1) the anisotropy of the photon distribution function at the moment of decoupling.
2) the high-redshift integrated Sachs-Wolfe effect generated by the change in growth rate of matter perturbations at the post-recombination epoch due to the non-negligible contribution of photons to the total energy density.
3) the contribution of the first adiabatic/Doppler peak to the low ℓ multipoles.

Table 2 summarizes the normalization amplitudes inferred from the $\otimes_{RMS}$ values derived from the 53 and 90 GHz (A+B) sky maps, excluding the quadrupole. The results obtained when the quadrupole is included are in excellent agreement with these once the bias introduced by the low observed quadrupole is accounted for (see Banday et al. 1994). The steeper power law approximation fits render lower amplitudes than the exactly computed CDM power spectra, and should not be considered as sufficiently good descriptors of this class of cosmological model (see Górski, Stompor & Banday 1995). The Q_{rms-PS} normalization for CDM models is also $\sim 19\ \mu$K.

Table 2: Q_{rms-PS} normalizations for specific cosmological models.

Model	Q_{rms-PS} (μK)
power law, $n = 1$	$19.4^{+2.3}_{-2.1}$
power law, $n = 1.1$	$18.2^{+2.1}_{-1.8}$
power law, $n = 1.15$	$17.9^{+2.0}_{-1.8}$
CDM ($h = 0.5$, $\Omega_b = 0.01$)	$18.8^{+2.3}_{-1.9}$
CDM ($h = 0.5$, $\Omega_b = 0.03$)	$18.9^{+2.2}_{-1.9}$
CDM ($h = 0.5$, $\Omega_b = 0.05$)	$18.8^{+2.2}_{-1.9}$
CDM ($h = 0.5$, $\Omega_b = 0.10$)	$18.7^{+2.3}_{-1.9}$
CDM ($h = 0.8$, $\Omega_b = 0.03$)	$18.8^{+2.2}_{-2.0}$

Acknowledegments: We acknowledge the efforts of those contributing to the *COBE*-DMR. *COBE* is supported by the Office of Space Sciences of NASA headquarters. We thank Radek Stompor for providing us with CDM anisotropy power spectrum coefficients.

REFERENCES

Banday, A.J., et al. 1994, ApJ Letters, 436, L99
Bennett, C.L., et al. 1994, ApJ, 436, 423
Górski, K., et al. 1994, ApJ Letters, 430, L89
Górski, K., Stompor, R., & Banday, A.J. 1995 ApJ Letters, in preparation
Wright, E.L., et al. 1994a, ApJ, 420, 1
Wright, E.L., Smoot, G.F., Bennett, C.L., & Lubin, P.M.1994b, ApJ, 436, 443

Simulated CMB Maps at 0°.5 Resolution: Basic Results

G. Hinshaw, C.L. Bennett & A. Kogut
Code 685, Laboratory for Astronomy and Solar Physics
NASA/Goddard Space Flight Center, Greenbelt MD 20771
hinshaw@stars.gsfc.nasa.gov

ABSTRACT

We have simulated full-sky maps of the Cosmic Microwave Background (CMB) anisotropy expected from Cold Dark Matter (CDM) models at 0°.5 and 1°.0 angular resolution. Statistical properties of the maps are presented as a function of sky coverage, angular resolution, and instrument noise, and the implications of these results for observability of the Doppler peak are discussed. We find that sensitivity to the Doppler peak depends primarily on the fraction of sky covered, and only secondarily on the angular resolution and noise level. A sample map simulation is presented to illustrate the anisotropies.

1. INTRODUCTION

Observations of the cosmic microwave background anisotropy are fundamental to understanding the formation and evolution of structure in the universe. On angular scales smaller than a few degrees, the CMB anisotropy probes the physical scales and causal mechanisms responsible for observed current structure (see eg., Efstathiou, Bond, & White 1992; Coulson et al. 1994). Several groups have reported detections of anisotropy on degree angular scales (Schuster et al. 1993; Wollack et al. 1994; Cheng et al. 1994; De Bernardis et al. 1994; Dragovan et al. 1994; Devlin et al. 1994; Clapp et al. 1994) but the implications of these experiments for models of structure formation are unclear. Interpretation of observational results in a specific patch of sky are complicated by the fact that cosmological models of structure formation are typically stochastic. Thus observable measures of anisotropy such as *rms* fluctuations or broad-band power are expected to deviate from the universal average dictated by cosmology. This intrinsic scatter, termed "cosmic variance", will be more severe if only a small portion of the sky is observed. The additional variance introduced by limited sky coverage is sometimes called "sample variance" (Scott, Srednicki, & White 1994). Questions of importance both to observers planning future instruments and to theorists analyzing existing data include the following: what is the role of sample variance in statistical analyses of small regions of the sky? What fraction of the sky must be observed to constrain models at specified confidence? What is the relation between limiting sources of uncertainty: the signal-to-noise ratio per pixel, cosmic variance, sample variance, and sky coverage? Are large-area scans with relatively greater instrument noise per pixel better than small-area deep observations? How important is it to produce a map of the microwave sky, as opposed to observing chopped scans? Are the answers to these questions strongly model-dependent?

In this paper, we describe Monte Carlo simulations of CMB anisotropy maps in the context of CDM models at 0°.5 and 1°.0 angular resolution, which we use to quantify the power of simple statistical tests that can be applied to the maps. As an example, we construct and evaluate a likelihood function for the universal baryon density, Ω_b, and show that the sensitivity to Ω_b is determined primarily by the fraction of sky covered, and, to a lesser extent, by the angular resolution, and the instrument sensitivity.

2. MAP SIMULATIONS

To generate a particular realization of the CMB temperature observed at an instrumental dispersion σ_b we evaluate the sum over spherical harmonics, $\Delta T(\theta,\phi) = \sum_{\ell,m} a_{\ell m} Y_{\ell m}(\theta,\phi) e^{-\frac{1}{2}\ell(\ell+1)\sigma_b^2}$, where the amplitudes $a_{\ell m}$ ($2 \le \ell \le 500$) are Gaussian random variables of zero mean and variance C_ℓ, ($\langle a_{\ell m} a^*_{\ell' m'} \rangle = C_\ell \delta_{\ell\ell'} \delta_{mm'}$) appropriate to the model under consideration (Stompor 1994; Bond & Efstathiou 1987). In the present work we adopt a scale-invariant primordial spectrum of fluctuations ($n = 1$) with a power spectrum normalization, expressed in terms of the quadrupole, of $Q_{rms-PS} = 20$ μK (Górski et al. 1994) based on the two-year *COBE*-DMR data (Bennett et al. 1994). A familiar feature in these spectra is the existence of the so-called Doppler peak at multipole order $\ell \simeq 220$. For a given primordial Sachs-Wolfe spectrum and a given Hubble constant, the amplitude of the first Doppler maximum increases with the baryon density Ω_b (thus Ω_b provides a convenient means to parameterize the height of the Doppler peak, as is done in §3.) It should also be noted that the location of the first Doppler maximum, which probes the angular size of the horizon at the epoch of last scattering, depends primarily on Ω_0: $\ell_{max} \sim 200/\Omega_0^{1/2}$ (Kamionkowski, Spergel & Sugiyama 1994). We restrict attention, in this paper, to flat models with $\Omega_0 = 1$.

Figure 1: Simulated full-sky CMB map at 0°.5 resolution. The temperature range is ± 300 μK. The map was generated using a CDM spectrum with baryon density $\Omega_b = 0.05$ and nominal values for the other parameters (see text).

As an illustration of the anisotropies expected in a CDM model, Figure 1 shows a "standard" simulation ($H_0 = 50$ km/s/Mpc, $\Omega_b = 0.05$) at 0°.5 angular resolution. The *rms* fluctuations in this map are 88 μK while the peak-peak amplitude is roughly ± 350 μK. Additional maps made with different values of Ω_b are presented in Figure 2 of Hinshaw, Bennett, & Kogut (1995). The increase of power in the high ℓ modes with increasing Ω_b is clearly apparent in those maps.

3. LIKELIHOOD ANALYSIS

The covariance matrix $M_{ij} = \langle t_i t_j \rangle = \frac{1}{4\pi} \sum_\ell (2\ell + 1)\, B_\ell^2\, C_\ell\, P_l(\cos\alpha_{ij})$ fully characterizes the statistical properties of a map for Gaussian fluctuations. Here t_i is the temperature in pixel i of the map, the angled brackets denote a universal ensemble average, $B_\ell^2 = e^{-\ell(\ell+1)\sigma_b^2}$ is the beam factor and C_ℓ is the power spectrum. $P_l(\cos\alpha_{ij})$ is the Legendre polynomial of order ℓ, and α_{ij} is the angle between pixels i and j. Assuming a uniform prior distribution of model parameters, the likelihood of a model spectrum, C_ℓ, given a map t, is $L(C_\ell) \propto [\det M(C_\ell)]^{-1/2} e^{-\frac{1}{2}\vec{t}^T \cdot M^{-1}(C_\ell) \cdot \vec{t}}$.

In the present work we parameterize the power spectrum C_ℓ by the baryon density Ω_b, leaving all other parameters fixed at nominal values. We then examine the *width* of the likelihood $L(\Omega_b)$ as a function of sky coverage, angular resolution, and instrument noise per pixel to see which of these experimental factors are most important in maximizing sensitivity to the power spectrum. Figure 2 shows the likelihood functions obtained from single, selected patches of the sky for each resolution and patch size considered. The width of the likelihood function, which measures the sensitivity to Ω_b, is nearly independent of the angular resolution of the map (in the small range considered), but strongly sensitive to the fraction of sky observed. This point is reiterated in Table 1 which gives the statistics of the recovered maximum likelihood values for the ensemble of patches. In each case, the median of the 100 determinations is within 2σ of the input, $\Omega_b = 0.05$, while the standard deviation ranges from 0.12 in the large, high-resolution, low-noise patches, to 0.41 in the small, high-noise, low-resolution patches. This table also demonstrates that for moderately high signal-to-noise ratios, $\sigma \lesssim 40$ μK per pixel, the sensitivity to Ω_b is nearly independent of noise level. We interpret these results as follows: a precise determination of the height of the Doppler peak requires a large number of "typical" high spatial frequency ($\ell \sim 200$) peaks and troughs to determine a "typical" amplitude. Moreover, given sufficient instrument sensitivity, a 1° beam will not dilute these features to the point of non-detection.

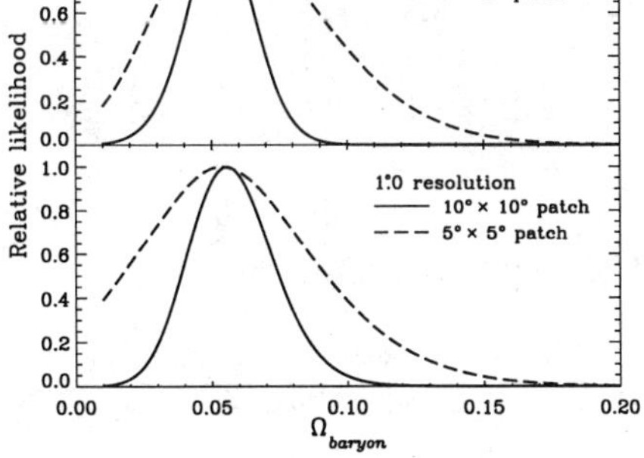

Figure 2: Likelihood functions for the baryon density Ω_b for selected sky patches. In each case the noise per pixel was 10 μK, and the input value of Ω_b was 0.05. Note that the width of the likelihood depends relatively strongly on the patch size, but not on the resolution.

Table 1 - Ω_b likelihood maxima from $\Omega_b = 5.0\%$ simulations (%)

Patch size	Noise per pixel (μK)			
	5	10	20	40
		0°.5 resolution		
5°.2 × 5°.2	5.1 ± 2.5	5.1 ± 2.7	5.3 ± 3.0	5.5 ± 3.2
10°.4 × 10°.4	4.8 ± 1.2	5.1 ± 1.4	5.1 ± 1.4	4.8 ± 1.7
		1°.0 resolution		
5°.2 × 5°.2	4.6 ± 2.7	4.7 ± 3.0	4.9 ± 3.7	4.2 ± 4.1
10°.4 × 10°.4	4.9 ± 1.5	4.8 ± 1.7	4.9 ± 1.8	4.6 ± 2.1

Of course the above analysis makes the optimistic assumption that all other cosmological parameters are known exactly and that there are no significant systematic effects (see, e.g., Wilkinson 1994) or foreground contaminants (see, e.g. Bennett et al. 1992). There is also considerable degeneracy amongst cosmological parameters with regard to the interpretation of the CMB anisotropy power spectrum in the sense that different parameter combinations can yield similar angular power spectra (Bond et al. 1994). The conclusions we draw about experimental sensitivities to Ω_b are subject to such considerations, however our conclusions regarding the experimental sensitivities to the CMB power spectrum itself are valid independent of theoretical models. Our general conclusion is that, given limited observing time, optimal sensitivity to the CMB power spectrum is obtained by covering as large a portion of the sky as possible consistent with obtaining a moderately high signal to noise ratio.

We are very grateful to Radek Stompor for providing us with CDM power spectra in advance of publication. We also acknowledge useful conversations with Tony Banday and Krys Górski.

REFERENCES

Bennett, C.L., et al. 1992, ApJ, 396, L7
Bennett, C.L., et al. 1994, ApJ, 436, 423
Bond, J.R., & Efstathiou, G. 1987, MNRAS, 226, 655
Bond, J.R., et al. 1994, Phys.Rev.Lett., 72, 13
Cheng, E.S., et al. 1994, ApJ, 422, L37
Clapp, A.C., et al. 1994, ApJ, 433, L57
Coulson, D., Ferreira, P., Graham, P., and Turok, N. 1994, Nature, 368, 27
De Bernardis, P., et al. 1994, ApJ, 422, L33
Devlin, M.J., et al. 1994, ApJ, 430, L1
Dragovan, M., et al. 1994, ApJ, 427, L67
Efstathiou, G., Bond, J.R., & White, S.D.M. 1992, MNRAS, 258, 1P
Górski, K., et al. 1994, ApJ, 430, L89
Hinshaw, G., Bennett, C.L., & Kogut, A. 1995, ApJ Letters, in press
Kamionkowski, M., Spergel, D.N., & Sugiyama, N. 1994, ApJ, 426, L57
Schuster, J., et al. 1993, ApJ, 412, L47
Scott, D., Srednicki, M., & White, M. 1994, ApJ, 421, L5
Stompor, R. 1994, A&A, 287, 693
Wilkinson, D. 1994, Lake Louise Winter Institute
Wollack, E.J., et al. 1994, ApJ, 419, L49

CMB ANISOTROPY AT 0°.5 ANGULAR SCALE II. ANALYSIS OF PEAKS

A. Kogut, G. Hinshaw, and C.L. Bennett
Code 685, NASA Goddard Space Flight Center, Greenbelt, MD 20771
kogut@stars.gsfc.nasa.gov

ABSTRACT

The recent MSAM detection of unresolved features in the CMB at 0°.5 angular resolution (Cheng et al. 1994) has been interpreted by some authors as possible evidence for a new population of mm-bright objects or as a signature of non-Gaussian processes in the CMB. We have simulated CMB maps using *COBE*-normalized cold dark matter (CDM) models and searched the resulting maps for features at 0°.5 to 1°.5 resolution. Contrary to expectation, the detection of isolated unresolved features ("point sources") is common in CDM models. We derive the number density of fitted peaks and compare the results to recent observations. The MSAM "sources" are in agreement with standard CDM ($n = 1$, $h = 0.5$, $\Omega_b = 0.05$) and are not necessarily evidence for processes beyond the standard model.

1. INTRODUCTION

Measurements of the angular distribution of the cosmic microwave background radiation are an effective probe of conditions in the early universe. Several groups have reported detections of anisotropy on degree angular scales: ACME (Schuster et al. 1993), ARGO (De Bernardis et al. 1994), Big Plate (Wollack et al. 1994), IAB (Piccirillo and Calisse 1993), MAX4 (Devlin et al. 1994), MSAM (Cheng et al. 1994), and Python (Dragovan et al. 1994). Particular interest has been generated by the MSAM collaboration's detection of bright unresolved features consistent with a CMB spectrum. Are bright unresolved sources evidence of foreground contamination, instrumental artifacts, new cosmological information, or are they an expected feature in the standard scale-invariant cold dark matter model? In this paper we address the question of unresolved features in the CMB, and attempt to quantify the number density per steradian of such sources as a function of intensity and cosmological model.

2. ANALYSIS

We simulate a series of fully-sampled CMB maps at resolution 0°.5 to 1°.5 FWHM using power spectra for CDM models normalized by the *COBE* detection of of anisotropy on large angular scales. The maps at 0°.5 consist of 4096 equal-area pixels within a 10.4° × 10.4° patch on the sky, while the maps at 1°.0 and

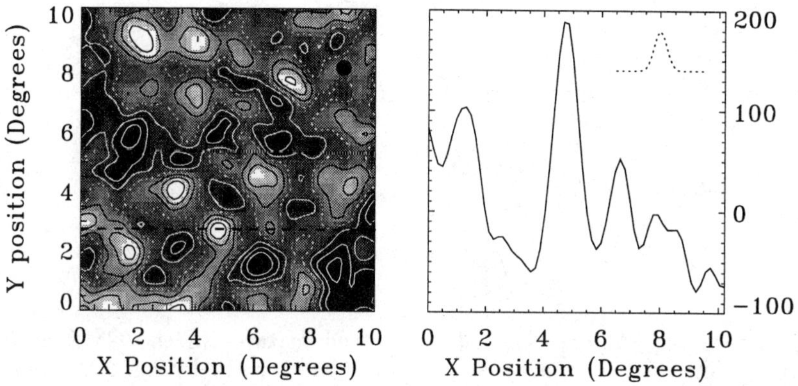

Figure 1: (left) simulated CMB anisotropy maps for standard CDM model (n=1, $h = 0.5$, Ω_b=0.05, $Q_{rms-PS} = 20\ \mu$K). Contours are -150μK to +150 μK in steps of 50 μK. The dashed line is the zero contour. The 0°.5 beam width is indicated at the upper right. (right) Temperature profile in one-dimensional slice indicated by the dotted line in the map. The beam shape is indicated at upper right.

0°.5 use a 20.8° × 20.8° patch. For details see the companion paper by Bennett et al. in this volume.

We fit each map to a set of unresolved sources by searching the map for the brightest pixel, calculating the mean of the 80 pixels in a 9 × 9 array surrounding that pixel, and fitting the mean-subtracted patch to a Gaussian profile with the same instrumental beam dispersion (*e.g.* 0°.5 FWHM) used to generate the map. The fitting algorithm searches the pixels in order of decreasing temperature, removing each fitted profile from the map before proceeding to the next pixel. We add no instrument noise to the maps, but estimate the significance of each fitted amplitude using the standard deviation of the 80 pixels surrounding each candidate source. Further analysis is restricted to those pixels in which the fitted amplitude is more than 3 times this local "confusion noise". Note that the Gaussian width is fixed: spatial structure larger than the beam will be decomposed into several neighboring "sources". Extended structure significantly larger than the 9 × 9 pixel patch will be removed as the fitted mean and will not affect the source decomposition. This closely parallels the beam-switching techniques used by current experimental groups, which also remove large-scale power: the 1°.5–3°.6 patch fitted to each source closely approximates the typical scale of beam throws used by observers.

Figure 1 shows a contour plot of a typical realization for spatially flat cold dark matter (CDM) models, with baryon density $\Omega_b = 0.05$, Hubble constant $H_0 = 50$ km s^{-1} Mpc^{-1} and *COBE* power-spectrum normalization $Q_{rms-PS} = 20\ \mu$K and $n = 1$ (Górski et al. 1994). Approximately 14% of the map area is covered by features brighter than 5 times the local "confusion noise". A one-dimensional slice through a bright candidate feature is also shown. The chopping strategies employed by medium-scale experiments act as a high-pass filter on the sky; consequently, these features would be detected as unresolved CMB sources by an experiment with sufficiently low instrumental noise. Enough of the map area lies within one beam width of a bright feature for their detection to be statistically not unlikely.

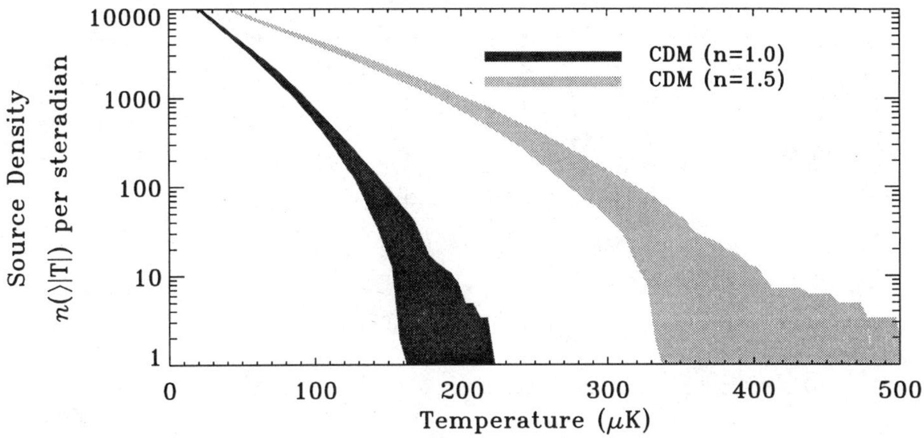

Figure 2: Integral source density per steradian $n(>|T|)$ for fitted source amplitudes brighter than threshold $|T|$ observed at instrumental resolution 0°.5 FWHM. The gray bands represent the 68% confidence interval from 100 noiseless realizations of $n = 1$ and $n = 1.5$ models normalized to match the *COBE* data at each value of n.

Figure 2 shows the integral number density per steradian $n(>|T|)$ of fitted amplitudes brighter than threshold $|T|$ for CDM models with n=1.0 and 1.5. For n=1, an experiment covering 0.01 sr of the sky would expect to observe on average one feature brighter than 135 μK, with less than 5% probability to observe a source brighter than 175 μK. An experiment with smaller sky coverage would observe correspondingly fewer features: at 10^{-3} sr, a typical experiment would observe one source brighter than 85 μK, with 5% probability to observe a source brighter than 150 μK. The source densities depend somewhat on the cosmological model: changing the index n from 1 to 1.5 increases the fitted amplitudes by a factor 2.2, while changing the baryon density Ω_b from 0.05 to 0.2 increases the amplitudes by 1.5.

The source curves in Figure 2 may be used to compare observations to theory. We define the brightest expected source in any experiment, $T_{\rm N=1}$, as the amplitude T for which $n(>|T|) = 1/\Omega$ so that an experiment observing a solid angle Ω would observe, on average, one source as bright or brighter. Figure 3 compares the brightest source in recent medium-scale observations to the brightest source $T_{\rm N=1}$ in the standard CDM model. Of these experiments, only the MSAM collaboration reports a detection of an unresolved feature consistent with a CMB spectrum. For the other experiments, Figure 3 shows either the brightest unresolved source regardless of its fitted spectrum (ACME) or the brightest pixel convolved with the point-source response function of the chop strategy employed by that instrument. Selecting the brightest pixel in place of the brightest fitted feature for groups not reporting source detections is clearly only an approximation, but our simulations indicate it is reasonable.

Three of the experiments (ACME, Big Plate, and MSAM) observe a brightest source close to the value expected for the standard $n = 1$ CDM model.

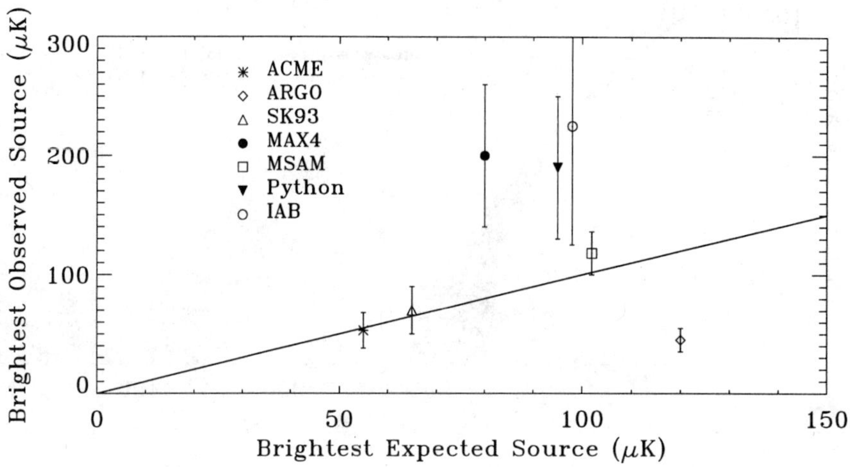

Figure 3: Comparison of the brightest observed feature in recent degree-scale experiments with the brightest feature expected given the angular resolution and sky coverage. The solid line is the prediction for *COBE*-normalized CDM.

ARGO observes a paucity of bright sources given the relatively large solid angle, while MAX4 and Python observe peak amplitudes brighter then expected, closer to the $n = 1.5$ model than $n = 1$ if the *COBE* normalization is retained. Similar results arise from analysis of the power spectra (e.g. Wright et al. 1994), which utilize the full data sets instead of the local "brightest pixel" test used here. Details of the chop strategy and scan pattern are not expected to significantly modify these values, provided that the beam throw does not significantly exceed the 3°.6 scale of the largest patch fitted in the simulations. Larger beam throws will increase the expected brightest source.

We conclude that the detection of isolated unresolved features in the CMB is not uncommon for standard cosmological models and is not necessarily evidence for non-Gaussian statistics or processes beyond the standard model. While observers are understandably cautious about isolated bright features in their data, these features are expected in standard models and should be included in statistical analyses.

REFERENCES

Cheng, E.S., ct al. 1994, ApJ, 422, L37
De Bernardis, P., et al. 1994, ApJ, 422, L33
Devlin, M.J. et al. 1994, ApJ Letters, 430, L1
Dragovan, M., et al. 1994, ApJ, 427, L67
Górski, K., et al. 1994, ApJ Letters, 430, L89
Piccirillo, L, & Calisse P., 1993, APJ, 411, 529
Schuster, J, et al., 1993, ApJ, 412, L47
Wollack, E.J., et al. 1994, ApJ, 419, L49
Wright, E.L., Smoot, G.F., Bennett, C.L., & Lubin, P.M.1994, ApJ, 436, 443

ALTERNATIVE APPROACHES TO THE DARK MATTER PROBLEM

ALTERNATIVES TO DARK MATTER

Demosthenes Kazanas
NASA Goddard Space Flight Center, Code 665
Greenbelt, MD 20771.
kazanas@lheavx.gsfc.nasa.gov

ABSTRACT

The deviation of the observed accelerations in extragalactic systems (galaxies and clusters of galaxies) has been interpreted, by a minority of researchers, as signal of the breakdown of Newtonian (and for that matter Einstein) gravity at these distance scales rather than the presence of large amounts of "dark matter". A number of such proposals is reviewed with particular emphasis on their predictions on the rotation curves of spiral galaxies, the most unequivocal measure of such deviations.

1. INTRODUCTION

1.1 General Background

The fact that the observed accelerations in extragalactic systems, and in particular in clusters of galaxies, are far greater than implied by the amount and distribution of their luminous mass (Zwicky 1933) has been with us for over 60 years. The advent of observation has further indicated that a problem of similar nature is encountered also in galaxies both spirals and ellipticals, (Rubin et al 1978; Bosma 1981), while at larger scales, if current prejudices that the Universe is at its critical density $\Omega = 1$, are correct, its global dynamics must also be dominated by matter which is non-luminous and moreover non-baryonic. All these issues are in fact discussed extensively in the present volume. On the other hand, there is a small minority of researchers in the field who believe that these observational conundrums are nothing more than the indication that some of our fundamental laws (most notably the law of gravity) fail on galactic or larger scales. There have therefore been a number of proposals motivated by these observations, most of them explicitly constructed so that they resolve the issue of the missing light problem at some or all scales. The present review will discuss a number of the most popular of them, outline their basic premises and examine their success in dealing with the problem of galactic dynamics and in particular the problem of the "flat" galactic rotation curves, the issue which is most unequivocally amenable to modeling and comparison with observation. This review draws heavily of the much more extended and complete one by Bekenstein (1987) which the interested reader is strongly advised to consult.

1.2 Measurement of accelerations - Phenomenology of spirals

Unlike the velocity (or at least its component along the line of sight) which is directly measurable, the dynamical measurement of mass involves the measurement of acceleration; the measurement of the latter, however, is fraught with uncertainties which make the eventual estimates rather model dependent. The chain of arguments for the dynamical estimate of ρ goes as follows: *(i.)* A gravitational theory relates the distribution of matter (its stress-energy tensor) to the gravitational field (this reduces Poisson's equation in the non-relativistic limit, i.e $\nabla^2\Phi \sim G\rho$). *(ii.)* The conservation of the matter stress-energy tensor yields the geodesic equation which relates the acceleration a with the gradient of the gravitational potential yielding Newton's second law (*i.e.* $a = \nabla\Phi$). *(iii.)* The assumption of (virial) equilibrium provides a relation between the acceleration and the velocity, $\langle v^2\rangle \simeq 2a\langle R\rangle$, where $\langle R\rangle$, $\langle v\rangle$ are the average dimension and velocity of the system. Measurement of $\langle v^2\rangle$ and $\langle R\rangle$ can then provide estimates for a and hence ρ, which can then be directly compared to the distribution and amount of light in the system.

The above discussion makes clear that the more involved and complicated the system under consideration and its dynamics, the more equivocal the comparison of the estimated to the observed values of ρ and hence their potential discrepancy will be. As such, the rotation curves of spiral galaxies, having the least observational uncertainty (the largest component of the velocity is by far the circular component v_c; the dynamics are dominated by the rotation and the geometry is rather well defined allowing the precise calculation of the acceleration) are the laboratory at which these theories can be tested with the highest degree of confidence. This class of galaxies appears to consist of flattened configurations, of transverse dimension roughly ten times smaller than their size, supported by rotation, with surface density which can be fit well by an exponential, $\Sigma = \Sigma_0 exp(-r/r_d)$, where r_d is the characteristic length of the light brightness distribution on the galactic disk. The mass of the luminous matter can then be easily calculated, $M = 2\pi\Sigma_0 r_d^2$, and the total matter content can be estimated by assuming a constant mass to light ratio. Of particular interest is the fact that the value of the central brightness Σ_0 is observationally found to be constant independent of the other attributes of the galaxy.

The assumption of equilibrium between the Newtonian force of gravity and the centrifugal force in the disk leads to $v^2/R \sim G\Sigma_0$ (R is a length scale of the order of the characteristic disk length r_d) or, eliminating R in favor of M, $v^4 \sim G^2\Sigma_0 M$, a relation which (for constant Σ_0 as indicated by the observations) can be recognized as the empirically derived Tully–Fisher (hereafter TF) relation for galaxies. It is generally considered that theories which account for the dynamics of spiral galaxies should be able to account for this empirical relation. One can see however that in as much as the dynamics of spirals are dominated at small radii by the Newtonian forces due to the observed matter and which do establish the magnitude of v as well as M, an account of the TF relation might simply involve physics not necessarily related to the deviation of the galactic rotation curves from their Newtonian form (in the above example where the Newtonian force is dominant, this relation simply reflects the unknown todate reason which forces Σ_0 to be constant).

The remaining of this review will therefore deal only with this particular aspect of the missing light problem, having in mind that these theories will have eventually to satisfy the constraints imposed by the other astrophysical systems (such as clusters of galaxies) whose dynamics imply the presence of dark matter.

1.3 Ways Out

The discrepancy between the accelerations estmated from the dynamics and the observed distribution of light in spiral galaxies could be reconciled by introducing modification in the chain of arguments given above by either: *(1.)* Changing the matter stress-energy tensor, i.e. introducing matter component(s) which do not radiate, thereby resetting the normalization between the estimated and computed density ρ. It is important to point out that since the equations are sensitive to the density rather than the total mass, one should prescribe not just a number but a function in order to reconcile the disagreement between observed and the estimated dynamics. *(2.)* Changing the gravitational tensor, i.e. the relation between the matter distribution and the resulting gravitational potential providing additional potential for the given matter density. *(3.)* Changing the relation between the acceleration a and $\nabla\Phi$, i.e. the form of Newton's second law (Milgrom 1983a,b,c). We discuss these possibilities in more detail:

(1.) It is generally thought that this is the simplest way out of this puzzle. Unfortunately, the nature of the dark matter is to this day unknown, but recent observations (see contributions by C. Alcock and P.D. Mannheim) indicate that it is unlikely to be of stellar origin (interstellar bricks are still allowed but they are very hard to make). Furthermore, its distribution is also unknown; what is certain is that it has to be different from that of the luminous matter, otherwise the dynamics of galaxies would be wrong. It is generally assumed that the dark matter is distributed in the haloes of normal spirals and extends to much larger distances than the optical light distribution. It is usually modeled as an isothermal sphere whose density at large distance is $\rho(r) \propto r^{-2}$. This distribution naturally produces asymptotically flat rotation curves for disk galaxies and for this reason it is considered as the distribution of choice. However, it should be born in mind that the flat rotation curves observed obtain not because the dynamics are dominated by this halo matter component, but because of a fine interplay between the contributions of the disk and halo components which conspire to produce a flat rotation curve (disk-halo conspiracy; Bachall & Casertano 1985). The extent of the halo component to distances much larger than the optical size of galaxies implies a non-dissipative form for the halo matter, making weakly interacting massive particles (undiscovered as yet) as the prime such candidate (see contribution by F. Wilcek). In the long run, the acceptance of this solution rests with the discovery of the necessary massive particle candidates in the laboratory.

(2.) Changing the gravitational field equations has been in general viewed unfavorably as a way of fixing the dark matter problem, in that it appears as ad hoc as populating a galactic halo with an unknown massive particle. There is also an additional general reluctance in considering alternatives to Einstein gravity. This reluctance stems, to a large extent, from the success of Einstein gravity in generalizing the Newtonian one and addressing successfully certain observational issues in the solar system and also the the decay of the orbit of the binary pulsar. However, it should be born in mind that: a. From the observational point of view, its validity has been tested only on scales of the order of the solar system, while the potential failures appear at scales 10^9 times larger. b. On the theoretical side, the fundamental principle instrumental in the development of General Relativity, namely general covariance, does not suffice to single Einstein gravity out of all theories compatible with it, hinting that the specific form of the field equations may not be as compelling as generally thought. Besides the above, there are several additional issues which the currently accepted theory of gravity has problems in addressing: When coupled to the Higgs field (introduced to account for the masses of elementary particles through spontaneous symmetry breaking) leads to a huge cosmological constant, in gross violation with observations, a situation quietly ignored, simply because its effects are too drastic to be dealt with. Einstein gravity (and all second order theories in general) is not renormalizable within the conventional schemes,

because of its dimensionful coupling constant.

(3.) This class of theories are indeed pathological, in the sense that they violate the stress-energy conservation law since they are not derivable from a Lagrangian. Bekenstein (1987) questions whether the epistemological price to be paid by agreeing to theories of the sort is worth the benefits of doing away with the issue of dark matter. He then proceeds to produce (Bekenstein & Milgrom 1984; Bekenstein 1993) a theory of type *(2.)* above which does not suffer from these problems and which reduces in the non-relativistic regime to a relation similar to that put forward by Milgrom (1983a,b,c).

In the remaining of this review the fundamentals of a number of such theories will be presented as well as their confrontation with the data of the rotation curves of spiral galaxies.

2. ALTERNATIVE GRAVITATIONAL THEORIES

2.1 Logarithmic Potentials

The simplest modification to the Newtonian potential consistent with the flatness of the galactic rotation curves at large distances is the addition of a logarithmic term which becomes important at galactic scales (Tohline 1982; Kuhn & Kruglyak 1987)

$$\Phi = -GM[r^{-1} - \lambda^{-1} ln(r/\lambda)]. \tag{1}$$

Asymptotically, the gradient of the above expression, $GM/r\lambda$, has to match the centrifugal acceleration v_c^2/r, where circular motion has been assumed and v_c is the magnitude of the "flat" galactic rotation curve velocity. The matching can be achieved by the appropriate choice for λ, which on dimensional grounds is expected to be of the order of kpc. What makes this potential additionally attractive is the fact that this form of the gravitational potential makes thin galactic disks stable and therefore the Ostriker-Peebles haloes unecessary (Christodoulou 1991). In general the fits to the galactic rotation curves are good. Fig. 1 gives the fit to the rotation curve of the spiral *NGC 3198*, the galaxy most studied in relation to the issue of flat galactic rotation curves because of the high quality of the data and because it is the galaxy for which the rotation curve data extend to the largest number of disk radii. Fits for other galaxies of similar mass are similarly good but it appears that this scheme fails for systems of larger mass for which a different value for the constant λ is needed. Bekenstein (1987) suggests that this discrepancy is related to the fact that the asymptotic form of the force law, when balanced by the centrifugal force, yields $M \propto v_c^2$ rather than the TF scaling which is observed in nature.

2.2 Finite Length Scale Antigravity (FLAG)

The basic idea for this scheme (Sanders 1984; 1986a) is found in supergravity theories according to which part of the gravitational interaction between bodies is carried by a massive vector particle. This leads to a gravitational potential enlarged by an additional Yukawa-type term

$$\Phi = \frac{G_\infty M}{r}\left[1 + ae^{-r/r_0}\right] \tag{2}$$

where a and r_0 are two constants which give the strength and the range of the part of gravity which is mediated by the massive vector field. Sanders chooses $a \simeq -0.9$ and $r_0 \simeq 36$ kpc so that the resulting potential can fit the observed rotation curves. In fact Sanders obtains the values for these constants by fitting the data of *NGC 3198* and uses these values to "predict" the rotation curves of additional galaxies from the observed distribution of light in the disk. The fits obtained are indeed good but they cover only a limited range of galactic masses. The very small and very large galaxies are difficult to fit with the same values of the parameters. Bekenstein (1987) argues that this discrepancy may have its origin in the linearity between the potential and the mass and that this theory, as that of the Tohline - Kuhn logarithmic potentials, leads to $M \propto v^2$ rather than the TF relation. As argued above, the latter relation obtains if one assumes that the centrifugal acceleration is balanced by the non-Newtonian law. If on the other hand the Newtonian and non-Newtonian terms are found to be comparable, the TF relation is reduced to the constant value of Σ_0.

2.3 Modified Newtonian Dynamics (MOND)

The empirical TF relation plays an important role in constraining the theories of galactic dynamics. It has been argued that it can be obtained only by a theory in which the potential - mass relation is non-linear (Bekenstein 1987). Milgrom (1983a,b,c) has indeed conjectured a modification of Newton's second law which does have this non-linear behavior; he has suggested that this law be replaced by

$$\boldsymbol{g} = \boldsymbol{a}\mu(|\boldsymbol{a}|/a_o) = \boldsymbol{a}\mu(x) \tag{3}$$

where μ is a function of the argument $x = |\boldsymbol{a}|/a_0$ and a_0 is a universal constant. Milgrom has given only the asymptotic values of the function $\mu(x)$

$$\mu(x) \rightarrow \begin{cases} 1 & x \gg 1 \\ x & x \ll 1 \end{cases} \tag{4}$$

thus for accelerations larger than a_0 the standard Newtonian law is recovered while for smaller accelerations one obtains

$$a^2/a_0 = g \quad \text{or} \quad v_c^4 = GMa_o\ . \tag{5}$$

Interestingly, it is found that the value of the arbitrary constant a_0 numerically need be of order $cH_o \simeq 10^{-8}$ cm s^{-2} in order to yield deviations from the Newtonian dynamics at the scales of galaxies as observed. The theory can easily account for the TF relation, yielding for the circular velocities in the asymptotic ($x \ll 1$) regime $v_c^4 \simeq GMa_0$ while at the same time producing an asymptotically flat rotation curve.

It was pointed out early-on (Felten 1984) that in its original form, this theory, not being derivable from a lagrangian, did violate the energy and momentum conservation laws. The situation has been remedied by Bekenstein & Milgrom (1984) and Bekenstein (1993) with the introduction of a theory of type *(2.)* above, which modifies instead the form of the gravitational tensor. This theory introduces a scalar field to the gravitational action and it is akin to the Brans–Dicke scalar tensor theories, with the difference that the coupling of the scalar field to gravity is not a constant but it depends on the gradient of the scalar field (see Sanders 1986b). The acceleration is related in a non-linear fashion to the gradient of the Newtonian potential $\nabla\Phi_N$

$$a = g = -\nu(|\nabla\Phi_N|/a_o)\nabla\Phi_N \tag{6}$$

$$\nu(x) \to \begin{cases} 1 & x \gg 1 \\ x^{-1/2} & x \ll 1 \end{cases}$$

This theory (called AQUAL because it involves a Lagrangian not quadratic in the scalar field gradient) gives the same phenomenology as *MOND* and does not suffer from its deep problems. Though the original version had a problem with the superluminal propagation of scalar waves, a more recent version (phase coupled gravity) seems to have overcome this problem. Despite the fact that a deeper justification for the choice of such a theory is currently lacking, it is of interest because of the ease with which it confronts the observations.

2.4 Conformal Gravity

The alternatives discussed above were all introduced with the explicit intent to address and (hopefully) resolve the issue of dynamics of galaxies or other extragalactic structures. They all have a common approach in that they aim in guessing the (presumably) correct gravitational theory, justifying the guess by its ability to account for the existing phenomenology.

In contrast to the above alternatives to the Einstein – Newton gravity stands the approach of Mannheim and Kazanas (Mannheim & Kazanas 1989; Mannheim 1990; Kazanas & Mannheim 1991; Mannheim & Kazanas 1991; Mannheim 1992; 1993a,b; Mannheim & Kazanas 1994) who set out to study, at the classical level, a conformally invariant theory of gravity, without the intent of addressing (much less resolving) from the outset any of the issues discussed herein. Rather, their study was motivated by the fact that (Einstein) gravity is alone, among the modern theories of fundamental interactions, with a dimensionful coupling constant in its action (a property which, incidentally, is directly responsible for the non-renormalizability of the quantized theory) and the fact that it is also the only theory of a fundamental interaction whose action is not fixed by a local invariance principle but rather by the fact that the resulting equations be no higher than second order. (Parenthetically it is pointed out that the actions of the other fundamental interactions have dimensionless coupling constants and derive all their scales (masses) from dynamics, i.e. from spontaneous symmetry breaking; moreover, since Lorentz invariance alone, the flat space analog of general covariance, does not suffice to uniquely fix the form of the fundamental actions, these theories then appeal to ***principles of local invariance***, the so-called gauge invariances, to precisely restrict the available freedom and uniquely determine the corresponding actions).

Motivated by the above considerations, Mannheim and Kazanas (see above references) have considered the analogous principle of local scale (conformal) invariance to restrict the form of the gravitational action, thus abandoning Einstein's demand of second order equations, while still retaining his principle of general covariance. This additional principle, which demands that the laws of physics be invariant under any and all local stretchings of the geometry $g_{\mu\nu}(x) \to \Omega^2(x)g_{\mu\nu}(x)$ leads indeed to a unique action for the gravitational field which has the form

$$I_W = -\alpha \int d^4x(-g)^{1/2}C_{\lambda\mu\nu\kappa}C^{\lambda\mu\nu\kappa} \tag{7}$$

where $C_{\lambda\mu\nu\kappa}$ is the conformal or Weyl tensor and α is a universal dimensionless coupling constant. The theory derived from the above action will hereafter be referred to as *Conformal* or *Weyl Gravity.*

The equations of motion for the gravitational field are derived from the action of eq. (5) by the standard methods of functional variation with respect to $g_{\mu\nu}$, and lead to equations of the form $4\alpha W_{\mu\nu} = T_{\mu\nu}$ where $W_{\mu\nu}$ is to replace the familiar Einstein tensor and $T_{\mu\nu}$ is the stress energy tensor. These equations are the analog of the Einstein equations in Weyl gravity. In general they are fourth order, non-linear equations for the metric coefficients $g_{\mu\nu}$. They consequently do not reduce to those of Einstein in any obvious limit and hence this theory does not trivially reduce to that of Einstein gravity, which immediately raises the question of its viability. However, inspection of these equations indicates that they involve powers of the Ricci tensor $R_{\mu\nu}$ and its second derivatives. Noting that the vacuum ($T_{\mu\nu} = 0$) Einstein equations reduce to $R_{\mu\nu} = 0$, any vacuum solution of Einstein gravity would then also be a vacuum solution to the equations of Weyl gravity, thus directly and immediately guaranteeing that the theory would pass the classic static tests of General Relativity. On the other hand, the vacuum equations of Weyl gravity, $W_{\mu\nu} = 0$, may be realized in ways other than by simply having the Ricci tensor vanish, thus leading to solutions of Weyl gravity more general than those of Einstein gravity (even while being compatible with them).

Conformal invariance and the tracelessness of $C_{\lambda\mu\nu\kappa}$ imply that $W_{\mu\nu}$ and hence $T_{\mu\nu}$ are also traceless, thus obviating the cosmological constant problem which plagues the Einstein theory as discussed in section 1.3 above. In addition, the condition $T^{\mu}{}_{\mu} = 0$ places severe constraints on the structure of gravitational sources which go contrary to our cherished notions and which will be discussed shortly.

Since a lot of insight is usually gained by examining the behavior of exact highly symmetric solutions to the equations of motion, Mannheim & Kazanas (1989) have studied the static, spherically symmetric geometry within Weyl gravity, the geometry given by the line element

$$ds^2 = B(r)dt^2 - A(r)dr^2 - r^2 d\Omega$$

with $B(r) = 1/A(r)$ (conformal invariance always allows for the latter relation). The exact, comlete solution to the 4th order equations is given by

$$-g_{00} = 1/g_{rr} = B(r) = 1 - \beta(2 - 3\beta\gamma)/r - 3\beta\gamma + \gamma r - kr^2 \tag{8}$$

where β, γ, and k are three appropriate dimensionful integration constants. It is apparent that the above expression reduces to the Schwarzschild solution when $\gamma = 0$, $k = 0$, in agreement with the arguments given above, while in the case that γ, $k \neq 0$ the solution differs markedly from that of the Scharzschild at large distances thus offering the possibility of addressing the issue of dynamics of extragalactic systems. However, addressing this issue requires the solution to the non-vacuum problem of the theory since the potential has to be coupled to the observed matter distribution. To this we turn right away.

Though the field equations are rather complicated 4th order (in general) differential equations for the function $B(r)$, the combination $(W^0{}_0 - W^r{}_r)/B$ of the components of $W_{\mu\nu}$, has a particularly simple form

$$\frac{3(W^0_{\ 0} - W^r_{\ r})}{B} = \nabla^4 B(r) = f(r) = \frac{3(T^0_{\ 0} - T^r_{\ r})}{4\alpha B} \tag{9}$$

i.e. it reduces *exactly* (rather than approximately) to a fourth order (rather than second) Poisson's equation. Considering that the Green's function of the operator ∇^4 is $-|\mathbf{r} - \mathbf{r}'|/4\pi$ the interior solution can therefore be obtained in terms of integrals over the source function $f(r)$ (Mannheim & Kazanas 1994)

$$B(r > R) = -\frac{1}{6} \int_0^R dr' f(r')[3r'^2 r + r'^4/r]\,. \tag{10}$$

In this solution one can identify the coeficients of the Newtonian and linear terms as moments of the interion mass distribution (the other terms can in fact be identified as integrals over the exterior matter distribution)

$$\beta(2 - 3\beta\gamma) = \frac{1}{6} \int_0^R dr' f(r') r'^4 \quad , \quad \gamma = -\frac{1}{2} \int_0^R dr' f(r') r'^2\,. \tag{11}$$

One should note the rather unexpected outcome, that even though the Green's function of the 4th order theory is linear in the distance r, after integration over a finite size source we recover not only the linear term r but also the Newtonian $1/r$ term even though the Laplacian operator ∇^2 is not present in our equations. Thus, while the second order Poisson equation is sufficient to generate a $1/r$ potential is not in fact necessary. The Newtonian piece is effectively what is left over after the leading contribution of the $|\mathbf{r} - \mathbf{r}'|$ Green's function of the ∇^4 operator is extracted out. It should therefore come as not too much of a surprise that the Newtonian term is itimately related to the internal structure of gravitational sources. The fact that the Newtonian potential is explicitly associated with the short distance behavior of the theory stands in sharp constrast to the type of behavior generally sought in 4th order theories, wherein it is attempted to reduce the 4th order equations to those of Einstein at large enough distances thereby obtaining Newton's law as a long distance aspect of the theory.

Before we turn to the discussion of galactic dynamics, one should not that the source function $f(r)$ above should not be considered as the uniform energy density distribution within a macroscopic body (this would lead to contradiction with observation according to eq. (11)) but rather as the distribution of energy density within an elementary gravitational source, which is this theory simply cannot be considered as a "billiard ball" *i.e.* pointlike and structureless. A model such form for $f(r)$ is given in Mannheim & Kazanas (1994) and involves both δ-functions and its derivatives. The potential of a macroscopic body should then be considered as the result of the sum of exterior potentials of such elementary sources, as in fact one implicitly does in the standard second order theories anyway, with the difference that in the latter case one can replace the sum of singularities with an averaged-out continuous density, a procedure not possible in the 4th order theory which probes deeper into the structure of these sources.

Armed with the notion that each star (or each proton) in a galaxy puts out a potential of the form $V(r) = -\beta/r + \gamma r/2$ one may integrate this over an exponential thin disk distribution of form $\Sigma = \Sigma_0 exp(-r/r_d)$. For this idealized distribution it can be shown that the circular rotational velocity has the form (Mannheim 1993a)

$$v^2(r) = \frac{N\beta r^2}{2r_d^3}\left[I_0(x)K_0(x) + \left(\frac{\gamma r_d^2}{\beta} - 1\right) I_1(x)K_1(x)\right] \quad (12)$$

where $x = r/2r_d$. The form of this function fit to the data of *NGC 3198* is shown in fig. 4 assuming $r_d = 2.72$ kpc and an effective value for $1/\gamma \simeq 3 \times 10^{29}$ cm, a value of the order of the Hubble radius. As seen in this figure, the flatness of the rotation curve is indeed achieved from the interplay between the linear and the Newtonian terms, but that involves no conspiracy in that they both are calculated using the observed light distribution and assuming a reasonable mass to light ratio. In this fit the value of γ is indeed allowed to be fixed by the data, however, the fits of other galaxies using the value obtained from the *NGC 3198* fit is also quite good (Mannheim 1993a).

Since the flatness of the rotation curve is obtained from the interplay of the linear and the Newtonian terms, while the magnitude of the velocity at small distances is indeed determined by the Newtonian one, the TF relation follows simply from the observational fact that the value of Σ_0 is constant and its eventual explanation has to be sought there. Considering the lack of freedom available, conformal gravity fares quite well in addressing the dynamics of spiral galaxies. Founded on very solid theoretical basis, it is in this respect ahead of the competitors, however the relevance or not of the theory rests entirely with its ability to address the observations not only of galaxies but also of their clusters as well as cosmological issues.

I would like to thank Philip Mannheim for numerous discussions, comments and suggestions. œ

REFERENCES

Bahcall J. N. & Cassertano, S. 1985, ApJL, 293, L7
Begeman, K. G., Broeils, A. H. & Sanders, R. H. 1991, MNRAS, 249, 523
Bekenstein, J. D. 1987, in Proc. 2nd Canadian Conf. on General Relativity and Rel. Astrop. ed. A. Coley, C. Dyer & T. Tupper (Singapore: World Scientific), 68
Bekenstein, J. D. 1993, Phys. Rev. D, 48 3641
Bekenstein, J. D. & Milgrom M. 1984, ApJ, 286, 7
Bosma, A. 1981, AJ, 86, 1791
Christodoulou, D. M. 1991, ApJ, 372, 471
Felten, J. E. 1984, ApJ, 286, 7
Kazanas, D. & Mannheim, P. D. 1991, ApJS, 76, 431
Kuhn, J.R. & Kruglyak, L. 1987, ApJ, 313, 1
Mannheim, P. D. 1990, Gen. Rel. Grav. 22, 289
Mannheim, P. D. 1992, ApJ, 391, 429
Mannheim, P. D. 1993a, ApJ, 419, 150
Mannheim, P. D. 1993b, Gen. Rel. Grav., 25, 697
Mannheim, P. D. & Kazanas, D. 1989, ApJ, 342, 635
Mannheim, P. D. & Kazanas, D. 1991, Phys. Rev. D, 44, 417
Mannheim, P. D. & Kazanas, D. 1994, Gen. Rel. Grav., 26, 337
Milgrom, M, 1983a, ApJ, 270, 365
Milgrom, M, 1983b, ApJ, 270, 371
Milgrom, M, 1983c, ApJ, 270, 384

Rubin, V.C. *et al.* 1980, ApJ, 238, 471
Sanders, R. H. 1984, AA, 136, L21
Sanders, R. H. 1986a, AA, 154, 135
Sanders, R. H. 1986b, MNRAS, 223, 559
Tohline, J. E. 1982, in Internal Kinematics and Dynamics of Galaxies ed. A. Athanassoula (Dordrecht: Reidel), 205

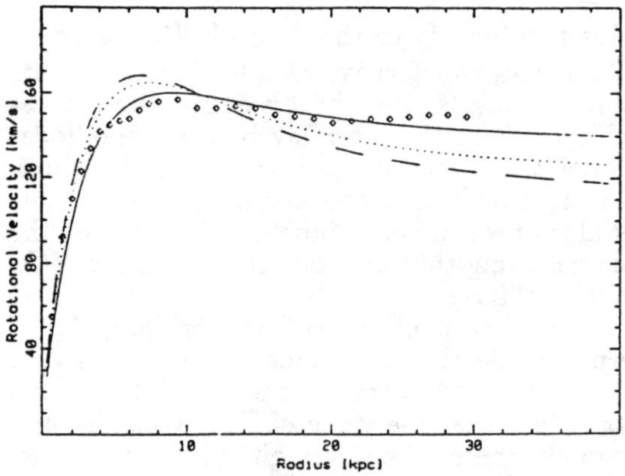

Fig. 1

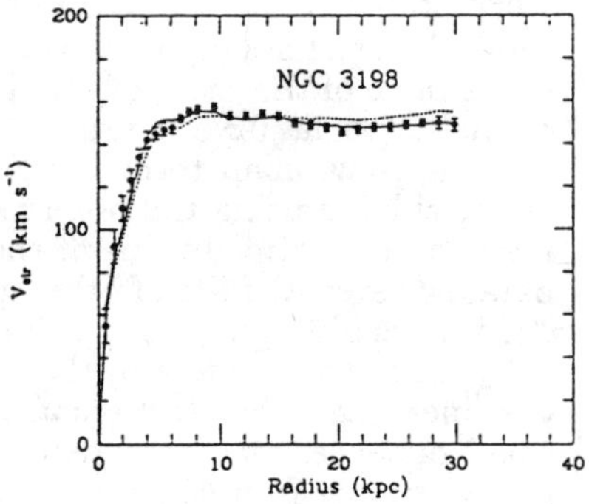

Fig. 2

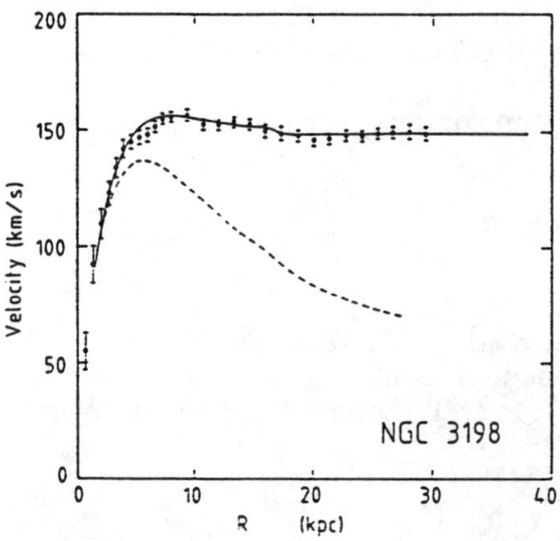

Fig. 3

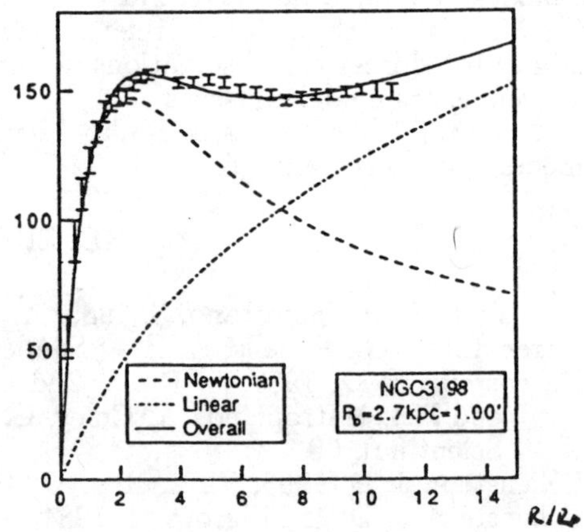

Fig. 4

Rotation curves of the spiral *NGC 3198* as computed by logarithmic potentials (Kuhn & Kruglyak 1987) (fig. 1), *MOND* (Begeman *et al.* 1991) (fig. 2), *FLAG* (Sanders 1986a) (fig. 3), and Conformal gravity (Mannheim 1993a) (fig. 4).

MICROLENSING, NEWTON-EINSTEIN GRAVITY, AND CONFORMAL GRAVITY

Philip D. Mannheim
Department of Physics, University of Connecticut, Storrs, CT 06269-3046
mannheim@uconnvm.uconn.edu

We discuss some implications of the current round of galactic dark matter searches for galactic rotation curve systematics and dynamics, and show that these new data do not invalidate the conformal gravity program of Mannheim and Kazanas which has been advanced as a candidate alternative to both the standard second order Newton-Einstein theory and the need for dark matter.

At the present time the search for galactic dark matter stands at an extremely critical juncture, and while it is still far too early to draw any permanent conclusions regarding the matter content of the Milky Way Galaxy, it would appear that to the extent that the Galaxy actually contains dark matter at all, not a great deal of it (and in all probability not as much of it as had been widely surmised) may turn out to be in the form of conventional astrophysical sources. Such standard baryonic sources essentially fall into two categories, namely sources which are never luminous such as brown dwarves with masses below the nuclear fusion burning threshold or strong gravity objects such as black holes, and sources which are in principle detectable optically but which are so faint (either by having low absolute luminosity or by being too far away from us) so as to simply fall below the available level of detectability. Searches for objects in both of these categories are now being vigorously and urgently pursued through the gravitational microlensing observations (which can detect either category) of the OGLE, MACHO and EROS collaborations (see e.g. the report of B. Paczynski, these proceedings) and through the use of the recently repaired Hubble Space Telescope (Bahcall, J. N., Flynn, C., Gould, A., & Kirhakos, S. 1994, ApJ. 435, L51) which now provides unprecedented optical sensitivity in the search for faint sources. During the last year the first results from both of these programs have been announced, with these early results turning out to not be overly encouraging for the standard galactic dark matter scenario. At the very minimum one can say that these results have certainly not yet achieved their intended goal of confirming the standard picture, while at the maximum one can say that they have actually thrown the entire picture into doubt.

In order to identify the potential implications of these searches for gravitational theory, it is useful to recall some general aspects of the standard picture, a picture which is driven by essentially two inputs, one being the theoretical demonstration by Ostriker and Peebles that disk shaped Newtonian spiral galaxies are intrinsically unstable, and the other being the subsequent observational discovery made by many groups that rotational velocity curves of spiral galaxies plotted as a function of distance R from the center of each galaxy remain tantalizingly flat far outside the optical disk region where most of the luminous galactic matter is concentrated. Indeed, typical observed galactic surface brightnesses behave as $\exp(-R/R_0)$ with the scale length R_0 providing a characteristic length for a given galaxy, with rotation curves then going out as far as $10R_0$ in some cases. Since an exponential Newtonian disk of matter gives a peak in the rotation curve at around $2R_0$, beyond that region a slow Keplerian fall off should then be expected in precisely the kinematic region where the curves are found to be flat. Thus if the matter distribution in fact is the same as that of the light (so that light traces mass) then no matter what the numerical value of the mass to light ratio M/L which then normalizes the matter distribution to the surface brightness, by itself the luminous Newtonian contribution then fails to fit the observed shape of the curves. (In passing we note that while the flatness of the curves is extremely striking, the most significant theoretical feature is the fact that the curves are actually non-Keplerian at all). As regards the actual normalization of M/L it is usually the case that the luminous contribution can be normalized to the Newtonian peak at $2R_0$ so that the entire inner region of the rotation curve is then fitted purely by the luminous disk in what are called maximum disk fits, with these fits actually being found to nicely accommodate the observed initial rise in the rotation curves. One of the virtues of these maximum disk fits is that the luminous distribution then naturally respects the Tully-Fisher relation which is an observed correlation between the fourth power of the peak velocity (and thus the average velocity if the rotation curve is flat) and the luminosity of each galaxy. However, in the maximum disk fits to rotation curves the M/L ratios inferred from the fitting are typically found to be a factor or so larger than that actually detected optically in the solar neighborhood. Thus independent of what may or may not occur in the non-Keplerian outer region of a galaxy, we see that already in the inner disk region there is a kind of dark matter problem for a maximal disk, since the light does seem to be tracing the mass just as desired but with the wrong apparent normalization. (On the other hand if the disk is less than maximal then something other than the disk would be needed in the inner region to again give an inner region dark matter problem, with it then not immediately being clear how this extra something would still manage to leave the Tully-Fisher relation intact). Now in order to explain both the stability of galaxies and the flatness of rotation curves in the outer region the most popular approach within the standard Newtonian context is invoke an ad hoc dark matter spherical halo which is asymptotically an isothermal sphere. (While such a halo is the most economical option, we note that flatness itself does not actually require a sphere - a disk with a $1/R$ surface brightness also yields a flat rotation curve - and a less than spheroidal halo could still yield stability, so that the case for an explicitly spheroidal, isothermal halo is actually somewhat tenuous. Moreover, since the parameters associated with the halo are completely unrelated to those of the disk, the halo is only able to enforce the Tully-Fisher relation by an artificial tuning of its parameters to those of the disk - the so-called conspiracy which currently lacks any explanation). With the spherical dark halo proposal we thus encounter a second, independent dark matter problem, this one in the outer region, so that both the inner and outer regions need to be explored observationally, with differing outcomes potentially being realizable in the two regions.

With the advent of microlensing it became possible to explicitly investigate the above dark matter scenario, with microlensing of the LMC in principle exploring the dark matter content of the halo and microlensing of the galactic bulge exploring the dark matter content of the disk. So far the initial searches are finding many more lenses off the bulge than anticipated and precious few in the halo, far fewer in fact than widely anticipated in advance of the observations. The data thus suggest that the disk of the Galaxy could actually be maximal with a large M/L ratio, and that not a great deal of matter of a conventional astrophysical nature actually resides in the halo (a position also supported by the failure to find copious amounts of faint stars in the recent HST optical survey). Thus in a sense the microlensing observations are nicely solving the inner region dark matter problem while simultaneously actually making the outer region dark matter problem more severe, since not only is the halo dark matter not being found, but as the known amount of matter in or near the plane of the Galaxy is found to increase, the spherical halo would have to contain all that much more matter again if it is indeed going to stabilize the now much heavier disk. Further, these data also create a potential new problem for the fitting of rotation curves of dwarf galaxies, since here the standard dark matter fits favor a dark halo in both the inner and outer regions with the stars having to have anomalously low M/L ratios compared to spirals (low by as much as a factor of 100 in some cases). The trend required in dwarfs is thus completely opposite to that in regular spirals. Now the astrophysics which might lead to such anomalously low M/L ratios for the stars in dwarfs was never in fact identified even before the microlensing observations (the detected stars themselves are not that different from those detected in spirals), and now that we see that large not small M/L ratios are actually being preferred for normal spiral star regions, the dwarfs thus present a rather acute puzzle. Moreover, not only did such low M/L ratios for the stellar disk components of dwarf galaxies not emerge from theory, they simply emerged from doing unconstrained fitting with an inner region dark halo giving better fitting than a standard, conventionally normalized inner region disk. However, the dwarf galaxies themselves appear to fall into two categories. There are some for which a maximal disk gives acceptable fitting (with a less than maximal one and a consequently bigger halo then just giving better fitting), and there are some for which maximal disks completely fail in both the shape and the normalization of the inner region rotation curves. Thus something has to give somewhere, and it would therefore appear to be worthwhile to again measure the surface brightnesses of dwarf galaxies, optimally in many filters (this would also be a good objective for HST), to see if some optical components have been missed or if perhaps some scale length values might change, so that the inner regions of dwarf galaxies might then in fact be fitted with typical spiral M/L ratios after all.

Now given the apparent smallness of the baryonic content of the spherical halo of the Galaxy, there is still one option left for the standard theory, namely that the halo be predominantly non-baryonic being built out of some of the exotic particles now being theorized in elementary particle physics. (While little can currently be said either for or against this possibility, it would be quite remarkable if such a non-baryonic halo would always be able to adjust itself to the baryonic content of each galaxy each and every time in order to ensure the fulfillment of the Tully-Fisher relation galaxy by galaxy). However, unless this exotic scenario is actually realized for galaxies, we apparently would then have nowhere left to turn but to say that there may not actually be much of a halo, and that rather than being missing, the missing mass may not in fact be there at all. Such a step would then oblige us to have to think the unthinkable and question the one remaining ingredient of the standard picture, namely the validity of the Newton-Einstein theory of gravity itself on galactic distance scales, with the problem then lying not with any lack of knowledge of the astrophysical makeup of galaxies but rather with the laws of gravity themselves.

To look beyond Newton-Einstein (as well as to see how it is even in principle possible to go beyond Einstein while retaining its tested features) constitutes the program entered into a few years ago by Mannheim and Kazanas with their development and promotion of fourth order conformal gravity (for a review of this program as well as of some other candidate alternatives to the conventional wisdom see the report of D. Kazanas, these proceedings). One of the main findings of their work is the recognition that the standard second order Poisson gravitational equation (and accordingly its second order covariant Einstein generalization) is only sufficient to give Newton's Law of Gravity but not in fact necessary, a result they demonstrated explicitly (Mannheim, P. D., and Kazanas, D. 1994, GRG 26, 337) simply by obtaining Newton's Law in another covariant theory. Specifically, they found the potential $V(r) = -\beta/r + \gamma r/2$ to be the solution to the fourth order Poisson equation which explicitly emerges in the conformal fourth order covariant theory, to thereby, in principle at least, divorce the Newtonian potential from the standard second order theory. Thus at the present time Einstein is only sufficient to yield Newton but not necessary, and the theoretical challenge that Mannheim and Kazanas threw out to the community was to either find some underlying fundamental principle which would make the standard theory necessary too (and thus actually force us to second order gravity), or to recognize that it is the very absence of any such principle (rather than the phenomenological issue of dark matter) which actually opens the door to candidate alternatives even at this late date.

Since the conformal gravity potential reduces to Newton on short enough distance scales and then first deviates from it galactically (for an appropriately chosen γ), the viewpoint espoused by Mannheim and Kazanas is that from the study of the solar system we only measure the first few terms in a perturbation series and that at larger distances the series may simply differ from that inferred from Newton-Einstein, i.e. that it may depart from the standard model in precisely the kinematic region where the conventional wisdom is currently having problems, with the apparent need for dark matter then simply stemming from having guessed the wrong series. To explore the conformal option fits have been made to a representative set of galactic rotation curves using the $V(r) = -\beta/r + \gamma r/2$ potential in conjunction with the luminous matter distribution alone (to thus yield a fully specified and hence falsifiable theory), and the fits are found (Mannheim, P. D. 1993, ApJ. 419, 150) to rival in quality standard dark matter fits. (The fits achieve flatness in the explored $2R_0$ to $10R_0$ region by an interplay between the falling Newtonian contribution and a rising linear contribution with the linear contribution mirroring that of a dark halo in this region. The fits then differ from the standard dark matter fits in the yet to be explored region beyond $10R_0$ with the conformal theory predicting a testable eventual rise in the rotation curves). Additionally, unlike the dark matter fits where the two components (luminous and dark) have independent normalizations, in the conformal theory the linear and Newtonian potential contributions must both be integrated over the same matter distribution and thus must both have the same normalization. Consequently, in order to avoid falling below the curve at all observed distances the disk must be noneother than maximal. Further, in the fits, the coefficient γ is found to be universal for the regular spirals considered, and intriguingly to take a value of order the inverse Hubble radius. This universality for galactic γ turns out to be equivalent to the Tully-Fisher relation (precisely because the linear and Newtonian contributions are both simultaneously normalized to the one luminosity). Moreover, despite having no halo, linear potential disks also turn out to be stable (Christodoulou, D. M. 1991, ApJ. 372, 471). The conformal linear potential theory thus appears capable of reproducing all the desirable aspects of galactic dark matter without needing the dark matter itself, and also appears to have survived the microlensing observations unscathed. Thus it must currently be regarded as viable. This work has been supported in part by the Department of Energy under grant No. DE-FG02-92ER40716.00.

MAGNETOGRAVITODYNAMICS MODEL of GALAXY FORMATION

Howard D. Greyber
Greyber Assoc., 10123 Falls Road, Potomac, MD 20854, U.S.A.
e-mail: hgreyber@cap.gwu.edu

ABSTRACT

Unless the tau neutrino is discovered to have an appropriate size mass, the theories of galaxy formation involving only gravity may not fit the observations. A model within the Big Bang hypothesis is described with processes occurring after "Breakout" that lead to the concentration of matter along thin spatially curved current sheets. When a critical density is reached, gravitational collapse occurs, forming galaxies in many places along thin spatially curved sheets surrounding huge voids, in a cellular structure,as in the observations by Geller, Huchra et al. The origin of a primordial magnetic field concentrated in the sheets is explained. The model predicts that much of the missing dark matter is located along the thin spatially curved sheets of galaxies.

1. INTRODUCTION

In this paper the suggestion is made that *cosmical magnetism*, together with *gravitation*, may be of basic importance in explaining three major astrophysics problems at once: the large-scale structure of galaxies, - - the location of much, if not all, of the missing dark matter, - - and the origin of galactic magnetic fields.

In what may be called "Induced Galaxy Formation", a series of processes, involving magnetic fields, plasma physics and gravitation, as the Universe expands in the Big Bang model right after "Breakout" (i.e. when the Universe is about 300,000 years old and the radiation escapes from equilibrium with the matter), is described.

The evidence that galaxies tend to lie in thin sheets around large voids in a cellular distribution has been steadily getting stronger with more observations by several groups, but mainly due to the careful, remarkable, detailed observations by Margaret Geller, John Huchra and collaborators. The Big Bang model for the origin of the Universe has considerable strong evidence in favor of it, summarized by P. J. E. Peebles et al (1991) in a recent article.

Yet, the superb Geller/Huchra results on the large-scale structure of galaxies are very impressive, especially when one views their video presentation which rotates the data, showing how *extraordinarily thin* the sheets of galaxies are. Obviously, the 2D figures of their work distort, hiding this important aspect. In the COBE satellite results there is not the slightest sign of embryonic sheets of galaxies and voids. Clearly new concepts are necessary if the Big Bang model, which already explains so well many of the light element isotope abundances, is to be retained. Some new concepts are as follows:

P. J. E. Peebles (1993) calculates a significant residual ionization, 6×10^{-5}, up to 10^{-3}, after "Breakout" under various assumptions. The COBE satellite results tell us that the overall mass density distribution is relatively smooth at "Breakout". Now, it must be emphasized that another distribution in the Universe, of comparatively very low mass, such as the electron space and velocity distribution, need not be smooth at all. In fact some fraction of the electrons *must* clump into electric currents, producing magnetic fields in some plasma volumes. The Debye shielding distance, h, in centimeters, is

$$h = 6.90 \ (T/n_e)^{0.5}$$

where T is the temperature in degrees Kelvin, and n is the number of electrons per cubic centimeter. Lyman Spitzer states that "If h is small compared to other lengths of interest, an ionized gas is called a plasma - -". Obviously, putting in numbers, one sees that for any reasonable lengths relevant to galactic formation, h is extremely small compared to those lengths, and so we are dealing with a Plasma Universe, and electromagnetic effects dominate.

Gigantic electric currents, and strong, long-range magnetic fields can exist for times long compared to the age of the Universe.Thus, by far the most reasonable assumption is that after "Breakout", there was a very low mass irregular distribution of electric currents and magnetic fields in this early Universe - while the overall mass density distribution was relatively smooth. Arguing for perfect homogeneity is completely untenable.

P. J. E. Peebles (1971) emphasized two different lines of thought: "One is that the Universe must have been very smooth and regular because it is unstable, even if mildly so, so the approximate homogeneity of the Universe now means that it used to be much more homogeneous. The opposing line of thought was that the early Universe was chaotic. - - A second reason for studying primeval turbulence is the thought that the galaxies may be the fossil remnants of primeval turbulence eddies."

It is important to notice that the first line of thought is satisfied if one believes in the Guth inflation hypothesis, yet the second can exist also as described below.

The noted physicist, George Gamow, strongly urged the second line of thought, emphasizing the possibility of *large-scale turbulence* in the early Universe. Greyber (1967, 1989) has analyzed the effect of dissipation with time on a bounded turbulent medium.

Clearly as the Universe expanded from "Breakout" , (a) turbulent energy of wavelengths smaller than the size of a typical volume of plasma is dissipated relatively rapidly, (b) volumes of high density plasma with strong magnetic field inside *jostled* and *pushed* against other volumes of low density plasma with little or no magnetic field inside, and (c) the average volume of each of these plasma volumes grew enormously with time.

2. Current Generation and Galaxy Formation in the Early Universe

From plasma physics, one finds that Maxwell's Equations require there to be an electric current at the boundary between two stationary volumes of plasma with widely different magnetic field inside. This current is proportional to the jump across the boundary of the component of the magnetic field parallel to the boundary on each side of the boundary. One might describe this as due to plasmas being diamagnetic. Notice that some of these boundaries, or spatial interfaces, will be *Rayleigh-Taylor stable* when one plasma volume, denser and with stronger magnetic field, compresses an adjacent plasma volume with less density and a weaker magnetic field. At these particular local boundaries, gigantic currents are generated over millions of years of time. The polytrope equation of state of matter is

$$P(r) = K \rho (r)^{n+1/n}$$

where ρ is the density, and P is the pressure. An exponent of 1 fits an isothermal gas. An exponent of 5/3 represents an adiabatic monatomic gas that is either completely ionized or completely neutral. An index of n = 3, or an exponent of 4/3 , represents a gas that is dominated by relativistic particles. For instance, an exponent of 4/3 represents accurately very massive stars that are radiation dominated, that is dominated by photons. At the time soon after "Breakout", the matter pressure in the Universe is probably represented by an exponent starting at 4/3 and then dropping quickly to 1.

Therefore, since the magnetic pressure goes as B^2 , as the "piston-like" mass in one volume compresses the other plasma volume, due to the remaining *large-scale* turbulence, the jump in the component of the magnetic field parallel to the interface will *grow* steadily, as long as the exponent in the equation of state of the matter is significantly less than 2.

Let us assume, for a quick estimate, that the exponent is 1, i.e. an isothermal gas. Then a thirty-fold doubling, i.e. a compression of roughly a billion times is quite possible over millions of years. Thus an initial magnetic field of, for instance, a microgauss would become something like 1000 gauss, and the surface current would be amplified in the same ratio. This means that the surface current in the interface, (which is proportional to the jump in $B_{//}$) will also *be amplified enormously.* It follows that the charged matter in the plasma will be *drawn* into the interface by the well-known "Pinch Effect" in plasma physics. Thus the charged matter density will *increase greatly* along the sheets.

In addition, both by gravitational attraction, and by the electron and ion scattering against neutral atoms, neutral matter will also be drawn in to the interface. Relatively rapidly (in millions of years), the density along the spatially curved sheets (which are the interfaces between plasma volumes) must increase *until critical density for self-gravitation* occurs.

Then *gravitational collapse will occur* at each such critical density place, forming galaxies and quasars on thin spatially curved sheets surrounding huge voids.

3. DISCUSSION AND CONCLUSIONS

An attractive feature of this magnetogravitodynamics (MGD) theory of galaxy formation is that, besides offering an explanation for the cellular structure of the Universe found by Geller and Huchra, it provides a clear, physical explanation for the *origin of galactic magnetic fields.* They are created due to the gigantic surface currents generated in the sheets. This *primordial magnetic field* obviously will not exist everywhere in the Universe, but is concentrated along the sheets. For decades, a small minority of astrophysicists (like this author) have opposed the popular "galactic dynamo" model, preferring a preexisting or "primordial" magnetic field. One strong member of this small minority has been Professor Russell M. Kulsrud at Princeton University. The surprisingly high magnetic fields, milligauss level, discovered in very young evolving galaxies, by observations in the Lyman alpha forest of lines up to Z of about 3, by Philipp P. Kronberg and collaborators, are highly significant, and put the galactic dynamo model in grave doubt.

Astronomer Arthur M. Wolfe commented on these results that "the prevalent dynamo model for the formation of magnetic fields leads one to expect that the fields do not form until much later, after a galaxy has completed several rotations". In MGD one *must expect significant magnetic fields to be observed as soon as the galaxy forms.*

Perhaps someday the electric currents predicted by MGD, which should be still circulating through the sheets of galaxies, will be observed by some technique. Some recent observations suggest galaxy formation occurring even earlier than an age of the Universe of only one billion years. It is very probable that MGD will permit such very early galaxy formation whereas the orthodox HDM/CDM models, based only on gravitation acting on the very tiny ripples observed by COBE, would find it extremely hard to do.

Following Gamow, one can ask, of course, why was the primordial gas in a supersonic turbulent state? Gamow suggests that turbulence may have been the result of some kind of internal instability in a mixture whose matter density and energy density are equal (at about the time of "Breakout"), and that this needs investigation. As the last sentence of his Proceedings of the National Academy of Sciences article, Gamow (1954) says, "In fact, mechanical and thermodynamic properties of such as a system have never been investigated, and further investigations are badly needed in that field".

That this prediction made by the famed George Gamow forty years ago is correct can be seen if we analyze the following qualitative scenario. Consider a large macroscopic volume of the Universe just before or around "Breakout", i.e. at an age of the Universe of about 300,000 years. By chance, one half of the volume has already combined, changing phase from full ionization to the small residual ionization calculated by Peebles (1993) of 10^{-3} or less, while the second half of the volume has not yet combined.

Clearly, the radiation energy produced by the combination in the first half will flow across the interface to the second half, heating it up and delaying its combination from the general cooling due to expansion of the Universe. Also, it is important to notice that shock waves must push out from the heated first half of the volume into the second half of the volume. If, as is very probable, *the very same process occurs at random at myriads of places* all over the very young Universe around this critical time of "Breakout", the previously very homogeneous and smooth Universe is thus prepared in a turbulent state with turbulence on all scales that George Gamow surmised, and is prepared for the previous processes described.

ACKNOWLEDGEMENTS

It is a pleasure to thank my friend Martin D. Kruskal for a valuable conversation, and Gart Westerhout for permission to use the U. S. Naval Observatory Library. This paper is dedicated to the memory of a truly great scientist, and a humorous friend, George Gamow.

REFERENCES

Bothun, G.D., Geller, M.J. et al, 1992, ApJ, **395**, 347
Gamow, G., 1952, Physical Review **86**, 251
Gamow, G., 1954, Proceedings of the National Academy of Sciences, **40**, 480
Greyber, H.D., 1967, in "Instabilite Gravitationelle et Formation des Etoiles, des Galaxies de Leurs Structures Caracteristique", Memoirs Royal Society of Sciences of Liege, **XV**, 189
Greyber, H.D., 1967, ibid., **XV**, 197
Greyber, H.D., 1989, Comments on Astrophysics, **13**, 201
Greyber, H.D., 1989, in "The Center of the Galaxy", ed. M. Morris, Kluwer Acad. Press, 360
Greyber, H.D., 1994, in "Cosmical Magnetism - Contributed Papers", published by the Institute of Astronomy, from the NATO Advanced Study Institute on "Cosmical Magnetism", held in honour of Professor L. Mestel, FRS, at the Institute of Astronomy, Cambridge University, England, July 5-9, 1993, page 110.
Kronberg, P. P.& Perry, J. J., 1982, ApJ **263**, 518
Kronberg, P. P., 1994, in Reports on Progress in Physics, v.**57** (in press)
Kronberg, P. P., 1987, in "Interstellar Magnetic Fields", eds. R. Beck & R. Grave, published by Springer-Verlag
Kulsrud, R.M. et al, 1992, Ap.J., **396**, 606
de Lapparent, V., Geller, M.J. & Huchra, J.P., 1986, ApJ, **302**, L1
Peebles, P.J.E. et al, 1991, Nature, **352**, 769
Peebles, P.J.E., 1993, in "Principles of Physical Cosmology", Princeton Univ. Press, Fig. 6.8
Peebles, P.J.E., 1971, in "Physical Cosmology", Princeton University Press, page 220
Spitzer, Jr., L., 1962 in "Physics of Fully Ionized Gases", Interscience Publishers, Chap. 2

EVIDENCE FOR A POSSIBLE COMMON ORIGIN OF BARYONIC AND NON- BARYONIC DARK MATTER

E. J. Sternglass
Radiation Physics Laboratory, Scaife Hall RC-406
University of Pittsburgh, Pittsburgh, PA 15261

ABSTRACT

A single "positronium" or integrally charged "quarkonium"-like model for matter originally developed for the spin 0 neutral pion is shown to explain the nature of cold dark matter in the early Universe when in a highly excited spin 1 vector boson state, as well as the properties of baryonic matter following a phase transition to form a three-component nucleon structure in its final step of decay to the ground-state accompanied by a large release of energy that represents the Big Bang. The model gives the masses and sizes of cosmological structures, resolves the problem of nucleosynthesis at the high present baryon density, and explains the origin of dark baryonic halos as well as the rising mass-luminosity ratio with distance.

1. INTRODUCTION

In a detailed review of the dark matter problem, Trimble (1987) expressed the hope that there might be a single type of particle that would explain the nature of non-luminous matter on many scales, calling a single candidate "the most elegant solution". It is the purpose of the present note to show that there does in fact exist such a single type of particle that begins as an extremely massive, dense and therefore "cold" or slowly moving form of matter at the maximum possible or Planck density which divides itself in a series of division processes by two and finally ends up in producing baryons in a phase transition that heats the Universe sufficiently to allow nucleosynthesis of the lightest elements. This particle surprisingly turns out to be the simplest of all known particles, namely the electron and its anti-particle the positron, in a highly relativistic spin 1 state whose spin 0 form was developed to explain the properties of the neutral pion or π_o (Sternglass 1961, hereafter S 61).

2. PHYSICAL MODEL

The principal feature of the relativistic pair model that differentiates it from ordinary low energy positronium is that by calculating the force between the rotating charges for an observer at rest relative to one of the charges, the ordinary Coulomb force is increased by the Lorentz factor $\gamma_{12} = (1- \beta^2_{12})^{-1/2}$ so that for $\gamma_{12} >> 1$ it equals the relativistic centrifugal force at a minimum approach distance, equal to a quarter of the classical electron radius $r_{cl} = e^2/2m_oc^2 = 1.41 \times 10^{-13}$ cm, while the electron itself shrinks to increasingly smaller size r_{cl} / γ_{12} . For all quantized angular momentum states L= nh one finds positive mass-energy levels $n \gamma_{12} m_oc^2 = n (2/\alpha) m_oc^2$, where $1/\alpha$ is $(e^2/\hbar c)$= 137.036 so that for n=1 one obtains a mass close to that of pions. Not only does this system have no singularity in the sense of a zero size and infinite mass, but it can also contain any amount of excitation without disintegrating up to the extremely high energy states needed to provide the seeds of cosmological systems, including the state associated with the highest possible density, namely the Planck density $\rho_{Pl} = c^5/\hbar G^2 = 5.18 \times 10^{93}$ g/cm^3.

Using the geometric mean radius of the $e^+ e^-$ orbit and $(2/\alpha)$ times this value in analogy to the case of the electron (Motz 1962), one finds that the field energy confined into this effective radius $<Ree> = 4r_{cl}$ with a non-Euclidean volume $V_{ee} = 2\pi^2 <Ree>^3$ closely equals the mass of the Universe M_u as given by the Dirac Large Number Relation $q^2 / M^2G = (M_u / M)^{1/2}$ for the case of $q = e$ and $M= m_o$. As shown in S84, this mass is $M_u = e^4/m_o^3G^2$ and the volume Vo into which it is confined is obtained by setting the Planck density equal to (M_u/V_o) so that $V_o = M_u\hbar G^2/c^5$ then eliminating G by using the Dirac relation to give $G = (e^2/m_o^2)(m_o/M_u)^{1/2}$ resulting in the expression for $V_o = \hbar e^4/m_oc^2 = (8/\alpha) r_{cl}^3 = 1096 r_{cl}^3$. This volume turns out to be almost exactly equal to $V_{ee} = 1398 r_{cl}^3$, thus indicating that the physical origin of the Dirac Large Number Relation appears to be that the electron with its purely electromagnetic mass is the basic building block of all luminous and dark matter. For a finite mass $M_u = 1.7362 \times 10^{85} m_o$, the theory of General Relativity gives a finite outer size equal to the Schwarzschild radius $R_s(Mu) = 2M_uG/c^2 = 2(e^2/m_oc^2)(M_u/m_o)^{1/2}$, thereby explaining the fact that the Dirac number $(e^2 / m_o^2 G) = 4.1667 \times 10^{42}$ also gives the size of a closed Universe in units of the electron size. Its value is 2.3484×10^{30} cm, or 7.612×10^4 Mpc, equal to 2.4823×10^{12} lt-yr, or much larger than the Hubble radius c/H_o of about $10^{-20} \times 10^9$ lt-yr so that space is very nearly "flat" as in the standard inflation model. Since the pair system is stabilized by the centrifugal force for any value of the mass, there is no problem of "fine-tuning" the initial conditions so as to obtain a final state that is stable and that will neither collapse into a singularity nor fly apart. Instead it will approach the final form of a static, rigidly rotating Einstein Universe (Ozsvath and Schucking 196). This rotation takes place at the angular velocity of the original pair system given by $\omega(M_u) = c/R_s(M_u) = 1.2767 \times 10^{-20}$ s^{-1} corresponding to a period $\tau_u = 2\pi/\omega\ (M_u)$ of 1.5597×10^{13} yr. In this model, the density is uniformly distributed in the Einstein radius $R_E\ (Mu) = (1/8\pi^2)\, R_s(M_u) = 31.438 \times 10^9$ lt-yr , giving a value of $\rho_E = 3.0454 \times 10^{-29}$ g/cm^3. This value of the density is close to the critical density for a non-rotating Universe $\rho_{cr} = 3H_o^2/8\pi G$ for $h_o = H_o/100 = 1$ given by $1.88 \times 10^{-29}\ h_o^2$ g/cm^3 , and even more closely equal to the critical density for a rotating Universe $\rho_{cr}' = 3H_o^2/4\pi G = 3.76 \times 10^{-29}\ h_o^2$g/cm^3 and exactly equal to the latter for $h_o = 90.1$ kms^{-1} Mpc^{-1}, a value close to the most recently measured values (Pierce 1994)(Freedman 1994).

As originally suggested by Lemaitre (1932), the initial "primeval atom" decays in a series of steps or divisions by 2 that lead to the "seeds" of the various cosmological structures, ending with the formation of ordinary matter accompanied by the release of large amounts of radiation in the last step. In the present model, if e, mo, c and h are assumed to remain constant throughout the evolution of the Universewhile increasingly pairs with decreasing mass Mee are created, the volume $V_o = he^4/m_oc^2$ must also remain constant, so that the Planck density $M_{ee}/V_o = c^5/hG^2$ at which space "tears" and pairs are produced must also decline.The only way that this process can continue is that G must increase within the physical space defined by its Schwarzschild radius R_s(Mee) so as to keep $M_{ee}G_{ee}^2$ constant. This means that the Dirac relation between the gravitational constant and e, mo and Mu must take on the form $G_{ee} = (e^2/m_o^2)(m_o/M_{ee})^{1/2}$, leading to $Rs_{(Mee)} = 2(e^2 / m_o c^2)\ (M_{ee} / m_o)^{1/2}$. Thus, the present model requires that the sizes or spacings of cosmological systems vary as the square-root of their masses as described in S92, in agreement with observations for the large, diffused structures. Another interesting result is that as the masses decline towards that of the elecctron, G_{ee} approaches (e^2 / m_o^2) so that the local gravitational force between the particles equals the strength of the Coulomb force, very close to the same large value used by Motz(1962) in his model for a gravitationally stabilized electron.

As shown in S 84, by taking the case where two pions of average mass $\langle M_\pi \rangle$ are formed in the ground state of mass $[(2/\alpha)-2]m_o = 272.072$mo so as to conserve momentum, by setting $M_u/2\langle M_\pi \rangle$ equal to 2^{N_L} where N_L is the number of divisions, it is found that $N_L = 274.072$, which is extremely close to the measured value of $2/\alpha$ within the experimental uncertainty. This leads to the surprising result that in the electron pair model of matter, the square of the Dirac number, or a pure number equal to the number of electron masses in the Universe(M_u/m_o) as first calculated by Eddington, is fixed by the value of another dimensionless number, namely the atomic fine-structure constant α. The expression for this number is $N_E = 2[(2/\alpha)-2]\, 2^{2/\alpha}$, with the result that a single pure number determines the masses and sizes of the nuclear particles as well as the properties of the large-scale Universe.

Moreover, α also determines the sizes, masses and motions of the large-scale Universe as laid down prior to the Big Bang during the period when the production of dark matter pairs takes place and the Universe undergoes an inflationary period, in the present case driven by the exponential growth in the number of pairs since they have a minimum center to-center separation of $(e^2/m_oc^2) = 1.41 \times 10^{-13}$ cm, which also determines the size of baryons and mesons as discussed in S 65a and 65b. This comes about as the result of the existence of the volume V_o within which the field strength is large enough to create new pairs by internal pair production, held together by the intense gravitational force acting between these massive particles as if they were confined to a "bag" of radius $4r_{cl}$, analogous to the MIT bag model (Chodos 1974). V_o is seen to have a size of $(8/\alpha)$ or 1096 $r_{cl}{}^3$, so it can hold about 2^{10} or 1024 pairs before they are forced out by the continuing pair production process to start formation of another cluster of 1024 pairs in a step-wise decay. Some of the pairs will be forced out immediately, but others will be retained for some time to form the nuclei of the various systems which we see today only to be gradually ejected in a series of violent processes as proposed by Ambartsumian (1958), Burbridge *et al.*(1963), Arp*et al.*(1992) and more recently by Hoyle *et al.*(1994).

to be gradually ejected in a series of violent processes as proposed by Ambartsumian (1958), Burbridge *et al.*(1963), Arp*et al.*(1992) and more recently by Hoyle *et al.*(1994).

As to the time required for a division process to take place, this will be of the order of the rotational period for the seed pairs given by $2\pi R_s(M_{ee})/c = 1.56 \times 10^{13} (M_{ee}/M_u)^{1/2}$ yr . Now nucleosynthesis takes place beginning a few minutes after baryon formation in the big bang, which in the present model occurs when the 27th major stage of decay is reached, at at time $T_u = [1-(2^{-1/2})]^{-1} = 3.414$ times the minimum time for the first pair to divide since the periods of rotation decrease as $M_{ee}^{1/2}$. Taking the shortest possible time for scission or division of the field energy of the pair as half a period, one obtains a time to baryogenesis of 2.45×10^{13} yr from the formation of the first pair. Thus the present model provides for a long period of slow inflation during which the structure of the universe is laid down by CDM particles on a microscopic scale..

At that stage, the total mass-energy in a single pair is 9151.68 m_oc^2 or 4676.54 Mev. With a neutron mass of 939.57 Mev, the formation of two neutrons takes 1879.14 Mev, leaving 2797.40 Mev for kinetic energy and the production of pions that decay to relativistic electron-positron jets, most of which goes eventually into photons and neutrinos, heating the Universe. Thus, of the total mass-energy available, 60% goes into radiation kinetic energy, and only 40% goes into baryon rest-mass. Assuming that about half the radiation ends up as photons, or about 700 Mev per baryon formed, the temperature suddenly rises to 8.1×10^{12} K, well above the energy needed for nucleosynthesis. For a typical black body temperature of 2700 to 3000 K at recombination, the number of photons of 0.25 ev average energy produced will be 2.8×10^9, corresponding to a baryon to photon ratio $\eta = 3.6 \times 10^{-10}$ in good agreement with the range of $3\text{-}10 \times 10^{-10}$ required to explain the observed abundances of ^4He, ^2D, ^3He and ^7Li (Kolb and Turner 1990). Because only baryons but not anti-baryons are formed in this model from the original spin 1 electron pair which can exist in either "matter" or "anti-matter" form as discussed in S 65a, there is no annihilation of baryons at the time of baryogenesis to add further energy to the radiation field, and all the baryons and photons created are preserved.

When baryogenesis took place just a few minutes before nucleosynthesis, only a fraction of the total mass of the Universe had escaped from the central clusters to form baryons since significant mass must have remained to form the nuclei around which they were organized at every step. By the present epoch, however, a much larger fraction of all cosmological systems will have ejected the massive CDM pairs from the central nucleus, so that the present baryon density Ω_{bp} would be expected to be much larger in agreement with observations (Trimble 1987), probably of the order of 0.1 to 0.3. From the observed fact that large luminous systems such as superclusters sometimes contain500 to 1000 galaxies, and stellar clusters a similar number, it would appear that the fraction x of the 1024 pairs originally in the central cluster which escapes at every step must be large. Knowing that nucleosynthesis requires $0.015 < \Omega_{bn} h_o^2 < 0.026$ (Colb and Turner 1990) with $h_o = 0.9$ and taking into account that 60% of the energy goes into kinetic energy and radiation, one can get an estimate for x by taking $\Omega_{bn} = 0.025$ and $\Omega_{bp} = 0.15\text{-}0.25$. The 11th step is the last one to trap baryons long enough so as to make them unavailable for nucleosythesis, while the three largest systems with rotation periods larger than the Hubble age continue to trap CDM particles in their nuclei at the present epoch. One therefore gets two expressions for Ω_{bn} and Ω_{bp} namely $x^{11} = 0.025/0.4$ and $x^3 = \Omega_{bp} / 0.4$, giving a value of $x = 0.777$ for the fraction of massive CDM pairs forced out of V_o at every major step by the production of new pairs and leading to a value for $\Omega_{bp} = 0.187$. Thus, the present model solves one of the major puzzles of the standard big-bang model, explaining how the observed fraction of ^4He can be synthesized despite much too large an observed baryonic density today if the Universe is to have $\Omega = 1$.

Since all the systems are rotating like galaxies while simultaneously expanding during their early development, it also becomes possible to explain two other puzzlingphenomena, namely the origin of the massive dark halos accompanied by a rise in M/L with increasing size of the systems, and the evidence for large-scale coherent streaming of galaxies and superclusters. When each of the systems goes through its expansion phase, initially no radiation or matter can escape beyond $R_s(M_{ee})$. But as G_{ee} returns to its global value that is much smaller, the massive seed pairs that have escaped from the central cluster out to R_E find themselves rotating at velocities as high as $(1/8\pi^2)$ c, or 3800km s^{-1}, much too large to remain bound except to the very largest systems. Some fraction f_c will have much smaller velocities and will therefore fall back towards the central cluster that contains 22.3% of the total mass, forming the seed of the system, such as the nucleus of a globular cluster, a galaxy a star, or a smaller object too low in mass to initiate a fusion reaction, while another fraction will escape to form a dark halo at a distance where it can be bound by the large mass of a galaxy or cluster internal to the radius.

One can obtain an estimate for the fraction that falls back and the fraction that escapes to large distances using the theoretical total "seed" mass for a stellar system, 6.58 $M_\odot$ and correct it for the escape of 60% of the energy to get a value for the total baryonic mass of $2.63M_\odot$, and comparing this with the typical main sequence mass of actual stars, 0.8 - 1.2 $M_\odot$, or to a first approximation $1M_\odot$. This leads to the estimate that $1.63M_\odot$ or 62% of a stellar mass and of every system escaped to form a distant halo, dispersing the various systems to such large distances in the course of the continuing expansion of the entire Universe that they become so faint that they cannot be readily seen. Thus, each step of decay gives a factor 2.63 to the mass-luminosity ratio M/L, so that one gets for this ratio the values 2.63 , 6.92 and 18.22 for stellar systems in the visible part of the galaxy in good agreement with the values of 10 in Table 2 of Trimble(1987).For galaxies and groups this becomes 48, compared with Trimble's estimate of 10-100. For rich clusters or superclusters this becomes 126-332, compared with Trimble's range of 100-300, while for the largest systems it becomes 879, compared with Trimble's value of 700±150.

Finally, the rotation of all systems, including superclusters and complexes of superclusters at equilibrium velocities of the order of 1000km s^{-1} for superclusters can explain the large scale coherent motions which stretch over large regions of the sky, too large to have been formed in the short time since the Hubble age. However, it is interesting to note that because the centrifugal forces associated with rotation balance the gravitational force and thus providing a physical explanation for Einstein's cosmological constant Λ, there is no deceleration at this time in the history of the Universe, so that the "age-crisis" is not as serious as in the standard model, where the Big Bang occurs at 2/3 the Hubble age.

3. CONCLUSION

In summary, the model for hadronic matter based on a relativistic electron-positron system appears to explain many of the presently known aspects of both baryonic and cold dark matter on all scales. Future observations will be able to test the predictions of the theory so it has the desirable aspect of being clearly falsifiable. Thus, the definite disovery of such objects as WIMPS or neutrinos with finite rest-mass would falsify the present model., which requires that the electron and its anti-particle are the basic particles from which all forms of matter are formed, and that all mass is electromagnetic in origin. It therefore provides a physical model for supersymmetry, and superstrings can now be identified with the highly relativistically contracted tubes of force between rotating charges of finite cross-section, eliminating the point charge infinities and the singularity in General Relativity, so that there is no breakdown of the laws of physics on any scale. As the images of quasars and distant galaxies become available, they will test the prediction of the model that galaxies begin as point-like objects from which stars are ejected to form small, spherical objects that become spiral galaxies as time advances.

REFERENCES

Ambartsumian, V. A. 1958. *Solvay Conference Reports*
Arp, H. C. *et al.* 1992 .*Nature* 357, 287
Burbridge, G.R., Burbridge, E. M. and Sandage, A. R. 1963. *Rev. Mod. Phys.* 35,951
Chodos, A. et al. 1974. *Phys.Rev.* D10, 2599
Freedman, W. L. et al. 1994. *Nature* 371, 757
Hoyle,F., Burbridge.G. and Narlikar, J.V.1993. *Ap.J.* 410, 437
Kolb, E. W. and Turner, M.S. 1990 *The Early Universe* , Addison-Wesley , New York
Lemaître, G. 1931.*Nature* 128, 701
Motz,L. 1962. *Nuovo Cimento* 26, 672
Ozsvath , I. and Schucking, E. 1962. *Nature* 193, 1168
Pierce, M. J. *et al.* 1994. 371, 385
Sternglass, E. J. 1961. *Phys Rev.* 123,391 (S 61)
------------------ 1965. *Nuovo Cimento* 35, 227 (S 65a)
-------------------1965. *Proc. 2nd Conf.on Resonant Particles, Ohio University*, B. A. Munir, ed., p33
-------------------1984. *Lett. Nuovo Cimento*, 41, 203
-------------------1992. *Testing the AGN Paradigm*, A.I.P. Conf. Proc. 254, Holt,S. Neff,S.G.Urry, C. M. eds. American Institute of Physics, New York, 105
Trimble, V.1987. *Ann. Rev. Astron. Astrophys.* 25, 425

DARK MATTER AND COSMIC PLASMA PHYSICAL PROCESSES

P.Ammiraju
Stevens Institute Of Technology, Hoboken, New Jersey 07030

INTRODUCTION

The classic evidence for dark matter comes from the flat rotation curves of spiral galaxies. An extensive search is under way in the astronomical laboratories around the world to decipher the nature of this dark matter across the varied large scale structure of the Cosmos. As early as 1955-56, the late Winston H.Bostick, in a series of seminal laboratory plasma physics experiments at the Stevens Institute of Technology, had recorded evidence of the formation of spiral structures from interacting "plasmoids" (spontaneously arising sharp boundaried entities),and noted the striking similarities to their galactic analogues. These led to speculations concerning the origin and structure of barred spiral galaxies. The latter were exhibited in computer simulations of galactic configurations having helical arms with initial boundary conditions that were similar to plasma physics experiments at Stevens Tech. Assuming that the large scale structure of the Universe is the result of the combined forces of gravity and electromagnetism, and the properties of plasma are the same everywhere from the micro-metered dimensional structures seen under plasma focus conditions to the Hubble distance of the Cosmos, then a dedicated facility of Dense Plasma Pinch Focus (DPPF) machine will go a long way in unraveling the patterns of the large scale structure of the galaxies and, in particular, throw light on the existence of dark matter, which is considered the number one problem in astronomy for the present decade

PROPERTIES OF PLASMA PHYSICAL PROCESSES

1. It is estimated that, at least by volume, more than 99% of the observable matter in the Universe is in the plasma state.
2. Plasmas are dominated by electromagnetic force fields. The Biot-Savirt law of force between current elements is very much stronger (~ 1/r) than the gravitational force between mases (~ $1/r^2$).
3. The cylindrical magnetic field formed around plasma electron current flow focusses that flow into a beam by the so-called "Bennet Pinch" effect. This gives rise to multiple stands of conducting current .
4. Magnetic-Field aligned "Birkeland Currents" are formed whenever magnetic and electric field lines in a plasma are made to run parallel to each other.
5 Electric Double Layers, due to localized charge separation, are formed within the Birkeland currents, This gives rise to particle acceleration, in particular the synchrotron radiation emission at radio, optical, and X-ray wavelengths.
6. Filamentary structure is a natural morphology exhibited by energetic plasmas.
7. Formation of current sheets gives space a "cellular structure".

PLASMA PHYSICS AND COSMOLOGICAL STRUCTURES

1. The observed radio emission in the shape of an arc some 120 light-years long at the center of the Milky Way, using very large Array radio telescope is an example of Birkeland currents on a galactic scale as shown in computer simulation calculations.
2. Various diagnostic instrumentation profiles often show filamentary magnetic "rope-like" structure from plasmas produced in multiterrawatt pulse power generators or dense plasma pinch focus machines. Double radio galaxies could be part of filamentary structure stretched across billions of light years.

CONCLUSIONS

The pinch effect in the dense plasmas at the cosmic level will assist gravitation in the formation of interstellar clouds. The interaction between pinched filaments leads to synchrotron radiation and the evolution of these plasmas into spiral, elliptic and irregular galaxies have been shown, in computer simulations, to exhibit the characteristics of the observed galactic patterns.

Filamentary plasmas on supergalactic scales are shown to exhibit a wide variety of galaxy shapes, and detailed comparison on spiral galaxy shapes have shown good agreement with actual observations and computer simulation involving pinch effect, thus reducing the need for large quantities of unknown invisible matter.

Scientific progress essentially rests on close interplay between theory, observation and experiment. We do not yet know how the universe generates all of its observed structures. So, it is prudent to allow all types of investigations. Unlike the huge and expensive astronomical telescopes - optical as well as radio - the DPPF is a simple, reliable and low cost instrument which is useful in other areas such as magnetic lens in focusing of TeV beams in particle acceleration, radiographic techniques in industrial materials testing, medical diagnostics in cancer therapy and, not but least, fusion pinch experiments that are being revived once again to find alternative sources of energy.

ACKNOWLEDGEMENTS

I thank the late W.H.Bostick, V.Nardi, T.Erber, J.Brzosko and C.Powell for discussion over the last few years on the application of Dense Plasma Pinch Focus with special reference to Astrophysical and Cosmological applications.

REFERENCES

W.H.Bostick, (1956) Phys.Rev. 104, 292
W.H.Bostick, (1958) Rev.Mod.Phys. 30, 1090
V.Nardi, W.H.Bostick, J.Feugeas and W.Prior (1980) Phys.Rev. 27A, 2217
A.L.Peratt, (1986) IEEE Transactions on Plasma Science PS-14, 639 and 763
H.Alfven, ibid., p.629 and 779

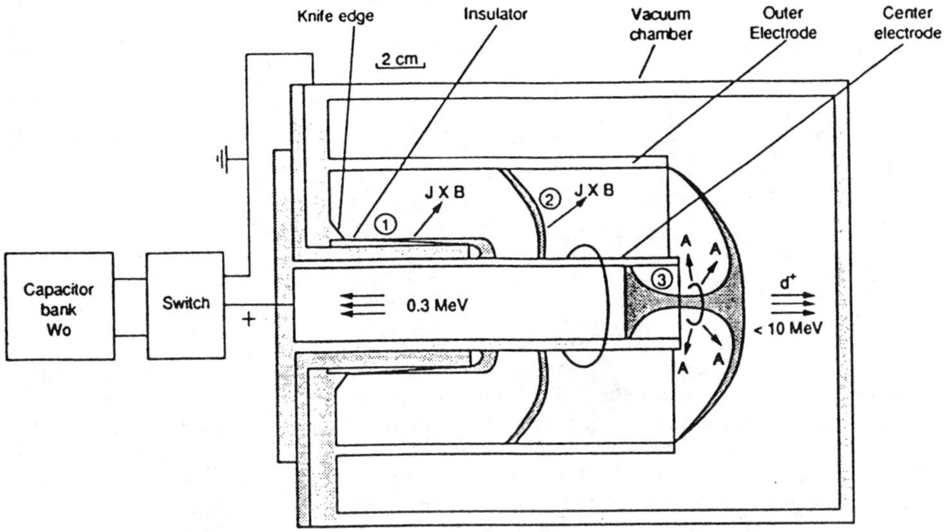

Fig. 1 Schematic view (axial cross section) of a Mather-type plasma focus machine with field distortion elements: APF (Advanced Plasma Focus) = PF with a field distortion element (here a circular knife edge of the breech).

The configuration of the current sheath, where the leading peak of the interelectrode current density is concentrated is outlined during the three main stages of the discharge:

1. Initial phase: gas breakdown, plasma-sheath formation and lift-off from the insulator.
2. Axial acceleration phase: the $\underline{J} \times \underline{B}$ force accelerates current sheath and accumulated mass.
3. Radial compression phase: typical pinch radius at maximum of radius compression $\cong 2$ mm.

Instabilities and high nuclear reactivity occur after phase 3 (Ref. 2).

BARYON FRACTION IN THE UNIVERSE, ANTIPROTON LIFETIME, AND THE DIFFUSE RADIATION

S.A. Stephens
Tata Institute of Fundamental Research
Homi Bhabha Road, Bombay 400005, India

and

N. Finetti
Particle Astrophysics Laboratory
New Mexio State University, Las Cruces, NM 88003

ABSTRACT

We have calculated the diffuse spectrum in the energy region between 0.1 KeV and 500 MeV, whivh arise from the decay of antiprotons in the baryon Symmetric Universe. Comparing this with the observation, we derived a life time of > $2.3x10^{24}$ sec for the antiproton, which is 17 orders better than the one determined from laboratory experiments. We predict a cut-off in the diffuse radiation beyond 460 MeV, and if seen, would provide evidence for the existence of antimatter domains in the universe.

1. INTRODUCTION

The question of whether the Universe is baryon symmetric or not is a long standing problem (eg: Sakharov, 1967; Stegman, 1976; Brown & Stecker, 1979; Turner & Kolb, 1990). Attempt has been made in the past to look for the signature of its existence in the diffuse radiation from the annihilation products of antimatter with matter at large red-shifts (Stecker et al, 1971). No conclusive evidence could be obtained from approach and Stegman (1976) concluded that the matter and antimatter must be separated on scales larger than clusters of galaxies. If matter and antimatter domains could exists in the baryon symmetric universe (Brown & Stecker, 1979; Sato, 1981), one could to predict possible annisotropies in the diffuse radiation (Gao et al, 1990) In this context, to search for antinuclei in cosmic rays is indeed a significant step towards establishing the existence of antimatter in the Universe. The current upper limits of < 10^{-4} on the fraction of antinuclei (Stephens, 1990; Ormes & Streitmatter, 1991) clearly rules out their presence in our galactic neighbourhood, but they may exist in clusters at distances well beyond 10 Mpc. Following the suggestion of O'Connor(1994), we consider the possibility that antiprotons decay to a neutral pion and an electron, which is expected to be the dominent decay mode. While the measured lifetime of proton is > 1.6×10^{40} s (Becker-Szendy et al, 1990), attempt made to measure the life time of antiproton has given only a lower limit of 8.8×10^{6} s (Gabrielse et al, 1990). The existence of antiprotons in cosmic radiation (see Stephens and Golden, 1987) clearly suggests that the life time of antiprotons has to be > 10^{14} s. If antimatter exists in the Universe, it is obvious that the decay products of antiprotons would produce diffuse radiation, whose character would be very different from the other radiations. In this paper we examine this aspect in detail.

2. PROCEDURE

When antiproton decays at rest, the energies of electron and pion are 459.4 and 478.8 MeV respectively. While electrons interact with the microwave and starlight photons, lifting their energies to X and gamma ray energy regime, neutral pions decay directly to 2 gamma rays. The flux of these radiation at energy E, as observed at the present epoch is given by (Stecker, 1991),

$$dJ(E)/dE = c/(4\pi.H_oh) \int Q\{E',z\}dz/\{(1+z)^4(1+\Omega z)^{0.5}\}$$

Here, H_oh is the Hubble constant, = 1.0 is the cosmic density term. The production spectrum Q of the radiation at a red-shift z with energy $E'=E(1+z)$ is written as,

$$Q(E') = f_b.(N_b/\tau_{\bar{p}}).\{[dN/dE']_{\pi o} + [dN/dE']_{comp}\} \int m(dn/dm)dm$$

where, $\tau_{\bar{p}}$ is the life time of antiprotons for the above mode of decay and f_b = 0.5 is the antimatter fraction in the Symmetric Universe. In the above equation, N_b is the number of baryons in a cluster of mass 10^{15} $M_\odot$. Study of X-rays in clusters show that the baryon mass fraction is about 15% of the total mass (Mushotzky, 1994) and hence we take N_b = 1.8×10^{71} baryons. We use the differential cluster mass distribution of Press and Schechter (1974), which is given by Perrenod (1980) for Ω = 1.0 as,

$$dn(m,z)/dm = A.(1+z)^4.(m/m_o)^{-(1+\alpha)}.\exp\{-(1+z)^2.(m/m_o)^{2(1-\alpha)}\}$$

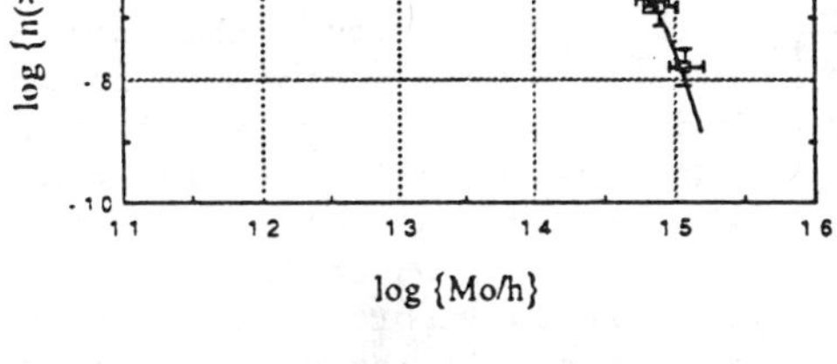

Fig.1 The calculated integral mass distribution is compared with the observations.

Using this expression we derived the integral cluster distribution upto z = 0.1, in the region between 10^{12} $(M_\odot/h)$ and 10^{15} $(M_\odot/h)$, and compared with the observation of Bachall and Cen(1993). This is shown in Fig.1. Notice the excellent fit to the data points except at $M_{12}(M_{12}=10^{12}M_\odot)$. From this comparison, we derived the values of A= 5.0×10^{-4} cluster.{$(Mpc/h)^{-3}$ per M_{15}}, α = 0.55 and m_o = 0.17 M_{15}/h. We use these parameters to determine the baryonic mass in clusters as a function of z.

The energy spectrum of photons from the decay of pions can be approximated using kinematics of decay and can be written as,

$$(dN/dE)_{\pi o} = 4.353\times10^{-3} \text{ photon/MeV} \qquad \text{for } 9.709 \text{ MeV} < E < 469.135 \text{ MeV}$$

= 0.0 otherwise

and the inverse Compton sepctrum can be written as,

$$(dN/dE)comp = \int n_{ph}[\varepsilon,z].\sigma(E,\varepsilon,E_e).c.t_{1/2}[z].d\varepsilon$$

where n_{ph} is the energy distribution of blackbody photons, $\sigma(E,\varepsilon,E_e)$ is the cross-section for the electron of energy E_e to scatter an ambient photon of energy ε to an energy E, and $t_{1/2}$ is the time taken for the electron to lose half its energy. The life time of electrons against energy loss is primarily determined by the inverse Compton scattering with the microwave background photons. As a result, the life time due to energy loss is much shorter than the cosmic time scale, specially at large red-shifts. In the case of starlight we assume a constant black body temperature of 10^4K and a dilution factor of $10^{16}/(1+z)^3$ in clusters.

3. RESULTS

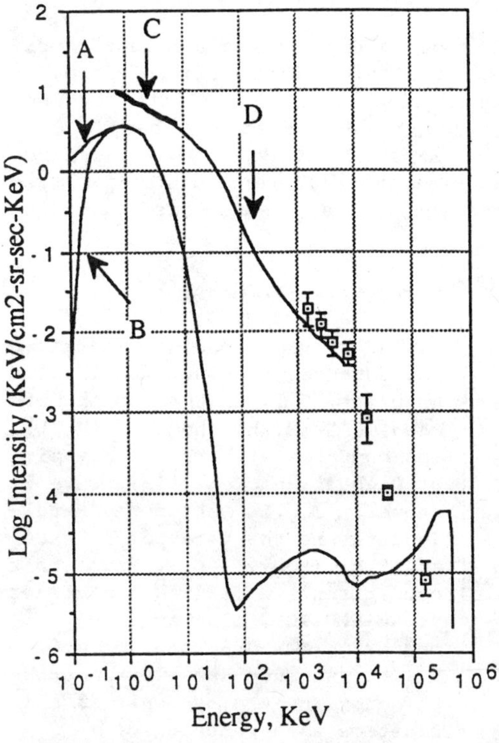

Fig.2. The calculated diffuse spectrum is compared with the observation as explained in the text.

Curve A in Fig.2 is the calculated radiation arising from the above processes from .1 KeV to 500 MeV upto a red-shift z = 5. Curve B is the expected spectrum by taking into account the absorption for a hydrogen column density of 10^{20} n_H cm^{-2} and using the cross-section of Morrison & McCammon (1983). The absorption is important only below 1 KeV. Four features can be seen in the spectrum. 1) The one KeV feature corresponds to the inverse Compton scattering with the blackbody photons. 2) The one around one MeV is from the star light. 3) The peak at about 300 MeV is due to pion decay. 4) the cut-off above 450 MeV is due to the kinematics of pion decay. We found that contributions from regions of large red shifts are very small. For instance, 90% of the radiation at 1 KeV comes from z < 1.0 and at 100 MeV, 90% comes from z < 3.0. It is important to notice that the calculated spectrum is divided by 10^{24} for convenience

We have also shown in Fig. 2, the observed diffuse radiation above 1 KeV. The spectrum below 10 KeV denoted by C is the ASCA results (Gendreau et al,

1994), Curve D is the best fit to the data upto 6 MeV as given by Gruber (1991) and the data points are from direct obsrvations (Schonfelder et al, 1980; observed flux at 150 MeV is due to the decay of antiprotons in the Symmetric Universe, we obtain an lower limit to the life time of antiproton as $4x10^{24}h^2$ s for this decay mode. A similar comparison made at about a few KeV gives a value of$4.3x10^{23}h^2$ s. If the gamma ray spectrum is extended to higher energies, one would able to look for the feature corresponding to the sharp cut-off in the spectrum above 400 MeV and to get a better limit for the life time of antiproton by about an order of magnitude.

The value of h varies from 0.4 to 1.0, and if we take a value of 0.75, we get a lower limit of $2.25x10^{24}$ s, which is about 17 orders of magnitude better than the one determined from laboratory experiments and is about 9 orders of magnitude better than the inference made from the cosmic ray life time. The spectral cut-off beyond 460 MeV, if seen provides an evidence to the existence of antimatter domains in the universe and would then set a strict limit on the life time of antprotons.

ACKNOWLEDGEMENT: We thank Dr. R.E. Streitmatter for the help and useful discussions. N. Fenetti acknowledges that her stay at the New Mexico State University is financed through the Fellowship from the University of Bari, Italy.

REFERENCES

Bahcall, N. and Cen, R, Ap. J., 407, L49 (1993).
Becker-Szendy, R., et al, Phys. Rev., D42, 2974 (1990).
Brown, R.W. ans Stecker, F.W., Phys. Rev. Lett., 43, 315 (1979)
Fichtel C.E., et al, Ap.J., 222, 833 (1978).
Gabrielse, G. et al, Phys. Rev. Lett., 65, 1317 (1990).
Geo, Y.T. et al., Ap. J., 361, L37 (1990)
Gendreau, K.C. et al, preprint (1994)
Mushotzky, R., Reported in this conference
Gruber, D.E., Proc. Int. Workshop on X-rays, ED. Barcons & Fabian, Laredo, Spain (1991).
Morrison, R. and McCammon, D., Ap. J., 270, 119 (1983).
O'Connor, D.J., preprint INFN PI/AE 94/06, 1994.
Ormes, J.F. and Streitmatter, R.E., Int. School of Astrophysics D Chalonge, 1991, ed. Sanchez & Zichichi, World Scientific, p.614
Perrenod, S.C., Ap. J., 236, 373 (1980).
Press, W. and Schechter, P., Ap. J., 187, 425 (1974).
Sato, K., Phys. Lett., 99B, 66 (1981).
Sakharov, A.D., JETP lett., 5, 24 (1967).
Schonfelder, V. et al, Ap. J., 240, 350 (1980).
Stecker, F.W., 1971, 'Cosmic Gamma Rays' NASA SP-249
Stecker, F.W. et al, Phys. Rev. Lett., 27, 1469 (1971).
Steigman, G., Ann. Rev. Astron. Astrophys., 14, 339 (1976)
Stephens, S.A., Proc. 21st Int. Cosmic Ray Conf., 11, 99 (1990)
Stephens. S.A. and Golden, R.L., Space Sci. Rev., 46, 31 (1987).
Turner, M. and Kolb, E., The Early Universe, Addison-Wesley (1990)

SUMMARY

SUMMARY

Joseph Silk
Departments of Astronomy and Physics, and Center for Particle Astrophysics
University of California, Berkeley, CA 94720, USA

This conference has presented an almost united front on the dark matter problem. In this summary talk, I will review the highlights of the meeting, taking full advantage of my temporary need to indulge in handwaving to a greater extent than I ordinarily do.

Historically, dark matter has had a venerable history as was eloquently outlined by TRIMBLE. In the solar system, orbital anomalies led to the discovery of the planet Neptune. Similar anomalies also led to the discovery of Pluto, despite the fact that this planet was subsequently realized to have insufficient mass to account for the anomalies, once its satellite Charon was utilized to give a direct mass determination. Orbital anomalies can also be misleading: recall the failure to find the hypothesized planet Vulcan. Instead, the explanation for the precession of the perihelion of Mercury was accounted for by Einstein in his theory of general relativity.

On a much larger scale, Zwicky drew attention in 1933 to a dark matter problem in galaxy clusters on a scale of 1 Mpc. The definitive study of the Coma cluster in 1970 confirmed the dark matter problem in galaxy clusters. From the mid 1960's, studies of binary galaxies by Page and others suggested that there was a dark matter problem on scales of order 100 kpc. Modern observations have not succeeded in producing more precise mass estimates for binary galaxies. However in the 1970's, the work of Rubin, Roberts, and others demonstrated from the rotation curves of galaxies that dark matter dominated galaxy halos on scales of 10 to 100 kpc. In 1975, two independent studies by Ostriker *et al* and Einasto *et al* argued that dark halos were required in order to guarantee that galactic disks are stable.

Interestingly enough, this theoretical argument has since been realized to be incorrect. Spheroids can provide much of the necessary stabilization. However, the combination of observations and theory had led by 1980 to a consensus among astronomers that at least 90 percent of the matter in the universe was non-luminous. In 1980, the situation took a turn for the worse when Guth argued convincingly that the very early universe underwent a short-lived phase of exponential inflation. A consequence of inflationary cosmology was the prediction that the universe should be at critical density to within one part in a hundred thousand. This meant that 99 percent of the matter in the universe had to be dark if the universe underwent inflation.

Dark matter is the most critical issue facing modern astronomers. How can one be content with a galaxy 90 or 99 percent of which is of unknown composition? Astronomers are rushing to meet the challenge, as became abundantly clear at this meeting. The introductory talks by REES, PACZYNSKI and WILCZEK reviewed possible candidates for the dark matter. These span the mass range from 10^4 solar masses to 10^{-5} eV. The mass range above the Planck mass consists of baryonic dark matter candidates, including black holes and degenerate dwarfs. Such objects are known to exist but their contribution to the dark matter density on any scale is unknown.

The complementary mass range to baryonic dark matter is that of particle dark matter, for which the most attractive candidate is the lightest supersymmetric particle. This particle has a mass of order 100 proton masses, is weakly interacting and its abundance can be calculated because such particles once were in thermal equilibrium in the early universe. Unfortunately, we have no guarantee that such particles actually exist, although there are intriguing hints from the experimentally inferred convergence of the electro-weak and strong nuclear interaction strengths that supersymmetry may be a valid description at sufficiently high energy.

The first item to be clarified is of course whether there indeed is a dark matter problem. KAZANAS eloquently defended the thesis that the law of gravitation may be suitably modified on large scales with respect to Einstein's theory so as to dispense entirely with dark matter. The most promising attempt in this direction utilizes a relativistic fourth order theory of gravitation based on conformal invariance. It remains to be seen whether such a theory can account for such diverse phenomena as gravitational lensing by galaxy clusters as well as rotation curves of galaxies.

I will divide my summary of the results presented at this meeting according to a sequence of scales. First, I consider the question of dark matter distributed on scales of up to 1 Mpc. On the smallest scales discussed, that of the disk of our galaxy, the amount of dark matter is still unresolved: a recent estimate attributed to Gould and quoted by ALCOCK was 75($\pm$25) solar masses per square parsec in the vicinity of the sun. Approximately 50 solar masses per square parsec can be accounted for in terms of known types of stars and interstellar matter.

The microlensing results described by ALCOCK, with some 45 events towards the galactic bulge in the MACHO experiment and 11 events in the OGLE experiment, indicated more lensing objects than standard disk models had predicted by at least a factor of 3. However, whether these are dark matter candidates or simply low mass stars is unclear. The mass estimates range between 0.01 and 1 solar mass. Conventional interpretations, in terms of known types of stars, require either a nearly aligned central bar (within 15 degrees), so that star-star lensing is important, or else a maximum disk, which would require a surface density of about 100 solar masses per square parsec near the sun. Interestingly enough, a corollary of the latter interpretation is that the halo scale length is increased by a factor of 2 or so, thereby reducing the predicted microlensing event rate towards the Large Magellanic Cloud by a similar factor. Highlights of the new bulge microlensing detections include a binary lens, which was seen as a beautiful double caustic light curve, and a long duration event with sufficient signal-to-noise that the correction for the earth's acceleration around the sun was necessary to model the light profile. This means that we are on the verge of being able to use parallax to probe the distance of the lensing object and the structure of the Einstein ring.

The local initial mass function was reviewed by LIEBERT, who showed that brown dwarfs do not make a significant contribution to the mass density in either population I or population II. In particular, LIEBERT presented the results of a new parallax survery of nearby population II stars, which showed clear evidence for a turn-down below about 0.3 solar masses in the luminosity function. This is similar to earlier results obtained for disk stars. The work of TINNEY, superceding the earlier study of Kroupa, Tout and Gilmore, finds evidence for a possible upturn in the disk IMF below $\sim 0.1 M_{\odot}$. However the mass contributed by the low mass stars is negligible for the mass budget. BLITZ showed recent infrared (2 micron) luminosity functions, in particular for the Trapezium cluster in the K-band, which reveal a similar turn-down.

Rotation curves were discussed by BOSMA, who argued that the wide diversity of rotation curves shapes ranging from rising to flat and to declining means that there need be no disk-dark halo conspiracy. Observations of dwarf spirals such as DDO 154 show that the rotation curves display nearly rigid body rotation within the inner scale length or two before flattening. This is in apparent contradiction to the predictions of cold dark matter collapse to form a dark halo. However, WHITE argued that the simulations were still controversial, citing his own recent results which appeared to give a significant cold dark matter core length and could fit the observations of inner rotation curves.

The question of how flat halos may be was addressed by BLITZ. He argued that the flaring of the atomic hydrogen layer in the Milky Way disk beyond about 2.5 scale lengths demonstrated that the dark matter could not be in a thin disk, but had to be either in a thick disk or spheroid. In a poster paper, SACKETT argued that observations of polar ring galaxies contrained the shape of the dark halo along both the minor and major axes, to have an ellipticity between E5 and E8. However, PFENNIGER presented a model for the dark halo as an extended flat disk consisting of cold extremely dense molecular hydrogen and gas distributed in density acording to a fractal distribution. He proposed circumventing the disk flaring issue by appealing to extremely non-circular motions for the halo gas in the disk. The case for dark matter in ellipticals was noted by MAKISHIMA. He showed ASCA observations of the elliptical galaxy NGC 4636, for which the mass-to-light ratio is approximately 150 at 100 kpc radius.

Moving on to 10 Mpc scales, galaxy clusters provide an ideal probe of the underlying fluctuation power spectrum as well as the dark matter content of the universe. Clusters of mass $10^{15}M_{15}\mathrm{M}_\odot$ have condensed from a comoving radius of $10M_{15}^{1/3}(\Omega h^2)^{1/3}$ Mpc. Moreover, the fluctuations in the galaxy number counts have unit amplitude when averaged over spheres of radius $8h^{-1}$ Mpc. The abundance and mass function of galaxy clusters accordingly probes the rms density fluctuations in the dark matter, conventionally written as σ_8 over this scale. KOCHANEK found that the cosmological constant-dominated flat model, a popular way to reconcile a high Hubble constant with inflationary cosmology, or at least a flat universe, was in severe difficulty, the large proper volume element resulting in overproduction of the number of wide-separation lensed quasars. A conservative bound is $\Lambda < 0.7$, in terms of the possible vacuum contribution to the cosmological density. Gravitational lensing by intervening clusters also enhances the frequency of splittings of double quasars, effectively probing σ_8 and preferring a value $\sigma_8 \approx 0.7$. N. BAHCALL used optical observations of the cluster mass function to favor a low density universe, cold dark matter in an $\Omega = 1$ universe overproducing the abundances of rich clusters by a large factor.

The most interesting new results on clusters were presented by the x-ray observers, who showed ASCA intensity and temperature maps of galaxy clusters. MAKISHIMA was able to spatially resolve the central cD galaxy in the Fornix cluster of galaxies and study the gas and metallicity distribution in the surrounding cluster. The iron abundance increased from one-third of the solar value in the outer cluster environment to forty percent within a core radius and fifty percent near the central galaxy. The gas fraction also increased with radius. At a distance of about 200 kpc, the gas fraction is approximately twenty percent of the total gravitating mass, and remains at this level out to as far as the mass profile can be traced at approximately 0.8 Mpc. MUSHOTZKY presented ASCA observations of three rich clusters chosen to have little or no evident substructure. Temperature determinations can be made to within an uncertainty of approximately 0.1 keV. These enable one to model the mass profiles and infer a ratio of dark-to-luminous matter that was between 3.1 and 3.6 for the three clusters. The mass-to-light ratios are about 150 at 1 Mpc and the gas fractions are approximately 20 percent. All of these numbers are for a Hubble constant of 50 km/s/Mpc. The measured x-ray temperature is equivalent to a gas velocity dispersion that is about 80 percent of the velocity dispersion observed for the galaxies.

Similar results were presented by OHASHI for a sample of poor galaxy clusters. He demonstrated that the metallicity was about 40 percent of the solar value, and the gas fraction is again about 20 percent of the total gravitating mass, for clusters in the temperature range 2-6 keV. Some clusters showed metallicity gradients, the most extreme being in the Centaurus cluster, where the iron abundance increases by a factor of 4 towards the core, while other poor clusters showed no metallicity gradient. These poor clusters present a marked contrast to the compact groups of galaxies studied by MULCHAEY. These groups typically have very few galaxies, low metallicity and low gas fraction compared to clusters. For example, in the case of NGC 2300, the iron abundance is 8($\pm$2)% of the solar value, and the gas fraction is 6% of the total mass. It would seem that compact groups have lost both gas and the accompanying metals that may have been produced by early supernovae. However, it remains somewhat of a mystery as to why poor clusters would have retained so much more of the early ejecta from galaxies.

Gravitational lensing of background galaxies by rich clusters provides an important new probe of the dark matter distribution in galaxy clusters. TYSON showed an image of the cluster Abell 1689 in which he has identified some 5,000 arclets. In a study of 28 clusters, he concluded from modelling the distortions of the background galaxies that the cores of the dark matter distributions were small, below 75 kpc. BROADHURST described a different technique for modelling the dark mass distribution, which utilized the difference in background galaxy number counts produced by cluster lensing. His method should give an absolute determination of the dark matter mass, whereas use of background image distortions measures the shear and only gives a differential measure of the gradient in the dark matter distribution. SQUIRES showed several dark matter maps produced by applying an inversion procedure to the sheared images of the background galaxies. In one cluster, he inferred a dark matter mass that was approximately three times that determined from x-ray and optical measurements, which rely on the virial theorem assumption of hydrostatic equilibrium. Remarkably, however, the analysis of Abell 2163, the hottest and one of the most luminous x-ray clusters, revealed no significant lensing detection.

The apparent discrepancy between dark matter cores, as determined by lensing relative to the x-ray and optical cores, amounts to a factor of at least 3 in terms of core size determinations. This was explained by WHITE, who showed that in realistic simulations of x-ray clusters, the dark matter profile was poorly approximated by a King profile with a constant core radius. Rather, the dark matter continuously and slowly flattened so that the relevant core radius for the x-ray measurements, determined by the gas emission measure, would be up to an order of magnitude larger than the core radius required by lensing measurements. These latter appeal to the surface density exceeding a critical value.

On larger scales, our knowledge of the dark matter distribution becomes progressively poorer. On scales of 10 to 100 Mpc, two techniques have been applied to infer the dark matter content and fluctuation level of the underlying density distribution. DAVIS showed that the bulk flow measurements which yield the product of σ_8 and $\Omega^{0.6}$ favor a value near unity, although there remain serious doubts about the detailed comparison between the reconstructed density field from the POTENT analysis and the IRAS density field. A second approach uses power spectrum reconstruction, assuming an underlying gaussian density field applied to different redshift surveys, to infer a similar quantity. Poster papers by HEAVENS argued that the degeneracy between σ_8 and Ω could be removed, favoring a value of Ω of about 0.3. However, a poster paper by BRAINERD suggested that halo selection would produce an apparent bias, possibly allowing a higher value of Ω.

The cosmic microwave background provides a powerful measure of fluctuations on scale upward of 100 Mpc, a comoving scale of 100 Mpc at high redshift subtending an angle in the sky of approximately $\Omega^{1/2}$ degrees. These scales, namely 0.5 to 10 degrees, are being probed by many experiments, following the COBE DMR measurement of larger scale fluctuations. GORSKI showed that the COBE DMR data provide a strong constraint on dark matter models. For example, the standard cold dark matter model, when normalized to COBE, requires a value of σ_8 equal to 1.4. Various variations on cold dark matter have been explored in order to try to reconcile the COBE normalization with the locally measured value for σ_8 of about 0.5. ZUREK argued that a certain amount of velocity bias was inevitable, and this effect, combined with a more careful treatment of clustering in redshift surveys, weakened the pairwise galaxy velocity constraint on σ_8 from 0.5 to about 0.7. Of course, this still presents a great challenge to CDM.

One possible solution is that of mixed dark matter (MDM), in which a small admixture of the dark matter is hot. KLYPIN described calculations of a model which contained two neutrinos of mass 2.5 eV and argued that this could meet all observational constraints. The most severe problem lies in accounting for the 20 percent baryon fraction in rich clusters, a difficulty for any model in which $\Omega = 1$ in view of the primordial nucleosynthesis result that $\Omega_B = 0.015h^{-2}$. The issue of structure formation at high redshift is also a challenge for the MDM model. MA found that a 20 percent admixture of hot dark matter significantly improved the confrontation with observations, earlier models with a 30 percent hot dark matter admixture failing to have sufficient small scale power. The fraction of baryons in condensed objects at redshift 3 or larger as measured by damped Lyman alpha clouds towards quasars can probably be explained with a 20 percent hot dark matter fraction. Of course a Hubble constant of 50 km s^{-1} Mpc^{-1} is required if the $\Omega = 1$ model is to survive.

No model, with or without cold dark matter, hot dark matter or baryonic dark matter, was able to account for the apparent dipole motion reported by POSTMAN for a frame defined by Abell clusters at a distance of about $150h^{-1}$Mpc. If these observations continue to hold up when improved distance estimators become available, more than a trivial adjustment of extant model parameters will be required. For the moment, however, one can simply hope that the significance of the Lauer-Postman bulk flow will diminish as new observations are obtained. This certainly has historically been the case for other apparently severe anomalies reported in previous decades that have challenged standard cosmology. These include the spectral distortions in the cosmic blackbody radiation, the apparent lack of fluctuations in the microwave background, and Hubble constant determinations that implied a cosmological age less than that of the earth.

But what if the anomalous large-scale Lauer-Postman dipole prevails? And what if the recent reports of a high Hubble constant determinations are vindicated? Not to mention other horrors lurking in the wings, such as the apparent refusal of the periodicity in pair counts reported for deep pencil beam surveys to vanish quietly. All of these are potentially severe difficulties for the standard cosmological model, and its many variants. Rather than pursue an entirely new theory, as some newspaper headlines occasionally urge, we should be open to more radical possibilities that however remain within the context of the big bang theory

Perhaps we will have to resort to a model such as the following. Imagine our galaxy being within a huge underdense "void", of order 100 Mpc across and at 50 percent of the global density. One could have a locally anomalous, and high, Hubble constant, a density of luminous galaxies locally measured to be half that inferred from deep redshift surveys, and, if we are 10 percent offcenter, a dipole motion as measured by Lauer and Postman. Sadly perhaps, such a vast "hole" is highly improbable in the context of any current theories. Despite the lack of theoretical support, one could point to the deep redshift surveys which reveal a huge underdensity of galaxies, bounded by the Perseus-Pisces supercluster 5500km s^{-1} to the north and its apparent extension, the Great Wall, some 7500km s^{-1} to the south. It is certainly premature to infer that such a hole, and presumably other holes, must definitively be present. However were such structures to be confirmed, they would even lead to anomalous correlations in the galaxy counts on scales of about 100 Mpc if the holes are compensated (in mass, so as to preserve uniform spatial curvature, and, incidentally, reduce any cosmic microwave background signature) by a surrounding shell of galaxies. No doubt theories could readily be devised, for example containing early first order inflationary phase transitions, to contain such voids, should the need arise.

Evidently dark matter is here to stay. On galactic scales, cluster scales, supercluster scales, and even the horizon scale, there are increasingly compelling arguments for the presence of dark matter in excess of that observed directly. It is fair to say, however, that only in galaxy halos is the case absolutely compelling. X-ray observations of clusters are hinting at far more gas than previously believed to be present within galaxy clusters. While these systems are conveniently massive and deep potential wells where the hot x-ray emitting gas can be studied, it is by no means so obvious that comparably large amounts of gas could not be present outside clusters, to bring Ω up to about 10 percent of the critical value, where the primpordial nucleosynthesis constraints would not be grossly violated. For that matter, it is not impossible that dense clouds of cold molecular gas constitute the dark matter of galaxy halos. We are now on the verge of undertaking sensitive searches for dark matter in many of its various guises. This undertaking will undoubtedly energize and stimulate the field of cosmology for the next decade.

APPENDIX A: CONFERENCE PROGRAM

Program of Oral Presentations

Monday 10 October 1994

0800	Registration		
0845	1. Introductory Overviews to Dark Matter		**S. Holt**, chair
0900	**M. Rees**	The Dark Matter Problem	
0945	**B. Paczynski**	Current Views	
1030	Coffee		
	2. Dark Matter and Fundamental Physics		**M. Turner**, chair
1100	**F. Wilczek**	Dark Matter and Particle Physics	
1130	**D. Kazanas**	Dark Matter or a New Theory of Gravity?	
1200	**D. Spergel**	Heresies: Omega<1, Lambda>0, and Topological Defects	
1230	Lunch		
	3. Dark Matter in the Milky Way		**J. Mather**, chair
1400	**C. Alcock**	Gravitational Lensing and MACHOS	
1430	**J. Liebert**	Brown Dwarfs as Dark Matter: What the IMF Tells Us	
1500	**L. Blitz**	Dark Matter in the Disk and Halo	
1530	**M. Valtonen**	The Little Bang: Peculiarities in the Local Group	
1540	Discussion		
1600	Tea		
	4. Dark Matter in External Galaxies		**M. Abramowicz**, chair
1630	**C. Canizares**	Constraints on Dark Matter in Clusters and Galaxies from the Comparison of X-ray and Optical Isophotes	
1700	**A. Bosma**	H I Observations	
1730	**K. Makishima**	X-Ray (ASCA) Measurements of Elliptical Galaxies	
1800	Discussion		
1830	Wine and cheese and poster perusal (to 2200)		

Tuesday 11 October 1994

5. Dark Matter in Rich Clusters — **N. Bahcall**, chair

0830	**T. Broadhurst**	Optical Evidence for Dark Matter in Rich Clusters
0900	**R. Mushotzky**	ASCA Measurements of X-Rays from Rich Clusters
0930	**D. Cannon**	Mapping the Properties of Galaxy Clusters with ROSAT
0940	**N. Bahcall**	Dark-Matter in Clusters and on Large Scales
0950	Discussion	
1000	Coffee	

6. Dark Matter in Groups of Galaxies — **C. Bennett**, chair

1030	**C. Bird**	New Optical Data on Poor Clusters
1100	**J. Mulchaey**	X-ray Measurements of Groups of Galaxies
1130	**T. Ohashi**	ASCA Measurements of X-Rays from Poor Clusters
1200	Discussion	
1230	Lunch	

7. Gravitational Lensing By Cluster-Scale Dark Matter — **L. Krauss**, chair

1400	**A. Tyson**	Dark Matter Mapping by Gravitational Lens Tomography
1435	**G. Squires**	Mapping the Dark Matter in Clusters by Weak Gravitational Lensing
1440	**C. Kochanek**	Gravitational Lensing Results
1510	**R. Narayan**	Interpretation of Gravitational Lens Results
1540	**A. Loeb**	Evidence from Gravitational Lensing for a Non-Thermal Pressure Support in the Cluster A2218
1545	Discussion	
1600	Tea	

8. Dark Matter on Large Scales — **A. Szalay**, chair

1630	**M. Davis**	Dark Matter Inferred from Large Scale
1700	**M. Postman**	Cluster Streaming
1730	**A. Klypin**	Cosmological Models with a Mixture of Cold and Hot Dark Matter
1800	Discussion	
1830	CONFERENCE BANQUET Featuring shadowy remarks from **M. Turner**	

Wednesday 12 October 1994

	9. The Nature of Dark Matter	**E. Wright**, chair
0830 **T. Bowles**	Neutrinos as Dark Matter	
0900 **S. White**	Discriminators for the Nature of Dark Matter	
0930 **D. Pfenniger**	Cold Molecular Gas as Dark Matter Candidate in Galaxies and Clusters	
0940 **C.-P. Ma**	Structure Formation in Cold+Hot Dark Matter Models	
0950	Discussion	
1000	Coffee	
	10. Dark Matter and the Universe	**V. Rubin**, chair
1030 **W. Zurek**	Second Coming of Cold Dark Matter?	
1100 **K. Gorski**	The CMB/Dark Matter Connection	
1130 **V. Trimble**	Looking Backward, Darkly	
1200	Discussion	
1230	Lunch	
	11. Conference Summary	**S. Holt**, chair
1330 **J. Silk**	Rapporteur	

APPENDIX B: LIST OF ATTENDEES

Attendees

NAME	AFFILIATION	E-MAIL ADDRESS
ABRAMOWICZ, MAREK	U GOTHEBORG, SWEDEN	MAREK@FY.CHALMERS.SE
AIZENMAN, MORRIS L.	NSF	MAIZENMAN@NSF.GOV
AJHAR, EDWARD	NOAO/KPNO	AJHAR@NOAO.EDU
ALCOCK, CHARLES	LLNL/IGPP	ALCOCK@SUNLIGHT.LLNL.GOV
ALLEN, RON	STScI	RJALLEN@STSCI.EDU
AMMIRAJU, P.	STEVENS INST. OF TECHNOLOGY	
ANDREWS, THOMAS B.		
ARNAUD, KEITH	NASA/GSFC	ARNAUD@LHEAVX.GSFC.NASA.GOV
AUDLEY, DAMIAN	NASA/GSFC	AUDLEY@LHEAVX.GSFC.NASA.GOV
BAHCALL, NETA A.	PRINCETON	NETA@ASTRO.PRINCETON.EDU
BANDAY, TONY	USRA/GSFC	BANDAY@STARS.GSFC.NASA.GOV
BARBIER, LOUIS M.	NASA/GSFC	BARBIER@LHEAVX.GSFC.NASA.GOV
BARRETT, P.	NASA/GSFC	BARRETT@LHEAVX.GSFC.NASA.GOV
BARTELMANN, MATTHIAS	CFA & MPA, GARCHING	MSB@CFA.HARVARD.EDU
BAYKAL, A.	NASA/GSFC	BAYKAL@LHEAVX.GSFC.NASA.GOV
BENDER, PETER	JILA/U COLORADO	PBENDER@JILA.COLORADO.EDU
BENNETT, CHUCK	NASA/GSFC	BENNETT@STARS.GSFC.NASA.GOV
BENNETT, DAVID	UC-BERKELEY	BENNETT@SUNLIGHT.LLNL.GOV
BERTSCH, D.	NASA/GSFC	DLB@SAGE0.GSFC.NASA.GOV
BIRD, CHRISTINA M.	U KANSAS	TBIRD@KULA.PHSX.UKANS.EDU
BLACK, K.	NASA/GSFC	BLACK@LHEAVX.GSFC.NASA.GOV
BLITZ, LEO	U MARYLAND	BLITZ@ASTRO.UMD.EDU
BOCK, JAMIE	UC-BERKELEY	JBOCK@PHYSICS.BERKELEY.EDU
BOESHAAR, PATRICIA C.	DREW UNIVERSITY	BOESHAA@DREW.DREW.EDU
BOLDT, ELIHU	NASA/GSFC	BOLDT@LHEAVX.GSFC.NASA.GOV
BONNELL, JERRY	NASA/GSFC	
BOSMA, ALBERT	OBS DE MARSEILLE	BOSMA@BATIS.CNRS-MRS.FR
BOWEN, TED	U ARIZ	MCGUIRE@SOLITON.PHYSICS.ARIZONA.EDU
BOWLES, THOMAS J.	LANL	TJB@LANL.GOV
BRAINERD, TEREASA	LANL	TGB@T6-SERV.LANL.GOV

BRANDT, JOHN C.	LASP/U COLORADO	LYRAE::BRANDT
BRIEU, PHILIPPE	PRINCETON OBS	PHILIPPE@ASTRO.PRINCETON.EDU
BROADHURST, THOMAS	JOHNS HOPKINS U	TJB@SKYSRV.PHA.JHU.EDU
BROEILS, ADRICK	CORNELL U	BROEILS@ASTROSUN.TN.CORNELL.EDU
BROMLEY, BEN	LANL	BROMLEY@EAGLE.LANL.GOV
BUOTE, DAVID	MIT	DBUOTE@SPACE.MIT.EDU
CANIZARES, CLAUDE	MIT	CRC@SPACE.MIT.EDU
CANNON, DAMIAN	U BIRMINGHAM	DBC@STAR.SR.BHAM.AC.UK
CHENG, ED	NASA/GSFC	EC@COBI.GSFC.NASA.GOV
CHEUNG, CYNTHIA	NASA/GSFC	CCHEUNG@NSSDC.GSFC.NASA.GOV
CHI, LAN KEH	U.S. NAVAL ACADEMY	CHI@SCS.USNA.NAVY.MIL
CHOUDHURY, LATIF	ELIZ C ST U	JUTTA.CHOUDHURY@LAUNCHPAD.UNC.EDU
CHRISTIAN, E.	NASA/GSFC	CRISTIAN@LHEAVX.GSFC.NASA.GOV
CLINE, THOMAS L.	NASA/GSFC	CLINE@LHEAVX.GSFC.NASA.GOV
COMER, GREG	PARKS COLLEGE ST LOUIS	COMER@NEWTON.SLU.EDU
CONTOPOULOS, J.	NASA/GSFC	CONTO@ROSSERV.GSFC.NASA.GOV
COWEN, RON	SCIENCE NEWS	
DAVE, ROMEEL	UC-SANTA CRUZ	RAD@LICK.UCSC.EDU
DAVIS, DAVID	NASA/GSFC & U MARYLAND	DDAVIS@ASTRO.UMD.EDU
DAVIS, MARC	UC-BERKELEY	MARC@BKYAST.BERKELEY.EDU
DAVIS, S.	NASA/GSFC	
DE LAIX, ANDREW A.	OHIO STATE U	DELAIX@MPS.OHIO-STATE.EDU
DE YOUNG, DAVID	NOAO/KPNO	DEYOUNG@NOAO.EDU
DEBATTISTA, VICTOR	RUTGERS U	DEBATTIS@PHYSICS.RUTGERS.EDU
DIXIT, VIJAI	PARKS COLLEGE ST. LOU	DIXITVV@SLUVCA.SLU.EDU
DOLAN, J. F.	NASA/GSFC	15658::TEJFD
DRACHMAN, RICHARD J.	NASA/GSFC	DRACHMAN@STARS.GSFC.NASA.GOV
DRESSEL, LINDA L.	APPLIED RESEARCH	DRESSEL@ARCFOS1.ARCLCH.COM
DWEK, ELI	NASA/GSFC	EDWEK@STARS.GSFC.NASA.GOV
ESPOSITO, J.	NASA/GSFC	ESPOSITO@LHEAVX.GSFC.NASA.GOV
ESTEBAN, ERNESTO	U. PUERTO RICO	E_ESTEBAN@UPR1.UPR.CLU.EDU
FAHEY, RICHARD P.	NASA/GSFC	FAHEY@STARS.GSFC.NASA.GOV
FELTEN, JAMES E.	NASA/GSFC	FELTEN@STARS.DNET.NASA.GOV
FICHTEL, CARL	NASA/GSFC	SCCEF@SCFVM.BITNET
FISCHER, PHIL	AT & T BELL LABS	PHILF@PHYSICS.ATT.COM

FOCKE, WARREN	NASA/GSFC	FOCKE@LHEAVX.GSFC.NASA.GOV
FONG, DICK	U DURHAM	RICHARD.FONG@DURHAM.AC.UK
FRAMPTON, PAUL H.	U. NORTH CAROLINA	FRAMPTON@PHYSICS.UNC.EDU
FROST, KENNETH J.	NASA/GSFC	FROST@NSSDCA.GSFC.NASA.GOV
FUCHS, BURKHARD	INST HEIDELBERG	S34@MVS.URZ.UNI-HEIDELBERG.DE
GAIDOS, ERIC J.	MIT	EJGAIDOS@SPACE.MIT.EDU
GARILLI, BIANCA	IFCTR-CNR MILANO	BIANCA@IFCTR.MI.CNR.IT
GEHRELS, NEIL	NASA/GSFC	GEHRELS@LHEAVX.GSFC.NASA.GOV
GERBAL, DANIEL	INSTITUT D'ASTROPHYSIQUE	GERBAL@IAP.FR
GIACCONI, RICCARDO	EUROPEAN SOUTHERN OBSERVATORY	
GILES, B.	NASA/GSFC	GILES@LHEAVX.GSFC.NASA.GOV
GORSKI, KRZYSZTOF	NASA/GSFC	GORSKI@STARS.GSFC.NASA.GOV
GRABER, JIM	LIBRARY OF CONGRESS	JGRA@LOC.GOV
GREYBER, HOWARD D.	GREYBER ASSOCIATES	HGREYBER@CAP.GWU.EDU
GROGIN, NORMAN A.	CFA	GROGIN@CFA.HARVARD.EDU
GUERTLER, HELENA		
GUHATHAKURTA, RAJA	STScI	RAJA@STSCI.EDU
GULL, THEODORE	NASA/GSFC	GULL@STARS.GSFC.NASA.GOV
HAIG, S. J.	LOYOLA COLLEGE	HAIG@LOYOLA.EDU
HANAMI, HITOSHI	IWATE U	HHANAMI@YEELEN.HSS.IWATE-U.AC.JP
HARDING, ALICE	NASA/GSFC	HARDING@LHEAVX.GSFC.NASA.GOV
HARTMAN, BOB	NASA/GSFC	RCH@SAGEO.GSFC.NASA.GOV
HASAN, HASHIMA	NASA HEADQUARTERS/JPL	HHASAN@HQ.NASA.GOV
HASSENBALG, FEDERICO	SO CAROLINA	PHYSICS@SCUCHE.PSC.SCAROLINA.EDU
HAUSER, MICHAEL	NASA/GSFC	HAUSER@STARS.GSFC.NASA.GOV
HAWKINS, MIKE	ROYAL OBS EDINBURGH	MRSH@STARLINK.ROE.AC.UK
HEAP, SALLY	NASA/GSFC	
HENRIKSEN, MARK	U ALABAMA	MARK@BAMA.ASTR.UA.EDU
HENRY, DICK	JOHNS HOPKINS UNIVERSITY	RCH@PHA.JHU.EDU
HINSHAW, GARY	HUGHES STX	HINSHAW@STARS.GSFC.NASA.GOV
HINTZEN, PAUL	NASA/GSFC & UNLV	HINTZEN@STARS.GSFC.NASA.GOV
HOLT, STEVE	NASA/GSFC	HOLT@LHEAVX.GSFC.NASA.GOV
JAHODA, KEITH	NASA/GSFC	JAHODA@LHEAVX.GSFC.NASA.GOV
JONES, FRANK C.	NASA/GSFC	JONES@LHEAVX.GSFC.NASA.GOV
JOSTIES, JERRY	U.S. NAVAL OBS	JOS@CASA.UNSO.NAVY.MIL

KAFATOS, MENAS	GEORGE MASON U	MKAFATOS@COMPTON.GMU.EDU
KAMIONKOWSKI, MARC	COLUMBIA U	KAMION@PHYS.COLUMBIA.EDU
KAPLAN, GEORGE	U.S. NAVAL OBS	GHK@CERES.USNO.NAVY.MIL
KAYSER, SUSAN	NATIONAL SCI FOUND	SKAYSER@NSF.GOV
KAZANAS, DEMOS	NASA/GSFC	KAZANAS@LHEAVX.GSFC.NASA.GOV
KELLEY, RICHARD	NASA/GSFC	KELLEY@LHEAVX.GSFC.NASA.GOV
KERNAN, PETE	CWRU	PETE@THEORY2.PHYS.CWRU.EDU
KERR, FRANK	U MARYLAND & USRA	FKERR@GVSP.USRA.EDU
KHAN, IRSHAD		IKHAN@MITLNS.MIT.EDU
KHERSONSKY, VALERY	U PITTSBURGH	VKK@PHYAST.PITT.EDU
KII, TSUNEO	ISAS, JAPAN	KII@ASTRO.ISAS.AC.JP
KOCHANEK, CHRISTOPHER	HARVARD UNIVERSITY	KOCHANEK@CFA.HARVARD.EDU
KOGUT, AL	HUGHES STX	KOGUT@STARS.GSFC.NASA.GOV
KONDO, YOJI	NASA/GSFC	KONDO@STARS.GSFC.NASA.GOV
KRAUSS, LAWRENCE	CWRU	KRAUSS@THEORY1.PHYS.CWRU.EDU
KRIZMANIC, J.	NASA/GSFC	KRIZMANIC@LHEAVX.GSFC.NASA.GOV
LAFLAMME, RAYMOND	LANL	LAF@TDO-SERV.LANL.GOV
LANDY, STEPHEN	CARNEGIE OBSERVATORIES	LANDY@OCIW.EDU
LEIGHLY, K.	NASA/GSFC	UPHSTKL@MTSUNIX1.BITNET
LEISAWITZ, DAVID	NASA/GSFC	LEISAWITZ@STARS.GSFC.NASA.GOV
LEONARD, PETER	U MARYLAND	PJTL@ASTRO.UMD.EDU
LEVENTHAL, MARV	U MARYLAND	ML@ASTRO.UMD.EDU
LIEBERT, JAMES	U ARIZONA	LIEBERT@AS.ARIZONA.EDU
LOEB, AVI	HARVARD U	LOEB@CFAT3.HARVARD.EDU
LOEWENSTEIN, MICHAEL	NASA/GSFC	LOEWENSTEIN@LHEAVX.GSFC.NASA.GOV
LUBIN, LORI M.	PRINCETON U	LML@ASTRO.PRINCETON.EDU
MA, CHUNG-PEI	CALTECH	CPMA@KARL.TAPIR.CALTECH.EDU
MADAU, PIERO	STSCI	MADAU@STSCI.EDU
MADEJSKI, G.	NASA/GSFC	MADEJSKI@LHEAVX.GSFC.NASA.GOV
MAKISHIMA, KAZUO	U TOKYO	MAXIMA@MIRANDA.PHYS.S.U-TOKYO.AC.JP
MANNHEIM, PHILIP	U CONNECTICUT	MANNHEIM@UCONNVM.UCONN.EDU
MARAN, STEPHEN P.	NASA/GSFC	HRSMARAN@STARS.GSFC.NASA.GOV
MARANI, GABRIELLA	GEORGE MASON U	GMARANI@SCIENCE.GMU.EDU
MARKEVITCH, MAXIM	ISAS	MAXIM@ASTRO.ISAS.AC.JP
MAROCHINIK, LEONID	CSC	LMAROCHN@EXPLORER.CSC.COM

MARSHALL, FRANK	NASA/GSFC	MARSHALL@LHEAVX.GSFC.NASA.GOV
MARTEL, HUGO	U TEXAS	HUGO@SAGREDO.AS.UTEXAS.EDU
MATHER, JOHN C.	NASA/GSFC	MATHER@STARS.GSFC.NASA.GOV
MCGUIRE, PATRICK	U ARIZ	MCGUIRE@SOLITON.PHYSICS.ARIZONA.EDU
MEURER, GERHARDT R.	JOHNS HOPKINS U	MEURER@POUTINE.PHA.JHU.EDU
MICHALITSIANOS, ANDREW	NASA/GSFC	MICHALTS@TORTE.GSFC.NASA.GOV
MITCHELL, J.	NASA/GSFC	
MIYAJI, TAKAMITSU	U MARYLAND & NASA/GSFC	
MOORE, BEN	U WASH	MOORE@HERMES.ASTRO.WASHINGTON.EDU
MORRISON, HEATHER	CASE WESTERN RESERVE U.	HEATHER@NOAO.EDU
MULCHAEY, JOHN	CARNEGIE INSTITUTE	MULCHAEY@PEGASUS.OCIW.EDU
MUSHOTZKY, RICHARD	NASA/GSFC	RICHARD@XTELAB.GSFC.NASA.GOV
NADEAU, RICHARD J.		
NARAYAN, RAMESH	HARVARD U/CfA	NARAYAN@CFA.HARVARD.EDU
NEFF, SUSAN	NASA/GSFC	NEFF@STARS.GSFC.NASA.GOV
NEMIROFF, ROBERT	GMU & NASA/GSFC	NEMIROFF@GROSSC.GSFC.NASA.GOV
NORRIS, JAY P.	NASA/GSFC	NORRIS@LHEAVX.GSFC.NASA.GOV
NOVICK, ROBERT	COLUMBIA U	RN@CUPHY3.PHYS.COLUMBIA.EDU
OHASHI, TAKAYA	TOKYO METRO U	OHASHI@PHYS.METRO-U.AC.JP
OLLING, ROB	COLUMBIA U	OLLING@TOSCA.PHYS.COLUMBIA.EDU
ORMES, JONATHAN	NASA/GSFC	ORMES@LHEAVX.GSFC.NASA.GOV
OZERNOY, LEONID	GMU & NASA/GSFC	OZERNOY@STARS.GSFC.NASA.GOV
PACZYNSKI, BOHDAN	PRINCETON U OBS	BP@ASTRO.PRINCETON.EDU
PALMER, D.	NASA/GSFC	PALMER@TGRS.GSFC.NASA.GOV
PERSIC, MASSIMO	SISSA-TRIESTE	PERSIC@TSMI19.SISSA.IT
PETRE, ROBERT	NASA/GSFC	PETRE@LHEAVX.GSFC.NASA.GOV
PFENNIGER, DANIEL	GENEVA OBS	PFENNIGER@SCSUN.UNIGE.CH
PIZZELLA, ALLESSANDRO	U PADOVA	PIZZELLA@ASTRPD.PD.ASTRO.IT
POOLA, R.	NASA/GSFC	
POSTMAN, MARC	STScI	POSTMAN@STSCI.EDU
PTAK, A.	NASA/GSFC	PTAK@LHEAVX.GSFC.NASA.GOV
QUINN, PETER J.	MT STROMLO & SS OBS	PJQ@MSO.ANU.EDU.AU
RAND, RICHARD	U MARYLAND	RJR@ASTRO.UMD.EDU
REES, MARTIN	INST of ASTRONOMY	MJR@MAIL.AST.CAM.AC.UK
REINES, FREDERICK	UC-IRVINE	FREINES@UCI.EDU

RHEE, GEORGE	U NEVADA-LAS VEGAS	GRHEE@NEVADA.EDU
RHO, JEONGHEE	NASA/GSFC & U MD	RHO@XRAY-1.GSFC.NASA.GOV
ROBERTS, MORTON	NRAO	MROBERTS@NRAO.EDU
RODRIGUES-WILLIAMS, LILIYA	U WASHINGTON	LLRW@ASTRO.WASHINGTON.EDU
ROOD, H. J.	INST FOR ADVANCED STUDY	ROOD@SNS.IAS.EDU
ROSENBAUM, DORIS	SMU	ROSENBAUM@PHYVMS.PHYSICS.SMU.EDU
RUBIN, VERA	CARNEGIE INST OF WASH	RUBIN@GAL.CIW.EDU
SACKETT, PENNY D.	INST FOR ADV STUDY	PSACKETT@GUINNESS.IAS.EDU
SALPETER, ED	CORNELL UN	TOMION@ASTROSUN.TN.CORNELL.EDU
SCHARF, CALEB	NASA/GSFC	CALEB@ROSSERV.GSFC.NASA.GOV
SCHILD, RUDY	CFA	RSCHILD@CFA.HARVARD.EDU
SCHLEGEL, E.	NASA/GSFC	ERIC@HEASFS.GSFC.NASA.GOV
SCOVILLE, ANTHONY	AEROVEST, INC	ACLVILLE@CAIS.COM
SECKEL, DAVID	BARTOL RESEARCH INST	SECKEL@BARTOL.UDCL.EDU
SERLEMITSOS, PETER	NASA/GSFC	SERLEMITSOS@LHEAVX.GSFC.NASA.GOV
SHAPIRO, PAUL	U TEXAS	SHAPIRO@ASTRO.AS.UTEXAS.EDU
SHAYA, ED	U MARYLAND	SHAYA@IMG.UMD.EDU
SHRADER, C.	NASA/GSFC	GROSSC::SHRADER
SICOTTE, HUGUES	PRINCETON U	SICOTTE@PUPGG.PRINCETON.EDU
SILK, JOSEPH	UC-BERKELEY	SILK@UCBAST.BERKELEY.EDU
SINGH, K. P.	NASA/GSFC	KPS@ROSSERV.GSFC.NASA.GOV
SINHA, RAMESH	TIFR (Pune, India)	SINHA@GMRT.ERNET.IN
SMITH, HOWARD A.	SI-NASM	HOWARD@NASM.EDU
SNOW, GEORGE	U MARYLAND	SNOW@UMDHEP.UMD.EDU
SPERGEL, DAVID	PRINCETON U OBS	DNS@ASTRO.PRINCETON.EDU
SQUIRES, GORDON	CITA/U TORONTO	SQUIRES@CITA.UTORONTO.CA
STAHLE, C.	NASA/GSFC	STAHLE@LHEAVX.GSFC.NASA.GOV
STARK, M.	NASA/GSFC	
STECHER, TED	NASA/GSFC	STECHER@UIT.GSFC.NASA.GOV
STECKER, F. W.	NASA/GSFC	STECKER@LHEAVX.GSFC.NASA.GOV
STEINMETZ, MATTHIAS	MPI ASTROPHYS	MHS@MPA-GARCHING.MPG.DE
STEPHENS, ALFRED	NASA/GSFC	ASTEPHENS@PALVX.NMSU.EDU
STERNGLASS, E. J.	U PITTSBURGH	
STILLER, BERTRAM	GEORGE WASHINGTON U	BSTILLER@CAP.GWU.EDU
STREEKUMAR, P.	NASA/GSFC	SREEKUMAR@LHEAVX.GSFC.NASA.GOV

STREITMATTER, ROBERT	NASA/GSFC	STREITMATTER@LHEAVX.GSFC.NASA.GOV
STROHMAYER, T.	NASA/GSFC	
SUBBARAO, MARK	JOHNS HOPKINS	SUBBARAO@ETA.PHA.JHU.EDU
SUMMERS, FRANK	PRINCETON U	SUMMERS@ASTRO.PRINCETON.EDU
SWANK, JEAN	NASA/GSFC	SWANK@LHEAVX.GSFC.NASA.GOV
SZALAY, ALEXANDER	JOHNS HOPKINS U	SZALAY@STSCI.BITNET
SZYMKOWIAK, ANDREW	NASA/GSFC	ANDREW.SZYMKOWIAK@GSFC.NASA.GOV
TAWARA, YUZURU	NAGOYA U	TAWARA@SATIO.PHYS.NAGOYA-U.AC.JP
TAYLOR, ANDY	U EDINBURGH	ANT@STAR.ROE.AC.UK
TEEGARDEN, B.	NASA/GSFC	TEEGARDEN@TGRS.GSFC.NASA.GOV
TEMKIN, A.	NASA/GSFC	TEMKIN@STARS.GSFC.NASA.GOV
THOMPSON, DAVE	NASA/GSFC	DJT@SAGE0.GSFC.NASA.GOV
THOMPSON, RODGER	U ARIZONA	RTHOMPSON@AS.ARIZONA.EDU
THOUL, ANNE	INST FOR ADV STUDY	THOUL@GUINNESS.IAS.EDU
TITARCHUK, LEV	NASA/GSFC	TITARCHUK@LHEAVX.GSFC.NASA.GOV
TRASCO, JOHN	U MARYLAND	JTRASCO@ASTRO.UMD.EDU
TRAVIS, JOHN	SCIENCE MAGAZINE	JSTRAVIS@WORLD.STD.COM
TREMBLAY, BENOIT	RUTGERS U	BENOIT@PHYSICS.RUTGERS.EDU
TRIMBLE, VIRGINIA	U MARYLAND	
TULLY, BRENT	U HAWAII	TULLY@IFA.HAWAII.EDU
TURNER, MICHAEL	FERMI NATIONAL ACCEL.	
TURNSHEK, DAVID	U PITTSBURGH	TURNSHEK@VM2.CIS.PITT.EDU
TYSON, TONY	AT&T BELL LABS	TYSON@PHYSICS.ATT.COM
VALTONEN, MAURI	TUORLA OBSERVATORY	MAVALTO@SARA.CC.UTU.FI
VAN DER MAREL, ROELAND P.	INST FOR ADV STUDY	MAREL@GUINESS.IAS.EDU
VAN STEENBERG, MICHAEL E.	NASA/GSFC	
VOGELEY, MICHAEL	JOHNS HOPKINS U	VOGELEY@PHA.JHU.EDU
WALSH, WILFRED	U NEW S WALES & ATNF	WWALSH@ATNF.CSIRO.AU
WAMBSGANSS, JOACHIM	ASTROPHYS INST POTSDAM	JWAMBSGANSS@AIP.DE
WAN, LIN	PRINCETON U	LINWAN@PHOENIX.PRINCETON.EDU
WARREN, MICHAEL S.	LANL	MSWARREN@LANL.GOV
WASSERMAN, IRA	CORNELL U	IRA@IRA.TN.CORNELL.EDU
WESNEY, JOSEPH C.	GREENWICH HIGH/SACRED HEART U	
WHITE, RICHARD A.	NASA/GSFC	RWHITE@AMARNA.GSFC.NASA.GOV
WHITE, SIMON	MPI ASTROPHYS	SWHITE@MPA-GARCHING.MPG.DE

WILCZEK, FRANK	INSTITUTE FOR ADVANCED STUDY	
WILLIAMS, HAROLD	MONTGOMERY COLLEGE	HAROLDW@UMD5.UMD.EDU
WILSON, ANDREW	U MARYLAND	WILSON@ASTRO.UMD.EDU
WILSON, GILLIAN	U DURHAM	GILLIAN.WILSON@DURHAM.AC.UK
WOOD, KENT S.	NAVAL RESEARCH LAB	WOOD@SSD0.NRL.NAVY.MIL
WOODGATE, BRUCE	NASA/GSFC	WOODGATE@UIT.GSFC.NASA.GSFC
WRIGHT, NED	UCLA	BONNIE::WRIGHT
WU, CHI-CHAO	CSC	WU@STSCI.EDU
YAQOOB, T.	NASA/GSFC	YAQOOB@LHEAVX.GSFC.NASA.GOV
ZABLUDOFF, ANN	CARNEGIE OBSERVATORIES	AIZ@LIBRA.OCIW.EDU
ZAPOLSKY, HAROLD	RUTGERS U	ZAPOLSKY@RUTHEP.RUTGERS.EDU
ZARITSKY, DENNIS	LICK OBSERVATORY	DENNIS@LICK.UCSC.EDU
ZEPF, STEVE	UC-BERKELEY	ZEPF@ASTRON.BERKELEY.EDU
ZHANG, W.	NASA/GSFC	ZHANG@LAMPF.BITNET
ZUREK, WOJCIECH	JET PROPULSION LAB	RWZUREK@NASAMAIL.NASA.GOV

APPENDIX C: PHYSICAL CONSTANTS

TABLE OF PHYSICAL CONSTANTS

CONSTANT	SYMBOL	MKS		CGS		OTHER
speed of light	c	$3.00 \cdot 10^{8}$	m/s	$3.00 \cdot 10^{10}$	cm/s	(2.997925)
electron charge	e	$1.60 \cdot 10^{-19}$	coul	$4.80 \cdot 10^{-10}$	esu	
Planck constant	h	$6.63 \cdot 10^{-34}$	J•s	$6.63 \cdot 10^{-27}$	erg•s	
	$\hbar$	$1.05 \cdot 10^{-34}$	J•s	$1.05 \cdot 10^{-27}$	erg•s	
	hc	$1.99 \cdot 10^{-25}$	J•m	$1.99 \cdot 10^{-16}$	erg•cm	
	$\hbar c$	$3.15 \cdot 10^{-26}$	J•m	$3.15 \cdot 10^{-17}$	erg•cm	200 MeV•fm
Boltzmann constant	k	$1.38 \cdot 10^{-23}$	J/K	$1.38 \cdot 10^{-16}$	erg/K	$8.6 \cdot 10^{-5}$ eV/K
	k/h	$2.08 \cdot 10^{10}$	s^{-1}/K	$2.08 \cdot 10^{10}$	s^{-1}/K	
	k/hc	69.5	m^{-1}/K	0.695	cm^{-1}/K	
Gravitational constant	G	$6.67 \cdot 10^{-11}$	$N \cdot m^2/kg^2$	$6.67 \cdot 10^{-8}$	$dy \cdot cm^2/gm^2$	
Gas constant	R	8.314	J/K•mole	$8.31 \cdot 10^{7}$	erg/K•mole	
Avogadro's number (= R/k)	N	$6.02 \cdot 10^{26}$	amu/kg	$6.02 \cdot 10^{23}$	amu/kg	$6 \cdot 10^{23}$ molecules/mole
electron mass	m_e	$9.11 \cdot 10^{-31}$	kg	$9.11 \cdot 10^{-28}$	gm	0.51 MeV
proton mass	M_p	$1.67 \cdot 10^{-27}$	kg	$1.67 \cdot 10^{-24}$	gm	938 MeV
neutron mass	M_n	$1.67 \cdot 10^{-27}$	kg	$1.67 \cdot 10^{-24}$	gm	939 MeV
pion mass ($=270 \cdot m_e$)	m_π	$2.46 \cdot 10^{-28}$	kg	$2.46 \cdot 10^{-25}$	gm	140 MeV
muon mass ($=207 \cdot m_e$)	m_μ	$1.89 \cdot 10^{-28}$	kg	$1.89 \cdot 10^{-25}$	gm	106 MeV
classical elect radius ($=e^2/mc^2$)	r_c	$2.82 \cdot 10^{-15}$	m	$2.82 \cdot 10^{-13}$	cm	
Compton wavelength ($=h/mc$)	λ_c	$2.43 \cdot 10^{-12}$	m	$2.43 \cdot 10^{-10}$	cm	0.02 Å

Thomson cross-section	σ_T	$6.65 \bullet 10^{-29}$	m^2	$6.65 \bullet 10^{-25}$	cm^2	
Planck length ($=\sqrt{\hbar G/c^3}$)	l_{Pl}	$1.61 \bullet 10^{-35}$	m	$1.61 \bullet 10^{-33}$	cm	
Planck time ($=\sqrt{\hbar G/c^5}$)	t_{Pl}	$5.39 \bullet 10^{-44}$	s	$5.39 \bullet 10^{-44}$	s	
Planck density ($=c^5/\hbar G^2$)	ρ_{Pl}	$5.16 \bullet 10^{96}$	kg/m^3	$5.16 \bullet 10^{93}$	gm/cm^3	
Bohr radius ($=\hbar^2/me^2$)	r_B	$0.53 \bullet 10^{-10}$	m	$0.53 \bullet 10^{-8}$	cm	0.5 Å
Fine structure constant ($=e^2/\hbar c$)	α	$7.30 \bullet 10^{-3}$		$7.30 \bullet 10^{-3}$		1/137
Bohr magneton ($=e\hbar/2m_ec$)	μ_B	$9.27 \bullet 10^{-24}$	J/T	$9.27 \bullet 10^{-21}$	erg/gauss	
Nuclear magneton ($=e\hbar/2M_pc$)	μ_N	$5.05 \bullet 10^{-27}$	J/T	$5.05 \bullet 10^{-24}$	erg/gauss	
Permittivity of vacuum	ε_o	$8.85 \bullet 10^{-12}$	fd/m			$1/4\pi\varepsilon_o = 9.0 \bullet 10^9$
Permeability in vacuum	μ_o	$4\pi \bullet 10^{-7}$	Hen/m			
Stefan-Boltzmann constant	σ	$5.67 \bullet 10^{-8}$	$W/m^2 \bullet K^4$	$5.67 \bullet 10^{-5}$	$erg/s \bullet cm^2 \bullet K^4$	
Rydberg ($=m_e e^4/2\hbar^2$)	R_∞	$2.18 \bullet 10^{-18}$	J	$2.18 \bullet 10^{-11}$	erg	13.6 eV
1 amu		$1.66 \bullet 10^{-27}$	kg	$1.66 \bullet 10^{-24}$	gm	931.5 MeV
1 calorie		4.19	J	$4.19 \bullet 10^{7}$	erg	
1 year		$3.16 \bullet 10^{7}$	s	$3.16 \bullet 10^{7}$	s	
1 atmosphere		$1.01 \bullet 10^{5}$	N/m^2	$1.01 \bullet 10^{6}$	$dyne/cm^2$	$14.2\ lbs/in^2$
		$1.01 \bullet 10^{5}$	Pascal			760 Torr
1 eV		$1.6 \bullet 10^{-19}$	J	$1.6 \bullet 10^{-12}$	erg	11,605 K
		$1.24 \bullet 10^{-6}$	m	$1.24 \bullet 10^{-4}$	cm	
		1	Tesla	10^4	gauss	

ASTROPHYSICAL CONSTANTS

CONSTANT	SYMBOL	MKS	CGS	OTHER
astronomical unit	AU	$1.50 \cdot 10^{11}$ m	$1.50 \cdot 10^{13}$ cm	
	AU/year			4.74 km/s
parsec	pc	$3.09 \cdot 10^{16}$ m	$3.09 \cdot 10^{18}$ cm	3.26 LY
solar mass	$M_\odot$	$1.99 \cdot 10^{24}$ kg	$1.99 \cdot 10^{33}$ gm	
solar luminosity	$L_\odot$	$3.90 \cdot 10^{26}$ J/s	$3.90 \cdot 10^{33}$ erg/s	
solar effective temperature	$T_{eff\odot}$	5780 K	5780 K	
solar radius	$R_\odot$	$6.96 \cdot 10^{8}$ m	$6.96 \cdot 10^{10}$ cm	
Earth radius	$R_\oplus$	$6.38 \cdot 10^{6}$ m	$6.38 \cdot 10^{8}$ cm	
Earth mass	$M_\oplus$	$5.98 \cdot 10^{24}$ kg	$5.98 \cdot 10^{27}$ gm	
Earth density	$\rho_\oplus$	5520 kg/m^3	5.52 gm/cm^3	
Jansky	Jy	$1.0 \cdot 10^{-26}$ $W/m^2 \cdot Hz$	$1 \cdot 10^{-23}$ $erg/s \cdot cm^2 \cdot Hz$	
Hubble constant	H_o	$3.24h \cdot 10^{-18}$ s^{-1}	$3.24h \cdot 10^{-18}$ s^{-1}	$100h$ km/s•Mpc
critical density ($=3H_o^2/8\pi G$)	ρ_o	$1.88h^2 \cdot 10^{-26}$ kg/cm^3	$1.88h^2 \cdot 10^{-29}$ gm/cm^3	
plasma frequency				$8.98\sqrt{n_e(cm^{-3})}$ kHz/gauss
radian				$57.29578° = 206,265''$
CMB photon density	n_γ	$4.15 \cdot 10^{5}$ m^{-3}	415 cm^{-3}	

AUTHOR INDEX

AUTHOR INDEX

A

Alcock, C., 77
Allsman, R. A., 77
Ammiraju, P., 517
Arnaud, K., 231
Awaki, H., 190
Axelrod, T. S., 77

B

Bahcall, N. A., 201
Banday, A. J., 477, 481
Barbier, L. M., 53
Barker, D. L., 53
Bartelmann, M., 307, 335, 355
Beers, T. C., 221
Bennett, C. L., 477, 481, 485, 489
Bennett, D. P., 77
Bernstein, G. M., 91, 339, 343
Bertola, F., 145
Bird, C. M., 221, 251
Biviano, A., 272
Blitz, L., 101
Boeshaar, P. C., 91
Boroson, T., 157
Bosma, A., 111
Bowen, T., 53
Bowles, T. J., 44
Brainerd, T. G., 407
Broadhurst, T., 320
Broeils, A., 125
Bromley, B. C., 397, 433
Buote, D. A., 182, 186
Byrd, G. G., 450

C

Caldwell, D. O., 411
Canizares, C. R., 182, 186
Cannon, D., 264
Carrasco, L., 212
Cen, R., 347
Christian, E. R., 53
Cole, S., 351
Combes, F., 161
Cook, K. H., 77

D

Davis, A. J., 53
Davis, M., 361
Doroshkevich, A. G., 429
Dunn, A. M., 425
Durret, F., 272

F

Finetti, N., 520
Fischer, P., 339, 343
Fong, R., 429
Freeman, K. C., 77
Frenk, C., 351
Friese, V., 129
Fuchs, B., 129
Fukazawa, Y., 231

G

Gaidos, E. J., 194
Garilli, B., 212
Gerbal, D., 272, 276
Golden, R. L., 53
Górski, K. M., 467, 477, 481
Greyber, H. D., 509
Griest, K., 77
Guhathakurta, P., 339

H

Halverson, P. G., 53
Harding, P., 157
Hawkins, M. R. S., 331
Heavens, A. F., 381, 386

Henriksen, M. J., 280
Hinshaw, G., 477, 481, 485, 489
Holt, S. S., xi
Holtzman, J., 411
Huchra, J. P., 221

I

Innanen, K. A., 450

K

Kazanas, D., 495
Kendall, K. R., 53
Khan, I., 133
Klypin, A., 411
Kochanek, C. S., 297
Kogut, A., 477, 481, 485, 489
Kriessler, J. R., 221
Krombel, K. E., 53

L

Labrador, A. W., 53
Laflamme, R., 425
Le Fevre, O., 272
Leibert, J., 71
Le Peintre, F., 272
Lineweaver, C. H., 481
Lobo, C., 272
Loeb, A., 330
Loewenstein, M., 190, 231
Lubin, L. M., 208

M

Ma, C.-P., 420
Maccagni, D., 212
Makishima, K., 172, 190
Mannheim, P. D., 505
Mao, S., 330
Marshall, S. L., 77
Martel, H., 441, 446
Matsumoto, H., 190
Matsushita, K., 190
Mazure, A., 272
McCall, M. L., 450
McGuire, P. C., 53
Merritt, D., 149
Metcalfe, T. S., 53
Mewaldt, R. A., 53
Mitchell, J. W., 53
Moore, B., 168
Morrison, H., 157
Mulchaey, J. S., 243
Mushotzky, R. F., 190, 231, 251, 268

N

Narayan, R., 307
Norton, R. S., 53
Nusser, A., 361

O

Ohashi, T., 255
Olling, R. P., 121
Ormes, J. F., 53
Ostriker, J. P., 347

P

Paczyński, B., 10
Peebles, P. J. E., 390
Perlmutter, S., 77
Persic, M., 137
Peterson, B. A., 77
Pfenniger, D., 161
Pifer, A. E., 53
Pizzella, A., 145
Pogge, R. W., 141
Ponman, T., 264
Postman, M., 371
Pratt, M. R., 77
Primack, J. R., 411

Q

Quinn, P. J., 77, 433

R

Recillas, E., 212
Rees, M. J., 3
Rhee, G., 343
Rodgers, A. W., 77
Rodrigues-Williams, L. L., 331

S

Sackett, P. D., 141, 157
Sakima, Y., 268
Salucci, P., 137
Schild, R., 95
Schindler, S. M., 53
Serna, A., 276
Shapiro, P. R., 441, 446
Sicotte, H., 390
Silk, J., 527
Slezak, E., 272
Smoot, G. F., 481
Spergel, D. N., 457
Steinmetz, M., 355
Stel, F., 137
Stephens, S. A., 520
Sternglass, E. J., 513
Stochaj, S. J., 53
Streitmatter, R. E., 53
Stubbs, C. W., 77
Sutherland, W., 77

T

Tawara, Y., 268
Taylor, A. N., 381, 386
Tenorio, L., 481
Thomson, D. J., 95
Thoul, A. A., 437
Tremblay, B., 149
Trimble, V., 57
Turchaninov, V. I., 429
Turner, E. L., 347
Tyson, J. A., 91, 287, 339, 343

V

Valinia, A., 441
Valtonen, M. J., 450
van Gorkom, J. H., 121
Villumsen, J. V., 407, 441

W

Wambsganss, J., 347
Warren, M. S., 397, 433
Webber, W. R., 53
Weiss, A., 355
Wielen, R., 129
Wilczek, F., 18
Williams, T. B., 149
Wilson, G., 351
Wright, E. L., 477

Y

Yamashita, K., 268

Z

Zabludoff, A. I., 216
Zaritsky, D., 165, 216
Zepf, S. E., 153
Zheng, J. Q., 450
Zurek, W. H., 397, 433

SUBJECT INDEX

SUBJECT INDEX*

A

ASCA 172, 190, 231, 243, 251, 255, 268, 527

B

Baryon density 3
BBXRT 172, 231
Black holes 3, 57, 129, 527
Brown dwarfs 3, 57, 71, 77, 101, 527
Bulk flows - *see Galaxy velocity fields*

C

CFHT 339, 343
Cold Dark Matter (CDM) model 3, 168, 182, 201, 208, 287, 297, 335, 347, 355, 371, 397, 407, 411, 420, 425, 429, 433, 441, 446, 467, 485, 489, 513, 527
Conformal gravity 495, 505
Cooling flows 172
Cosmic Background Explorer (COBE) 3, 101, 201, 297, 347, 355, 361, 371, 381, 386, 397, 407, 411, 433, 441, 457, 467, 477, 481, 485, 489, 509, 527
Cosmic microwave background 133, 297, 361, 371, 381, 386, 390, 397, 457, 467, 477, 481, 485, 489, 527
Cosmological constant 10, 133, 201, 297, 371, 386, 411, 457, 467, 495, 513, 527
CTIO 91, 141, 149, 339

D

Deuterium abundance 3
Disk stability 101, 111
Dust 165, 331
Dwarf spiral galaxies 129, 168, 190

E

Einstein Observatory 186, 190, 212, 221, 243, 251
Elliptical galaxy dark matter 145, 153, 172, 190, 194, 208, 221, 255
EROS 10, 77, 505, 527
Exosat 251

F

Faint blue galaxies 287, 307, 343

G

Galaxy dynamics in clusters 221, 231
Galaxy velocity fields 361, 371, 381, 397, 425, 429, 450, 467, 527
Galileo 57
Giant molecular clouds 101
Ginga 172, 251, 255
Globular cluster dynamics 153
Gluinos 18
Gravitational macrolensing 10, 201, 216, 231, 287, 297, 307, 320, 331, 335, 339, 343, 347, 351, 355, 527
Gravitational microlensing 10, 77, 95, 505, 527
Gravity theory 495, 505
Gravity waves 457

H

Harrison-Zeldovich spectrum 3
Helium abundance 3
Higgs particle 18, 495
Hot dark matter (HDM) model *-see Mixed dark matter model*
Hubble constant determination 343
Hubble Space Telescope (HST) 3, 71, 77, 91, 153, 287, 320, 429, 505

*Page numbers refer to the *first* page of the contribution in which the subject appears.

Huchra's lens 10
Hybrid dark matter model *-see Mixed dark matter model*
Hydrogen clouds 101, 161

I

IMAX telescope 53
Initial mass function 101, 527
Intergalactic hydrogen 57, 446
Intracluster medium (ICM) 172, 201, 221, 243, 251, 255, 264, 268, 280, 330
Ionizing massive particles 53
IRAS 361, 371, 381, 386, 407, 527

J

Jupiters *-see Stars, low mass*

K

Keck telescope 3
KPNO 91, 157, 339, 343

L

LSND 44
Luminosity function *-see Stellar luminosity function*

M

MACHO 10, 77, 505, 527
McGraw-Hill Telescope 194
Metallicity 153, 182, 243, 255, 268, 527
Mexican National Astronomical Observatory 212
MSW neutrino oscillations 44
Mixed Dark Matter (MDM) model 3, 201, 335, 371, 411, 420, 441, 457, 467, 527
Mt. Stromlo Observatory 77

N

Neutrinos 3, 44, 53, 411, 420
New Technology Telescope (NTT) 320
Newton 57

O

OGLE 10, 77, 505, 527
Oort limit 57

P

Planetary nebulae dynamics 149
Polar ring galaxies 141, 161, 190, 527
POTENT 361, 386, 411, 527
Primordial Baryon Isocurvature (PBI) model 201, 339, 371

Q

QCD 18
Quark Nuggets 53

R

Reionization 446, 457
Rich clusters, velocity dispersion 57
ROSAT 172, 182, 186, 190, 194, 216, 231, 243, 264, 268, 272
Rotation curves 57, 101, 111, 121, 125, 133, 137, 157, 168, 495, 527
Rutgers Fabry-Perot 149

S

Standard model 18, 44
Stars, low mass, dwarfs 71, 91, 101, 157
Stellar luminosity function 71, 91
Stellar remnants 3, 57, 71, 527
Sunyaev-ZXeldovich effect 57, 330
Supernovae, Type 1a 10, 361, 371
Supersymmetry (SUSY) 3, 44

Swope telescope 77

U

Unification 18, 44

V

VLA 125

W

Weyl gravity 495, 505
WSRT 125

X

X-ray cluster isophotes 186, 194, 216, 231, 243, 255, 264
X-ray cluster boundary 255

AIP Conference Proceedings

		L.C. Number	ISBN
No. 244	Radio Frequency Power in Plasmas (Charleston, SC, 1991)	91-77853	0-88318-937-2
No. 245	Basic Space Science (Bangalore, India, 1991)	91-78379	0-88318-951-8
No. 246	Space Nuclear Power Systems (Albuquerque, NM, 1992)	91-58793	1-56396-027-3 1-56396-026-5 (pbk.)
No. 247	Global Warming: Physics and Facts (Washington, DC, 1991)	91-78423	0-88318-932-1
No. 248	Computer-Aided Statistical Physics (Taipei, Taiwan, 1991)	91-78378	0-88318-942-9
No. 249	The Physics of Particle Accelerators (Upton, NY, 1989, 1990)	92-52843	0-88318-789-2
No. 250	Towards a Unified Picture of Nuclear Dynamics (Nikko, Japan, 1991)	92-70143	0-88318-951-8
No. 251	Superconductivity and its Applications (Buffalo, NY, 1991)	92-52726	1-56396-016-8
No. 252	Accelerator Instrumentation (Newport News, VA, 1991)	92-70356	0-88318-934-8
No. 253	High-Brightness Beams for Advanced Accelerator Applications (College Park, MD, 1991)	92-52705	0-88318-947-X
No. 254	Testing the AGN Paradigm (College Park, MD, 1991)	92-52780	1-56396-009-5
No. 255	Advanced Beam Dynamics Workshop on Effects of Errors in Accelerators, Their Diagnosis and Corrections (Corpus Christi, TX, 1991)	92-52842	1-56396-006-0
No. 256	Slow Dynamics in Condensed Matter (Fukuoka, Japan, 1991)	92-53120	0-88318-938-0
No. 257	Atomic Processes in Plasmas (Portland, ME, 1991)	91-08105	0-88318-939-9
No. 258	Synchrotron Radiation and Dynamic Phenomena (Grenoble, France, 1991)	92-53790	1-56396-008-7
No. 259	Future Directions in Nuclear Physics with 4π Gamma Detection Systems of the New Generation (Strasbourg, France, 1991)	92-53222	0-88318-952-6

No.	Title	LCCN	ISBN
No. 260	Computational Quantum Physics (Nashville, TN, 1991)	92-71777	0-88318-933-X
No. 261	Rare and Exclusive B&K Decays and Novel Flavor Factories (Santa Monica, CA, 1991)	92-71873	1-56396-055-9
No. 262	Molecular Electronics—Science and Technology (St. Thomas, Virgin Islands, 1991)	92-72210	1-56396-041-9
No. 263	Stress-Induced Phenomena in Metallization: First International Workshop (Ithaca, NY, 1991)	92-72292	1-56396-082-6
No. 264	Particle Acceleration in Cosmic Plasmas (Newark, DE, 1991)	92-73316	0-88318-948-8
No. 265	Gamma-Ray Bursts (Huntsville, AL, 1991)	92-73456	1-56396-018-4
No. 266	Group Theory in Physics (Cocoyoc, Morelos, Mexico, 1991)	92-73457	1-56396-101-6
No. 267	Electromechanical Coupling of the Solar Atmosphere (Capri, Italy, 1991)	92-82717	1-56396-110-5
No. 268	Photovoltaic Advanced Research & Development Project (Denver, CO, 1992)	92-74159	1-56396-056-7
No. 269	CEBAF 1992 Summer Workshop (Newport News, VA, 1992)	92-75403	1-56396-067-2
No. 270	Time Reversal—The Arthur Rich Memorial Symposium (Ann Arbor, MI, 1991)	92-83852	1-56396-105-9
No. 271	Tenth Symposium Space Nuclear Power and Propulsion (Vols. I–III) (Albuquerque, NM, 1993)	92-75162	1-56396-137-7 (set)
No. 272	Proceedings of the XXVI International Conference on High Energy Physics (Vols. I and II) (Dallas, TX, 1992)	93-70412	1-56396-127-X (set)
No. 273	Superconductivity and Its Applications (Buffalo, NY, 1992)	93-70502	1-56396-189-X
No. 274	VIth International Conference on the Physics of Highly Charged Ions (Manhattan, KS, 1992)	93-70577	1-56396-102-4
No. 275	Atomic Physics 13 (Munich, Germany, 1992)	93-70826	1-56396-057-5

No. 276	Very High Energy Cosmic-Ray Interactions: VIIth International Symposium (Ann Arbor, MI, 1992)	93-71342	1-56396-038-9
No. 277	The World at Risk: Natural Hazards and Climate Change (Cambridge, MA, 1992)	93-71333	1-56396-066-4
No. 278	Back to the Galaxy (College Park, MD, 1992)	93-71543	1-56396-227-6
No. 279	Advanced Accelerator Concepts (Port Jefferson, NY, 1992)	93-71773	1-56396-191-1
No. 280	Compton Gamma-Ray Observatory (St. Louis, MO, 1992)	93-71830	1-56396-104-0
No. 281	Accelerator Instrumentation Fourth Annual Workshop (Berkeley, CA, 1992)	93-072110	1-56396-190-3
No. 282	Quantum 1/f Noise & Other Low Frequency Fluctuations in Electronic Devices (St. Louis, MO, 1992)	93-072366	1-56396-252-7
No. 283	Earth and Space Science Information Systems (Pasadena, CA, 1992)	93-072360	1-56396-094-X
No. 284	US-Japan Workshop on Ion Temperature Gradient-Driven Turbulent Transport (Austin, TX, 1993)	93-72460	1-56396-221-7
No. 285	Noise in Physical Systems and 1/f Fluctuations (St. Louis, MO, 1993)	93-72575	1-56396-270-5
No. 286	Ordering Disorder: Prospect and Retrospect in Condensed Matter Physics: Proceedings of the Indo-U.S. Workshop (Hyderabad, India, 1993)	93-072549	1-56396-255-1
No. 287	Production and Neutralization of Negative Ions and Beams: Sixth International Symposium (Upton, NY, 1992)	93-72821	1-56396-103-2
No. 288	Laser Ablation: Mechanismas and Applications-II: Second International Conference (Knoxville, TN, 1993)	93-73040	1-56396-226-8
No. 289	Radio Frequency Power in Plasmas: Tenth Topical Conference (Boston, MA, 1993)	93-72964	1-56396-264-0
No. 290	Laser Spectroscopy: XIth International Conference (Hot Springs, VA, 1993)	93-73050	1-56396-262-4
No. 291	Prairie View Summer Science Academy (Prairie View, TX, 1992)	93-73081	1-56396-133-4

No. 292	Stability of Particle Motion in Storage Rings (Upton, NY, 1992)	93-73534	1-56396-225-X
No. 293	Polarized Ion Sources and Polarized Gas Targets (Madison, WI, 1993)	93-74102	1-56396-220-9
No. 294	High-Energy Solar Phenomena A New Era of Spacecraft Measurements (Waterville Valley, NH, 1993)	93-74147	1-56396-291-8
No. 295	The Physics of Electronic and Atomic Collisions: XVIII International Conference (Aarhus, Denmark, 1993)	93-74103	1-56396-290-X
No. 296	The Chaos Paradigm: Developments an Applications in Engineering and Science (Mystic, CT, 1993)	93-74146	1-56396-254-3
No. 297	Computational Accelerator Physics (Los Alamos, NM, 1993)	93-74205	1-56396-222-5
No. 298	Ultrafast Reaction Dynamics and Solvent Effects (Royaumont, France, 1993)	93-074354	1-56396-280-2
No. 299	Dense Z-Pinches: Third International Conference (London, 1993)	93-074569	1-56396-297-7
No. 300	Discovery of Weak Neutral Currents: The Weak Interaction Before and After (Santa Monica, CA, 1993)	94-70515	1-56396-306-X
No. 301	Eleventh Symposium Space Nuclear Power and Propulsion (3 Vols.) (Albuquerque, NM, 1994)	92-75162	1-56396-305-1 (Set) 156396-301-9 (pbk. set)
No. 302	Lepton and Photon Interactions/ XVI International Symposium (Ithaca, NY, 1993)	94-70079	1-56396-106-7
No. 303	Slow Positron Beam Techniques for Solids and Surfaces Fifth International Workshop (Jackson Hole, WY 1992)	94-71036	1-56396-267-5
No. 304	The Second Compton Symposium (College Park, MD, 1993)	94-70742	1-56396-261-6
No. 305	Stress-Induced Phenomena in Metallization Second International Workshop (Austin, TX, 1993)	94-70650	1-56396-251-9
No. 306	12th NREL Photovoltaic Program Review (Denver, CO, 1993)	94-70748	1-56396-315-9

No.	Title	LC No.	ISBN
No. 323	Atomic Physics 14 Fourteenth International Conference on Atomic Physics (Boulder, CO 1994)	94-73219	1-56396-348-5
No. 324	Twelfth Symposium on Space Nuclear Power and Propulsion (Albuquerque, NM 1995)	94-73603	1-56396-427-9
No. 325	Conference on NASA Centers for Commercial Development of Space (Albuquerque, NM 1995)	94-73604	1-56396-431-7
No. 326	Accelerator Physics at the Superconducting Super Collider (Dallas, TX 1992-1993)	94-73609	1-56396-354-X
No. 327	Nuclei in the Cosmos III Third International Symposium on Nuclear Astrophysics (Assergi, Italy 1994)	95-75492	1-56396-436-8
No. 328	Spectral Line Shapes, Volume 8 12th ICSLS (Toronto, Canada 1994)	94-74309	1-56396-326-4
No. 329	Resonance Ionization Spectroscopy 1994 Seventh International Symposium (Bernkastel-Kues, Germany 1994)	95-75077	1-56396-437-6
No. 330	Computational Chemistry F.E.C.S. Conference (Nancy, France 1994)	95-75843	1-56396-457-0
No. 331	Non-Neutral Plasma Physics II (Berkeley, CA 1994)	95-79630	1-56396-441-4
No. 332	X-Ray Lasers 1994 Fourth International Colloquium (Williamsburg, VA 1994)	95-76067	1-56396-375-2
No. 333	Beam Instrumentation Workshop (Vancouver, B. C., Canada 1994)	95-79635	1-56396-352-3
No. 334	Few-Body Problems in Physics (Williamsburg, VA 1994)	95-76481	1-56396-325-6
No. 336	Dark Matter (College Park, MD 1994)	95-76538	1-56396-438-4
No. 338	Intersections Between Particle and Nuclear Physics 5th Conference (St. Petersburg, FL 1994)	95-77076	1-56396-335-3
No. 339	Polarization Phenomena in Nuclear Physics Eighth International Symposium (Bloomington, IN 1994)	95-77216	1-56396-482-1

No. 307	Gamma-Ray Bursts Second Workshop (Huntsville, AL 1993)	94-71317	1-56396-336-1
No. 308	The Evolution of X-Ray Binaries (College Park, MD 1993)	94-76853	1-56396-329-9
No. 309	High-Pressure Science and Technology—1993 (Colorado Springs, CO 1993)	93-72821	1-56396-219-5 (Set)
No. 310	Analysis of Interplanetary Dust (Houston, TX 1993)	94-71292	1-56396-341-8
No. 311	Physics of High Energy Particles in Toroidal Systems (Irvine, CA 1993)	94-72098	1-56396-364-7
No. 312	Molecules and Grains in Space (Mont Sainte-Odile, France 1993)	94-72615	1-56396-355-8
No. 313	The Soft X-Ray Cosmos ROSAT Science Symposium (College Park, MD 1993)	94-72499	1-56396-327-2
No. 314	Advances in Plasma Physics Thomas H. Stix Symposium (Princeton, NJ 1992)	94-72721	1-56396-372-8
No. 315	Orbit Correction and Analysis in Circular Accelerators (Upton, NY 1993)	94-72257	1-56396-373-6
No. 316	Thirteenth International Conference on Thermoelectrics (Kansas City, Missouri 1994)	95-75634	1-56396-444-9
No. 317	Fifth Mexican School of Particles and Fields (Guanajuato, Mexico 1992)	94-72720	1-56396-378-7
No. 318	Laser Interaction and Related Plasma Phenomena 11th International Workshop (Monterey, CA 1993)	94-78097	1-56396-324-8
No. 319	Beam Instrumentation Workshop (Santa Fe, NM 1993)	94-78279	1-56396-389-2
No. 320	Basic Space Science (Lagos, Nigeria 1993)	94-79350	1-56396-328-0
No. 321	The First NREL Conference on Thermophotovoltaic Generation of Electricity (Copper Mountain, CO 1994)	94-72792	1-56396-353-1
No. 322	Atomic Processes in Plasmas Ninth APS Topical Conference (San Antonio, TX)	94-72923	1-56396-411-2